Biocatalysis and Nanotechnology

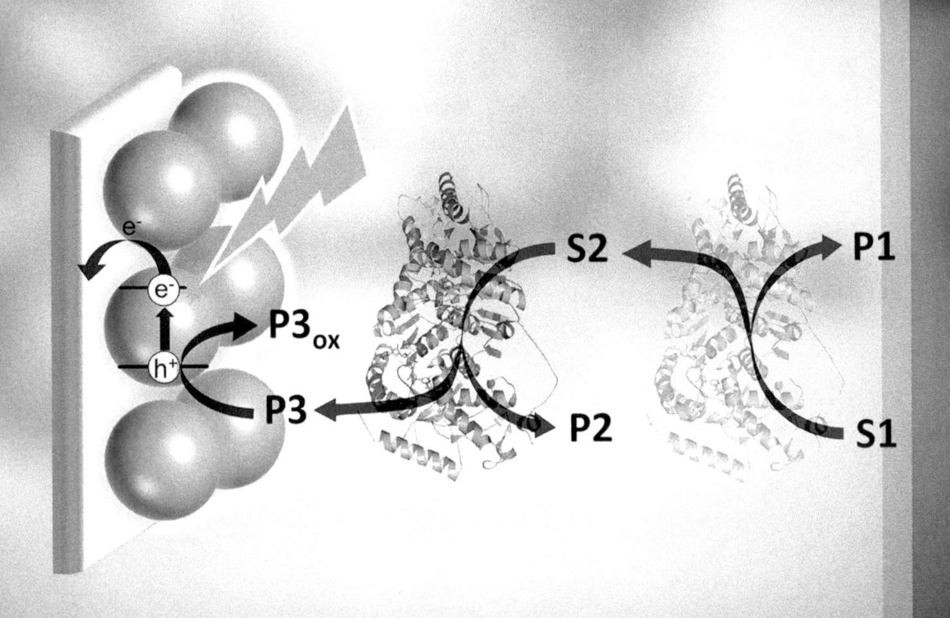

Pan Stanford Series on Biocatalysis

Series Editor
Peter Grunwald

Titles in the Series

Published

Vol. 1
Industrial Biocatalysis
Peter Grunwald, ed.
2015
978-981-4463-88-1 (Hardcover)
978-981-4463-89-8 (eBook)

Vol. 2
Handbook of Carbohydrate-Modifying Biocatalysts
Peter Grunwald, ed.
2016
978-981-4669-78-8 (Hardcover)
978-981-4669-79-5 (eBook)

Vol. 3
Biocatalysis and Nanotechnology
Peter Grunwald, ed.
2017
978-981-4613-69-9 (Hardcover)
978-1-315-19660-2 (eBook)

Forthcoming

Vol. 4
Pharmaceutical Biocatalysis
Peter Grunwald, ed.
2018

Pan Stanford Series
on Biocatalysis
Volume 3

Biocatalysis and Nanotechnology

edited by Peter Grunwald

Pan Stanford Publishing

Published by

Pan Stanford Publishing Pte. Ltd.
Penthouse Level, Suntec Tower 3
8 Temasek Boulevard
Singapore 038988

Email: editorial@panstanford.com
Web: www.panstanford.com

British Library Cataloguing-in-Publication Data
A catalogue record for this book is available from the British Library.

Biocatalysis and Nanotechnology

Cover image: Courtesy of Mark Riedel, Daniel Schäfer, and Fred Lisdat, Technical University of Wildau, Germany.

ISBN 978-981-4613-69-9 (Hardcover)
ISBN 978-1-315-19660-2 (eBook)

Printed in the USA

Contents

Preface

The 23 chapters of this book provide an actual overview of the various relations between two scientific fields: nanotechnology and biocatalysis. In the recent past, nanobiocatalysis has rapidly developed into a subarea of enzyme biotechnology. It combines the advances in nanotechnology, which have generated nanoscale materials of different sizes, shapes, and physicochemical properties by various methods, including the employment of microbes as nanofactories, with the excellent characteristics of biocatalysts to an innovative technology.

The first chapter, written by Torben Kodanek, Sara Sánchez Paradinas, Franziska Lübkemann, Dirk Dorfs, and Nadja C. Bigall, introduces readers to the fundamentals of nanotechnology/nanochemistry discussing physical effects occurring at the nanometer scale such as electronic properties, surface-related properties, superparamagnetism, and localized surface plasmon resonance together with the different methods of nanofabrication. This is followed by an introduction to biocatalysis that—in addition to discussing general aspects such as protein biosynthesis and architecture—provides a brief overview of different designs of biotransformations, including a section on hydrogenase mimics (important in connection with the transition from a fossil-based to a hydrogen-based economy) and methods to improve the properties of biocatalysts, which result in enhanced robustness (tolerance to higher temperatures and pH values), yields, and productivity. In their chapter about an environmentally benign nanomaterial synthesis, Lihong Liu, Fang Xie, Xiuxia Meng, Vishnu Parek, and Shaomin Liu present an alternative to the traditional physicochemical methods of the production of nanoparticles, the microorganism culture broth-mediated technology; the topic is discussed in the context of other biological processes for NP synthesis, including the use of microbes as nanofactories.

In Chapter 4, Omkar V. Zore, Rajeswari M. Kasi, and Challa V. Kumar show how enzymes and water-soluble polymers interact and how synthetic polymers can enhance the physical, chemical, and biological properties of enzymes as demonstrated with the example of core/shell nanoparticles prepared from various enzymes and polyacrylic acid of different molecular weight. The approach is based on the hypothesis that the physical confinement of the enzyme within a polymeric shell will decrease the enzyme's conformational entropy, thus contributing to the stabilization of the structure of the enzyme. Biological strategies in the nanobiocatalyst assembly are reviewed by Ian Dominic F. Tabañag and Shen-Long Tsai in Chapter 5. This strategy to produce immobilized enzyme systems by mimicking assembly systems found in nature has several advantages over traditional immobilization methods, including a limited denaturation caused by the immobilization process, proper orientation of the enzyme's active site, reduction of costly purification techniques, or possibility to fine-tune the binding affinity via suited receptor–ligand pairs. This is followed by a chapter on graphene-based nanobiocatalytic systems, authored by Michaela Patila, George Orfanakis, Angeliki C. Polydera, Ioannis V. Pavlidis, and Haralambos Stamatis. Different procedures to functionalize graphene-based nanomaterials (mono- and few-layer graphene, ultrathin graphite, graphene quantum dots, graphene oxide, and reduced graphene oxide) and the applications of these nanomaterials conjugated with hydrolases and oxidoreductases in various biocatalytic processes of industrial interest, biofuel production, degradation of pollutants and biosensing are summarized. In Chapter 7, Alan S. Campbell, Andrew J. Maloney, Chenbo Dong, and Cerasela Z. Dinu discuss the application of nanosupports for the immobilization of biocatalysts as it relates to green chemistry practices as well as the development of green technology principles and devices. Numerous examples for the application of enzyme immobilization are given, including the production of goods and chemicals, wastewater treatment, production of biofuels, fabrication of biosensors and biofuel cells, and development of antifouling and active-surface decontaminating coatings.

Chapter 8, authored by Peng-Cheng Chen, Xue-Yan Zhu, Jin Li, and Xiao-Jun Huang, deals with enzyme immobilization on membranes and its application in bioreactors. Enzymatic membrane bioreactors combine the product separation process with enzymatic catalysis in a continuous operation. The chapter focuses on the analysis of the research in enzyme immobilizations on membranes from the perspectives of immobilizations on mesoporous and nanofiber membranes, and the structure and application of enzyme-immobilized membrane bioreactors. As outlined by Avinesh Byreddy and Munish Puri in Chapter 9, various nanosupports, due to their high surface area and high mechanical stability allowing for effective enzyme binding with minimum diffusion limitation, are considered for potential application for the immobilization of lipases in connection with industrial biodiesel production. In the following chapter, Ananiy Kohuta, Scott W. Pryor, Andriy Voronov, and Sergiy Minko describe a method of encapsulation of exoenzymes by nanoparticles with grafted polymer brushes, which they term enzymogel nanoparticles, as a novel type of phase-boundary biocatalyst. The brush-decorated nanoparticles are based on a superparamagnetic particle core to facilitate recycling. The enzymogel nanoparticles imitate catabolism of either intracapsular or external substrates and are suited for a broad range of applications as demonstrated with cleavage of insoluble substrates like cellulose by cellulases. Chapter 11, contributed by Zheng Liu, Jun Ge, Diannan Lu, Guoqiang Jiang, and Jianzhong Wu, deals with the molecular simulation-mediated design and synthesis of nanostructured enzyme catalysts that exhibit high stability and activity in organic solvents at high temperature. This is achieved by an aqueous-based, two-step in situ polymerization procedure to synthesize enzyme nanogels and by nano-conjugates consisting of enzymes and Pluronic, a co-polymer with a hydrophobic backbone (poly(propylene oxide)) flanked by two hydrophilic side chains.

Naiqian Zhan, Goutam Palui, Wentao Wang, and Hedi Mattoussi report in Chapter 12 on probing enzymatic activity in vitro and in vivo by combining luminescent quantum dots, gold nanoparticles, and energy transfer. The examples presented comprise fluorescence resonance energy transfer between quantum dot donors and dye or fluorescent protein acceptors, bioluminescent resonance energy transfer between luminescent biological and chemical substrates and quantum dot acceptors, and energy transfer quenching of quantum dots and dye emission by AuNPs. In Chapter 13, Jing Mu, Hao Lun Cheong, and Bengang Xing focus on FRET reporter molecules for the identification of enzyme functions; they discuss the recent progress in the design of FRET-based reporters for the visualization and sensing of enzymatic activities in living systems and non-invasively monitoring cellular events in vitro and in vivo and discuss some typical enzymes in both physiological and pathological conditions. The authors emphasize the importance of such studies for the disease diagnosis and efficient therapy in the future. In Chapter 14, Mark Riedel, Daniel Schäfer, and Fred Lisdat report on photoelectrochemical sensors representing a rather new area of research, which has developed fast due to the progress in nanoparticle synthesis, light-directed read-out methods, and high sensitivities. As a rather simple and cost-effective technique, photoelectrochemical sensing based on quantum dots in combination with electrodes provides an interesting approach for the direct analysis of redox-active small molecules or the detection of enzymatically formed products or consumed co-substrates. The application includes the analysis of biospecific binding events using recognition elements such as DNA, aptamers, antibodies, and molecular imprinted polymers.

In Chapter 15, Ruben Ragg, Karsten Korschelt, Karoline Herget, Filipe Natalio, Muhammad Nawaz Tahir, and Wolfgang Tremel review efforts that have been made to synthesize inorganic nanomaterials capable of mimicking natural enzymes termed as "artificial enzymes," that are more stable and cost efficient compared with their natural counterpart. The overwhelming number of publications are on peroxidase mimics and their applications. One of the so far rare exceptions are molybdenum oxide nanoparticles, which exhibit an intrinsic sulfite oxidase-like activity and may be used as a replacement to treat sulfite oxidase deficiency leading to severe neurological damage. Another medical field where enzyme nanomaterials find application is diabetes care. The recent development in this field has been summarized by Wanyi Tai and Zhen Gu in Chapter 16 with a focus on glucose oxidase–anchored nanomaterials that are used for glucose detection and advanced noninvasive insulin delivery. The report includes non-enzyme nanosensors constructed from metal oxides and a discussion of the closed-loop system for insulin delivery, which mimics the pancreatic function of healthy people and releases insulin in response to the blood glucose level.

The following two chapters deal with the use of nanostructured materials for the fabrication of enzymatic biofuel cells. Takanori Tamaki (Chapter 17) focuses on "current density" and summarizes the nanostructured electrode materials (carbon black, carbon nanotubes, three-dimensional monolithic carbonaceous foams, mesoporous carbon materials) used in high-current-density enzymatic biofuel cells. The chapter considers the aspects concerning the importance of different gas-diffusion biocathodes on the current density, and of the enhanced stability of enzymes immobilized on nanostructured materials compared with the immobilization on electrodes without nanostructures. The chapter contributed by Dan Wen and Alexander Eychmüller (Chapter 18) describes recent developments in the application of porous nanostructures for the fabrication of enzymatic biofuel cells, which are apart from carbon nanotubes engineered into highly porous nanoarchitectures, graphene, porous carbon–derived 3D nanostructures, porous nanostructured noble metal surfaces (porous template–assisted metal nanoparticle nanoarchitectures, porous structures from metal nanoparticles such as noble metal aerogels), and porous nanostructures from polymeric materials. The performance of the different biofuel cells are quantified by the parameters open-circuit voltage, maximum power density, and operational stability.

The monitoring of biological phenomena through surface plasmon resonance or localized surface plasmon resonance is discussed by Bruno P. Crulhas, Caroline R. Basso, and Valber A. Pedrosa in their contribution about nanoplasmonic biosensors (Chapter 19). In addition to an introduction to the theoretical fundamentals, applications are treated such as cell analysis and drug delivery, the monitoring of cell environment, biomedical diagnostics, as well as aptamer-based localized surface plasmon resonance, and aptamer-based surface-enhanced Raman scattering biosensors. Chapter 20, by Marcos Pita, analyzes how enzyme biocomputing paradigms have been developed and how their complexity has grown from single logic operations to networks; furthermore, it discusses the evolution from model to applied systems and their interface with different materials to achieve applied systems and devices. Chapter 21, about functional nano-bioconjugates for targeted cellular uptake and specific nanoparticle–protein interactions, has been written by Sanjay Mathur, Shaista Ilyas, Laura Wortmann, Jasleen Kaur, and Isabel Gessner. The authors provide a concise overview and scope of "click chemistry" and their application for the conjugation of macro-biomolecules such as proteins and antibodies to nanoparticulate carriers with focus on gold and iron oxide nanoparticles as the most investigated examples for highly functional protein-conjugated nanoprobes. The authors of Chapter 22, Apurva Atak and Sanjeeva Srivastava, present an overview of the different approaches of cell-free expression–based microarrays, where surface plasmon resonance or carbon nanotubes are used as label-free detection techniques. Research in this field is among others of importance for studying the interaction of proteins with other proteins or small molecules and in connection with microarray-based functional nanoproteomics.

The concluding chapter, contributed by Ali Kermanizadeh, Kim Jantzen, Astrid Skovmand, Ana C. D. Gouveia, Nicklas R. Jacobsen, Vicki Stone, and Martin J. D. Clift, provides an overview of the current knowledge and challenges associated with human exposure to nanomaterials. It discusses nanomaterial-induced human health outcomes related to the respiratory, cardiovascular, central nervous, hepatic, and renal systems and the gastrointestinal tract.

Peter Grunwald

Chapter 1

Fundamentals of Nanotechnology

Torben Kodanek, Sara Sánchez Paradinas, Franziska Lübkemann, Dirk Dorfs, and Nadja C. Bigall

Leibniz Universität Hannover,
Institute of Physical Chemistry and Electrochemistry,
Callinstr. 3A, D-30167 Hannover, Germany

nadja.bigall@pci.uni-hannover.de, dirk.dorfs@pci.uni-hannover.de

1.1 Introduction

Nano—from the Greek word nannos meaning dwarf—originally is simply a prefix replacing the factor of 10^{-9} for SI units. However, in the last half of the twentieth century, this prefix has developed to become nearly a term for itself, now being frequently used when objects with nanometer-sized dimensions are meant. This is reasoned by the huge amount of novel materials and effects discovered on the nanometer scale (which is usually in the range of 0.2 to 100 nm), since objects of that size are close to the size of molecules, so that already molecular effects can come into play. While there are surely many valid definitions available nowadays for the terms nanotechnology and nanoscience, the ones from the UK Royal Society and Royal Academy of engineering of 2004 are widely recognized by many institutions, such as by the scientific committee on emerging and newly identified health

Biocatalysis and Nanotechnology
Edited by Peter Grunwald

ISBN 978-981-4613-69-9 (Hardcover), 978-1-315-19660-2 (eBook)
www.panstanford.com

risks (SCENIHR) of the European Commission. In these definitions, nanoscience is "the study of phenomena and manipulation of materials at atomic, molecular, and macromolecular scales, where properties differ significantly from those at a larger scale." Instead, the term nanotechnology is defined as "the design, characterization, production and application of structures, devices and systems by controlling shape and size at the nanoscale," and a nanoscaled object is considered to be one "having one or more dimensions of the order of 100 nm or less." Feynman's lecture "there is plenty of room at the bottom" is probably the most famous one about nanotechnology, and the last decades have proven that Feynman was absolutely right in that sense, since almost every day novel materials with exciting new properties are developed by scientists all over the world, and we are by far not yet at the end.

Taking the above-cited definitions into account, a huge variety of different effects and techniques can be thought of already, and the amount of examples increases daily. This chapter presents an introduction to nanoscience with a special emphasis on nanochemistry. In the first subchapter, an introduction to a variety of nanoscale properties will be given. Especially the large specific surface of nanomaterials, which can lead to reduced melting points of nanoscopic objects or also to enhanced catalytic activity, is described, as well as special physical effects such as the occurrence of superparamagnetism, localized surface plasmon resonances (LSPRs) or size quantization for magnetic, metallic, and semiconducting nanomaterials, respectively. Also, when nanosized objects are close to each other, interparticle interactions can occur, which can make the design of multifunctional nanoobjects a challenging task. However, we will show that this is not always just a drawback, but it can again lead to fascinating new physical effects. In the second subchapter, an introduction in the most common fabrication methods from nanotechnology will be given, such as colloidal, hydrothermal and sol-gel synthesis, ball milling, laser ablation, and a variety of lithographic techniques, including classical lithographic ones and nanochemistry-based ones such as polymer lithography. Generally, when getting in touch with nanotechnology, one frequently meets two different fabrication principles called "top-down" and "bottom-up." This classification dates back to several decades and means that one can either obtain nanoscaled objects by disintegrating a larger one (top-down), or by assembling smaller objects, atoms or molecules, e.g., by chemical reactions (bottom-up). However, a variety of techniques can be seen as a mixture between top-down and bottom-up. Here we will divide the strategies depending on the medium (environment) in which the nanoparticles are formed, having wet chemical, solid-state and gas-phase processes. After these two subchapters, a brief summary will be given.

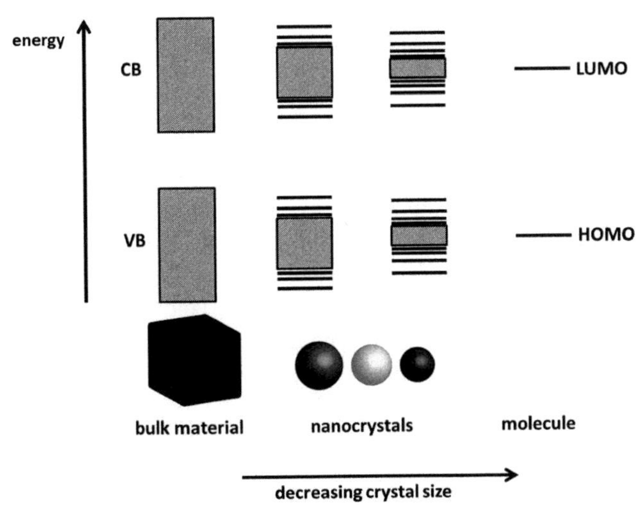

Figure 1.1 Schematic diagram of the size dependency of the energy levels in a semiconductor material. Note that the changes in conduction band (CB) and valence band (VB) are in most cases not symmetric but depend on the relative effective masses of electron and hole (see text for details).

1.2 Physical Effects on the Nanometer Scale

In the following, the most prominent special (mainly physical) effects occurring at the nanoscale will be briefly explained. Also, crucial references will be given so that the interested reader is assisted in finding important related reviews.

1.2.1 Electronic Properties

As it has been already mentioned in the introduction, nanocrystals represent a transition state from molecules to solid-state bodies.

This of course is also valid for their electronic behavior. How and to what extend the particle size influences the electronic structure of a nanocrystal of a given material first of all depends on whether this material is a metal, a (narrow or wide band gap) semiconductor or an insulating material.

As nanocrystals can be seen as intermediates between molecules and bulk materials, it is only logical that there are two fundamental different theoretical approaches to describe the electronic behavior of nanocrystals. They can either be described as large molecules, e.g., via (quantum mechanical) tight binding calculations or they can be described as small solid-state bodies using solid-state physics concepts like, e.g., the effective mass approximation. The first ones, though powerful approaches, are quite complicated and the interested reader is referred to the literature (see, e.g., Ramaniah et al., 1993, for tight binding calculations of CdSe quantum dots). Here we will focus mainly on solid-state physical approaches and discuss quantum mechanical effects only qualitatively and briefly summarized.

In solid-state physics, the charge carriers generated (e.g., by photo excitation) in a bulk crystal, namely electrons and defect electrons or holes, can be assigned a so-called effective mass which is nothing else but a measure for their acceleration in an electric field.

Analogue to a hydrogen atom, the positively charged hole and the negatively charged electron form a quasi-particle within a bulk crystal which is called exciton. The equilibrium radius of an exciton can be calculated also completely analogously to the Bohr radius of the hydrogen atom with the only difference that the effective masses of the charge carriers and the dielectric constant of the material instead of that of the vacuum have to be used:

$$a_0 = \frac{4\pi\varepsilon_0\hbar^2}{m_e e^2}$$

$$r_{\text{Bohr-Exciton}} = \frac{4\pi\varepsilon_0\varepsilon_r\hbar^2}{e^2}\left(\frac{1}{m_e^*}+\frac{1}{m_h^*}\right) = a_0\cdot\varepsilon_r\cdot m_e\left(\frac{1}{m_e^*}+\frac{1}{m_h^*}\right)$$

Equation 1.1 Bohr radius of the hydrogen atom a_0 and Bohr radius of an exciton $r_{\text{Bohr-Exciton}}$. m_e^* and m_h^* are the effective masses of the electron and the hole in the respective material, ε_r is the relative permittivity of the material.

It can be intuitively understood, that the electronic properties of the crystal change dramatically, when the radius of the whole crystal reaches the same order of magnitude as the exciton Bohr radius (which is larger for materials with low effective masses of the charge carriers and high relative permittivity according to Eq. 1.1 and is for typical semiconductor materials in the range of 2 to 30 nm).

Combining the solid-state physical approximation of the effective masses of the charge carriers with the quantum mechanical model of the particle in the box, the change in energy for each charge carrier can be derived as a function of the particle radius. This most simple way of a quantitative description of the size quantization effect is called the Brus equation (Brus, 1984).

$$E_{\text{gap}}(r) = E_{\text{gap(bulk)}} + \frac{h^2}{8r^2} \cdot \left(\frac{1}{m_{\text{e}}^*} + \frac{1}{m_{\text{h}}^*} \right) - \frac{1.8e^2}{4\pi\varepsilon_{\text{r}}\varepsilon_0 r}$$

Equation 1.2 Brus equation for calculating the energy band gap E_{gap} of a quantum dot as a function of its radius (r). Necessary material constants are the effective masses of the charge carriers (m_{e}^* and m_{h}^*) in the material, the relative permittivity (ε_{r}) of the material and the bulk band gap of the material ($E_{\text{gap(bulk)}}$).

The Brus equation in this form contains one term derived from the particle in the box model (for both charge carriers) and an additional term which is a correction term for the attractive Coulomb interaction between the two charge carriers.

Figure 1.1 schematically shows the changes in the energy levels from a bulk semiconductor towards a nanometer-sized semiconductor nanocrystal (a quantum dot). We see two main observations: (i) the band gap increases with decreasing particle size (note that the shift in valence band edge and conduction band edge are not necessarily symmetric but depend on the relative effective masses of the two charge carriers) and (ii) the continuous bands develop more and more into well-defined isolated states (this effect is especially pronounced for states near the band gap).

The aforementioned thoughts are mainly valid for semiconducting materials but can be easily transferred also to insulating materials. In insulating materials typically the Bohr exciton radii are much smaller than in semiconductor materials and can even reach down to typical interatomic distances.

Therefore size quantization is much weaker pronounced or even absent in typical insulating materials, and their electronic structure stays more or less unaltered when compared to the bulk material (nevertheless, their physical properties can differ significantly from the bulk material, e.g., due to surface effects). Metals, on the other hand, do not have any band gap in their bulk modification; nevertheless, especially very small metal clusters (typically less than 50 atoms) can show drastic size quantization and undergo a so-called metal-to-semiconductor-transition, which means that these small clusters lose their metallic properties and become semiconducting (some of them even show fluorescence, which is normally a typical phenomenon for semiconducting nanoparticles but not for metallic particles).

Our considerations so far were derived for the approximation of spherical nanocrystals. Even though an in-depth discussion of the electronic properties of anisotropic shapes of nanocrystals would exceed the scope of this chapter, we, nevertheless, would like to give a very short overview and some nomenclature for nanocrystals with an anisotropic shape.

For semiconductor materials, 0D (that is an object with quantum confinement in all three dimensions) quasi-spherical nanocrystals are commonly referred to as quantum dots. For 1D (meaning objects with quantum confinement in only two dimensions) structures, the terms quantum wire and quantum rod can be found. The term quantum wire typically refers to structures which are in 1D that long that no quantization effect in this direction can be observed, while the term quantum rod is typically used for structures which are much longer in one dimension than in the other two dimensions but still show slight quantum confinement in all three dimensions. 2D structures (that is objects with quantum confinement in only one dimension) are typically referred to as quantum wells.

One interesting finding on all these structures is that the band gap (as calculated from the Brus equation) is mainly dependent on the strongest confined dimension. The exact electronic structure (degeneracy at the band edges as well as density of states within the bands) however can become a complicated function of the particle dimensions. The electronic properties of differently sized quantum dots can be directly seen, e.g., in their absorption spectra (see Fig. 1.2).

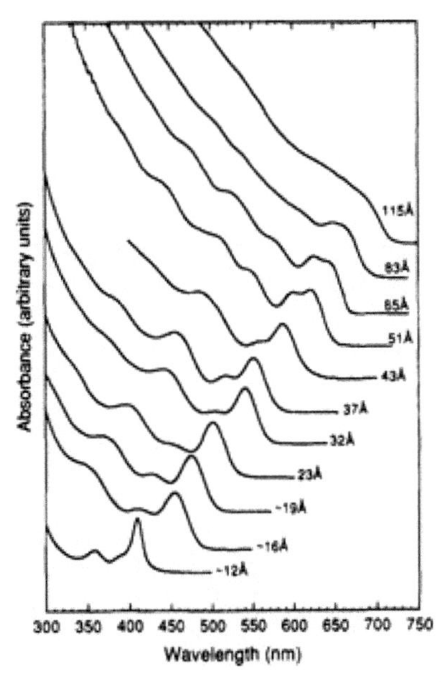

Figure 1.2 Absorption spectra of differently sized CdSe quantum dots (1.2–11.5 nm). Reprinted (adapted) with permission from *J. Am. Chem. Soc.*, **115**, (1993), pp. 8706–8715. Copyright American Chemical Society (Murray et al., 1993).

One intriguing property of semiconductor nanocrystals which is not observed in their bulk counterparts is an efficient fluorescence (see Fig. 1.3). This is caused by the strong confinement of the charge carriers to one spatial region given by the size of the quantum dot which significantly enhances the probability of a radiative recombination of the charge carriers (opposed to the behavior in bulk in which charge carriers can easily separate, and therefore radiative recombination is usually so slow that non-radiative processes occur with a much higher probability). In quantum dots, on the other hand, the fluorescence properties are mainly dominated by the particle size and by the surface properties of the nanocrystal.

Especially non-saturated surface atoms with free electron pairs or electron vacancies can act as very efficient electron or hole "traps" which means nothing else but that they represent spatially localized states within the band gap which can trap the free charge carriers generated via photoexcitation at the particle surface. Usually the relaxation of trapped charge carriers occurs non-radiatively. Therefore, the so-called fluorescence quantum yield (which is the ratio of emitted to absorbed photons) is the higher, the better the particle surface is passivated either with molecular ligands or with an insulating inorganic shell material.

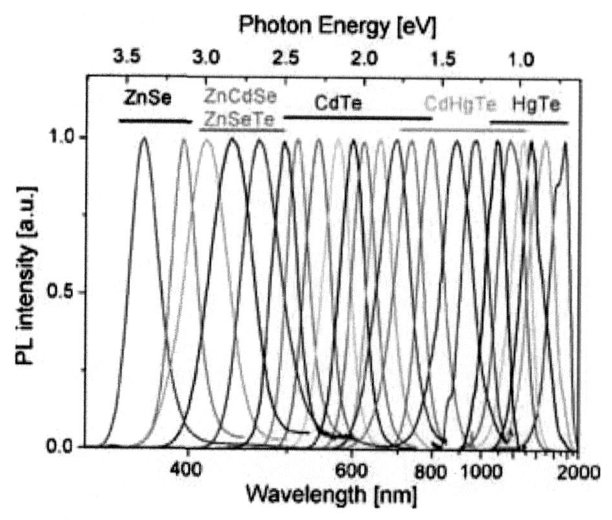

Figure 1.3 Emission spectra of nanocrystals of different sizes and different materials. The emission range covers the full spectral range from the UV/Vis throughout the visible to the near infrared part of the spectrum. Reproduced ("in part") from Gaponik (2010), with permission of The Royal Society of Chemistry.

1.2.2 Surface-Related Properties of Nanocrystals

Another property which is inherent to any nanoparticle is the very high surface to volume ratio compared to bulk materials. Depending on the way a specific nanocrystal is produced, its surface structure can be quite different. While particles obtained from gas-phase methods like molecular beam epitaxy (MBE) or alike typically have ligand free surface, particles synthesized via solution based chemistry often carry some sort of passivating ligands on the particle surface. Both constellations can have specific advantages and disadvantages (also depending on the actual nanomaterial of course). Ligand free particles are typically more sensitive, e.g., to surface oxidation under ambient conditions and might have so-called "dangling bonds" (e.g., lone electron pairs, unpaired electrons or electron deficiencies, see also "trap states" in the paragraph about fluorescence above) which might significantly affect their reactivity. In contrast, ligand covered particles naturally have a better protected and passivated surface, which can of course create other problems like a higher electronic resistance to the ligand shell or simply a chemically not accessible surface reducing the catalytic activity. Due to high surface energy of nanocrystals, these crystals also typically show a strongly size dependent melting point which decreases with decreasing particle size. The melting point of, e.g., gold nanoparticles drops drastically from the bulk value of 1064°C when particles are getting smaller than 10 nm in diameters. For particles around 3 nm diameter, the melting temperature can be as low as ca. 400°C (Buffat and Borel, 1976).

1.2.3 Magnetism-Superparamagnetism

Superparamagnetism is an effect which can occur for ferromagnetic objects at the nanoscale. An overview about this phenomenon will be given in this paragraph. When decreasing a ferromagnetic particle's size more and more (see Fig. 1.4), the number of magnetic domains also decreases until the particle contains a single domain only (that is, the small particle is a fully magnetized permanent magnet with its magnetization being the saturation magnetization all the time). Permanent magnetic particles are known to assemble into chains since each particle interacts with the field of the neighboring particles. This can easily lead to agglomeration effects when happening with freely mobile particles, e.g., in suspensions. For such a single domain permanent magnet particle, in order to change the direction of magnetization, which is usually oriented along the so-called easy axis of the crystal, the anisotropy energy of the magnetic particle has to be overcome (since the magnetization direction cannot be changed by moving domain walls as it could be if more than one domain were present). The anisotropy energy depends both on the crystal type (material) and on the volume of the particle. Therefore, when further decreasing the particle size, less and less energy is needed to reorient the magnetization direction inside the small crystal. When the nanoparticle becomes small enough, the anisotropy energy is that small, that it can be overcome by the thermal energy. If that happens for example at room temperature, the magnetization direction is being statistically reoriented permanently, and that means that the particle would be no permanent magnet anymore. Such a nanoparticle is called superparamagnetic, since its magnetization direction can be reoriented by external magnetic fields, and since in absence of an external magnetic field no magnetization of such a superparamagnetic nanoparticle ensemble can be detected externally. However, when looking at the saturation magnetization of a superparamagnetic particle in an external magnetic field, its order of magnitude is similar to the saturation magnetization of a ferromagnetic particle and hence much larger than that of a paramagnetic particle. However, similarly to paramagnetism, each particle acts independently from the ensemble. Since superparamagnetic nanoparticles do not exhibit permanent magnetic fields, they can display stable suspensions, a phenomenon which can for example be observed in ferrofluids. To state it very simple, superparamagnetic particles behave similar to ferromagnetic particles as long as an external magnetic field is dominant, however they never become permanent magnets themselves.

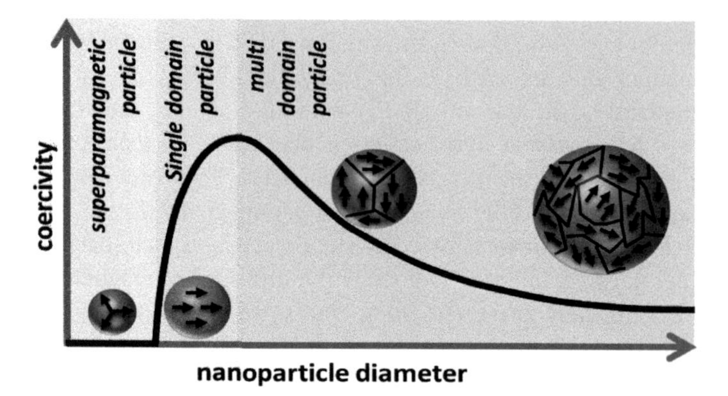

Figure 1.4 Schematic drawing of the coercivity of ferromagnetic particles of different sizes.

While the thermal instability of the magnetization direction in magnetic nanoparticles can be problematic for data storage units, superparamagnetic nanoparticles can be employed for a variety of other applications and display an important branch of nanomaterials nowadays. Especially for solutions and dispersions of nanoparticles, it can be very advantageous that by employing external magnetic fields the colloidal nanoparticles can be manipulated (reoriented, accelerated, or reoriented manifold by alternating fields so that heat is released to the surroundings). Examples are ferrofluidic ones (Odenbach, 2004) (e.g., in loud speakers), and in biomedicine, such as magnetofection, magnetically guided drug delivery, cell sorting, hyperthermia and as contrast agents for magnetic resonance imaging. (Gupta and Gupta, 2005; Pankhurst et al., 2003). Currently, many researchers work on combining superparamagnetism with other physical and chemical properties to yield, e.g., multifunctional nanoobjects (nanovectors) suitable for a variety of diagnosis, treatment and signaling methods simultaneously, which, e.g., is highly desirable for cancer therapy. Furthermore, a variety of superparamagnetic nanoparticles exhibit also catalytic properties, and the combination of both can also lead to interesting applications, e.g., if the catalyst can be separated from the reaction solution by external magnetic fields. Examples for such nanocatalytic and at the same time magnetic materials are Fe_xO_y (Lesin et al., 2012), Co, (Lewis et al., 2012), and FePt (Khalid et al., 2011).

When superparamagnetic nanoparticles approach, the magnetization of each particle can be influenced more and more by that of its neighboring nanoparticles so that interparticle interactions can occur. These interparticle interactions are strongly dependent on the distance between the nanoparticles and can lead to interesting physical coupling phenomena, as was shown for a variety of supracrystals (ordered three-dimensional assemblies of nanoparticles) already (Bigall et al., 2013).

1.2.4 Localized Surface Plasmon Resonances

Nanoparticles with a high charge carrier density are frequently found to enable the presence of so-called LSPRs energetically in the ultraviolet, visible or near infrared range of the electromagnetic spectrum. Since nanoparticles can be significantly smaller than the wavelength of light, these LSPRs can be excited upon irradiation. Briefly, such a LSPR can be seen as a resonance of a collective vibration of quasi free charge carriers. Also in a bulk material, under certain circumstances light can excite a surface plasmon; in this case the plasmon propagates along the surface of the bulk until dissipation. Instead, for nanoparticles, two more effects happen: firstly, since the nanoparticle size is smaller or comparable to the penetration depth of such a surface plasmon, the LSPR penetrates through the whole particle. Secondly, due to the limited size of the nanoparticles, standing waves occur. Most notably, in most of the cases such a LSPR can be described classically by combining the Drude model for free charge carriers and the Mie theory for the interactions of an electromagnetic field with a particle. This means, the occurrence of LSPRs in a first approximation is not a quantum mechanical effect, even though it can be only observed for nanoscaled materials. Nevertheless, the occurrence of LSPRs is one of the most exciting properties of nanotechnology, since it leads to many interesting effects, techniques and applications. The probably most prominent example is the typical red color of old glasses and church windows, which is frequently caused by the absorbance of visible light of gold nanoparticles which are included in the glasses. In the year 1857, Faraday was the first to synthesize such gold nanoparticles in aqueous solution, and most notably this solution is (more than 150 years later!) still colloidally stable to date and can be marveled in the Faraday museum in London (Thompson, 2007). Originally, LSPRs are famous in metal nanoparticles. Here, gold and silver are the most prominent materials since already small nanoparticles of them exhibit the typical red and yellow colors in ensemble, respectively. It has also been shown that for larger noble metal nanoparticle sizes, also other metals such as platinum exhibit LSPRs in the UV/Vis regime of the electromagnetic spectrum (Bigall et al., 2013). Furthermore, in the recent past, LSPRs have been observed to also occur in degenerately doped semiconductor systems. Since the resonance frequency strongly depends on the charge carrier densities, in such materials the LSPR wavelength is in the near infrared of the spectrum. Prominent examples for such plasmonic semiconductor materials are $Cu_{2-x}Se$, CuS, and ITO, among many more (Dorfs et al., 2011; Xie et al., 2013).

With plasmonic nanoparticles, many interesting physical effects can be observed. First, as long as the nanoparticles are small in comparison to the wavelength of light, no significant dependency of the energetic position of the LSPR occurs (in this effect, the so-called quasi static approximation can be employed for calculations). Deviations from this rule are observed mainly for very tiny nanoparticles, for which the dielectric function becomes a function of the size, and for particles much larger than the wavelength of the irradiating light, for which charge carriers distributed within the NP volume experience different field accelerations (which means that the quasi static approximation cannot be employed any more). Second, instead, the LSPR position is strongly influenced by the dielectric function of the environment. For example, if the surrounding of the nanoparticles is changed towards higher refractive indexes, their absorption and scattering spectra are shifted towards larger wavelengths. This effect is so strong, that plasmonic nanoparticles have many times been suggested as nanoscopic sensor units for the detection of a variety of molecules. Third, the absorption and scattering spectra of plasmonic nanoparticles are critically influenced by their shape. While as mentioned above, the absolute size of the nanoparticles plays an inferior role to the shape of their absorption and scattering spectra, instead, factors such as aspect ratios (see Fig. 1.5) and tip curvatures are very important for them. Hence, for nanoparticles with complex shapes, frequently more than one LSPR are observed, which can be correlated, e.g., to longitudinal and transversal vibration modes in addition to multipole vibrations in some cases. As a result, the extinction spectra of nanorods are inherently different from those of spherical nanoparticles, nanoprisms, nanocubes, or nanostars, etc. (Nikoobakht and El-Sayed, 2003).

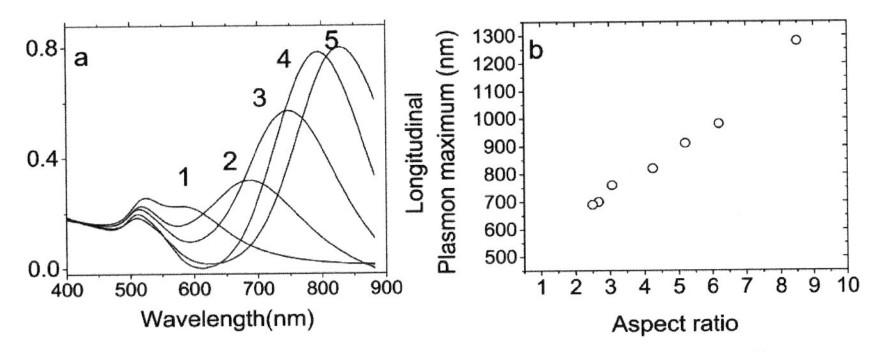

Figure 1.5 (left) Absorption spectra of gold nanorods with five different aspect ratios. (right) Dependence of the longitudinal plasmon modes spectral position on the aspect ratio. Note that spherical gold particles would always exhibit only one LSPR at around 520 nm (not shown). Reprinted (adapted) with permission from *Chem. Mater.*, 2003, 15(10), pp. 1957–1962. Copyright (2003) American Chemical Society (Nikoobakht and El-Sayed, 2003).

1.2.5 Interactions between Two Nanoparticles

Of course, all of the above-mentioned nanoparticles (metallic, magnetic, semiconducting) can interact with one another when two particles come close to each other (e.g., in all sorts of particle assemblies, like films or gels). Examples of the most important interactions are discussed very briefly in the following.

The main interactions between fluorescent quantum dots in close proximity is the so-called Förster (or Fluorescence) resonant energy transfer (FRET) which occurs between a donor and an acceptor in very close proximity (few nm). Its main characteristic is the very strong distance dependence making it the dominant mechanism for energy transfer for very short distances of nanoparticles but basically negligible for distances larger than 10 nm. For FRET to occur the most important necessity is an overlap of the emission spectrum of the donor and the absorption spectrum of the acceptor (nevertheless, it is a non-radiative energy transfer mediated purely by dipol/dipol interactions). According to the previously explained size quantization effect, this condition is, e.g., always fulfilled for FRET from a smaller quantum dot to a larger quantum dot made of the same material. For details, the interested reader is referred to the literature (Franzl et al., 2004).

Interactions between two metallic nanoparticles can also be of complicated demeanor especially when looking at plasmonic particles in close proximity to one another. In this case the coupling between the two plasmon resonances becomes that strong that the LSPR frequency shifts dramatically with respect to that of the non-interacting particles. Fascinatingly, this shift of the LSPR frequency is that distance sensitive that it can be used, e.g., as "molecular ruler" to determine the distance of, e.g., Ag or Au nanoparticle with sub-nanometer resolution by simple measuring scattering spectra in a confocal microscope (Sönnichsen et al., 2005).

Interactions between plasmonic and semiconducting nanoparticles are also very complicated and can result in almost any result between strong fluorescence quenching and extreme fluorescence enhancement, mainly depending on the actual distance and geometry of the nanoparticle assembly. Again the interested reader is referred to the literature (Bigall et al., 2012).

1.2.6 Interactions within Nanoheterostructures

A so-called nanoheterostructure is a particle comprising two or more compartments of different materials (e.g., core–shell-type structures but also janus-like or more complicated geometries like seeded rods, etc.). In structures comprising two semiconductor materials, the effects observed mainly depend on the size and relative positions (relative to vacuum level) of the band gaps of the two materials to one another. Depending on these two parameters, photoexcited charge carriers are either confined in one compartment of the heterostructure (type-I heterostructure) or they are separated in the two compartments (type-II heterostructure). The fluorescence properties change accordingly (e.g., higher fluorescence quantum yields in type-I structure, high fluorescence lifetimes in type-II structures) (Gaponik et al., 2010).

Another fundamentally important process occurs in semiconductor metal/hybrid nanoparticles. In this type of heterostructures in the photoexcited state, charge carriers (electrons or holes, depending on the relative alignment of the fermi level of the two components) generated in the semiconductor part of the heterostructure can be very quickly (ps time scale) transferred to the metal compartment. Hence such type of metal/semiconductor heterostructure on the one hand typically loses all its fluorescence, but instead can be highly beneficial in applications where separated charge carriers are required (e.g., photocatalytic water splitting and other photocatalytic reactions) (Costi et al., 2010).

1.2.7 Summary (Physical Effects on the Nanometer Scale)

Summarizing, the variety of nanoscopic effects ranges from enhanced catalytic activities due to the extremely large specific surfaces, to strongly reduced melting points of special materials. Localized surface plasmon resonances, superparamagnetism, and size quantization display fascinating examples for the manifold novel phyiscochemical effects on the nanoscale. While interparticle interactions, especially in nanoparticle assembly architectures, many times need to be controlled, their proximity can lead to further interesting phenomena. A careful exploitation of all those fascinating properties may help developing advantageous next generation devices such as sensors, solar cells and LEDs and biocatalyst platforms. In the next subchapter, common techniques for the synthesis/fabrication of nanoscopic structures will be introduced.

1.3 Nanofabrication

Over the past years, a large number of different techniques to produce nanoparticles ranging from simple mechanical milling procedures to complex chemical reaction routes have been developed. According to the nanoparticle requirements concerning their properties, an appropriate method has to be chosen. For example, if only the particle size matters, then milling approaches present an easy and fast way to obtain nanocrystals. On the other hand, crystalline nanoparticles of high quality can be achieved by solvothermal synthesis. Hence, every synthesis method has its legitimacy. Nonetheless, the production of nanoparticles is a challenging area, since several factors have to be considered to obtain the desired nanoparticles. In case of oxidizable materials, for instance, working under inert conditions is necessary to prevent the oxidation of the nanoparticles during their formation.

In the early stage of nanotechnology, several books have subdivided the synthesis techniques into two classes, namely "bottom-up" and "top-down" approaches. In bottom-up methods, nanoparticles are synthesized starting from atoms, molecules or clusters. As opposed to this, top-down approaches describe the preparation of nanoparticles from a bulk piece of the corresponding material. Nowadays, this subdivision is obsolete, since some techniques, e.g., lithography, represent a mixture of both approaches. The synthetic methods can also be grouped according to the growth media (Cao and Wang, 2004). In this manner, it can be distinguished between

- vapor-phase growth, including, e.g., laser pyrolysis and atomic layer deposition;
- liquid-phase growth, basically colloidal synthesis and electrochemical methods;
- solid-phase formation, such as ball milling;
- hybrid growth, for example, liquid-vapor-solid (VLS) growth of nanowires.

Another possible classification of the synthetic approach is as chemical and physical processes. Here, chemical methods would mostly overlap with the bottom-up approaches, while most of the physical methods would be based on top-down techniques. Physical methods include laser ablation, pulsed wire discharge and mechanical milling. In contrast, solvothermal synthesis, sol-gel methods as well as hot-injection approaches, for example, belong to the chemical route.

The design of nanoparticles for specific applications with the desired nanoscopic properties requires close monitoring of the synthesis conditions. At present, a huge variety of methods for producing uniform nanoparticles with narrow size distribution has been developed. However, reproducibility and further upscaling, parameters of high importance for envisioned applications, are not always easy to implement. By carefully choosing the reaction conditions for the respective synthesis technique, nanoparticles of high quality can be obtained with a high degree of control over their properties, which will determine their scope: catalysis (Tao, 2014), energy storage (Arico et al., 2005; Zhang et al., 2013a), imaging (Bogart et al., 2014), drug delivery (Tang et al., 2012), etc. This chapter will highlight common methodologies for the synthesis of nanoparticles and their advantages and disadvantages, pointing out some of the most recent examples.

1.3.1 Wet Chemical Preparation

1.3.1.1 Hydro-/solvothermal synthesis

The first preparation technique, which will be introduced in this chapter, is the hydro-/solvothermal approach. Generally, this method includes all synthesis procedures, which are carried out at high temperatures and pressures (Demazeau, 2008). Here, a reaction mixture consisting of a solvent, precursors as well as other additives is tightly sealed in a pressure-resistant vessel called autoclave and heated above the boiling point of the employed solvent (Cushing et al., 2004). For instance, in case of aqueous solutions those syntheses are usually performed at temperatures above 100°C (Daou et al., 2006; Chiu and Yeh, 2007). As a result of the heating, the pressure rises inside the autoclave. In this regard, the autogenous pressure can be adjusted, in addition to the temperature, by the filling level of the autoclave by working under supercritical conditions. The benefit of these reaction conditions becomes apparent in the fact that the precursors dissolve more easily than under ambient conditions. In turn, this results in a speed-up of the reaction. Depending on the employed solvent, it can be distinguished between hydrothermal and solvothermal synthesis. While, in the first case water serves as solvent, the term solvothermal refers to all other media (Cushing et al., 2004).

Besides the preparation of spherical nanoparticles (Daou et al., 2006; Yang et al., 2007), the hydro-/solvothermal method also offers the possibility to produce anisotropic nanostructures. By adjusting the reaction conditions (e.g., temperature, addition of surfactants, etc.), the growth kinetics can be controlled leading to different shapes, e.g., nanowires (Tang et al., 2003), nanocubes (Cheng et al., 2009) or nanotubes (Zhuo et al., 2009). For instance, Yang et al. (2007) have synthesized various silver nanostructures by changing the temperature, the reaction time and/or the precursor to surfactant ratio. Furthermore, it has been shown, that the solvent is able to influence the morphology of the nanoparticles as well. In this context, CdS nanorods or nanodiscs have been obtained by using either ethylenediamine or pyridine as solvent (Tang et al., 2003). However, general drawbacks of this approach become noticeable in the poor monitoring of the synthesis as well as the limited flexibility regarding precursor additions during the reaction. Moreover, difficulties often appear in upscaling the synthesis.

1.3.1.2 Microwave-assisted synthesis

Over the past years, the significance of microwave-assisted synthesis has increased continuously. Thus, it is not surprising, that this technique describes a well-established method to produce different kinds of nanoparticles these days, which is summarized in a review of Bilecka and Niederberger (2010). The fundamental principle is based on the resonant interaction between the oscillating electric field induced by microwaves and an ionic or dipolar component. As a consequence of this interaction, dipoles or ions permanently attempt to reorient themselves according to the oscillating electric field leading to heating due to dielectric losses or collisions (Bilecka and Niederberger, 2010). Since microwaves induce an efficient as well as fast internal heating reducing the reaction time, this technique has been used as an attractive alternative for the synthesis of nanoscale materials. Due to non-homogeneous heating when using microwaves, temperature gradients in the reaction medium can be avoided by, e.g., rotating the reaction flask and thus a more homogeneous environment for nucleation and growth processes is provided. Microwave-assisted syntheses are able to produce nanoparticles with narrow size distribution, although it is not easy to precisely control their shape. In this way, silver nanoparticles (Fuku et al., 2013), carbon nanomaterials (Schwenke et al., 2015), iron oxide nanoparticles (Wang et al., 2007), among others have been prepared. In contrast to conventional heating methods, this technique additionally offers the possibility to selectively heat specific species in the reaction flask, which gives access to different kinds of heterostructures. For instance, core–shell nanostructures can be produced in this manner using preformed nanoparticles as seeds (Ziegler et al., 2007).

Besides, microwaves are often combined with hydro-/solvothermal synthesis, which are gaining a lot of interest due to several advantages like rapidity, efficient and volumetric heating, homogeneity, fast kinetics, less energy requirements, etc. (Bhosale et al., 2015; Zou et al., 2015). Nonetheless, similar drawbacks as in the hydro-/solvothermal approach regarding the monitoring and the upscaling occur. Some insights in microwave-assisted techniques for the synthesis of well-defined nanomaterials and nanocatalysts are listed in a recent review of Gawande et al. (2014).

1.3.1.3 Aqueous colloidal synthesis

A distinction is made between two types of synthesis: chemical reduction and chemical precipitation to produce colloidal nanoparticles in aqueous solution.

In 1857, M. Faraday reported the first method for the synthesis of metallic gold nanoparticles by chemical reduction (Faraday, 1857). Nowadays, reduction of metal precursors in solution is still a frequently employed route to produce colloidal metal nanoparticles. Typically, this method consists of dissolving a salt from the precursor, a reducing agent and a stabilizing agent in the same liquid phase. The mean size, the size distribution and the shape of the resulting nanoparticles can be controlled by varying the concentration and the type of the reagents, the reducing agent and the stabilizing agent, as well as the nature of the dispersing medium. Furthermore, this method provides dispersions stable for long periods of time (see also the gold colloid of Faraday mentioned in the introduction of this subchapter, dispersion which remains stable until today). In the early 50s of the last century, Turkevich et al. (1951) reported a method for the synthesis of metallic colloids. It is a standard and reproducible method for the preparation of spherical 20 nm gold nanoparticles based on the reduction of chloroauric acid ($AuCl_4$) with sodium citrate. In this case, the citrate ions act as both the reducing and capping agent. Additionally, in this work a mechanism for the formation of nanoclusters based on nucleation and growth was proposed. According to the commonly accepted mechanism (Goia and Matijevic, 1998), the metal atoms generated during the reduction process aggregate forming clusters (also termed embryos) which further grow becoming nuclei. The most frequently used metal precursors in the aqueous chemical reduction are nitrates, acetates, chlorates and citrates, prevailing in each case the lower cost and more chemically stable metal salts or complexes. Among the reducing agents, sodium borohydride is one of the most often used inorganic compounds. For the synthesis of transition metals which do not crystallize well at room temperature (e.g., Co and Ni), an alternative method at high temperatures can be applied. Here, the reduction of metal salts in the presence of stabilizing agents takes places at high temperatures by the injection of a strong reducing agent (e.g., $LiHB(CH_2CH_3)_3$) (Sun and Murray, 1999). More information about metal colloids and the advantages of metal colloid synthesis can be found in several reviews and books (Schmid, 1992, 1994; Alexander and Shlomo, 2010).

Chemical precipitation can also be used for the synthesis of colloidal nanoparticles in aqueous solution. The precipitation usually takes place by changing the solubility of a substance in the solution by adding a precipitation agent, pressure- and temperature change, solvent evaporation or changing the polarity of the solvent by changing the pH value. After precipitation the solid phase is separated by centrifugation. These methodologies are commonly used to synthesize semiconductor nanoparticles like CdS (Spanhel et al., 1987), CdTe (Gaponik et al., 2002), ZnSe (Shavel et al., 2004), and many more.

1.3.1.4 Organic colloidal synthesis

The most common synthesis routes in organic media are the hot-injection and heating-up approaches, which will be discussed in more detail.

The hot-injection approach represents a clear change with respect to water-based synthetic methods. Here, generally, coordinating or non-coordinating mixtures of long chained hydrocarbons with high boiling points and organometallic compounds as precursors are used. Typically, hot-injection methods are characterized by a rapid nucleation, quenched by the immediate cooling of the reaction mixture. Moreover, nucleation and growth take place in separate periods of time, so that nanocrystals with narrow size distribution can be achieved (Yin and Alivisatos, 2005). The general procedure of this method consists in the rapid injection of a solution containing the organometallic precursor into a hot solvent. Due to the high temperatures, the precursors convert into highly reactive species called monomers, whose transient supersaturation leads to the instantaneous formation of nuclei. Accordingly, this process is called nucleation. The decrease of the monomer concentration as well as the reduced temperature, resulting from the injection of a "cold" solution, subsequently prevents a further nucleation. Instead, the nuclei start to grow into mature nanoparticles by the incorporation of monomers, which are still present in the reaction media (Yin and Alivisatos, 2005; de Mello Donegá et al., 2005). In this context, the growth kinetics depends on several factors such as temperature, type of the coordinating solvent and co-surfactants. Since the monomers exhibit high reactivities, reactions are usually carried out under inert gas atmosphere using Schlenk techniques (see Fig. 1.6A). After the synthesis, the nanoparticles can be precipitated by the addition of a non-solvent changing the polarity and redissolved in an appropriate organic solvent. Typically, long-chain alkylphosphines R_3P (e.g., TOP), alkylphosphine oxides R_3PO (e.g., TOPO), alkylamines (e.g., hexadecylamine), or their mixtures are employed as high-boiling coordinating solvents for the hot-injection approach (Murray et al., 1993). The resulting nanoparticles form a stable colloidal suspension sterically stabilized by ligand molecules attached to their surface. In 1993, Murray, Norris and Bawendi published a pioneer work about the chemical synthesis of semiconductor nanoparticles of cadmium chalcogenides based on the use of organometallic precursor (Cd, Se) in a high-boiling apolar coordinating solvent (TOPO) (Murray et al., 1993). Since that time, the hot-injection organometallic synthesis method has evolved and has been adapted to prepare nanoparticles from II-VI, IV-VI and III-V semiconductors (Čapek et al., 2009; Manna et al., 2000; Carbone et al., 2007) and was also carried out in non-coordinating solvent like 1-octadecence with added coordinating ligands (Yu and Peng, 2002). Some of these modifications, concerning the nucleation and growth temperature, the use of organic capping agents, the growth of inorganic shell structures, etc., aim to vary the size and shape of the nanoparticles, while improving the surface properties, the size distribution and luminescent properties (see Fig. 1.6 C–E). As previously mentioned, the nanocrystals obtained after such hot-injection approaches are stabilized with organic ligands, which is why these particles are only soluble in organic media. Hence, several post-preparative methodologies have been developed to render them water-soluble or biocompatible, such as ligand exchange (Bagaria et al., 2006), polymer coating (Pellegrino et al., 2004) or encapsulation into hydrophilic shells (Jana et al., 2007). Summarizing, hot-injection methods represent a versatile way to prepare defect-free, well-passivated nanocrystals with highly luminescent optical properties and narrow size and shape distributions. Among the disadvantages, the use of expensive and sometimes highly toxic organic precursors as well as the hydrophobicity of the as-synthesized particles can be mentioned.

The heating-up method (also known as "non-injection" method) represents an alternative to the aforementioned hot-injection technique. This method allows the synthesis of high quality nanomaterials on large scale with small or no variation between the batches. The general synthesis route consists of mixing all reagents into a reaction vessel at room temperature and increasing its temperature. The reaction solution contains the precursors, stabilizer and the solvent. When increasing the reaction temperature the precursors experience an increased thermodynamic driving force to form monomers, which leads to the instantaneous formation of nuclei. At higher temperature the nuclei form nanocrystals. The underlying mechanism is, in principle, similar to the hot-injection methods. Heating-up methods allow for the synthesis of many nanocrystals (see Fig. 1.6 B, F, G), such as binary semiconductor nanocrystals, ternary and quaternary nanocrystals, metal and metal oxide nanocrystals, all of them collected in literature in a recent review (van Embden et al., 2015).

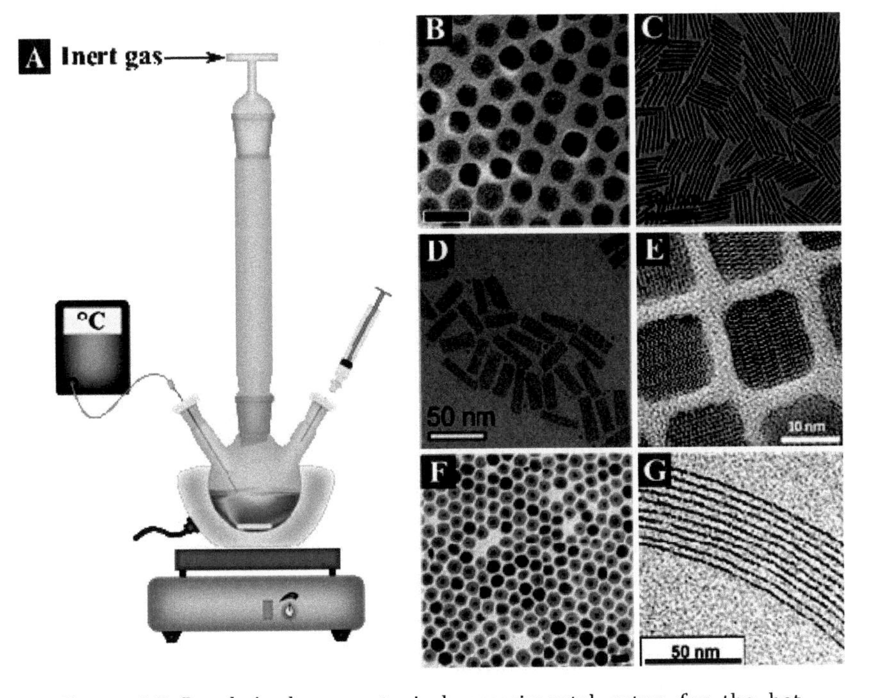

Figure 1.6 Panel A shows a typical experimental setup for the hot-injection approach. Generally, a precursor is rapidly injected into an organic solution at high temperature. Panels B-G illustrate transmission electron microscope images of nanoparticles prepared by the hot-injection as well as heating-up method: iron oxide nanoparticles (B) [adapted with permission from Soon et al., 2007. Copyright 2007 American Chemical Society], CdSe@CdS dot-in-rod nanocrystals (C) [adapted with permission from Hill et al., 2014. Published by The Royal Society of Chemistry], CdSe nanoplatelets (D) [adapted with permission from Bouet et al., 2013. Copyright 2013 American Chemical Society], CuTe nanocubes (E) [adapted with permission from Li et al., 2013. Copyright 2013 American Chemical Society], Au@Fe_3O_4 core–shell nanocrystals (F) [adapted with permission from Shi et al., 2006. Copyright 2006 American Chemical Society], Au nanowires (G) [adapted with permission from Pazos-Pérez et al., 2008. Copyright 2008 American Chemical Society].

1.3.1.5 Sol-gel methods

The sol-gel method has been used in recent years to prepare a wide variety of nanostructured materials. The method is attractive because it involves low temperature processes. The sol-gel method is a suitable wet chemical route for the synthesis of nanostructured metal oxides or polymers. Aqueous and organic solvents can be used to dissolve the precursors (metal alkoxides and metal chlorides) which after hydrolysis and condensation reactions give rise to the final product with sizes ranging commonly from 1 to 100 nm.

Two stages can be distinguished in the sol-gel process; the first one is the formation of a colloidal suspension of particles in a liquid medium that is referred to as the sol. In a second step, the particles react with each other forming a cross-linked 3D network with sub-micrometer-sized pores and polymeric chains known as a gel. More details about the process, advantages and disadvantages, as well as applications can be found in a book from A. C. Pierre (1998). Silica nanoparticles are the most famous example of nanoparticles prepared by sol-gel methods (Singh et al., 2014; Kessler et al., 2006). And among the routes used for the synthesis of silica nanoparticles, the most relevant one is the process developed by Werner Stöber in 1968 (Stöber et al., 1968). Examples of other materials can be found in the literature for semiconductor CuO nanoparticles (Jayaprakash et al., 2015), magnetic-silica composites (Mao et al., 2014), and many more.

Sol-gel methodologies can also be extended to the fabrication of non-ordered assemblies named gels. As defined by Hüsing and Schubert (1998), *a gel consists of a sponge like, three-dimensional solid network whose pores are filled with another substance.* After the two stages already mentioned of the sol-gel process (hydrolysis and condensation of, e.g., metal or semimetal alkoxides), the pores in the gel are filled with water and/or alcohols and therefore, the gels are called aquagels, hydrogels, or alcogels. Going beyond, the liquid in the pores can be replaced by air, obtaining aerogels (when the pore liquid is removed by supercritical drying), xerogels (when the pore liquid evaporates under normal atmospheric conditions) or lyophilized aerogels (when the pore liquid is removed by freeze-drying). The design of the properties and applications of this type of assemblies is a field of growing interest in which nanocrystals are playing an increasing role. Nanocrystals can be used in the fabrication of gels and they may conserve their nanoscopic properties when assembled, obtaining solid-state macroscopic architectures that present, e.g., the quantum size effect (Bigall et al., 2009; Sánchez-Paradinas et al., 2015; Hendel et al., 2013; Mohanan et al., 2005).

1.3.1.6 Dendrimers and micelles

Dendrimers have been used as nanoreactors for the preparation of nanoparticles with defined sizes and shapes. Dendrimers are highly branched macromolecules with nanometer-scale dimensions that consist basically of a central core, intermediary repetitive units (the branches), and functional surface groups (Vögtle et al., 2009; Fischer and Vögtle, 1999). Dendrimers have attracted attention because of their well-defined structures and chemical versatility, which make them an interesting instrument for the preparation of nanoparticles. Such three-dimensional macromolecules have been reported to act as suitable templates for the synthesis of metallic and bimetallic nanoparticles. Monodisperse Au nanoparticles of 1–2 nm were prepared by a template-based approach using poly-(amidoamine) (PAMAM) dendrimers (Kim et al., 2004). Bimetallic Pd-Pt (Anderson et al., 2013), Pt-Ni (Aranishi et al., 2013) and Au-Pt dendrimer-encapsulated nanoparticles (DENs) (Iyyamperumal et al., 2013) have been successfully prepared in more recent publications with promising applications in catalysis and electrocatalysis. Dendrimer encapsulated CdS nanoparticles (DE-CdS) were also prepared by means of these nanoreactors which sequester Cd^{2+} and S^{2-} ions and, after their reaction, stabilize the resulting CdS nanoparticles by preventing agglomeration (Lemon and Crooks, 2000).

Micelles have also been used for the synthesis of colloidal nanoparticles. For example, bis(2-ethylhexyl) sulfosuccinate sodium salt (AOT) reverse micelles were used either as nanoreactors or as stabilizing agents for the synthesis of various spherical copper nanocrystals and magnetic cobalt nanocrystals in a work of Lisiecki (2005). In the same way like the dendrimers, micelles represent confined compartments with nanometer size which determine the nanoparticle growth.

Micelles were firstly used for the synthesis of CdS nanoparticles in 1986 (Lianos and Thomas, 1986). Cadmium and sulfur precursors were dissolved in heptane solution containing AOT micelles.

Although the work demonstrated the utility of inverted micelles for the preparation of small CdS nanoparticles, some big nanoparticles and aggregates were also observed. This method was lately used by several authors. In 1997, Gacoin et al. reported an inverse micelle route to prepare CdSe nanoparticles with 4-fluorophenylthiolate molecules as ligands (see Fig. 1.7). Arrested precipitation approaches based on water-in-oil microemulsions (inverse micelles) are normally conducted at room temperature. The as-prepared nanoparticles are finally stabilized in solution by the electrostatic interaction between surface Cd^{2+} ions and the anionic AOT surfactant. Upon treatment with thiol molecules (e.g., mercaptoundecanoic acid) the original ligands can be replaced under appropriate conditions. Although very promising, compared to high temperature synthesis approaches, micellar methods are reported to produce nanoparticles with lower crystallinity, high defect densities and thus, in case of photoluminescent particles lower emission quantum yields.

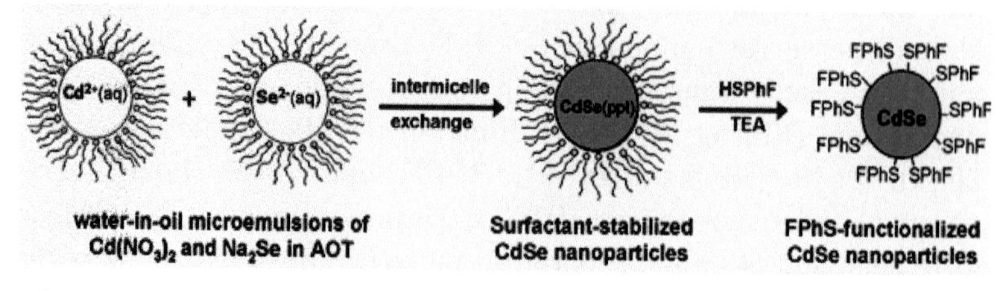

Figure 1.7 Inverse-micellar arrested precipitation approach for nanoparticle synthesis (e.g. CdSe) and the surface functionalization with thiolate molecules. AOT molecules in heptane with a small fraction of water are employed as the surfactant. After treatment with triethylamine (TEA) and 4-fuorophenylthiol (FPhSH), thiolate-caped CdSe nanoparticles are obtained. [Reprinted with permission from Arachchige and Brock (2007). Copyright (2007) American Chemical Society].

1.3.2 Solid-State Preparation

1.3.2.1 Milling

Mechanical milling is a very convenient and promising method to produce nanoparticle powders and alloys from bulk materials (see Fig. 1.8B) (Damonte et al., 2004; Chin et al., 2005; Ban et al., 2011). Grinding mills typically used in the process include the attrition jet, planetary, oscillating and vibration mills, all of which are classified as high-energy mills. Among these high-energy mills, the planetary mill is an efficient as well as suitable device for the production of nanoparticles. As its name implies, the milling container is rotated about two separate parallel axes in a manner analogous to the revolution of a planet around the sun (see Fig. 1.8A). In this connection, high rotational speeds ensure powerful collisions between the milling balls inside the container being responsible for the grinding process. Indeed, the motion of the milling balls exhibits a complex behavior and depends on the revolution speed, which occasionally affects the performance of the grinding. In addition to the rotational speed, milling time, ball to powder ratio as well as filling level of the milling vessel must also be taken into account with regard to the grinding efficiency (Burmeister and Kwade, 2013). However, an essential problem of this technique concerns the crystallinity of the final nanoparticles. Due to the collisions between the milling balls, high impact energies are released, which induces the formation of lattice defects such as dislocations or grain boundaries in materials (Düvel et al., 2011). The accumulation of these defects during the milling process often results in a complete destruction of the crystal structure called amorphization. Nonetheless, different kinds of nanostructures such as oxides (Lin and Nadiv, 1979) have been prepared by grinding.

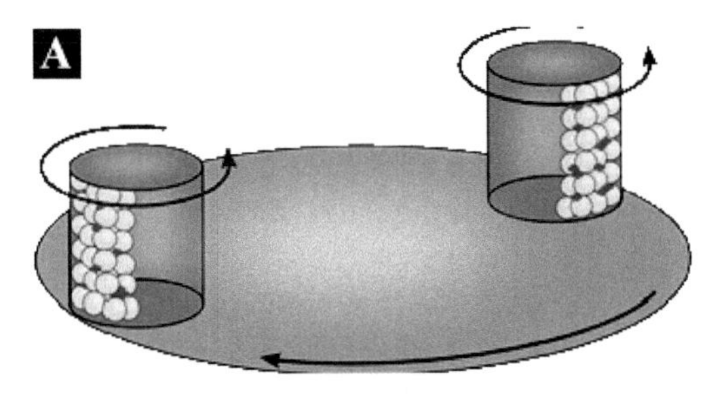

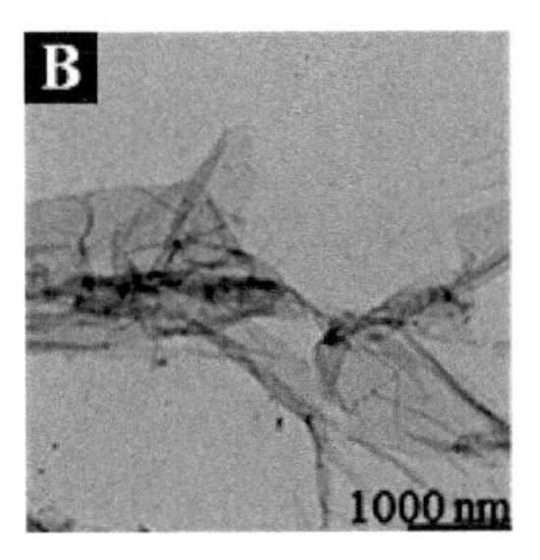

Figure 1.8 Panel A shows a scheme of the operation principle of a planetary ball mill. The containers with the ball mills (yellow) and the sample (green) inside rotate opposite to the main axes. Panel B illustrates a transmission electron microscope image of graphene synthesized by high-energy ball milling [adapted with permission from León et al., 2014, copyright 2014 American Chemical Society].

Besides the mechanical milling of bulk materials, high-energy mills are frequently applied for mechanochemical synthesis.

Here, the high impact energies released by the collisions of the milling balls are used for the reaction between the precursor powders. Starting in the 1980s, many works have focused on the preparation of nanoparticles by mechanochemistry till now, which finds expression in a great number of different compounds. This covers, among others, alloys (Nasu et al., 2001), oxides (Lin and Nadiv, 1979), sulfides (Short and Steward, 1957) as well as nitrides (Calka, 1991). Furthermore, this approach provides the possibility to synthesize metastable materials (Düvel et al., 2011) being difficult to access by other methods. Indeed, the mechanism is not completely understood yet, but it is assumed that metastable structures result from local hot spots induced by the collisions. Since these hot spots are restricted to small areas, the temperature can decrease quickly so that metastable states are "frozen" (Suryanarayana, 2001). For a detailed discussion about mechanochemistry, the interested readers are referred to a recent review of Baláž et al. (2013). In the past view years the use of additional surfactants enjoys great popularity. The function of the surfactants consists in the adsorption on freshly formed nanoparticle surfaces to prevent agglomeration as well as cold welding during the milling process (Guérard, 2008). In this way, nanostructures of rare earth elements (Zheng et al., 2012) and calcium carbonate nanoparticles (Cho et al., 2009), for example, have been developed.

1.3.2.2 Lithography

This technique allows the miniaturization of devices in many fields such as solar cells, microelectronic components, or biosensors. The methodologies used to produce nanostructured materials are grouped in top-down and bottom-up. In the top-down approaches thin films or bulk materials are scaled down to produce nanoscale devices (Colson et al., 2013; Gates et al., 2005). These techniques include optical lithographic processes such as photolithography, electron beam lithography (Pease, 2006) or X-ray lithography (Smith and Flanders, 1980). There are limitations to these approaches such as the high cost for acquisition and operating cost and also in the nanoscale dimensions. Bottom-up techniques are based on interactions between molecules or colloidal particles to form two- or three-dimensional nanostructures. These methodologies allow building up nanoscale devices by the use of small building blocks such as atoms, molecules or nanoparticles. There are also techniques which combine the top-down and bottom-up approaches. These techniques include self-assembly (Gates et al., 2005), scanning probe lithography (Krämer et al., 2003), nanosphere lithography (Colson et al., 2013) and procedures which are divided in molding and embossing. Molding includes curing a precursor against a topographically patterned substrate which is used for example by techniques like step-and-flash imprinting lithography (Colburn et al., 1999) or replica molding (Xia et al., 1996). Embossing approaches such as nanoimprinting lithography (Kooy et al., 2014) transfer a structured mold into a flat polymer. Typically, lithographic techniques involve two basic fabrication steps. Initially, it is necessary to create or built up a mold, mask or template which can give the final nanostructure of the thin films. According to the type of lithography, the mask will be created in different ways. For the mask deposition different methodologies like solvent evaporation (Denkov et al., 1992), dip coating (Dimitrov and Nagayama, 1996), spin coating, and many others can be used. As example, optical lithography uses a photoresist and light of different wavelength. Techniques like molding, embossing or printing use molds or stamps to transfer a patterned topography into a substrate. The second step is the fabrication of the nanostructure from the required material by using the stamp, mask or template. Certain techniques include as a third step the removal of the mask by using sonication in an adequate solution or etching.

1.3.3 Gas-Phase Methods

Most of the gas-phase methods are based on evaporation of a source material and homogeneous nucleation in the gas phase followed by condensation of the nanoparticles in the cooler parts of the system. Different techniques (based on either physical or chemical methods) can be used to obtain a supersaturated vapor phase. Typical gas-phase methods include, among others, furnace flow reactors, laser pyrolysis, flame reactors, plasma reactors, sputtering as well as inert gas condensation (Swihart, 2003; Kruis et al., 1998). In order not to overextend this part of the chapter, we have selected some of the mentioned methods that will be introduced in more detail.

1.3.3.1 Furnace flow reactors

The furnace flow reactors represent the simplest system to produce a saturated vapor using an oven. Here, an important requirement of the samples to which this method is applied, is that these are substances having a sufficient high vapor pressure at intermediate temperatures up to about 1700°C. In this regard, the choice of source material will determine the operating temperature. Initially, the starting material is placed in a heated flow of inert gas carrier and nanoparticles will be formed by cooling in a last step. Using furnace flow reactors, nanoparticles such as Ag (Scheibel and Porstendörfer, 1983), Ga (Deppert et al., 1996), ZrO_2 (Widiyastuti et al., 2010), ZnS (Lenggoro et al., 2000) as well as Bi_2WO_6 (Mann et al., 2011) have been obtained.

1.3.3.2 Laser pyrolysis

In the case of laser pyrolysis, the formation of nanoparticles is reached by a gas-phase decomposition of the reactants in a flowing gas that is heated with an infrared laser. In principle, the laser beam causes the molecules from the source material to be heated selectively, resulting in a chemical reaction. The carrier gas is, however, only indirectly heated by collisions with the reactant molecules. Recent works have dealt with, among others, Si nanoparticles (Huisken et al., 2002; Malumbres et al., 2015), SiC nanopowders (Réau et al. 2012) or multiwalled carbon nanotube-TiO_2 nanocomposite (Wang et al., 2015) synthesized by irradiation with carbon dioxide lasers. Instead of infrared lasers, nanoscopic structures, e.g., iron nanoparticles, can also be fabricated by photochemical dissociation of precursors by using an ultraviolet laser (Eremin et al., 2013). Furthermore, lasers provide the possibility to evaporate a sample target in an inert gas flow reactor. The generated vapor is then cooled by collisions with the inert gas molecules, causing supersaturation and thus, inducing nanoparticle formation (Pithawalla et al., 2001; Kobayashi et al., 2013).

1.3.3.3 Laser ablation of solids

Generally, laser ablation is based on a selective and rapid heating of a solid target using a pulsed laser beam, resulting in the formation of an energetic plasma above the target. Afterwards, nanoparticles are formed by condensation processes of the generated active species. By varying the pulse duration and energy, the relative amounts of ablated atoms and particles can be modified. Typically, laser ablation syntheses are performed in reaction chambers, providing the opportunity to work under vacuum or alternatively under different atmospheres, or in liquid media (Yan and Chrisey, 2012; Zeng et al., 2012; Amendola and Meneghetti, 2009). With the help of laser ablation, colloidal metal nanoparticles solutions (Mafuné et al., 2000, 2003) or nanoscale oxides (Amarilio-Burshtein et al., 2010) for example, have been prepared. Reactive laser ablation, with reactive gases like oxygen acting as a reactant, is also used. For instance, Murali et al. (2001) ablated an indium target in an air atmosphere producing indium oxide nanoparticles. In addition, this technique provides access to nanoscale films, in which case it is named pulsed laser deposition (Pergolesi et al., 2010; Infortuna et al., 2008).

1.3.3.4 Inert gas condensation

This method consists of heating a solid to evaporate it into a carrier gas; then the vapor is mixed with a cold gas to reduce the temperature. Inert gas condensation is already used on a commercial scale for a wide range of materials, for example metal nanoparticles (Pérez-Tijerina et al., 2010; Callini et al., 2010). An alternative to achieve supersaturation and induce the nucleation of nanoparticles is by means of chemical reactions in the gas phase. These approaches are called chemical vapor condensation in analogy to the chemical vapor deposition (CVD) processes used to deposit thin solid films on surfaces. There are many good examples of the application of this method in the literature, in which the synthesis of magnetic nanoparticles is highlighted (Lee et al., 2008a; Kim and Lee, 2013). The number of impurities in this type of synthesis processes can be reduced significantly by using purified inert gas and high-quality source material. Among the disadvantages, one can mention that sometimes the particle size distributions are relatively wide. As can be derived up to here, a large number of synthesis methods of nanoparticles in the gas phase have been developed. Examples of all these methods (and many more), types of nanoparticles that can be synthesized and their applications can be found in several reviews (Kruis et al., 1998; Swihart, 2003).

1.3.3.5 Electrospraying

Electrospraying describes an effective approach for liquid atomization via electrical forces, which not only enjoys wide currency in analytical techniques like, e.g., mass spectrometry, but also can be employed to produce nanoscale materials or films (Jaworek and Sobczyk, 2008). Generally, the concept is based on the disintegration of a liquid into fine droplets by applying high voltage to a thin capillary, through which the solution flows. The applied voltage causes an accumulation of charges in the liquid, resulting in a deformation of its meniscus at the capillary nozzle. This deformation occurs, in simple terms, due to the combination of liquid surface tension and electrostatic attraction of the counter electrode acting in opposite directions. Indeed, the behavior of the meniscus is more complex and depends on several parameters such as liquid flow rate, applied voltage as well as conductivity of the liquid (Jaworek and Sobczyk, 2008; Cloupeau and Prunet-Foch, 1990). Thus, the control over the meniscus plays an essential role in electrospraying, since it determines the size and the homogeneity of the droplets, which, in turn, influences the uniformity of nanoparticles. Here, different spraying modes can be distinguished according to the shape of the meniscus. In this regard, the so-called "cone jet" mode is one of the most widely used modes due to the good control over the droplet size. In this case, the meniscus is elongated into a thin jet, whose tip disintegrates into droplets due to instabilities (Cloupeau and Prunet-Foch, 1990). In this way, deposition of films, preparation of nanoparticles as well as their encapsulation can be achieved. In principle, the preparation of films is carried out by spraying either a colloidal suspension or a solution containing precursors onto a substrate. Afterwards, the evaporation of the liquid induces a solidification of the dissolved material resulting in the formation of a tight layer. In case of precursor solutions, the fine droplets act additionally as reaction vessel, where the generation of nanoparticles takes place. Moreover, the substrate is usually heated to accelerate the evaporation of the solvent as well as to induce reactions, if precursors are employed (Jaworek and Sobczyk, 2008). In general, electrospraying allows the deposition of thin and homogeneous films, which are required for several applications, e.g., thin-film solar cells or nanoelectronic devices (Ghimbeu et al., 2007; Chaparro et al., 2007). The preparation of nanoparticles by means of electrospraying includes different chemical reaction routes (Lenggoro et al., 2000; Choi, 2001; Salata, 2005). In contrast to the previously mentioned methods, these reactions occur in the droplets instead of conventional vessels, e.g., autoclaves or flasks, whereby the reaction space is restricted by the droplet dimension. Thus, it is logical that the droplet size with the corresponding size distribution plays a crucial role for the homogeneity of the nanoparticles. For this reason, events which broaden the size distribution such as uncontrolled droplet explosions have to be avoided. As already mentioned, electrospraying provides also the possibility to encapsulate nanoscale materials. For instance, the dispersion of nanoparticles into a solution containing the precursors for the shell growth describes the simplest way. During the evaporation, the shell can solidify around the nanoparticles leading to core–shell structures (Jaworek and Sobczyk, 2008).

1.3.4 Deposition Methods

In the last part of this chapter, a brief introduction of common deposition techniques providing access to nanostructured films will be given. As already mentioned, some gas-phase methods, e.g., electrospraying or laser ablation are also suitable for the deposition of thin films. Particularly with regard to prospective application such as solar cell or electronic devices, homogeneous and reproducible films are of great importance to ensure a constant quality. Here, we will focus on CVD, physical vapor deposition (PVD) as well as MBE and their basic concepts.

Broadly, CVD techniques consist of applying solid thin-film coatings to surfaces, but it is also used to produce high-purity bulk materials and powders (Dobkin and Zuraw, 2003). In typical CVD, a substrate is exposed to one or more volatile precursors, which react and/or decompose on the substrate surface to produce the desired deposit. Frequently, a gas flow through the reaction chamber serves to remove volatile by-products. Nowadays, a huge variety of CVD formats exist, of which some of them will be pointed out. Depending on the operating pressure, three formats can be distinguished, named *atmospheric-pressure CVD*, *low-pressure CVD*, and *ultrahigh-vacuum CVD*. The techniques can also be classified depending on how the precursor are introduced in the system, having *aerosol-assisted CVD*, when the precursors are transported to the substrate by means of a liquid/gas aerosol, or *direct-liquid-injection CVD*, when the liquid precursor is injected in the vaporization chamber and then transported to the substrate as in classical CVD. A good alternative for depositing a variety of thin films at lower temperatures is *plasma-enhanced CVD* (PECVD). A scheme for the setup in PECVD is shown in Fig. 1.9. Here, a plasma is generated to enhance chemical reaction rates of the precursors. Since the formation of the reactive species occurs by collision in the gas phase, the substrate (and thus, the film formation) can be maintained at a low temperature. CVD can also be used to alternatively deposit successive monolayers of different elements on a substrate to produce layered films. Nowadays, this technique is also known as *atomic layer epitaxy* or *atomic layer deposition*.

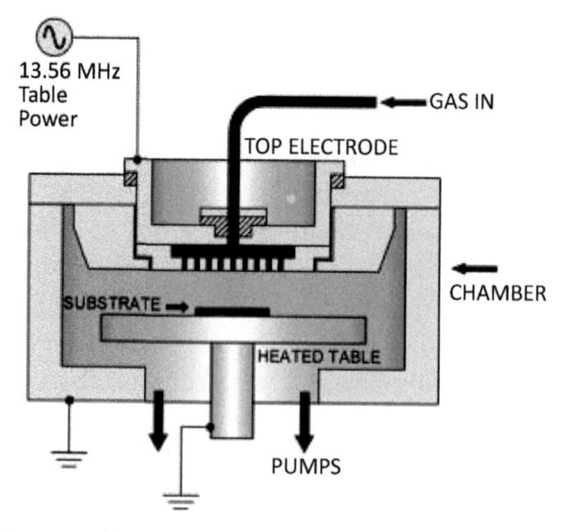

Figure 1.9 Plasma-enhanced chemical vapor deposition (PECVD) setup. Copyright © 2014 Oxford Instruments plc, Tubney Woods, Abingdon, Oxon OX13 5QX, UK. All Rights Reserved.

Chemical vapor deposition has emerged as the most versatile, promising and commercially viable technique to fabricate graphene films (Hofmann et al., 2015; Zhang et al., 2013b). Several recent examples can be found in literature (Liu et al., 2014; Sun et al., 2015). Silicon and derivatives are other widely deposited materials in films by means of CVD, which gain great importance for applications in transistors and solar cells (Meyerson, 1986; Leach et al., 2002). Relative to other approaches (evaporation, sputtering, growth from a liquid phase, etc.), CVD is complex and expensive. Moreover, the use of precursor chemicals often introduces impurities in the solid films. Nonetheless, CVD is the deposition technique of choice for most of the films which aim to be applied, for example, in microelectronics, becoming an industrial standard process technique for the manufacturing of 2D materials of "electronic-grade" (Hofmann et al., 2015). The main advantage of CVD is to accomplish coating of complex substrate surfaces in a very uniform way.

Physical Vapor Deposition includes deposition processes in which a material is vaporized from a solid or liquid source in the form of atoms or molecules, transported in the form of a vapor through a vacuum or low-pressure gaseous environment to the substrate where it condenses (Mattox, 2010). The coating method in PVD involves purely physical processes rather than involving a chemical reaction at the surface to be coated as in CVD. Thus, PVD techniques can be divided in three main categories, named vacuum evaporation, sputter deposition and arc vapor deposition. In vacuum deposition, the vacuum environment allows the thermally vaporized source to reach the surface, with no collisions with the gas molecules in the space between the source and the substrate. By physical sputtering process, the surface atoms of a solid source are physically ejected by atomic-sized accelerated particles. In the case of arc vapor deposition, a high current, low-voltage arc is used to vaporize an electrode and to deposit the vaporized material on a substrate. PVD has been applied successfully for the fabrication of nanowires from different materials such as gold (Cross et al., 2007) and ZnO (Kong et al., 2001), and other interesting configurations such as Y-shaped nanorods (Wang et al., 2005).

Generally, MBE is based on the epitaxial growth of layers on a heated substrate (Franchi et al., 2003). To achieve this kind of growth, the lattice mismatch between the epilayer and substrate has to be sufficiently small. For the deposition, pure elements, of which the film is to consist, are vaporized under ultra-high vacuum, resulting in a directed molecular beam. In this regard, the term "molecular beam" refers to the non-interacting atoms in the gas phase. Afterwards, the atoms condense on a heated substrate, where potential reactions between different gaseous species as well as the formation of a nanostructured film take place. Moreover, the ultra-high vacuum as well as the heating of the substrate during the deposition promotes an undisturbed epitaxial growth without the inclusion of any contaminants. Thus, MBE offers the opportunity to produce films of high quality and purity. However, this technique is time consuming and requires complex apparatus. Nowadays, MBE is widely used and covers, for example, the deposition of nanowires (Guo et al., 2010), carbon nanostructures (Park et al., 2010) as well as tight films (Lee et al., 2008b; Yu et al., 2000).

1.4 Summary

In this chapter we focused on mainly two points. First we introduced a manifold of physical properties of nanoparticles with a special focus on such properties which are either very different than in bulk, or which are even absent in the corresponding bulk material. In the second part of the chapter we introduced a large variety of fabrication methods for nanomaterials, ranging from solid-phase methods over liquid-phase methods to gas-phase methods each time trying to point the major advantages and disadvantages of each method. Throughout the whole chapter we tried to emphasize that nanoscience is an interdisciplinary field which involves basically all natural sciences and which also can offer benefits to virtually all fields of natural sciences and engineering. Of course it is not possible to cover any of the mentioned properties or fabrication methods in depth in the framework of one book chapter. The interested reader is therefore again referred to the review articles and original papers mentioned in the relevant subchapters.

References

Alexander K., Shlomo M., Aqueous Dispersions of Metallic Nanoparticles, in *Nanoscience Colloidal and Interfacial Aspects*, Starov V. M., ed., CRC Press, Boca Raton (2010).

Amarilio-Burshtein I., Tamir S., Lifshitz Y., *Appl. Phys. Lett.*, **96** (2010), 103104.

Amendola V., Meneghetti M., *Phys. Chem. Chem. Phys.*, **11** (2009), 3805–3821.

Anderson R. M., Zhang L., Loussaert J. A., Frenkel A. I., Henkelman G., Crooks R. M., *ACS Nano*, **7** (2013), 9345–9353.

Arachchige I. U., Brock S. L., *Acc. Chem. Res.*, **40** (2007), 801–809.

Aranishi K., Singh A. K., Xu Q., *Chemcatchem*, **5** (2013), 2248–2252.

Aricò A. S., Bruce P., Scrosati B., Tarascon J. M., van Schalkwijk W., *Nat. Mater.*, **4** (2005), 366–377.

Bagaria H. G., Ada E. T., Shamsuzzoha M., Nikles D. E., Johnson D. T., *Langmuir*, **22** (2006), 7732–7737.

Baláž P., Achimovičová M., Baláž M., Billik P., Cherkezova-Zheleva Z., Criado J. M., Delogu F., Dutková E., Gaffet E., Gotor F. J., Kumar R., Mitov I., Rojac T., Senna M., Streletskii A., Wieczorek-Ciurowa K., *Chem. Soc. Rev.*, **42** (2013), 7571–7637.

Ban I., Stergar J., Drofenik M., Ferk G., Makovec D., *J. Magn. Magn. Mater.*, **323** (2011), 2254–2258.

Bhosale M. A., Chenna D. R., Ahire J. P., Bhanage B. M., *RSC Adv.*, **5** (2015), 52817–52823.

Bigall N. C., Herrmann A.-K., Vogel M., Rose M., Simon P., Carrillo-Cabrera W., Dorfs D., Kaskel S., Gaponik N., Eychmüller A., *Angew. Chem. Int. Ed.*, **48** (2009), 9731–9734.

Bigall N. C., Parak W. J., Dorfs D., *Nano Today*, **7** (2012), 282–296.

Bigall N. C., Wilhelm C., Beoutis M. L., Garcia-Hernandez M., Khan A. A., Giannini C., Sanchez-Ferrer A., Mezzenga R., Materia M. E., Garcia M. A., Gazeau F., Bittner A. M., Manna L., Pellegrino T., *Chem. Mater.*, **25** (2013), 1055–1062.

Bilecka I., Niederberger M., *Nanoscale*, **2** (2010), 1358–1374.

Bogart L. K., Pourroy G., Murphy C. J., Puntes V., Pellegrino T., Rosenblum D., Peer D., Lévy R., *ACS Nano*, **8** (2014), 3107–3122.

Bouet C., Mahler B., Nadal B., Abecassis B., Tessier M. D., Ithurria S., Xu X., Dubertret B., *Chem. Mater.*, **25** (2013), 639–645.

Brus L. E., *J. Chem. Phys.*, **80** (1984), 4403–4409.

Buffat P., Borel J. P., *Phys. Rev. A*, **13** (1976), 2287–2298.

Burmeister C. F., Kwade A., *Chem. Soc. Rev.*, **42** (2013), 7660–7667.

Calka A., *Appl. Phys. Lett.*, **59** (1991), 1568–1569.

Callini E., Pasquini L., Rude L. H., Nielsen T. K., Jensen T. R., Bonetti E., *J. Appl. Phys.*, **108** (2010), 73513.

Cao G., Wang Y. (eds.), *Nanostructures and Nanomaterials*, 2nd. ed., World Scientific Publishing, Singapore (2004).

Čapek R. K., Lambert K., Dorfs D., Smet P. F., Poelman D., Eychmüller A., Hens Z., *Chem. Mater.*, **21** (2009), 1743–1749.

Carbone L., Nobile C., Giorgi M. de, Della Sala F., Morello G., Pompa P., Hytch M., Snoeck E., Fiore A., Franchini I. R., Nadasan M., Silvestre A. F., Chiodo L., Kudera S., Cingolani R., Krahne R., Manna L., *Nano Lett.*, **7** (2007), 2942–2950.

Chaparro A., Benítez R., Gubler L., Scherer G., Daza L., *J. Power Sources*, **169** (2007), 77–84.

Cheng C., Xu G., Zhang H., *J. Cryst. Growth*, **311** (2009), 1285–1290.

Chin P. P., Ding J., Yi J. B., Liu B. H., *J. Alloy. Compounds*, **390** (2005), 255–260.

Chiu H.-C., Yeh C.-S., *J. Phys. Chem. C*, **111** (2007), 7256–7259.

Cho K., Chang H., Kil D. S., Kim B. G., Jang H. D., *J. Ind. Eng. Chem.*, **15** (2009), 243–246.

Choy K., *Mater. Sci. Eng., C*, **16** (2001), 139–145.

Cloupeau M., Prunet-Foch B., *J. Electrostatics*, **25** (1990), 165–184.

Colburn M., Johnson S. C., Stewart M. D., Damle S., Bailey T. C., Choi B., Wedlake M., Michaelson T. B., Sreenivasan S. V., Ekerdt J. G., Willson C. G., *Proc. SPIE Int. Soc. Opt. Eng.*, **3676** (1999), 379–389.

Colson P., Henrist C., Cloots R., *J. Nanomater.*, **2013** (2013), 1–19.

Costi R., Saunders A. E., Banin U., *Angew. Chem. Int. Ed.*, **49** (2010), 4878–4897.

Cross C. E., Hemminger J. C., Penner R. M., *Langmuir*, **23** (2007), 10372–10379.

Cushing B. L., Kolesnichenko V. L., O'Connor C. J., *Chem. Rev.*, **104** (2004), 3893–3946.

Damonte L. C., Zelis L. A. M., Soucase B. M., Fenollosa M. A. H., *Powder Technol.*, **148** (2004), 15–19.

Daou T. J., Pourroy G., Bégin-Colin S., Grenèche J. M., Ulhaq-Bouillet C., Legaré P., Bernhardt P., Leuvrey C., Rogez G., *Chem. Mater.*, **18** (2006), 4399–4404.

de Mello Donegá C., Liljeroth P., Vanmaekelbergh D., *Small*, **1** (2005), 1152–1162.

Demazeau G., *J. Mater. Sci.*, **43** (2008), 2104–2114.

Denkov N., Velev O., Kralchevski P., Ivanov I., Yoshimura H., Nagayama K., *Langmuir*, **8** (1992), 3183–3190.

Deppert K., Bovin J.-O., Malm J.-O., Samuelson L., *J. Cryst. Growth*, **169** (1996), 13–19.

Dimitrov A. S., Nagayama K., *Langmuir*, **12** (1996), 1303–1311.

Dobkin D., Zuraw M. K. (eds.), *Principles of Chemical Vapor Deposition*, 1st. ed., Springer Science+Business Media, New York (2003).

Dorfs D., Hartling T., Miszta K., Bigall N. C., Kim M. R., Genovese A., Falqui A., Povia M., Manna L., *J. Am. Chem. Soc.*, **133** (2011), 11175–11180.

Düvel A., Wegner S., Efimov K., Feldhoff A., Heitjans P., Wilkening M., *J. Mater. Chem.*, **21** (2011), 6238–6250.

Eremin A. V., Gurentsov E. V., Priemchenko K. Y., *J. Nanopart. Res.*, **15** (2013), 1–15.

Faraday M., *Philos. Trans. R. S. London*, **147** (1857), 145–181.

Fischer M., Vögtle F., *Angew. Chem. Int. Ed.*, **38** (1999), 884–905.

Franchi S., Trevisi G., Seravalli L., Frigeri P., *Prog. Cryst. Growth Charact. Mater.*, **47** (2003), 166–195.

Franzl T., Klar T. A., Schietinger S., Rogach A. L., Feldmann J., *Nano Lett.*, **4** (2004), 1599–1603.

Fuku K., Hayashi R., Takakura S., Kamegawa T., Mori K., Yamashita H., *Angew. Chem. Int. Ed.*, **52** (2013), 7446–7450.

Gacoin T., Malier L., Boilot J. P., *J. Mater. Chem.*, **7** (1997), 859–860.

Gaponik N., *J. Mater. Chem.*, **20** (2010), 5174–5181.

Gaponik N., Hickey S. G., Dorfs D., Rogach A. L., Eychmuller A., *Small*, **6** (2010), 1364–1378.

Gaponik N., Talapin D. V., Rogach A. L., Hoppe K., Shevchenko E. V., Kornowski A., Eychmüller A., Weller H., *J. Phys. Chem. B.*, **106** (2002), 7177–7185.

Gates B. D., Xu Q., Stewart M., Ryan D., Willson C. G., Whitesides G. M. *Chem. Rev.*, **105** (2005), 1171–1196.

Gawande M. B., Shelke S. N., Zboril R., Varma R. S., *Acc. Chem. Res.*, **47** (2014), 1338–1348.

Ghimbeu C. M., van Landschoot R., Schoonman J., Lumbreras M., *J. Eur. Ceram. Soc.*, **27** (2007), 207–213.

Goia D. V., Matijević E., *New J. Chem.*, **22** (1998), 1203–1215.

Guérard D., *Rev. Adv. Mater. Sci.*, **18** (2008), 225–230.

Guo W., Zhang M., Banerjee A., Bhattacharya P., *Nano Lett.*, **10** (2010), 3355–3359.

Gupta A. K., Gupta M., *Biomaterials*, **26** (2005), 3995–4021.

Hendel T., Lesnyak V., Kühn L., Herrmann A.-K., Bigall N. C., Borchardt L., Kaskel S., Gaponik N., Eychmüller A., *Adv. Funct. Mater.*, **23** (2013), 1903–1911.

Hill L. J., Richey N. E., Sung Y., Dirlam P. T., Griebel J. J., Shim I.-B., Pinna N., Willinger M.-G., Vogel W., Char K., Pyun J., *CrystEngComm*, **16** (2014), 9461–9468.

Hofmann S., Braeuninger-Weimer P., Weatherup R. S., *J. Phys. Chem. Lett.*, **6** (2015), 2714–2721.

Huisken F., Ledoux G., Guillois O., Reynaud C., *Adv. Mater.*, **14** (2002), 1861–1865.

Hüsing N., Schubert U., *Angew. Chem. Int. Ed.*, **37** (1998), 22–45.

Infortuna A., Harvey A. S., Gauckler L. J., *Adv. Funct. Mater.*, **18** (2008), 127–135.

Iyyamperumal R., Zhang L., Henkelman G., Crooks R. M., *J. Am. Chem. Soc.*, **135** (2013), 5521–5524.

Jana N. R., Earhart C., Ying J. Y., *Chem. Mater.*, **19** (2007), 5074–5082.

Jaworek A., Sobczyk A., *J. Electrostatics*, **66** (2008), 197–219.

Jayaprakash J., Srinivasan N., Chandrasekaran P., Girija E., *Spectrochim. Acta Part A*, **136** (2015), 1803–1806.

Kessler V., Spijksma G., Seisenbaeva G., Håkansson S., Blank D. A., Bouwmeester H. M., *J. Sol-Gel. Sci. Technol.*, **40** (2006), 163–179.

Khalid W., El Helou M., Murböck T., Yue Z., Montenegro J. M., Schubert K., Göbel G., Lisdat F., Witte G., Parak W. J., *ACS Nano*, **5** (2011), 9870–9876.

Kim W.-B., Lee J.-S., *J. Nanosci. Nanotechnol.*, **13** (2013), 4622–4626.

Kim Y.-G., Oh S.-K., Crooks R. M., *Chem. Mater.*, **16** (2004), 167–172.

Kobayashi K., Kokai F., Sakurai N., Yasuda H., *J. Phys. Chem. C*, **117** (2013), 25169–25174.

Kong Y. C., Yu D. P., Zhang B., Fang W., Feng S. Q., *Appl. Phys. Lett.*, **78** (2001), 407–409.

Kooy N., Mohamed K., Pin L. T., Guan O. S., *Nanoscale Res. Lett.*, **9** (2014), 320.

Krämer S., Fuierer R. R., Gorman C. B., *Chem. Rev.*, **103** (2003), 4367–4418.

Kruis F. E., Fissan H., Peled A., *J. Aerosol Sci.*, **29** (1998), 511–535.

Leach W. T., Zhu J., Ekerdt J. G., *J. Cryst. Growth*, **240** (2002), 415–422.

Lee D. W., Yu J. H., Kim B. K., Jang T. S., *ChemInform*, **39** (2008a).

Lee Y. J., Lee W. C., Nieh C. W., Yang Z. K., Kortan A. R., Hong M., Kwo J., Hsu C.-H., *J. Vac. Sci. Technol. B*, **26** (2008b), 1124.

Lemon B. I., Crooks R. M., *J. Am. Chem. Soc.*, **122** (2000), 12886–12887.

Lenggoro I., Okuyama K., Fernández de la Mora J., Tohge N., *J. Aerosol Sci.*, **31** (2000), 121–136.

León V., Rodriguez A. M., Prieto P., Prato M., Vázquez E., *ACS Nano*, **8** (2014), 563–571.

Lesin V. I., Pisarenko L. M., Kasaikina O. T., *Colloid J.*, **74** (2012), 85–90.

Lewis E. A., Jewell A. D., Kyriakou G., Sykes E. C. H., *Phys. Chem. Chem. Phys.*, **14** (2012), 7215–7224.

Li W., Zamani R., Rivera Gil P., Pelaz B., Ibáñez M., Cadavid D., Shavel A., Alvarez-Puebla R. A., Parak W. J., Arbiol J., Cabot A., *J. Am. Chem. Soc.*, **135** (2013), 7098–7101.

Lianos P., Thomas J. K., *Chem. Phys. Lett.*, **125** (1986), 299–302.

Lin I. J., Nadiv S., *Mater. Sci. Eng.*, **39** (1979), 193–209.

Lisiecki I., *J. Phys. Chem. B*, **109** (2005), 12231–12244.

Liu R., Chi Y., Fang L., Tang Z., Yi X., *J. Nanosci. Nanotechnol.*, **14** (2014), 1647–1657.

Mafuné F., Kohno J.-Y., Takeda Y., Kondow T., *J. Phys. Chem. B*, **107** (2003), 4218–4223.

Mafuné F., Kohno J.-Y., Takeda Y., Kondow T., Sawabe H., *J. Phys. Chem. B*, **104** (2000), 9111–9117.

Malumbres A., Martínez G., Hueso J. L., Gracia J., Mallada R., Ibarra A., Santamaría J., *Nanoscale*, **7** (2015), 8566–8573.

Mann, Amanda K. P., Skrabalak S. E., *Chem. Mater.*, **23** (2011), 1017–1022.

Manna L., Scher E. C., Alivisatos A. P., *J. Am. Chem. Soc.*, **122** (2000), 12700–12706.

Mao H., Liu X., Yang J., Li B., Yao C., Kong Y., *Mater. Sci. Eng. C*, **40** (2014), 102–108.

Mattox D. M. (ed.), *Handbook of Physical Vapor Deposition (PVD) Processing*, 2nd ed., Norwich, Elsevier Science, Oxford (2010).

Meyerson B. S., *Appl. Phys. Lett.*, **48** (1986), 797–799.

Murali A., Barve A., Leppert V. J., Risbud S. H., Kennedy I. M., Lee, Howard W. H., *Nano Lett.*, **1** (2001), 287–289.

Murray C. B., Norris D. J., Bawendi M. G., *J. Am. Chem. Soc.*, **115** (1993), 8706–8715.

Nasu T., Araki F., Uemura O., Usuki T., Kameda Y., Takahashi S., Tokumitsu K., *J. Metastable Nanocrystalline Mater.*, **10** (2001), 203–210.

Nikoobakht B., El-Sayed M. A., *Chem. Mater.*, **15** (2003), 1957–1962.

Odenbach S., *J. Phys. Condensed Matter*, **16** (2004), R1135–R1150.

Pankhurst Q. A., Connolly J., Jones S. K., Dobson J., *J. Phys. D Appl. Phys.*, **36** (2003), R167–R181.

Park J., Mitchel W. C., Grazulis L., Smith H. E., Eyink K. G., Boeckl J. J., Tomich D. H., Pacley S. D., Hoelscher J. E., *Adv. Mater.*, **22** (2010), 4140–4145.

Pazos-Pérez N., Baranov D., Irsen S., Hilgendorff M., Liz-Marzán L. M., Giersig M., *Langmuir*, **24** (2008), 9855–9860.

Pease R. F. W., *Contemp. Phys.*, **22** (2006), 265–290.

Pellegrino T., Manna L., Kudera S., Liedl T., Koktysh D., Rogach A. L., Keller S., Rädler J., Natile G., Parak W. J., *Nano Lett.*, **4** (2004), 703–707.

Peìrez-Tijerina E., Mejiìa-Rosales S., Inada H., Joseì-Yacamaìn M., *J. Phys. Chem. C*, **114** (2010), 6999–7003.

Pergolesi D., Fabbri E., D'Epifanio A., Di Bartolomeo E., Tebano A., Sanna S., Licoccia S., Balestrino G., Traversa E., *Nat. Mater.*, **9** (2010), 846–852.

Pierre A. C., *Introduction to Sol-Gel Processing*, 1st ed., Springer Science+ Business Media, New York (1998).

Pileni M. P., *Acc. Chem. Res.*, **41** (2008), 1799–1809.

Pithawalla Y. B., El-Shall M. S., Deevi S. C., Ström V., Rao K. V., *J. Phys. Chem. B*, **105** (2001), 2085–2090.

Ramaniah L. M., Nair S. V., *Phys. Rev. B*, **47** (1993), 7132–7139.

Réau A., Guizard B., Canel J., Galy J., Ténégal F., Danforth S., *J. Am. Ceram. Soc.*, **95** (2012), 153–158.

Salata O. V., *Curr. Nanosci.*, **1** (2005), 25–33.

Sánchez-Paradinas S., Dorfs D., Friebe S., Freytag A., Wolf A., Bigall N. C., *Adv. Mater.*, **27** (2015), 6152–6156.

Scheibel H. G., Porstendörfer J., *J. Aerosol Sci.*, **14** (1983), 113–126.

Schmid G., *Chem. Rev.*, **92** (1992), 1709–1727.

Schmid G. (ed.), *Clusters and Colloids*; WILEY-VCH, Weinheim (1994).

Schwenke A. M., Hoeppener S., Schubert U. S., *Adv. Mater.*, **27** (2015), 4113–4141.

Shavel A., Gaponik N., Eychmüller A., *J. Phys. Chem. B.*, **108** (2004), 5905–5908.

Shi W., Zeng H., Sahoo Y., Ohulchanskyy T. Y., Ding Y., Wang Z. L., Swihart M., Prasad P. N., *Nano Lett.*, **6** (2006), 875–881.

Short M. A., Steward E. G., *Z. Phys. Chem.*, **13** (1957), 298–315.

Singh L. P., Bhattacharyya S. K., Kumar R., Mishra G., Sharma U., Singh G., Ahalawat S., *Adv. Colloid Interface Sci.*, **214** (2014), 17–37.

Smith H. I., Flanders D. C., *J. Vac. Sci. Technol.*, **17** (1980), 533–535.

Sönnichsen C., Reinhard B. M., Liphardt J., Alivisatos A. P., *Nat. Biotechnol.*, **23** (2005), 741–745.

Soon G. K., Piao Y., Park J., Angappane S., Jo Y., Hwang N. M., Park J. G., Hyeon T., *J. Am. Chem. Soc.*, **129** (2007), 12571–12584.

Spanhel L., Haase M., Weller H., Henglein A., *J. Am. Chem. Soc.*, **109** (1987), 5649–5655.

Stöber W., Fink A., Bohn E., *J. Colloid Interface Sci.*, **26** (1968), 62–69.

Sun J., Chen Y., Priydarshi M. K., Chen Z., Bachmatiuk A., Zou Z., Chen Z., Song X., Gao Y., Rümmeli M. H., Zhang Y., Liu Z., *Nano Lett.*, **15** (2015), 5846–5854.

Sun S., Murray C. B., *J. Appl. Phys.*, **85** (1999), 4325–4330.

Suryanarayana C., *Prog. Mater. Sci.*, **46** (2001), 1–184.

Swihart M. T., *Curr. Opin. Colloid Interface Sci.*, **8** (2003), 127–133.

Tang F., Li L., Chen D., *Adv. Mater.*, **24** (2012), 1504–1534.

Tang K.-B., Qian Y.-T., Zeng J.-H., Yang X.-G., *Adv. Mater.*, **15** (2003), 448–450.

Tao F. (ed.), *Metal Nanoparticles for Catalysis Advances and Applications*, RSC, Cambridge (2014).

Thompson D., *Gold Bull.*, **40** (2007), 267–269.

Turkevich J., Stevenson P. C., Hillier J., *Discuss. Faraday Soc.*, **11** (1951), 55–75.

van Embden J., Chesman, Anthony S. R., Jasieniak J. J., *Chem. Mater.*, **27** (2015), 2246–2285.

Vögtle F., Richardt G., Werner N. (eds.), *Dendrimer Chemistry*, WILEY-VCH Verlag, Weinheim (2009).

Wang J., Huang H., Kesapragada S. V., Gall D., *Nano Lett.*, **5** (2005), 2505–2508.

Wang J., Lin Y., Pinault M., Filoramo A., Fabert M., Ratier B., Bouclé J., Herlin-Boime N., *ACS Appl. Mater. Interfaces*, **7** (2015), 51–56.

Wang W.-W., Zhu Y.-J., Ruan M.-L., *J. Nanopart. Res.*, **9** (2007), 419–426.

Widiyastuti W., Balgis R., Iskandar F., Okuyama K., *Chem. Eng. Sci.*, **65** (2010), 1846–1854.

Xia Y., Kim E., Zhao X.-M., Rogers J. A., Prentiss M., Whitesides G. M., *Science*, **273** (1996), 347–349.

Xie Y., Riedinger A., Prato M., Casu A., Genovese A., Guardia P., Sottini S., Sangregorio C., Miszta K., Ghosh S., Pellegrino T., Manna L., *J. Am. Chem. Soc.*, **135** (2013), 17630–17637.

Yan Z., Chrisey D. B., *J. Photochem. Photobiol. C*, **13** (2012), 204–223.

Yang Y., Matsubara S., Xiong L., Hayakawa T., Nogami M., *J. Phys. Chem. C*, **111** (2007), 9095–9104.

Yin Y., Alivisatos A. P., *Nature*, **437** (2005), 664–670.

Yu W. W., Peng X. G., *Angew. Chem. Int. Ed.*, **41** (2002), 2368–2371.

Yu Z., Ramdani J., Curless J. A., Finder J. M., Overgaard C. D., Droopad R., Eisenbeiser K. W., Hallmark J. A., Ooms W. J., Conner J. R., Kaushik V. S., *J. Vac. Sci. Technol. B*, **18** (2000), 1653.

Zeng H., Du X.-W., Singh S. C., Kulinich S. A., Yang S., He J., Cai W., *Adv. Funct. Mater.*, **22** (2012), 1333–1353.

Zhang Q. F., Uchaker E., Candelaria S. L., Cao G. Z., *Chem. Soc. Rev.*, **42** (2013a), 3127–3171.

Zhang Y., Zhang L., Zhou C., *Acc. Chem. Res.*, **46** (2013b), 2329–2339.

Zheng L., Cui B., Zhao L., Li W., Hadjipanayis G. C., *J. Alloy. Compound*, **539** (2012), 69–73.

Zhuo L., Ge J., Cao L., Tang B., *Cryst. Growth Des.*, **9** (2009), 1–6.

Ziegler J., Merkulov A., Grabolle M., Resch-Genger U., Nann T., *Langmuir*, **23** (2007), 7751–7759.

Zou X., Fan H., Tian Y., Zhang M., Yan X., *Dalton Trans.*, **44** (2015), 7811–7821.

Chapter 2

Biocatalysis: An Introduction

Peter Grunwald

Department of Physical Chemistry,
University of Hamburg, D-20146 Hamburg, Germany

grunwald@chemie.uni-hamburg.de

2.1 Introduction

Biocatalysis, a subfield of biotechnology, is the use of enzymes or whole cells as an alternative to chemical catalysts in connection with the production of goods. Making beer (and wine) belongs to the oldest of all "biotechnological" processes. The Sumerians brewed beer already about 7,000 years ago, and in China, the preparation of alcoholic drinks even has a 9,000-year-old tradition. Another historic application of biocatalysis is the use of chymosin, a protease from the stomachs of, e.g., ruminant animals for the cheese-making process.

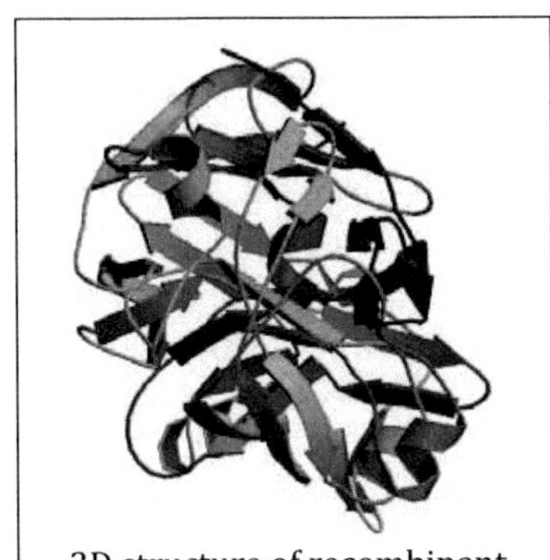

3D structure of recombinant bovine chymosin at 2.3 Å resolution (Gilliland et al., 1990).

Biocatalysis and Nanotechnology
Edited by Peter Grunwald

ISBN 978-981-4613-69-9 (Hardcover), 978-1-315-19660-2 (eBook)
www.panstanford.com

Millions of years of evolution have generated an unimaginable multiplicity of organisms (microorganisms, plants, animals) where biocatalysts regulate and control all metabolic reactions in a very selective way, thereby enabling high reaction rates under mild reaction conditions, even though microorganisms have been found to live under unusual environmental conditions. This means that they adapted to their respective surroundings by developing an enzyme equipment that enabled them to survive in soils spoiled by toxic chemicals as they are found near to disused chemical plants, in deep sea habitats where they grow at temperatures just above 0°C and at a pressure of more than 1000 bar, or at even lower temperatures, e.g., under the conditions of the Siberian permafrost. The upper extreme are microorganisms existing in so-called ecological niches, first found in the hot springs of the Yellow Stone National Park about 60 years ago. They produce enzymes that function at temperatures up to 130°C and extreme pH-values. Such robust enzymes are of great interest for various biotechnological processes.

Not long ago enzymes had to be employed as provided by Nature and were often available in only limited amounts. Meanwhile, due to an increasing understanding of enzyme structure–function relationship, the development of gene and recombinant technologies, and the advent of protein design (rational design and directed evolution) methods, it has become possible to generate high quantities of enzymes from a suited host organisms, and to tailor the properties (stability, activity, selectivity, substrate specificity) of a biocatalyst to meet the requirements of a given process. These advances paved the way for a sustainable production of compounds for applications in almost every field of daily life, i.e., among others food, animal feed, agriculture, clothing, hygiene, pharmaceuticals, and for analytical and diagnostic purposes.

2.2 Biocatalysts: General Aspects

In the following subsections, some properties of enzymes are recalled (see also Grunwald, 2009), including their classification together with a brief historical background, structural aspects, basics of enzyme kinetics, and the phenomenon of promiscuity.

Furthermore, the design of biotransformations and methods to improve biocatalysts are discussed.

2.2.1 Classification of Biocatalysts

At the beginning of the 19th century, G. S. Constantin Kirchhoff (1764–1833) found that starch is degraded to simple sugars by means of diluted inorganic acids. In 1835 the Swedish chemist Jöns Jakob Berzelius (1779–1848) proposed the name catalysis (from the Greek: kata: wholly, and lyein: to loosen) for the conversion of substances under mild conditions in presence of compounds seemingly not involved into the reaction, without being able to explain what really happens during catalysis. Further findings during those times were the decomposition of starch to sugars by a compound denoted as "diastase" by the French chemist Anselme Payen (1795–1878) or the digestion of flesh in presence of pepsin prepared from animal tissues by Theodor Schwann in 1836. These compounds were first named ferments until Wilhelm Kühne (1837–1900) neoterized in 1887 the term enzyme (from the Greek: en: in, and zyme: sourdough) which should be confined to ferments not needing the surroundings of a cell to develop activity. On the other hand, Louis Pasteur (1822–1895) held the view that fermentation is not possible in absence of life; however, this opinion was disproved by the discovery made by Eduard Buchner (1860–1917) during his time in the chemical laboratory of Hans von Pechmann at the University of Tübingen, and that he published in 1897 under the title "On Alcoholic Fermentation Without Yeast Cells" (original German title: "Über die alkoholische Gärung ohne Zellen"); these findings proved that all enzymes are chemical compounds not requiring any vital force to be effective. Buchner was awarded the Noble Prize for chemistry in 1907.

In parallel to the increasing number of enzymes discovered, the situation with respect to their denotation became more and more confusing because in the absence of rules for naming enzymes, researchers themselves invented names that often did not refer to the type of reaction catalyzed, or used the same name for different enzymes. To overcome this unsatisfying situation the International Union of Biochemistry (IUB)—now the International Union of Biochemistry and Molecular Biology (IUBMB)—appointed an International Enzyme Commission in 1955. The assignment of the Enzyme Commission was "to consider the classification and nomenclature of enzymes and coenzymes, their units of activity and standard methods of assay, together with the symbols used in the description of enzyme kinetics." This ambitious work was done in cooperation with the Biological Nomenclature Commission of the IUPAC (International Union of Pure and Applied Chemistry), and both organizations work together in the Joint IUPAC-IUB Commission on Biochemical Nomenclature (JCBN).

Table 2.1 The general classes of the enzyme classification system developed by the Enzyme Commission (EC)

S.No.	Class	Reaction type	Remarks
1.	Oxidoreductases	Oxidation/reduction	Two-substrate reactions; cofactors are required
2.	Transferases	Group transfer	Two-substrate reactions
3.	Hydrolases (transferases; transfer of groups to water)	Hydrolysis reactions	water is involved as one of two substrates (in most cases in excess) playing the part of an acceptor
4.	Lyases	Non-hydrolytic bond breaking (a), bond formation (b)	(a) One-substrate reaction (b) Two-substrate reaction
5.	Isomerases	Isomerization reactions	Intramolecular group transfer
6.	Ligases	Bond formation	ATP-dependent two-substrate reaction

The enzyme classification adopted by the Enzyme Commission of the IUB is based on a hierarchically ordered sequence of numbers EC w.x.y.z: The first number denotes the general class (e.g., transferase, "2"). The second and the third number refer to the mechanism of the catalyzed reaction but allow to distinguish between different activities (e.g., transferring C1-groups, "2.1" together with methyltransferase, "2.1.1"). The forth number is a serial one within the sub-subclass. It is in the nature of such classification that it makes no distinction between enzymes catalyzing the same type of reaction but being of different sources, so that they may differ from each other with respect to chemical or physical properties due to which they can be separated from each other by physical methods. A similar situation exists in case of isoforms of enzymes that underwent different post-translational modifications. Enzymes belonging to the first enzyme class are oxidoreductases (EC 1.-.-.-) catalyzing oxidation-reduction reactions. If the catalyzed reaction is the oxidation of an alcohol (EC 1.1.-.-), the enzyme acts on the –CHR–OH group of the donor, being oxidized by donating hydrogen or an electron to the acceptor, e.g., NAD^+ (EC 1.1.1.-). The fourth digit is a running number; the resulting systematic name is alcohol: NAD oxidoreductase with the EC classification 1.1.1.1 and the common name alcohol dehydrogenase. All enzymes have a systematic and a common name and are subdivided into six general classes as presented in Table 2.1.

2.2.2 Protein Biosynthesis and Protein Architecture

The central dogma of the flow of information in biological systems is

$$\textbf{DNA} \xrightarrow{\textit{transcription}} \textbf{RNA} \xrightarrow{\textit{transcription}} \textbf{enzyme/protein}$$

The sense strand of the double-stranded structural gene (DNA) carries the information about the order of amino acids within the protein to be synthesized in the form of the sequence of four nucleotides, the bases adenine (A), guanine (G), which are purines, and the pyrimidines thymine (T) and cytosine (C). Different amino acids are defined by a trinucleotide sequence, the (sense) codon; the possible base triplets result in $4^3 = 64$ codon combinations for the 20 canonical amino acids (codon degeneracy; for most amino acids several codons exist), three stop (non-sense) codons and one start codon (also specifying for methionine (formylmethionine in bacteria). Different cells make use of different codons for one and the same amino acid. Some organisms also encode the non-canonical amino acids selenocysteine and pyrolysine known as 21st and 22nd amino acid, respectively.

The sense strand is first transcribed into a messenger RNA (mRNA; RNA contains A, G, C, and uracil (U)) catalyzed by DNA-dependent RNA polymerases in presence of ribonucleoside triphosphates before the information is translated into the respective amino acid sequence constituting the primary structure of the polypeptide chain. The site of the protein biosynthesis, always starting with (N-formyl) methionine (fMet or Met), is the ribosome, a protein/RNA complex. Translation proceeds by linking amino acids to their respective transfer RNAs (tRNA) for transfer to the ribosome. tRNA in its three-dimensional compact structure resembles an "L" containing four unpaired loops. As a consequence of the L-like tertiary structure, the molecule is dipolar. One pole is the anticodon in a variable loop; the three nucleobases of the anticodon are complementary to an amino acid codon of the mRNA. The second pole is located at a distance of about 8 nm from the anticodon. It is characterized by the invariant sequence 5′-CCA-3′ identical in all tRNAs, and serves to bind the respective aa. Aminoacyl-tRNA-synthetases (AARSs) charge the different tRNAs in an ATP-dependent reaction. The AARSs are specific for one amino acid each, and for the corresponding tRNA. The latter is recognized by the synthetase mainly via characteristic structural elements near to the anticodon of the tRNA, and its acceptor 3′-end, respectively. Linking the different activated amino acids via peptide bond formation occurs at the ribosome consisting of a small 30 S subunit and a large 50 S subunit (prokaryotic ribosome; S: Svedberg units). The small subunit made up by about 20 proteins serves to stabilize the position of the mRNA and a 16S-rRNA. This rRNA recognizes the so-called Shine-Delgarno sequence (AGGAGGU) of an mRNA and supports mRNA binding by complimentary base pairing. The initiation phase is followed by the elongation process. The large subunit, possessing peptidyl transferase (PT) activity, has two binding sites, the P (peptide) and the A (acceptor) site, with tRNAfMet (tRNAMet) bound to the P-site. The elongation process requires elongation factors (EF) and starts with binding the second tRNA—matching the next codon of the mRNA—at the A-site. After the second aa residue has been transferred to the initiation aa fMet, tRNAfMet is released and the mRNA with the bound second tRNA moves from the A- to the P-site so that the next aminoacyl-tRNA can bind. This process is continued until the A-site presents one of the stop codons having no complimentary tRNAs. Under these conditions, releasing factors come into action and catalyze the final step, the hydrolysis of the ester bond between the last tRNA and the C-terminus of the peptide chain. A third site—the E (exit)-site is required for releasing the deacetylated tRNAs after transferring its amino acids to the growing peptide chain.

The next structural level of proteins is characterized by *secondary structure* elements, mainly the α-helix and the β-sheet. Their formation occurs spontaneously through the interaction between neighbored amino acid residues. These non-covalent interactions are stabilized by hydrogen bonds (binding energy ~40 kJ/mol) and fold into domains to yield a characteristic 3D or *tertiary structure*, maintained by forces acting between charged and/or hydrophobic amino acid residues, and dipole-dipole interactions. Different domains within a protein may be linked by oligopeptide loops. Another aspect contributing significantly to the stability of the protein molecule is the hydrophobic effect; it results from the fact that folding in an aqueous surrounding leads to a more or less water-free hydrophobic core whereas polar and charged amino acid residues are preferentially located on the protein surface. Finally, a variety of enzymes form a *quaternary structure* through the non-covalent interaction of subunits. The correct folding of proteins is assisted by ATP-dependent folding catalysts known as chaperones. They play, e.g., an important role in the endoplasmic reticulum and its associated degradation machinery in connection with the synthesis of glycoproteins (e.g., Ninagawa et al., 2014; Lederkremer, 2007).

Many proteins are co- or posttranslationally modified, e.g., by (reversible) phosphorylation, *N*- and *O*-glycosylation (both types occur in human erythropoietin (EPO)), fatty acid acylation of membrane-anchored proteins, or hydroxylation of lysine and proline (collagen). A visualization of protein-specific glycosylation in living cells has recently been shown to be possible by means of methods that find wide application in Nanotechnology too. As reported by Doll et al. (2016) a sugar molecule modified in a way to be able to enter a cell where it is then metabolically incorporated is labeled inside the cell with a fluorophore as an acceptor via biorthogonal ligation chemistry. To the proteins of interest enhanced green fluorescent protein (EGFP) is attached, serving as a donor. A detection of Förster Resonance Energy Transfer (FRET;

Chapter 1) resulting from protein-specific glycosylation is possible, presupposed EGFP-labeling is achieved within spitting distance to the site of glycosylation.

Pro-proteins are biosynthesized as inactive precursors and require proteolytic processing for maturation; examples are serine proteases (e.g., chymotrypsin) or the hormone insulin, synthesized as pro-insulin. Proteins designated for the export out of the cytoplasm normally contain an N-terminal signal peptide; hydrolysis of such pre-pro-proteins yields the respective pro-proteins that are activated by splitting of one of few peptides.

Another strategy used by bacteria and fungi to produce peptides as secondary metabolites among them antibiotics (vancomycin, bacitracin), immunosuppressives (cyclosporine) or biosurfactants (surfactin) which are in part of high pharmacological/medical relevance are generated nonribosomally by large multi-enzyme complexes the so-called nonribosomal peptide synthases (NRPS) as well as the polyketide synthases (PKS) that, e.g., synthesize erythromycin, an alternative to β-lactam antibiotics (*Streptomyces erythreus*) or the blood cholesterol lowering lovastatin (*Aspergillus terreus*). In this kind of peptide synthesis, the role of an mRNA as matrix is taken over by single modules of the enzyme complex enabling a stepwise elongation of the respective metabolite. The accepted building blocks are not restricted to the canonical amino acids but D-amino acids and modified amino acids may be also incorporated (Sieber and Marahiel, 2005).

2.2.3 Many Biocatalysts Require Cofactors

Frequently, enzymes develop their catalytic activity solely on the basis of a specific arrangement of some few amino acid residues. However, there are many others that need the assistance of **cofactors** of organic or inorganic origin for being (fully) active. The latter are metal ions, mainly 3d transition metal ions and those from the alkaline and earth alkaline group such as Na^+, K^+, or Mg^{2+}, and Ca^{2+}, respectively. They are either tightly attached to the active site or bound only loosely. Some metal ions are more associated with the substrate than with the enzyme, as in the case of Mg^{2+}-ATP. There are also enzymes where more than one metal ion is essential, e.g., iron-molybdenum clusters in nitrogenases and xanthine oxidase containing in each of its two subunits one

Mo and two [Fe_2S_2] centers. A metal ion or another non-protein group that is bound covalently to an enzyme's active site is termed prostetic group. These are usually distinguished from so-called coenzymes that are organic molecules of low molecular weight. Many coenzymes are in dissociation/association equilibrium with the catalysts and are modified during the reaction so that they rather have the function of a co-substrate; examples are NAD^+ and NADH (the oxidized and the reduced form are termed nicotinamide adenine dinucleotide) that transfer hydride (H^-) ions. However, other coenzymes as the electron transferring cofactors FAD (flavin adenine dinucleotide) and FMN (flavin mononucleotide) are bound to the respective enzyme rather tightly—sometimes even covalently as in case of the flavoprotein succinate dehydrogenase (EC 1.3.99.1). Coenzymes in addition may contain a metal ion that is coordinated by electron pair donating atoms; in some rare cases, real metalorganic M^{z+}–carbon bonds are formed.

Many cofactors/coenzymes such as NAD^+ and NADH involved in multiple redox reactions are still rather expensive; on the other hand, depending on the direction of the redox reaction to be catalyzed NADH or NAD^+ must be present in stoichiometric concentrations. For running such a process continuously under economically feasible conditions (Chapter 2, Vol. 1), i.e., in presence of catalytic amounts of cofactors, they have to be regenerated in situ, which is possible by different methods, including chemical, electrochemical, photochemical, and enzymatic ones, out of which only the enzymatic methods have been proven until now. Three different enzyme-mediated cofactor regeneration methods are known, the *coupled-substrate-recycling*, the *coupled enzyme recycling*, and the *coupled system recycling*. The first one is the simplest of the three methods; it merely needs a co-substrate accepted by the enzyme used for the transformation in question. The reaction sequence is as follows: reduction of the substrate in presence of NADH leads to the formation of NAD^+ that on its part oxidizes the co-substrate and is reduced back to NADH; for more information about this topic and recent developments, see Lauterbach et al. (2013) or Uppada et al. (2014).

An interesting example of how it is possible to electrically wire the active site of a redox enzyme has been demonstrated already in 2003 by Xiao et al. with a "plugging into glucose oxidase"

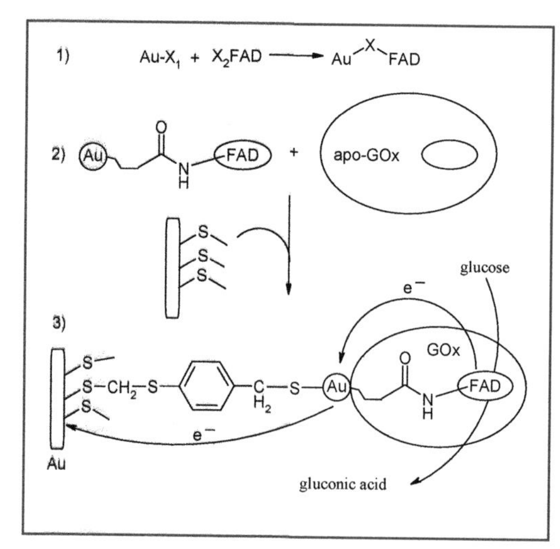

(GOx). As is to be seen from the opposite scheme (redrawn from Xiao et al. with modifications) they first functionalized gold nanoparticles (Au nanocrystals with a diameter of 1.4 nm) with *N*-hydroxy-succinimide; the modified Au-nanoparticles were then reacted with N^6-(2-aminoethyl)-FAD. Thereafter, this Au-X-FAD adduct was used to reconstitute an apo-glucose oxidase. Finally, the Au-X-FAD-GOx construct was assembled on an Au-electrode (0.2 cm^2) the surface of which was modified with a 1,4-dimercaptoxylene monolayer. The surface coverage with reconstituted GOx was determined as 10^{-12} mole/cm^2, corresponding to a 60% coverage by a protein monolayer. Electron transfer was characterized by cyclic voltammograms, showing an increase of the electro oxidant current with increasing glucose concentration. The unimolecular electron transfer rate constant calculated from the saturation current and the coverage of the electrode surface with enzyme was 5,000 s^{-1}, about seven times more than in case of native GOx with oxygen as electron acceptor. The Au-nanoparticle serves as an electrical relaying element between the surface of the macroelectrode and the FAD redox site within GOx.

2.2.4 Enzyme Kinetics

The term enzyme kinetics is somewhat misleading as one may draw the conclusion that basic principles of chemical kinetics are not valid in this area, which is of course not the case; enzymes accelerate a reaction as other catalysts do by reducing the (Gibbs free) activation energy—the activation energy between the transition state (TS) and the substrate (S)

$$\Delta G^{\#} = G_{TS} - G_S$$

but without changing the reaction Gibbs energy ΔG of the reaction which is 0 when the equilibrium is reached. As $\Delta G^{\#}_{uncat} > \Delta G^{\#}_{cat}$, the interaction between an enzyme (E) and a substrate (S) generates and stabilizes a new reaction path with a transition state having a lower activation energy.

2.2.4.1 The active site

The catalyzed conversion of a substrate to a product requires its interaction with the catalyst—in this case the enzyme. The site of encounter is the enzyme's active site where the substrate—and if required the cofactor—bind. Of the many amino acid residues an enzyme normally consists of (> 100 amino acids; ~10 kd, diameter: ~2.5 nm) only a few make up the three-dimensional active site; all the others serve among others to stabilize its structure in the environment where it acts.

The amino acids contributing to the formation of the active site, the catalytic groups, are usually located at the surface of the enzyme within a predominantly polar cavity. They originate from rather different parts of the enzyme's sequence, and are in closer contact than immediately neighbored ones. Their interaction with the substrate is in most cases highly specific and established mainly by many weak non-covalent interactions and directed hydrogen bonds forming short-ranged forces.

Already in 1894 Emil Fischer, who was awarded the Nobel Prize in Chemistry 1902, gave an explanation for the substrate specificity with the analogy between a lock and a key. In one of his publications, Fischer put his conclusions in the words: "*To use a picture I want to say that enzyme and glucoside must go together like* ***key and lock*** *in order to enable a chemical effect.*" In this analogy, the substrate represents the correctly sized key with correctly arranged and sized teeth on it, mimicking the functional groups, and the lock is the enzyme itself containing the active site as the key hole. The key and lock (or template) theory is still valid and of great use. However, not all experimental results that emerged from progress in enzymology could be explained by this rigid model. Protein dynamics provide, e.g., a link between hydrogen tunneling and the catalysis of C–H bond cleavage. Protein motions are also associated with substrate/ligand binding and release. An advancement of Fischer's ideas is the *induced fit theory* proposed by Daniel E. Koshland Jr. (1958), which is based on the assumption of structural flexibility of the enzyme molecule with the consequence that free enzyme (often named the open form) and enzyme-substrate complex (closed form) are different in their three-dimensional shape—in other words enzymes may adopt a complementary form to the substrate only after binding to it.

The activity of most of all enzymes acting in solution is pH-dependent due to the fact that they contain carboxylate-, amino-, or other functional catalytic groups. If such a group can be protonated, the activity of the enzyme will increase until to a maximum, the so-called pH-optimum which lies at pH = 7.5 for the majority of enzymes, beyond which the protonation of a second active site group will result in a stepwise decrease of activity.

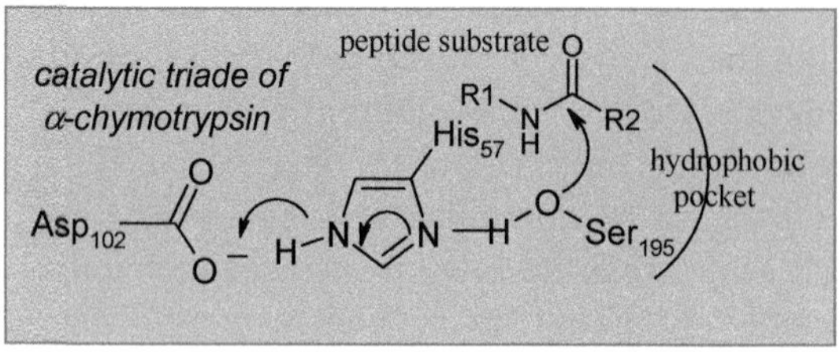

As an example of the mechanism of an enzyme, the action of a peptidase on the hydrolysis of peptide bonds is briefly discussed. Peptidases are divided into different groups of which one is the serine endopeptidase α-chymotrypsin that cleaves peptide bonds after bulky amino acid residues such as tyrosine, phenylalanine, or tryptophan. The active site of this enzyme is made up by the *catalytic triad* Ser195, His 57, and Asp 102. A proton transfer from Ser to His and from there to Asp initiates a nucleophilic attack of the Ser oxygen on the carbonyl bond of the substrate as depicted in the scheme opposite. The following reaction steps (not shown) are the separation of the bond substrate by base catalysis of His 57 into a leaving group (binding the H atom transferred from Ser to His) and a second part that remains bond to the Ser as an acyl-enzyme intermediate; it is released by a nucleophilic attack of water supported by His 57 and Asp 102 (for more details see Berg et al., 2002).

2.2.4.2 The Michaelis–Menten equation

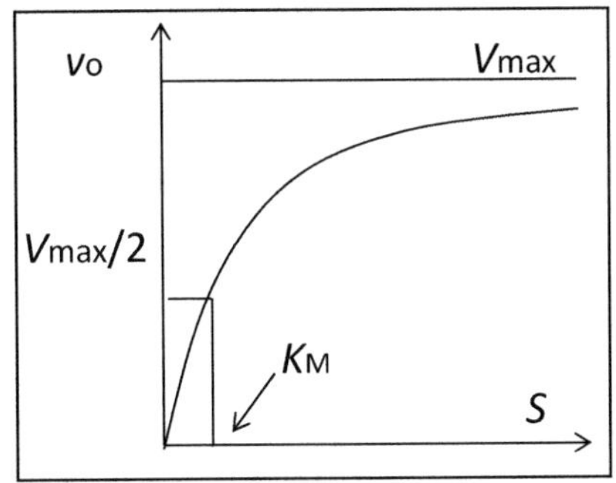

A characteristic of many enzyme-catalyzed reactions is that the reaction rate (V) increases with increasing substrate concentration (S); the increase is first linear but goes down successively to finally reach a maximum (V_{max}). In 1913, Leonard Michaelis (1875–1945) and Maud Menten (1879–1960) published an equation by which the kinetics of many enzyme-catalyzed reactions can be described based on the steady-state approximation. The concept is based on the decay of an enzyme-substrate complex (Victor Henri, 1902) to the free enzyme and products as the rate-determining step.

$$E + S \rightleftharpoons ES \rightarrow E + P$$

The Michaelis–Menten equation can be then written as

$$V = \frac{k_1 k_{cat}(S)}{k_1(S) + k_{-1} + k_{cat}}$$

or

$$V = \frac{k_{cat}(S)}{K_M + (S)},$$

where k_{-1} and k_1 are the rate constants for the association/dissociation equilibrium between E and S and the ES complex and the Michaelis–Menten constant is given by $(k_{-1} + k_{cat})/k_1$. From V_{max} in the above scheme, which is the product of k_{cat} and the total enzyme concentration (E_t) present, k_{cat} results as $V_{max}/(E_t)$, which is the total turnover number, the number of substrate molecules converted to the product P presupposed the enzyme molecules are fully saturated with the substrate.

The ratio of $\boldsymbol{k}_{cat}/K_M$ (in $s^{-1}\cdot M^{-1}$) is taken as a measure of the catalytic efficiency η. For example, it has been mentioned above that chymotrypsin favors the conversion of bulky peptide substrates;

for those with phenylalanine η is about 10^5, for glycine this value is just 1.3×10^{-1} (Ferch, 1999).

A question of interest is the possible maximum value of k_{cat}/K_M, i.e., what is the maximal efficiency of an enzyme. This may be answered by substituting K_M for the rate constants that it consists of

$$\frac{k_{cat}}{K_M} = \frac{k_2}{\frac{k_{-1} + k_2}{k_1}}.$$

Hence with $k_2 >> k_{-1}$, k_{cat}/K_M approaches k_1. As k_1 is now the rate constant for the formation of the ES-complex this process is rate determining, and the maximal reaction rate or efficiency of the biocatalyst is characterized by the situation that each encounter between the enzyme and the substrate molecule is successful with respect to product formation. This encounter in solution is diffusion-controlled so that the maximal value of k_1 or k_{cat}/K_M, respectively is in the order of 10^8 to 10^9 $L \cdot mol^{-1} \cdot s^{-1}$. It is said that enzymes with such k_{cat}/K_M-values catalyze a reaction at the upper limit, or that they act as perfect catalysts.

2.2.4.3 Temperature dependence

The temperature dependence of many enzyme-catalyzed reactions can be described by the well-known Arrhenius equation. However, the activity decreases after exceeding a certain temperature range due to the thermal instability of enzymes. This temperature range of maximum activity is often named "temperature optimum," a term that has obviously been chosen in analogy to the term pH-optimum, which could be misunderstood as the phenomenon pH-optimum is solely explained by thermodynamics, based on well-described equilibria. However, the effect of temperature on enzyme-catalyzed reactions is governed by two different properties of an enzyme. Firstly, activity increases with increasing temperature as normally expected, but, secondly, the stability of the enzyme decreases

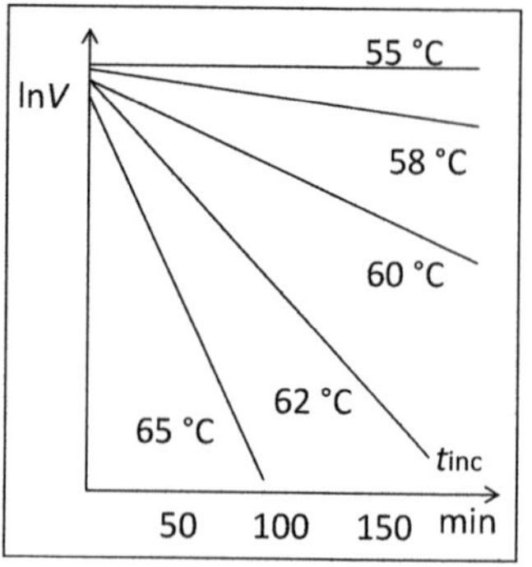

producing a decline in activity due to heat denaturation, a process that proceeds with increasing incubation time. Thus, a description of the experimental conditions for measuring temperature dependence of such reactions must include the incubation time t_{inc} of the enzyme at the respective reaction temperature. The data given oppositely in a schematic way for the enzyme β-galactosidase demonstrate that the heat inactivation occurs within a rather narrow temperature range—in this case between 58°C and 65°C. As all measuring points can be fitted rather satisfactorily to a linear equation, the conclusion to be drawn is that the transition of an enzyme from its active to its denaturated state can simply be described by first-order kinetics. A plot of the logarithmized slopes from these data versus the respective reciprocal absolute incubation temperatures yields an activation energy for the denaturation of β-galactosidase of E_a = 740 kJ/mol, indicating again the strong temperature dependence of this process.

2.2.4.4 Promiscuous enzymes

The occurrence of promiscuity of enzymes, treated comprehensively by Wu and Lin (2014) is assumed to be due to gene duplication events during evolution as a starting point for divergent evolution of new enzyme families (Thoden et al., 2004, and literature cited therein). The result might have been an enzyme that makes use of the original chemistry for transformation but is able to apply this chemistry to alternative substrates due to some mutations in the amino acid sequence; such enzymes exhibit substrate promiscuity and are characterized by a broad substrate specificity. Alternatively, an enzyme is evolved by mutation in a way that the active site developed the capability of catalyzing reactions based on different reaction mechanisms; these enzymes are functionally promiscuous (catalytic promiscuity). It was shown by Khersonsky et al. (2011) through screening the *Escherichia coli* proteome with a series of xenobiotics as substrates that substrate promiscuity occurred significantly more frequent than catalytic promiscuity.

A remarkable example of a catalytically promiscuous enzyme is the *Candida antarctica* lipase B (CAL-B; EC 3.1.1.3) that belongs to the hydrolases. Branneby et al. in 2003 reported that the enzyme is also capable of catalyzing C–C bond formation, normally done by aldolases where a Zn^{2+} ion or a lysine residue is involved in the underlying catalytic mechanism. The CAL-B catalyzed aldol addition of hexanal performed in quasi anhydrous cyclohexane at room temperature is depicted in the above scheme.

2.3 Design of Biotransformations and Enzymatically Catalyzed Reactions

This chapter deals with different ways of designing enzyme-catalyzed reactions. Apart from transferring the biocatalyst to an insoluble state by immobilization the reaction in question may be performed in solvents other than water which are organic solvents, ionic liquids, or supercritical fluids. In addition the application of active site mimics is briefly discussed.

2.3.1 Immobilization of Biocatalysts

Enzymes and whole cells are both used as dissolved or suspended catalysts, respectively, or are transferred into an immobilized—water insoluble—state through attachment to the surface of a water-insoluble carrier of organic or inorganic origin or by entrapment within the porous matrix of such materials. The pro and cons of micromagnetic porous and nonporous biocatalyst carriers have been discussed theoretically and with practical examples by Stolarow et al. (2014); the state of the art of robust enzyme preparations for industrial applications is the topic of an article provided by Thum et al. (2014).

The attachment of a biocatalyst via a covalent linkage, an adsorptive or ionic binding, etc., may require the decoration of the carrier's surface with suited spacers containing terminal functional groups such as $-SH$, $-NH_2$, aldehyde, epoxide, *N*-hydroxysuccinimide esters and others for reaction with exposed functional groups of the biocatalyst (e.g., $-NH_2$ of Lys), leading in the case of covalent attachment to its chemoligation in a non-specific way. Chemically inert polymers (polyethylene, polystyrene, etc.) can be treated by means of various plasma technologies (Siow et al., 2006) resulting in carboxylated (CO_2-plasma), hydroxylated (O_2-plasma) or aminated (H_2/N_2-plasma) polymer surfaces with enhanced hydrophilic character; for successful application examples, see Ghasemi et al. (2011) or Vorhaben et al. (2010). The advantages of immobilization are an easy separation of the catalyst from the reaction mixture, the possibility to reuse the catalysts several times, e.g., in a batch or fed-batch reactor, and an often higher stability which may enable to carry out the reaction at higher temperatures. To the possible disadvantages of immobilization belong a sometimes lower activity of the biocatalyst compared to its free counterpart, mainly due to mass transfer limitations and/or the chemistry used for immobilization, loss of the biocatalyst through leakage, and additional costs for the carrier material. Diffusion limitations may be reduced by the shape of the carriers (e.g., the lens-shaped Lenticat® beads prepared from polyvinyl alcohol; Vorlop and Jekel, 1998) and/or the choice of the material employed (Schoenfeld et al., 2013). Carrier-free immobilization is achieved by cross-linking enzyme molecules—mostly glutaraldehyde is used as bifunctional reagent—either in their crystalline state (cross-linked enzyme crystals, CLECs) or based on precipitated enzyme aggregates (cross-linked enzyme aggregates, CLEAs; Sheldon, 2011). The catalysts obtained by both these procedures are characterized by a largely preserved native structure, higher stability towards organic solvents and heat, and an often high storage and operational stability. Because the production of CLECs requires purified enzymes and their laborious crystallization, the CLEA® technology (also in the form of Combi-CLEAs where different enzymes are co-precipitated and co-cross-linked; Mateo et al., 2006) is preferred for industrial applications, among others due to the fact that they may be prepared from crude enzyme preparations (Sheldon and van Pelt, 2013, and literature cited therein).

Immobilization procedures should be thoroughly documented, e.g., according to the guidelines published by the Working Party on Immobilized Biocatalysts within the European Federation of Biotechnology (1983), illustrated in more detail with the immobilization of *Nitrosomonaseuropaea* by entrapment through ionotropic gelation within calcium alginate (van Ginkel et al., 1983). Compared to enzymes, immobilization is less common in cases where cells are utilized as catalysts. However, an approach of some biotechnological relevance is an immobilization of whole cells via cell surface expression of carbohydrate-binding modules such as the cellulose-binding domain (Wang et al., 2001), or the chitin-binding domain for the stable immobilization of cells onto the surface of chitin beads (Wang and Chao, 2006). The different immobilization methods and kinds of carrier materials with their pros and cons have been described, e.g., by Cantone et al. (2013), Liese and Hilterhaus (2013), Datta et al. (2013), DiCosimo et al. (2013), Buchholz et al. (2012), Grunwald (2009), Spahn and Minteer (2008), Blickerstaff (1997), and others. Franssen et al. (2013) in a comprehensive publication treat the topic of enzyme immobilization in connection with the production of biorenewables. For a variety of biocatalytic processes, immobilization strategies have been developed for application in a membrane bioreactor (MBR). Chakrabortyace et al. (2016) have summarized the advantages of MBR over conventional (e.g., packed bed etc.) ones such as reduced energy consumption and simultaneous process intensification caused by large surface of the immobilization matrix, reuse of the enzymes and improved product recovery. The authors review various immobilization techniques with the respective reactor setups as they are actually applied in the fields of sugar, starch, drinks, in the pharmaceutical industry, and for energy conversion.

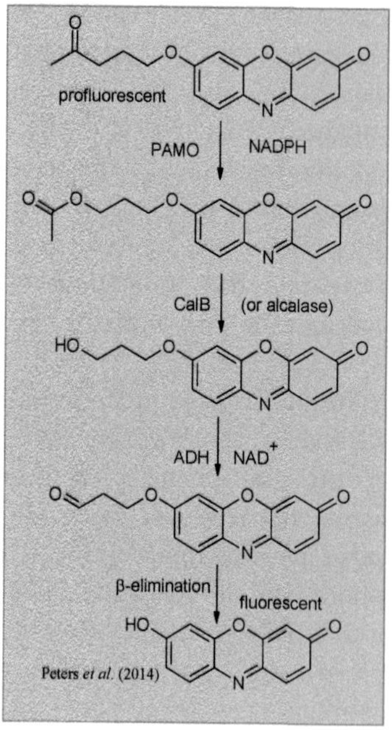

An immobilization with a focus on the selective/directed covalent attachment of proteins onto solid supports plays an important role in the fabrication of biosensors, or protein microarrays in connection with high-throughput methods for studying protein-protein or protein-small molecule and other interactions, a topic excellently reviewed, e.g., by Wong et al. (2009), Jonkheijm et al. 2008, and also in Chapter 22.

A particular immobilization strategy employed by Nature in connection with cascade reactions shall be mentioned where the sequential steps of the respective pathways are catalyzed by enzymes involved in countless chemical reactions for different purposes, including energy storage and conversion or communication with the cellular environment (García-Junceda et al., 2015). To circumvent possible incompatibilities between different catalysts, substrates, intermediates, products, reaction conditions, or whole pathways, Nature uses compartmentalization, i.e., different reactions are run in different organelles, a strategy developed to perfection during evolution of eukaryotic cells. Peters et al. (2014) described cascade reactions in multicompartmentalized polymersomes (artificial vesicles surrounded by a membrane synthesized from amphiphilic block copolymers). They encapsulated different enzymes (CAL-B and ADH) in polystyrene-b-poly(3-(isocyano-L-alanyl-amino-ethyl)-thiophene) (PS-b-PIAT) nanoreactors serving as organelle mimics, which were then mixed with a "cytosolic" enzyme phenylacetone monooxygenase (PAMO) and the required reagents; the mixture was subsequently encapsulated within micrometer-sized polybutadiene-b-poly(ethylene oxide) (PB-b-PEO) vesicles to generate a functional cell mimic. The average sizes of the porous enzyme nanoreactors retaining the large biopolymers but enabling substrate diffusion across the membrane were 187 nm (CAL-B) and 318 nm (ADH), respectively. The cascade reaction studied is shown in the opposite scheme. The profluorescent resorufin derivative is oxidized by NADPH-dependent PAMO; the resulting ester is hydrolyzed by the lipase CAL-B to give an alcohol that is oxidized in presence of an alcohol dehydrogenase (ADH) and NAD^+ to the respective aldehyde converted into the fluorescent product resofurin via a spontaneous

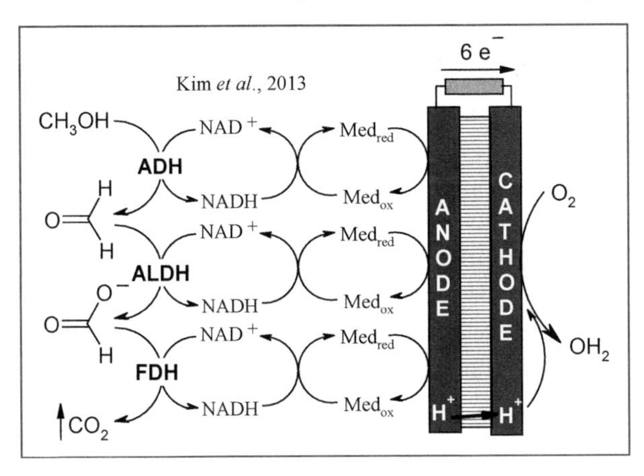

That this multienzyme and multicompartmentalized β-elimination cascade design fulfills one of the main characteristics of compartmentalization, the spatial separation of incompatible enzymes, was shown substituting CAL-B by the protease alcalase. In presence of free alcalase, the cascade reaction was severely inhibited due to degradation of the other two enzymes present. When, however, the reaction was performed with alcalase separated from the reaction mixture by encapsulating in a PS-b-PIAT nanoreactor, formation of the product within the cell mimics took place as confirmed by confocal fluorescence microscopy. Such multicompartmentalized systems may be among others of great use to study cellular processes in more detail (Peters et al., 2014).

As a final field of high research activities, the fabrication of biobatteries shall be mentioned. In 2013, Kim et al. reported about a new approach to combine the three enzymes alcohol dehydrogenase (ADH), an aldehyde DH (ALDH) and a formate DH (FDH)—all NAD(H)-dependent—into a synthetic metabolic pathway to convert methanol to CO_2. They made use of findings published by Petka et al. (1998), who reported about hydrogel forming engineered proteins. The self-assembling artificial proteins were constructed as triblocks containing a central polyelectrolyte water-soluble domain devoid of any regular secondary structure and flanked by terminal basic region leucine zipper (bZIP) domains. Characteristic of the bZIP motif is a periodic repetition of leucine (L) residues within an α-helix up to 80 aa residues in length at every seventh position, known as heptad repeat (Landschulz et al., 1988). The gelation of the artificial proteins synthesized by Petka et al. (1998) is driven by the aggregation of the two ZIP domains through coiled-coil hydrophobic interactions whereas the central Ala-Gly-rich polyelectrolyte part contributes to the highly swollen character of the obtained hydrogels. The group of Banta had demonstrated that this synthesis strategy can be transferred to globular proteins with a quaternary structure, such as a thermostable ADH (Wheeldon et al., 2009); here the hydrogel-like properties materialize as a consequence of the attached leucine zipper domains and additional protein-protein interactions due to the quaternary structure of the proteins involved, the catalytic activity of which is not affected by this procedure. For their methanol biobattery (Kim et al., 2013) they genetically appended a ZIP domain and a randomly structured soluble peptide domain to the N-termini of the three enzymes employed, a tetrameric ADH (*Bacillus stearothermophilus*), a tetrameric human ALDH, and a dimeric FDH (*S. cerevisiae*); all three enzymes obey an ordered bi-bi mechanism. Remarkably, two of the modified enzymes showed significantly improved catalytic efficiencies; the k_{cat}/K_M-values for ADH and FDH increased by six orders and by two orders of magnitude compared to those of the w.t. enzymes. With this device a maximum current density of >26 $mA \cdot cm^{-2}$ was obtained with methanol as fuel which is significantly more than achieved earlier with other biobatteries.

2.3.2 Non-Conventional Reaction Media

Most biotransformations are carried out in aqueous solutions; however, non-conventional media such as organic solvents may also be employed, and in principle ionic liquids, or supercritical solvents, looked at as "green answers" to organic solvents (Anastas and Eghbali, 2010), too. A substitution of water by unconventional solvents is in many cases associated with a variety of advantages with the main one being the possibility to use substrates insoluble in water. Compared with water as reaction medium, enhanced stability, and catalytic activity of biocatalysts together with a change of properties, including, e.g., an inversion of the enantioselectivity have been reported. Whether or not such advantages apply to a given biotransformation system cannot be predicted and non-conventional reaction media employed since the 1980s still represent an active field of research. In any case at least a small amount of residual water is indispensable for biocatalytic activity.

2.3.2.1 Organic solvents

The use of organic solvents as reaction medium for enzyme-catalyzed reactions is known since many decades (Sym, 1936) but the findings published obviously went unnoticed by the scientific community. About 30 years ago, Zaks and Klibanov (1984) reported that porcine pancreatic lipase catalyzes the transesterification reaction between tributyrin and various alcohols in quasi water-free organic media at temperatures above 100°C, and already some years earlier Klibanov et al. (1977) proposed "a new approach to preparative organic synthesis" where the enzymatic process is performed in a system containing water together with a water-immiscible organic solvent as a second phase. Biocompatible organic solvents have several advantages:

- they offer the possibility to enzymatically convert nonpolar substrates and to increase the solubility of substrates and products with a positive effect on space/time yield and volumetric productivity;
- the outcome of the reaction in question is changed towards synthesis instead of hydrolysis;
- in addition a solvent may change the catalyst's properties—sometimes termed "solvent mediated enzyme engineering"—such as its enantioselectivity (Tawaki and Klibanov, 1992; see also Chen and Sih, 1989).

However, the presence of an organic solvent is also often associated with a significant reduction of catalytic activity (Zaks and Klibanov, 1988). Nonpolar solvents (log $P > 2$), although often contributing to significantly increased temperature stability (Grunwald et al., 1988, 1993), may impair the hydrophobic core of the enzyme molecules; more hydrophilic solvents, on the other hand, remove essential water molecules from the enzyme surface and its active site, or disrupt hydrogen bond interactions between subunits, thereby affecting the kinetic values (V_{max}, K_M). Hence the polarity of the reaction medium has to be thoroughly balanced with respect to enzyme stabilization and inactivation (Karan et al., 2012; and literature cited therein). Investigations into the so-called "water activity, a_W" by Halling (1984, 1994) and Goderis et al. (1987) led to new insights concerning the relation between the amount of water in a non-aqueous reaction medium and the activity of enzymes. For a recent systematic study on the activity and enantioselectivity of a (*S*-) selective hydroxynitrile lyase in organic solvents (MTBE, toluene, octane) as a function of the water concentration, see Paravidino et al. (2010). Reviews about biocatalysis in organic solvents have been published, e.g., by Carrea and Riva (2000), van Rantwijk and Sheldon (2007), Lozano (2010), Moniruzzaman et al. (2010), and for examples of industrial biotransformations in organic solvents or biphasic systems see Liese et al. (2006), Krishna (2002), Milner and Maguire (2012).

Three additional application examples are mentioned here in order to underline the diversity of this approach. The synthesis of nonionic surfactants from renewable resources for application in food, cosmetics, and the pharmaceutical industry is an actual biosynthetic topic. Lin et al. (2016) employed vinyl laurate and glucose as starting materials. They tested among others a lipase from *Aspergillus oryzae* for its capability to catalyze this conversion and found that productivity results were comparable to those published for, e.g., Novozyme 435 when the conversion was performed in anhydrous 2-methyl-2-butanol, at 60°C, and a vinyl laurate to glucose ratio of 2:1.

A completely new strategy was recently presented by Mukherjee and Gupta (2016) for the biodiesel synthesis (Chapter 9) from soybean oil using a lipase from *Thermomyceslanuginosus* that they precipitated over Fe_3O_4 nanoparticles. Under optimum reaction conditions they achieved a complete conversion within 3 h at 40°C in a nearly solvent-free medium.

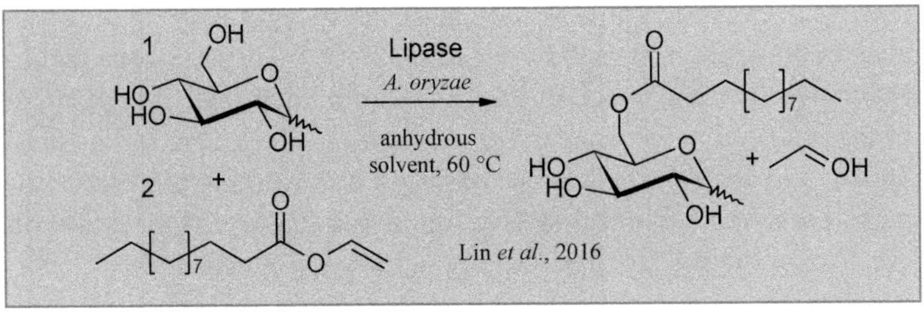

In connection with the search of substrates for transacylation reactions catalyzed by the enzyme *Candida antarctica* lipase A (CAL-A) that compared to CAL-B has not found as much application in this field so far, Wikmark et al. (2015) modified natural Cal-A through directed evolution. A combinatorial gene library was generated with seven positions of the active site targeted simultaneously. The different variants of the enzyme were isolated on a microliter plate via a His_6-tag. A screening for transacylation of 1-phenylethanol in isooctane revealed the highest enantioselectivity for the double mutant Y93L/L367I with an E-value of 100 (*R*) compared to the wild-type CAL-A (E=3).

2.3.2.2 Ionic liquids

Organic solvents represent a high amount of hazardous industrial waste with the well-known negative effects concerning air and soil pollution, climate change, etc. An alternative is the use of ionic liquids (ILs) with a melting point below 100°C. Their polarity can be tailored by appropriate combination of anions and cations. They are normally of high thermal and storage stability and able to dissolve a large variety of inorganic, organic, and polymeric compounds, including cellulosic material (e.g., Abdulkhani et al., 2013). ILs have a negligible vapor pressure, are inert, and due to their pronounced biocompatibility do not inactivate enzymes as polar organic solvents often do. Typical "first generation" ILs were prepared from, e.g., 1-butyl-3-methylimidazolium (BMIM) as cation with tetrafluoroborate or hexafluorophosphate as anion and exert polarities similar to methanol; the first reports about successful (two-phase) biotransformations in such solvents date back to the year 2000 (Cull et al, 2000; Lau et al., 2000). These ILs are rather expensive, which in part is due to the fact that they have to be purified as a requirement for successful application (Park and Kazlauskas, 2003), and their "greenness" is limited due to their high stability, partial water-solubility and low biodegradability; for a review on the environmental fate and toxicity of ILs see Thi et al. (2010).

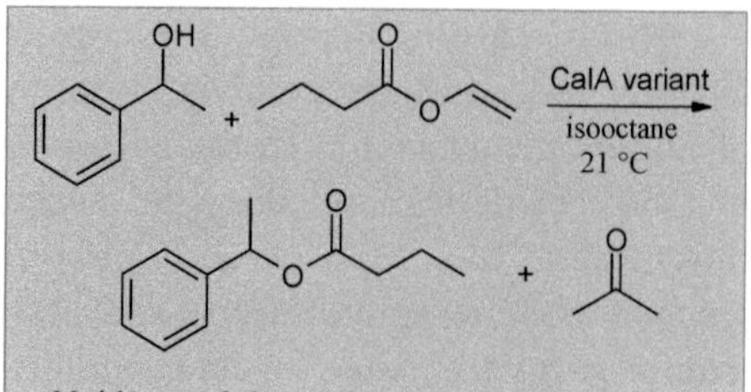

CalA variant
isooctane
21 °C

Model transacylation reaction; Wikmark et al., 2015

Several examples of alternative easy to prepare and biodegradable ILs, so-called advanced ILs, have been reported during the recent years, among them protic ionic liquids (PILs) of the type $[R^1R^2R^3NH^+][X^-]$, simply synthesized by mixing tertiary amines with carboxylic acids, so that X^- is an acetate, propionate, or hexanoate anion. They were successfully tested by de los Rios et al. (2012) with the kinetic resolution of 1-phenyl ethanol by *Candida antarctica* lipase (not soluble in the PILs). Other examples of comparatively easy to synthesize ILs with improved biodegradability contain choline as cation combined with anions derived from amino acids, sugars, or alkylsulfates; they may be used as cosolvents with water to support the solubility of hydrophobic substrates. Furthermore, the rather new in most cases water-soluble and very promising deep eutectic solvents (DESs) are among the advanced ILs; they consist of a salt such as choline chloride mixed with an uncharged hydrogen bond donor, e.g., glycerol or urea, and the eutectics are liquid at room temperature. The advantages and limitations of using cholinium-based DESs and ILs have been studied by Bubalao et al. (2015) with immobilized CAL-B as catalyst to synthesize butyl acetate from 1-butanol and acetic anhydride. The addition of water as protic solvent to the DESs enhanced enzymatic activity and product yield.

One of the most advantageous properties of ILs—apart from being more or less green—is that they can replace polar organic solvents without destroying the enzyme's catalytic activity and hence extend biotransformations to a polarity range that was inaccessible before the advent of ILs (Park and Kazlauskas, 2003). Disadvantages are apart from being expensive problems related to the separation of products from ILs and their recovery and reuse. Reviews about ionic liquids in biocatalysis have been published, e.g., by Kragl et al. (2002), Park and Kazlauskas (2001, 2003), Yang and Pan (2005), van Rantwijk and Sheldon (2007), Cantone et al. (2007), Gorke et al. (2010), Habulin et al. (2011), Tavares et al. (2013). For a new class of deep eutectic solvents derived from natural products see Dai et al. (2013), and for reports focusing on properties and applications of different kinds of deep eutectic solvents Zhang et al. (2012), Hayyan et al. (2013), and Guo et al. (2013). It was also reported that imidazolium-based ILs such as 3-alkyl-1-methyl imidazolium ILs may serve as solvent as well as catalyst as demonstrated with the esterification between α-tocopherol and succinic acid (Tao et al., 2016).

2.3.2.3 Supercritical fluids

Supercritical fluids (SCFs) are used as solvents as well as extraction media, e.g., in combination with ionic liquids (see below). They are generated from compounds such as CO_2 heated and compressed above their critical point as described in Chapter 16, Vol. 1. The critical volume V_c of CO_2 is distinctly lower than the one of normal gases, indicating that gases in the critical state are considerably denser and explains the designation "supercritical fluid." Hence, SCFs can be used as reaction media with a variety of advantages: the diffusivity of dissolved molecules is higher than in real solvents meaning that mass transfer limitations are low, as is the viscosity of SCFs.

Apart from CO_2, water, methanol, ethane, ethylene, SF_6 (nearly as non-polar as xenon) acetone, and others have been used as SCFs. The reason why $scCO_2$ is often given preference over other SCFs is that it lacks all the hazardous properties organic solvents may have, and upon degassing the reaction system, the solvent vanishes completely (e.g., Anastas and Eghbali, 2010). In addition, the T_c-value is rather low so that temperature-labile compounds are less impaired. On the other hand there are a variety of reports according to which $scCO_2$ may show inhibitory effects. For example, Kamat et al. (1992, 1993, 1995) found fluoroform instead of $scCO_2$ to be an ideal solvent for the transesterification of methylmethacrylate with 2-ethylhexanol catalyzed by *Candida cylindracea* lipase. Problems arising from the use of $scCO_2$ come from the possible formation of carbamates from CO_2 and basic amino acid side chains and the formation of carbonic acid, lowering the pH value of the enzyme's microenvironment. A further disadvantage of SCFs in general is that their application—although a mature technique—requires a sophisticated apparatus equipment.

The influence of pressure on the reactions in SCFs is governed by two effects, that of pressure on the reaction rate itself, and changes in the reaction behavior by pressure dependent parameters such as solvating properties, partitioning coefficient, and the dielectric constant; an increase of the latter with increasing pressure means that the solvent becomes more hydrophilic and may, e.g., lead to a change in enantioselectivity (Kamat et al., 1993). The water activity is a key parameter of enzyme kinetics in organic solvents or ILs, which also holds for SCFs. The solubility of H_2O in the rather nonpolar CO_2 is low but may be adjusted to some extent by the reaction temperature (Jackson et al., 1995). The low polarity of $scCO_2$ is also the reason why this dense gas is preferentially employed for biotransformations with hydrophobic compounds. Kasche et al. (1988) reported that α-chymotrypsin, trypsin, and penicillin acylase partially unfold in $scCO_2$ during the depressurization step, particularly in humid $scCO_2$. A review about supercritical fluid technologies as an alternative to conventional processes for preparing biodiesel has been provided by Bernal et al. (2012). For a review of enzyme stability and activity in non-aqueous reaction media see Wang et al. (2016).

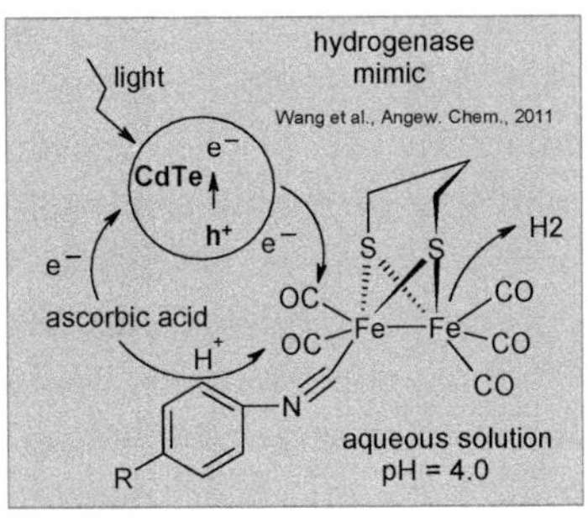

Several instances of the combined use of ILs and SCFs have meanwhile been described, among them the CAL-B catalyzed acylation of octan-1-ol (Reetz et al., 2002; see also Lozano et al. (2002) for a similar approach). In this continuously working process, the substrates were introduced into the CO_2 stream, and after passing the reactor containing the IL/CAL-B suspension, the $scCO_2$ was depressurized, followed by product collection and analysis. In such reaction systems with $scCO_2$ as mobile (and extractive) phase the employed ILs may also serve to protect the enzyme against the sometimes deleterious effect of CO_2 (Garcia et al., 2004).

2.3.3 Hydrogenase Mimics for Hydrogen Production

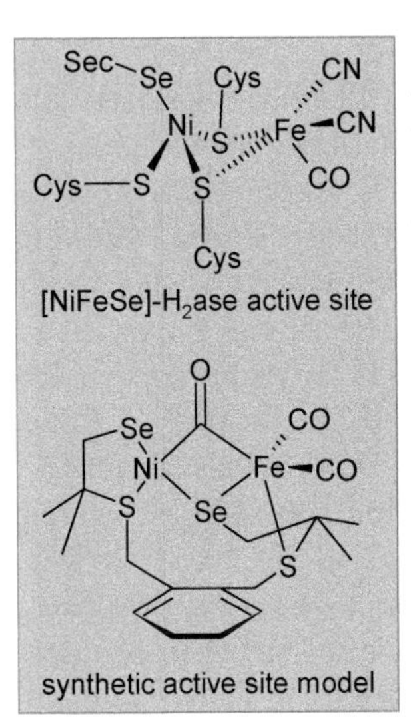

Chapter 15, provided by Tremel and coworkers, deals with inorganic nanoparticles that mimic enzymatic activity; these nanoparticles are also termed nanozymes. An alternative are enzyme mimics that rely on synthetic replications of the active site of enzymes. In connection with a transfer from a fossil-based to a hydrogen-based economy hydrogenase ([Fe]-hydrogenases, [FeFe]-hydrogenases, [FeNi]-hydrogenases) mimics are of particular relevance. For the photocatalytic H_2-production various inorganic hybrid photocatalysts have been developed were a hydrogenase or a hydrogenase mimic are combined with an inorganic semiconductor.

An efficient photocatalytic system for hydrogen production with a hydrogenase [FeFe]–H_2ase mimic (opposite scheme) in an aqueous solution at room temperature by irradiation with visible light has been described by Wang et al. (2011). To the artificial active site of the [FeFe]–H_2ase mimic hydrophilic ether groups (not shown) were attached via a cyanide group to enhance the solubility. As a photosensitizer CdTe Qdots, stabilized by 3-mercapto-propionic acid were used. Ascorbic acid served as a proton source as well as an electron donor for the hole formed in the CdTe Qdots after electron transfer. Optimal conditions for the light-driven H_2 production were 1.56×10^{-4} M [FeFe]–H_2ase catalyst, 1×10^{-3} M CdTe Qdot, and 8.52×10^{-2} M ascorbic acid at pH = 4, which led to a total of 17.6 mL H_2 after 10 h of irradiation ($\lambda > 400$ nm). An improved version consisting of a polyethylene imine-grafted [FeFe]-H_2ase mimic working in a broad pH-range due to the high buffering capacity and stabilizing ability provided by the branched PEI secondary coordination sphere has been reported by Liang et al. (2015). Becker et al. (2016) reported the first [FeFe]-H_2ase mimic that is equipped with a ligand acting as an electron reservoir. The redox-active ligand donates an electron to the active site during the catalytic cycle. The oxygen-tolerant catalyst has a turnover frequency of 7.0×10^4 s^{-1} at an overpotential of 0.66 V in dilute sulfuric acid. [NiFeSe]-hydrogenases are a subclass of [NiFe]-hydrogenases with a selenocysteine residue coordinated to the active site nickel center in place of a cysteine (opposite scheme), displaying a high H_2 evolution rate and O_2 tolerance. Wombwell et al. (2015) utilized [NiFeSe]-H_2ase from Desulfomicrobiumbaculatum for the design of various photocatalytic H_2-production devices, e.g., the [NiFeSe]-H_2ase on a dye-sensitized TiO_2 nanoparticle for visible light-driven H_2-generation. The active site model [$NiFeS_2Se_2(CO)_3$] shown oppositely was synthesized by Wombwell and Reisner (2015). Becker et al. (2016) reported the first [FeFe]-H_2ase mimic that is equipped with a ligand acting as an electron reservoir. The redox-active ligand donates an electron to the active site during the catalytic cycle. The oxygen-tolerant catalyst has a turnover frequency of 7.0×10^4 s^{-1} at an overpotential of 0.66 V in dilute sulfuric acid. Honda et al. (2016) described photocatalytic H_2-production with a recombinant strain of *E. coli*, expressing genes encoding [FeFe]-hydrogenase and maturases from *Clostridium acetobutylicum* in combination with methyl viologen as redox mediator and TiO_2; the maturases are involved in maturing hydrogenases into their catalytically active form (Kuchenreuther et al., 2012). For a review about electrobiocatalysis by hydrogenases, see Reisner (2011) or McPherson and Vincent (2014). Another example of hydrogen production by a nanophoto-biocatalyst has been reported by Balasubramanian et al. (2013). They made use of the light-harvesting proton pump bacteriorhodopsin that self-assembled on the surface of Pt/TiO_2 (anatase) NPs through the interaction between carboxylate groups and exposed oxygen atoms on the TiO_2 surface. Bacteriorhodopsin served as a photosensitizer and transferred protons to the Pt nanocatalyst for subsequent reduction to hydrogen. This hybrid system produced 5275 μmole of H_2 per μmole protein and hour at pH 7 in presence of methanol as electron donor when illuminated with white light.

2.4 Improving the Properties of Enzymes

To make use of biocatalysts in chemical processes as substitutes for traditional chemical catalysts an adaption of their properties to the needs of the process in question is required, including enhanced robustness (e.g., tolerance to higher temperatures and pH values), yields and productivity. During the last two decades, a range of molecular tools have been developed, enabling considerable improvements of enzyme functions. Some of them are briefly discussed here.

2.4.1 Making Use of Biodiversity

A possibility to obtain better enzymes is to exploit the enormous bio-diversity of living organisms (e.g., bacterial communities) by screening them either for a desired enzyme or for the corresponding gene(s) (scheme opposite). The latter is the area of environmental "metagenomics" permitting the investigation into the (chemical) diversity of organisms living in a given environment as a resource of novel functional sequence space (Lorenz et al., 2002), and touching on the problems associated with "not-yet-culturable" organisms (Lorenz et al., 2003; Streit et al., 2004). It is estimated that up to now more than 99% of all microorganisms have not yet been cultivated; metagenomics means to generate DNA libraries from environmental samples after isolation and purification of DNA and cloning; after sequencing the genes (homology-driven or functional/activity-based); the enzymes they code for may be expressed in a suited host and characterized (Ferrer et al., 2009).

2.4.2 Enzyme Engineering Approaches

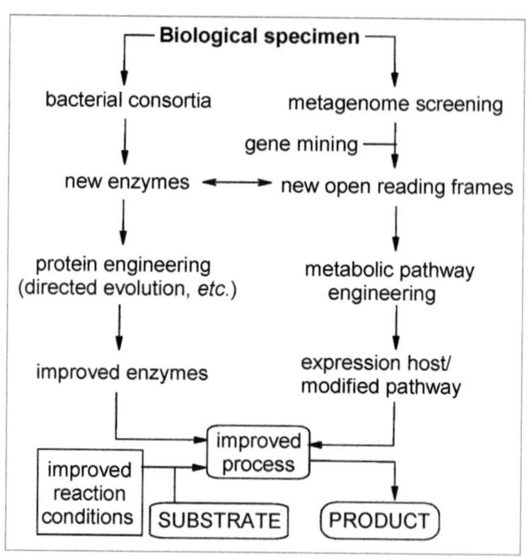

Enzyme-engineering approaches may be roughly divided apart from rational design, in directed evolution techniques, semi-rational design, and de novo design of new enzymes by means of computational biology (Quin and Schmidt-Dannert, 2011). Directed evolution is completely different from, and an alternative to rational design, relying on structural (and mechanistic) information of the respective enzyme/protein or at least on data obtained from homology modeling. *Rational design* uses sequence, structural, and mechanistic information mainly for targeted mutations of active site residues. Although most of the alterations within the active site lead to inactivation, the method may provide insight into details of the catalytic mechanism, and has further led to the design of numerous enzyme variants that not only showed improved properties (catalytic activity, solvent and temperature tolerance, etc.) but were also capable of catalyzing alternative types of reactions even in cases where only one amino acid residue was exchanged. This may have happened also during natural evolution and could explain the phenomenon of "promiscuity" (Section 2.2.4.4). Rational design (for an excellent review see Toscano et al., 2007) normally does not involve changes of the global protein fold. It seems, however, not to be clear whether active site mutation or random mutation of the whole catalyst is more efficient, and in many cases altered properties of an enzyme are caused by a change of conformational dynamics of the entire protein molecule. A frequent experience is that techniques successfully employed in one experimental study may fail when transferred to a seemingly analogous situation (Parikh et al., 2005).

A main difference in rational design and directed evolution is that in the first case amino acid residues within or near to the active site are targeted for mutation, whereas the generally employable directed evolution techniques include mutations distant from the active site. A disadvantage, however, is that the generation of a mutant with the desired properties is accompanied by the creation of an extremely large library, requiring elaborate screening methods. To circumvent this problem, several strategies based on the *semi-rational approach* have been developed in the recent past: considering some knowledge about mechanism, sequence, and crystal structure of the enzyme in question it is possible to manage with a significantly smaller library—albeit at the expense of general applicability.

2.4.3 Directed Evolution

Among the technologies employed for library creation, one distinguishes between non-recombining (error prone PCR, saturation mutagenesis, CAST, etc.) and recombining (DNA shuffling, staggered extension process, ITCHY, SCRATCHY) mutagenesis, depending on whether one parent gene or several parent genes are used in directed evolution approaches. This results in variants with point mutations, or in libraries, consisting of chimeras.

3DM analysis is an approach to create "smart" libraries, characterized by their significantly reduced size together with a comparatively high probability to identify hits; it is based on a software tool that can generate superfamily-specific databases, and has been developed by Jochens and Bornscheuer (2010). The applicability of this concept was demonstrated with the improvement of the enantioselectivity of an esterase from *P. fluorescens* (PFE) with the sterically demanding 3-phenyl butyric acid *p*-nitrophenyl ester (3-PB-*p*NP) as substrate. The enantioselectivity of PFE is influenced by the four positions W28, V121, F198, and V225 (Park et al., 2005). This amino acid distribution was determined by means of a structural alignment of 1751 sequences of α/β-hydrolase fold enzymes; frequently occurring residues/"mutations" that obviously have already proven during natural evolution to contribute to the integrity of the protein were defined as allowed (**A**) and rarely occurring ones at not allowed (NA). Based on this analysis an appropriate codon choice was possible. In this approach site-saturation mutagenesis was performed simultaneously at all four targeted positions in order to allow possible cooperative effects. With 3-PB-*p*NP as substrate the w-t PFE exhibits only very low activity and an enantioselectivity of $E = 3.2$, whereas in library **A**a significant number of mutants with improved reaction rates towards either enantiomer of up to 240-fold and enantioselectivities up to $E = 80$ could be identified (as in similar cases the activity was determined by following the release of *p*-nitrophenol through measuring the absorbance at 410 nm).

2.4.4 HotSpot Wizard

HotSpot Wizard is a web server for identification of hot spots in protein engineering (http://loschmidt.chemi.muni.cz/hotspotwizard/) developed by the group of Damborsky (Pavelka et al., 2009). HotSpot Wizard is able to automatically identify sites suited for engineering by integration of structural functional and evolutionary information resulting from databases such as RCSB (Research Collaboratory for Structural Bio-informatics), PDB (Protein Data Bank), UniProt (Universal Protein Database), PDBSWS—mapping PDB chains to UniProt knowledgebase entries (Martin, 2005), the Catalytic Site Atlas (CAS; Furnham et al., 2014), and nr (non-redundant) nucleotide database maintained by NCBI (National Center for Biotechnology Information), together with the application of various tools: CASTp (Computer Atlas of Surface Topology of Proteins), CAVER (a software tool for analysis and visualization of tunnels and channels in proteins), BLAST (Basic Local Alignment Search Tool; Altschul et al., 1997), CD-HIT (Cluster database at High Identity and Tolerance), MUSCLE (Multiple Sequence Comparison by Log-Expectation), and Rate4Site: an algorithmic tool for the identification of functional regions in proteins by surface mapping of evolutionary determinants within their homologues (Pupko et al., 2002). The only required input is a structure (PDB code/file). As one result of the calculations the mutability of residues (derived from evolutionary conservation) is provided ranging on a scale from one (low mutability) to nine. For the haloalkane dehalogenase Dha A (*Rhodococcus* sp.), HotSpot Wizard identified 17 hot spots with a mutability between six and nine, lining the active site pocket and two tunnels. With the exception of one amino acid located next to a catalytic residue, all these hot spots showed a high degree of mutability. The analysis of gene site saturation mutagenesis successfully applied by Kretz et al. (2004) to improve the thermostability of a haloalkane dehalogenase revealed that 30% of the residues identified for possible mutation could not be replaced without impairing the catalytic activity of the enzyme.

2.4.5 De Novo Computational Protein Design

As a final and rather new method to get access to improved enzymes or to those for which a natural counterpart is not known (Diels-Alderase, Kemp eliminase) *de-novo* protein design is briefly discussed here. It has been pointed out elsewhere that a catalyst stabilizes the transition state (TS, see, e.g., activated complex theory), resulting in a lower activation energy E_a of the respective reaction. In the case of biocatalysts this principle is supported by an interaction of the TS with distinct amino acid residues leading to rate accelerations of up to 10^{15} and more under mild reaction conditions. Hence the first step in computational enzyme design is the in silico modeling of the TS of a target reaction together with residues arranged so that they stabilize the TS; these residues include functional side chains acting as, e.g., hydrogen acceptors (for proton abstraction), H-donors (H^+-donation) or residues carrying a positive charge to stabilize an emerging negative one in the TS. The outcome of this procedure is known as theozyme; it bears the catalytic residues in an optimized geometry and is subsequently computationally inserted into the binding pocket of a protein scaffold selected from the Protein Data Bank; this

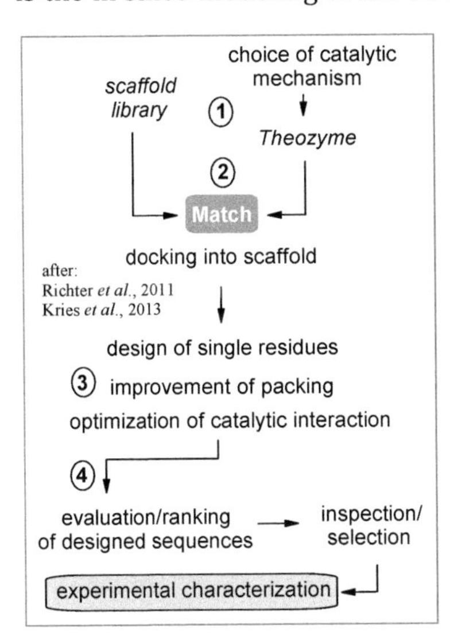

part is termed "matching": the identification of sites within scaffolds suited to harbor the theozyme. The next steps serve to further improve the packing of the theozyme and the surrounding side chains are varied for optimal accommodation within the binding pocket through enhancing the binding affinity to the TS by means of computational modeling tools such as RosettaMatch (Zanghellini et al., 2006); this algorithm enables a rapid elimination of catalytically unfavorable active-site geometries. The different designs are then produced and kinetically characterized, and may subsequently be further optimized with respect to their catalytic efficiencies, e.g., by directed evolution (Quin and Schmidt-Dannert, 2011; Li et al., 2012; Kries et al., 2013). It should be emphasized here that this procedure—unlike as in the case of directed evolution—requires a detailed understanding of the catalytic mechanism of an enzyme, or in other words, reliable structure predictions together with provisions for protein stability and specifications of inter-molecule interactions are essential for a successful de novo design of enzymes (e.g., Pantazes et al., 2011). For a recent review, see Woolfson et al. (2015).

2.4.6 Systems Metabolic Engineering Approaches

Today the so-called "*classical metabolic engineering*" is part of what is referred to as "systems metabolic engineering" that also includes the research areas "systems biology," "synthetic biology," and "evolutionary engineering." The integrated *systems metabolic engineering* design process and the different domains involved, together with some characteristics, are schematically shown below. The driver for the considerable progress in these overlapping areas of systems metabolic engineering is the need for a sustainable production of fine and bulk chemicals, and (alternative) biofuels. The classical experience-based engineering of strains and screening for overproduction is supported by the tools provided by the other research areas, and in addition, enzyme engineering treated in the previous chapter is of high importance as a means to improve properties of key enzymes involved in a production pathway (Liang et al., 2014). The potential of *adaptive evolution* as a possible alternative to rational design in bioengineering has been demonstrated by Hong et al. (2011). A *Saccharomyces cerevisiae* strain was subjected to adaptive evolution for improving galactose utilization; adaptive evolution strains were obtained from three different serial dilutions over 62 days (≈400 generations) on glucose minimal media and led to the isolation of three evolved mutants showing the desired faster glucose uptake and higher growth rates. An analysis at the systems level (changes in transcriptome, metabolome, and genome sequence) compared to the parent strain revealed that the phenotype changes are not caused by mutations of genes related to galactose and reserve carbohydrate (trehalose, glycogen) metabolism, but due to mutations of genes encoding for proteins involved in the Ras/PKA signaling pathway, as derived from transcriptome and metabolome data.

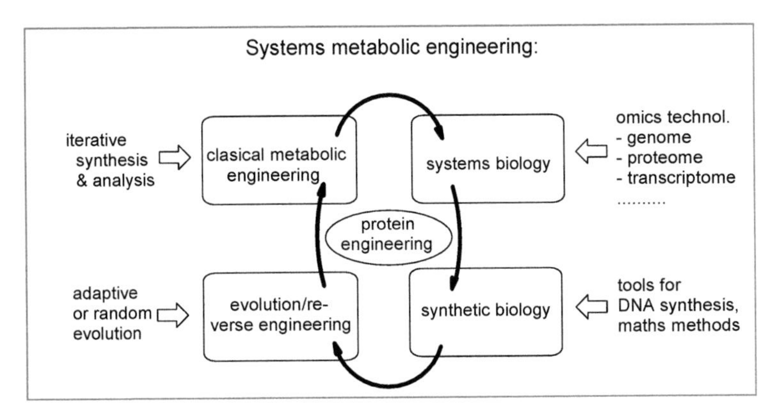

Advances in directed evolution at the pathway and genome scale, and new strategies for use in synthetic biology have been reviewed, e.g., by Cobb et al. (2012), Yadav et al., (2012), Liang et al. (2014), or Jones et al., 2015). Nucleotide sequences targeted in a biosynthetic gene for directed evolution are the promoter (binding of RNA polymerase mediated by transcription factors to initiate the polymerization of an mRNA), ribosomal binding sites (translation of the base sequence into an amino acid sequence requires binding of a ribosome to facilitate tRNA binding to the mRNA codons), intergenic regions, and the order of pathway genes.

2.5 Application of Enzymes and Outlook

It is well known that the rapid industrial development with the chemical industry being predominantly based on the exploitation of fossil resources, despite many obvious advantages, goes along with a variety of negative environmental impacts. Industrial biotechnology is based on the exploitation of renewable resources and contributes significantly to what is termed Green Chemistry. Unlike traditional chemical technology, biotechnology makes use of enzymes and whole cells as catalysts for producing fuels, bulk and fine chemicals and other materials. Biocatalysts work under mild reaction conditions with often high efficiency and selectivity to yield pure products. These properties enable production processes characterized by lower energy consumption and less waste generation as compared to their chemical counterparts and are prerequisites for making the different areas of the chemical industry increasingly more sustainable, resulting among others in reduced greenhouse gas emissions and in overcoming problems associated with the depletion of fossil resources.

According to a recent report by BCC (Business Communications Company) Research about Enzymes in Industrial Applications (2011), the global enzyme market is estimated at US$ 3.3 billion in 2010 and forecasted to reach US$ 4.4 billion by 2015, which means a compound annual growth rate (CAGR) of about 6%. The respective numbers for so-called technical enzymes that in contrast to speciality enzymes are normally used as bulk enzymes are US$ 1 billion for 2010 and 1.5 billion (2015) with the principal customer being the leather industry, followed by companies engaged in bioethanol production. The market for enzymes employed in the food and beverage industry is expected to increase within this 5-year period from US$ 975 to US$ 1.3 (CAGR ~5%); in 2009 nearly 50% of these enzyme sales were allotted to processing of milk and dairy products.

Most of all enzymes employed for production on industrial scale are hydrolases, which, however, may change because enzymes belonging to other classes such as lyases—e.g., aldolases, hydroxynitrile lyases—or oxidoreductases are meanwhile successfully used for various processes. Of the applied hydrolases, esterases, lipases (EC 3.1.-.-), glycoside hydrolases (EC 3.2.-.-) and peptidases (EC 3.4.-.-, acting on peptide bonds) make up about 70%. The products resulting from fermentation processes range from comparatively inexpensive bulk products (bioethanol: world production 70,000,000 $t \cdot a^{-1}$; biodiesel: 13,000,000 $t \cdot a^{-1}$;

L-glutamic acid: 1,500,000 t·a^{-1}; citric acid: 1,800,000 t·a^{-1}; L-lysin: 1,500,000 t·a^{-1}) to highly expensive fine chemicals such as vitamins (vitamin B_{12}: 3 t·a^{-1}) or antibiotics (bulk products: 30,000 t·a^{-1}; specialities 5,000 t·a^{-1}). The different areas where enzymes are used in industry are the detergent industry, starch and fuel, food and dairy, baking, animal feed, beverage, textile, pulp and paper, fats and oils, the leather industry, personal care and biofuels. In this case the focus is on cellulose, hemicellulose and lignin degradation for the generation of biomass derived (second generation) chemicals.

If applied as "processing aid" these enzymes must match the specific conditions of the respective process, such as enhanced temperatures, pH stability, or tolerance to organic solvents. Methods to improve the properties of enzymes with respect to process parameters are, e.g., recombinant DNA-technology and pathway engineering. Such technologies allow for manipulating the genetic material of microorganisms used in fermentation processes or to insert genes from higher organisms as plants or animals into industrial microorganisms in order to generate the desired function by expression.

To sum up, the global enzyme market will grow steadily, a development driven by the demand for biocatalysts required as processing aids or for the sustainable production of fine and bulk chemicals, pharmaceuticals, bioplastics, energy, etc., from renewable feedstock as an alternative to petroleum-based chemical processes. Further development of advanced biotechnological tools such as industrial systems biology combined with high-throughput screening not only will lead to enzymes with an improved adaption to the respective process requirements but may also pave the way for the discovery of so far unknown enzymes, including plant-derived biocatalysts, with novel properties for new applications in industrial biotechnology. Furthermore, enzymes emerging from the drawing board probably gain increasing importance although the field of computational enzyme design is still in its infancy.

References

Abdulkhani A., Marvast E. H., Ashori A., Karimi A. N., *Carbohydr. Polym.*, **95** (2013), 57–63.

Altschul S. F., Madden T. L., Schäffer A. A., Zhang J., Zhang Z., Miller W., Lipman D. J., *Nucleic Acids Res.*, **25** (1997), 3389–3402.

Anastas P., Eghbali N., *Chem. Soc. Rev.*, **39** (2010), 301–312.

Balasubramanian S., Wang P., Schaller R. D., Rajh T., Rozhkova E. A., *Nano Lett.*, **13** (2013), 3365–3371.

Becker R., Amirjalayer S., Li P., Woutersen S., Reek J. N. H., *Sci. Adv.*, **2** (2016), e1501014.

Berg J. M., Tymoczko J. L., Stryer L., *Biochemistry*, 5th ed., W. H. Freeman, New York (2002).

Bernal J. A., Lozana P., García-Verdugo E., Burguette M. I., Sánchez-Gómez G., López-López G., Pucheault M., Vaultier M., Luis S. V., *Molecules*, **17** (2012), 8696–8719.

Blickerstaff G. F., Immobilization of Enzymes and Cells, Humana Press, Totowa, 1997.

Branneby C., Carlqvist P., Magnusson A., Hult K., Brinck T., Berglund P., *J. Am. Chem. Soc.*, **125** (2003), 874–875.

Bubaloa M. C., Tušeka A. J., Vinkovićb M., Radoševića K., Srčeka V. G., Redovnikovića, I. R., *J. Mol. Catal. B Enzymatic*, **122** (2015), 188–198.

Buchholz K., Kasche V., Bornscheuer U. T., *Biocatalysts and Enzyme Technology*, 2nd. Ed., Wiley-Blackwell (2012).

Cantone S., Hanefeld U., Basso A., *Green Chem.*, **9** (2007), 954–971.

Cantone S., Ferrario V., Corici L., Ebert C., Fattor D., Spizzo P., Gardossi L., *Chem. Soc. Rev.*, **42** (2013), 6262–6276.

Carrea G., Riva S., *Angew. Chem.*, **112** (2000), 2312–2341; *Angew. Chem. Int. Ed.*, **39** (2000), 2226–2254.

Chakrabortyace S., Handajaya Ruslib H., Nath A., Sikder J., Bhattacharjee C., Curcio S., Driolia E, *Crit. Rev. Biotechnol.*, **1** (2016), 43–58.

Chen C. S., Sih C. J., *Angew. Chem.*, **101** (1989), 711–724; *Angew. Chem. Int. Ed.*, **28** (1989), 695–707.

Cobb R. E., Si Tong, Zhao H., *Curr. Opin. Chem. Biol.*, **16** (2012), 285–291.

Cull S. G., Holbrey J. D., Vargas-Mora V., Seddon K. R., Ley G. J., *Biotechnol. Bioeng.*, **69** (2000), 227–233.

Dai Y., van Spronsen J., Witkamp G.-J., Verpoorte R., Choi Y. H., *Anal. Chim. Acta*, **766** (2013), 61–68.

Datta S., Christena L. R., Rajaram Y. R. S., *3 Biotech.*, **3** (2013), 1–9.

DiCosimo R., McAuliffe J., Poulose A. J., Bohlmann G., *Chem. Soc. Rev.*, **42** (2013), 6437–6474.

De los Rios A. P., van Rantwijk F., Sheldon R. A., *Green Chem.*, **14** (2012), 1584–1588.

Doll F., Buntz A., Späte A.-K., Schart V. F., Timper A., Schrimpf W., Hauck C. R., Zumbusch A., Wittmann V., *Angew. Chem.*, **128** (2016), 2303–2308; *Angew. Chem. Int Ed.*, **55** (2016), 2262–2266.

Ferch A. in *Structure and Mechanism in Protein Science: A Guide to Enzyme Catalysis and Protein Folding*; W. H. Freeman &Company, New York (1999).

Ferrer M., Beloqui A., Timmis K. N., Golyshin P. N., *J. Mol. Microbiol. Biotechnol.*, **16** (2009), 109–123.

Fischer E., *Ber. Dt. Chem. Ges.*, **27** (1894), 2895–2993.

Franssen M. C. R., Steunenberg P., Scott E. L., Zuilhof H., Sanders J. P. M., *Chem. Soc. Rev.*, **42** (2013), 6491–6533.

Furnham N., Holliday G. L., de Beer T. A., Jacobsen J. O., Pearson W. R., Thornton J. M., *Nucleic Acids Res.*, **42** (2014), D485–D489.

Garcia S., Lourenço N. M. T., Lousa D., Sequeira A. F., Mimoso P., Cabral J. M. S., Afonso C. A. M., Barreiros S., *Green Chem.*, **6** (2004), 466–470.

García-Junceda E., Lavandera I., Rother D., Schrittwieser J. H., *J. Mol. Catal. B. Enzymatic*, **114** (2015), 1–6.

Ghasemi M., Minier M. J. G., Tatoulian M., Chehimi M. M., Arefi-Khonsari F., *J. Phys. Chem. B*, **115** (2011), 10228–10238.

Goderis H. L., Ampe G., Feyten M. P., Fouwé B. L., Guffens W. M., van Cauwenbergh S. M., Tobback P. P., *Biotechnol. Bioeng.*, **30** (1987), 258–266.

Gorke J., Srienc F., Kazlauskas R., *Biotechnol. Bioprocess Eng.*, **15** (2010), 40–53.

Grunwald P., *Biocatalysis: Biochemical Fundamentals and Applications*, Imperial College Press (2009).

Grunwald P., Hansen K., Remus M., *Ind. J. Chem.*, **32B** (1993), 70–72.

Grunwald P., Hansen K., Stollhans M., *Mededel. Faculteit Landbouww. Rijksuniv. Gent*, **53** (1988), 2057–2064.

Guo W., Hou Y., Ren S., Tian S., Wu W., *J. Chem. Eng. Data*, **58** (2013), 866–872.

Habulin M., Primožič M., Knez Z., Application of ionic liquids in biocatalysis, in *Ionic Liquids: Applications and Perspectives* (Kokorin A., ed.), InTech, available from http://www.intechopen (2011).

Halling P. J., *Enzyme Microb. Technol.*, **16** (1994), 178–206.

Halling P. J., *Enzyme Microb. Technol.*, **6** (1984), 513–516.

Hayyan A., Mjalli F. S., AlNashef I. M., Al-Wahaibi Y. M., Al-Wahaibi T., Hashim A., *J. Mol. Liquids*, **178** (2013), 137–141.

Henri V., *Compt. Rend Acad. Sci.*, **135** (1902), 916–919.

Honda Y., Hagiwara H., Ida S., Ishihara T., *Angew. Chem.*, **128** (2016), 8177–8180; *Angew. Chem. Int. Ed.*, **55** (2016), 8045–8048.

Hong K.-K., Vongsangnak W., Vemuri G. N., Nielsen J., *Proc. Natl. Acad. Sci. U. S. A.*, **108** (2011), 12179–12184.

Jackson K., Bowman L. E., Fulton J. L., *Anal. Chem.*, **67**(1995), 2368–2372.

Jochens H., Bornscheuer U. T., *ChemBioChem*, **11** (2010), 1861–1866.

Jones J. A., Toparlak Ö. D., Koffas M. A. G., *Curr. Opin. Biotechnol.*, **33** (2015), 52–59.

Jonkheijm P., Weinrich D., Schröder H., Niemeyer C. M., Waldmann H., *Angew. Chem.*, **120** (2008), 9762–9792; *Angew. Chem. Int. Ed.*, **47** (2008), 9618–9647.

Kamat S. V., Barrera J., Beckmann E. J., Russel A. J., *Biotechnol. Bioeng.*, **40** (1992), 158–166.

Kamat S. V., Critchley G., Beckman E. J., Russel A. J., *Biotechnol. Bioeng.*, **46** (1995), 610–620.

Kamat S. V., Iwaskewycz B. Beckman E. J., Russel A. J., *Proc. Natl. Acad. Sci. U. S. A.*, **90** (1993), 2940–2944.

Karan R., Capes M. D., DasSarma S., *Aquatic Biosystems* (2012), 8:4.

Kasche H., Schlothauer R., Brunner G., *Biotechnol. Lett.*, **8** (1988), 569–574.

Khersonsky O., Malitsky S., Rogachev I., Tawfik D. S., *Biochemistry*, **50** (2011), 2683–2690.

Kim Y. H., Campbell E., Yu J., Minteer S. D., Banta S., *Angew. Chem.*, **125** (2013), 1477–1480; *Angew. Chem. Int. Ed.*, **52** (2013), 1437–1440.

Klibanov A. M., Samokhin G. P., Martinek K., Berizin I. V., *Biotechnol. Bioeng.*, **19** (1977), 1351–1361.

Koshland D. E., *Proc. Natl. Acad. Sci. U. S. A.*, **44** (1958), 98–104.

Kragl U., Eckstein M., Kaftzik N., *Curr. Opin. Biotechnol.*, **13** (2002), 565–571.

Kretz K. A., Richardson T. H., Gray K. A., Robertson D. E., Tan X., Short J. M., *Methods Enzymol.*, **388** (2004), 3–11.

Kries H., Blomberg R., Hilvert D., *Curr. Opin. Chem. Biol.*, **17** (2013), 1–8.

Krishna S. H., *Biotechnol. Adv.*, **20** (2002), 238–267.

Kuchenreuther J. M., Britt R. D., Swartz J. R., *PLoS ONE*, **7** (2012), e45850. (Maturases).

Landschulz W. H., Johnson P. F., McKnight S. L., *Science*, **240** (1988), 1759–1764.

Lau R. M., van Rantwijk F., Seddon K. R., Sheldon R. A., *Org. Lett.*, **2** (2000), 4189–4191.

Lauterbach L., Lenz O., Vincent K. A., *FEBS J.*, **280** (2013), 3058–3068.

Lederkremer G. Z., *Intracellular Lectin Involvement in Glycoprotein Maturation and Quality Control in the Secretory Pathway* in Glycobiology (Sansom C., Markman O., eds.), Scion Publishing Ltd (2007).

Li X., Zhang Z., Song J., *Comput. Struct. Biotechnol. J.* (2012), e201209007.

Liang W.-J., Wang F., Wen M., Jian J.-X., Wang X.-Z., Chen B., Tung C. H., Wu L.-Z., *Chem. Eur. J.*, **21** (2015), 3187–3192.

Liang Y., Ang E. L., Zhao H., Directed evolution of enzymes for industrial biocatalysis, in *Industrial Biocatalysis* (Grunwald P., ed.), Pan Stanford Publishing Pte. Ltd., Singapore (2014).

Liese A., Hilterhaus L., *Chem. Soc. Rev.*, **42** (2013), 6236–6249.

Liese A., Seelbach K., Wandrey Ch. (eds.), *Industrial Biotransformations*, WILEY-VCH, 2nd ed. (2006).

Lin X. S., Zhao K.-H., Zhou Q.-L., Xie K.-Q., Halling P. J., Yang Z., *Bioresour. Process*, **3** (2016), 2.

Lorenz P., Liebeton K., Niehaus F., Eck J., *Curr. Opin. Biotechnol.*, **13** (2002), 572–577.

Lorenz P., Liebeton K., Niehaus F., Schleper C., Eck J., *Biocat. Biotransformation*, **21** (2003), 87–91.

Lozano P., *Green Chem.*, **12** (2010), 555–569.

Lozano P., de Diego T., Carrié D., Vaultier M., Iborra J. L., *Chem. Commun.*, (2002), 692–693.

Martin A. C. R., *Bioinformatics*, **21** (2005), 4297–4301.

Mateo C., Chmura A., Rustler S., van Rantwijk F., Stolz A., Sheldon R. A., *Tetrahedron Asymmetry*, **17** (2006), 320–323.

Meyer D., Buescher J. M., Eck J., *Use of Newly Discovered Enzymes and Pathways: Reaction and Process Development Strategies for Synthetic Applications with Recombinant Whole-Cell Biocatalysts and Metabolically Engineered Production Strains* in Industrial Biocatalysis, Pan Stanford Series on Biocatalysis, Vol. 1., Pan Stanford Publishing (2014).

Michaelis, L. Menten M. L., *Biochem. Zeitschrift*, **49** (1913), 333–369.

Milner S. E., Maguire A. R., *Arkivoc* (2012), 321–382.

Moniruzzaman M., Nakashima K., Kamiya N., Goto M., *Biochem. Eng. J.*, **48** (2010), 295–314.

Mukherjee J., Gupta M. N., *Bioresource Biotechnol.*, **209** (2016), 166–171.

Ninagawa S., Okada Y., Sumitomo Y., Kamiya Y., Kato K., Horimoto S., Ishikawa T., Takeda S., Sakuma T., Yamamoto T., Mori K., *J. Cell Biol.*, **206** (2014), 347–356.

Pantazes R. J., Grisewood M. J., Maranas C. D., *Curr. Opin. Struct. Biol.*, **21** (2011), 1–6.

Paravidino M., Sorgedrager M. J., Orru R. V. A., Hanefeld U., *Chem. Eur. J.*, **16** (2010), 7596–7604.

Parikh M. R., Matsumura I., *J. Mol. Biol.*, **352** (2005), 621–628.

Park S., Kazlauskas R. J., *J. Org. Chem.*, **55** (2001), 8395–8401.

Park S., Kazlauskas R. J., *Curr. Opin. Biotechnol.*, **14** (2003), 432–437.

Park S., Morley K. L., Horsman G. P., Holmquist M., Hult K., Kazlauskas R. J., *Chem. Biol.*, **12** (2005), 45–54.

Pavelka A., Chovankova E., Damborsky J., *Nucleic Acids Res.*, **37** (2009), W376–W383.

Peters R. J. R. W., Marguet M., Marais S., Fraaije M. W., van Hest J. C. M., Lecommandoux S., *Angew. Chem.*, **126** (2014); 150–154; *Angew. Chem. Int. Ed.*, **53** (2014), 146–150.

Petka W. A., Harden J. L., McGrath K. P., Wirtz D., Tirell D. D., *Science*, **281** (1998), 389–392.

Pupko T., Bell R. E., Mayrose I., Glaser F., Ben-Tal N., *Bioinformatics*, **18** (2002), Suppl 1, S71–S77.

Quin M. B., Schmidt-Dannert C., *ACS Catal.*, **1** (2011), 1017–1021.

Reetz M. T., Wiesenhöfer W., Franciò G., Leitner W., *Chem. Commun.* (2002), 992–993.

Richter F., Leaver-Fay A., Khare S. D., Bjelic S., Baker D., *PLoS ONE*, **6** (2011), e19230.

Reisner E., *Eur. J. Inorg. Chem.* (2011), 1005–1016.

Schoenfeld I., Dech S., Ryabenky B., Daniel B., Glowacki B., Ladish R., Tiller J. C., *Biotechnol. Bioeng.* (2013), DOI: 10.1002/bit.24906.

Schwann T., *Archiv für Anatomie, Physiologie und wissenschaftliche Medicin*, (1836), 90–138; *Ann. Phys.*, **114** (1936), 358–364.

Sheldon R. A., *Org. Proc. Res. Develop.*, **15** (2011), 213–232.

Sheldon R. A., van Pelt S., *Chem. Soc. Rev.* (2013), DOI: 10.1039/c3cs60075k.

Sieber S. A., Marahiel M. A., *Chem. Rev.*, **105** (2005), 715–738.

Siow K. S., Britcher L., Kumar S., Griesser H. J., *Plasma Process. Polym.*, **3** (2006), 392–418.

Spahn C., Minteer S. D., *Recent Patents in Engineering*, **2** (2008), 195–200.

Stepankova V., Damborsky J., Chaloupkova R. *Hydrolases in Non-Conventional Media: Implications for Industrial Biocatalysis* in Industrial Biocatalysis, Pan Stanford Series on Biocatalysis, Vol. 1., Pan Stanford Publishing (2014).

Stolarow J., Gerçe B., Syldatk C., Magario I., Morhardt C., Franzreb M., Hausmann R., Micromagneticporous and nonporousbiocatalyst carriers, in *Industrial Biocatalysis*, Pan Stanford Series on Biocatalysis, vol. 1, pp. 521–552 (Grunwald P., ed.), Pan Stanford Publishing Pte. Ltd., Singapore (2014).

Straathof A. J. J., Panke S., Schmid A., *Curr. Opin. Biotechnol.*, **13** (2002), 549–556.

Streit W. R., Daniel R., Jaeger K. E., *Curr. Opin. Biotechnol.*, **15** (2004), 285.

Sym E. A., *Enzymologia*, **1** (1936), 156–160.

Tao Y., Dong R., Pavlidis I. V., Chen B., Tan T., *Green Chem.*, **18** (2016), 1240–1248.

Tavares A. P., Rodríguez O., Macedo E. A., New Generations of Ionic Liquids Applied to Enzymatic Biocatalysis, in *Ionic Liquids: New Aspects for the Future* (Kadokawa J.-I., ed.), InTech, available from http://www.intechopen (2013).

Tawaki S., Kilibanov A. M., *J. Am. Chem. Soc.*, **114** (1992), 1882–1884.

Thi P. T. P., Cho C.-W., Yun Y.-S., *Water Res.*, **44** (2010), 352–372.

Thoden J. B., Taylor Ringia E. A., Garrett J. B., Gerlt J. A., Holden H. M., Rayment I., *Biochemistry*, **43** (2004), 5716–5727.

Thum O., Hellmers F., Ansorge-Schuhmacher M., Robust enzyme preparations for industrial applications, in *Industrial Biocatalysis*, Pan Stanford Series on Biocatalysis, vol. 1, pp. 553–582 (Grunwald P., ed.), Pan Stanford Publishing Pte. Ltd., Singapore (2014).

Toscano M. D., Woycechowsky K. J., Hilvert D., *Angew. Chem.* (2007); *Angew. Chem. Int. Ed.*, **46** (2007), 3212–3236.

Uppada V., Bhaduri S., Noronha S. B., *Curr. Sci.*, **106** (2014), 946–957.

Van Ginkel C. G., Tramper J., Luyben K. Ch. A. M., Klapwijk A., *Enzyme Microb. Technol.*, **5** (1983), 297–304.

Van Rantwijk F., Sheldon R. A., *Chem. Rev.*, **107** (2007), 2757–2785.

Vorhaben T., Böttcher D., Jasinski D., Menyes U., Brüser V., Schröder K., Bornscheuer U. T., *ChemCatChem.*, **2** (2010), 992–996.

Vorlop K.-D., Jekel M., German Patent DE 19827552 C1 (1998).

Wang A. A., Mulchandani A., Chen W., *Biotechnol. Prog.*, **17** (2001), 407–411.

Wang F., Wang W.-G., Wang X.-J., Wang H. Y., Tung C.-H., Wu L.-Z., *Angew. Chem.*, **123** (2011), 3251–3255; *Angew. Chem. Int. Ed.*, **50** (2011), 3193–3197.

Wang J.-Y., Chao Y.-P., *Appl. Environ. Microbiol.*, **72** (2006), 927–931.

Wang S., Meng X., Zhou H., Liu Y., Secundo F., Liu Y., *Catalysts*, **6** (2016), 32.

Wheeldon I. R., Campbell E., Banta S., *J. Mol. Biol.*, **392** (2009), 129–142.

Wikmark Y., Svedendahl Humble M., Bäckvall J. E., *Angew. Chem.*, **127** (2015), 4358–4362; *Angew. Chem. Int. Ed.*, **54** (2015), 4284–4288.

Wombwell C., Caputo C. A., Reisner E., *Acc. Chem. Res.*, **48** (2015), 2858–2865.

Wombwell C., Reisner E., *Chem. Eur. J.*, **21** (2015), 8096–8105.

Wong L. S., Khan F., Micklefield J., *Chem. Rev.*, **109** (2009), 4025–4053.

Woolfson D. N., Bartlett G. J., Burton A. J., Heal J. W., Niitsu A., Thomson A. R., Wood C. W., *Curr. Opin. Struct. Biol.*, **33** (2015), 16–26.

Wu Q., Lin X.-F., Promiscuous biocatalysts: Applications for synthesis from laboratory to industrial scale in *Industrial Biocatalysis* (Grunwald P., ed.), Pan Stanford Publishing Pte. Ltd., Singapore (2014).

Xiao Y., Patolsky F., Katz E., Hainfeld J. F., Willner I., *Science*, **299** (2003), 1877–1981.

Yadav V. G., De Mey M., Lim C. G., Ajikumar P. K., Stephanopoulos G., *Metab. Eng.*, **14** (2012), 233–241.

Yang Z., Pan W., *Enzyme Microb. Technol.*, **37** (2005), 19–28.

Zahnghellini A., Jiang L., Wollacott A. M., Cheng G., Meiler J., Althoff E. A., Rothlisberger D., Baker D., *Protein Sci.*, **15** (2006), 2785–2794.

Zaks A., Klibanov A. M., *Science*, **224** (1984), 1249–1251.

Zaks A., Klibanov A. M., *J. Biol. Chem.*, **263** (1988), 3194–3201.

Zhang Q., de Oliveira Vigier K., Royer S., Jérôme F., *Chem. Soc. Rev.*, **41** (2012), 7108–7146.

Chapter 3

Environmentally Benign Nanomaterial Synthesis Mediated by Culture Broths

Lihong Liu,[a] Fang Xie,[b] Xiuxia Meng,[c] Vishnu Parek,[a] and Shaomin Liu[a]

[a]*Department of Chemical Engineering, Curtin University, Perth, Western Australia 6845 Australia*
[b]*Department of Materials, Imperial College London, United Kingdom*
[c]*School of Chemical Engineering, Shandong University of Technology P. R. China*

Shaomin.Liu@curtin.edu.au

3.1 Introduction

Owing to their unique optical, electrical, chemical and physical properties, nanoscale materials have shown remarkable potential for widespread applications in electronics, catalysis, sensing and nanomedicine, among other fields (Kamyshny and Magdassi, 2014; Corain et al., 2008; Kumar et al., 2015; Etheridge et al., 2013). A multitude of protocols have been well established to synthesize nanoparticles (NPs), including chemical reduction (Flores et al., 2013), physical vapor deposition (Wang et al., 2007) and irradiation routes (Shin et al., 2004; Dhand et al., 2015). Unfortunately, conventional physicochemical methods result, in most cases, in high environmental and economic costs (Dahl et al., 2007). Hence, there is a significant benefit in developing

Biocatalysis and Nanotechnology
Edited by Peter Grunwald

ISBN 978-981-4613-69-9 (Hardcover), 978-1-315-19660-2 (eBook)
www.panstanford.com

nontoxic and environmentally benign biological processes for NP synthesis (Faramarzi and Sadighi, 2013).

To advance the sustainability of nanosynthesis, biological reactants including amino acids/peptides/proteins, enzymes, nucleotides, plant extracts and microorganisms have been introduced to regulate the formation of NPs under aqueous and ambient conditions (Durán et al., 2011; Liu et al., 2013). There are numerous reviews available on biogenic NPs. Particularly, NP fabrication using microbes as nanofactories has been extensively investigated. We will briefly discuss each of these technologies below to place the microorganism culture broth-mediated method into context.

Liquid broth culture is one of the most common ways to multiply microorganisms. A nutrient broth typically contains water, yeast/meat extract, dextrose, peptone, other digested proteins, sodium chloride, etc. Some of these ingredients (like yeast extract and dextrose) are able to produce Ag and Au NPs (Roy et al., 2015; Liu et al., 2006; Badwaik et al., 2011). Despite the success achieved with these readily available "green" reactants, traditional microbial NP synthesis heavily relies on manipulating the specific microbes rather than exploiting the broth compositions, and therefore, the potential of microbial nanoparticle synthesis is yet to be fully realized. Research in our group focuses on sustainable synthesis of functional nanomaterials. This work aims to review our efforts in further improving the efficiency of microbial nanosynthesis. Metallic Au and Ag NPs are considered as models to explore the possibility of broth-mediated technology (Liu et al., 2014a; Liu et al., 2014b). NPs produced by broths are more stable, monodisperse and the synthesis rate is faster than in the case of microorganisms. Dextrose, yeast extract, peptone, or other digested proteins are found to play pivotal roles, not only in reducing metal ions, but also in preventing the agglomeration of NPs. Although the biochemical mechanism seems straightforward, the results of our study highlight the important synchronous reduction by various broths that have been ignored for a long time. Combining with a hydrothermal green process and using renewable sun light, we further developed high-quality ZnO nanomicrostructures enwrapped with a thin carbon layer (ZnO@C, Shen et al., 2015a) and Ag NP-embedded AgCl nanocomposites (Ag@AgCl, Shen et al., 2015b), respectively.

Both particles show high photocatalytic activity towards the decomposition of organic dye solutions under solar light, with excellent photocatalytic stability.

3.2 Biosynthesis of NPs

By investigating the biosynthetic processes inside microorganisms, plants or their parts and extracts, an interesting observation is that biomolecules are always associated with the formation of nanomaterials when challenged with the corresponding precursor ions. Biomolecules are organic molecules that exist naturally in living organisms. Large biomolecules such as polypeptides, proteins, carbohydrates and nucleic acids, with typical sizes in the range of ~5 to 200 nm, i.e., close to the size range of nanomaterials, have been used to fabricate complicated nanomaterials at the molecular level (Lu et al., 2004; Xiong et al., 2013). Small molecules such as lipids and vitamins are not reviewed in this chapter. As for DNA-based NPs, Baumann et al. (2015) have published excellent review. For details on NP synthesis using plants, the reader is referred to several reviews (Makarov et al., 2014; Park, 2014; Kharissova et al., 2013). Moreover, in-depth elucidation of the mechanisms behind phyto-synthesis (using plant or plant extracts) reveals that NP growth occurs via the same biomolecule route as with micro-organisms, which is tightly linked to amino acid residues in proteins and polysaccharides (Akhtar et al., 2013). Therefore, we focus on the role of these two main biomolecule types in guiding NPs formation.

3.3 Production of NPs Using Proteins and Polysaccharides

The use of amino acids, peptides and proteins as reducing and capping agents for the synthesis of NPs has catalytically provided alternative routes to nanostructures that are prepared under biological conditions (Maruyama et al., 2015). Metal ions may form salts with $-COO^-$ groups in aspartic acid and glutamic acid, or complex with nucleophilic groups, such as imidazole in histidine, $-SH$ in cysteine, $-SCH_3$ in methionine, and $-NH_2$ in lysine (Xie et al., 2007). The binding affinity exhibited by these and other amino acids (e.g., serine, threonine, asparagine, glutamine, and arginine) has been hypothesized to provide reaction sites and to stabilize the respective NPs (Faramarzi and Sadighi, 2013; Gruen, 1975). Tyrosine, tryptophan, and phenylalanine have often been identified as reductants (Wangoo et al., 2014). Under alkaline conditions, the phenolic group of tyrosine and the indole group of tryptophan have been reported as reducing functional groups, leading to the reduction of Ag^+ ions into Ag NPs. During this reduction process, oxidized tyrosine and tryptophan molecules also act as capping agents to stabilize AgNPs in aqueous solutions (Shankar and Rhim, 2015; Selvakannan et al., 2004).

A short chain of amino acid units linked by amide bonds, called a peptide, has been intensively studied for the preparation of NPs via the combination of reducing and capping properties of amino acid residues in the peptide structure. For example, Naik et al. (2002) found that phage display-selected silver-binding peptides can be used to synthesize a variety of Ag NPs. Si et al. (2007) prepared Au and Ag NPs using tryptophan-based peptides as both the reducing and stabilizing agent. Tan and co-workers (2010) examined different amino acid sequences for the right combination of reducing, capping, and morphogenic functions.

Proteins are larger polymeric compounds of amino acids exhibiting a tertiary and possibly quaternary structure. A few proteins such as bovine serum albumin (BSA), lysozyme, and casein have been reported to produce metallic nanoclusters (Ashraf et al., 2013). BSA-assisted synthesis of highly stable dispersions of water-soluble Au nanoplates has been exploited by Xie et al. (2007). Similarly, several amino acid residues of BSA containing sulfur–, oxygen–, and nitrogen-bearing groups have been suggested to reduce $AuCl_4^-$ and to direct the plate growth. Lysozyme, another widely available protein in plants, fungi, and bacteria, has been used to direct the growth of AuNPs and experimental results confirmed that histidine residues, as well as other metal-binding ligands, play critical roles in tuning the growth process (Wei et al., 2011).

Recently, tremendous effort has been devoted toward synthesizing NPs with various sugars, to reduce or eliminate the use and generation of hazardous substances. Monosaccharides, such as glucose, are capable of directly participating in the synthesis of NPs. It has been suggested that the aldehyde group of glucose is oxidized into a carboxyl group via the nucleophilic addition of OH^-, which in succession leads to the reduction of metal ions (Liu et al., 2006; Badwaik et al., 2011; Makarov et al., 2014). Panigrahi et al. (2004) reported a general method for the synthesis of different metal (Au, Ag, Pt, Pd) NPs using commonly available sugars, e.g., glucose and fructose as reducing agents. In an exhaustive study, Pettegrew et al. (2014) employed other monosaccharide sugars, including ribose, fructose, sorbose, xylose, and galactose to obtain AgNPs, and tested their respective antimicrobial performance. It is important to mention that disaccharide sucrose is a typical non-reducing sugar. However, hydrothermal decomposition of sucrose to glucose has led to Ag, Pd, and magnetite (Fe_3O_4) NP formation (Sun et al., 2009). Many naturally occurring polysaccharides, especially those from plant and algal origins, are highly economical, stable, non-toxic and biodegradable reagents to produce NPs. Using polysaccharides as templates for AgNP assembly has been reviewed by Emam et al. (2016). To date, the most widely used polysaccharide is plant-derived cellulose and its derivatives. Starch is another popular benign reagent employed in sweet nanochemistry. A complete green synthesis of AgNPs has been realized by Raveendran and co-workers (2003), in which glucose and starch served as the reducing and stabilizing agents, respectively. A buffered starch-glucose solution has been developed by Engelbrekt et al. (2009, 2010) for monodisperse Au and Pt NP synthesis. Contrarily, there has been very limited research on microbial polysaccharide-facilitated NP synthesis, owing mainly to the difficulty in competing with low cost plant polysaccharides. Interest in microbial cellulose became intense after a great deal of research performed on high yield bacterial cellulose (BC) production. For example, BC has been reported for the biosynthesis of CdS (Li et al., 2009). In situ synthesis of antimicrobial AgCl and AgNPs was successfully attained by Hu et al. (2009) and Wu et al. (2014). An anionic extracellular bacterial polysaccharide, xanthan gum (consisted of β–1,4-linked D-glucose backbone, substituted alternately with a trisaccharide side chain linked to every second glucose residue) was able to reduce Ag^+ ions to Ag^0 (Xu et al., 2014). Generally, the oxidation of hydroxyl groups to carbonyl groups plays an important role in the reduction of metal ions. Other polysaccharides containing amino reducing end groups also provide complexing and stabilizing function (Sathiyanarayanana et al., 2013).

3.4 Production of NPs by Various Microbes

It is well known that minerals at nano- and microlength scales can be accumulated by either uni- or multi-cellular microorganisms owing to a biologically controlled mineralization process (Lovley et al., 1987; Kröger et al., 2002; Bäuerlein, 2003). The inherent ability of these microbes to produce biominerals in such an eco-friendly way has served as the driving force behind the development of research focused on the microbial synthesis of nanomaterials.

AgNPs possess excellent antibacterial activities and their potential application in nanomedicine has promoted intensive studies on the microbial synthesis of these particles, in order to overcome toxicity (Roy et al., 2013). A wide diversity of biological entities has successfully been used to produce AgNPs. Although detailed description of these biochemical pathways falls outside the scope of this review, biogenic metallic NPs and quantum dots are frequently linked to the microbial resistance mechanism towards toxic ions (Johnston et al., 2013; Mukherjee et al., 2001a; Ramanathan et al., 2013; Bao et al., 2010). Microbes, including bacteria, actinomycetes, fungi and yeasts known to produce AgNPs are listed in Table 3.1. In a typical whole organism biosynthesis protocol, isolated or genetically engineered microbes were first cultivated in a nutritional liquid, such as lysogeny broth (LB) and nutrient broth (NB) media (Suresh et al., 2010; Babu et al., 2009). After that, cells were washed to remove broth components and redispersed in an aqueous silver salt solution to synthesize NPs intracellularly or extracellularly. Alternatively, cells were removed by centrifugation and only the spent culture supernatants (Saravanan et al., 2011; Nanda and Saravanan, 2009; El-Shanshoury et al., 2011) or cell-free filtrates were employed for extracellular NP synthesis. This approach would benefit the downstream particle recovery. It is generally presumed that the secretion of enzymes, such as nitrate reductase, by specific microbes is one of the significant factors in inducing nanoparticle formation (Kumar et al., 2007).

Table 3.1 List of various species of microorganisms reported to synthesize silver nanoparticles (AgNPs)

Organism	Reducing agent	NPs synthesis conditions	Organism culture broth and conditions	References
Bacteria				
Pseudomonas stutzeri AG259	H_2O-washed cells	30°C, 48 h, dark	Lennox L (LB)	Klaus et al. (1999)
Shewanella oneidensis MR-1	H_2O-washed cells	30°C, 48 h	Luria-Bertani (LB), 30°C, 200 rpm, 24 h	Suresh et al. (2010)
Plectonema boryanum UTEX 485	H_2O-washed cells	25, 65, 100°C, 28 d, dark	BG-11 medium, 29°C, 6–8 w	Lengke et al. (2007)
Aeromonas SH10	H_2O-washed cells	60°C, 12–144 h, dark	Soya peptone and beef extract, 30°C, 24 h	Zhang et al. (2007)
Bacillus licheniformis	Phosphate buffer washed cells	37°C, 24 h	Nutrient broth (NB)	Kalimuthu et al. (2008)
Bacillus cereus	H_2O-washed cells	37°C, 120 h	Nutrient broth (NB)	Babu et al. (2009)
Bacillus megaterium NCIM 2326	Cell-free spent MH broth	RT, 48 h, dark	Muller Hinton (MH) broth, 37°C, 15 h	Saravanan et al. (2011)
Morganella sp.	Cells and LB broth	37°C, 20 h, dark	LB without NaCl, 37°C, 15 h	Parikh et al. (2008)

(*Continued*)

Table 3.1 *(Continued)*

Organism	Reducing agent	NPs synthesis conditions	Organism culture broth and conditions	References
Morganella psychrotolerans	Cells and LB broth	20 h, 24 h, 5 days, and 14 days at 25, 20, 15, and 4°C, dark	Luria-Bertani without NaCl, at 15, 20, and 25°C for 24 h and at 4°C for 5 days	Ramanathan et al. (2011)
Staphylococcus aureus	Cell-free spent MH broth	5 m, bright	Muller Hinton broth, 37°C, 24 h	Nanda and Saravanan (2009)
Escherichia Coli ATCC8739 *Bacillus subtilis* ATCC6633 *Streptococcus thermophilus* ESh1	Cell-free spent LB broth Cell-free spent LB broth Cell-free spent ST broth	5 m, bright	Luria and Burrous (LB) broth, 35°C, 24 h Luria and Burrous (LB) broth, 35°C, 24 h ST broth, 35°C, 24 h	El-Shanshoury et al. (2011)
Actinomycetes				
Brevibacterium casei	H_2O-washed cells	37°C, 24 h	Nutrient broth containing beef-extract , NaCl and peptone, 24 h	Kalishwaralal et al. (2010)
Streptomyces hygroscopicus	Cell-free spent NB broth	30°C, 96 h, dark	Nutrient broth (NB), 25–28°C, 96 h	Sadhasivam et al. (2010)
Streptomyces albidoflavus	Cells and fresh broth	RT, 72 h	Broth containing yeast extract and dextrose, 72 h	Prakasham et al. (2012)

Organism	Reducing agent	NPs synthesis conditions	Organism culture broth and conditions	References
Fungi				
Fusarium oxysporum	Cells in H_2O or cell-free H_2O filtrate	28°C, dark	Broth containing yeast extract, 28°C, 6 d	Durán et al. (2005)
Aspergillus fumigatus	Cell free filtrate in H_2O	25°C, 72 h, dark	Broth containing yeast extract and glucose, 25°C, 72 h	Bhainsa and D'Souza (2006)
Phaenerochaete	H_2O-washed mycelium	37°C, 72 h, dark	Malt extract broth, 5 d	Vigneshwaran et al. (2006)
Aspergillus flavus	H_2O-washed mycelium	37°C, 72 h, dark	Yeast malt broth, 37°C, 5 d	Vigneshwaran et al. (2007)
Penicillium fellutanum	Cell-free filtrate in H_2O	25°C, 24 h, dark	Broth containing yeast extract and glucose, 25°C, 72 h	Kathiresan et al. (2009)
Cladosporium cladosporioides	Cell-free Spent broth	27°C, 78 h	Broth containing glucose, 27°C, 1 w	Balaji et al. (2009)
Coriolis versicolor	Washed mycelium or cell-free spent broth	37°C, dark, 1–72 h depends on pH values	Broth containing glucose, malt extract and peptone, 25–30°C, 7–8 d	Sanghi and Verma (2009)

(*Continued*)

Table 3.1 (*Continued*)

Organism	Reducing agent	NPs synthesis conditions	Organism culture broth and conditions	References
Trichoderma viride	Cell-free filtrate in H_2O	27°C, dark, 24 h	72 h in broth containing yeast extract and glucose, 27°C, followed by 48 h in H_2O	Fayaz et al. (2010)
Amylomyces rouxii KSU-09	Cell-free filtrate in H_2O	28 ± 0.5°C, 72 h	Potato dextrose broth, 4 d in broth, 3 d in H_2O	Musarrat et al. (2010)
Aspergillus flavus NJP08	Cell-free filtrate in H_2O	28°C, 72 h, dark	MGYP broth, 28°C, 4 d in broth, 3 d in H_2O	Jain et al. (2011)
Rhizopus stolonifer	Inactivated cell-free filtrate	25°C, 72 h, dark	Broth containing yeast extract and glucose, 25°C, 72 h; autoclaved then in water for72 h	Binupriva et al. (2010)
Yeasts				
Yeast strain MKY3	Cells in spent broth	24 h, dark	Broth containing tryptone, yeast extract and glucose	Kowshik et al. (2003)
Saccharomyces Cerevisiae	Cells in glucose and phosphate buffer	25°C	Broth containing peptone, yeast extract and dextrose, 37°C, 48 h	Korbekandi et al. (2014)

Apparently, traditional microbiological methods suffer from several drawbacks: (1) isolating and screening potential NP producing strains is complex, time consuming, and labor intensive (Malhotra et al., 2013); (2) microbial synthesis is a slow process, especially when fungi are employed and when the reactions are carried out in dark conditions (Bhainsa and D'Souza, 2006); (3) a variable cell culture process may cause reproducibility concerns; (4) as the living microorganisms can tolerate a narrow window of pH, temperature or toxic ion concentration, it is difficult to optimize reaction conditions to provide sufficient control over particle size, morphology and monodispersity; (5) potential risks to human health are posed when pathogenic microbes are used (Nanda and Saravanan, 2009; Mokhtari et al., 2009); (6) in order to release the resultant intracellular nanoparticles, downstream operations like ultrasonic breakdown, detergent disintegration, high temperature calcination and freeze-thawing processes are normally required (Park et al., 2010); these costly downstream processes are highly energy intensive and time consuming, which presents a significant barrier when scaling up.

3.5 Microorganism Culture Broth-Mediated NP Synthesis

Greener biosynthesis through microorganism culture broth might circumvent these limitations. The individual ingredient concentrations of commonly used broths are listed in Table 3.2. Yeast mold broth (YMB) is used for cultivating yeasts, molds, and other aciduric microorganisms. Lysogeny broth (LB) is one of the most common media for cultivating recombinant strains of *E. coli*. Tryptic soy broth (TSB) and nutrient broth (NB) are two other general-purpose media for the cultivation of a wide variety of microorganisms. Their common components are found to be saccharides, yeast extract, peptone, and other digested proteins, etc. As discussed before, various saccharides are well-known agents for the green synthesis of NPs. Yeast extract (YE) is concentrated from autolyzed *Saccharomyces cerevisiae*. *S. cerevisiae* cells have been employed to synthesize Au, Ag, TiO_2, and CdTe NPs in high yields (Lim et al., 2011; Jha et al., 2009; Bao et al., 2010). With its high content of B vitamins, tyrosine

(1.2% total amino acids) and other unknown peptides, the YE itself has the potential to reduce both gold and silver ions (Nadagouda and Varma, 2006; Swami et al., 2004; Yamal et al., 2013). The general scheme of microbe culture broth-mediated NP formation is shown in Fig. 3.1.

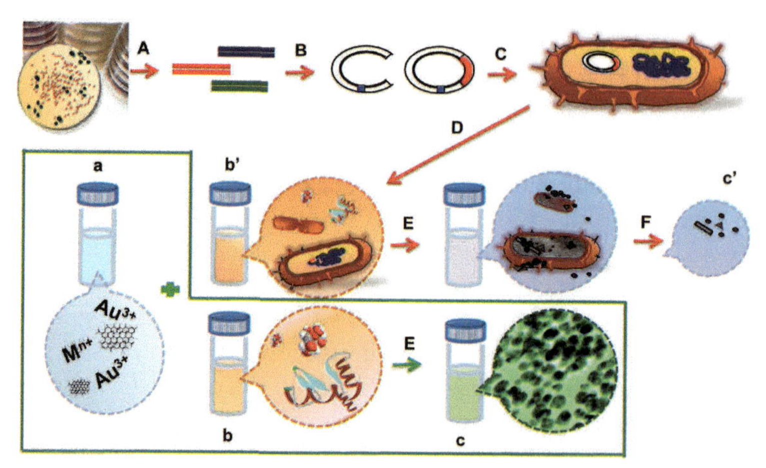

Figure 3.1 Schematic representation of the broth-mediated greener synthetic strategies versus the conventional microbial nanomaterial synthesis. Nanoparticles (c) can be produced by simply mixing the precursors in water (a) and the microbe culture broth (b), followed by incubation (E) under mild conditions. The microbial process is dramatically simplified by eliminating cell screening (A), gene identification (B), DNA recombination (C), DNA insertion (D) and particle separation (F) steps. The abundant reducing agent like dextrose (atoms are represented as spheres) and digested proteins (represented as ribbons) in the broth might contribute to the particle formation and stabilization. Reproduced from *RSC Adv.*, **4** (2014), 14564–14568 with permission from the Royal Society of Chemistry.

In the following sections, we summarize our recent efforts in developing novel broth-mediated procedures in preparing NPs. We begin the discussion with the synthesis of AuNPs in YMB with tunable particle size and morphology. This is followed by the description of the preparation of AgNPs in LB, NB, TSB and YMB. More complex nanostructures, including nano-Ag embedded AgCl and carbon/silver-modified hierarchical ZnO photocatalysts, are prepared in LB and YMB, respectively. Their excellent performance in organic dye degradation has also been investigated.

Table 3.2 The individual ingredient concentrations (g L^{-1} in water) of common microorganism culture broths

Broth name	[1]Glucose [2]Sucrose [3]Starch [4]Mannitol	Yeast extract	Peptone	Malt extract	Casein	Soybean meal	Potato extract	Beef extract	NaCl	K_2HPO_4	$MgSO_4$	$CaCO_3$
Yeast mold broth	[1]10.0	3.0	5.0	3.0	—	—	—	—	—	—	—	—
Tryptic soy broth	[1]2.5	—	—	—	17.0	3.0	—	—	5.0	2.5	—	—
LB Broth MILLER	—	5.0	10.0	—	—	—	—	—	10.0	—	—	—
Nutrient broth	—	—	5.0	—	—	—	—	3.0	—	—	—	—
MGYP broth	[1]10.0	3.0	5.0	3.0	—	—	—	—	—	—	—	—
ST broth	[2]10.0	5.0	—	—	10.0	—	—	—	—	2.0	—	—
Mueller Hinton broth	[3]1.5		—	—	1.75	—	—	2.0	—	—	—	—
Yeast extract mannitol	[4]10.0	1.0	—	—	—	—	—	—	0.1	0.5	0.2	1.0
Potato dextrose broth	[1]20.0		—	—	—	—	4.0	—	—	—	—	—
Streptomyces broth	[1]4.0	4.0	—	10.0	—	—	—	—	—	—	—	2.0

3.5.1 Au NPs

Au^{3+} has the largest positive standard reduction potential (E_0 = 1.5 V) therefore has the greatest affinity to be reduced. We first evaluated the capability of the YMB in producing AuNPs. The broth was autoclaved at 121°C for 15 min prior to mixing with $HAuCl_4$ solution. In a series of experiments, a 5 mL chloroauric acid solution was combined with 5 mL of YMB (at different pH values). The vials were agitated at 100 rpm on an orbital shaker in an incubator at 37°C. It was found that the color of the solution changed gradually from light yellow to pink within 2 h. Adjusting the pH values of YMB resulted in controlling not only the NP growth rate, but also the NP stability. At pH 12, the reaction took place fast and a highly stable AuNP formation was evidenced by the UV-vis absorption peak centered at 523 nm (Fig. 3.2a). The highly concentrated particles were spherical in shape with an average diameter of 20 nm and a polydispersity index (PDI) of 0.15. The AuNPs exhibited excellent stability for at least 6 months under ambient conditions.

It was furthermore shown that Au^{3+} concentration had a profound influence on the size and morphology of the NPs. By fixing the pH value at 12 but increasing the [Au^{3+}] from 0.5 to 5.0 mM, we found that higher concentrations generally led to larger particle sizes as well as distinctively different colors of Au hydrosols (Fig. 3.2c inset). The AuNPs could be easily extracted by centrifugation and be readily re-dispersed in deionized water by hand-shaking, without obvious irreversible aggregation. A number of Bragg reflections can be seen in Fig. 3.2b which correspond to the (111), (200), (220), (311) and (222) reflections of face-centered cubic metallic gold (Mukherjee et al., 2001b). In terms of multibranched AuNPs, several strategies focusing on multistep (Lu et al., 2008) or template-mediated synthesis (Wang et al., 2010; Sau et al., 2011; Mohanty et al., 2010) have been reported to synthesize AuNPs with spiky morphology. A facile, one-pot, template-free synthesis of flower-like AuNPs (AuNFs) was reported by Xie's group (2008). Noticeably, AuNFs were observed by adjusting the initial [Au^{3+}] at 2.5 mM in this work. Figure 3.2f depicts the transmission electron microscopy (TEM) image of AuNFs after 4 h reaction at room temperature. The particulates have a solid core with short spikes. The dimension of the AuNFs branch was 10 nm in diameter and 15 nm in length. The z-average diameter was 54.4, 70.6, and 86.3 nm at 4, 8, and 24 h, respectively, with the SPR peak at 556 nm.

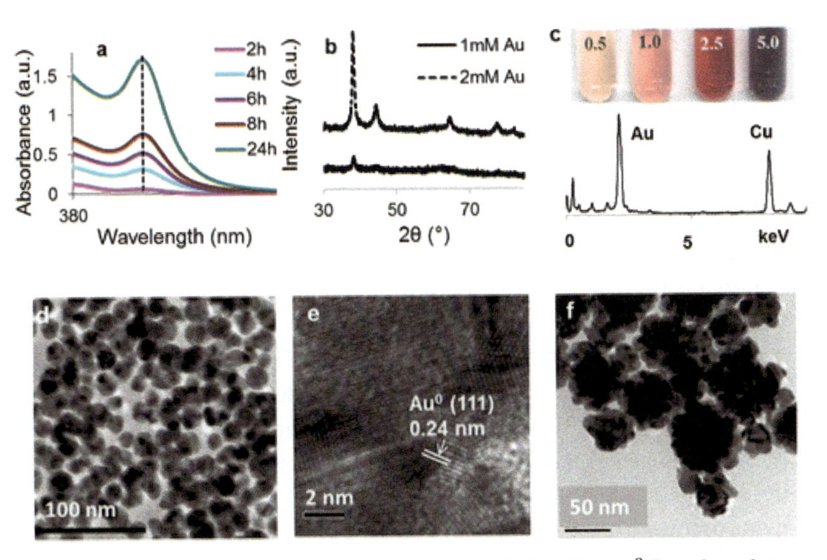

Figure 3.2 (a) UV-vis absorbance spectra of 1 mM [Au^{3+}] reduced in pH 12 yeast mold broth at 37°C. (b) XRD pattern of AuNPs formed in the system. (c) Typical EDS spectrum of the AuNPs. The inset shows four samples with increasing [Au^{3+}] from 0.5 to 5.0 mM after a 4 h reaction at 37°C. (d) TEM micrograph of AuNPs prepared with 1 mM [Au^{3+}] at 37°C. (e) High-resolution TEM image from a region in (d). (f) TEM image of Au nanoflowers formed after centrifugation of AuNPs with initial [Au^{3+}] concentration at 2.5 mM. Reproduced from *RSC Adv.*, **4** (2014), 14564–14568 with permission from the Royal Society of Chemistry.

The growth of the AuNPs in YMB at room temperature was relatively slow in comparison with the facilitated kinetics of the particle formation at 37°C. A characteristic peak at 2.195 keV in the energy dispersive X-Ray spectroscopy (EDS) spectrum (Fig. 3.2c) provides evidence that AuNPs can be obtained under milder conditions. Facile particle synthesis at ambient temperature would be more favorable for large-scale applications thanks to easier operation and lower energy consumption.

Several enzymes, including NADH-dependent nitrate/sulfite reductase, hydrogenase, α-amylase, etc., have been suggested to involve AuNP formation together with other electron shuttles

(Pereira et al., 2015). However, their fundamental roles continue to be debated within the scientific community. For instance, no AuNPs were obtained in heat-denatured cell-free extract of *Rhizopus oryzae* (Das et al., 2010). Contrarily, Maliszewska et al. (2009) reported Au^0 formation at appreciable rates in the heat-treated cell-free filtrate of *Trichoderma koningii*. Based on the fact that YM broths were still active in producing NPs despite being autoclaved at 121°C for 15 min, we believe that the native enzyme structure is not indispensable in the biogenic process. Dextrose solution at pH 12 promoted AuNP formation in a very rapid manner (few seconds). This fast production rate is highly beneficial to scale up the NP synthesis. The AuNP growth in YE was quite slow. After stirring the mixture at room temperature for 3 days, the average AuNP size was 44 nm when YE concentration was 0.5%. A black gold colloidal solution was formed in 1% YE solution after a 24 h reaction. A similar phenomenon was reported by Johnston et al. (2013) when a secondary metabolite secreted by *Delftia acidovorans* was incubated with $AuCl_3$. The authors demonstrated that a nonribosomal peptide acted to generate AuNPs. In our experiments, FTIR spectroscopic measurements disclosed that the major functional groups of yeast extract at 1579 cm^{-1} (Amide II, N-H, and C-H vibrations of the peptide bond in different protein conformations) and 1399 cm^{-1} (C=O of COO^- symmetric stretching in proteins) were attached on AuNPs. These results suggested a new formation mechanism of AuNPs in pH 12 YMB, as follows. Similar to the reduction of Au^{3+} by $NaBH_4$ and the particle stabilization by sodium citrate, the growth process may start with a rapid reduction by dextrose, followed by a nucleation process of gold clusters which undergo coalescence processes until the final size is reached, and stabilized by peptides/proteins available in the broth.

The formation of AuNPs in other typical microbial culture broths has confirmed the proposed formation mechanism. Although the experimental conditions were the same for all the different broths, the gold sol formation was drastically slow in LB and TSB, in striking contrast with YMB. The broth dependence of AuNP formation rate might be explained by the decreasing dextrose concentration in YMB, TSB and LB (1%, 0.25% and 0%, respectively).

Without noticing the effects of pH, precursor concentration and dextrose content on AuNP nucleation, the very slow reduction rate in microbe-mediated control experiments could mislead the researchers to ignore the critical role of broth, while overestimating the importance of bacteria or yeasts on the NP synthesis.

3.5.2 AgNPs

To investigate whether or not this strategy is applicable to other metallic nanoparticle synthesis, silver nitrate was added to the broth. Due to the larger energy barrier for Ag^+ reduction than Au^{3+} (E^0 of Ag^+ and Au^{3+}: 0.8 vs. 1.5 V), AgNPs in YMB were not observed in the dark even after one month of incubation. In previous microbial Ag bioproductive studies, certain control experiments were conducted in the dark and particles were not formed as compared to samples exposed to microbes (Ramanathan et al., 2011). This is likely the second reason why the broth-reducing capabilities have often been ignored and too much effort has been directed toward engineering microbial strains with a special protein expression. In contrast, in the presence of ambient light, stable Ag nanoparticulates were formed in all broths. In broths containing NaCl, AgCl NPs were identified as the major intermediate, and could be converted to AgNPs via one-pot photo-reduction (Zhang et al., 2014a; Liu et al., 2014). Figure 3.3a is a photograph of AgNPs formed with different initial concentrations of Ag^+ in YMB and TSB, respectively. The Ps had been stored inside the tubes for one month without any aggregation. The importance of ambient light in reducing Ag^+ has been reported by Nam et al. (2008). In their study, the authors firstly transformed the *S. cerevisiae* strain EBY100 which overexpressed the hexaglutamic acid (E_6) peptide, then employed the yeast cells as biological scaffolds for Ag^+ reduction. No AgNPs were found in cells that were kept in the dark. As proposed by Nam et al., we also believe that photochemical reduction of Ag is facilitated after binding of Ag ions to reducing peptides in the broth. SEM-EDS spectrum (Fig. 3.3b) showed the occurrence of silver with sulfur, mirroring the important role of metallothioneins in the reduction and binding of Ag^+ (Lengke et al., 2006).

To further verify that our greener process is microbe independent, silver nitrate was incubated with native *E. coli* (ATCC25922) instead of the recombinant one. After exposing *E. coli* and Ag^+-containing TSB to natural light, AgNPs were synthesized extracellularly as indicated by the increased 430 nm absorbance intensity. To check whether AgNPs were formed inside the cells, the cell pellets were centrifuged and washed for ultrastructural investigation by TEM. In contrast to silver-resistant *Pseudomonas stutzeri* AG259 cultured in the presence of 50 mM $AgNO_3$, for which the majority of the AgNPs were deposited in the periplasmic space of the cells, the size of the AgNPs in our study was smaller and distributed evenly all over the cell body (Fig. 3.3f). The control experiments without Ag^+ showed no AgNP formation (Fig. 3.3e). Importantly, extracellular AgNPs were formed in TSB containing either *E. coli* or *B. subtilis*, thereby echoing that the broth-only strategy is indeed microbe independent.

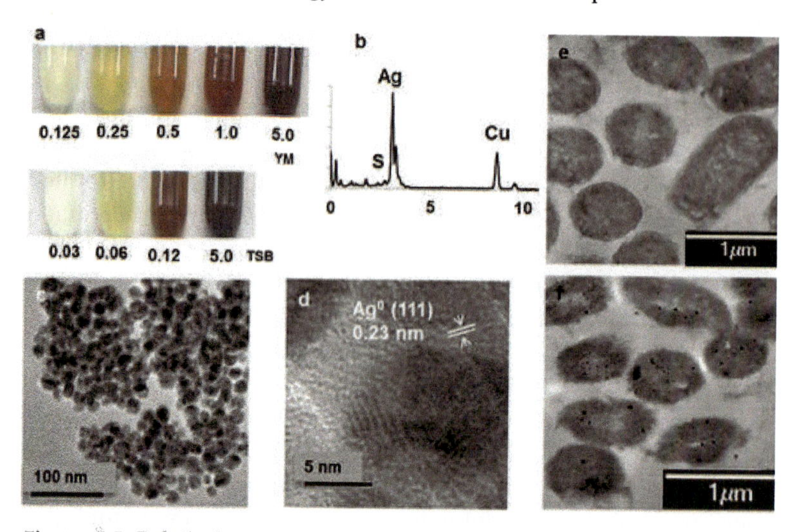

Figure 3.3 Relatively monodisperse Ag nanoparticles (AgNPs) prepared in broths versus irregular AgNPs accumulated in *E coli.* (a) Color photographs of samples synthesized with increasing [Ag^+] in yeast mold broth (YMB) and tryptic soy broth (TSB), respectively, after one month storage at room temperature. (b) Typical EDS spectrum of the AgNPs reduced with YMB. (c) TEM micrograph of AgNPs prepared with 1 mM [Ag^+] at 37°C in pH 12 YMB. (d) High-resolution TEM image from a region in c. (e) TEM images of *E. coli* and AgNPs accumulated inside the bacterial cells (f). Adapted from *Enzyme Microb. Technol.*, **67** (2014) 53–58 with permission from the Elsevier.

3.5.2.1 nano-Ag embedded AgCl

Ag@AgX (X = Cl, Br) plasmonic photocatalysts are expected to play important roles in environmental remediation. In the context of the current drive to conserve resources and protect the environment, Liu et al. (2012a) have demonstrated that Gluconacetobacter xylinum enabled the fabrication of antimicrobial Ag@AgCl composite. In their study, the G. xylinum was cultured in broth at 30°C for 5 days followed by an $AgNO_3$ challenge for 2 to 3 days to obtain Ag@AgCl nanoparticles. Finally, the solution containing Ag/AgCl particles and biomass was filtered through membranes to remove bacterial cell bodies. Apparently, there are disadvantages in the use of microbes, e.g., time-consuming procedures in microbe culture and downstream purification. Thus, the importance of cell-free biosystems with higher product yields and faster reaction rates has been attempted by our group. Colloidal AgCl formation was observed in TSB and LB (Zhang et al., 2014). LB has a higher content of NaCl than TSB. It also contains yeast extract and peptone, that are rich in amino acids and peptides. Inspired by the latest results on the electron transfer (ET) peptide-mediated Ag NP formation by laser irradiation on AgCl (Kracht et al, 2015), we concluded that the biogenic AgCl may be partially reduced under sunlight to produce small plasmonic Ag@AgCl (Fig. 3.4).

Additionally, an in situ photoreduction strategy is expected to introduce midgap defect states in AgCl to enhance photocatalytic activities (Ma et al., 2014). In this study, stable AgCl colloidal spheres with a size of 50 nm were formed first through a precipitation reaction of $AgNO_3$ and NaCl in LB Miller broth. By rationally changing the precursor concentration and light exposure time, we were able to fine-tune the bandgap of the resultant composites. After a 5 min solar light activation at room temperature, the as-synthesized Ag@AgCl NPs demonstrated even better performance than commercially available P25 (TiO_2) in the degradation of Rhodamine 6G. The excellent activities could be attributed to the embedded AgNPs associated with stronger plasmon resonance of Ag@AgCl. Herein, we developed for the first time a facile one-pot biological process that is superior to most conventional biomimetic methods in preparing Ag@AgCl NPs with a unique configuration.

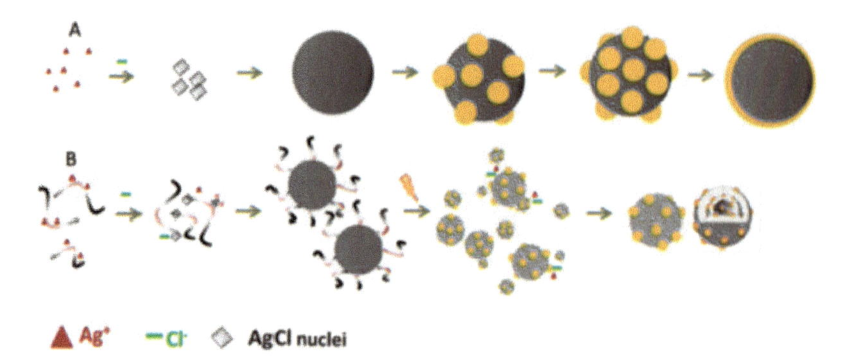

Figure 3.4 Mechanisms proposed for biogenic Ag@AgCl formation through photo-induced and accelerated ripening process: (A) the growth of Ag@AgCl particles during a typical chemical reduction process, and (B) the formation of Ag@AgCl plasmonic nanohybrid within LB broth. Reproduced from *RSC Adv.*, **5** (2015), 80488–80495 with permission from the Royal Society of Chemistry.

3.5.3 FeMn NPs

For almost 40 years, the fastidiousness and the sensitivity of most magnetotactic bacteria to even low concentrations of oxygen have limited the progress in obtaining biogenic nano-sized magnetic particles. Taking full advantage of a microfluidic droplet generator, Jung et al. (2012) successfully employed a recombinant E. coli, expressing both Arabidopsis thaliana phytochelatin synthase (AtPCS) and Pseudomonas putida metallothionein (PpMT), for crystalline FeMn magnetic nanoparticle fabrication. By simply mixing an aqueous solution of FeCl2·4H2O and MnCl2·4H2O with YM broth (pH 12), we also obtained relatively homogeneous FeMn paramagnetic NPs. Exceptionally, these particles demonstrated an amorphous nature, as evidenced by X-Ray diffraction (XRD) and selected-area electron diffraction (SAED) (Fig. 3.5). The dangling bands and high surface area have rendered the amorphous iron oxides superior to their crystalline counterparts in arsenite (As^{3+}) removal (Shan and Tong, 2013). Amorphous magnetic nanoparticles have been usually prepared by sonochemistry. This research provides an alternative way for obtaining FeMn binary oxides with potential applications in water splitting and purification.

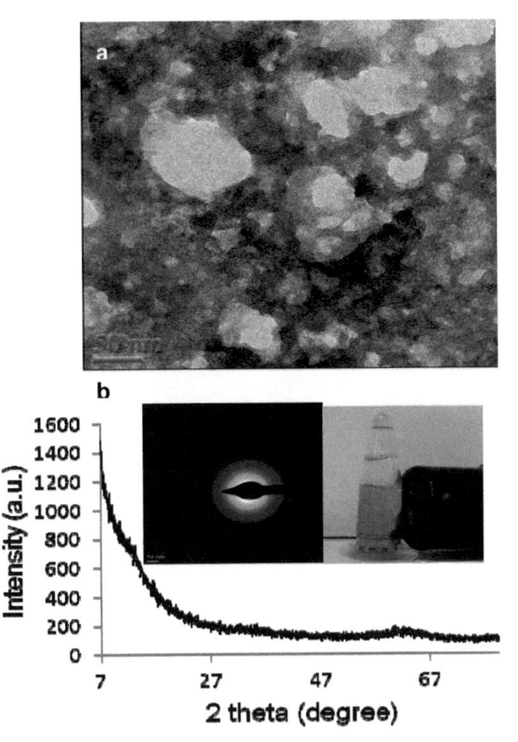

Figure 3.5 FeMn amorphous nanoparticles. (a) TEM micrograph of FeMn nanoparticles. (b) XRD and SAED patterns of amorphous FeMn particles formed in the system. Inset, particles were attracted to the right side by an external magnet.

3.5.4 Reduced Graphene Oxide (rGO)

In order to extend the application of graphene from electronic devices to biomedical fields, a few green strategies have been recently attempted to reduce graphene oxide (GO) with less contaminants produced (Gao et al., 2010). *Shewanella* species represent an important family of GO-reducing bacteria and their extracellular electron transfer (EET) network was suggested to be involved with the GO reduction (Salas et al., 2010; Wang et al., 2011). In our own work, we have demonstrated that pH12 YMB alone could reduce GO starting from room temperature, although a higher reduction rate was achieved in 15 min at 121°C in an autoclave (Fig. 3.6a–d).

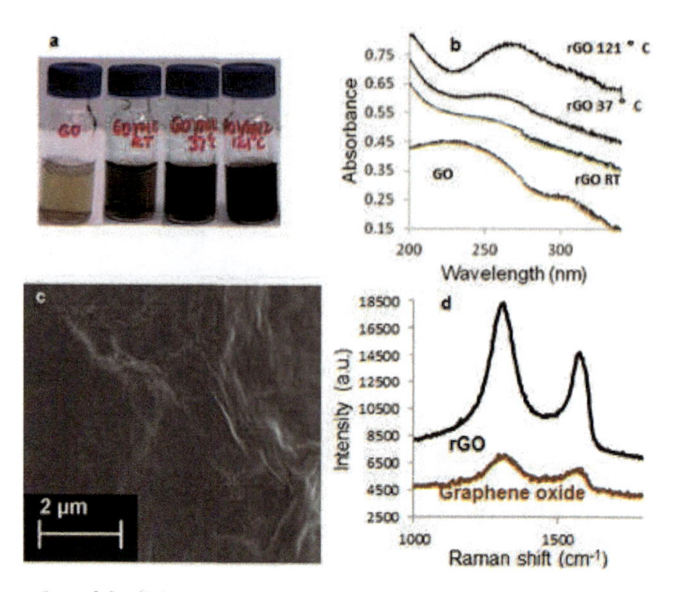

Figure 3.6 (a) Color photographs show a temperature dependence of graphene oxide (GO) reduction by YMB. (b) UV-vis spectra of rGO prepared at room temperature (18 h), 37°C (8 h) and 121°C (15 m), respectively. (c) SEM micrograph of the reduced graphene oxide. (d) Raman spectra of GO and rGO. Adapted from *RSC Adv.*, **4** (2014), 14564–14568 with permission from the Royal Society of Chemistry.

So far, we have presented a new concept for the greener synthesis of metal, metal oxides and semiconductor nanomaterials by using microbe-free broths. The capability to flexibly tune the particle size and morphology adds an extra advantage to this methodology. The general approach demonstrated so far could be extended to fabricate complex nanomaterials, for example, carbon and silver co-modified zinc oxide nanocomposites.

3.5.5 ZnO Nano/Micro Structures

Advances in green nanobiotechnology have led to numerous biomimetic approaches towards the synthesis of ZnO nanocrystals with a wide range of reactants, including pure amino acids (Subramanian and Ghaferi, 2014; Brif et al., 2014; Ramani et al., 2014), artificial peptides with an affinity for ZnO (Umetsu et al., 2005), gelatin (Tseng et al., 2012) or microorganisms like bacteria and fungi (Jain et al., 2013; Hussein et al., 2009). However, developing a biological process to supersede conventional physicochemical nanosynthesis remains a great challenge due to the fact that certain properties of ZnO are strongly dependent on its extrinsic characteristics, such as morphology, size, and exposed facets. Most of biogenic ZnO NPs were spherical in nature and enclosed by the less-reactive {10–10} facets, thus having no obvious photocatalytic activity (Jayaseelan et al., 2012). Additionally, pure ZnO for organic pollutant degradation suffers from both low solar light utilization (4%) and serious photocorrosion under long-term light irradiation. To solve these problems, a hetero-nanostructure of noble metal or a residual carbon coating is often introduced to a ZnO catalyst. For instance, AgNPs incorporated in ZnO may response to the visible light, trap photogenerated electrons, and delay the electron–hole recombination process, resulting in higher photocatalytic activity (Height et al., 2006). Other research groups have attempted coating ZnO with hydrothermal carbon (HTC) originating from biomass and poly- to monosaccharides to suppress the photocorrosion of ZnO (Zhang et al., 2009). Recently, Zhang's group (2014b) reported the synthesis of ZnO@C gemel hexagonal microrods by a facile one-step hydrothermal method with furfural as the carbon precursor. A number of ZnO assemblies have been mediated with natural polysaccharides, such as pectin (Wang et al., 2012), hyaluronic acid and chondroitin-6-sulfate (Waltz et al., 2012). Gum arabic, a complex mixture of glycoproteins and highly branched polysaccharides, has been successfully employed to guide the evolution of twin-brush ZnO mesocrystals (Liu et al., 2012b). As listed in Table 3.2, YMB powder is sugar rich, as it is composed of 47.6% dextrose, 14.3% yeast extract (abundant in yeast polysaccharide mannan), 23.8% peptone and 14.3% malt extract. We hypothesize that polysaccharide mannan should be able to promote secondary ZnO aggregation. Combined with a hydrothermal process, the dextrose and mannan may follow a chemical reaction path to form 5-hydroxymethyl-furfural-aldehyde (HMF) and subsequently transform to hydrothermal carbon (Titirici et al., 2008).

By optimizing the pH value, precursor concentration, yeast mold powder concentration, and hydrothermal reaction time, we finally obtained high quality thin hydrothermal carbon (HTC)-coated ZnO mesocrystals. The FTIR data indicated the involvement of proteins and mannans in the synthesis process. The Ag-modified ZnO@C samples, denoted as ZnO@C-Ag, were synthesized by mixing $AgNO_3$ and ZnO@C solutions under a sun simulator for 5 min. After centrifugation, the products were annealed at 400°C in air for 30 min. The weight percentage of carbon in the ZnO@C-Ag sample was 0.9%. The particle morphology, size, and crystalline phase are displayed in Fig. 3.7. All major diffraction peaks match with those of a typical wurtzite ZnO. The peaks at 2θ values of 38.1°, 44.4°, and 64.4° were attributed to crystal planes of metallic Ag. The presence of Ag, carbon and ZnO was further revealed by EDS analysis (Fig. 3.7.D).

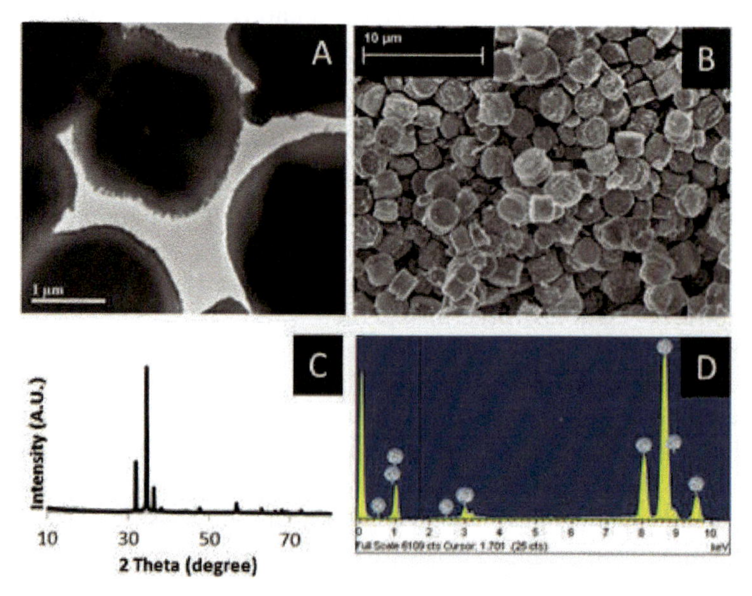

Figure 3.7 (A) TEM images showing the micro-hamburger morphology of the ZnO@C-Ag particles; (B) SEM images; (C) XRD pattern of ZnO and Ag NPs; (D) EDX analysis of ZnO@C-Ag after annealing at 400°C for 30 min. Reprinted with permission from *ACS Sustainable Chem. Eng.*, **3** (2015), 1010–1016. Copyright (2015) American Chemical Society.

Under solar light, the ZnO@C-Ag samples degrade methylene blue in 15 min. The photocatalytic performance depends strongly on efficient light absorption and charge transportation. AgNPs have been reported as "electron reservoirs" that effectively prolong the lifetime of photogenerated electron-hole pairs, thus leading to a remarkably improved photoactivity. The recycled samples have been evaluated for their photostability and it was found that the efficiency of ZnO@C-Ag remained at 96% after three cycling runs, indicating no serious corrosion of ZnO due to the hybridized HTC coating layer. In this study, YMB provides the carbon precursor as well as directing both ZnO and Ag nucleation.

3.6 Conclusions and Outlook

In summary, the preliminary studies of broth-mediated NP synthesis are reviewed in this chapter. By optimizing reaction parameters, such as microbial culture broth, pH value of broth, light intensity, temperature, and precursor concentration, we are able to realize the implementation of environmentally benign methods to fabricate technologically important nanomaterials. The broth-only greener approach has many advantages, for example the nanoparticles can be produced in a larger scale, over shorter periods of time and without the involvement of living microorganisms. This approach also has a significantly lower cost, as it relies in less expensive starting materials compared to pure peptides, amino acids, or enzymes. Hopefully, researchers in the microbial nanosynthesis field could be inspired to carefully design control experiments to avoid the misunderstanding of the microbial synthesis mechanisms and realize the implementation of complete green methods to fabricate technologically important nanomaterials.

Despite the encouraging results, much work is needed to extend the innovation to synthesize a broad range of different nanomaterials (e.g., CdSe nanocrystal and copper-containing nanocomposites).

A combination of the process with other physiochemical methods (e.g., hydrothermal, laser ablation and microwave) to enable nanosynthesis of other materials could make microbe-free synthesis more economically feasible.

Acknowledgment

L. Liu is supported by the Department of Chemical Engineering, Curtin University. This work was partially funded by the Australian Research Council (DP110104599). The authors wish to thank Dr Ioannis Theodorou for reviewing and proofreading the manuscript.

References

Akhtar M. S., Panwar J., Yun Y., *ACS Sustainable Chem. Eng.*, **1** (2013), 591–602.

Ashraf S., Abbasi A. Z., Pfeiffer C., Hussain S. Z., Khalid Z. M., Gil P. R., Parak W. J., Hussain I., *Colloids Surf. B*, **102** (2013), 511–518.

Babu M. M. G., Gunasekaran P., *Colloids Surf. B*, **74** (2009), 191–195.

Badwaik V. D., Bartonojo J. J., Evans J. W., Sahi S. V., Willis C. B., Dakshinamurthy R., *Langmuir*, **27** (2011), 5549–5554.

Balaji D. S., Basavarajaa S., Deshpande R., Maheshb D. B., Prabhakar B. K., Venkataraman A., *Colloids Surf. B*, **68** (2009), 88–92.

Bao H., Hao N., Yang Y., Zhao D., *Nano Res.*, **3** (2010), 481–489.

Bäuerlein E., *Angew. Chem. Int. Ed.*, **42** (2003), 614–641.

Baumann V., Muhammed M. A. H., Blanch A. J., Dey P., Rodríguez-Fernández J., DOI: 10.1002/ijch.201500031, *Isr. J. Chem.*, (2015), 1–20.

Bhainsa K. C., D'Souza S. F., *Colloids Surf. B*, **47** (2006), 160–164.

Binupriva A. R., Sathishkumar M., Yun S. I., *Colloids Surf. B*, **79** (2010), 531–534.

Brif A., Ankonina G., Drathen C., Pokroy B., *Adv. Mater.*, **26** (2014), 477–481.

Corain B., Schmid G., Toshima N. (eds.), *Metal Nanoclusters in Catalysis and Materials Science*, Elsevier, Amsterdam (2008).

Dahl J. A., Maddux B. L., Hutchison J. E., *Chem. Rev.*, **107** (2007), 2228–2269.

Das S. K., Das A. R., Guha A. K., *Small*, **6** (2010), 1012–1021.

Dhand C., Dwivedi N., Loh X. J., Ng A., Verma N. K., Beuerman R. W., Lakshminarayanan R., Ramakrishna S. R., *RSC Adv.*, **5** (2015), 105003–105037.

Durán N., Marcato P. D., Alves O. L., De Souza G. I., Esposito E., *J. Nanobiotechnol.*, **3** (2005), 8, doi:10.1186/1477-3155-3-8.

Durán N., Marcato P. D., Durán M., Yadav A., Gade A., Rai M., *Appl. Microbiol. Biotechnol.*, **90** (2011), 1609–1624.

El-Shanshoury A. E. R., ElSilk S. E., Ebeid M. E., *ISRN Nanotechnol.*, **2011** (2011), doi:10.5402/2011/385480.

Emam H. E., Ahmed H. B., *Carbohydr. Polym.*, **135** (2016), 300–307.

Engelbrekt C., Sørensen K. H., Lübcke T., Zhang J., Li Q., Pan C., Bjerrum N. J., Ulstrup J., *Chemphyschem.*, **11** (2010), 2844–2853.

Engelbrekt C., Sørensen K. H., Zhang J., Welinder A. C., Jensen P. S., Ulstrup J., *J. Mater. Chem.*, **19** (2009), 7839–7847.

Etheridge M. L., Campbell S. A., Erdman A. G., Haynes C. L., Wolf S. M., McCullough J., *Nanomedicine*, **9** (2013), 1–14.

Faramarzi M. A., Sadighi A., *Adv. Colloid Interface Sci.*, **189–190** (2013), 1–20.

Fayaz M., Tiwary C. S., Kalaichelvana P. T., Venkatesan R., *Colloids Surf. B*, **75** (2010), 175–178.

Flores G. Y., Miñán A., Grillo C. A., Salvarezza R. C., Vericat C., Schilardi P. L., *ACS Appl. Mater. Interfaces*, **5** (2013), 3149–3159.

Gao J., Liu F., Ma N., Wang Z., Zhang X., *Chem. Mater.*, **22** (2010), 2213–2218.

Gruen L. C., *Biochim. Biophys. Acta*, **386** (1975), 270–274.

Height M., Pratsinis S., Mekasuwandumrong O., Praserthdam P., *Appl. Catal. B*, **63** (2006), 305–312.

Hu W., Chen S., Li X., Shi S., Shen W., Zhang X., Wang H., *Mater. Sci. Eng. C*, **29** (2009), 1216–1219.

Hussein M. Z., Azmin W. H. W. N., Mustafa M., Yahaya A. H., *J. Inorg. Biochem.*, **103** (2009), 1145–1150.

Jain N., Bhargava A., Majumdar S., Tarafdar J. C., Panwar J., *Nanoscale*, **3** (2011), 635–641.

Jain N., Bhargava A., Tarafdar J. C., Singh S. K., Panwar J., *Appl. Microbiol. Biotechnol.*, **97** (2013), 859–869.

Jayaseelan C., Abdul Rahuman A., Kirthi A. V., Marimuthu S., Santhoshkumar T., Bagavan A., Gaurav K., Karthik L., Rao K. V. B., *Spectrochim. Acta A*, **90** (2012), 78–84.

Jha K., Prasad K., Kulkarni A. R., *Colloids Surf. B*, **71** (2009), 226–229.

Johnston C. W., Wyatt M. A., Li X., Ibrahim A., Shuster J., Southam G., Magarvey N. A., *Nat. Chem. Biol.*, **9** (2013), 241–243.

Jung J. H., Park T. J., Lee S. Y., Seo T. S., *Angew. Chem. Int. Ed.*, **51** (2012), 5634–5637.

Kalimuthu K., Babu R. S., Venkataraman D., Bilal M., Gurunathan S., *Colloids Surf. B*, **65** (2008), 150–153.

Kalishwaralal K., Deepak V., Pandiana S. R. K., Kottaisamy M., BarathManiKantha S., Kartikeyana B., Gurunathan S., *Colloids Surf. B*, **77** (2010), 257–262.

Kamyshny A., Madgsssi S., *Small*, **10** (2014), 3515–3535.

Kathiresan K., Manivannan S., Nabeel M. A., Dhivya B., *Colloids Surf. B*, **71** (2009), 133–137.

Kharissova O. V., Dias R. H. V., Kharisov B. I., Pérez B. O., Pérez V. M. J., *Trends Biotechnol.*, **31** (2013), 240–248.

Klaus T., Joerger R., Olsson E., Granqvist C. G., *PNAS,* **96** (1999), 13611–13614.

Korbekandi H., Mohseni S., Jouneghani R. M., Pourhossein M., Iravani S. *Artif. Cells Nanomed. Biotechnol.* (2014), 1–5, DOI: 10.3109/21691401.2014.937870.

Kowshik M., Ashtaputre S., Kharrazi S., Vogel W., Urban J., Kulkarni S. K., Paknikar K. M., *Nanotechnology,* **14** (2003), 95–100.

Kracht S., Messerer M., Lang M., Eckhardt S., Lauz M., Grobety B., Fromm K. M., Giese B., *Angew. Chem., Int. Ed.*, **54** (2015), 2912–2916.

Kröger N., Lorenz S., Brunner E., Sumper M., *Science,* **298** (2002), 584–586.

Kumar S. A., Abyaneh M. M., Gosavi S. W., Kulkarni S. K., Pasricha R., Ahmad A., Khan M. I., *Biotechnol. Lett.*, **29** (2007), 439–445.

Kumar S., Ahlawat W., Kumar R., Dilbaghi N., *Biosens. Bioelectron.*, **70** (2015), 498–503.

Lengke M. F., Fleet M. E., Southam G., *Langmuir,* **15** (2006), 7318–7323.

Lengke M. F., Fleet M. E., Southam G., *Langmuir,* **23** (2007), 2694–2699.

Li X., Chen S., Hu W., Shi S., Shen W., Zhang X., Wang, W., *Carbohydr. Polym.*, **76** (2009), 509–512.

Lim H., Mishra A., Yun S., *J. Nanosci. Nanotechnol.*, **11** (2011), 518–522.

Liu L., Liu T., Tade M. O., Wang S., Li X., Liu S., *Enzyme Microb. Technol.*, **67** (2014b), 53–58.

Liu J., Qin G., Raveendran P., Ikushima Y., *Chem. Eur. J.*, **12** (2006), 2131–2138.

Liu L., Shao Z., Ang H. M., Tade M. O., Liu S., *RSC Adv.*, **4** (2014a), 14564–14568.

Liu C., Yang D., Wang Y. G., Shi J. F., Jiang Z. Y., *J. Nanopart. Res.*, **14** (2012a), 1084–1095.

Liu M., Tseng Y., Greer H. F., Zhou W., Mou C., *Chem. Eur. J.*, **18** (2012b), 16104–16113.

Liu L., Yang J., Xie J., Luo Z., Jiang J., Yang Y., Liu S., *Nanoscale,* **5** (2013), 3834–3840.

Lovley D. R., Stolz J. F., Nord Jr G. L., Phillips E. J. P., *Nature,* **330** (1987), 252–254.

Lu L., Ai K., Ozaki Y., *Langmuir,* **24** (2008), 1058–1063.

Lu Q., Gao F., Komarneni S., *Adv. Mater.*, **18** (2004), 1629–1632.

Ma X., Dai Y., Yu L., Lou Z., Huang B., Whangbo M., *J. Phys. Chem. C,* **118** (2014), 12133–12140.

Makarov V. V., Love A. J., Sinitsyna O. V., Makarova S. S., Yaminsky, I. V., Taliansky, M. E., Kalinina N. O., *Acta Nat.*, **6** (2014), 35–44.

Malhotra A., Dolma K., Kaur N., Rathore Y. S., Mayilraj A. S., Choudhury A. R., *Bioresour. Technol.*, **142** (2013), 727–731.

Maliszewska I., Aniszkiewicz Ł., Sadowski Z., Proceedings of the III National Conference on Nanotechnology NANO (2009).

Maruyama T., Fujimoto Y., Maekawa T., *J. Colloid Interface Sci.*, **447** (2015), 254–257.

Mohanty A., Garg N., Jin R., *Angew. Chem. Int. Ed.*, **49** (2010), 4962–4966.

Mokhtari N., Daneshpajouh S., Seyedbagheri S., Atashdehghan R., Abdi K., Sarkar S., Minaian S., Shahverdi H. R., Shahverdi A. R., *Mater. Res. Bull.*, **44** (2009), 1415–1421.

Mukherjee P., Ahmad A., Mandal D., Senapati S., Sainkar S. R., Khan M. I., Parischa R., Ajayakumar P. V., Alam M., Kumar R., Sastry M., *Nano Lett.*, **1** (2001a), 515–519.

Mukherjee P., Ahmad A., Mandal D., Senapati S., Sainkar S. R., Khan M. I., Ramani R., Parischa R., Ajayakumar P. V., Alam M., Sastry M., Kumar R., *Angew. Chem. Int. Ed.*, **40** (2001b), 3585–3588.

Musarrat J., Dwivedi S., Singh B. R., Al-Khedhairy A. A., Azam A., Naqvi A., *Bioresour. Technol.*, **101** (2010), 8772–8776.

Nadagouda M. N., Varma R. S., *Green Chem.*, **8** (2006), 516–518.

Naik R. R., Stringer S. J., Agarwal G., Jones S. E., Stone M. O., *Nat. Mater.*, **1** (2002), 169–172.

Nam K. T., Lee Y. J., Krauland E. M., Kottmann S. T., Belcher A. M., *ACS Nano*, **2** (2008), 1480–1486.

Nanda A., Saravanan M., *Nanomedicine*, **5** (2009), 452–456.

Panigrahi S., Kundu S., Ghosh S. K., Nath S., Pal T., *J. Nanopart. Res.*, **6** (2004), 411–414.

Park T. J., Lee S. Y., Heo N. S., Seo T. S., *Angew. Chem., Int. Ed.*, **49** (2010), 7019–7024.

Park Y., *Toxicol. Res.*, **30** (2014), 169–178.

Parikh R. Y., Singh S., Prasad B. L. V., Patole M. S., Sastry M., Shouche Y. S., *ChemBioChem*, **9** (2008), 1415–1422.

Pereira L., Mehboob F., Stams A. J. M., Mota M. M., Rijnaarts H. H. M., Alves M. M., *Crit. Rev. Biotechnol.*, **35** (2015), 114–128.

Pettegrew C., Dong Z., Muhi M. Z., Pease S., Mottaleb M. A., Islam M. R., *ISRN Nanotechnol.*, **2014** (2014), Article ID 480284, 8 pages.

Prakasham, S. R., Kumar B. S., Kumar Y. S., Shankar G. G., *J. Microbiol. Biotechnol.*, **22** (2012), 614–621.

Ramanathan R., P. O'Mullane A., Parikh R. Y., Smooker P. M., Bhargava S. K., Bansal V., *Langmuir*, **27** (2011), 714–719.

Ramanathan R., Field M. R., O'Mullane A. P., Smooker P. M., Bhargava S. K., Bansal V., *Nanoscale*, **5** (2013), 2300–2306.

Ramani M., Ponnusamy S., Muthamizhchelvan C., Marsili E., *Colloids Surf. B*, **117** (2014), 233–239.

Raveendran P., Fu J., Wallen S. L., *J. Am. Chem. Soc.*, **125** (2003) 13940–13941.

Roy N., Gaur A., Jain A., Bhattacharya S., Rani V., *Environ. Toxicol. Pharmacol.*, **36** (2013), 807–812.

Roy K., Sarkar C. K., Ghosh C. K., *Appl. Nanosci.*, **5** (2015), 953–959.

Sadhasivam S., Shanmugam P., Yun K., *Colloids Surf. B*, **81** (2010), 358–362.

Salas E. C., Sun Z., Lüttge A., Tour J. M., *ACS Nano*, **4** (2010), 4852–4856.

Sanghi R., Verma P., *Bioresour. Technol.*, **100** (2009), 501–504.

Saravanan M., Vemu A. K., Barik S. K., *Colloids Surf. B*, **88** (2011), 325–331.

Sathiyanarayanana G., Kiran G. S., Selvin J., *Colloids Surf. B*, **102** (2013), 13–20.

Sau T. K., Rogach A. L., Döblinger M., Feldmann J., *Small*, **2011**, **7**, 2188–2194.

Selvakannan P., Mandal S., Phadtare S., Gole A., Pasricha R., Adyanthaya S. D., Sastry M., *J Colloid Interface Sci.*, **269** (2004), 97–102.

Shan C., Tong M., *Water Res.*, **47** (2013), 3411–3421.

Shankar S., Rhim J., *Carbohydr. Polym.*, **130** (2015), 353–363.

Shen Z., Liang P., Wang S., Liu L., Liu S., *ACS Sustainable Chem. Eng.*, **3** (2015a), 1010–1016.

Shen Z., Liu B., Pareek V., Wang S., Li X., Liu L., Liu S., *RSC Adv.*, **5** (2015b), 80488–80495.

Shin H. S., Yang H. J., Kim S. B., Lee S. S., *J. Colloid Interface Sci.*, **274** (2004), 89–94.

Si S., Mandal T. K., *Chemistry*, **13** (2007), 3160–3168.

Subramanian, N., Ghaferi, A. A., *RSC Adv.*, **4** (2014), 4371–4378.

Sun X., Zheng C., Zhang F., Yang Y., Wu G., Yu A., Guan N., *J. Phys. Chem. C*, **113** (2009), 16002–16008.

Suresh A. K., Pelletier D. A., Wang W., Moon J., Gu B., Mortensen N. P., Allison D. P., Joy D. C., Phelps T. J., Doktycz M. J., *Environ. Sci. Technol.*, **44** (2010), 5210–5215.

Swami A., Kumar A., D'Costa M., Pasricha R., Sastry M., *J. Mater. Chem.*, **14** (2004), 2696–2702.

Tan Y. N., Lee J. Y., Wang D. I., *J. Am. Chem. Soc.*, **132** (2010), 5677–5686.

Titirici M., Antoniettia M., Baccile N., *Green Chem.*, **10** (2008), 1204–1212.

Tseng Y., Liu M., Kuo Y., Chen P., Chen C., Chen Y., Mou C., *Chem. Commun.*, **48** (2012), 3215–3217.

Umetsu M., Mizuta M., Tsumoto K., Ohara S., Takami S., Watanabe H., Kumagai I., Adschiri T., *Adv. Mater.*, **17** (2005), 2571–2575.

Vigneshwaran N., Ashtaputre N. M., Varadarajan P. V., Nachane R. P., Paralikar K. M., Balasubramanya R. H., *Mater. Lett.*, **61** (2007), 1413–1418.

Vigneshwaran N., Kathe A. A., Varadarajan P. V., Nachane R. P., Balasubramanya R. H., *Colloids Surf. B*, **53** (2006), 55–59.

Waltz F., Wißmann G., Lippke J., Schneider A. M., Schwarz H., Feldhoff A., Eiden S., Behrens P., *Cryst. Growth Des.*, **12** (2012), 3066–3075.

Wang A., Liao Q., Feng J., Zhang P., Lia A., Wang J., *CrystEngComm*, **14** (2012), 256.

Wang G., Qian F., Saltikov C. W., Jiao Y., Li Y., *Nano Res.*, **4** (2011), 563–570.

Wang Z., Zhang J., Ekman J. M., Kenis P. J. A., Lu Y., *Nano Lett.*, **10** (2010), 1886–1891.

Wang X., Zuo J., Keil P., Grundmeier G., *Nanotechnology*, **18** (2007), 265303.

Wangoo N., Kaur S., Bajaj M., Jain D. V. S., Sharma R. K., *Nanotechnology*, **25** (2014), 435608 (7p.).

Wei H., Wang Z., Zhang J., House S., Gao Y., Yang L., Robinson H., Tan L. H., Xing H., Hou C., Robertson I. M., Zuo J., Lu Y., *Nat. Nanotechnol.*, **6** (2011), 93–97.

Wu J., Zheng Y., Song W., Luan J., Wen X., Wu Z., Chen X., Wang Q., Guo S., *Carbohydr. Polym.*, **102** (2014), 762–771.

Xie J., Lee J. Y., Wang D. I. C., *J. Phys. Chem. C*, **111** (2007), 10226–10232.

Xie J., Zhang Q., Lee J. Y., Wang D. I. C., *ACS Nano*, **2** (2008), 2473–2480.

Xiong J., Wu X., Xue Q., *J. Colloid Interface Sci.*, **390** (2013), 41–46.

Xu W., Jin W., Lin L., Zhang C., Li Z., Li Y., Song R., Li B., *Carbohydr. Polym.*, **101** (2014), 961–967.

Yamal G., Sharmila P., Rao K. S., Pardha-Saradhi P., *PLOS ONE*, **8** (2013), DOI: 10.1371/journal.pone.0061750.

Zhang L., Cheng H., Zong R., Zhu, Y., *J. Phys. Chem. C*, **113** (2009), 2368–2374.

Zhang S., Liu L., Pareek V., Becker T., Liang J., Liu S., *J. Microbiol. Methods*, **105** (2014a), 42–46.

Zhang P., Li B., Zhao Z., Yu C., Hu C., Wu S., Qiu J., *ACS Appl. Mater. Interfaces*, **6** (2014b), 8560–8566.

Chapter 4

Rational Design of Enzyme–Polymer Biocatalysts

Omkar V. Zore,[a,b] Rajeswari M. Kasi,[a,b] and Challa V. Kumar[a,b,c,d]

[a]*Department of Chemistry, U-3060, University of Connecticut Storrs, Connecticut 06269-3060, USA*
[b]*Institute of Materials Science, U-3136, University of Connecticut Storrs, Connecticut 06269-3069, USA*
[c]*Department of Molecular and Cell Biology, University of Connecticut Storrs, Connecticut 06269-3125, USA*
[d]*Department of Inorganic and Physical Chemistry, Indian Institute of Science, Bengaluru, Karnataka 560012, India*

kasi@ims.uconn.edu, challa.kumar@uconn.edu

4.1 Introduction

Enzymes are biological catalysts, which accelerate chemical or physical reactions in vivo and in vitro with very high selectivity, specificity, and efficiency. In contrast to other man-made catalysts, enzymes have several advantages, for example, they operate under physiological conditions at atmospheric pressures. Most importantly, enzymes are green and eco-friendly biocatalysts derived from earth abundant elements such as carbon, hydrogen and oxygen and are produced by sustainable approaches from renewable resources (Aehle W., *Enzymes in Industry*, Wiley, 2008; Campbell M K., Farrel S.O., *Biochemistry*, Brooks Cole Publishing

Biocatalysis and Nanotechnology
Edited by Peter Grunwald

ISBN 978-981-4613-69-9 (Hardcover), 978-1-315-19660-2 (eBook)
www.panstanford.com

Company; 2011). Unlike enzymes, most synthetic catalysts such as noble metals are not produced by sustainable or green methods and they are not often biocompatible. Also, either during the synthesis or in operating conditions of such non-biological catalysts, hazardous reaction conditions are used or many non-biodegradable byproducts are produced. Therefore, one of the modern industrial goals is to manufacture sustainable and degradable biocatalysts for production of clean materials and clean energy by green processes to reduce waste, conserve energy, and discover appropriate and functional replacements for hazardous chemicals.

Most enzymes also function with very high substrate specificity, which implies that even a minor change in the substrate structure impedes catalysis and a new enzyme is required to catalyze the corresponding reaction. One such example is glucose oxidase, which selectively catalyzes the oxidation of D-glucose by molecular oxygen at high rates. However, a different enzyme is required to oxidize other sugars like amylose or arabinose and glucose oxidase is not useful for the oxidation of these other sugars. This limitation of enzymes is not easily rectified, unless one isolates the corresponding enzyme that works on the desired substrate. One approach could be to modify the enzyme structure in a subtle manner, via adsorption onto a polymer or nanoparticles, such that the enzyme active site is distorted to broaden its selectivity without losing enzymatic activity. However, this is difficult to broaden the selectivity of any given enzyme and still remains to be a challenge in biocatalysis. Therefore, greater understanding of how enzymes and non-biological surfaces interact might provide some rational approaches for such systematic manipulation of enzyme structure in the near future.

In this chapter, we will focus on how enzymes and water-soluble polymers interact and how the synthetic polymers can enhance the physical, chemical, and biological properties of enzymes. To harness the catalytic properties of enzymes under non-biological conditions, enzymes need to be stabilized against denaturants that are common under laboratory conditions and properly designed synthetic polymers are useful to achieve this goal. The thermal stability, organic solvent stability, pH stability, reusability, enhanced operational life, resistance to poisoning or inhibition, for example, are to be enhanced for laboratory or industrial applications. One approach taken by our laboratory to address these issues is to stabilize enzymes by wrapping them with simple organic polymers such that the polymer chain would hold and stabilize the delicate enzyme structure against these denaturants.

While wrapping the enzyme with the polymer, the redox and electrochemical activity and catalytic activity of enzymes should be retained fully, and should allow for full diffusional transport of the substrate to the active site of the enzyme and release of the product. While simple wrapping of the enzyme without any covalent conjugation can also result in un-wrapping, covalent conjugation of the polymer chains onto the side chains of the enzyme via suitable linkers has been proposed as a method to stabilize 3D structures of enzymes under these denaturing conditions (Abad et al., 2005). However, the choice of the polymer and the type and the number of linkages to be made between the polymer and the enzyme surface are important decisions, and these can have a strong influence on the properties of the resulting biocatalysts. The covalent chemistry should not interfere with the active site chemistry and should leave sensitive residues untouched. Thus, there is an urgent need to design rational strategies to stabilize enzymes under ordinary laboratory conditions but with predictable properties of the enzyme–polymer conjugate. Water-soluble polymers are generally the preferred soft materials for wrapping enzymes and this is due to several important factors such as the availability of these polymers in large quantities with different functional groups which would allow for different conjugation strategies between the enzyme and the polymer; water-soluble polymers are more likely to produce a hydration shell around the enzyme and promote the retention of native-like structure of the enzyme; and the hydrophilic groups of the polymer are more likely to interact favorably with the polar groups on the enzyme surface and stabilize it. Thus, this strategy of wrapping enzymes with polymers allows for the combination of the biological properties of enzymes with the material properties of synthetic polymers and produce unique biocatalysts that are otherwise not possible to make either in the biological world or in the synthetic world. Such unique hybrids enjoy special status and their design can profit from the progress made in both these vastly different disciplines.

The use of enzymes in several applications is mired under non-physiological conditions such as high temperature, extreme pHs and organic solvents due to loss of their 3D structure, denaturation and consequent loss of catalytic activities under these non-biological conditions (Jaenicke and Bohm, 1998; Polizzi et al., 2007). These issues have been overcome by modification of the enzymes, linking enzymes to the surfaces or layered materials, etc. (Bhambani and Kumar, 2006; Kumar and Chaudhari, 2000; Kumarand Chaudhari, 2001; Kumar and Chaudhari, 2002; Kim and Grate, 2003; Novick and Dordick, 1998; Khalaf et al., 1996; Xu and Klibanov, 1996; Woodward and Kaufman, 1996; Khmelnitsky et al., 1991; Antonini et al., 1981).

Our general hypothesis is that the physical confinement of the enzyme within a polymeric shell will decrease the enzyme's conformational entropy, which will directly and positively impact the retention and stability of the 3D hierarchical structure of the enzyme. This is because the confined space around the enzyme would only permit access to limited amount of space and hence, the conformational distribution is severely limited when compared to free space around the enzyme. Decreased conformational entropy (ΔS) results in the increases of the free energy (ΔG) required to denature the enzyme ($\Delta G = \Delta H - T\Delta S$), where the enthalpy term (ΔH) is essentially fixed by the primary sequence of the enzyme. Therefore, a decrease in ΔS produces an increase in the ΔG term and more vigorous conditions are required to denature the enzyme under these modified conditions. At room temperature, ΔG for enzyme denaturation is large and positive, and at the denaturation temperature the native and denatured states are at equilibrium, and ΔG equals zero. This simple strategy of entropy control of the enzyme denaturation process has been widely used in our laboratory to control the thermo dynamic stabilities of enzymes by a variety of different approaches. For example, when the enzyme surface is wrapped with a synthetic polymer such that ΔH is unchanged, the increased conformational entropy of the enzyme is expected to increase the ΔG needed to denature the enzyme and hence should enhance enzyme stability. This concept is schematically illustrated in Scheme 4.1.

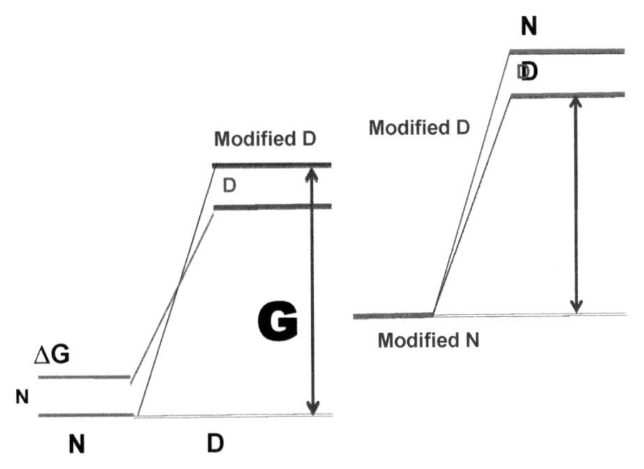

Scheme 4.1 Stabilization of the native state (N) and destabilization of the denatured state (D) would enhance the thermodynamic stability of the enzyme (Left). When the conformational entropy of the enzyme is lowered by confinement of the enzyme with a thin polymer shell (right), there will be a corresponding increase in ΔG, when ΔH is kept constant. The latter is thought to depend on the primary sequence of the enzyme and hence, unaltered in this scheme. The net result is an increase in the thermodynamic stability of the polymer-encased enzyme.

Our group has used hemoglobin (Hb), horseradish peroxidase (HRP), catalase, myoglobin, and glucose oxidase (GOx) as model enzymes to prepare and evaluate enzyme–polymer interactions. Both physical measurements as well as chemical conjugation methods were used to gain insight into these interactions. May research groups have selected these and other enzymes due to their widespread use in enzymology with applications in organic synthesis, multi-enzyme cascade catalysis as well as biofuel cells (Dunn et al., 2015; Hebbel and Shalev, 1996; Loewen and Hillar, 1995). For example, some of these heme-based enzymes have peroxidase activity, and this chemistry is extensively being used industrially for wastewater treatment in the oxidation of polycyclic aromatic hydrocarbons (Gao et al., 2014). We also chose these enzymes so that we can examine how an increase in the enzymes' molecular weights influences the enzyme–polymer interactions. The van der Waals attractive interactions are expected to increase with increase in the molar mass and hence, above certain mass, the polymer and the enzyme are expected to spontaneously combine, unless there are repulsive interactions impeding them. The non-ionic intermolecular forces are known to scale as a product of the radii of the partners ($U = c\ [R_1R_2/(R_1 + R_2)]$) where R_1 and R_2 are the radii of the two interacting spheres at a fixed distance of separation of their surfaces, and c being the proportionality constant (Parsegian, 2006). Thus, the intermolecular forces will steadily strengthen with molecular size.

This set of enzymes we chose for our modeling studies shows a variety of three-dimensional structures, consisting of alpha-helices, beta sheets and random coils. It is intriguing to ask if one kind of secondary structure is better suited for favorable interactions with synthetic polymers over the others. Will different structures show different extents of stabilization with a given polymer molecule? Such questions are not being addressed but would be of general interest in this area of enzyme–polymer interactions. In addition to the differing 3D structures, this set of enzymes also has increasing numbers of lysine residues in their primary sequence. This detail is important because the NH_2 groups of these amino acid side chains can be used for ligation with the COOH groups of polymers, and hence, the number of covalent links between the polymer and the enzyme can be controlled in a systematic manner by taking advantage of the natural distribution of this amino acid on enzyme surfaces. Key properties of the above enzymes (column 1) such as molecular weight (column 2), total number of residues (column 3) and total number of lysine residues (column 4) are presented in Table 4.1.

Our group has used polyacrylic acid (PAA) to covalently conjugate with enzymes (Ghimire et al., 2015; Thilakarathne et al., 2011; Mudhivarthi et al., 2012, Briand et al., 2012; Thilakarathne et al., 2012; Ghimire et al., 2014). PAA is a hydrophilic polymer with one COOH group per monomer being available for conjugation with the amine groups on enzyme surfaces. The hydrophilic, water-soluble polymers such as PAA are expected to keep the enzymes hydrated upon polymer conjugation, and these COOH groups may also serve as a local buffer to keep the local pH constant, and these polar groups would enhance the solubilities of certain hydrophobic enzymes in the aqueous media. For covalent chemistry between the amine groups of the enzyme and the acid groups of PAA, we use 1-ethyl-3-(3-dimethylamino propyl)carbodiimide (EDC), which is often used to activate the

COOH groups for conjugation to lysine amino groups (NH_2) of enzymes (Zore et al., 2014). Each newly formed amide bond is biocompatible and stable to aqueous media, while holding the enzyme and the polymer together. One can imagine multiple such links between the enzyme and the polymer would form a hydrophilic thin film around the enzyme and protect it from denaturation.

Table 4.1 Key properties of hemoglobin, horseradish peroxidase, and glucose oxidase

Enzyme	Molecular weight (kDa)	Total number of residues	Number of Lys residues
Hemoglobin 	64.5	574	48
Glucose oxidase 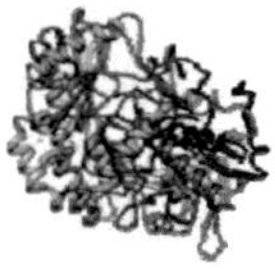	160	1166	30
Horseradish peroxidase 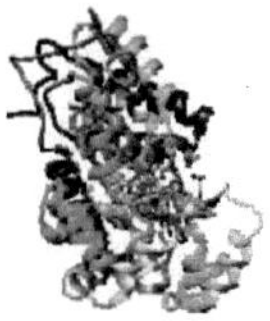	40	306	6

On the other hand, the artificially created peptide bond can be hydrolyzed by peptidases and the biocatalyst can be biodegraded when once the biocatalyst is released into the environment, an important criterion for low environmental impact of the biocatalyst. In addition, PAA with molecular weights of 5000 or less are also biodegradable and hence, complete degradation of the biocatalyst in the environment over reasonable time period is anticipated. Moreover, the rate of degradation may be pre-programmed to control its disintegration under specific environmental cues. Noble metal catalysts, on the other hand, suffer from toxic release and adverse environmental impact unless care is taken not to release the catalyst and ensure that it is fully recycled, which is difficult to practice in the real world.

Our strategy for the synthesis of enzyme-PAA conjugates is explained further, below. The polymer molecular weight is varied strategically such that it wraps around the desired enzymes and provides protection against the external stresses such as heat, temperature, organic solvents, denaturants or pH. The larger the enzyme, the longer the length of the polymer required to fully wrap the enzyme, as derived from simple geometry considerations. The polymer sheath around the enzyme may provide the microenvironment conducive for the retention of biological activity even under harsh conditions. If not, the functional groups that are remaining on the polymer after conjugation with the enzyme may be chemically modified to tailor the microenvironment around the enzyme or even improve upon enzyme's inherent characteristics. For example, the polymer network around the enzyme could impede access to microbes or peptidases and protect the enzyme against these deleterious agents (Riccardi et al., 2014). The electrical charge of the polymer may further be used to control diffusion of the substrates or the inhibitors to the enzyme surface, thus further providing additional selectivity over the enzyme's catalytic cycle. Thus, the polymer coating on the enzyme surface could be used for additional protection of the enzyme as well as control over its behavior in a rational and systematic manner.

These enzyme–polymer biocatalysts are made with approaches by adhering to the green chemistry principles and the resulting enzymatic products may be considered as "clean materials" as these are produced by green catalysts. The biocatalysts reported here are mostly non-toxic to biological systems, and they may be biodegradable or biocompatible depending on the molecular weight and nature of the polymer. Rational design of these biocatalysts with complete control over their molecular properties require fundamental understanding of protein- polymer interactions at the atomic level and methodologies need to be refined to tune the enzyme's environment as dictated by the intended applications.

4.2 Strategy for the Multi-Site Covalent Attachment of PAA to Enzyme Surfaces

Multi-site attachment on the polymer as well as the enzyme is expected to lower the conformational entropy of the enzyme the most, when compared to single site attachment. Hence, we focus here only on the multi-site attachment approach for maximum impact on enhancing the enzyme stability to the fullest possible extent. Therefore, by definition, one needs multiple functional groups on both the enzyme and the polymer to accomplish the multi-site attachment. Most enzymes have multiple numbers of lysine for multiple conjugation with PAA. In addition, one can use elegant biotechnology approaches to create multiple reactive residues at specific sites on the enzyme (Heredia and Maynard, 2007). Regardless of the origin of the reactive sites on the enzyme surface, our approach is to attach these functional groups on the enzymes with the COOH groups of PAA. In general terms, polymers with reactive groups such as COOH groups can be conjugated to lysines, while polymers with amine groups can be conjugated to the aspartic acid or glutamic acid residues on the enzymes (Pasut, 2014). Thus, the polymer is to be chosen such that it has functional groups that complement those present abundantly on the enzyme surface.

The COOH/NH_2 approach is used for covalent conjugation of the enzymes with PAA (Scheme 4.2) where the multi-site covalent conjugation of PAA to enzymes is illustrated, in general. The carbodiimide, EDC, is used to activate the carboxyl groups on the PAA (blue string), and then the activated PAA is reacted with the enzyme (red circle) through formation of amide bonds to form enzyme-PAA multi-site conjugate. In the following sections, specific examples of enzymes containing surface lysine groups will be conjugated to EDC activated PAA and the resulting conjugates discussed in detail.

Our hypothesis is that in the enzyme–polymer multi-site conjugates, the polymer will effectively confine the enzyme in its space such that these enzyme molecules will be isolated from each other by the intricate polymer network surrounding them. The resulting constraints on the enzyme conformations will limit the motions of the enzyme peptide chains and the decrease in the entropy of the states will stabilize the confined enzyme. This hypothesis is tested by (1) synthesizing enzyme–polymer conjugates using different enzymes and PAA as model systems; (2) elucidating the 3D structure of the enzyme in the conjugate and evaluating the morphology of conjugate, and (3) biocatalytic applications of these stabilized enzyme–polymer conjugates are tested under denaturing conditions. In this book chapter, we will exploit our ability to stabilize the enzyme for biocatalysis in organic solvents and in multi-enzyme cascade catalysis and this will be one of the main points of this chapter. The material is divided in terms of the enzyme or the protein used to form the multi-site conjugate with PAA.

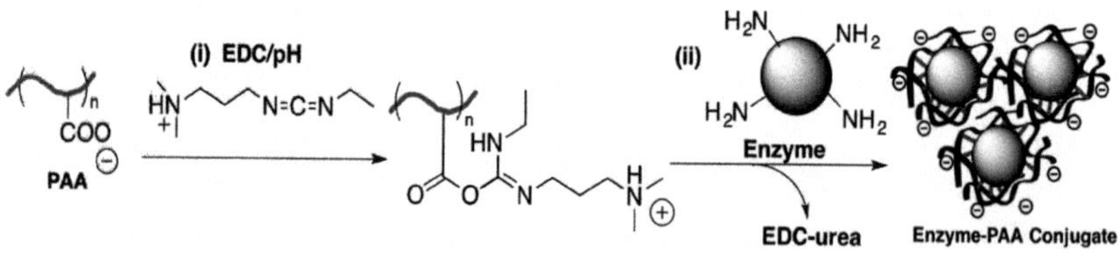

Scheme 4.2 General scheme of synthesis of enzyme–polymer conjugate. Polyacrylic acid (PAA) is used as a model polymer. Red circle is shown as an enzyme, blue string as PAA, EDC is used as a cross-linker. After EDC activation –COOH groups of PAA get covalently attached to $-NH_2$ groups on enzyme through amide bond formation.

4.3 Hemoglobin

Hemoglobin is a tetrameric globular protein that is responsible for oxygen transport in the blood, and several approaches are being tested to use Hb as an artificial blood substitute. It is composed of four different polypeptide chains which self-assemble into a single functional unit, without any covalent linkages between them. The heme prosthetic group is bound to each polypeptide chain, thus, maintaining four hemes per one Hb molecule, and the heme is also non-covalently bound to the polypeptide chain. Many attempts were made to stabilize Hb by using covalent attachment and entrapment chemistries. Besides oxygen, Hb also has a high binding affinity for toxic gases such as nitric oxide and carbon monoxide, and Hbis used as a sensor platform but suffers from short shelf life and lack of long-term stability.

We first reported the use of high molecular weight PAA (450 kDa) that is EDC activated and then multi-site cross-linked with Hb to produce Hb-PAA nanogels that presented retention of its biological activity and secondary structure. The Hb-PAA multi-site conjugate has a long shelf life when compared to unmodified hemoglobin (Mudhivrthi et al., 2012; Briand et al., 2011; Zore et al., 2014). We used PAA of different molecular weights to prepare Hb-PAA core–shell nanoparticle conjugates which can be further cross-linked using the –COOH groups of PAA and these confirmed retention of biological activity of Hb, and its secondary structure. Further these conjugates also showed exceptional thermal stability and stability against steam sterilization conditions (Mudhivarthi et al., 2012; Ghimire et al., 2015). However, these examples showed that Hb-based enzyme catalysis is limited to aqueous medium since most organic solvents denature this enzyme.

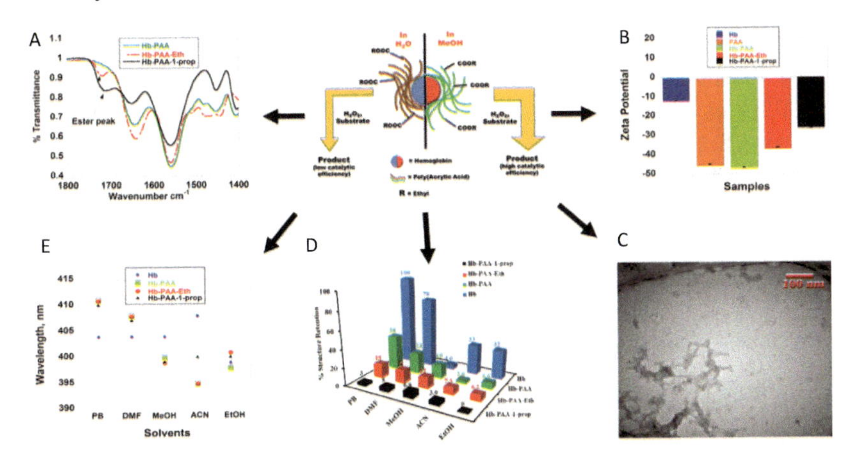

Figure 4.1 (A) IR spectra of Hb-PAA-prop (black), Hb-PAA-Eth (red) and Hb-PAA (green) (B) Zeta potential of Hb-PAA-prop (black), Hb-PAA-Eth (red), Hb-PAA (green), PAA (orange) and Hb (blue) (C) TEM study of Hb-PAA-Eth nanogels (D) Structure retention of Hb-PAA-prop (black), Hb-PAA-Eth (red), Hb-PAA (green), and Hb (blue) in various solvents. (E) UV study of Hb-PAA-prop (black), Hb-PAA-Eth (red), Hb-PAA (green) and Hb (blue) showing peak positions at different wavelengths and effect of organic solvent on peak shift.

In the literature, numerous attempts were made to use Hb as a biocatalyst in organic solvents with very limited success (Shi et al., 2004). We wished to explore methods to alter

Hb/solvent interface systematically by covalent conjugation with PAA, and post-functionalization of –COOH groups of the conjugate manipulate the microenvironment of Hb further, which is described in this Chapter. Conjugates of Hb-PAA were prepared using EDC activated chemistry with subsequent modification of –COOH groups in PAA to form esters. This modification protocol could confer some benefits for solubility in organic media and provide a systematic handle to control the Hb/solvent interface in a predictable manner (Fig. 4.1). When stabilized against organic solvent denaturation, Hb can be used for a variety of organic transformation in organic solvents among them acetonitrile (ACN) or methanol, which has not been easy to date (Klibanov and Klyachko, 1992).

4.3.1 Synthesis of Hb-PAA Conjugates

The solution of Hb was prepared by dissolving in phosphate buffer (PB 10 mM, pH 7.4). A solution of PAA was made at 2 wt% by dissolving in deionized water (DI) and adjusted to pH 7.4. EDC was added to this PAA solution such that EDC: COOH ratio was 1.5:1. This mixture was stirred for 15 min. Hb solution was slowly added to EDC activated PAA solution over the period of 10 min. After 6–8 h the samples were quenched or reacted further with alcohols. This conjugate, in which the mole ratio of Hb:PAA was 1.3:1, was referred to as Hb-PAA.

4.3.2 Synthesis of Hb-PAA-Ester Conjugates

Hb-PAA-ester conjugates were made using Hb-PAA. Unreacted PAA from Hb-PAA was activated using EDC such that EDC: COOH ratio was 1.5:1. This solution was stirred for 15 min and 30% v/v EtOH was added to it and stirred for 6–8 h. The mole ratio of EtOH: COOH was kept at 132:1. The synthesis of Hb-PAA-ester conjugates was a two-pot process and this conjugate was named as Hb-PAA-Eth. Similarly, synthesis was carried out using 1-propanol and conjugate was named as Hb-PAA-1-prop.

4.3.3 Infrared Spectroscopy (IR)

IR studies were carried out to evaluate the successful conjugation of PAA to Hb and alcohols to PAA. The IR bands at

1710–1720 cm^{-1} are clearly noted for Hb-PAA (green curve), Hb-PAA-Eth, (red broken line) and Hb-PAA-1-prop, (black line). This ester stretch was at higher frequencies than C=O stretch of PAA. The presence of ester stretch is in direct support of the ester formation. Also, intensity of ester stretch for Hb-PAA-1 -prop was nearly twice as high as that of Hb-PAA-Eth. This was in accordance with the electrostatic charges observed in zeta potential.

4.3.4 Zeta Potential Studies

Conjugation of Hb with PAA and following esterification of the COOH groups from PAA transformed the net charge that was evaluated by zeta potential studies (Fig. 4.1B). For example, Hb-PAA had a net negative charge of –45 mV, which was nearly the same as that of PAA. Conversion of carboxyl groups (PAA) into the corresponding ester (Hb-PAA and Hb-PAA-ester) is anticipated to lessen the negative charge on the conjugate. Thus, as expected the charge on Hb-PAA-Eth (red bar) and Hb-PAA-1-prop was reduced to –35 and –25 mV, respectively. Both the ester conjugates were negatively charged, though the reduction of the net charge compared to that of Hb-PAA confirmed the esterification. From the net charges of the conjugates it was determined that higher numbers of COOH from PAA were esterified and relative values of charge reduction were 45% and 22%, respectively. Thus, esterification is not complete but its formation was supported by IR spectroscopy.

4.3.5 TEM and DLS

The average hydrodynamic size and morphology of the esters were studied by means of dynamic light scattering (DLS) and transmission electron microscopy (TEM) studies, respectively. The TEM images (Fig. 4.1C) displayed that Hb-PAA-Eth and Hb-PAA formed cross-linked, water-swollen nanogels and Hb exhibited discrete aggregation into nanoparticles. Estimated sizes of the conjugates were gauged from the TEM images to be 100 (Hb-PAA-Eth) and 80 nm (Hb-PAA). These estimations were also confirmed by DLS, which showed average hydrodynamic diameter of 113 and 80 nm for Hb-PAA-Eth and Hb-PAA, respectively.

4.3.6 Protein Structure within the Conjugates

Changes in secondary structure of Hbin phosphate buffer (PB) are reflected by the changes in its Soret absorption and Soret circular dichroism (CD) bands. The heme environment was examined by Soret CD spectra and displayed diminished intensities of the conjugates when compared to that of Hb. Though the Soret absorption band positions were not considerably altered, the CD band intensities reduced rather severely, and the band intensities were used to estimate the range of retention of chirality of the prosthetic group (Kelly et al., 2005). The intensities followed the order Hb>Hb-PAA>Hb-PAA-Eth>Hb-PAA-1-prop and the chiral nature of the environment was distressed to a substantial extent in the conjugates, even though the chemical environment of the heme seemed to be nearly unchanged. Thus, the absorption and CD spectra gave different aspects of the heme environment.

Then, we tested the absorption and CD spectra of the conjugates in organic solvents such as, DMF, MeOH, ACN and EtOH. The order of the organic solvents correlates with their decreasing Log *P* (logarithm of octanol/water partition coefficient) values. DMF is most hydrophilic and EtOH is the least. The soret absorption band positions and soret CD intensities of the conjugates were tested in organic solvents and PB and are compared in Fig. 4.1D,E. The soret maximum of the conjugates serially shifted to shorter wavelengths in the following order of solvents: PB>DMF>MeOH>ACN>EtOH. The amount of secondary structure retention was evaluated using CD spectra of the samples and compared to Hb in PB. The extent of structure retention was plotted as a function of solvent polarity (Fig. 4.1D). The trend of structure retention was Hb>Hb-PAA>Hb-PAA-Eth> Hb-PAA-1-prop with respect to the conjugate and solvent dependency was PB>DMF>MeOH>ACN>EtOH. The reference was set at 100% for CD spectra intensity of Hb in PB. The lowest secondary structure retention was observed for Hb-PAA-1-prop in EtOH at 3%.

In DMF (20% v/v), CD spectra for Hb-PAA-1-prop, Hb-PAA-Eth, Hb-PAA and Hb showed 14%, 5%, 23%, and 78% secondary structure retention, respectively. The absorbance peak for Hb-PAA-Eth, Hb-PAA-1-prop, Hb-PAA and Hb were noted at 408, 407, 408 and 404 nm, respectively. In MeOH (60% v/v), CD spectra showed 12%, 6.8%, 15%, and 4.6% structure retention for these conjugates. The absorbance spectra showed peaks for Hb-PAA-Eth, Hb-PAA-1-prop, Hb-PAA and Hb at 399, 399, 400 and 404 nm, respectively. Hb-PAA-Eth and Hb-PAA showed higher secondary structure retention compared to Hb-PAA-1-prop and Hb in MeOH. MeOH showed better characteristics for structure retention and absorbance peak positions.

In ACN (60% v/v), the absorbance peaks were observed at 395, 400, 395 and 408 nm for Hb-PAA-Eth, Hb-PAA-1-prop, Hb-PAA, and Hb, respectively, whereas the corresponding CD spectra demonstrated 7.3%, 3.9%, 2.6%, and 33% structure retention. The absorbance peaks in EtOH (60% v/v) for Hb-PAA-Eth, Hb-PAA-1-prop, Hb-PAA, and Hb were at 401, 400, 398 and 399 nm and corresponding CD spectra showed 6.2%, 0%, 6.6% and 32% structure retention. In EtOH, the absorbance peak positions overlapped well and showed distinctly less difference in peak positions which indicates an improved preservation of the heme environment in this solvent. However, the CD spectra indicated low chirality of the 1-propanol derivative. This decrease in % of structure retention in EtOH could be attributed to increased hydrophobicity of the polymer coating around the protein (Fig. 4.1D).

According to UV studies, all samples retained the chemical environment of the heme-binding pocket to a large extent. PAA acted as a protective sheath around Hb and further esterification of PAA improved solubility in organic solvents. CD studies were performed and showed highest retention of chiral environment in buffer and all organic solvents for Hb, except in MeOH. All ester conjugates showed a low structure retention (<20%) in organic solvents. This loss of structural retention could be due to solubility of polymer in the organic solvents and the loss of structurally important water from protein surface. All conjugates showed higher % structure retention in DMF (most hydrophobic) compared to all other solvents.

The chirality retention increased in relation to increase in solvent polarity and hydrophobicity of the Hb-solvent interface. Hb-PAA showed a better structure retention than Hb-PAA-1-eth and Hb-PAA-1-prop. To summarize, modification of Hb-PAA using esterification with alcohols can be used as a modular approach to control solubility in organic solvents. Next, we tested the effect of the structures of the Hb-conjugates on the peroxidase like activity of Hb in organic solvents to measure the influence of the Hb-solvent interface on catalytic behavior.

4.3.7 Activity Studies

Hb does not function as an enzyme in biological environments but its peroxidase-like activity is widely used in biosensors, and the oxidation of polycyclic aromatic hydrocarbons (Jagannadham et al., 2006). In this study, oxidation of 2-methoxyphenol with hydrogen peroxide catalyzed by Hb is used for activity monitoring (Maehly et al., 1954). The oxidation product of 2-methoxyphenol absorbs at λ = 470 nm. Therefore, a colorimetric method is used to track the product formation and kinetic traces of absorbance at 470 nm versus reaction time at certain catalyst concentrations to determine initial reaction rates.

Activity studies were performed in presence of 2.5% to 80% v/v concentrations of organic solvents (the residual 20% v/v contained aqueous solutions of the conjugate, *o*-methoxyphenol and H_2O_2). The initial rates were used to estimate specific activities for comparison with the initial activities of Hb in PB as a reference. In all the plots, Hb-PAA-Eth, Hb-PAA-1-prop, Hb-PAA and Hb are represented as red, black, green, and blue bars, respectively.

The maximum activities were outstanding for all conjugates at 60% v/v MeOH, and Hb-PAA-Eth, Hb-PAA-1-prop and Hb-PAA had activities of 62%, 40% and 55%, respectively (Fig. 4.2A). Hb exhibited the lowest activity compared with all the conjugates under the chosen reaction conditions. Surprisingly at upper MeOH concentrations (60–80% v/v), Hb displayed enhanced activities when compared to lower MeOH concentrations (2.5–50% v/v) a behavior not shown by the conjugates.

In ACN, Hb-PAA-Eth exhibited high activities (40–60%) when scanned over a concentration range of 2.5–80% v/v of ACN, while, Hb-PAA-1-prop showed activities of 10–25% up to 40% v/v ACN. Between 40% and 70% v/v ACN, the activity values for Hb-PAA-1-prop were 16–20%. Hb-PAA showed minimum activity at 40% v/v ACN and the highest one at 70% v/v, while Hb in ACN revealed a decrease in activity from 2.5% to 50% v/v ACN. Unexpectedly, the activity of Hb increased again upon extra addition of ACN and presented highest activity at 80% v/v ACN unlike the other conjugates under similar conditions (Fig. 4.2B).

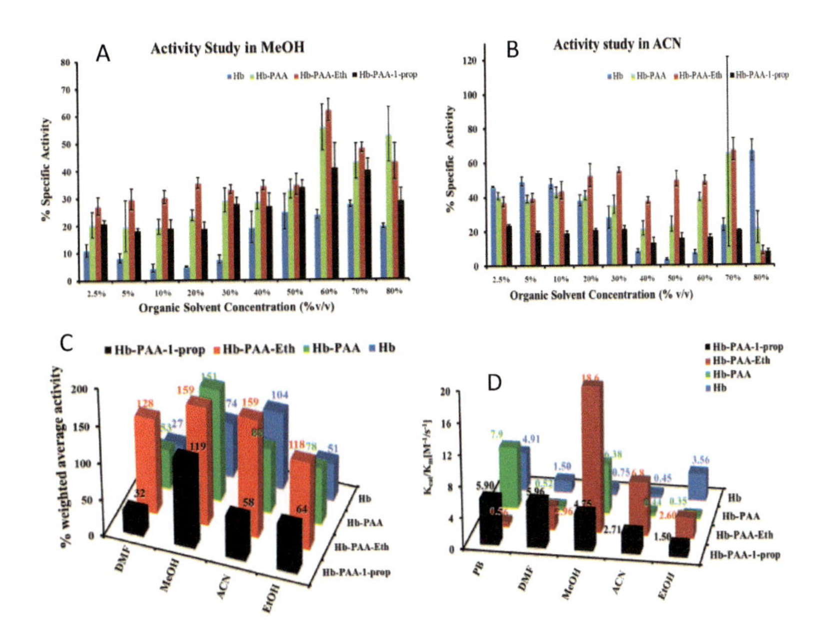

Figure 4.2 Specific activities of Hb-PAA-Eth (red), Hb-PAA-1-prop (black), Hb-PAA (green) and Hb (blue) with respect to that of Hb in PB (10 mM, pH 7.4) (100%), as a function of increasing concentrations of (A) MeOH, and (B) ACN. (C) Averages of the specific activities of Hb-PAA-Eth (red), Hb-PAA-1-prop (black), Hb-PAA (green) and Hb (blue) bar in DMF, MeOH, ACN and EtOH. Solvents are arranged in increasing log *P* values from DMF to EtOH. Catalytic efficiencies (D) of Hb-PAA-Eth (red), Hb-PAA-1-prop (black), Hb-PAA (green) and Hb (blue) in PB (10 mM, pH 7.4), DMF (20% v/v), MeOH (60% v/v), ACN (60% v/v) and EtOH (60% v/v). Specific solvent compositions corresponding to highest specific activities were used. Solvents are arranged in the order of increasing log (*P*).

These activity trends in the four organic solvents were inspected in the context of the solvent log *P* values, a measure of lipophilicity. Log *P* values that may be also taken as a rough measure of solvent polarity are −1.01 (DMF), −0.77 (MeOH), −0.34 (ACN), and −0.30 (EtOH) (Sangster, 1989) and show that the polarity increases from EtOH to DMF. Increased solvent hydrophobicity is known to reduce its ability to strip off the water surrounding the protein and to improve protein stability (Gorman and Dordick, 1992; Laane et al., 1987).

For a better understanding of the above trends as a function of log(*P*) of the solvent, the average activities of each conjugate in the respective solvent were calculated. The average activities of the conjugates were calculated using the following:

$$\% \text{ weighted average activity} = \frac{\sum(\% \text{ organic solvent} \times \% \text{ activity})}{100} \tag{4.1}$$

and compared to Hb at equivalent conditions; the specific activities and the corresponding solvent compositions were averaged out. It is assumed here that activities follow a smooth function of solvent composition for each solvent and weighted averages can be compared among different solvents. The resulting activity data plots are shown in Fig. 4.2C. Overall, Hb-PAA conjugates showed higher % values for weighted average activities when compared to Hb. Thus, polymer conjugation to Hb helped to enhance the activity of Hb in different organic solvents with the exception of ACN. Also, ester modification approach assisted to further improve the activity in different organic solvents in contrast to Hb-PAA. For example, Hb-PAA-Eth showed higher catalytic activity than both Hb-PAA and Hb, in all solvents. However, further increase in hydrophobicity of the conjugates from Hb-PAA-Eth to Hb-PAA-1-prop resulted in reduced activity owing to a discrepancy in the polarity of the solvent and the polymer shell around the protein. Therefore, the equilibrium of lipophilicity of the conjugates and the solvent chosen as the reaction medium for catalysis is enormously significant and supports as general parameter the design of Hb-PAA conjugates and successive polymer modification by esterification.

4.3.7.1 K_M and V_{max} Studies

The rates of several steps in the catalytic cycle play crucial role in enzymatic activity, and insight into these steps was gained by determining the Michaelis constant (K_M) and maximum reaction velocity (V_{max}) (see also Chapter 2). Using the initial rates determined for different substrate concentrations, Lineweaver–Burk plots (plot of inverse of initial rate versus inverse of substrate concentration) and Michaelis–Menten plots were set up via the equation

$$\frac{1}{V} = \frac{KM}{V_{max}} \cdot \frac{1}{[S]} + \frac{1}{V},$$

where $[S]$ = substrate concentration, V = initial reaction rate, K_M = Michaelis constant and V_{max} = maximum reaction velocity. Solvent concentrations were precisely adjusted as all the conjugates showed optimal catalytic activities under these conditions and the corresponding K_M, V_{max}, k_{cat}, and k_{cat}/K_M obtained from the experimental data were collected. Lower K_M values designate increased affinity of the substrate for the enzyme active site; k_{cat} is the turnover number, i.e. the number of moles of substrate molecules converted to the product molecules per unit time, while k_{cat}/K_M is the catalytic efficiency. Highest catalytic efficiency is the utmost desired.

Hb-PAA-1-prop exhibited higher affinity for the substrate in DMF but it varied very little with solvent polarity. Hb-PAA-Eth indicated higher affinity for substrate in MeOH and varied greatly with polarity; in contrast, Hb-PAA and Hb showed favored affinity in PB (10 mM, pH 7.4) and EtOH, respectively, and Hb-PAA indicated maximum sensitivity to solvent polarity. Also, Hb-PAA-Eth and Hb-PAA-1-prop showed highest V_{max} and k_{cat} in EtOH and PB (10 mM, pH 7.4), respectively.

Hb-PAA-Eth and Hb-PAA-1-prop had augmented affinity for the substrate in MeOH and DMF, respectively, and both ester conjugates exhibited peak catalytic efficiencies in MeOH and DMF, while Hb-PAA and Hb showed highest catalytic efficiency in PB (10 mM, pH 7.4). Due to the unmodified COOH groups, Hb-PAA is more hydrophilic than the ester conjugates, and therefore it displayed improved affinity for substrate in PB (10 mM, pH 7.4) and not in organic solvents. There were no specific trends in K_M as a function of solvent polarity for a given biocatalyst or the polarity of the polymer shell around the protein while in presence of MeOH the lowest average K_M values were obtained.

The catalytic efficiencies (k_{cat}/K_M) are plotted (Fig. 4.2D) to understand how the solvent affected the overall catalytic efficiency of the modified biocatalyst. The catalytic efficiencies showed a systematic relationship with solvent polarity, and the efficiencies peaked smoothly with polarity for each of the conjugates. While Hb showed the highest efficiency in PB, two of the conjugates (Hb-PAA and Hb-PAA-1-prop) showed higher efficiencies in PB than Hb itself. Two of the conjugates (Hb-PAA-1-Eth and Hb-PAA-1-prop) showed higher efficiencies than Hb in DMF, MeOH, and ACN, while Hb showed the highest efficiency in EtOH. Therefore, conjugates can be synthesized depending on the log P value of the reaction medium to obtain the highest catalytic efficiency. The data prove that the conjugates can exhibit many-fold higher catalytic efficiencies in organic solvents than the native catalysts have in PB. Based on kinetic data such as the catalytic efficiency we may be able to forecast which enzyme conjugate will demonstrate high efficiency in a particular solvent, but more in depth studies are essential to map out the complex relationships between the solvent, the substrate, the interface, the product and the solvent.

4.4 Glucose Oxidase (GOx) and Horseradish Peroxidase (HRP) Bienzyme

Glucose oxidase (GOx) is an oxidoreductase with a molecular weight of ~160 kDa. It contains 1166 amino acid residues of which 30 are lysine residues (Carlsson et al., 2005). Horseradish peroxidase has a molecular weight of 40 kDa and comprises 6 lysine amino acids totaling 306 amino acid residues (Kimmoju et al., 2011). Both GOx and HRP are industrially important enzymes and used in various applications such as sensing of glucose, biofuel cell applications and biocatalysis (Ferri et al., 2011). Both these enzymes get deactivated at temperatures above 50°C and have rather narrow pH range 4.6 to 6. While the stabilization of single enzyme systems is well understood, stabilization of multi-enzyme cascade systems has never been explored. If GOx and HRP could be stabilized together such that they perform at higher temperatures and within a broader pH range, then these two biocatalysts are applicable in cascade reactions at higher temperatures with improved activities. Stabilized and catalytically active bienzyme catalysts will not only enhance the productivity but will also reduce the cost of enzyme-catalyzed bioprocesses (Dunn et al., 2012).

To achieve this, we first developed bienzyme covalent conjugation within a PAA matrix. Furthermore, these conjugated bienzymes were assembled physically on graphene oxide (GO) sheets. Stabilized bienzyme hybrid materials with topologically distinct host materials for the first time achieved remarkable pH, chemical-denaturant, and high-temperature stability. The bienzyme conjugate and the dual-hybrid system are presented in Fig. 4.3 (Scheme 4.1), where GOx and HRP are shown and green lines represent PAA. GOx and HRP are conjugated to PAA by EDC chemistry to form GOx-HRP-PAA conjugates, which are then assembled onto the GO sheets to form the GOx-HRP-PAA/GO cascade biohybrid. GOx-HRP-PAA/GO is viewed as a nanogel of GOx-HRP-PAA assembled onto the GO sheets (Zore et al., 2015).

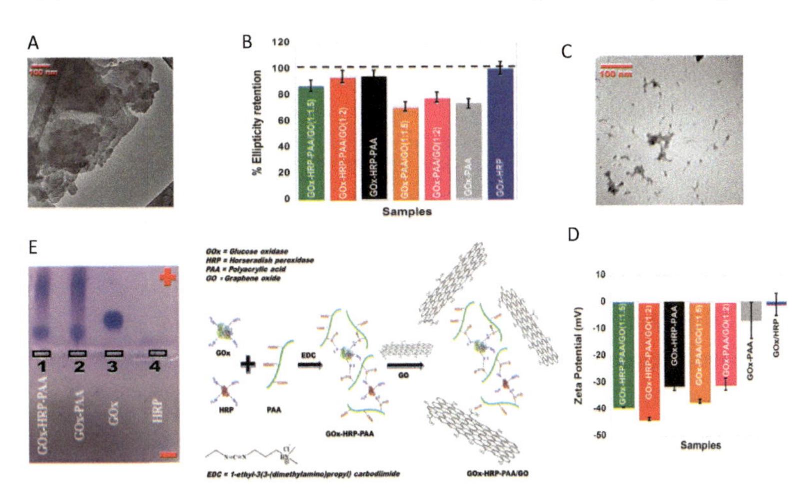

Figure 4.3 Synthesis and characterization of GOx-HRP-PAA conjugates and GOx-HRP-PAA/GO conjugate hybrid materials. (A) TEM micrograph of GOx-HRP-PAA/GO showing nanogels assembled on GO sheets. (B) Structure retention studies using CD spectroscopy showing all the samples having structure retention > 70% (data referenced to 1 μM GOx/HRP). (C) TEM micrograph of GOx-HRP-PAA conjugate showing formation of nanogels. (D) Zeta potential study of GOx/HRP, and other PAA single enzyme conjugates. (E) Agarose gel electrophoresis of GOx-HRP-PAA, GOx-PAA, GOx and HRP showing successful conjugation of the enzymes to PAA.

4.4.1 Synthesis of Bienzyme-PAA and Single Enzyme Conjugates

To synthesize GOx-HRP-PAA (bienzyme conjugate) and GOx-PAA (single-enzyme conjugate), EDC chemistry was used and all the syntheses were carried out in pH 7.4 phosphate buffer. In case of GOx-PAA and GOx-HRP-PAA, the total enzyme to PAA mole ratio was kept at 1.2:1. Effect of GO on hybrid formation was assayed from 0–31% (mass percent of GO). Higher activities were observed at higher GO loadings onto single and bienzyme-PAA conjugates. Therefore, in this study optimal GO loadings were used (25 and 31 mass %). The final mass (in %) of each component (100 mg scale) in the hybrid and single/bienzyme conjugate is given in Table 4.2. Purification of each sample was carried out using a 25 kDa dialysis membrane for 3 cycles (6 h each). Agarose gel electrophoresis and zeta potential studies were performed to test the success of the synthesis.

Table 4.2 Different conjugates and the mass percent of enzymes, PAA, and GO used for synthesis

Samples	Mass percent				Description
	GOx	HRP	PAA	GO	
GOx-HRP-PAA/GO(1:1.5)	13	3.3	58	25	Bienzyme conjugate hybrids
GOx-HRP-PAA/GO(1:2)	12	3.0	54	31	
GOx-HRP-PAA	18	4.5	78	0	Bienzyme conjugate
GOx-PAA/GO(1:1.5)	17	0	58	25	Single enzyme conjugate hybrids
GOx-PAA/GO(1:2)	15	0	54	31	
GOx-PAA	22	0	78	0	Single enzyme conjugate
GOx/HRP	80	20	0	0	Unbound Bienzyme

4.4.2 Agarose Gel Electrophoresis

Conjugation of enzymes to PAA and its further non-covalent interactions with GO was tested using agarose gel electrophoresis. The agarose gel can be viewed in Fig. 4.3E; Lane 1, 2, 3, and 4 were loaded with GOx-HRP-PAA, GOx-PAA, GOx and HRP, respectively. The conjugation of PAA to enzymes should increase the negative charge on the enzymes which results in increasing the electrophoretic mobilities. Although the PAA conjugation will also increase the molecular weight and decrease the electrophoretic mobility. This is responsible for the streaking in lane 1 and 2 for GOx-HRP-PAA and GOx-PAA, respectively, confirming the formation of enzyme polymer nanogels with variable molecular weights. Since HRP is an isoenzyme, it was not observed in the lane 4. Its isoelectric points range from 3.0 to 9.0. Therefore, HRP spreads across both sides of the agarose gel and was invisible after staining. GOx-HRP-PAA/GO conjugates showed same mobility as

GOx-HRP-PAA and confirms that GOx-HRP-PAA to GO interactions are non-covalent.

4.4.3 Zeta Potential Studies

The conjugation of enzyme to PAA was further tested using zeta potential measurement for establishing the net charge before and after modification of the enzymes with PAA and GO (all measurements done at pH 7.0), (Fig. 4.3D). The modification of enzyme(s) by PAA (covalently) and GO (non-covalently) should result in increased negative charge on the conjugate because of carboxyl groups present on GO and PAA. GOx-HRP-PAA (black bar) showed a zeta potential of –31 mV, which was very low compared to bienzyme (–1.1 mV, blue bar) and is due to the –COOH residues from PAA (pKa = 4.2). Successive GO modification decreased the zeta potential of GOx-HRP-PAA/GO (1:1.5) and GOx-HRP-PAA/GO (1:2) to –39.5 mV (green) and –43.8 mV (red). Similar results were seen for the single enzyme conjugate GOx-PAA, (–7.1 mV, gray bar). After GO conjugation the zeta potential decreased drastically to –37.2 (orange) and –30.8 mV (pink), respectively, for GOx-PAA/GO (1:1.5) and (1:2). Further, we tested the secondary structure of bienzyme, single enzyme conjugates and conjugate hybrids.

4.4.4 Circular Dichroism Studies

Enzyme(s) secondary structure was examined using far UV-CD studies. The % relative ellipticity retention of bienzyme/PAA conjugates, GO conjugates and conjugate hybrids were calculated taking bienzyme as a reference. Figure 4.3B gives the relative % ellipticity retention of all samples. Conjugation to PAA and GO is expected to reduce its secondary structure due to strain put on the enzymes by conjugation to PAA and GO. The studies were carried out in PB and ellipticity at 222 nm of GOx/HRP is referenced to 100%. The secondary structure of enzymes is required for their biological function and biological activity. Conjugation to PAA and GO resulted in maximum of 30% loss in secondary structure that can be related to unfavorable hydrophobic interaction of the enzymes interior with GO. GOx-HRP-PAA/GO (1:1.5) (green bar) and (1:2) (red bar) showed more than 85%

of ellipticity retention. GOx-PAA/GO (1:1.5) (orange) and (1:2) (pink) showed more than 71–78% of ellipticity retention. Apart from this, enzymes largely retained their secondary structure. Next, the effect of reduction in secondary structure on activity of enzyme was examined.

4.4.5 Transmission Electron Microscopy

Morphology of distorted structure of GOx-HRP-PAA and GOx/HRP conjugates was scanned using TEM. As seen from Fig. 4.3C, showed nanogel morphology for GOx-HRP-PAA, whereas Fig. 4.3A displayed nanogels of GOx-HRP-PAA assembled on GO plates. Further, the activity of enzymes was tested at various conditions.

4.4.6 Activity Studies at Room Temperature and pH 7.4

Standard activity studies were executed using unbound bienzyme, conjugates, and conjugate hybrids in pH 7.4 PB at room temperature. For all activity studies, activity of GOx/HRP at these conditions was referenced at 100%. Activity studies at pH 7.4 and room temperature for all samples are presented in Fig. 4.4A; all samples (GOx-HRP-PAA/GO(1:1.5), GOx-HRP-PAA/GO(1:2) and GOx-HRP-PAA) showed an activity of 144%, 157%, and 141%, respectively. As can be seen, the activity of conjugates and conjugate hybrids were very high compared to GOx/HRP. This encouraged us to test these samples at harsher conditions for activity studies.

4.4.6.1 Activity studies at different pHs in presence and absence of a chemical denaturant

Enzymatic activities of bio-hybrids were tested in presence and absence of SDS, a denaturant, at various pHs. SDS reduces the activity of GOx and HRP at a concentration above 2 mM. Therefore we tested the activity of enzymes ranging from pH 2.5 to 7.4 without and in presence of SDS (Fig. 4.4A,B). Bienzyme conjugate hybrids and conjugates showed high activities (>85%) at pH 2.5–7.4, whereas bienzyme (GOx/HRP) showed as little as 14% retention of activity at pH 2.5. In presence of 4 mM SDS (Fig. 4.4B) bienzyme conjugate hybrids exhibited up to about 90% activity retention at pH 5.5. Bienzyme conjugate at this pH showed 47% retention in activity. At lower pHs bienzyme (GOx/HRP) completely lost its activity (pH 2.5 and 4). At pH 7.4 bienzyme conjugate hybrids and bienzyme conjugate displayed an activity above 100%. This can be attributed to substrate channeling of H_2O_2 across the catalyst matrix by internal diffusion. H_2O_2, generated during GOx catalysis and used as substrate by HRP, gets recycled in the GO and PAA matrix resulting in higher turnover to the product ultimately resulting in an increase in activity.

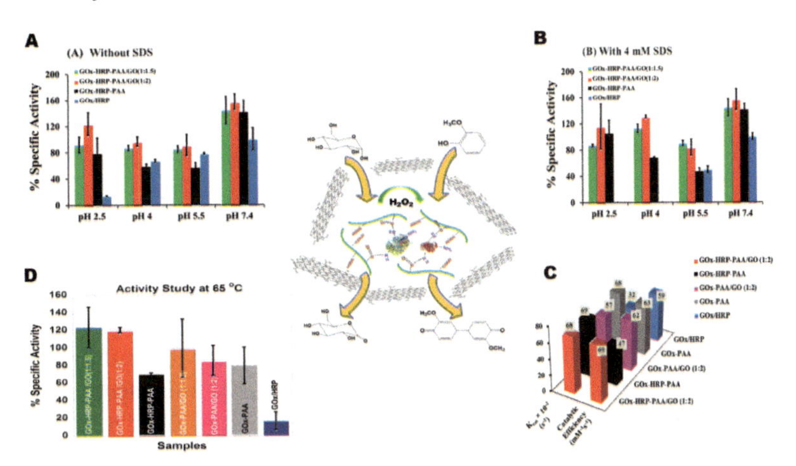

Figure 4.4 (A) Activity study as a function of increasing pH without SDS. (B) Activity study at increasing pHs with 4.0 mM SDS (denaturant). (C) Catalytic efficiency and K_{cat} of conjugates and conjugate hybrids (D). Activity study at 65°C for the samples of dual and single enzyme systems. Activity of GOx/HRP at room temperature in PB is referenced at 100%. GOx-HRP-PAA/GO (1:1.5) (green), GOx-HRP-PAA/GO (1:2) (red), GOx-HRP-PAA (black), GOx-PAA/GO(1:1.5) (orange), GOx-PAA/GO (1:2) (pink) GOx-PAA (gray) and GOx/HRP (blue).

4.4.6.2 High-temperature activity

The maximum temperature for biological activity of GOx is ~60°C. Therefore, activity studies were tested at higher temperature of 65°C to check the strength and efficacy of the bienzyme conjugate hybrids and conjugates. Activity retention in % for bienzyme conjugate hybrids and conjugates are given in Fig. 4.4D. The activity of GOx/HRP at room temperature and pH 7.4 was used as the reference (100%). GOx-HRP-PAA/GO exhibited ~120% activity which was 40% above that of single enzyme conjugate and hybrids. Single enzyme conjugate hybrids presented 70–80% activity. GOx/HRP exhibited ~12% activity while bienzyme and single enzyme conjugates had 60–80% activity.

4.4.6.3 Kinetics studies

Catalytic efficiency (k_{cat}/K_M) and k_{cat} values of all the conjugates are presented in Fig. 4.4C in a 3D diagram. Kinetic plots of GOx-HRP-PAA/GO(1:1.5) and GOx-PAA/GO(1:1.5) were not calculated because conjugate hybrid, GOx-HRP-PAA/GO(1:2) and GOx-PAA/GO(1:2) presented improved catalytic activity at higher temperature and at lower pHs. GOx-HRP-PAA displayed maximum k_{cat} of 69 × 10^{-2} s^{-1} followed by GOx-HRP-PAA/GO(1:2), GOx-PAA, GOx-PAA/GO(1:2) and GOx-HRP with values of 68 × 10^{-2}, 68 × 10^{-2}, 57 × 10^{-2}, and 32 ×10^{-2} s^{-1}, respectively. Catalytic efficiency was highest for GOx-HRP-PAA/GO(1:2) (69 $mM^{-1}s^{-1}$) and lowest in the case of GOx-HRP-PAA (47 $mM^{-1}s^{-1}$). GOx/HRP, GOx-PAA and GOx-PAA/GO (1:2) exhibited catalytic efficiency of 59, 63 and 63 $mM^{-1}s^{-1}$, correspondingly. K_M values mirror the affinity of the enzyme within the biohybrid/conjugate towards the substrate. The lesser the K_M value the greater the affinity and vice versa. GOx/HRP exhibited the lowermost K_M and V_{max} of 0.55 mM and 0.016 $\mu M^{-1}s^{-1}$, respectively, whereas GOx-HRP-PAA presented the uppermost K_M and V_{max} values of 1.46 mM and 0.035 $\mu M^{-1}s^{-1}$ in that order. K_M and k_{cat} values of native GOx were reported as 1.34 mM (Shin et al., 1993) and 16 × 10^{-2} s^{-1} (Witt et al., 2000), deviating considerably from our values (0.55 mM and 32 × 10^{-2} s^{-1}), a finding predictable for cascade catalysis. The diffusional features of substrates for free enzymes versus bound enzymes contribute to these differences.

4.5 Conclusions

We developed an easy, modular, and general synthetic strategy to covalently confine one or more enzymes within polyacrylic acid chains, and further assembled on to graphene oxide sheets. Reduction of conformational entropy by confinement strategy was used to directly stabilize the enzymes when used under non-physiological harsh conditions, which has previously not been possible. The extent of covalent conjugation and retention of the three dimensional structure of the enzyme was quantified by a variety of techniques. The morphology of these conjugates and conjugate hybrid materials was established by microscopic analysis. More importantly, these enzyme–polymer conjugates or enzyme–polymer conjugate GO hybrids could be used as biocatalysts at higher temperatures, broad pH range, with surfactants, and in organic solvents. The synthetic strategy expanded our ability to tailor the structure and properties of multi-enzyme conjugate hybrids that are used in cascade reactions at higher temperatures with enhanced catalytic activities compared to native enzymes. Work in progress includes investigation of cooperative effects of several enzymes when covalently bonded to PAA and confined with GO as well as electrochemical studies of these materials in biofuel cells. Thus, using a simple strategy, we are able to synthesize and stabilize enzyme based biocatalysts from mostly renewable and degradable resources for use in green processes and technologies.

References

Abad, J. M., Mertens, S. F. L., Pita, M., Fernandez, V. M., Schiffrin, D. J., *J. Am. Chem. Soc.,* **127** (2005), 5689–5694.

Aehle, W., *Enzymes in Industry*. Wiley (2008).

Antonini, E., Carrea, G., Cremonesi, P., *Enzyme Microb. Technol.,* **3** (1981), 291–296.

Bhambhani, A., Kumar, C. V., *Adv. Mater.,* **18** (2006), 939–942.

Briand, V. A., Thilakarathne, V. K., Kasi, R. M., Kumar, C. V., *Talanta,* **99** (2012), 113–118.

Briand, V. A., Thilakarathne, V., Kumar, C. V., Kasi, R. M., Protein-polymer conjugates for use as hybrid functional materials, *PMSE Preprints* (2011).

Campbell, M. K., Farrell, S. O., *Biochemistry*. Brooks Cole Publishing Company (2011).

Carlsson, G. H., Nicholls, P., Svistunenko, D., Berglund, G. I., Hajdu, J., *Biochemistry,* **44** (2005), 635–642.

Dunn, M. F., *Arch. Biochem. Biophys.,* **519** (2012), 154–166.

Dunn, B., Huang, Y., Lan, E., Blaik, R. A., *ACS Nano* (2015), DOI: 10.1021/acsnano.5b04580.

Ferri, S., Kojima, K., Sode, K., *J. Diabetes Sci. Technol.*, **5** (2011), 1068–1076.

Gao, Y., Peng, A., Li, H., Chen, Z., *Environ. Sci. Pollut. Res. Int.*, **21** (2014), 10696–10705.

Ghimire, A., Kasi, R. M., Kumar, C. V., *J. Phys. Chem. B*, **118** (2014), DOI: 10.1021/jp500310w.

Ghimire, A., Zore, O. V., Thilakarathne, V. K., Briand, V. A., Lenehan, P. J., Lei, Y., Kasi, R. M., Kumar, C. V., *Sensors*, **15** (2015), 23868–23885.

Gorman, L. A. S., Dordick, J. S., *Biotechnol. Bioeng.*, **39** (1992), 392–397.

Hebbel, R. P., Shalev, O., *Blood*, **87** (1996), 3948–3952.

Heredia, K. L., Maynard, D. H., *Org. Biomol. Chem.*, **5** (2007), 45–53.

Jaenicke, R., Bohm, G., *Curr. Opin. Struct. Biol.*, **8** (1998), 738–748.

Jagannadham, V., Bhambhani, A., Kumar, C. V., *Microporous Mesoporous Mater.*, **88** (2006), 275–282.

Kelly, S. M., Jess, T. J., Price, N. C., *Biochim. Biophys. Acta*, **1751** (2005), 119–139.

Khalaf, N., Govardhan, C. P., Lalonde, J. J., Persichetti, R. A., Wang, Y.-F., Margolin, A. L., *J. Am. Chem. Soc.*, **118** (1996), 5494–5495.

Khmelnitsky, Y. L., Belova, A. B., Levashov, A. V., Mozhaev, V. V., *FEBS Lett.*, **284** (1991), 267–269.

Kim, J., Grate, J. W., *Nano Lett.*, **3** (2003), 1219–1222.

Kimmoju, P. R., Chen, Z. W., Bruckner, R. C., Mathews, F. S., Jorns, M. S. *Biochem.*, **50** (2011), 5521–5534.

Klibanov, A. M., Klyachko, N. L., *Appl. Biochem. Biotechnol.*, **37** (1992), 53–68.

Kumar, C. V., Chaudhari, A., *J. Am. Chem. Soc.*, **122** (2000), 830–837.

Kumar, C. V., Chaudhari, A., *Chem. Mater.*, **13** (2001), 238–240.

Kumar, C. V., Chaudhari, A., *Chem. Commun.* (2002), 2382–2383.

Laane, C., Boeren, S., Vos, K., Veeger, C., *Biotechnol. Bioeng.*, **30** (1987), 81–87.

Loewen, P. C., Hillar, A., *Arch. Biochem. Biophys.*, **323** (1995), 438–436.

Maehly, A. F., Chance, B., *Methods Biochem. Anal.*, **1** (1954), 357–424.

Mudhivarthi, V. K., Cole, K. S., Novak, M. J., Kipphut, W., Deshapriya, I. K., Zhou, Y., Kasi, R. M., Kumar, C. V., *J. Mater. Chem.*, **22** (2012), 20423–20433.

Novick, S. J., Dordick, J. S., *Chem. Mater.*, **10** (1998), 955–958.

Parsegian, V. A., *Van der Waals Forces: A Handbook for Biologists, Chemists, Engineers, and Physicists*, Cambridge University Press (2006).

Pasut, G., *Polymers*, **6** (2014), 160–178.

Polizzi, K. M., Bommarius, A. S., Broering, J. M., Chaparro-Riggers, J. F., *Curr. Opin. Chem. Biol.*, **11** (2007), 220–225.

Riccardi, C. M., Cole, K. S., Benson, K. R., Ward, J., Bassett, K., Zhang, Y., Zore, O., Stromer, B., Kasi, R. M., Kumar, C. V., Toward 'Stable-on-the-Table' Enzymes: Improving key properties of catalase by simple covalent conjugation with poly(acrylic acid), *Bioconj. Chem.*, **25** (2014), 1501–1510.

Sangster, J., *J. Phys. Chem. Ref. Data.*, **18** (1989), 1111–1227.

Shi, J., Gao, Q., Wang, Q., *J. Am. Chem. Soc.*, **126** (2004), 14346–14347.

Shin, K. S., Youn, H. D., Han, Y. H., Kang, S. O., Hah, Y. C., *Eur. J. Boichem.*, **215** (1993), 747–752.

Thilakarathne, V. K., Briand, V. A., Kasi, R. M., Kumar, C. V., *J. Phys. Chem. B*, **116** (2012), 12783–12792.

Thilakarathne, V. K., Briand, V. A., Zhou, Y., Kasi, R. M., Kumar, C. V., *Langmuir*, **27** (2011), 7663–7671.

Thilakarathne, V., Briand, V. A., Zhou, Y., Kasi, R. M., Kumar, C. V., *Langmuir*, **27**, (2011), 7663–7671.

Witt, S., Wohlfahrt, G., Schomburg, G., Hecht, H. J., Kalisz, H. M., Boichem, J., **347** (2000), 553–559.

Woodward, C. A., Kaufman, E. N., *Biotechnol. Bioeng.*, **52** (1996), 423–428.

Xu, K., Klibanov, A. M., *J. Am. Chem. Soc.*, **118** (1996), 9815–9819.

Zore, O. V., Lenehan, P. J., Kumar, C. V., Kasi, R. M., *Langmuir*, **30** (2014), 5176–5184.

Zore, O., Pattamattel, A., Gnanaguru, S., Kumar, C. V., Kasi, R., *ACS Catal.*, **5** (2015), 4979–4988.

Chapter 5

Biological Strategies in Nanobiocatalyst Assembly

Ian Dominic F. Tabañag and Shen-Long Tsai

National Taiwan University of Science and Technology, No. 43, Sec. 4, Keelung Road, Da'an District, Taipei City 10697, Taiwan (R.O.C.)

stsai@mail.ntust.edu.tw

5.1 Introduction

Enzymes are a group of proteins that promote the transformation of chemical species in biological systems (Berg et al., 2007). They are able to catalyze a multitude of different simple as well as complex chemical reactions found in nature. Because of their excellent functional properties (e.g., activity, selectivity, and specificity), they have a vast potential when applied as industrial catalysts.

Since the early application of enzymes in crude form, the advances in industrial-scale purification, and recent progress in recombinant DNA-technology have increasingly paved the way for replacing conventional chemical catalysts by enzymes for the industrial production of fine chemicals, food additives, pharmaceuticals, etc. (Bornscheuer and Buchholz, 2005).

Biocatalysis and Nanotechnology
Edited by Peter Grunwald

ISBN 978-981-4613-69-9 (Hardcover), 978-1-315-19660-2 (eBook)
www.panstanford.com

An attractive advantage of enzymes over conventional catalysts is their benefit to the environment. The application of enzymes reduces the complexity of the process (less reaction steps); vice versa, there is less production of waste (by-products), hence making it economically attractive. Depending on the type of reaction catalyzed, enzymes may even replace rare transition metal catalysts (e.g., gold, palladium, platinum, etc.) which are a problem to the operating cost of a process (Sheldon and van Rantwijk, 2004). Thus, enzymes are referred to as "green" catalysts because they are both environmentally sustainable and economically competitive.

The development of methods enabling enzyme production on a large scale has been a major highlight in its potential as an industrial catalyst. Although enzymes have been optimized for bulk production, their properties are often not optimized for working inside industrial reactors (Guisan, 2006). Despite their excellent catalytic properties, disadvantages in connection with their industrial applications are among others that enzymes are water-soluble catalysts, susceptible to inhibition by substrates and/or products, and that their activity is sensitive to small changes in pH and temperature, which may result in denaturation and loss of activity. When multiple enzymes are involved in a process, the transport of substrate is also a factor that could limit the rate of reaction (Kim et al., 2012).

The utilization of enzymes on an industrial scale requires their reusability for a very long time since most chemical processes catalyzed by enzymes require its continuous use for both technical and economic reasons (Katchalski-Katzir, 1993; Bickerstaff Jr, 1997). To address the concerns on the enzyme instability characteristics and reusability, it has been found out that enzyme immobilization provides one of the best solutions to overcome these problems. The term "immobilized enzymes" refers to the "enzymes physically confined or localized in a certain defined region of space with retention of their catalytic abilities, and which can be used repeatedly and continuously" (Katchalski-Katzir, 1993). Furthermore, enzyme immobilization has been used to increase the overall activity and stability by maintaining a proper orientation of the enzyme molecules towards incoming substrates, and it may also function as an additional protection of the enzyme structure against inhibiting molecules and denaturation. The introduction of immobilization techniques has greatly improved the technical performance and economy of industrial processes (Bornscheuer and Buchholz, 2005).

The development of enzyme immobilization protocols and their industrial applications has occurred in the 1960s. Research ranged from single enzyme immobilization, multiple enzyme immobilization including cofactor immobilization and cell immobilization (around 1980s to early 1990s), to the ever-expanding multidisciplinary developments and applications to different fields such as bioaffinity chromatography, biosensors and others (Hartmeier, 1988; Khan and Alzohairy, 2010). Along with the establishment of enzyme immobilization protocols, different types of immobilization matrices have also been developed ranging from inorganic supports (e.g., bentonite, silica), natural (e.g., polysaccharides, proteins) and synthetic (e.g., polyacrylate, polyamide) polymers, processed materials (e.g., glass), and nanomaterials (e.g., carbon nanotubes) (Brena et al., 2013).

The improvement of nanomaterials via nanotechnology and its application in the field of biocatalysis led to the emergence of nanobiocatalysis, which is an application of nanobiotechnology, a rapidly growing research area these days. Nanobiocatalysis refers to the immobilization of enzymes onto nanomaterials (nanobiocatalysts). The use of nanobiocatalysts is quite attractive because their effective enzyme concentration (enzyme loading) and enzyme activity (per unit mass or volume of nanomaterial) are substantially improved due to the higher surface area of the nanomaterial onto which the enzyme is immobilized, compared to conventional materials (inorganic supports, polymers, etc.) (Kim et al., 2008).

There are presently four general methods reported for synthesizing nanobiocatalysts. These methods are (1) physical adsorption onto nanomaterials, (2) encapsulation and entrapment, (3) chemical conjugation, and (4) biological assembly. Among these methods, this chapter focuses on the discussion about the different strategies of biological assembly of nanobiocatalysts, because this method offers several advantages compared to the other approaches. The advantages of biological assembly over the methods are (a) the denaturation caused by the immobilization process is limited because properly folded enzymes are used, (b) a proper orientation of the enzyme's active site maximizes its interaction with the substrate, (c) reduces costly purification techniques for enzymes are reduced because specific interactions and similar working environments have the potential to directly immobilize the enzymes from the cell lysates, (d) fine-tuning of the binding affinity to guarantee highly efficient immobilization can be easily achieved by choosing suited receptor–ligand pairs (Kim et al., 2012).

The different methods of biological assembly together with the theoretical background and the characteristics of the resulting nanobiocatalysts are discussed in the following sections.

5.2 DNA-Directed Immobilization

Initially described in the early 1990s, DNA-directed immobilization (DDI is a chemically mild immobilization technique for proteins (Niemeyer et al., 1994). It is an efficient, reversible, and site selective process, which exploits the ability of oligonucleotides to selectively bind proteins tagged with complementary oligomers. This selectivity is due to the specificity of Watson–Crick base pairing interactions of DNA (Niemeyer et al., 1999). A scheme of the DDI process is shown in Fig. 5.1.

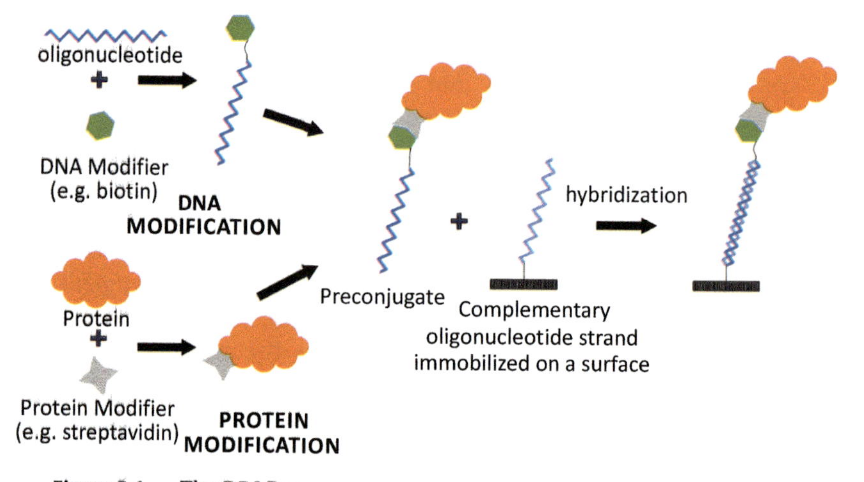

Figure 5.1 The DDI Process.

Several years after its initial description, DDI method has shown its efficiency, reversibility, and selectivity (Niemeyer et al., 1999) in the immobilization of proteins (Boozer et al., 2006; Lee et al., 2010; Wong et al., 2011), especially enzymes (Fruk et al., 2007; Muller and Niemeyer, 2008; Wilner et al., 2009), and nanomaterials (Aldaye and Sleiman, 2006; Zheng et al., 2006). It has proven its robustness and versatility in numerous applications ranging from biosensing and biomedical diagnostics, to fundamental studies in biology and medicine on the single-cell level (Meyer et al., 2014). In the establishment and implementation of DDI methods for particular applications, this brings along concerns in two primary directions: the synthesis of the protein-of-interest and short synthetic oligonucleotides by chemical means and the surface substrates which are to be functionalized with the capture oligonucleotides complementary to the DNA-protein conjugate (Niemeyer, 2010). This implies that the suitability of DDI methods in particular applications is dependent on the conjugation techniques used for the coupling of oligonucleotides to the protein of interest. A summary of the different conjugation techniques that have been utilized in the synthesis of DNA-protein conjugates along with the advances in DDI research is presented in the review of Meyer et al. (2014).

The application of DDI methods in the field of biocatalysis has been proven efficient, reversible, and selective (Niemeyer et al., 1999), and has paved the way for the assembly of artificial multi-enzyme complexes (large polypeptides with defined tertiary and quaternary structure comprising multiple catalytic centers) (Muller and Niemeyer, 2008). Multi-enzyme complexes display distinct advantages over isolated enzymes during a multi-step transformation of a substrate; the close proximity of the active sites of the multi-enzyme complex accelerates the rate of diffusional transport of the substrate to the active sites and displays "substrate channeling" (the coupling of two or more enzymatic reactions in which the common intermediate is transferred in between the enzymes without escaping to the bulk phase) which avoids side reactions or protects the unstable intermediate (Spivey and Ovadi, 1999).

The application of DNA-directed immobilization of different enzymes on one-dimensional (1D) DNA templates has been achieved by using biotinylated enzymes and DNA-streptavidin (STV) conjugates as the adaptor (DNA-STV adaptors) (Niemeyer et al., 1999). In the following, some studies conducted for the assembly of artificial enzyme complexes using DDI on 1D-DNA templates are discussed.

Niemeyer et al. (2002) have managed to assemble an artificial bienzymic complex by direct assembly of two enzymes, NAD(P)H: FMN oxidoreductase (NFOR) and luciferase (Luc), onto a single DNA-template. In this study, enzyme (or protein) modification by biotinylation was done in vivo to avoid enzyme denaturation by chemical means; a biotin molecule was attached to the N-terminus biotin carboxy carrier protein tail of two recombinant enzymes, NAD(P)H:FMN oxidoreductase (NFOR) and luciferase (Luc) using biotin ligase. These biotinylated enzymes were coupled to the DNA-STV adaptors to form single-stranded DNA-tagged enzymes and then immobilized to surface-bound capture oligonucleotides by means of the unique specificity of Watson–Crick base pairing to form the artificial bienzymic complex which displayed higher activity (measured in terms of the light released by the enzymatic product of the Luc enzyme and reported in terms of relative light units, RLU) compared with the same enzymes immobilized through random hybridization (Niemeyer et al., 2002; Kim et al., 2012).

Muller and Niemeyer (2008) also managed to assemble heterodimeric enzyme complexes (composed of glucose oxidase (GOX) and horseradish peroxidase (HRP)) by means of DDI. This artificial bienzymic complex of GOX-HRP combines the oxidation of glucose to gluconolactone and hydrogen peroxide (H_2O_2) and the fluorogenic dye Amplex Red to form a highly fluorescent resorufin by HRP. The GOX-HRP system has been widely used as a reporter system in the field of biosensing. This was the first time that the GOX-HRP system has been assembled by means of DDI and these two enzymes have been spatially arranged in such a way that they are in immediate proximity on the same DNA carrier strand. The results of this study provided experimental evidence of the channeling processes occurring in these complexes and also suggests that intercomplex diffusion plays an important role on the overall activity (Muller and Niemeyer, 2008).

The assembly of enzymes on a two dimensional (2D) DNA scaffold, as shown in Fig. 5.2B,C, has also been explored for enzyme immobilization and organization (Wilner et al., 2009); some of the results regarding the assembly of multi-enzyme complexes onto 2D DNA are as follows:

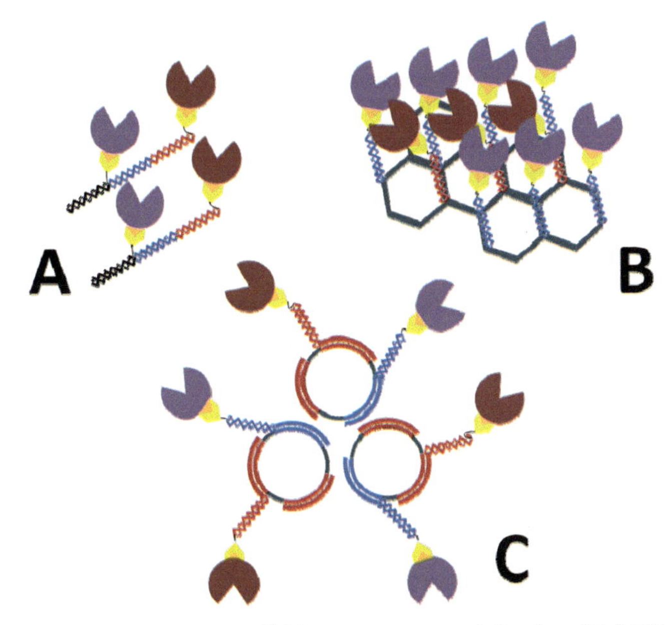

Figure 5.2 1D and 2D DNA scaffolds. Enzymes immobilized on 1D dsDNA [A], 2D hexagonal DNA [B], and 2D circular DNA [C].

Wilner et al. (2009) successfully assembled the enzymes GOX and HRP onto 2D DNA hexagonal scaffolds. A schematic representation of this enzyme complex is shown in Fig 5.1B. A set of single-stranded DNA oligos that are partially complementary to each other was used to form either two hexagon or four hexagon-like structures with each hexagon containing a 10 bp overhanging DNA tether for hybridization. The enzymes GOX and HRP were chemically functionalized with different DNA oligos and were attached onto two separate hexagons on the DNA scaffolds by hybridization with the complementary DNA tether in the hexagon structure. The overall activity of this enzyme cascade can be optimized by modifying the relative position of these enzymes on the DNA scaffolds. In this study, the flexibility of creating rigid 2D DNA hexagonal scaffolds offers the possibility of site-specific immobilization of multiple enzymes for the assembly of complex enzyme cascades (Wilner et al., 2009; Kim et al., 2012).

Wang et al. (2009) also have assembled an enzyme cascade consisting of GOX and HRP on 2D circular DNA scaffolds. In this study, two kinds of circular DNA components were generated by the hybridization of short single-stranded DNA chains for the respective attachment of GOX and HRP preconjugates. These circular DNA components each consist of, complementary domains for the anticocaine aptamer subunits, and sequence specific domains for the auxiliary hybridization of programmed nucleic acid-functionalized enzymes. The circular DNA scaffolds were linked to each other in the presence of cocaine via the cocaine-anticocaine aptamer linking. The GOX-HRP enzyme cascade was formed via hybridization of the ssDNA tagged enzymes to their respective complementary components in the assembled scaffold. The obtained cascade exhibited a 6-fold enhanced enzymatic activity over the free GOX-HRP which was attributed to the high local concentrations of the reacting components (in the vicinity of each enzyme which resulted to "substrate channeling" or intercomplex diffusion), achieved through immobilization of the biocatalysts in immediate vicinity to each other.

Fu et al. (2012) organized discrete GOX/HRP pairs on specific DNA origami tiles with controlled interenzyme spacing and position. In the DNA-directed assembly of GOX and HRP on DNA origami tiles, the DNA-conjugated enzymes, GOx-poly$(T)_{22}$ (5'-HS-TTTTTTTTTTTTTTTTTTTTTT-3') and HRP-poly$(GGT)_6$ (5'-HSTTGGTGGTGGTGGTGGTGGT-3'), were assembled on rectangular DNA origami tiles by hybridizing with the corresponding complementary strands displayed on the surface of the origami scaffolds. In this study, inter-enzyme substrate diffusion was investigated by varying the spacing between these two enzymes and it was found that enhanced activities were observed when enzymes were closely spaced; however, a drastic drop in the activity was observed when the enzymes were spaced as little as 20 nm apart from each other. The results of this study further suggests that the Brownian diffusion of intermediates in solution governed the variations in activity for more distant enzyme pairs, while dimensionally limited diffusion of intermediates across the connected protein surfaces contributed to the enhancement of the activity for the closely spaced GOX/HRP assembly (Fu et al., 2012).

As for the immobilization of enzymes onto complex three dimensional (3D) DNA structures, studies are yet to be conducted in this field; DNA has been used to create structures such as tetrahedrons, octahedrons and icosahedrons (He et al., 2008). Immobilization of enzymes onto a 3D structured DNA not only increases the stability and rigidity of the complex, but also allows complex arrangements of enzymes, which in turn could affect the overall activity of enzyme cascades (Kim et al., 2012).

5.3 Affinity Tag or Binding Peptides

Immobilization of enzymes onto solid surfaces can be either non-specific or site-specific (Wong et al., 2008). Non-specific immobilization methods (a.k.a. conventional immobilization methods) generally include adsorption (Besteman et al., 2003), non-specific covalent bonding (Letant et al., 2004; Pierre et al., 2006), entrapment (Kim et al., 2007), and encapsulation (Luckarift et al., 2004). Although widely applied in bioconversions, these conventional approaches often encounter some problems such as enzyme leaching (Manyar et al., 2008), loss of enzyme activity upon immobilization (due to unfolding or denaturation) (Hong et al., 2007), and strong diffusion resistance (Hudson et al., 2008). Site-specific immobilization strategies on the other hand, have been increasingly utilized to address the limitations of the conventional immobilization methods on maintaining the enzyme conformation, and retaining the biological activity on selected solid surfaces (Kumada et al., 2006; Kwon et al., 2006; Park et al., 2006).

One of the most popular site-specific enzyme immobilization strategies that have been exploited recently is via the use of synthetic protein scaffolds (popularly known as affinity tags or binding peptides) (Kim et al., 2012). Originally, these synthetic protein scaffolds were developed mainly for the purification of recombinant proteins with the following features: one-step adsorption purification, minimal effect on the structure and activity, easy and specific removal in the production of the native protein, simplicity and accuracy for assaying the recombinant protein during purification, and its vast applicability to a number of proteins (Terpe, 2003). However, the features provided by synthetic protein scaffolds are correspondingly applicable to the immobilization of proteins (especially enzymes) (Liu et al., 2012).

Synthetic protein scaffolds are usually incorporated genetically onto the enzyme and directed towards a solid surface or support (self-assembly) without the need of further chemistry on the surface or on the protein. Thus, these synthetic protein scaffolds have the potential to control the enzyme orientation and spacing, be highly specific, and provide reversibility (Whaley et al., 2000; Slocik and Naik, 2006; Krauland et al., 2007). One major advantage of this technique is the possibility to generate highly specific protein scaffolds or binding peptides for many materials of interest (Sarikaya et al., 2003). A wide variety of techniques ranging from phage display to mRNA display have been used to develop binding peptides which have the required affinity to a material (Brown, 1997; Whaley et al., 2000; Wilson et al., 2001).

In the immobilization of enzymes onto solid surfaces, incorporation of the synthetic protein scaffold is done via the recombinant DNA strategy. The gene of the enzyme of interest, along with the gene of the synthetic protein scaffold (attached to either the N or C-terminus of the enzyme) are cloned into an expression vector (e.g., a plasmid or a virus designed for protein expression in cells) to produce the recombinant vector which is then transformed into a host cell to express (either intracellular or extracellular expression) the enzyme of interest with the synthetic protein scaffold attached to it. After the protein expression process, the protein extraction is done by either cell lysis followed by physical separation (e.g., centrifugation, or sedimentation methods if the proteins were expressed intracellularly) or by direct physical separation of the proteins from the cell (if the proteins are expressed extracellularly) and then the mixture of proteins obtained from the protein extraction process are introduced onto the surface of the functionalized supporting material; only the enzymes with the synthetic protein scaffold would be immobilized onto the surface and then other proteins can be separated from the surface with the immobilized enzymes by physical separation.

This immobilization process is advantageous compared to conventional immobilization methods because after the protein expression and extraction of the transformed cells, there is no need for additional protein purification steps before immobilization due to the fact that the surface is selective (functionalized surface to synthetic protein scaffold interaction) to the enzyme to be immobilized. A schematic diagram presenting the general enzyme production and immobilization strategy of this process is shown in Fig. 5.3. The following sub-sections discuss the different types of synthetic protein scaffolds which have been developed and applied to enzyme immobilization.

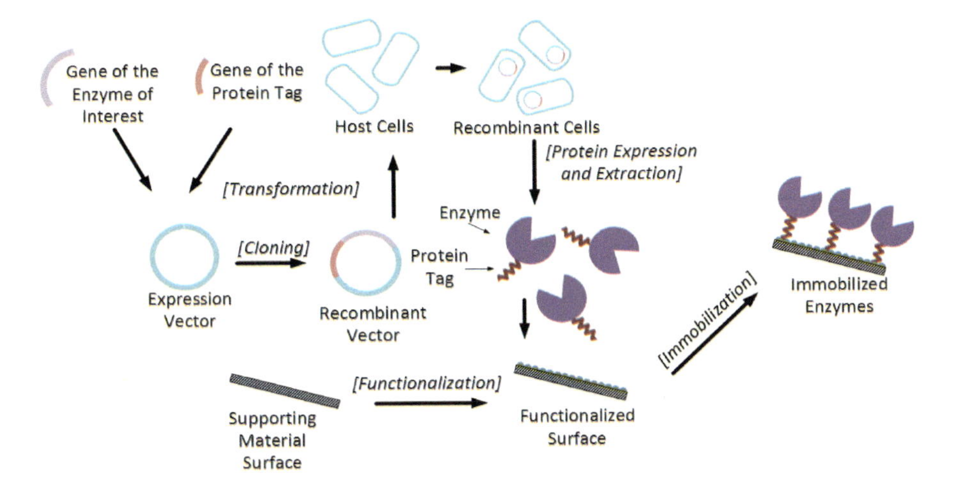

Figure 5.3 General strategy for production and immobilization of enzymes onto a functionalized supporting material via affinity tags (protein tags).

5.3.1 Polyhistidine Tag (His-Tag)

Developed for immobilized metal affinity chromatography (IMAC mainly for purification of recombinant proteins, this short affinity-tag consists of polyhistidine residues. The specific interaction between a transition metal (e.g., Ni^{2+}, Co^{2+}, etc.) which is immobilized on a matrix and specific amino acid chains is the main theory behind IMAC (Porath et al., 1975). It has been found out that the amino acid histidine exhibits the strongest interaction with immobilized metal ion matrices because the imadizole ring of the histidine residue is able to coordinate with the metal ion to form a stable conjugate. Thus, peptides containing consecutive histidine residues have higher efficiency to bind to the immobilized metal ion matrix compared to a single histidine residue (Khan et al., 2006; Gupta et al., 2008). This synthetic protein scaffold system is not only popular for purification of recombinant proteins but it is also a widely used system for enzyme immobilization. However, in order to successfully immobilize the enzyme, the polyhistidine tag must either be conjugated at either the N or C-terminus of the enzyme of interest; this mainly depends on the location of the active site of the enzyme (Liu et al., 2012). The following examples provide a brief overview of different enzymes that have been immobilized into different supporting materials using the polyhistidine tag system.

Horseradish Peroxidase, an enzyme widely used in immunochemistry for molecular labeling and detection, was immobilized on amino-nitrilotriacetic-Co(II) (ANTA) functionalized thiol-derivatized gold nanoparticles as described by Abad et al. (2005). It has been reported that the immobilized HRP using the His-tag immobilization strategy exhibited 93% relative enzymatic activity compared to the free HRP synthesized by recombinant engineering strategy.

D-Amino acid oxidase (DAO) is an enzyme widely used for the oxidation of cephalosporin C in the food and pharmaceutical industry, and it is also used as a biosensor for the detection of D-amino acids. Since this enzyme is widely used in the industry for its important applications, a variety of physical and chemical immobilization strategies have been applied to immobilize this enzyme. However, these said physical and chemical immobilization strategies are quite complex (in terms of synthesis) and this complexity gave rise to the need for finding alternatives in DAO immobilization. Kuan et al. (2008) have demonstrated the immobilization of DAO on Ni-NTA functionalized magnetic beads. The advantage of magnetic particles as enzyme supports results from the ease of separation of the immobilized enzyme from the reaction mixture by means of an applied magnetic field. It has been reported that the highest relative enzymatic activity of immobilized DAO compared to the free DAO was 75%. This reduced activity has been attributed to the following: (i) binding of DAO to its innate amino acid resides, (ii) multi-point attachment of DAO on the magnetic beads led to conformational changes (since a histidine residue is found on its active site). Although a reduced activity was observed for the immobilized DAO, its thermal activity and stability has been improved when compared to the free enzyme which was completely inactivated after incubation at 50°C for 1 h.

NADH oxidase (NOX), an oxidoreductase which catalyzes the reaction between oxygen and NADH to hydrogen peroxide or water and is industrially utilized for in-situ cofactor regeneration, has been immobilized on a variety of ANTA-Co(II) functionalized carbon nanomaterials such as single-walled carbon nanotubes (SWCNTs) (Wang et al., 2010), multi-walled carbon nanotubes (MWCNTs), and carbon nanospheres (CNS, fullerenes) (Wang et al., 2011). The preparation of the said carbon nanomaterials was done by formation of carboxylic acid groups on the nanomaterials (by acid digestion) prior to functionalization with ANTA-Co (II). The group of Wang et al. reported that the immobilization of NOX on SWCNTs had the highest relative enzymatic activity (~92% of the free enzyme activity was retained), and enzyme loading (~0.47 mg enzyme/mg material) when compared to NOX immobilized on MWCNTs and CNTs. The difference in the relative enzymatic activity and enzyme loading were attributed to the surface area and ligand density; the product of these two parameters can determine the number of ligands available for enzyme immobilization with respect to the amount of nanomaterial. SWCNTs have the highest number of available ligands (due to its high surface area) after functionalization when compared to MWCNTs and CNTs. In addition, it has been demonstrated that the immobilization of NOX is reversible and able to retain its activity for a couple of loading cycles. This implies that the functionalized carbon nanomaterial can be reused indefinitely in the process which is economically beneficial as the cost of the nanomaterial for enzyme immobilization would be a one-off investment.

Catechol 1,2-dioxygenase (CatA), a ring cleavage enzyme which is utilized mainly in remediation of toxic aromatic hydrocarbons such as catechol, has been immobilized on Ni/NiO decorated magnetic nanoparticles as demonstrated by Lee et al. (2011). It has been reported that the immobilized CatA has a relative enzyme activity of ~85%, an enzyme loading of 36 μg enzyme/mg nanoparticle. Furthermore, the immobilized CatA retained 72% of its catalytic activity after recycling for 6 times.

The cellulases (Cel48F, and Cel9G) from *Clostridium cellulolyticum*, are widely used in pretreatment of cellulose in lignocellulosic biomass to liberate reducing sugars for the ethanol fermentation processes, have been immobilized on iminodiacetic (IDA)-Ni^{2+}-functionalized magnetic core microspheres. Zhang et al. (2013) demonstrated a simplistic approach in the synthesis of the enzyme supporting material, Fe_3O_4/PMG (poly *N,N′*-methylenebisacrylamideco-glycidyl methacrylate) core/shell microspheres, by using the distillation-precipitation polymerization method. The F_3O_4/PMG microspheres are then functionalized by treatment of iminodiacetic acid (IDA) and Ni^{2+}. It has been reported that the immobilization of the cellulases Cel48F, and Cel9G had high relative enzymatic activities of ~99%, and 94%, respectively. Furthermore, the cellulases remain immobilized after recycling for 5 times without compromising its retained activity.

Ling et al. (2014) described a general strategy for enzyme immobilization using the His-tag system on nickel-nitriloacetic acid (Ni-NTA) functionalized mesoporous silica (SBA-15). This strategy involves the immobilization of the model protease (human rhinovirus 3C, PreScission protease (PSP)) which fused with a His-tagged enhanced green fluorescent protein (EGFP) on the functionalized SBA-15. The fusion of EGFP in between the His-tag and PSP serves as a spacer which distances the PSP active site away from the surface of the SBA-15 to avoid non-specific interactions with the functionalized surface to enable maximum exposure of the PSP active site to the solution while the His-tagged EGFP interacts with the pores on the surface of the functionalized SBA-15. The activity for this immobilized enzyme system is 1.6 times higher than the free enzyme, and has an enzyme loading of 50 μg enzyme/mg functionalized SBA-15. In addition, the immobilized enzyme exhibits increased stability and can retain ~90% of its initial activity after recycling for 5 times.

5.3.2 Some Other Commonly Used Affinity Tags

There are quite a number of affinity tags developed using techniques like phage and mRNA display for the purpose of generating specific interactions (affinity) with inorganic materials for protein purification or immobilization (Sarikaya et al., 2003). However, only a few of these affinity tags such as gold binding peptide (GBP), iron oxide affinity peptide (FeAP), silica affinity peptide (SiAP), and polystyrene affinity tag (PS-tag) have been utilized for biocatalysis because of their low affinity to nanomaterials, although this concern can be overcome by using multiple repeat tags or surface coated polymers to enhance their affinities (Sarikaya et al., 2003; Johnson et al., 2008). Some examples of enzymes of interest immobilized onto nanomaterials using the said affinity tags, and their characteristics are discussed in the following paragraphs.

Glutathione S-transferase (GST), is well known for catalyzing the conjugation of the reduced form of glutathione (GSH) to xenobiotic substrates for the purpose of detoxification and it is widely known for its clinical significance in cancer development and chemotherapeutic drug resistance (Sheehan et al., 2001). Kumada et al. (2006) demonstrated the immobilization of this enzyme on polystyrene (PS) supports (PS beads) by developing peptide tags with high affinity towards polystyrene surfaces. From the results of the screening and characterization of the said research group, the PS-tags with the sequences of RAFIASRRIKRP (PS-19), and AGLRLKKAAIHR (PS-23) fused with GST at its C-terminus showed the highest affinities (around >80% adsorption of the tagged proteins) when compared to the wild-type GST immobilized on the PS surface without a peptide tag. However, the fusion of the PS-tag to GST decreased its enzymatic activity when compared to the wild-type (free enzyme) because of the hydrophobic nature of the fused tag (Sakiyama et al., 2004). Thus, the relative enzymatic activities of the immobilized GST on polystyrene supports using PS-19 and PS-23 tags (~23% and 17%, respectively), are relatively low when compared to the wild-type (free enzyme) due to its decreased activity upon the fusion of the PS-tags to GST. Nonetheless, compared to the direct immobilization of wild-type GST, the activities of the immobilized PS-tag–GST fusion are 10-fold higher than those of the wild type. This result indicates that the immobilization of the PS-tag–GST fusion on the polystyrene support enhanced the exposure of the active site of GST to the substrate phase.

Haloalkane dehalogenase (dhlA), an enzyme involved in the degradation of environmentally toxic halogenated compounds (present in industrial by-products) and even the suspected carcinogenic compound 1,2-dichloroethane (DCA), has been used by Johnson et al. (2008) as a model enzyme to demonstrate the feasibility of the immobilization of enzymes on magnetic particles (MNPs) via the affinity peptides such as the iron oxide (FeAP) and silica (SiAP) affinity peptides. The MNPs used as supports in this study were the supermagnetic nanoparticles (consisting of an iron core surrounded by a shell of ferric oxide) and silica coated iron oxide MNPs (prepared using the SOL-DLC method). The affinity peptides were fused to the C-terminus of dhlA to yield fusion constructs of dhlA fused to (i) $(FeAP)_3$–His_6–three repeats of an iron oxide affinity peptide with the peptide sequence of RRTVKHHVN with an additional His-tag, and (ii) $(SiAP)_2$–His_6—two repeats of a silica affinity peptide with the peptide sequence of MSPHPHPRHHHT with an additional His-tag. An evaluation revealed that the fusion proteins (both the SiAP and FeAP tagged dhlAs) were immobilized by affinity binding on the surface of the MNPs (both Si-coated and uncoated). However, in terms of the relative enzymatic activity, the immobilized dhlA on both the Si-coated and uncoated magnetic particles were relatively low; this was attributed to the strong electrostatic or hydrophobic interactions of the solid inorganic surfaces with the proteins' native conformation. Thus, to prevent the undesired interactions, the surface of the MNPs (Si-coated and uncoated) were treated with organic silanes which act as barriers between the enzyme and the particle surface. It was found that coating the MNPs with alkoxysilanes such as 3-glycidoxypropyltrimethoxysilane (GPS) on iron oxide, and 3- aminopropyltriethoxysilane (APS) on Si-coated MNPs enhanced the immobilized enzyme activities of the dhlA-$(FeAP)_3$–His_6, and dhlA-$(SiAP)_2$–His_6, respectively. The relative enzyme activities of dhlA-$(FeAP)_3$–His_6 on GPS-coated MNPs, and dhlA-$(SiAP)_2$–His_6 on APS-coated MNPs were increased to 47% and 37% with an enzyme loading of 30% (51% of theoretical) and 36% (61% of the theoretical) μg per mg MNP, respectively. These results indicated that the coating of alkoxysilanes helped to preserve the enzyme activity although, it may have interfered with the specificity and binding strength of the affinity peptides due to the observed reduction of the amount of dhlA-AP adsorbed on the surface of the coated MNPs.

Alkaline phosphatase (AP), an enzyme widely used for immunoassays which is responsible for removing inorganic phosphate groups from various biomolecules (e.g., proteins, nucleic acids), was fused with several repeats of a well characterized gold binding peptide (GBP1) with the amino acid sequence of MHGKTQATSGTIQS and has been immobilized onto the surface of gold patterned silicon wafers (prepared via micro-contact printing) as demonstrated by Kacar et al. (2009). It has been shown that the fusion of a 5-tandem repeat of GP1 on AP provided the highest binding activity and specific enzyme activity per unit area which has been attributed to the proper orientation of the enzyme catalytic site to the solution.

Organophosphorus hydrolase (OPH), an enzyme capable of rapidly degrading organophosphorus toxins, such as phosphorus-based nerve agents and pesticides and frequently used for the electrochemical detection of the said compounds, was fused with the same gold binding peptide (GBP1) characterized by Kacar et al. (2009), immobilized onto gold nanoparticle-coated chemically modified graphene (Au-CMG), and then evaluated for its applicability in a flow-injection electrochemical biosensor system by Yang et al. (2011). The immobilized GBP1-OPH on Au-CMG hybrids exhibited a high enzymatic ability due to the well-oriented enzyme morphology (based on the evaluation of its electrochemical properties) and good stability (due to the binding specificity of the GBP). In addition, the GBP1-OPH/Au-CMG provided simultaneous functions of a biocatalysis, electron transport, and mass transfer. As a biosensor, the GBP1-OPH/Au-CMG has successfully demonstrated its efficiency in the detection of paraoxon (a model organophosphorus compound) with a sensitivity, limit of detection (LOD), and response time of 55.54 nA μM^{-1}, 9.54×10^{-8} M, and <3 s, respectively.

5.4 Enzyme-Assisted Covalent Immobilization

This method of biological assembly utilizes specific conjugating enzymes to achieve a strong covalent bonding between the enzyme of interest and the supporting material (e.g., nanomaterial). This is desirable in some applications which require improved stability of the enzyme-nanomaterial conjugates. Since this method utilizes the enzyme catalyzed covalent conjugation, the conjugation process is more specific (due to the conjugating enzyme's specificity) and milder (in terms of process conditions) when compared to the conjugation processes (e.g., conventional "click" chemistry) which uses chemical catalysts. Furthermore, the addition of ligands to the enzyme of interest (in the chemical covalent conjugation process) is no longer necessary because the conjugating enzyme in this method is responsible for this step. A schematic diagram of the enzyme-assisted covalent immobilization method is shown in Fig. 5.4. Several conjugation enzymes have been exploited for the site-specific covalent immobilization of industrially useful enzymes on particles. These are discussed in the following sub-sections.

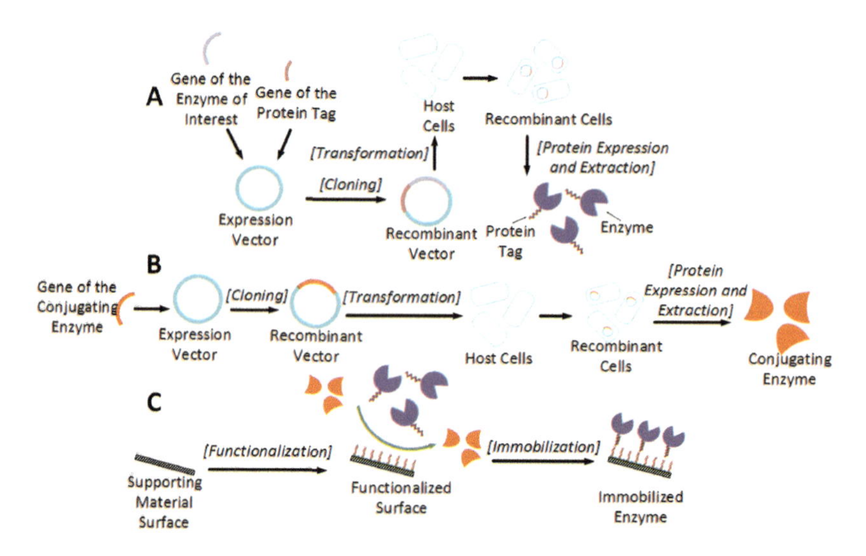

Figure 5.4 General strategy of enzyme-assisted covalent immobilization. The strategy involves the production of the tagged enzyme of interest [A], production of the conjugating enzyme [B], and enzyme mediated covalent conjugation yielding immobilized enzymes of interest [C].

5.4.1 Sortase A-Mediated Immobilization

Sortase A is an enzyme used by Gram-positive bacteria to anchor surface proteins to the cell wall through a condensation reaction between a C-terminal LPXTG (where X is any amino acid) tag on the former and a polyglycine bridge in the latter such that the enzyme cleaves the LPXTG sequence at the amide bond between the threonine and the glycine to form an acyl-protein (transpeptidase reaction) complex (Ton-That et al., 1997). The LPXTG recognition motif of Sortase A in the enzymatic ligation is highly selective which makes it attractive for protein modification. It has been demonstrated that Sortase A from *Staphylococcus aureus* provides a robust and mild approach for selective enzyme-mediated immobilization of proteins onto solid surfaces (Parthasarathy et al., 2007; Chan et al., 2007; Clow et al., 2008).

Ito et al. (2010) have successfully demonstrated a Sortase-A mediated covalent immobilization of glycosyltransferases (β-1,4-galactosyltransferase (hGalT) and α-1,3-fucosyltransferase (hFucT)) onto alkylamine displaying sepharose beads. These enzymes, fused with a LPETG-His_6 hybrid tag at their C-terminus, were introduced into a mixture with EAH sepharose beads (with reactive amino groups displayed on its surface) in the presence of Sortase A (recombinantly expressed and purified from *E. coli*) for the immobilization to proceed. The results showed that both enzymes were immobilized on the EAH sepharose beads by Sortase A-mediated specific transpeptidation of the C-terminal signal peptide (LPTEG tag) without an effect on their intact conformation which led to a retention of their enzymatic activity, improved stability, and reusability (Ito et al., 2010). Furthermore, the relative enzymatic activities of the immobilized hGalT and hFucT are 90% and 94%; with the relative activity after 10 reuses of 95% and 55%, respectively. It was also reported before that the sortase-catalyzed reaction is reversible and might result in low ligation efficiency. This could be overcome by addition of β-hairpin structures around the LPXTG tag, but the β-hairpin structures influence on enzyme activity was not mentioned in the report (Yamamura et al., 2011).

5.4.2 Transglutaminase-Mediated Immobilization

Transglutaminase is an enzyme that catalyzes an acyl transfer reaction by forming amide bonds (which exhibit high resistance to proteolytic degradation) between γ-carboxyl groups of glutamine (Q) residues and the primary amino groups of a variety of amines or the ε-amino groups of lysine (*K*) residues in proteins (Folk and Finlayson, 1977; Folk, 1980). It is a widely used tool in protein chemistry for the attachment of chemical labels, for probes to specific proteins, etc. In addition, it is used in the commercial food processing industry to bond proteins together (meat glue). This enzyme can even recognize specific amino acid sequences such as LLQG and MLAQGS. In enzyme immobilization applications, a number of enzymes have been immobilized onto different supporting materials (especially nanomaterials). Examples of transglutaminase mediated enzyme immobilization are discussed as follows.

Sugimura et al. (2007) have successfully immobilized the enzyme glutathione S transferase (GST) on an octanediamine-fixed NHS Sepharose gel (prepared by chemical modification of NHS-activated sepharose) via the conjugating enzyme transglutaminase 2 (TGase 2). A T26 peptide tag (with the amino acid sequence of HQSYVDPWMLDH), determined to be the most potent and preferred glutamine substrate for TGase2, was fused to the N-terminus of GST and then immobilized on the functionalized NHS-sepharose gel via TGase 2 catalytic reaction. Fascinatingly, the immobilized GST exhibited a higher enzymatic activity (~1.22 times higher) compared to the free GST, attributed to the increased localized concentration of GST on the solid phase due to site-specific conjugation. This site-specific and covalent immobilization of GST is much better in terms of enzymatic activity when compared to immobilization of GST via affinity tag (as discussed in Section 5.3.2).

The research group of Komiya et al., have demonstrated the immobilization of alkaline phosphatase (AP) on β-casein modified polyacrylic resin (Tominaga et al., 2004) and agarose gel beads (Tominaga et al., 2005), and magnetic particles (Moriyama et al., 2011) using microbial transglutaminase from *Streptomyces mobaraensis* (MTG). A variety of peptide tags containing reactive lysine residues recognized by MTG have been fused at the N-terminal of AP and then immobilized on the supporting materials (functionalized with reactive glutamine residues) via the catalytic action of MTG. The immobilization of AP on these supporting materials exhibited high relative enzymatic activities ranging from 86–100% and enhanced operational stability where there was 90–98% activity retention upon recycling for 10 times.

5.4.3 Phosphopantetheinyl Transferase-Mediated Immobilization

Sfp is one type of phosphopantetheinyl transferase, an enzyme that covalently transfers 4′-phosphopantetheinyl (Ppant) groups from coenzyme A (CoA) to conserved serine residues on PCP and acyl carrier protein (ACP) domains in nonribosomal peptide synthetases (NRPSs) and polyketide synthases (PKSs) in *Bacillus subtilis* (Lambalot et al., 1996). This enzyme is utilized in the notable protein labeling technology developed by Walsh et al. (Yin et al., 2006; Yin et al., 2005). In the said technology, the Sfp, or a related phosphopantetheinyl transferase (PPTase) enzyme catalyzes a reaction between CoA covalently immobilized on a solid surface and the protein of interest which contains a natural or engineered phosphopantetheinylation site. Yin et al. (2005) have developed a peptide tag "ybbR" (with the amino acid sequence DSLEFIASKLA), found to be an efficient substrate for Sfp-catalyzed protein labeling which serves as the engineered phosphopantetheinylation site. In the immobilization of enzymes, the Sfp-mediated immobilization attaches the enzyme of interest through a Ser residue of the fused ybbR-tag to the phosphopantetheine moiety of CoA attached to supporting materials.

Wong et al. (2008) first demonstrated the immobilization of luciferase and glutathione S-transferase (GST) on PEGA resin. It has been reported that the relative enzymatic activity of the immobilized luciferase was ~90%. Additionally, GST immobilized via Sfp-catalyzed conjugation exhibited a higher enzymatic activity when compared to the GST randomly attached to the PEG resin due to the site-specific immobilization of GST.

Furthermore, the same research group demonstrated the immobilization of the enzymes arylmalonate decarboxylase (AMDase) and glutamate racemase (GluR) on CoA functionalized polystyrene nanoparticles (Wong et al., 2010). These enzymes catalyze the decarboxylation of malonates. Prior to the Sfp-catalyzed immobilization, the ybbR tag was fused to the N-terminus of both enzymes and surprisingly, the activity of the ybbR-GluR fusion was higher when compared to the wild type GluR while the activity of AMDase was not altered; the increase in GluR activity was attributed to the improved stabilization and solubility of the tagged GluR. It was reported that the immobilized AMDase and GluR have lower enzymatic activities when compared to the free ybbR-tagged enzymes, which implied that there is a loss in enzymatic activity after the immobilization process. This result was attributed to the electrostatic repulsion of the negatively charged substrate attempting to approach the nanoparticles that were also negatively charged due to the unreacted surface carboxylates, and also to the possible steric hindrance of the immobilized enzymes thus stressing the importance of the surface groups on the supporting material in this method. Although the immobilized enzymes exhibited low enzymatic activities, the immobilized AMDase retained around 93% of its initial activity after recycling for 4 times while the immobilized GluR remained thermostable at temperatures up to 55°C.

In addition, the shortcomings of using this method in enzyme immobilization would be the instability of the immobilization of CoA on the supporting material, and the interference of the endogenous CoA from the cell lysate upon immobilization (when the enzyme of interest is obtained directly from the cell lysate) (Wong et al., 2008; Wong et al., 2009).

5.4.4 O^6-Alkyl Guanine Transferase-Mediated Immobilization

The human DNA repair protein O^6-alkyl guanine transferase (hAGT) is an enzyme which brings about the repair of O^6-alkylguanine in double-stranded DNA by transferring the alkyl group from the DNA to an internal cysteine residue in the AGT protein; this residue is located at position 145 in the human AGT (Pegg, 2000). This enzyme was utilized in one of the notable protein labeling technologies developed by Johnsson et al., where the hAGT is fused with the protein of interest, followed by the reaction of hAGT with the alkyl group present in the substrate (O^6-benzylguanine (BG) derivatives) that is fused to the label to form the protein-label conjugate (Keppler et al., 2003).

Unlike the enzyme-assisted covalent immobilization methods discussed in sub-sections 4.1 to 4.3, where the role of the enzyme is to form the covalent bond/s between the peptide tag (fused to the enzyme of interest) and the ligands displayed on the surface of the supporting material, this method utilizes the enzyme (hAGT) itself as a peptide tag which reacts to the ligands displayed on the supporting material. Kindermann et al. (2003) demonstrated the immobilization of the enzyme glutathione S-transferase on O^6—benzylguanine-functionalized agarose beads (FAB). In this study, the hAGT was fused to the C-terminal of GST and then immobilized on functionalized agarose beads via the demethylation of the O^6—benzylguanine displayed on its surface by the hAGT, forming the FAB-hAGT-GST conjugate. It has been reported that the GST retained 62% of its activity upon immobilization.

The advantage of the hAGT mediated immobilization method is the site-specific covalent attachment of the hAGT-fused enzyme to the benzyl-guanine (BG) functionalized surface where the BG is unreactive against other proteins than hAGT. However, since the size of hAGT is quite large when compared to the peptide tags discussed in the previous sections, there is a possibility that the fusion of hAGT to the enzyme of interest would have an effect on its conformation and activity as observed from examples from the study of discussed in Section 5.3.2 where the fusion of PS tags had an effect on the GST activity.

5.4.5 Enzyme Immobilization via Tyrosine Cross-Linking

Tyrosine cross-linking is typically observed in structural proteins such as elastin, silk, and even sea urchin eggs (Raven et al., 1971; Downie et al., 1973; Foerder and Shapiro, 1977). However, tyrosine residues are normally unreactive but they can be converted into free radicals that react with each other to form dityrosines. In nature, the oxidative cross-linking between tyrosine residues are catalyzed by a wide variety of metalloenzymes that use oxygen or hydrogen peroxide as terminal electron acceptors (Eickhoff et al., 2001; Oudgenoeg et al., 2002; Fukuoka et al., 2002; Chen et al., 2003; Takasaki et al., 2005; Steffensen et al., 2008). Dityrosine cross-linking can also be achieved by the catalytic action of metal oxidants conjugated to metal binding peptides (Brown et al., 1998; Brown et al., 1995). Endrizzi et al. (2006) even developed a method for the site-specific and covalent surface immobilization of proteins via dityrosine cross-linking inspired from the natural metal-catalyzed cross-linking process. The following examples from the research group of Komiya et al., discusses the immobilization of the enzyme alkaline phosphatase (AP) into water-oil-water microspheres and the synthesis of cross-linked AP aggregates using the dityrosine cross-linking method.

Minamihata et al. (2009) demonstrated the immobilization of Y-tagged (GGYYY) alkaline phosphatase (AP) onto water-in-oil-in-water (W/O/W) type microcapsules. The AP-GGYYY fusion was immobilized onto the W/O/W microcapsules via dityrosine cross-linking through a radical polymerization reaction in the oil phase and the resulting immobilized AP exhibited around 2.5 times enzymatic activity than that of the wild type. Additionally, the immobilized AP showed an enhanced operational stability because it showed around 95% of activity retention after being recycled for 5 times which may be due to the strong covalent cross-linking of the enzymes on the W/O/W microcapsules.

Another approach in the utilization of dityrosine cross-linking in the synthesis of cross-linked enzyme aggregates (CLEA is by enzyme mediated oxidative cross-linking as discussed above. Minamihata et al. (2011b) first demonstrated the use of the site-specific peroxidases-mediated cross-linking of a Y-tagged (GGGGY) alkaline phosphatase (AP). They utilized the enzyme horseradish peroxidase (HRP) to catalyze the dityrosine cross-linking between the tyrosine residues of the Y-tagged AP and reported that the relative enzyme activity of the cross-linked AP was 95% when compared to the free AP (wild type). In a subsequent study by the said research group (Minamihata et al., 2011a), they demonstrated the immobilization of AP onto biotin-coated plates by cross-linking the Y-tagged AP with a Y-tagged streptavidin (SA) via HRP-mediated dityrosine cross-linking. It was also reported that the AP-SA conjugate (using the GGGGY as the Y-tag) immobilized on the biotin-coated plate (the mode of immobilization in this case is not a covalent type but via the streptavidin-biotin affinity binding) exhibited a high enzymatic activity because of the higher heteroconjugation efficiency.

5.5 Other Affinity Tags: Unnatural Amino Acids

One major shortcoming of the different enzyme immobilization methods discussed in this chapter is the consideration of active site orientation of the enzyme of interest upon immobilization. All of the discussed methods in this chapter (Sections 5.2–5.4 and 5.6–5.8) are limited to the addition of tags to the N and C terminals of the enzymes of interest to control their orientation. However, there are some enzymes, where their active sites are located near their N and C terminal, and immobilization of these types of enzymes would directly lead to a great loss in enzymatic activity upon immobilization due to the disorientation of the active site (where the substrate cannot interact with the enzyme's active site). The best approach to achieve proper active site orientation of these enzyme types is to conjugate the ligands (present at the surface of the functionalized supporting material) to the specific amino acid residue location in the enzyme (which can be determined by studying its detailed crystal structure) that provides the optimum active site orientation.

To address this problem, the incorporation of unnatural amino acids (uAA) at the specific amino acid residue location (which provides the optimum active site orientation upon immobilization) of the enzyme of interest offers unique functional groups (which are not found in the 20 natural amino acids) which can be exploited for different types of site-specific conjugation. The incorporation of uAA's into the synthesis of proteins was first implemented independently by the Schultz and Chamberlin research groups in the late 1980s (Noren and Science, 1989; Bain et al., 1989), and then this method was further refined in the succeeding years (Wang et al., 2001; Xie and Schultz, 2005; Xie and Schultz, 2006; Ryu and Schultz, 2006). A good review on the development of the technologies in incorporating uAA's has been presented by Liu and Schultz (2010). The technology widely used in site-specifically incorporating a single uAA to a predetermined residue is the in vivo *Escherichia coli*-based system developed in the Schultz lab. This system employs the orthogonal tRNA and aminoacyl-tRNA synthetase derived from *Methanocaldococcus jannaschii* to incorporate the uAA only to the location encoded by the stop codon (UAG) on the mRNA.

The two commonly used uAA's in this method are the *p*-propargyloxyphenylalanine (*p*Pa) and *p*-azidophenylalanine (*p*Az) which are usually applied to conjugations via the highly selective Huisgen copper(I)-catalyzed azide-alkyne [3 + 2] cycloaddition click reaction (Chin et al., 2002; Bundy and Swartz, 2010). A schematic illustration of the site-specific incorporation of uAA's and its immobilization on a functionalized surface via the click reaction is shown in Fig. 5.5.

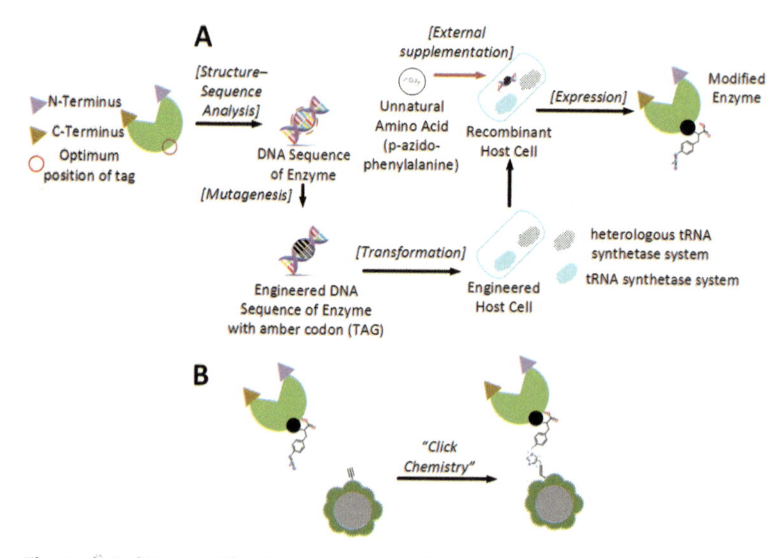

Figure 5.5 Site-specific incorporation of uAA. The steps involved in uAA incorporation [A], and the immobilization of the modified enzyme to the functionalized supporting material via click chemistry [B].

Seo et al. (2011) demonstrated the utilization of the said method by specifically incorporating *p*Az to a pre-determined site in the protein DrrA (a pathogenic protein that binds to human Rab1) and clicked an alkyne-biotin fusion with the *p*Az. The resulting conjugate (DrrA-biotin) was then immobilized via avidin-biotin interaction on a neutravidin functionalized surface and showed that the site-specifically immobilized DrrA (using the uAA tag) exhibited higher activity when compared to a randomly oriented immobilized DrrA. However, this non-covalent linking (avidin-biotin) of the DrrA on the avidin-functionalized surface can present a limitation to the sturdiness of the immobilized DrrA which could give an impact on its operational stability.

In the immobilization of proteins (especially enzymes) on supporting materials, the research group of Bundy, et al., introduced the Protein Residue-Explicit Covalent Immobilization for Stability Enhancement system (PRECISE) which enables the targeted immobilization of proteins to enhance protein stability and durability (Smith et al., 2013b). This PRECISE system utilizes the site-specific uAA (in this case the photo-stable *p*Pa, which is unlike the photochemically unstable *p*Az) incorporation by employing an *E. coli*-based cell-free protein synthesis (CFPS) approach that eliminates the membrane-transport limitations and has a 27-fold increase in production yields of proteins containing *p*Pa (Bundy and Swartz, 2010); and lastly, the system also includes a direct covalent immobilization strategy to eliminate the need of a non-covalent peptide linker. Additionally, the said CFPS approach has other process advantages which include the direct access and optimization of the synthesis environment, simplified purification, and high-throughput automation potential (Kim et al., 2011b; Bundy and Swartz, 2011; Shrestha et al., 2012).

Wu (2014) demonstrated the immobilization of *Candida antarctica* lipase B (CalB), an enzyme widely used in many industrial applications, on azide functionalized superparamagnetic beads employing the PRECISE system. The incorporation of the unnatural amino acid *p*Pa was done at the N98 position of CalB and some slight decrease in the enzymatic activity was observed when compared to the wild type CalB. Upon immobilization of the engineered CalB-N98*p*Pa (via click reaction with the azide functional groups displayed on the surface of the magnetic beads), the enzymatic activity is 1.23 times higher than the free CalB (wild type) suggesting that CalB might be a positive candidate for covalent conjugation as it retained a significant amount of activity post-immobilization. However, it was also reported that the CalB immobilization is not that feasible since the process requires the synthesis and purification of CalB in large amounts, which in turn is costly and time consuming.

In a succeeding study by the same research group of Wu et al. (2015) utilizing the same PRECISE system in enzyme immobilization on azide-functionalized superparamagnetic beads, the T4 bacteriophage lysozyme (TBL) (which is responsible for degrading bacterial cell walls) was chosen as the model enzyme for its well-characterized folding and denaturation pathways. The incorporation of the unnatural amino acid *p*Pa at positions away from the active site of the lysozyme exhibited around 25–30% loss in enzymatic activity when compared to the standard lysozyme (unmodified). In addition, a dramatic loss in enzyme activity (up to 90%) occurred upon the incorporation of *p*PA at positions near the active site. These results stress the importance in the selection of the uAA insertion site (which preferably would be away from the active site while taking into consideration the surface of the enzyme being exposed to the substrate phase); the optimum uAA insertion site for TBL determined in this study was at L91. The relative enzyme activities of the immobilized lysozyme is around 64.3% (with respect to the standard unmodified enzyme) and ~81% (with respect to the free engineered enzyme L91*p*Pa-TBL), and a slight decrease in the immobilized activity was detected in post-immobilization testing; these values exhibit the advantages of immobilization orientation control. Furthermore, the immobilized enzyme obtained via uAA incorporation exhibited improved operational stability (around 50–73%) when compared to immobilization via non-specific covalent interactions.

In another study by Deepankumar et al. (2015), it was demonstrated that a modified ω-transaminase (ω-TA) from *Sphaerobacter thermophilus* (an industrially relevant enzyme utilized for the preparation of optically pure amine compounds) can be immobilized on chitosan and polystyrene beads by using the residue-specific and site-specific strategies in uAA incorporation. Interestingly, the ω-transaminase utilized in this study was modified to incorporate the uAA, (4*R*)-fluoroproline [(4*R*)-FP], into its structure by replacing its own proline residues (residue-specific multiple uAA incorporation). It has been reported that the incorporation of (4*R*)-FP enhanced the solubility, and thermal stability (~2-fold increase) of the ω-TA:[(4*R*)-FP] when compared to the unmodified ω-TA. Moreover, the modified ω-TA: [(4*R*)-FP] with enhanced properties was incorporated with another uAA l-3,4-dihydroxyphenylalanine (DOPA) at its R23 position (ω-TA:DOPA:[(4*R*)-FP]) and evaluated for immobilization. The incorporation of DOPA enables the enzyme of interest to form a covalent bond between nucleophilic amino groups present on the supporting material with the electrophilic *ortho*-quinone group of DOPA (Ayyadurai et al., 2011). In the immobilization of ω-TA:DOPA:[(4*R*)-FP] onto chitosan, ~96% activity retention (with respect to ω-TA activity) with a 42% immobilization efficiency (amount immobilized enzyme per initial amount of enzyme introduced) were observed, implying that the immobilized ω-TA:DOPA:[(4*R*)-FP] retained the functional active form of the modified enzyme due to the site-specific incorporation of DOPA. In addition to the immobilization of ω-TA:DOPA:[(4*R*)-FP] on chitosan, the immobilization of ω-TA:DOPA:[(4*R*)-FP] on polystyrene beads showed that the immobilized ω-TA:DOPA:[(4*R*)-FP] can still be reused for 10 cycles suggesting that this immobilized enzyme system is suitable for industrial applications.

5.6 Leucine Zippers

The leucine zipper domain (simply leucine zipper) is a structural motif which functions as a dimerization domain as well as its presence can generate adhesion forces between parallel α-helices; these are commonly found in transcription factors (e.g., DNA binding proteins). These α-helices consist of multiple leucine residues at approximately 7-residue intervals, which form an amphipathic alpha helix with a hydrophobic region running along one side as shown in Fig. 5.6. The structure of the leucine zipper was first hypothesized in the late 1980s by Landschulz et al. (1988) and has been further elucidated in the study of the structures of the fos and jun oncogenes, and the GCN4 transcription factor (Kouzarides and Ziff, 1989; Busch and Sassone-Corsi, 1990; Alber, 1992; O'Shea et al., 1993; Lumb and Kim, 1995).

The assembly of the coiled-coil domains is reversible with the changes in its environment (e.g., temperature, pH, and ionic strength) and this reversibility makes the leucine zipper domain an ideal interaction candidate for cross-linking in responsive hydrogel structures (Petka et al., 1998; Wang et al., 1999). The reversibility and specificity of the coiled-coil domains of leucine zippers inspired the development of a wide variety of tools which utilize coiled-coil interactions in the field of protein engineering and the design of this coiled-coil motif designer zippers which can be utilized as probes, tethers (tags), molecule rulers, affinity matrices, or components of new material

(Woolfson, 2005; Gunasekar et al., 2008; Banwell et al., 2009). The discussed attributes of the coiled-coil interactions (especially leucine zippers) are also well suited for site-specific, reversible, non-covalent protein (especially enzymes) immobilization (Zhang et al., 2005) and an image shown in Fig. 5.7 depicts the utilization of designer zippers (K-coil and E-coil) in enzyme immobilization. In addition, the following examples represent a discussion on how leucine zippers (or other designer zippers) are applied in enzyme immobilization.

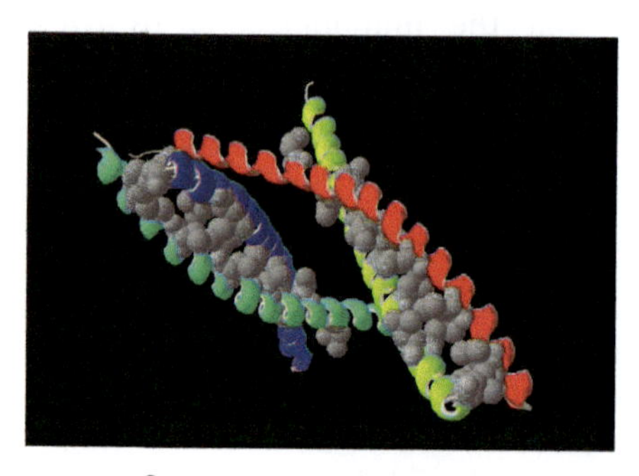

Figure 5.6 A Deepview® (Guex and Peitsch, 1997) generated image of c-fos [PDB ID: 1FOS; (Glover and Harrison, 1995)] and c-jun [PDB ID: 1JUN (Junius et al., 1996)] leucine zippers where leucine residues are shown in gray atoms.

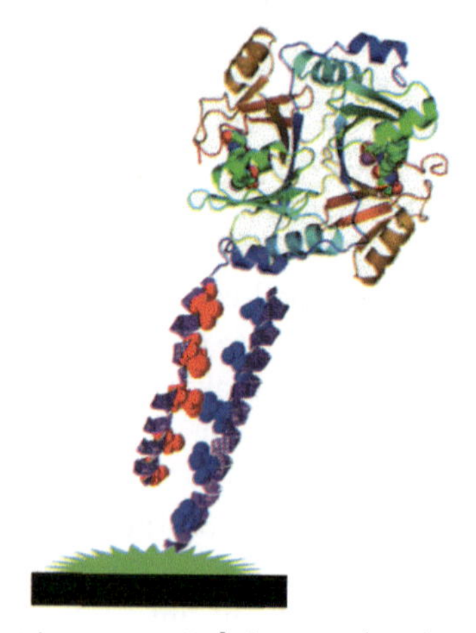

Figure 5.7 A computer generated image showing the immobilization of an enzyme using designer zippers. The K-coil tagged enzymes (K residues represented by red molecules) interacting with the E-coil (E residues represented by blue molecules) functionalized surface.

The research group of Banta et al., developed a general strategy of utilizing α-helical leucine zipper appendages as protein cross-linkers to create functional hydrogels (Wheeldon et al., 2007). This strategy involves the genetic fusion of these helices to both the N and C termini of proteins which enables them to self-assemble into cross-linked aggregates (e.g., hydrogels). Also, the application of this strategy to modify enzymes such as laccase (polyphenol oxidase) (Wheeldon et al., 2008), alcohol dehydrogenase (Wheeldon et al., 2009), and organophosphorus hydrolase (Lu et al., 2010) led to the creation of enzymatic hydrogels which can be employed as a biofuel cell cathode and anode, and an environmental organophosphorus biosensor, respectively. Recently, a cascade of dehydrogenase enzymes, namely alcohol dehydrogenase (responsible for methanol to formaldehyde oxidation), aldehyde dehydrogenase (responsible for formaldehyde to formate oxidation), and the dimeric formate dehydrogenase (responsible for formate to CO_2 oxidation), have been cross-linked to form a hydrogel that can produce a synthetic metabolic pathway for direct methanol to CO_2 oxidation (Kim et al., 2013). These enzymes that were modified via protein engineering cross-linking strategy enabled them to self-assemble into enzymatic hydrogels while either retaining or enhancing their catalytic performance in the 3D-structure of the hydrogel. As to why the enzymatic hydrogels have their activities retained or enhanced, a possible explanation would be that the water present in the hydrogel structure takes part in the transport of the substrate. In addition, the 3D-hydrogel structure can also be modified by controlling the cross-linking distance of the enzyme via its α-helix appendages, which creates an increased local enzyme concentration in the hydrogel.

Steinmann et al. (2010) have illustrated a method for intracellular production of enzyme-decorated particles which depends on the simultaneous synthesis of an insoluble protein inclusion body (also discussed in Section 5.7.3) and an enzyme of interest where the enzyme of interest is tightly attached to the particle surface in multiple copies. In this study, the microorganism of interest, *Escherichia coli*, was engineered to express polyhydroxybutyrate synthase PhaC from *Cupriavidus necator*, tagged with a designed negatively charged α-helical coil, Ecoil, (glutamic acid–rich coil) at its N-terminus. This forms inclusion bodies upon high-level expression. Conversely, this *E. coli* was also designed to coexpress galactose oxidase (GOase) from *Fusarium* spp. tagged with a designed positively charged coil, Kcoil (lysine rich coil) at its C-terminus. The enzyme is directly immobilized on the inclusion bodies via the coil-coil interactions. These coil-coil interactions are held together by both the hydrophobic interactions of a leucine zipper motif and by electrostatic interaction of the lysine and glutamate residues of the helical backbone (Chao et al., 1998). Moreover, the designed coils used in enzyme immobilization contain cysteine residues at the C-terminus of each coil which allows the formation of an intermolecular disulfide bond in an environment where oxidizing conditions prevail (Zhou et al., 1993). It has been observed that there was a slight reduction of cellular GOase activity upon the coexpression of the PhaC inclusion bodies which implies that the enzyme is produced in its active form in the presence of the inclusion bodies. Upon purification of the functional inclusion bodies (immobilized GOase) it was also found that the immobilized GOase retained 65% of the free GOase activity. The loading capacity of the proteinaceous inclusion bodies used in this study for GOase exceeds 200 mg (wet weight) of inclusion body per gram. Thus, the formation of the particle-immobilized enzymes may be limited by the expression yield of the enzyme moiety itself rather than by the binding capacity of the inclusion bodies. This method shows an in vivo enzyme production and immobilization approach which can be a cost-efficient alternative to whole-cell biotransformation.

The preceding sections discuss on different strategies of site-specific immobilization of enzymes on supporting materials and their attributes, while the succeeding Sections 5.7 and 5.8 on the other hand, will tackle on discussing in general the in vivo approach on enzyme-scaffold production, and immobilization. The advantage of the in vivo enzyme-scaffold production and immobilization approach is that it eliminates the need of a supporting material along with the functionalization steps required for the conjugation of the recombinant enzyme to its surface.

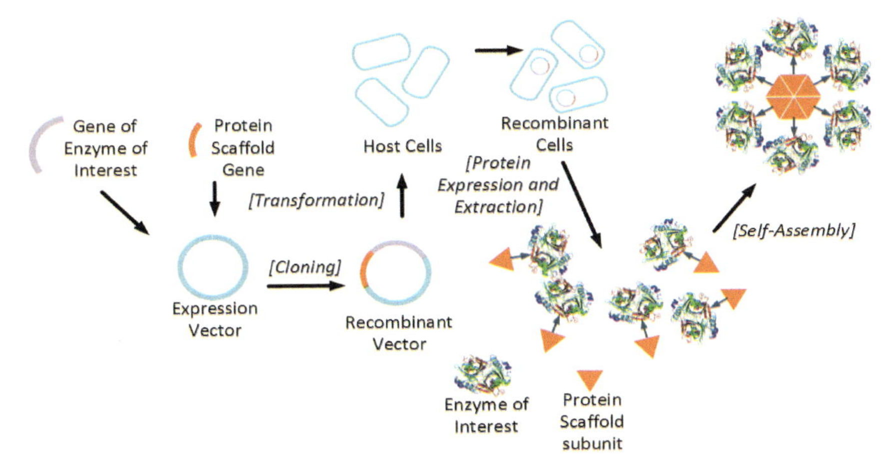

Figure 5.8 A general approach in enzyme immobilization on self-assembling protein scaffolds.

5.7 Self-Assembling Protein Scaffolds

A variety of proteins found in nature have an innate attribute to self-assemble specifically into a diverse range of 2D or 3D structures or even nanostructures via their different structural protein motifs. This self-assembly attribute of these types of proteins makes them attractive scaffolds that can be conjugated with enzymes to create a functional nanostructure with catalytic properties. These self-assembling enzymatic nanostructures can provide an ease in downstream processing especially the separation and purification steps. The self-assembling protein scaffolds covered in this section are S-layer proteins, stable boiling protein 1 (SBP1), polyhydroxyalkanoate (PHA) systems, oleosins (artificial oil bodies), and virus-like particles (VLPs).

5.7.1 S-Layer Proteins

Crystalline bacterial cell surface layer proteins (a.k.a. S-layer proteins as coined by Sleytr (1976)), are found in every taxonomic group of walled eubacteria and represent an almost universal feature in archaebacteria. Typically, S-layers comprise around 10–15% of the total protein found in the cell (Pum et al., 1991;

Messner and Sleytr, 1992; Sleytr et al., 1993). These S-layer proteins function as framework for determining and maintaining shape of the cell, protective coating with antifouling properties, molecular sieves, promoters for cell adhesion and surface recognition, and as an adhesion zone for the cell's exoenzymes (Sleytr et al., 2007). The lattice structures of S-layers vary with the species of the microorganism; the most generally exhibited structures are the oblique (p1, p2), square (p4), or the hexagonal space group symmetry. S-layers also exhibit the capability to assemble into 2D arrays both in vivo and in vitro and this self-assembly is one of the key properties which can be exploited for nanobiotechnological applications especially in the field of biocatalysis. The research group of Sleytr et al., has compiled vast information on S-layers and their biotechnological applications and a more detailed discussion on this matter is found in this cited review article (Sleytr et al., 2014). In this sub-section, several examples are presented along with a brief discussion regarding the utilization of S-layer proteins (SLP) in engineering self-assembling biocatalysts.

The enzymes invertase (an enzyme widely used in the food industry for breaking down sucrose into simple sugars), and glucose oxidase (GOX, responsible for breaking down glucose into hydrogen peroxide and D-glucono-δ-lactone, an important enzyme used in a lot of biotechnological applications, e.g., biosensors), were fused with hexagonal latticed S-layer proteins from *Bacillus stearothermophilus* (PV72) and *Thermoanaerobacter thermohydrosulfuricus* (L111-69), and then formed into a SLP microparticles in solution as demonstrated by Sára et al. (1993). On these S-layer lattices around 2–3 enzyme molecules (corresponding to 1000 μg invertase and 650 μg GOX per mg SLP) were immobilized per hexametric unit cell of SLP indicating that the immobilized enzymes formed a monolayer of enzymes on the surface of the S-layer lattice. The corresponding relative enzymatic activities of the immobilized invertase and GOX were reported to be 70% and 35%, respectively. This low relative enzymatic activity of GOX was due to the fact that its size was small enough for being immobilized entirely inside the pores or having some part of their structure penetrated into the pore openings of the S-layers. Because of this shortcoming, GOX was then immobilized onto SLP in the presence of spacers such as a 4-amino butyric acid or a 6-amino caproic acid, and an increase in relative enzyme activity (~1.7 fold) was observed.

In a following study conducted by Küpcü et al. (1995), small enzymes such as β-glucosidase (BGL) were fused onto hexagonal latticed SLP from *T. thermohydrosulfuricus* (L111-69) and formed into SLP microparticles in solution. It was reported that the immobilized BGL exhibited 16% of its free enzyme activity after direct fusion to the SLP and a tenfold increase in activity (~160% of the free enzyme) was observed when spacers were introduced into the SLP lattice which prevented the multipoint attachment and enzyme contact with the SLP pores.

In another study by Duncan et al. (2005), the enzyme β-1,4-glycanase (Cex), a kind of cellulase-xyalanase which degrades cellulose or xylan to its monosaccharide form, was fused and formed into functional microparticles via self-assembly with the SLP from *Caulobacter crescentus* RsaA having a hexagonal lattice type. Although, the microparticles did not exhibit a high enzymatic activities when compared to the free enzyme, it has been demonstrated that these SLP microparticles containing the fusion RsaA and Cex can be rapidly purified to 90–99% purity by simple coarse filtration; it can also be separated easily from the end product after the reaction, and lastly, it maintains 50% of its initial activity after being recycled for 10 times.

Schaffer et al. (2007) have evaluated different types of biocatalyst assembly exhibited by the fusion of the enzyme glucose-1-phosphate thymidylyltransferase (RmlA, a type of transferase found in the sugar metabolic pathway), with the oblique-latticed SLP of *Geobacillus stearothermophilus NRS2004/3a* (rSgsE) via self-assembly into soluble monomeric form, self-assembly in aqueous solution, and recrystallization on negatively charged liposomes. Of these three configurations, the monomeric assembly of RmlA-rSgsE fusion exhibited the highest relative enzymatic activity of 99.7% while 3% and 65.4% were found for the self-assembly in aqueous solution and liposomes, respectively. These results imply that the preparation procedure for the monomeric assembly did not impose conformational changes to the active site of RmlA, unlike the assembly in aqueous solution where multiple-layer formation introduces juxtaposition among the RmlA units which explains its significant activity reduction. For the assembly of RmlA-rSgsE on the liposome surface, the decrease in enzymatic activity was attributed to the fact that the active sites of RmlA were misoriented on the surface as a consequence of double or multi-layer formation. Overall, S-layer fusions conferred an increased stability (in terms of storage) without loss of enzymatic activity for the monomeric assembly, and reusability of the biocatalysts.

Tschiggerl et al. (2008) have exploited the self-assembly system via recrystallization of square latticed S-layer proteins of *Bacillus sphaericus CCM2177* (SbpA) onto different materials (silicon wafers, glass slides and different types of polymer membranes) for the site-directed immobilization of the cellulase, laminarinase (LamA). The recrystallization of the LamA-SbpA fusion on planar surfaces of different material types led to the formation (dependent on the calcium concentration of the medium) of an orderly monomolecular array which then exhibited relatively high enzymatic activities. This S-layer fusion approach and its recrystallization assembly strategy can then be applied to supports with low binding capacity for an enzyme because regardless of the type of support, the recrystallization of the S-layer fusions would still assemble into ordered monolayer lattices displaying the catalytic site of the enzyme. Additionally, this approach provides enhanced operational stability along with enhanced storage and temperature resistance.

5.7.2 Boiling-Stable Protein 1

The boiling-stable protein 1 (SP1) is a homo-oligomeric self-assembling protein (~148 kDa) with a ring-like structure (having a diameter of 11 nm and an internal cavity of 3 nm) around a pseudo six-fold axis composed of 12 subunits (dodecameric) firmly bound with each other (Wang et al., 2006). Further, the SP1 rings even tend to self-assemble themselves into a nanotube-like structure (Wang et al., 2003). This protein was first isolated from aspen trees (*Populus tremula*) and it exhibited a high resistance towards extreme environmental conditions (e.g., presence of proteases, detergents, and organic solvents, varying pH's, and even boiling temperatures of 107°C) (Wang et al., 2002; Wang et al., 2003). A 3D model of this protein is shown in Fig. 5.9. The self-assembly and stability attribute of the SP1 makes it a good choice for an enzyme scaffold.

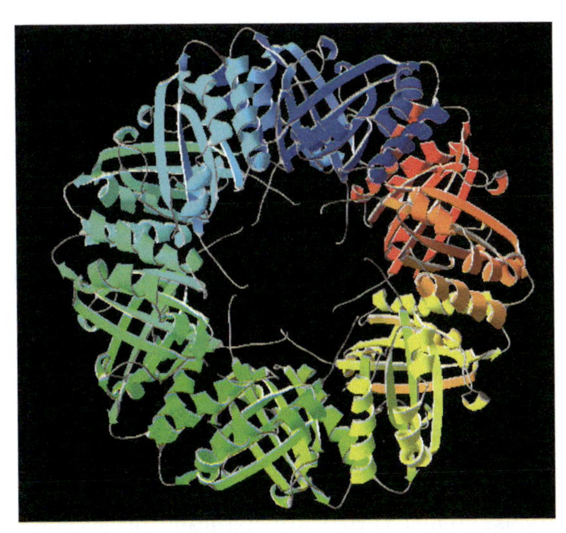

Figure 5.9 A Deepview® (Guex and Peitsch, 1997) generated image of the SP1 protein (PDB ID: 1TR0; Dgany et al., 2004).

The research group of Shoseyov et al., have utilized the SP1 as enzyme scaffolds for glucose oxidase (GOx) (Heyman et al., 2007b), and the cellulases such as an endoglucanase Cel5A (Heyman et al., 2007a), an exoglucanase Cel6B (Morais et al., 2010) both from *Thermobifida fusca* and have demonstrated that the immobilized enzymes generally exhibited higher retained activity (due to the proximity and the cooperative effects of the enzymes displayed on the SP1 scaffold via cohesin–dockerin interaction) and enhanced operational stability (e.g., thermal). Moreover, the incorporated enzymes that were bound on the SP1 subunit nearly correspond to the theoretical 1 unit enzyme per 1 SP1 subunit binding capacity (or almost a dozen enzyme units can be displayed to on the dodecameric SP1). In terms of the downstream processing (e.g., separation and purification), both the enzyme and the SP1 scaffold are expressed as insoluble protein aggregates either in their free or immobilized form thus, making them easier to handle during their preparation and their use in a catalytic reaction.

Interestingly, although the SP1 is a good choice for an enzyme scaffold, it can also be modified to function as an artificial enzyme that can perform a catalysis reaction at extremely high temperatures because of its inherent stability. Miao et al.

(2014) demonstrated the incorporation of selenocysteine (Sec), an amino acid present in the active center of the enzyme glutathione peroxidase (GPx), to the structure of the SP1 monomer surface by means of computational design and genetic engineering. The modified SP1 monomer displaying an artificial GPx active site on its surface self-assembled into a ring-shaped SP1 with homododecamer catalytic selenium centers (12 active sites) exhibiting an enzymatic activity that is ~1.30 times higher than that of the GPx from human plasma. Furthermore, the modified SP1 exhibits high GPx activity and thermostability at a broad range of operation temperatures (20–85°C).

5.7.3 Inclusion Body Display (PHA Systems)

Polyhydroxyalkanoates (PHA) are naturally occurring polyesters widely used in the industry as biopolymers to replace petrochemical-based polymers. These PHAs are composed of *R*-3-hydroxy fatty acids that are produced and accumulated by many bacteria as water-insoluble inclusion bodies (sometimes referred as PHA beads) under excess carbon source and nitrogen-limited conditions (Steinbüchel and Hein, 2001). The PHAs stored in bacteria function as an energy storage unit in order for the cells to survive nutrient-deprived conditions. The formation and accumulation of short-chain length PHAs (e.g., polyhydroxybutyrate) from sugars in cells are facilitated by three key enzymes in the sugar catabolism pathway. These three enzymes along with a description of their function are (i) β-ketothiolase (PhaA) which is responsible for the condensation of acetyl-CoA (produced from the conversion of sugar via the glycolysis and pyruvate dehydrogenase oxidation pathway) to acetoacetyl-CoA, (ii) acetoacetyl-CoA reductase (PhaB) reduces the acetoacetyl-CoA to the PHA precursor molecule *R*-3-hydroxybutyryl-CoA), and lastly, (iii) PHA synthase (PhaC) which polymerizes the precursor molecule to PHA in the form of granules (Rehm, 2003; Rehm, 2006; Grage et al., 2009). The industrial production of PHAs is mainly done via microbial fermentation and the comprehensive reviews by Gumel et al. (2013) and Tan et al. (2014) provide a good discussion on the recent advances in the production and recovery of PHAs.

Since it has been discussed in the previous paragraph that PHAs are mostly produced intracellularly as inclusion bodies (or insoluble granules) facilitated by the three key enzymes (PhaA, PhaB, and PhaC), it has been observed from the PHA biosynthesis pathway (from the PHA inclusion model organism *Ralstonia eutropha*) expressed in recombinant systems such as *Escherichia coli* (Gram negative), *Corynebacterium glutamicum* (Gram positive), and *Lactococcus lactis* (Gram positive), that out of the three key enzymes produced in the system, only PhaC remains covalently attached to the surface of the growing PHA granule (Hezayen et al., 2002; Jo et al., 2006; Valappil et al., 2007; Mifune et al., 2009). As the PhaC is attached to the surface of the PHA granule, it has become a suitable tag target for conjugating proteins especially enzymes on the surface of the PHA granules to produce functional granules in vivo. This in vivo approach when applied in enzyme production and immobilization offers the advantage of performing intracellularly both production and immobilization simultaneously which then provides the ease of downstream processing (e.g., separation and purification) of the functionalized PHA particles.

Rehm and coworkers have exploited the use of N or C terminal PhaC fusions to immobilize in vivo by means of an engineered *E. coli* PHA-producing system (a genetically modified *E. coli* which contains the PHA biosynthesis pathway as described in the previous paragraph), a variety of enzymes such as β-galactosidase (Peters and Rehm, 2006), α-amylase (Rasiah and Rehm, 2009), organophosphorus hydrolase (Blatchford et al., 2012), and *N*-ethylamide reductase (Robins et al., 2013) on PHA granules. Moreover, they have also demonstrated the immobilization of the two enzymes, *N*-acetylglucosamine 2-epimerase and *N*-acetylneuraminic acid aldolase, fused to both the N and C terminal of PhaC onto PHA granules which allowed them to create a functional PHA granule containing a multiple enzyme complex that can catalyze the synthesis of *N*-acetylneuraminic acid (Hooks et al., 2013). In terms of the performance of the immobilized enzymes using the PhaC fusion system, the synthesized functional PHA beads generally exhibited a moderate relative enzymatic activity (although for the case of the immobilized OPH, its activity increased ~1.6-fold with respect to the free enzyme), an increased substrate affinity when compared to the free enzyme, and increased stability.

The PHA bead display system exploited by Rehm et al. is versatile to simultaneous functional fusions on the N and C terminals of PhaC, even with two different enzymes, or any protein functionality of choice thus providing a bifunctional display system on PHA granules which could be used for a lot of biotechnological applications (Hooks et al., 2014).

5.7.4 Oleosins (Enzyme Immobilization on Artificial Oil Bodies)

Oil bodies, also referred to as lipid bodies, lipid droplets, oleosomes, and oil globules, are intracellular organelles that contain energy storage lipids which typically found in large quantities in oleaginous biomass (Chapman et al., 2012). These oil bodies generally have a spherical structure composed of a triacylglyceride (TAG) core surrounded by a phospholipid monolayer embedded in unique integral proteins such as oleosins, caleosins and steroleosins (Tzen, 2011; Murphy, 2012). The integral proteins serve as the membrane proteins which provide the structural integrity and stability of the oil bodies. A major component of the said integral proteins is oleosin (which comprises 80–90% of the integral proteins) and it is considered as the protein responsible for the integrity and stability of the oil bodies because of the steric hindrance and electronegative repulsion exhibited by its structure (Tzen and Huang, 1992; Tzen et al., 1992; Tzen et al., 2003). Oleosins are small proteins with a molecular mass of 15–24 kDa which have three structural domains consisting of the N and C terminal ampiphatic domains (located outside the oil body), and a central hydrophobic domain (which anchors to the TAG core of the oil body) (Huang, 1992; Murphy, 1993; Loer and Herman, 1993). A schematic illustration of an oil body and its components is shown in Fig. 5.10.

Mixing oleosins with TAGs and phospholipids in vitro in the same proportions as in native oil bodies can form artificial oil bodies (AOB) by virtue of self-assembly (Tai et al., 2002). Because of this unique self-assembly mechanism to form AOBs there have been quite a number of biotechnological applications that were developed in the past decade. These biotechnological applications of artificial oil bodies are extensively discussed in the reviews of Bhatla et al. (2010), Roberts et al. (2008), and

Tzen (2012). This sub-section however focuses on the application of the oleosin-fusions to immobilize enzymes via self-assembly on artificial oil bodies as presented in the following paragraphs.

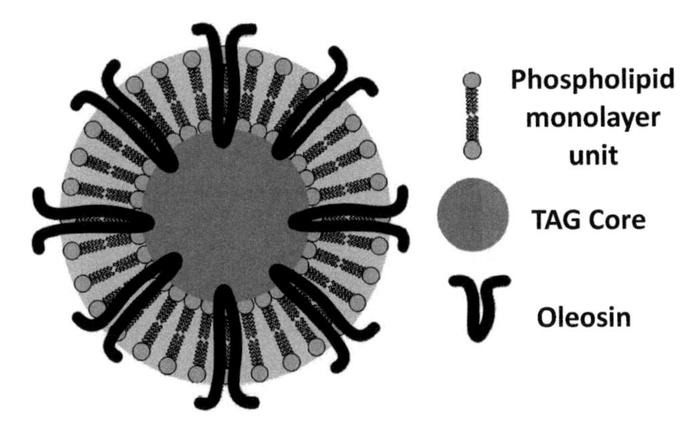

Figure 5.10 A schematic illustration of an oil body and its components.

Liu et al. (1997) first demonstrated the in vivo production and immobilization of *Neocallimastix patriciarum* xylanase (XynC) onto canola plant (*Brassica napus*) seed oil-bodies by fusing the XynC to the C-terminal of oleosin, which was then expressed in the transgenic plant. The obtained functional oil-bodies (which contained the immobilized XynC) exhibited retained enzymatic activity and characteristics similar to the free XynC produced via microbial fermentation. Moreover, the xylanase immobilized on the oil bodies were easily separated from the reaction mixture via flotation centrifugation and recycling for 5 times showed no loss of enzymatic activity; instead an increase of 50% in enzyme activity was observed during the second round of recycling and this increase was attributed to optimal enzyme folding that occurred during the incubation period or upon the loss of an inhibitory component by virtue of dilution.

Tzen and coworkers have immobilized in vitro a wide variety of enzymes such as D-hydantoinase (an enzyme widely used in the pharmaceutical industry) (Chiang et al., 2006), xylanase from *N. patriciarum* (Hung et al., 2008), Endo-1,4-β-D-glucanase from *Fibrobacter succinogenes* (Zeng et al., 2009), and D-psicose 3-epimerase (responsible for conversion of D-fructose to D-psicose) (Tseng et al., 2014) on AOBs. This in vitro approach of immobilizing enzymes on AOBs was done by expressing the oleosin-fused enzymes as inclusion bodies in *Escherichia coli*, adding plant oil to the *E. coli* containing the insoluble fusion protein, followed by sonicating the mixture to form the functional AOBs via self-assembly. The immobilized enzymes on AOBs generally exhibited enhanced operational stability (in terms of pH and temperature) under their optimal conditions, and reusability.

Chiang et al. (2013a), have successfully performed a multi-enzyme display on AOBs using the in vitro immobilization approach described by Chiang et al. (2006). The heterologous endoglucanase (celA), cellobiohydrolase (celK), and β-glucosidase (BGL) were individually fused with oleosin and then expressed in *E. coli* and recovered. After recovery, the insoluble proteins, containing the enzyme-oleosin fusion were mixed together along with the plant oil, and sonicated to form an artificial cellulosome (as discussed in detail in Section 5.8) on the surface of the AOBs. The cellulases co-immobilized on AOBs exhibited cellulose hydrolyzing activity (rate of release of glucose during the enzymatic reaction reported as 1 µmol of reducing sugar equivalent per min.) and can retain 80% of its initial activity after four consecutive reaction cycles. The advantage of using AOBs in the immobilization of cellulases is that they can be easily recovered and reconstituted for repeated use. Especially, in the case of a solid substrate (e.g., cellulosic biomass) for the reaction, the enzymes immobilized from AOB can be easily separated from the solid phase by flotation, thus reducing the complexity of the downstream processing.

5.7.5 Virus-Like Particles

Viruses are considered as a class of biological supramolecular entities (infectious agents) that cannot metabolize on their own but are capable of sabotaging the machinery of the cell for their own replication; because of these qualities, they have been described as organisms at the border of living matter. In their most basic form, viruses are made up of virions (virus particles) consisting the genetic material (either DNA or RNA) packaged inside the capsid (known as the protein coat that protects the genetic material) (Chapman and Liljas, 2003; Koonin et al., 2006; Aniagyei et al., 2008). When a virus infects a cell, its genetic material is replicated and translated into the protein subunits which self-assemble into capsids. This self-assembly process of capsids has been exploited to create non-infectious virus-like particles (VLPs) that resemble viruses but lack the viral genetic material necessary for replication. VLPs are nanometer-sized virus capsid-derived symmetric protein shells (containing a hollow space interior) with robust chemical and physical properties that can be engineered for a wide variety of applications such as vaccines, drug delivery vehicles, imaging agents, nanoelectronics, and specially biocatalysis (Smith et al., 2013a). Specifically in the field of biocatalysis, VLPs are good candidates for enzyme nanocarriers due to their self-assembly and reproduction properties. As enzyme nanocarriers, VLPs can be chemically or genetically modified such that enzymes are either placed/ conjugated inside or outside the VLPs (Cardinale et al., 2012). The different types of VLPs that have been used as enzyme nanocarriers are shown in Fig. 5.11. This sub-section presents the different examples of enzymes immobilized inside or outside the VLPs.

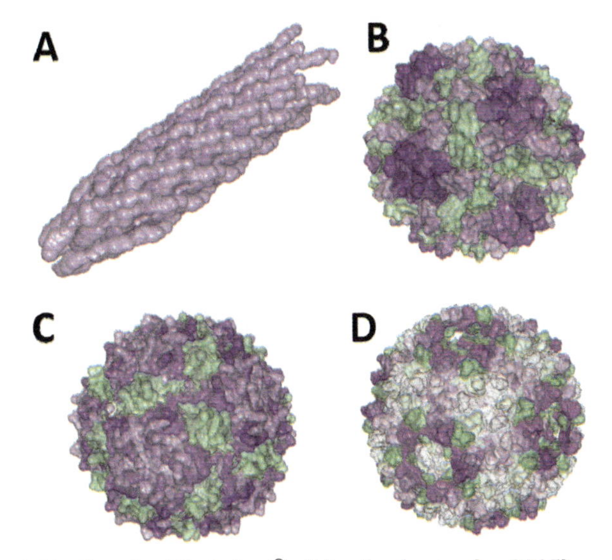

Figure 5.11 Protein Workshop® (Moreland et al., 2005) generated images of VLPs derived from: fd Bacteriophage [PDB ID: 2C0W; (Marvin et al., 2006)] [A], CCMV [PDB ID:1CWP; (Speir et al., 1995)] [B], $Q\beta$ Bacteriophage [PDB ID: 1QBE; (Golmohammadi et al., 1996)] [C], and P22 Bacteriophage [PDB ID: 3IYH; (Parent et al., 2010)] [D].

5.7.5.1 Phage display for enzyme improvement and selection

VLPs (commonly from bacteriophages) genetically fused with proteins are popularly used in phage display for the study of protein-protein, protein-peptide, and protein-DNA interactions (Smith, 1985). In this technique, the gene encoding the protein of interest is fused with a phage capsid gene thus, displaying the protein of interest outside the capsid while packing the gene inside and the displayed protein of interest can be screened in a process called biopanning against other proteins, peptides, or DNA sequences to detect interactions (Smith and Petrenko, 1997). In another perspective, phage display is also widely applied to select enzymes with improved properties (e.g., stability, activity, or substrate specificity). When both the enzyme and the substrate are displayed on the surface of the phage particle with close proximity in such a way that the substrate can minimize diffusion problems to the enzyme active site, the enzymatic reaction on the surface of the phage particle can be exploited to facilitate the isolation of different mutants of that enzyme from the bulk (Fernandez-Gacio et al., 2003).

McCafferty et al. (1991) demonstrated that fusing an enzyme, alkaline phosphatase (AP), on the surface of a minor coat protein of fd bacteriophage retained its enzymatic activity. The expressed AP that was fused to the N-terminus of the gene 3 (the minor coat protein of fd bacteriophage) formed an enzymatically active phage particle with similar kinetic properties like the free AP implying that the fusion or the immobilization of enzymes on phage particles does not affect the activity of the immobilized enzyme. These phage particles expressing AP on their surface were purified by binding the phage particles to an immobilized AP inhibitor system (arsenate–Sepharose) and then eluted with an inorganic phosphate; this demonstrates that the functional phage particles can be co-purified along with the DNA encoding it by using an affinity chromatography approach.

In a similar approach, Belien et al. (2005) showed that multiple enzymes displayed on the surface of phage particles can still exhibit their functionality after immobilization. Two family 11 endoxylanases such as endo-1,4-xylanase I from *Aspergillus niger* (ExlA) and endo-1,4-xylanase A from *Bacillus subtilis* (XynA) were fused to a gene 3 minor coat protein of M13 bacteriophage. The phage particles displaying both enzymes exhibited a clear hydrolytic activity towards an insoluble, chromogenic arabinoxylan substrate and could be recovered by affinity binding with immobilized endoxylanase inhibitors such as *Tritium aestivum* xylanase inhibitor (TAXI) and xylanase inhibitor protein (XIP).

Shukla and Krag (2010) developed a phage display system for identifying peptides that recognize specific receptors on the cancer cell surface by fusing a random peptide library upstream of the β-lactamase enzyme (which serves as a reporter in tracking peptides that bind to cells) to be displayed on the surface of M13 bacteriophage (pIII). The advantage of using this peptide-β-lactamase functionalized phage particle is that the cost and tedious process of normal phage-ELISA used for selection can be avoided just by direct measurement of β-lactamase activity towards reporter substrates (e.g., fluorescent, or chromogenic).

Sunbul et al. (2011) utilized the Sfp phosphopantetheinyl transferase (see Section 5.4.3)-catalyzed conjugation in displaying both the substrate and the enzyme to be selected on the same phage particle. In this method, the enzyme of interest, biotin ligase (BirA), and the ybbR peptide tag that were fused with the pIII of M13 bacteriophage allowed the display of the enzyme near the ybbR on the same phage particle. The CoA-functionalized substrate was then coupled in vitro to the ybbR tag expressed on the phage particle by the Sfp. This substrate attached phage prepared via Sfp-catalyzed conjugation was shown to be a useful screening platform to select enzymes based on the enzymatic activities in a phage library.

The previous paragraphs describe the utilization of VLPs in phage display to establish an enzyme selection and improvement scheme, and an affinity binding approach for purification of immobilized enzymes with retained catalytic activities on the surface of phage particles. The following examples deal with the utilization of VLPs as nanomaterials, serving as enzyme scaffolds (enzymes displayed on the surface of VLPs) or nanocages (the enzymes expressed inside the VLPs) in the construction of self-assembled catalytic systems that can be produced either in vivo or in vitro.

5.7.5.2 VLPs as enzyme scaffolds

A patented process developed by Straffon et al. (2006) describes the utilization of two immobilized enzymes, α-amylase from

Bacillus licheniformis and xylanase A from *Bacillus halodurans*, on the surface of the T7 bacteriophage capsid in a high-temperature starch liquefaction process or in de-inking mixed office waste. Both genes encoding the said enzymes were fused in tandem downstream the 10B gene of the T7 capsid (capsid–E1–E2) and produced phage particles with both enzymes displayed on their surface upon expression. The phage particles displaying two enzymes exhibit enzymatic activities toward the hydrolysis of starch and xylan. Further, the bacterial culture producing the functional phage particles can be directly incorporated in the process thus avoiding the costly step of enzyme purification.

Collins and coworkers have patented a process where engineered bacteriophages displaying at least one biofilm degrading enzyme could be utilized to complement the use of antibiotics as treatment of diseases caused by pathogenic bacteria with increased resistance due to the biofilms secreted by bacteria. These engineered bacteriophages can render a "two-pronged attack strategy" to the pathogenic bacteria and their biofilms by (i) degrading the extracellular matrix of the biofilm via the enzymatic activity of the enzyme displayed on the phage which exposes the bacteria to phage infection, and (ii) rapid phage replication with subsequent bacterial lysis once the bacteria are infected with the phage. The biofilm degrading enzymes that have been displayed on the T7 phage are endoglucanase (celA), which degrades the polysaccharide (e.g., cellulose) component of the biofilm (Collins and Lu, 2012), and dispersin B (DspB), which hydrolyzes the β-1,6-*N*-acetyl-D-glucosamine, an adhesion molecule needed for biofilm formation and integrity (Collins et al., 2012). In addition, the display of multiple enzymes with complementary biofilm degrading functions on the phage particle surface can render extended biofilm lysis (Sutherland et al., 2004).

Carette et al. (2007) constructed a self-assembled catalytic system based on a plant virus where its capsid is genetically modified to display lipase B (CalB) from *Candida antarctica*. The enzyme support utilized in this study was the VLP from Potato Virus X (PVX) which has a flexible rodlike structure, similar to that of the fd bacteriophage VLP (refer to Fig. 5.8A), with a length of 500 nm and diameter of 13 nm (Parker et al., 2002). The CalB was genetically introduced upstream of the

PVX capsid protein monomer (CPm; however, the CalB-CPm fusion is sterically hampered during self-assembly upon expression because CalB has a larger size compared to the CPm. To overcome the steric hindrance when expressing the enzyme on the PVX surface, a genetic sequence known as 2A (derived from the Foot and Mouth Disease Virus, FMDV) was genetically introduced between the CalB and CPm producing a CalB-2A-CPm fusion (Cruz et al., 1996; Donnelly et al., 2001; Uhde-Holzem et al., 2010). The obtained CalB functionalized PVX virus particle with a 1:3 CalB-2A-CPm to CPm ratio exhibited a reduced catalytic activity (about 45 times reduction) when compared to the free CalB. Further, the yield of the decorated PVX particles was low due to the rapid necrosis of the host plant *Nicotiana benthamiana* (Cardinale et al., 2012).

5.7.5.3 VLPs as enzyme nanocages

Comellas-Aragones et al. (2007) first demonstrated the in vitro encapsulation of the enzyme horseradish peroxidase (HRP) in a VLP. The VLP utilized in this study was the cowpea chlorotic mottle virus (CCMV) which has an icosahedral shape with an outer diameter of 28 nm and a well-defined inner cavity with diameter of 18 nm (refer to Fig. 5.11B). The CCMV VLPs have a unique reversible pH-dependent assembly/disassembly property which provides control for the mechanism of containing or releasing an entrapped material (Verduin, 1974; Verduin, 1978; Speir et al., 1995). The inclusion of HRP in the CCMV capsid involves the disassembly of CCMV capsid into dimers at pH 7.5 releasing the HRP into the solution containing the CCMV dimers; the CCMV capsid is reassembled again by decreasing the pH to 5.0. However, this simple reconstruction method by varying the pH is a statistical process and could not provide the precise control of the number of enzymes encapsulated per CCMV particle thus, there is a distribution of enzymes inside and outside the VLP.

To address the shortcoming of the method described by Comellas-Aragones et al. (2007), Minten et al. (2011) applied a direct assembly approach based on a heterodimeric coiled-coil linker (K and E coil designer zipper as discussed in Section 5.6). With this approach, a K-coil was attached downstream the CCMV capsid and on the other hand, an E-coil was attached to the enzyme lipase B from *Pseudozyma antarctica* (PalB). Directed assembly was done by adding the E-coiled PalB to the K-coiled CCMV dimer (disassembled at pH 7.5) in solution to form the CCMV dimer–PalB conjugate via the K and E coil interaction; lowering the pH of the solution to 5.0 reassembles the CCMV particles containing PalB. Using this approach, the number of PalB encapsulated in the VLP can be controlled by varying the PalB–E-coil and capsid–K-coil proportions. Interestingly, the encapsulation of PalB in CCMV exhibited an increased enzymatic activity when compared to the free PalB, and the increased number of PalB in the VLP did not increase the overall reaction rate. These effects were explained by the local concentration (confinement concentration) of the enzyme within the VLPs combined with the increased collision of the substrate and enzyme due to spatial confinement. It was reported for this system that an encapsulating of 1.3 PalB enzymes per CCMV particle (with an equivalent confinement molarity of 1 mM which is 25 times higher than the bulk enzyme concentration) led to a 4.55-fold increase in enzymatic activity.

Fiedler et al. (2010) developed an in vivo method of encapsulation of enzymes inside VLPs via RNA-directed encapsidation using a two-plasmid system to express the tagged VLP capsids and the tagged enzyme. In this method, the VLPs from Q^{β} bacteriophage, which forms an icosahedral structure similar to that of the CCMV, were utilized and two binding domains were introduced into the capsid (CP) mRNA of Q^{β} bacteriophage, carried on a first plasmid. An RNA aptamer (α-Rev) which binds to an arginine-rich peptide (Rev) was inserted upstream the ribosome binding site of the CP. The other binding domain, the sequence of a Q^{β} packaging hairpin (Q^{β}hp; this hairpin interacts with the interior-facing residues of the CP) was fused downstream of the CP stop codon and the resulting first plasmid contains the α-Rev–CP–Q^{β}hp. The second plasmid on the other hand, contains the Rev-tagged enzyme. Upon expression of these two plasmids in *E. coli*, the Rev-tagged enzyme will attach to the α-Rev–Q^{β}hp to form an enzyme–Q^{β}hp conjugate and the enzyme–Q^{β}hp conjugate will attach itself into the CP. The enzyme–Q^{β}hp–CP then assembles inside the Q^{β} VLPs with the enzyme of interest encapsulated within. Using this method, two enzymes such as peptidase E and firefly luciferase were packaged into the Q^{β} VLPs and these enzyme-containing Q^{β} VLPs exhibited retained enzymatic activities of 45% and 92% for peptidase E and luciferase, respectively, when compared to their free form. Also, from the kinetic analysis of these enzyme-containing Q^{β} VLPs, it was found out for the case of luciferase that upon encapsulation the substrate specificity (k_{cat}/K_m) was significantly decreased due to the increase in K_m which implied that the substrate encountered diffusivity problems.

The research group of Douglas et al. has engineered an in vivo one-pot heterologous expression-assembly system for the programmed encapsulation of fusion proteins in a high copy number utilizing the capsid and scaffold machinery of the VLPs derived from *Salmonella typhimurium* P22 bacteriophage (O'Neil et al., 2011). The P22 bacteriophage VLPs is composed of a coat protein that assembles into an icosahedral capsid with the aid of a scaffolding protein (SP) which is incorporated on its interior surface via non-covalent association. Moreover, the P22 VLPs undergo irreversible structural changes upon heating that lead to a significant increase of its internal volume and porosity and in turn increases the external accessibility to its hollow interior (Parker et al., 1998; Teschke et al., 2003; Parent et al., 2010). The engineered heterologous expression system contains the SP_{141} gene (truncated form of SP encoding amino acids 141–303 located at the C-terminal region (Parker et al., 1998; Weigele et al., 2005)) fused upstream to the P22 coat protein (CP). The protein of interest is then fused upstream of the SP_{141} to form the protein-SP_{141}–CP fusion, and upon expression and self-assembly, a VLP containing the protein of interest is obtained. Using this expression system, enzymes such as alcohol dehydrogenase (AdhD), β-glucosidase (CelB), and even a multiple enzyme cascade of β-glucosidase (CelB), galactokinase (GALK), and glucokinase (GLUK) which catalyze sequential reactions in sugar metabolism, have been encapsulated inside P22 VLPs (Patterson et al., 2012a; Patterson et al., 2012b; Patterson et al., 2014). For the case of the encapsulated AdhD, the functional VLPs exhibited a very low retained enzymatic activity (around 15% of the free AdhD) due to the high enzymatic loading (~250 enzymes present in the capsid corresponding to a confinement molarity of 7.6 mM) inside the VLPs which caused a crowding effect (Patterson et al., 2012a). However, for the case of CelB encapsulation, the encapsulated CelB retained 94% of the free enzyme activity with an enzyme loading of ~84 enzymes per capsid (corresponding to a confinement molarity of 2.4 mM) which suggests that the quaternary structure of the enzyme was not hampered by the assembly and encapsulation (Patterson et al., 2012b). Lastly, the encapsulation of a multiple enzyme cascade consisting of CelB, GALK, and GLUK in P22 VLPs showed that the arrangement of the enzymes within the enzyme-SP_{141} fusion had a significant effect on the overall enzymatic activity. The kinetic studies have validated the presence of substrate channeling and it was reported that channeling between sequential enzymes is dependent on the inter-enzyme distance and the balance between their kinetic parameters (Patterson et al., 2014).

5.8 Cellulosomes

The cellulosome is a discrete, cellulose binding, multi-enzyme complex found on the cell wall of several anaerobic cellulose-consuming microorganisms (e.g., clostridia, ruminal bacteria) that is responsible for degradation of cellulosic substrates (Lamed et al., 1983; Doi and Kosugi, 2004). The various enzymes in this elegant machinery act synergistically to efficiently hydrolyze the polysaccharide components of lignocellulosic biomass (e.g., cellulose, hemicellulose) (Murashima et al., 2003). The cellulosome is on one hand made up of a fibrillar scaffolding protein called scaffoldin, which contains binding domains known as cohesins that are positioned periodically along the fibrils, and on the other hand, the cellulosomal enzyme subunits which have various functions that invariably contain a cohesin binding domain called a dockerin. In addition, a unique feature found in the cellulosome scaffoldin is another binding domain known as the cellulose binding domain (CBD which is responsible for the attachment/binding of the whole cellulosome machinery to its cellulosic substrate, overcoming the substrate diffusion problems by decreasing the proximity of the substrate and enzyme for them to interact. The dockerin containing enzyme subunits interact via self-assembly with their respective cohesin binding domains to form the cellulosome. The cohesion–dockerin interaction is a high affinity, species specific interaction (analogous to a plug-and-socket assembly) that plays a crucial role in the unique assembly of cellulosomes in different bacterial species. Regarding the intricate nature of the cellulosome, the following reviews by Bayer and coworkers, Doi and Kusugi, Fontes and Gilbert provide a comprehensive discussion on its discovery, function, assembly, up to its significant biotechnological applications (Bayer et al., 2004; Bayer et al., 2008; Smith and Bayer, 2013; Doi and Kosugi, 2004; Fontes and Gilbert, 2010).

Exploiting the cellulosome's nature as a reconfigurable platform for immobilization of multiple enzymes including an enzyme cascade configuration, along with the advancement of molecular biology and conjugation techniques, led to the creation of "designer cellulosomes." These designer cellulosomes have been developed to recreate the natural multi-enzymatic complex nature of the native cellulosomes in fulfilling the long-standing objective of efficient biomass degradation. Furthermore, with the enhanced capabilities of designer cellulosomes in biomass degradation, it has become an essential tool for the consolidated bioprocessing (CBP) strategy (Bayer et al., 1994; Olson et al., 2012; Hyeon et al., 2013; Kricka et al., 2014; Goncalves et al., 2014). This section presents the different designer cellulosomes that have been developed and they have been categorized into two types: (i) cohesion–dockerin based and (ii) artificial cellulosomes.

5.8.1 Cohesin-Dockerin–Based Designer Cellulosomes

This type of designer cellulosome is a biomimicry of the native cellulosome which utilizes the self-assembly of the dockerin tagged enzymes (e.g., cellulases) to a configurable multi-cohesin fused protein scaffold or a scaffoldin fused to the cell wall of the microorganism (e.g., cell surface display of minicellulosomes). The assembly of this type of cellulosome occurs in vitro while the components, such as the enzymes and the configurable scaffoldin are produced in vivo. A schematic representation of the assembly of this type of designer cellulosome is shown in Fig. 5.9.

5.8.1.1 Protein scaffold-based assembly of designer cellulosomes and hemicellulosomes (a.k.a. xylanosomes)

The term "minicellulosomes" is often used to distinguish the designer cellulosomes that contain engineered scaffoldins (a.k.a. miniscaffoldin to be consistent with the minicellulosome) with non-native cohesin and cellulose-binding domains, in contrast with the definition of a scaffoldin (with reference to the native scaffolding protein) as discussed at the beginning of this section. Other researchers even use the term "truncated cellulosome" that is synonymous to the minicellulosome to even give it a clearer distinction with the cellulosome. The examples discussed in this sub-section that use the term chimeric cellulosomes refer to the minicellulosomes. This term is applied in this section to provide a distinction to designer cellulosomes that utilize specifically the miniscaffoldin and designer cellulosomes that utilize other protein scaffolds as presented in some examples (Doi and Kosugi, 2004; Mitsuzawa et al., 2009).

Bayer and coworkers, first demonstrated the production of active cellulosome chimeras (can be considered as minicellulosomes) wherein dockerin-tagged cellulases (CelA and CelF from *Clostridium cellulolyticum*) were assembled onto a chimeric scaffoldin (termed as a miniscaffoldin) designed to contain two divergent cohesin modules (derived from *C. cellulolyticum* and *C. thermocellum*) and a cellulose-binding domain (CBD). The resulting designer cellulosome containing the said cellulases exhibited enhanced synergistic action (around 2 to 3-fold increase in enzymatic activity when compared with free cellulases) when tested for its hydrolytic activity on crystalline cellulose (Fierobe et al., 2001). In a succeeding study, the chimeric scaffoldin was modified to contain an additional cohesin module (derived from *Ruminococcus flavefaciens*) where three dockerin-tagged enzymes: two cellulases (Cel48F and Cel9G from *C. cellulolyticum*) and a hemicellulase (XynZt from *C. thermocellum*) were assembled onto. The resulting trifunctional chimeric cellulosome containing two cellulases and a hemicellulase was tested for its hydrolytic activity on the complex substrate hatch straw (biomass containing cellulose and hemicellulose) and interestingly, the trifunctional cellulosome exhibited up to

4-fold hydrolytic activity when compared to the free enzymes showing the designer cellulosomes' configurability (Fierobe et al., 2005). In addition, the trifunctional designer cellulosomes has been upgraded into a tetra-functional designer cellulosome having different geometries/configurations containing two additional pivotal cellulases (which corresponds to four catalytic domains of 2 Cel48F and Cel9G pairs) along with additional CBDs. These developed multi-functional complexes having different geometries exhibited reduced activity on crystalline cellulose when compared to the conventional designer cellulosomes. This reduced activity was attributed to the increased number of protein-protein interactions which lead to the hindrance in the mobility of the enzyme complex. Nonetheless, the construction of high-order functional cellulosomes possessing different geometries still exhibited higher hydrolytic activities (around 2 to 3 times higher) than free enzymes (Mingardon et al., 2007).

The research group of Doi et al. engineered the host cell *Bacillus subtilis* to express, assemble in vivo, and secrete the minicellulosome of *Clostridium cellulovorans* containing its native scaffoldin protein CbpA (which consists of a CBD, a hydrophilic domain, and a cohesin domain) along with its dockerin-tagged endoglucanase EngB. This system involves co-expression of both the CbpA and EngB genes *from C. cellulovorans* which leads to the assembly of the minicellulosome in vivo and subsequent secretion of the minicellulosome outside the host cell. A yield of 0.2 mg minicellulosomes/L culture was obtained from this system and it was shown that the minicellulosome retained its hydrolytic activity when tested with carboxymethyl cellulose. However, the obtained minicellulosome could not degrade crystalline cellulose. This system for the production of *C. cellulovorans* minicellulosome has an advantage in the aspect of simple downstream processing (Cho et al., 2004). In another approach, two strains of *B. subtilis*, one expressing the gene of the scaffolding protein and the other the dockerin-tagged endoglucanase EngB or xylanase XynB, were developed to secrete the respective minicellulosome components and assemble them in vitro. The co-culture of the strain expressing the CbpA with the strain expressing either EngB or XynB was observed to secrete and assemble a functional minicellulosome containing EngB or XynB in the solution. This cooperative action of the co-culture in the assembly of the secreted components of the microorganisms was termed as "intracellular complementation (Arai et al., 2007)."

The construction of designer cellulosomes that utilize the cohesion–dockerin interaction with other scaffolding proteins aside from the native scaffoldin has also been exploited. The examples presented in Section 5.7.2 by Heyman et al. (2007a) and Morais et al. (2010) led to a discussion on the engineered functional cellulosomes with enhanced hydrolytic properties and thermal stability that were assembled via the attachment of the dockerin-tagged cellulases such as the endo- and exo-glucanases from *Thermobifida fusca* onto the cohesin functionalized on the boiling-stable protein 1 (SP1). Mitsuzawa et al. (2009) exploited a protein which they termed as a rossettasome, specifically a group II chaperonin derived from *Sulfolobus shibatae* that self assembles into a double-ring-like structure comprising 18 subunits in the presence of ATP/Mg^{2+}. Functionalizing a cohesin on a rossettasome subunit can assemble into a thermostable rossettasome with 18 cohesin binding domains. Fusion of four different cohesin domains on rossettasome subunits enabled the attachment of four appropriately dockerin-tagged cellulases (CelF, Cel9K, Cel9R, and Cel48S) from *C. thermocellum* onto the rossettasome via in vitro self-assembly. The engineered multi-enzyme structure was termed as rossettazyme. The rossettazyme containing four different types of cellulases (with a corresponding enzyme loading of 18 enzymes per rossettasome) exhibited 2.3–2.4-fold enhanced hydrolytic activity when compared to the free enzymes in solution.

As cellulosomes are utilized in the hydrolysis of cellulosic substrates, another machinery of natural lignocellulosic biomass-consuming microorganisms analogous to cellulosomes known as hemicellulosomes is employed for the hydrolysis of hemicellulosic components present in lignocellulosic biomass. The hemicellulosomes (sometimes termed as xylanosomes) are analogous to the cellulosomes in that they contain cohesin domains and a hemicellulose/xylan-binding domain (HBD or XBD). In a similar trend as the development of the minicellulosome, the exploitation of the hemicellulosomes lead to the development of minihemicellulosomes (sometimes termed as minixylanosome). Bayer and coworkers also exploited the use of minihemicellulosomes in degrading the hemicellulose component (of which xylan is the main component) in a complex substrate wheat straw. The designed minihemicellulosome contains a tetravalent enzyme complex of three endoxylanases (Xyn10A, Xyn10B, and Xyn11A) and a β-xylosidase (Xyl43A) from *Thermobifida fusca* along with a CBD which binds to the cellulose component of the complex substrate. The functionally assembled hemicellulosomes exhibited a ~1.7-fold increased hydrolytic activity when compared to their free enzyme counterparts, tested for their activity in the hydrolysis of hemicellulose in wheat straw (Morais et al., 2011). In another study βy McClendon et al. (2012), the design of a trifunctional minihemicellulosome containing two enzymes units (a combination of different xylanases and bi-functional arabinofuranosidase/xylosidases) was applied in the hydrolysis of hemicellulose in the complex substrates wheat and destarched corn bran. The resulting trifunctional minihemicellulosomes having different configurations exhibited enhanced hydrolytic activities of up to 1.3-fold increase compared to the free enzymes when tested for their hydrolytic capabilities on the said complex substrates.

5.8.1.2 Cell surface display of minicellulosomes and minihemicellulosomes: a whole-cell biocatalyst approach for the consolidated bio-processing strategy

The utilization of renewable materials and energy are in a booming pace in the present time as the fossil-based fuels and its raw materials contribute to the climate change. This transition is not a luxury; it has become a necessity. Thus, there is a need for establishing bioprocessing strategies in the green production of platform chemicals to answer our increasing demand of chemical products in a renewable and sustainable fashion. The most attractive renewable material that can be utilized in bioprocessing is the lignocellulosic biomass. It is the most abundant biomass (usually plants) found in nature which can be used in bioethanol production via fermentation its lignocellulose, and because of its abundance, this raw material can be obtained in fact at a very low cost; it may be the best substrate with the highest potential in bioprocessing. In fact, lignocellulose comprises 50–90% of all plant matter and it is composed of three major components which are cellulose, hemicellulose (which is composed mainly of xylan), and lignin. These major components of lignocellulose are present in their polysaccharide forms and hydrolysis of these polysaccharides yield fermentable sugars except for lignin (which produces phenolic compounds that actively inhibit fermentation). The major fermentable sugars produced from lignocellulose are glucose (abundant in cellulose), and xylose (abundant in hemicellulose). However, lignocellulosic biomass is not directly fermentable to produce the desired platform chemical product and it requires expensive upstream processing to release the fermentable sugars (Lynd et al., 2002; Lynd et al., 2005).

A recently developed solution for the upstream processing of lignocellulosic biomass which has a great potential of eliminating the expensive upstream processing is the consolidated bioprocessing (CBP) strategy. This strategy for the complete utilization of lignocellulosic biomass substrates involves the integration of four biologically mediated bioprocesses namely enzyme production (e.g., cellulases), hydrolysis of the polysaccharide components in the substrate, hexose and pentose fermentation. The goal of this strategy is to engineer CBP microorganism(s) capable of performing the said four bioprocesses in one bioreactor setting. With the development of minicellulosomes coupled with the cell surface display technology to mimic the native cellulosomes of lignocellulose-consuming microorganisms, recombinant strategies can be employed in the introduction of the minicellulosome to the microorganism(s) suited for fermentation (e.g., ethanologenic strains such as yeast for the case of ethanol production from sugars). Thus a suitable CBP microorganism(s) can be engineered (Kondo and Ueda, 2004; Olson et al., 2012; Kricka et al., 2014).

A general methodology of introducing the minicellulosome on the cell surface of a microorganism is shown in Fig. 5.12. The minicellulosome for this case is composed of the multi-enzyme complex with the corresponding cohesion–dockerin interactions on the miniscaffoldin. Additionally, the miniscaffoldin consists of a cellulose-binding domain (CBD) and the surface anchor protein which is responsible for the attachment of the whole minicellulosome assembly on the cell wall of the microorganism. This minicellulosome assembly strategy involves the in vivo production of the dockerin-tagged cellulases and the miniscaffoldin, followed by their secretion to the medium while the miniscaffoldin is anchored to the cell wall upon its secretion. The minicellulosome is constructed in vitro via the precise and specific self-assembly of the dockerin-tagged cellulases in the medium onto the miniscaffoldin attached on the cell surface.

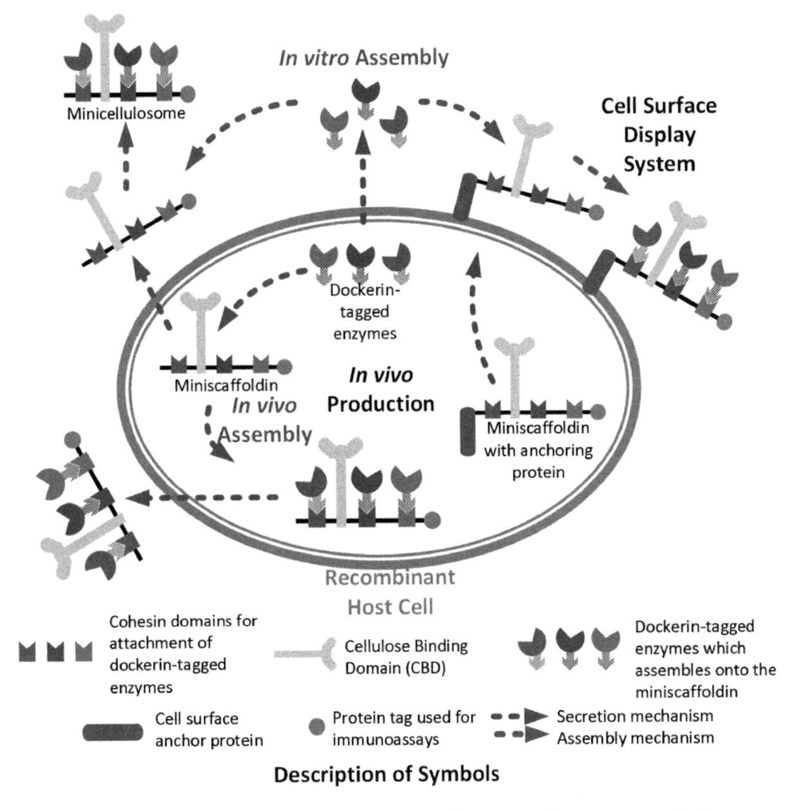

Figure 5.12 Schematic illustration of the different modes in minicellulosome assembly.

Chen and colleagues have demonstrated the in vitro functional assembly of a minicellulosome containing three cellulases, namely endoglucanase, exoglucanase, and a β-glucosidase, along with a CBD on the surface of yeast using the agglutinin (both Aga1 and Agα1) anchoring systems. The engineered yeast displaying the minicellulosome exhibited up to 3-fold increase in cellulose hydrolysis when compared to free enzymes while simultaneously producing ethanol from phosphoric acid-swollen cellulose (PASC) (Tsai et al., 2009). Furthermore, minicellulosome assembly using the intercellular complementation approach was exploited to enhance the ethanol production. This approach in the minicellulosome assembly involves the consortium of four different yeast strains of which one strain is responsible for expressing the miniscaffoldin using the two agglutinin anchor system under different promoters on one hand, and the remaining three strains each are responsible for the secretion of the dockerin-tagged cellulases (as mentioned in the previous example). This consortium mediated assembly approach produced more ethanol when compared to the previous example since there is an increased yeast concentration in this system that produces ethanol when compared to the previous example mentioned (Tsai et al., 2010; Goyal et al., 2011). In a succeeding study with the aim of enhancing cellulose hydrolysis, the functional display of a tetravalent designer cellulosome on the yeast surface was done via an adaptive assembly approach. The assembly of this designer cellulosome involves an anchoring scaffoldin containing two divergent cohesin domains displayed on the yeast surface where two dockerin-tagged bifunctional adapter scaffoldins are present, containing a CBD and two cohesin subunits each for the enzymes to assemble onto. The dockerin-tagged enzymes comprise two units of the endoglucanase and β-glucosidase pair which corresponds to a total of four enzymes displayed on the whole designer cellulosome assembly. This tetravalent designer cellulosome exhibited a 4.2-fold enhancement on the hydrolysis of PASC when compared to their free enzyme counterparts showing the feasibility of engineering a more complex cellulosome configuration (Tsai et al., 2013a). The minicellulosome assembly approaches described in the previous four examples presented in this paragraph have been incorporated in a patent by Chen and Tsai on engineering yeast for cellulosic ethanol production (Chen and Tsai, 2014). In another perspective, the minicellulosome assembly on yeast cell surface has also been exploited to develop minihemicellulosomes for the efficient hemicellulose hydrolysis. The minihemicellulosome functional assembly approach was similar to that of the first example by Tsai et al. (2009). The minihemicellulosome contains a trifunctional multi-enzyme complex consisting of the enzymes endoxylanase, a β-xylosidase, and an acetylxylan transferase, and a miniscaffoldin with a xylan binding domain (XBD). This minicellulosome assembly exhibited a ~3.3-fold increase in hydrolytic activity over free enzymes when tested for hydrolysis on xylan demonstrating that minihemicellulosomes can also be functionally expressed on the yeast surface Srikrishnan et al. (2013).

Wen et al. (2010) also have successfully assembled in vivo a trifunctional minicellulosome containing the enzymes endoglucanase, exoglucanase, and β-glucosidase along with a CBD on the surface of yeast. This in vivo assembly of the minicellulosome involved the co-expression of the miniscaffoldin CbpA3 (derived from *C. cellulovorans* CbpA) containing three cohesin domains and the agglutinin (Agα1) anchoring system along with the dockerin tagged cellulases and upon assembly, the minicellulosome was secreted and anchored to the yeast cell wall. The resulting minicellulosome displayed on yeast surface exhibited a ~8.8-fold enhancement in hydrolytic activity over the free enzymes and produced a high ethanol titer of ~1.8 g/L from the utilization of PASC.

Fan et al. (2012) have also utilized the adaptive assembly approach to construct in vitro the functional designer cellulosome that was displayed on the surface of yeast. The designer cellulosome system comprised of an anchoring scaffoldin with two cohesin modules which utilized the agglutinin (Aga1) anchoring system for yeast surface display, a trifunctional adaptive scaffold which contains a dockerin domain that attaches to the anchoring scaffold, and three cohesin domains onto which the dockerin-tagged cellulases (endoglucanase, exoglucanase, and β-glucosidase) along with a CBD are attached. The genes encoding the scaffold and the dockerin tagged cellulases were genetically introduced into one yeast strain under a co-expression system. The co-expression system expresses both the anchoring and adaptive scaffold under the influence of the GAL1 promoter and the dockerin-tagged cellulases under the TEF1 promoter on the other hand. Upon the co-expression of the scaffold genes and dockerin-tagged cellulase genes, these components are then secreted while the anchoring scaffold attaches itself to the yeast cell wall. Once the adaptive scaffold and its respective cellulases are secreted, the designer cellulosome will then construct itself via the precise self-assembly mechanism of the cohesins and dockerins. Overall, the functionally assembled designer cellulosome on the yeast surface contains 6 enzymes per functionally assembled cellulosome and consequently exhibited 1.3-fold enhanced hydrolytic activity over free enzymes and produced a high ethanol titer of 1.4 g/L from PASC.

The cell surface display of designer cellulosomes has a great potential for consolidated bioprocessing and by employing the strategies discussed in this sub-section along with a little bit of creativity, we can then develop a wide range of complex cellulosome configurations that can be integrated to the bioprocess that we want to design. As an example, our laboratory is currently working on the engineering of a microbial consortium that employs the functional assembly of both the minicellulosome and minihemicellulosome on the cell surface via the intercellular complementation approach. The goal of this research work is to efficiently utilize the cellulose and hemicellulose components of lignocellulosic biomass for the production of industrially relevant platform chemicals. The assembly scheme of these designer cellulosomes on the cell surface is shown in Fig. 5.13.

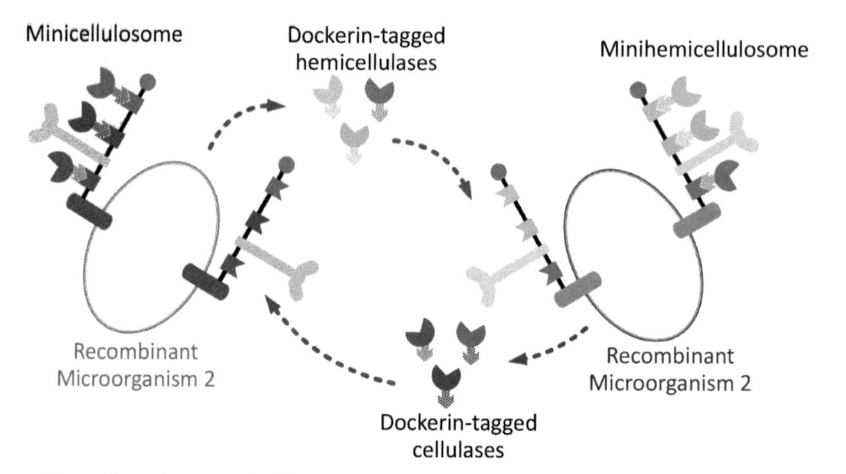

Figure 5.13 An example functional assembly of two "designer cellulosomes" via intercellular complementation approach for the microbial consortium of two recombinant microorganisms.

5.8.2 Artificial Cellulosomes

These are designer cellulosomes that were developed by employing the different bioconjugation technologies in attaching cellulases onto novel scaffold materials such as nanoparticles,

DNA, and even novel protein scaffolds, e.g., ankryn, staphylococcal protein A (SPA) as to be presented in the following paragraphs. Artificial cellulosomes can also be considered as unconventional cellulosomes in such a way that they do not utilize the crucial cohesion–dockerin interaction which is a functional component that defines a cellulosome.

Eklund et al. (2004) demonstrated the feasibility of a novel protein scaffold staphylococcal protein A (SPA) for the assembly of an artificial cellulosome. It has been reported that fusing a CBD to the SPA has served as a platform for docking reporter proteins that have been attached to a SPA-CBD fusion via the IgG-affibody conjugation, on the cellulose surface.

The immobilization of cellulases on nanoparticles using affinity binding has been presented in Section 5.3.1 referring to the example by Zhang et al. (2013). Kim et al. (2011a), utilized the streptavidin-biotin interaction to attach a CBD with the endoglucanase A (EglA) from *Aspergillus niger* both fused with a His_6-biotin acceptor peptide–IgA-hinge linker to streptavidin to form a CBD-EglA fusion. The CBD-EglA fusion was then immobilized on the functionalized CdSe nanoparticles via His-affinity interaction. The immobilized CBD-EglA exhibited a 7.2-fold increase in hydrolytic activity over the native free enzyme when tested with PASC demonstrating the functionality of nanoparticle-based artificial cellulosomes. Blanchette et al. (2012) immobilized the endoglucanase from *Trichoderma viride* onto polystyrene nanospheres via EDC and Sulfo-NHS coupling chemistry and reported a 1.1-fold increase in hydrolytic activity over the free enzyme demonstrating the efficiency of the cellulase-nanosphere complex. Tsai et al. (2013b), demonstrated the immobilization of two cellulases (CelA and CelE) onto quantum dots via metal-affinity binding (polyhistidine tag mediated) and investigated the effect of clustering on the synergistic activities exhibited by the bifunctional quantum dots by varying the size of the nanoparticle. It has been reported that the clustering effect which is attributed to the amount of enzymes being loaded onto the nanoparticle [similar to the crowding effect discussed in Section 7.5.3 in reference to the example of Patterson et al. (2012a)] incurs a more significant decrease in the synergistic activity of the immobilized bifunctional enzyme system than caused by the particle size effect. In other words, enzyme proximity or the enzyme loading (amount of enzymes immobilized onto the nanoparticle) is more crucial in controlling the enzyme synergism than the change in nanoparticle size.

Cunha et al. (2013) utilized the novel protein scaffold, ankryn, a repeat protein with roughly linear arrays of 33-residue repeating units possessing an α-helical consensus. This protein scaffold is hyperstable and has a denaturation temperature of 90°C. This feature made ankryn an ideal scaffold for thermostable cellulases such as endoglucanases CelA and CelE from *C. thermocellum* and *C. cellulolyticum*, respectively. Fusing the endoglucanases on this self-assembling ankryn scaffold exhibited retained hydrolytic activity for the CelA-ankryn fusion and a 2-fold enhancement for the CelE-ankryn fusion when tested with amorphous cellulose hydrolysis. These results demonstrated that the ankyrin arrays can be a promising scaffold for constructing artificial cellulosomes, which preserve or enhance the enzymatic activity while being thermostable at the same time.

The use of DNA as scaffolds for cellulases have also been exploited to create artificial cellulosomes. Mori et al. (2013) first demonstrated the covalent conjugation of the CBD-endoglucanase fusion on functionalized double stranded DNA scaffolds (dsDNA) mediated by microbial transglutaminase (MTG; discussed in Section 5.4.2). This strategy involves the modification of DNA scaffolds to contain a functional group (analogous to the Q-residue) that is recognized by the MTG. After modification of the dsDNA, the K-tagged CBD-endoglucanase fusion was mixed along with the MTG for the conjugation to occur at the analogous-Q and K residues in the scaffold, and the CBD-endoglucanase (Cel5A). The obtained artificial cellulosome consisting of multiple CBD-endoglucanase units on dsDNA exhibited a 5.7-fold improvement in hydrolytic activity on crystalline cellulose (Avicel). Another approach of using dsDNA scaffolds for artificial cellulosome construction is the use of the site-specific, non-covalent zinc-finger protein (ZFPs-guided) assembly as demonstrated by Sun et al. (2014). Two ZFPs, Zif268 and PE1A, each with binding affinity to a unique 9bp DNA sequence, have been prepared in such a way that its binding function is preserved, and fused with an elastin-like polypeptide (ELP) to act as a linker yielding ELP-Zif268, and ELP-PE1A fusions. Cel5A and CBD were fused upstream of the ELP to yield Cel5A-ELP-Zif268 and CBD-ELP-PE1A conjugates. These Cel5A-ELP-Zif268 and CBD-ELP-PE1A conjugates were then introduced onto the dsDNA scaffold and by virtue of site-specific non-covalent self-assembly, Cel5A and CBD were attached onto the DNA scaffold creating a bifunctional artificial cellulosome. This artificial cellulosome exhibited a 1.7-fold enhancement in avicel hydrolysis. One major feature of this artificial cellulosome is that can be easily engineered to control the number, and spacing of functional units displayed on the dsDNA scaffold by designing the sequence of the DNA scaffold.

5.9 Evaluation of the Different Immobilized Enzyme Systems as Industrial Biocatalysts

In an industrial setting, immobilized enzyme systems are usually favored because they can be recovered easily from the reaction media and can then be reused for another reaction cycle. These features make the utilization of immobilized enzyme systems economically feasible. Moreover, incorporating an immobilized enzyme in a process provides simplicity to the design of reactor systems (Mateo et al., 2007). In other words, the features of an immobilized enzyme system that can be used in an industrial scale operation are the activity, stability, and in some cases, the selectivity is included. Throughout this chapter, the immobilized enzyme systems that have been presented as examples were evaluated for their relative enzymatic activity (REA; sometimes referred to as the retained activity), enzyme loading (EL), and reusability (R). The relative enzyme activity is the activity of the immobilized enzyme with respect to its free counterpart (presented in this chapter as the percentage of the immobilized enzyme activity to the free enzyme activity) while the enzyme loading is the amount of enzyme being immobilized on the supporting material, and the reusability is the retained activity after being reused in a number of reaction cycles. The values for the REA, EL, and R for all the examples presented in this chapter are summarized in Table 5.1 in the supplementary data section.

Table 5.1 Summary of the Immobilized Enzyme Systems presented in this chapter

Biological assembly strategy	Enzyme of interest	Supporting material	Relative enzymatic activity	Enzyme loading	Reusability REA:RC	Reference
DNA directed immobilization	NAD(P)H:FMN oxidoreductase (NFOR) –luciferase (Luc)	1D DNA	~300%	—	—	Niemeyer (2002)
	Glucose oxidase (GOX) –Horseradish peroxidase (HRP)	1D DNA	~200%	—	—	Muller and Niemeyer (2008)
		2D circular DNA	~600%	—	—	Wang et al. (2009)
		2D Hexagonal DNA		—	—	Wilner et al. (2009)
		2D DNA origami tiles	~1800%	—	—	Fu et al. (2012)
Affinity tags (polyhistidine tag)	Horseradish peroxidase (HRP)	Gold nanoparticle	93%	—	—	Abad et al. (2005)
	Ferrodoxin-NADP$^+$ reductase (FNR)	Gold nanoparticle	97%	—	—	
	T4 DNA ligase	γ-nanoparticle	—	—	—	Herdt et al. (2007)
	D-amino acid oxidase	Magnetic beads	75%	12 μg enzyme per μL beads	37%: 20	Kuan et al. (2008)
	NADH oxidase	Single-walled carbon Nanotubes (SWCNTs)	92%	0.47 mg enzyme per mg CNP	70%: 10	Wang et al. (2010)
		Multi-walled carbon Nanotubes (MWCNTs)	87%	0.196 mg enzyme per mg CNP	65%: 10	Wang et al. (2011)

Biological assembly strategy	Enzyme of interest	Supporting material	Relative enzymatic activity	Enzyme loading	Reusability REA:RC	Reference
		Carbon nanospheres (CNS)	78%	0.052 mg enzyme per mg CNP	55%: 10	Wang et al. (2011)
	Catechol 1,2-deioxygenase	NiO decorated magnetic nanoparticle	85%	36 μg enzyme per μg CNP	72%: 6	Lee et al. (2011)
	Glycerol dehydrogenase	Single-walled carbon nanotubes (SWCNTs)	—	—	—	Wang et al. (2012)
	Cellulases (Cel48F and Cel9G)	Magnetic nanoparticles of	99% (Cel48F) 94% (Cel9G)	0.2 mmol enzyme per mL suspension	81%: 5*	Zhang et al. (2013)
	Human rhinovirus 3C (a.k.a. PreScission protease (PSP))	NiO nanoparticle decorated mesoporous silica	160%	50 μg enzyme per mg material	90%: 5	Ling et al. (2014)
Affinity tags (other affinity tag)	Glutathione S-transferases (GSTs)	Polystyrene beads	23% (PS-19) 17% (PS-23)	—	—	Kumada et al. (2006)
	Haloalkane dehalogenase (Dh1A)	Magnetic nanoparticle	47%	79 μg protein per mg MNP	—	Johnson et al. (2008)
		Sol-gel and dense liquid coated magnetic nanoparticle	37%	59 μg protein per mg MNP	—	
	Alkaline phosphatase (AP)	Micro-patterned gold substrates	~8%	—	—	Kacar et al. (2009)

(Continued)

Table 5.1 (*Continued*)

Biological assembly strategy	Enzyme of interest	Supporting material	Relative enzymatic activity	Enzyme loading	Reusability REA:RC	Reference
	Organophosphorus hydrolase (OPH)	Gold nanoparticle coated graphene sheet	—	—	—	Yang et al. (2011)
Enzyme-assisted covalent immobilization	Arylmalonate decarboxylase	Coenzyme A-derivatized polystyrene NPs	0.6% (?)	86 μg enzyme per μL suspension	93%: 4	Wong et al. (2010)
	Glutamate racemase		12.5% (?)	15 μg enzyme per μL suspension	—	
	–Galactosyltransferase	EAH sepharose beads	90%	—	~95%: 10	Ito et al. (2010)
	–Fucosyltransferase		94%	—	~55%: 10	
	Glutathione S-transferase	Coenzyme A-derivatized PEGA beads	90%	—	—	Wong et al. (2008)
		Octanediamine-fixed NHS sepharose gel	122%	0.62 ng enzyme per μL gel	—	Sugimura et al. (2007)
		O^6—benzylguanine-functionalized agarose beads	~62%	—	—	Kindermann et al. (2003)
	Alkaline phosphatase	–casein modified Polyacrylic resin	~100%	0.32 mg enzyme per g PAR	~98%: 10	Tominaga et al. (2004)

Biological assembly strategy	Enzyme of interest	Supporting material	Relative enzymatic activity	Enzyme loading	Reusability REA:RC	Reference
	Alkaline phosphatase	–casein modified agarose gel beads	86%	~0.89 μg enzyme per mL beads	~98%: 20d [continuous reactor]	Tominaga et al. (2005)
		Magnetic nanoparticles	~100%	~0.66 μmol enzyme per g MNP	90%: 10	Moriyama et al. (2011)
		W/O/W microcapsules	~250%	—	95%: 5	Minamihata et al. (2009)
		—	~95%	—	—	Minamihata et al. (2011b)
		Biotin-coated plate	—	—	—	Minamihata et al. (2011a)
Leucine zippers	Laccase (Polyphenol oxidase)	CLEA	—	—	—	Wheeldon et al. (2008)
	Alcohol dehydrogenase (AdhD)	CLEA	—	—	—	Wheeldon et al. (2009)
	Organophosphorus hydrolase (OPH)	CLEA	—	—	—	Lu et al. (2010)
	Galactose oxidase	CLEA	65%	200 mg inclusion bodies/g	—	Steinmann et al. (2010)
uAA tag	*Candida antarctica* lipase B	Azide functionalized magnetic beads	~123%	—	—	Wu (2014)

(*Continued*)

Table 5.1 (*Continued*)

Biological assembly strategy	Enzyme of interest	Supporting material	Relative enzymatic activity	Enzyme loading	Reusability REA:RC	Reference
	Lysozyme	Azide functionalized magnetic beads	64.3% (wrt std) ~81% (wrt free)	—	—	Wu et al. (2015)
	Modified *Sphaerobacter thermophilus*-transaminase	Chitosan and polystyrene beads	95% (for chitosan immobilized-TA)	—	—	Deepankumar et al. (2015)
Protein scaffoldin (S-layer proteins)	Invertase	S-layer protein	70%	950 µg enzyme per mg protein	—	Sara and Sleytr (1996)
		S-layer protein	70%	1000 µg enzyme per mg protein	—	
	Glucose oxidase	S-layer protein	35%	650 µg enzyme per mg protein	—	
		S-layer protein	35%	550 µg enzyme per mg protein	—	
		S-layer protein	35%	600 µg enzyme per mg protein	—	
	–Glucosidase	S-layer protein	160%	—	—	
	–Glycanase	S-layer protein	~78%	35 mg immobilized enzyme per L sol'n	50%: 10	Duncan et al. (2005)

Biological assembly strategy	Enzyme of interest	Supporting material	Relative enzymatic activity	Enzyme loading	Reusability REA:RC	Reference
	Glucose 1-phosphate thymidylyltransferase	S-layer protein	99.75	—	—	Schaffer et al. (2007)
		S-layer protein	3%	—	—	
		S-layer protein	65.4%	1.004 mol enzyme per mol liposome	61%: 1	
	Laminarinase	Silicon wafers	106.46%	0.5678 mg enzyme per cm^2 supporting material	31%: 8	Tschiggerl et al. (2008)
		Epoxy-activated glass slide [1]	101.90%	—	—	
		Glass slides PES membrane	111.76%	—	—	
		Cellulose Acetate membrane	78.89%	—	—	
		Sartobind® Epoxy 75 membrane [1]	271.19%	—	—	
		Cellulose membrane	190.66%	—	—	
		Polypropylene membrane	52.59%	—	—	

(*Continued*)

Table 5.1 (*Continued*)

Biological assembly strategy	Enzyme of interest	Supporting material	Relative enzymatic activity	Enzyme loading	Reusability REA:RC	Reference
Protein scaffoldin (SP1 scaffold)	Cellulase (*Thermobifida fusca* Cel5A)	SP1	94.44%	1.3743 g enzyme per g SP1 protein	—	Heyman et al. (2007a)
	Glucose oxidase	SP1	55% (?)	6.4516 g enzyme per g SP1 protein	—	Heyman et al. (2007b)
	Exoglucanase (*Thermobifida fusca*)	SP1	162.96%	1.7888 g enzyme per g SP1 protein	—	Morais et al. (2010)
	Artificial glutathione peroxidase	SP1	129.34%	—	—	Miao et al. (2014)
Protein scaffoldin (PHA systems)	–Galactosidase	PhAC	–(68,000 Miller Units)	—	—	Peters and Rehm (2006)
	–Amylase	PhAC	101.20%	—	78%: 3	Rasiah and Rehm (2009)
	Organophosphohydrolase (OpdA)	PhAC	160%	—	—	Blatchford et al. (2012)
	N-ethylmaleimide reductase (nemA)	PhAC	—	—	—	(Robins et al., 2013)
	N-acetylglucosamine 2-epimerase (Slr1975) N-acetylneuraminic acid aldolase (NanA)	PhAC	[MEC]			Hooks et al. (2013)

Biological assembly strategy	Enzyme of interest	Supporting material	Relative enzymatic activity	Enzyme loading	Reusability REA:RC	Reference
Protein scaffoldin (oleosins)	D-hydantoinase	AOBs	100% (pH 7–7.5) ~100% (25 C)	—	Active for 7 cycles	Chiang et al. (2006)
	Xylanase (*Neocallimastix patriciarum*)	AOBs	100%	—	150%: 4	Liu et al. (1997)
	Xylanase (*Neocallimastix patriciarum*)	AOBs	—	—	60%: 8	Hung et al. (2008)
	1,3-1,4-D-glucanase	AOBs	—	—	80%: 8	Zeng et al. (2009)
	Endo-1,4-D-glucanase	AOBs	—		~80%: 5	Chiang et al. (2013a)
	Endoglucanase (celA)-cellobiohydrolase (celK) system	AOBs	[MEC]	—	~80%: 4	Chiang et al. (2013b)
	D-psicose 3-epimerase	AOBs	—	—	>50%: 5	Tseng et al. (2014)
	Lipase (*Pseudomonas aeruginosa*)	AOBs	—	—	72%: 4	Bai et al. (2014)

(*Continued*)

Table 5.1 (*Continued*)

Biological assembly strategy	Enzyme of interest	Supporting material	Relative enzymatic activity	Enzyme loading	Reusability REA:RC	Reference
Protein scaffoldin (VLPs)	Alkaline phosphatase	fd bacteriophage VLPs	~100%	—	—	McCafferty et al. (1991)
	Endo 1,4-xylanase I (*Aspergillus niger*) + Endo 1,4-xylanase A (*Bacillus subtilis*)	M13 bacteriophage VLPs	—	—	—	Belien et al. (2005)
	α-amylase (*Bacillus licheniformis*) + xylanase A (*Bacillus halodurans*)	T7 bacteriophage VLPs	—	—	—	Straffon et al. (2006)
	Lipase B	Potato virus X (PVX) VLPs	~2.22% [compared to free CALB]	0.33 enzymes/capsid	—	Carette et al. (2007)
	Lipase B	Cowpea chlorotic mottle virus (CCMV) VLPs	~455%	1.3 enzymes/capsid	—	Minten et al. (2011)
	Horseradish peroxidase (HRP)	Cowpea chlorotic mottle virus (CCMV) VLPs	—	—	—	Comellas-Aragones et al. (2007)
	Peptidase E	Q bacteriophage VLPs	~45%	—	—	Fiedler et al. (2010)

Biological assembly strategy	Enzyme of interest	Supporting material	Relative enzymatic activity	Enzyme loading	Reusability REA:RC	Reference
	Luciferase	M13 bacteriophage VLPs	~92%	—	—	
	β-lactamase	M13 bacteriophage VLPs	—	—	—	Shukla and Krag (2010)
	Biotin ligase (BirA)	T7 bacteriophage VLPs	—	—	—	Sunbul et al. (2011)
	Endoglucanase (CelA, *clostridium thermocellum*),	T7 bacteriophage VLPs	—	—	—	Sutherland et al. (2004); Collins and Lu (2012)
	Dispersin B	P22 bacteriophage VLPs	—	—	—	Collins et al. (2012)
	Alcohol dehydrogenase (AdhD)	P22 bacteriophage VLPs	~15%	~250 enzymes/ capsid	—	Patterson et al. (2012a)
	β-glucosidase (CelB)	P22 bacteriophage VLPs	94%	~84 enzymes/ capsid	—	Patterson et al. (2012b)
	β-glucosidase (CelB) + galactokinase (GALK) + glucokinase (GLUK)	P22 bacteriophage VLPs	[MEC]			Patterson et al. (2014)
Cellulosomes (cohesin–dockerin based)	Cellulases CelA + CelF	Minicellulosome	~300%	2 enzymes per scaffold	—	Fierobe et al. (2001)

(*Continued*)

Table 5.1 (*Continued*)

Biological assembly strategy	Enzyme of interest	Supporting material	Relative enzymatic activity	Enzyme loading	Reusability REA:RC	Reference
	Cellulases Cel 48F + Cel 9G and hemicellulase XynZt	Minicellulosome	~400%	3 enzymes per scaffold	—	Fierobe et al. (2005)
	Multiple cellulases Cel 48F + Cel 9G	Minicellulosome	~200–300%	2–4 enzymes per scaffold	—	Mingardon et al. (2007)
	Endoglucanase (EngB)	Minicellulosome	—	—	—	Cho et al. (2004)
	Endoglucanase (EngB) and xylanase (XynB)	Minicellulosome	—	—	—	Arai et al. (2007)
	4 cellulases (CelF, Cel9K, Cel9R, and Cel48S) from *C. thermocellum*	Rossettasome	~230–240%	18 enzymes per rossettasome	—	Mitsuzawa et al. (2009)
	Endoxylanases (Xyn10A, Xyn10B, and Xyn11A) and a β-xylosidase (Xyl43A) from Thermobifida fusca	Minihemicellulosome	~170%	4 enzymes per scaffold	—	Morais et al. (2011)
	Xylanase and bi-functional arabinofuranosidase/xylosidase	Minihemicellulosome	~95–130%	2 enzymes per scaffold	—	McClendon et al. (2012)
	Endoglucanase, exoglucanase, and β-glucosidase	Yeast	~880%	3 enzymes per miniscaffoldin	—	Bai et al. (2014) Wen et al. (2010)

Biological assembly strategy	Enzyme of interest	Supporting material	Relative enzymatic activity	Enzyme loading	Reusability REA:RC	Reference
	Endoglucanase, exoglucanase, and β-glucosidase	Yeast	~300%	3 enzymes per miniscaffoldin	—	Tsai et al. (2009); Goyal et al. (2011); Tsai et al. (2010)
	2 units of endoglucanase and β-glucosidase pairs	Yeast	~420%	4 enzymes per assembly	—	Tsai et al. (2013a)
	Endoxylanases, a β-xylosidase, and an acetylxylan transferase	Yeast	~330%	3 enzymes per miniscaffoldin	—	Srikrishnan et al. (2013)
	2 units of the endoglucanase, exoglucanase, and β-glucosidase	Yeast	130%	6 enzymes per assembly	—	Fan et al. (2012)
Cellulosomes (Artificial Cellulosomes)	Endoglucanase A *Aspergillus niger*	Functionalized CdSe	~720%	—	—	Kim et al. (2011a)
	Endoglucanase (CelA) and exoglucanase (CelE)	CdSe	~62 to 197%	5:1 to 70:1 enzyme to QD ratio	—	Tsai et al. (2013b)
	Endoglucanases (CelA and CelE)	Ankryn scaffold	~100% (CelA) ~200% (CelE)	1 enzyme per ankryn scaffold	—	Cunha et al. (2013)
	Endoglucanase (Cel5A)	dsDNA	~570%	—	—	Mori et al. (2013)
	Endoglucanase (CelA)	dsDNA	~170%	1 enzyme per scaffold	—	Sun et al. (2014)

Using the evaluation scheme discussed in the previous paragraph along with the collated data, a semi-quantitative comparison approach adapted from the review paper of Roessl et al. (2010) was utilized to develop a rating system that can be used as a basis to compare the different biological assembly strategies presented in this chapter for the production of efficient and stable immobilized enzyme systems. The rating system evaluation matrices are shown in Table 5.2 and Table 5.3 at the supplementary data section. This rating system involves the evaluation of the activity, stability (operational, thermal, and storage), and complexity of the synthesis process (also termed as immobilized enzyme (I.E.) preparation) of the immobilized enzyme system being produced employing the assumed biological assembly strategy. The results of the comparison of the different biological assembly strategies for enzyme immobilization are shown in Fig. 5.14.

Table 5.2 Evaluation matrix for the different biological assembly strategies

	Performance criteria*				
Rating scale (description)	**REA**	**R[1]**	**Thermal stability**	**Storage stability**	**Immobilized enzyme preparation[2]**
5 (Exceptional Performance)	>100%; [MEC]	>100%	Significantly enhanced		One-pot synthesis
4 (Above average)	90–100%	90–100%	Slightly enhanced		No enzyme modification step
3 (Average)	70–89%	70–89%	No change		Conventional
2 (Needs improvement)	50–69%	50–69%	Slight deterioration		Complex material functionalization procedure
1 (Poor performance)	<50%	<50%	Detrimental		Complex

[1]The basis of three reaction cycles is used to evaluate for the reusability.
[2]Based on the number of preparation stages as generalized in Table 5.3.

Table 5.3 Basis for evaluation of the immobilized enzyme preparation criterion

Description	Synthesis scheme
One-pot synthesis	*In situ* enzyme-scaffold production and immobilization
No enzyme modification step	1-1 → 1-2 → 3; 2-1 → 3
Conventional	1-1 → 1-2 → 3; 2-1 → 2-2 → 3
Complex nanomaterial/scaffold preparation (1-x) procedure	
Complex	

Note: Each stage/step is represented as one reactor unit when the synthesis is at the industrial scale.*

Legend:

1-1—Nanomaterial/scaffold synthesis

1-2—Nanomaterial/scaffold functionalization

2-1—Enzyme production

2-2—Enzyme modification or another extra step

3—Immobilization

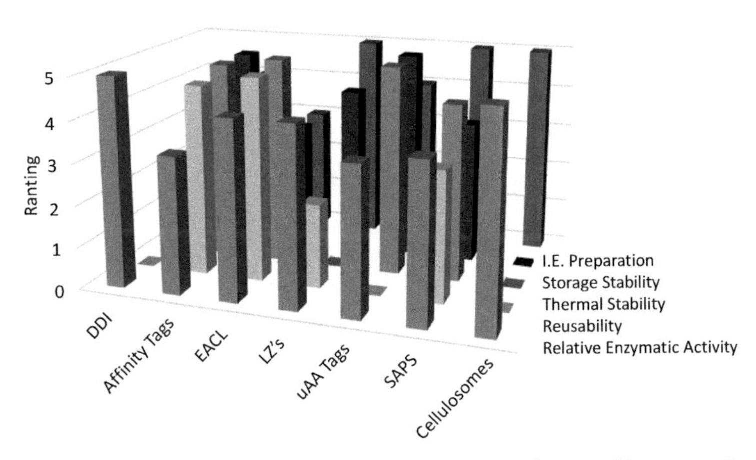

Figure 5.14 Comparison of the different biological assembly strategies for enzyme immobilization.

5.10 Concluding Remarks

The various biological assembly strategies that have been presented and discussed in this chapter along with their respective examples have produced immobilized enzyme systems having preserved or enhanced catalytic activity, reusability and stability. In Section 5.9, the immobilized enzyme systems synthesized with the biological assembly strategies presented throughout Sections 5.2–5.8 have been evaluated for their performance using a semi-quantitative approach. The results of the evaluation can be used as a guide in the selection of the appropriate strategy to be incorporated in the design of industrial bioprocesses.

The biological assembly strategies discussed in this chapter, which have been derived from mimicking assembly systems found in nature, are still subjected to innovation and further development. Furthermore, the application of these biological assembly strategies is not only limited to the "bottom-up" approach in the production of nanobiocatalyst systems but can also be exploited to assemble a wide range of technological tools which can be utilized in many research areas.

References

Abad, J. M., Mertens, S. F., Pita, M., Fernandez, V. M., and Schiffrin, D. J., 2005, *J. Am. Chem. Soc.,* **127,** 5689–5694.

Alber, T., 1992, *Curr. Opin. Genet. Dev.,* **2,** 205–210.

Aldaye, F. A., and Sleiman, H. F., 2006, *Angew. Chem. Int. Ed. Engl.,* **45,** 2204–2209.

Aniagyei, S. E., Dufort, C., Kao, C. C., and Dragnea, B., 2008, *J. Mater. Chem.,* **18,** 3763–3764.

Arai, T., Matsuoka, S., Cho, H. Y., Yukawa, H., Inui, M., Wong, S. L., and Doi, R. H., 2007, *Proc. Natl. Acad. Sci. U. S. A.,* **104,** 1456–1460.

Ayyadurai, N., Prabhu, N. S., Deepankumar, K., Jang, Y. J., Chitrapriya, N., Song, E., Lee, N., Kim, S. K., Kim, B. G., Soundrarajan, N., Lee, S., Cha, H. J., Budisa, N., and Yun, H., 2011, *Bioconjug. Chem.,* **22,** 551–555.

Bai, F. W., Yan, W., Zhang, S. J., Yu, D., and Bai, L. H., 2014, *Fuel,* **128,** 340–346.

Bain, J. D., Diala, E. S., Glabe, C. G., Dix, T. A., and Chamberlin, A. R., 1989, *J. Am. Chem. Soc.,* **111,** 8013–8014.

Banwell, E. F., Abelardo, E. S., Adams, D. J., Birchall, M. A., Corrigan, A., Donald, A. M., Kirkland, M., Serpell, L. C., Butler, M. F., and Woolfson, D. N., 2009, *Nat. Mater.,* **8,** 596–600.

Bayer, E. A., Belaich, J. P., Shoham, Y., and Lamed, R., 2004, *Annu. Rev. Microbiol.,* **58,** 521–554.

Bayer, E. A., Lamed, R., White, B. A., and Flint, H. J., 2008, *Chem. Rec.,* **8,** 364–377.

Bayer, E. A., Morag, E., and Lamed, R., 1994, *Trends Biotechnol.,* **12,** 379–386.

Belien, T., Hertveldt, K., Van den Brande, K., Robben, J., Van Campenhout, S., and Volckaert, G., 2005, *J. Biotechnol.,* **115,** 249–260.

Berg, J. M., Tymoczko, J. L., and Stryer, L., 2007, *Biochemistry,* W. H. Freeman.

Besteman, K., Lee, J. O., Wiertz, F. G. M., Heering, H. A., and Dekker, C., 2003, *Nano Lett.,* **3,** 727–730.

Bhatla, S. C., Kaushik, V., and Yadav, M. K., 2010, *Biotechnol. Adv.,* **28,** 293–300.

Bickerstaff Jr, G. F., 1997, *Immobilization of Enzymes and Cells,* Springer.

Blanchette, C., Lacayo, C. I., Fischer, N. O., Hwang, M., and Thelen, M. P., 2012, *PLoS One,* **7,** e42116.

Blatchford, P. A., Scott, C., French, N., and Rehm, B. H., 2012, *Biotechnol. Bioeng.,* **109,** 1101–1108.

Boozer, C., Ladd, J., Chen, S., and Jiang, S., 2006, *Anal. Chem.,* **78,** 1515–1519.

Bornscheuer, U. T., and Buchholz, K., 2005, *Eng. Life Sci.,* **5,** 309–323.

Brena, B., González-Pombo, P., and Batista-Viera, F., 2013, *Immobilization of Enzymes and Cells,* (Guisan, J. M., ed.), Humana Press, vol. 1051, pp. 15–31.

Brown, K. C., Yang, S. H., and Kodadek, T., 1995, *Biochemistry,* **34,** 4733–4739.

Brown, K. C., Yu, Z., Burlingame, A. L., and Craik, C. S., 1998, *Biochemistry,* **37,** 4397–4406.

Brown, S., 1997, *Nat. Biotechnol.,* **15,** 269–272.

Bundy, B. C., and Swartz, J. R., 2010, *Bioconjug. Chem.,* **21,** 255–263.

Bundy, B. C., and Swartz, J. R., 2011, *J. Biotechnol.,* **154,** 230–239.

Busch, S. J., and Sassone-Corsi, P., 1990, *Trends Genet.,* **6,** 36–40.

Cardinale, D., Carette, N., and Michon, T., 2012, *Trends Biotechnol.,* **30,** 369–376.

Carette, N., Engelkamp, H., Akpa, E., Pierre, S. J., Cameron, N. R., Christianen, P. C., Maan, J. C., Thies, J. C., Weberskirch, R., Rowan, A. E., Nolte, R. J., Michon, T., and Van Hest, J. C., 2007, *Nat. Nanotechnol.,* **2,** 226–229.

Chan, L., Cross, H. F., She, J. K., Cavalli, G., Martins, H. F., and Neylon, C., 2007, *PLoS One,* **2,** e1164.

Chao, H. M., Bautista, D. L., Litowski, J., Irvin, R. T., and Hodges, R. S., 1998, *J. Chromatogr. B,* **715,** 307–329.

Chapman, K. D., Dyer, J. M., and Mullen, R. T., 2012, *J. Lipid Res.,* **53,** 215–226.

Chapman, M. S., and Liljas, L., 2003, *Adv. Protein Chem.* (Wah, C., and John, E. J., eds.), Academic Press, vol. 64, pp. 125–196.

Chen, T., Small, D. A., Wu, L.-Q., Rubloff, G. W., Ghodssi, R., Vazquez-Duhalt, R., Bentley, W. E., and Payne, G. F., 2003, *Langmuir,* **19,** 9382–9386.

Chen, W., and Tsai, S., 2014, Google Patents.

Chiang, C. J., Chen, H. C., Kuo, H. F., Chao, Y. P., and Tzen, J. T. C., 2006, *Enzyme Microb. Technol.,* **39,** 1152–1158.

Chiang, C. J., Chen, P. T., Yeh, C. Y., and Chao, Y. P., 2013a, *Process Biochem.,* **48,** 1886–1892.

Chiang, C. J., Chen, P. T., Yeh, C. Y., Wang, Z. W., and Chao, Y. P., 2013b, *Appl. Microbiol. Biotechnol.,* **97,** 9185–9192.

Chin, J. W., Santoro, S. W., Martin, A. B., King, D. S., Wang, L., and Schultz, P. G., 2002, *J. Am. Chem. Soc.,* **124,** 9026–9027.

Cho, H. Y., Yukawa, H., Inui, M., Doi, R. H., and Wong, S. L., 2004, *Appl. Environ. Microbiol.,* **70,** 5704–5707.

Clow, F., Fraser, J. D., and Proft, T., 2008, *Biotechnol. Lett.,* **30,** 1603–1607.

Collins, J. J., Kobayashi, H., Kearn, M., Araki, M., Friedland, A., and Lu, T. K. T., 2012, Google Patents.

Collins, J. J., and Lu, T. K. T., 2012, Google Patents.

Comellas-Aragones, M., Engelkamp, H., Claessen, V. I., Sommerdijk, N. A., Rowan, A. E., Christianen, P. C., Maan, J. C., Verduin, B. J., Cornelissen, J. J., and Nolte, R. J., 2007, *Nat. Nanotechnol.,* **2,** 635–639.

Cruz, S. S., Chapman, S., Roberts, A. G., Roberts, I. M., Prior, D. A., and Oparka, K. J., 1996, *Proc. Natl. Acad. Sci. U. S. A.,* **93,** 6286–6290.

Cunha, E. S., Hatem, C. L., and Barrick, D., 2013, *Appl. Environ. Microbiol.,* **79,** 6684–6696.

Deepankumar, K., Nadarajan, S. P., Mathew, S., Lee, S. G., Yoo, T. H., Hong, E. Y., Kim, B. G., and Yun, H., 2015, *Chemcatchem,* **7,** 417–421.

Dgany, O., Gonzalez, A., Sofer, O., Wang, W., Zolotnitsky, G., Wolf, A., Shoham, Y., Altman, A., Wolf, S. G., Shoseyov, O., and Almog, O., 2004, *J. Biol. Chem.,* **279,** 51516–51523.

Doi, R. H., and Kosugi, A., 2004, *Nat. Rev. Microbiol.,* **2,** 541–551.

Donnelly, M. L. L., Luke, G., Mehrotra, A., Li, X., Hughes, L. E., Gani, D., and Ryan, M. D., 2001, *J. Gen. Virol.,* **82,** 1013–1025.

Downie, J. W., Labella, F. S., and Whitaker, S., 1973, *Connect. Tissue Res.,* **2,** 37–48.

Duncan, G., Tarling, C. A., Bingle, W. H., Nomellini, J. F., Yamage, M., Dorocicz, I. R., Withers, S. G., and Smit, J., 2005, *Appl. Biochem. Biotechnol.,* **127,** 95–110.

Eickhoff, H., Jung, G., and Rieker, A., 2001, *Tetrahedron,* **57,** 353–364.

Eklund, M., Sandstrom, K., Teeri, T. T., and Nygren, P. A., 2004, *J. Biotechnol.,* **109,** 277–286.

Endrizzi, B. J., Huang, G., Kiser, P. F., and Stewart, R. J., 2006, *Langmuir,* **22,** 11305–11310.

Fan, L. H., Zhang, Z. J., Yu, X. Y., Xue, Y. X., and Tan, T. W., 2012, *Proc. Natl. Acad. Sci. U. S. A.,* **109,** 13260–13265.

Fernandez-Gacio, A., Uguen, M., and Fastrez, J., 2003, *Trends Biotechnol.,* **21,** 408–414.

Fiedler, J. D., Brown, S. D., Lau, J. L., and Finn, M. G., 2010, *Angew. Chem. Int. Ed. Engl.,* **49,** 9648–9651.

Fierobe, H. P., Mechaly, A., Tardif, C., Belaich, A., Lamed, R., Shoham, Y., Belaich, J. P., and Bayer, E. A., 2001, *J. Biol. Chem.,* **276,** 21257–21261.

Fierobe, H. P., Mingardon, F., Mechaly, A., Belaich, A., Rincon, M. T., Pages, S., Lamed, R., Tardif, C., Belaich, J. P., and Bayer, E. A., 2005, *J. Biol. Chem.,* **280,** 16325–16334.

Foerder, C. A., and Shapiro, B. M., 1977, *Proc. Natl. Acad. Sci. U. S. A.,* **74,** 4214–4218.

Folk, J. E., 1980, *Annu. Rev. Biochem.,* **49,** 517–531.

Folk, J. E., and Finlayson, J. S., 1977, *Adv. Protein Chem.* (C. B. Anfinsen, J. T. E., and Frederic, M. R., eds.), Academic Press, vol. 31, pp. 1–133.

Fontes, C. M., and Gilbert, H. J., 2010, *Annu. Rev. Biochem.,* **79,** 655–681.

Fruk, L., Muller, J., Weber, G., Narvaez, A., Dominguez, E., and Niemeyer, C. M., 2007, *Chemistry,* **13,** 5223–5231.

Fu, J., Liu, M., Liu, Y., Woodbury, N. W., and Yan, H., 2012, *J. Am. Chem. Soc.*, **134**, 5516–5519.

Fukuoka, T., Tachibana, Y., Tonami, H., Uyama, H., and Kobayashi, S., 2002, *Biomacromolecules*, **3**, 768–774.

Glover, J. N., and Harrison, S. C., 1995, *Nature*, **373**, 257–261.

Golmohammadi, R., Fridborg, K., Bundule, M., Valegard, K., and Liljas, L., 1996, *Structure*, **4**, 543–554.

Goncalves, G., Mori, Y., and Kamiya, N., 2014, *Sustainable Chem. Process.*, **2**, 19.

Goyal, G., Tsai, S. L., Madan, B., DaSilva, N. A., and Chen, W., 2011, *Microb Cell Fact*, **10**, 89.

Grage, K., Jahns, A. C., Parlane, N., Palanisamy, R., Rasiah, I. A., Atwood, J. A., and Rehm, B. H., 2009, *Biomacromolecules*, **10**, 660–669.

Guex, N., and Peitsch, M. C., 1997, *Electrophoresis*, **18**, 2714–2723.

Guisan, J. M., 2006, *Immobilization of Enzymes and Cells*, Springer.

Gumel, A. M., Annuar, M. S. M., and Chisti, Y., 2013, *J. Polym. Environ.*, **21**, 580–605.

Gunasekar, S. K., Haghpanah, J. S., and Montclare, J. K., 2008, *Polym. Adv. Technol.*, **19**, 454–468.

Gupta, M., Caniard, A., Touceda-Varela, A., Campopiano, D. J., and Mareque-Rivas, J. C., 2008, *Bioconjug. Chem.*, **19**, 1964–1967.

Hartmeier, W., 1988, *Immobilized Biocatalysts: An Introduction*, Springer-Verlag.

He, Y., Ye, T., Su, M., Zhang, C., Ribbe, A. E., Jiang, W., and Mao, C., 2008, *Nature*, **452**, 198–201.

Herdt, A. R., Kim, B. S., and Taton, T. A., 2007, *Bioconjug. Chem.*, **18**, 183–189.

Heyman, A., Barak, Y., Caspi, J., Wilson, D. B., Altman, A., Bayer, E. A., and Shoseyov, O., 2007a, *J. Biotechnol.*, **131**, 433–439.

Heyman, A., Levy, I., Altman, A., and Shoseyov, O., 2007b, *Nano Lett.*, **7**, 1575–1579.

Hezayen, F. F., Steinbuchel, A., and Rehm, B. H., 2002, *Arch. Biochem. Biophys.*, **403**, 284–291.

Hong, J., Gong, P., Xu, D., Dong, L., and Yao, S., 2007, *J. Biotechnol.*, **128**, 597–605.

Hooks, D. O., Blatchford, P. A., and Rehm, B. H., 2013, *Appl. Environ. Microbiol.*, **79**, 3116–3121.

Hooks, D. O., Venning-Slater, M., Du, J. P., and Rehm, B. H. A., 2014, *Molecules,* **19,** 8629–8643.

Huang, A. H. C., 1992, *Annu. Rev. Plant Physiol. Plant Mol. Biol.,* **43,** 177–200.

Hudson, S., Cooney, J., and Magner, E., 2008, *Angew. Chem. Int. Ed. Engl.,* **47,** 8582–8594.

Hung, Y. J., Peng, C. C., Tzen, J. T., Chen, M. J., and Liu, J. R., 2008, *Bioresour. Technol.,* **99,** 8662–8666.

Hyeon, J. E., Jeon, S. D., and Han, S. O., 2013, *Biotechnol. Adv.,* **31,** 936–944.

Ito, T., Sadamoto, R., Naruchi, K., Togame, H., Takemoto, H., Kondo, H., and Nishimura, S., 2010, *Biochemistry,* **49,** 2604–2614.

Jo, S. J., Maeda, M., Ooi, T., and Taguchi, S., 2006, *J. Biosci. Bioeng.,* **102,** 233–236.

Johnson, A. K., Zawadzka, A. M., Deobald, L. A., Crawford, R. L., and Paszczynski, A. J., 2008, *J. Nanoparticle Res.,* **10,** 1009–1025.

Junius, F. K., O'Donoghue, S. I., Nilges, M., Weiss, A. S., and King, G. F., 1996, *J. Biol. Chem.,* **271,** 13663–13667.

Küpcü, S., Mader, C., and Sára, M., 1995, *Biotechnol. Appl. Biochem.,* **21,** 275–286.

Kacar, T., Zin, M. T., So, C., Wilson, B., Ma, H., Gul-Karaguler, N., Jen, A. K., Sarikaya, M., and Tamerler, C., 2009, *Biotechnol. Bioeng.,* **103,** 696–705.

Katchalski-Katzir, E., 1993, *Trends Biotechnol.,* **11,** 471–478.

Keppler, A., Gendreizig, S., Gronemeyer, T., Pick, H., Vogel, H., and Johnsson, K., 2003, *Nat. Biotechnol.,* **21,** 86–89.

Khan, A. A., and Alzohairy, M. A., 2010, *Res. J. Biol. Sci,* **5,** 565–575.

Khan, F., He, M., and Taussig, M. J., 2006, *Anal. Chem.,* **78,** 3072–3079.

Kim, Y. H., Campbell, E., Yu, J., Minteer, S. D., and Banta, S., 2013, *Angew. Chem. Int. Ed. Engl.,* **52,** 1437–1440.

Kim, T. W., Chokhawala, H. A., Hess, M., Dana, C. M., Baer, Z., Sczyrba, A., Rubin, E. M., Blanch, H. W., and Clark, D. S., 2011b, *Angew. Chem. Int. Ed. Engl.,* **50,** 11215–11218.

Kim, J., Grate, J. W., and Wang, P., 2008, *Trends Biotechnol.,* **26,** 639–646.

Kim, M. I., Kim, J., Lee, J., Jia, H., Na, H. B., Youn, J. K., Kwak, J. H., Dohnalkova, A., Grate, J. W., Wang, P., Hyeon, T., Park, H. G., and Chang, H. N., 2007, *Biotechnol. Bioeng.,* **96,** 210–218.

Kim, H., Sun, Q., Liu, F., Tsai, S. L., and Chen, W., 2012, *Top. Catal.,* **55,** 1138–1145.

Kim, D. M., Umetsu, M., Takai, K., Matsuyama, T., Ishida, N., Takahashi, H., Asano, R., and Kumagai, I., 2011a, *Small,* **7,** 656–664.

Kindermann, M., George, N., Johnsson, N., and Johnsson, K., 2003, *J. Am. Chem. Soc.,* **125,** 7810–7811.

Kondo, A., and Ueda, M., 2004, *Appl. Microbiol. Biotechnol.,* **64,** 28–40.

Koonin, E. V., Senkevich, T. G., and Dolja, V. V., 2006, *Biol. Direct,* **1,** 29.

Kouzarides, T., and Ziff, E., 1989, *Nature,* **340,** 568–571.

Krauland, E. M., Peelle, B. R., Wittrup, K. D., and Belcher, A. M., 2007, *Biotechnol. Bioeng.,* **97,** 1009–1020.

Kricka, W., Fitzpatrick, J., and Bond, U., 2014, *Front. Microbiol.,* **5,** 174.

Kuan, I., Liao, R., Hsieh, H., Chen, K., and Yu, C., 2008, *J. Biosci. Bioeng.,* **105,** 110–115.

Kumada, Y., Tokunaga, Y., Imanaka, H., Imamura, K., Sakiyama, T., Katoh, S., and Nakanishi, K., 2006, *Biotechnol. Prog.,* **22,** 401–405.

Kwon, Y., Coleman, M. A., and Camarero, J. A., 2006, *Angew. Chem. Int. Ed. Engl.,* **45,** 1726–1729.

Lambalot, R. H., Gehring, A. M., Flugel, R. S., Zuber, P., LaCelle, M., Marahiel, M. A., Reid, R., Khosla, C., and Walsh, C. T., 1996, *Chem. Biol.,* **3,** 923–936.

Lamed, R., Setter, E., and Bayer, E. A., 1983, *J. Bacteriol.,* **156,** 828–836.

Landschulz, W. H., Johnson, P. F., and McKnight, S. L., 1988, *Science,* **240,** 1759–1764.

Lee, S. Y., Lee, S., Kho, I. H., Lee, J. H., Kim, J. H., and Chang, J. H., 2011, *Chem. Commun. (Camb.),* **47,** 9989–9991.

Lee, J. H., Wong, N. Y., Tan, L. H., Wang, Z., and Lu, Y., 2010, *J. Am. Chem. Soc.,* **132,** 8906–8908.

Letant, S. E., Hart, B. R., Kane, S. R., Hadi, M. Z., Shields, S. J., and Reynolds, J. G., 2004, *Adv. Mater.,* **16,** 689–693.

Ling, D., Gao, L., Wang, J., Shokouhimehr, M., Liu, J., Yu, Y., Hackett, M. J., So, P. K., Zheng, B., Yao, Z., Xia, J., and Hyeon, T., 2014, *Chemistry,* **20,** 7916–7921.

Liu, C. C., and Schultz, P. G., 2010, *Annu. Rev. Biochem.,* **79,** 413–444.

Liu, J. H., Selinger, L. B., Cheng, K. J., Beauchemin, K. A., and Moloney, M. M., 1997, *Mol. Breed.,* **3,** 463–470.

Liu, W. S., Wang, L., and Jiang, R. R., 2012, *Top. Catal.,* **55,** 1146–1156.

Loer, D. S., and Herman, E. M., 1993, *Plant Physiol.,* **101,** 993–998.

Lu, H. D., Wheeldon, I. R., and Banta, S., 2010, *Protein Eng. Des. Sel.,* **23,** 559–566.

Luckarift, H. R., Spain, J. C., Naik, R. R., and Stone, M. O., 2004, *Nat. Biotechnol.,* **22,** 211–213.

Lumb, K. J., and Kim, P. S., 1995, *Science,* **268,** 436–439.

Lynd, L. R., van Zyl, W. H., McBride, J. E., and Laser, M., 2005, *Curr. Opin. Biotechnol.,* **16,** 577–583.

Lynd, L. R., Weimer, P. J., van Zyl, W. H., and Pretorius, I. S., 2002, *Microbiol. Mol. Biol. Rev.,* **66,** 506–577.

Manyar, H. G., Gianotti, E., Sakamoto, Y., Terasaki, O., Coluccia, S., and Tumbiolo, S., 2008, *J. Phys. Chem. C,* **112,** 18110–18116.

Marvin, D. A., Welsh, L. C., Symmons, M. F., Scott, W. R., and Straus, S. K., 2006, *J. Mol. Biol.,* **355,** 294–309.

Mateo, C., Palomo, J. M., Fernandez-Lorente, G., Guisan, J. M., and Fernandez-Lafuente, R., 2007, *Enzyme Microb. Technol.,* **40,** 1451–1463.

McCafferty, J., Jackson, R. H., and Chiswell, D. J., 1991, *Protein Eng.,* **4,** 955–961.

McClendon, S., Mao, Z., Shin, H.-D., Wagschal, K., and Chen, R., 2012, *Appl. Biochem. Biotechnol.,* **167,** 395–411.

Messner, P., and Sleytr, U. B., 1992, *Adv. Microb. Physiol.* (Rose, A. H., ed.), Academic Press, vol. 33, pp. 213–275.

Meyer, R., Giselbrecht, S., Rapp, B. E., Hirtz, M., and Niemeyer, C. M., 2014, *Curr. Opin. Chem. Biol.,* **18,** 8–15.

Miao, L., Zhang, X., Si, C., Gao, Y., Zhao, L., Hou, C., Shoseyov, O., Luo, Q., and Liu, J., 2014, *Org. Biomol. Chem.,* **12,** 362–369.

Mifune, J., Grage, K., and Rehm, B. H., 2009, *Appl. Environ. Microbiol.,* **75,** 4668–4675.

Minamihata, K., Goto, M., and Kamiya, N., 2011a, *Bioconjug. Chem.,* **22,** 2332.

Minamihata, K., Goto, M., and Kamiya, N., 2011b, *Bioconjug. Chem.,* **22,** 74–81.

Minamihata, K., Tokunaga, M., Kamiya, N., Kiyoyama, S., Sakuraba, H., Ohshima, T., and Goto, M., 2009, *Biotechnol. Lett.,* **31,** 1037–1041.

Mingardon, F., Chanal, A., Tardif, C., Bayer, E. A., and Fierobe, H. P., 2007, *Appl. Environ. Microbiol.,* **73,** 7138–7149.

Minten, I. J., Claessen, V. I., Blank, K., Rowan, A. E., Nolte, R. J. M., and Cornelissen, J. J. L. M., 2011, *Chem. Sci.,* **2,** 358–362.

Mitsuzawa, S., Kagawa, H., Li, Y., Chan, S. L., Paavola, C. D., and Trent, J. D., 2009, *J. Biotechnol.,* **143,** 139–144.

Morais, S., Barak, Y., Hadar, Y., Wilson, D. B., Shoham, Y., Lamed, R., and Bayer, E. A., 2011, *Mbio.,* 2, pii: e00233-11.

Morais, S., Heyman, A., Barak, Y., Caspi, J., Wilson, D. B., Lamed, R., Shoseyov, O., and Bayer, E. A., 2010, *J. Biotechnol.*, **147**, 205–211.

Moreland, J. L., Gramada, A., Buzko, O. V., Zhang, Q., and Bourne, P. E., 2005, *BMC Bioinformatics*, **6**, 21.

Mori, Y., Ozasa, S., Kitaoka, M., Noda, S., Tanaka, T., Ichinose, H., and Kamiya, N., 2013, *Chem. Commun. (Camb.)*, **49**, 6971–6973.

Moriyama, K., Sung, K., Goto, M., and Kamiya, N., 2011, *J. Biosci. Bioeng.*, **111**, 650–653.

Muller, J., and Niemeyer, C. M., 2008, *Biochem. Biophys. Res. Commun.*, **377**, 62–67.

Murashima, K., Kosugi, A., and Doi, R. H., 2003, *J. Bacteriol.*, **185**, 1518–1524.

Murphy, D. J., 1993, *Prog. Lipid Res.*, **32**, 247–280.

Murphy, D. J., 2012, *Protoplasma*, **249**, 541–585.

Niemeyer, C. M., 2002, *Trends Biotechnol.*, **20**, 395–401.

Niemeyer, C. M., 2010, *Angew. Chem. Int. Ed. Engl.*, **49**, 1200–1216.

Niemeyer, C. M., Boldt, L., Ceyhan, B., and Blohm, D., 1999, *Anal. Biochem.*, **268**, 54–63.

Niemeyer, C. M., Koehler, J., and Wuerdemann, C., 2002, *ChemBioChem*, **3**, 242–245.

Niemeyer, C. M., Sano, T., Smith, C. L., and Cantor, C. R., 1994, *Nucleic Acids Res.*, **22**, 5530–5539.

Noren, C. J., Anthony-Cahill, S. J., Griffith, M. C., Schultz, P. G., 1989, *Science*, **244**, 182–188.

O'Neil, A., Reichhardt, C., Johnson, B., Prevelige, P. E., and Douglas, T., 2011, *Angew. Chem. Int. Ed. Engl.*, **50**, 7425–7428.

O'Shea, E. K., Lumb, K. J., and Kim, P. S., 1993, *Curr. Biol.*, **3**, 658–667.

Olson, D. G., McBride, J. E., Shaw, A. J., and Lynd, L. R., 2012, *Curr. Opin. Biotechnol.*, **23**, 396–405.

Oudgenoeg, G., Dirksen, E., Ingemann, S., Hilhorst, R., Gruppen, H., Boeriu, C. G., Piersma, S. R., van Berkel, W. J., Laane, C., and Voragen, A. G., 2002, *J. Biol. Chem.*, **277**, 21332–21340.

Parent, K. N., Khayat, R., Tu, L. H., Suhanovsky, M. M., Cortines, J. R., Teschke, C. M., Johnson, J. E., and Baker, T. S., 2010, *Structure*, **18**, 390–401.

Park, T. J., Lee, S. Y., Lee, S. J., Park, J. P., Yang, K. S., Lee, K. B., Ko, S., Park, J. B., Kim, T., Kim, S. K., Shin, Y. B., Chung, B. H., Ku, S. J., Kim do, H., and Choi, I. S., 2006, *Anal. Chem.*, **78**, 7197–7205.

Parker, M. H., Casjens, S., and Prevelige, P. E., Jr, 1998, *J. Mol. Biol.,* **281,** 69–79.

Parker, L., Kendall, A., and Stubbs, G., 2002, *Virology,* **300,** 291–305.

Parthasarathy, R., Subramanian, S., and Boder, E. T., 2007, *Bioconjug. Chem.,* **18,** 469–476.

Patterson, D. P., Prevelige, P. E., and Douglas, T., 2012a, *ACS Nano,* **6,** 5000.

Patterson, D. P., Schwarz, B., El-Boubbou, K., van der Oost, J., Prevelige, P. E., and Douglas, T., 2012b, *Soft Matter,* **8,** 10158–10166.

Patterson, D. P., Schwarz, B., Waters, R. S., Gedeon, T., and Douglas, T., 2014, *ACS Chem. Biol.,* **9,** 359–365.

Pegg, A. E., 2000, *Mutat. Res./Rev. Mutat. Res.,* **462,** 83–100.

Peters, V., and Rehm, B. H., 2006, *Appl. Environ. Microbiol.,* **72,** 1777–1783.

Petka, W. A., Harden, J. L., McGrath, K. P., Wirtz, D., and Tirrell, D. A., 1998, *Science,* **281,** 389–392.

Pierre, S. J., Thies, J. C., Dureault, A., Cameron, N. R., van Hest, J. C. M., Carette, N., Michon, T., and Weberskirch, R., 2006, *Adv. Mater.,* **18,** 1822–1826.

Porath, J., Carlsson, J., Olsson, I., and Belfrage, G., 1975, *Nature,* **258,** 598–599.

Pum, D., Messner, P., and Sleytr, U. B., 1991, *J. Bacteriol.,* **173,** 6865–6873.

Rasiah, I. A., and Rehm, B. H., 2009, *Appl. Environ. Microbiol.,* **75,** 2012–2016.

Raven, D. J., Earland, C., and Little, M., 1971, *Biochim. Biophys. Acta,* **251,** 96–99.

Rehm, B. H., 2003, *Biochem. J.,* **376,** 15–33.

Rehm, B. H., 2006, *Biotechnol. Lett.,* **28,** 207–213.

Roberts, N., Scott, R., and Tzen, J., 2008, *Open Biotechnol. J.,* **2,** 13–21.

Robins, K. J., Hooks, D. O., Rehm, B. H., and Ackerley, D. F. 2013 *PLoS One,* **8,** e59200.

Roessl, U., Nahalka, J., and Nidetzky, B., 2010, *Biotechnol. Lett.,* **32,** 341–350.

Ryu, Y. H., and Schultz, P. G. 2006 *Nat. Methods,* **3,** 263.

Sakiyama, T., Ueno, S., Imamura, K., and Nakanishi, K., 2004, *J. Mol. Catal. B Enzymatic,* **28,** 207–214.

Sára, M., Küpcü, S., Weiner, C., Weigert, S., and Sleytr, U. B., 1993, *Immobilised Macromolecules: Application Potentials* (Sleytr, U. B., Messner, P., Pum, D., and Sára, M., eds.), Springer London, pp. 71–86.

Sara, M., and Sleytr, U. B., 1996, *Micron,* **27,** 141–156.

Sarikaya, M., Tamerler, C., Jen, A. K., Schulten, K., and Baneyx, F., 2003, *Nat. Mater.,* **2,** 577–585.

Schaffer, C., Novotny, R., Kupcu, S., Zayni, S., Scheberl, A., Friedmann, J., Sleytr, U. B., and Messner, P., 2007, *Small,* **3,** 1549–1559.

Seo, M. H., Han, J., Jin, Z., Lee, D. W., Park, H. S., and Kim, H. S., 2011, *Anal. Chem.*, **83**, 2841–2845.

Sheehan, D., Meade, G., Foley, V. M., and Dowd, C. A., 2001, *Biochem. J.*, **360**, 1–16.

Sheldon, R. A., and van Rantwijk, F., 2004, *Aust. J. Chem.*, **57**, 281.

Shrestha, P., Holland, T. M., and Bundy, B. C., 2012, *BioTechniques*, **53**, 163–174.

Shukla, G. S., and Krag, D. N., 2010, *J. Drug Target.*, **18**, 115–124.

Sleytr, U. B., 1976, *J. Ultrastruct. Res.*, **55**, 360–377.

Sleytr, U. B., Huber, C., Ilk, N., Pum, D., Schuster, B., and Egelseer, E. M., 2007, *FEMS Microbiol. Lett.*, **267**, 131–144.

Sleytr, U. B., Messner, P., Pum, D., and Sara, M., 1993, *Mol. Microbiol.*, **10**, 911–916.

Sleytr, U. B., Schuster, B., Egelseer, E. M., and Pum, D., 2014, *FEMS Microbiol. Rev.*, **38**, 823–864.

Slocik, J. M., and Naik, R. R., 2006, *Adv. Mater.*, **18**, 1988–1992.

Smith, G., 1985, *Science*, **228**, 1315–1317.

Smith, S. P., and Bayer, E. A., 2013, *Curr. Opin. Struct. Biol.*, **23**, 686–694.

Smith, M. T., Hawes, A. K., and Bundy, B. C., 2013a, *Curr. Opin. Biotechnol.*, **24**, 620–626.

Smith, G. P., and Petrenko, V. A., 1997, *Chem. Rev.*, **97**, 391–410.

Smith, M. T., Wu, J. C., Varner, C. T., and Bundy, B. C., 2013b, *Biotechnol. Prog.*, **29**, 247–254.

Speir, J. A., Munshi, S., Wang, G., Baker, T. S., and Johnson, J. E., 1995, *Structure*, **3**, 63–78.

Spivey, H. O., and Ovadi, J., 1999, *Methods*, **19**, 306–321.

Srikrishnan, S., Chen, W., and Da Silva, N. A., 2013, *Biotechnol. Bioeng.*, **110**, 275–285.

Steffensen, C. L., Mattinen, M. L., Andersen, H. J., Kruus, K., Buchert, J., and Nielsen, J. H., 2008, *Eur. Food Res. Technol.*, **227**, 57–67.

Steinbüchel, A., and Hein, S., 2001, *Biopolyesters*, Springer, pp. 81–123.

Steinmann, B., Christmann, A., Heiseler, T., Fritz, J., and Kolmar, H., 2010, *Appl. Environ. Microbiol.*, **76**, 5563–5569.

Straffon, M. J., Widdowson, C. A., and Zachariou, M., 2006, Google Patents.

Sugimura, Y., Ueda, H., Maki, M., and Hitomi, K., 2007, *J. Biotechnol.*, **131**, 121–127.

Sun, Q., Madan, B., Tsai, S. L., DeLisa, M. P., and Chen, W., 2014, *Chem. Commun. (Camb.)*, **50,** 1423–1425.

Sunbul, M., Emerson, N., and Yin, J., 2011, *ChemBioChem*, **12,** 380–386.

Sutherland, I. W., Hughes, K. A., Skillman, L. C., and Tait, K., 2004, *FEMS Microbiol. Lett.*, **232,** 1–6.

Tai, S. S., Chen, M. C., Peng, C. C., and Tzen, J. T., 2002, *Biosci. Biotechnol. Biochem.*, **66,** 2146–2153.

Takasaki, S., Kato, Y., Murata, M., Homma, S., and Kawakishi, S., 2005, *Biosci. Biotechnol. Biochem.*, **69,** 1686–1692.

Tan, G. Y. A., Chen, C. L., Li, L., Ge, L., Wang, L., Razaad, I. M. N., Li, Y., Zhao, L., Mo, Y., and Wang, J. Y., 2014, *Polymers*, **6,** 706–754.

Terpe, K., 2003, *Appl. Microbiol. Biotechnol.*, **60,** 523–533.

Teschke, C. M., McGough, A., and Thuman-Commike, P. A., 2003, *Biophys. J.*, **84,** 2585–2592.

Tominaga, J., Kamiya, N., Doi, S., Ichinose, H., and Goto, M., 2004, *Enzyme Microb. Technol.*, **35,** 613–618.

Tominaga, J., Kamiya, N., Doi, S., Ichinose, H., Maruyama, T., and Goto, M., 2005, *Biomacromolecules*, **6,** 2299–2304.

Ton-That, H., Faull, K. F., and Schneewind, O., 1997, *J. Biol. Chem.*, **272,** 22285–22292.

Tsai, S. L., DaSilva, N. A., and Chen, W., 2013a, *ACS Synth. Biol.*, **2,** 14–21.

Tsai, S.-L., Goyal, G., and Chen, W., 2010, *Appl. Environ. Microbiol.*, **76,** 7514–7520.

Tsai, S. L., Oh, J., Singh, S., Chen, R., and Chen, W., 2009, *Appl. Environ. Microbiol.*, **75,** 6087–6093.

Tsai, S. L., Park, M., and Chen, W., 2013b, *Biotechnol. J.*, **8,** 257–261.

Tschiggerl, H., Breitwieser, A., de Roo, G., Verwoerd, T., Schaffer, C., and Sleytr, U. B., 2008, *J. Biotechnol.*, **133,** 403–411.

Tseng, C. W., Liao, C. Y., Sun, Y., Peng, C. C., Tzen, J. T., Guo, R. T., and Liu, J. R., 2014, *J. Agric. Food Chem.*, **62,** 6771–6776.

Tzen, J., 2011, *Sesame Gen. Sesamum*, **48,** 187.

Tzen, J. T. C., 2012, *ISRN Botany*, **2012,** 1–7.

Tzen, J. T., and Huang, A. H., 1992, *J. Cell Biol.*, **117,** 327.

Tzen, J. T., Lie, G. C., and Huang, A. H., 1992, *J. Biol. Chem.*, **267,** 15626–15634.

Tzen, J., Wang, M., Tai, S., Lee, T., and Peng, C., 2003, *Adv. Plant Physiol.*, **6,** 93–105.

Uhde-Holzem, K., Schlosser, V., Viazov, S., Fischer, R., and Commandeur, U., 2010, *J. Virol. Methods,* **166,** 12–20.

Valappil, S. P., Boccaccini, A. R., Bucke, C., and Roy, I., 2007, *Antonie Van Leeuwenhoek,* **91,** 1–17.

Verduin, B. J. M., 1974, *FEBS Lett.,* **45,** 50–54.

Verduin, B. J. M., 1978, *J. Gen. Virol.,* **39,** 131–147.

Wang, L., Brock, A., Herberich, B., and Schultz, P. G., 2001, *Science,* **292,** 498–500.

Wang, W., Dgany, O., Dym, O., Altman, A., Shoseyov, O., and Almog, O., 2003, *Acta Crystallogr. D Biol. Crystallogr.,* **59,** 512–514.

Wang, W. X., Dgany, O., Wolf, S. G., Levy, I., Algom, R., Pouny, Y., Wolf, A., Marton, I., Altman, A., and Shoseyov, O., 2006, *Biotechnol. Bioeng.,* **95,** 161–168.

Wang, W. X., Pelah, D., Alergand, T., Shoseyov, O., and Altman, A., 2002, *Plant Physiol.,* **130,** 865–875.

Wang, C., Stewart, R. J., and Kopecek, J., 1999, *Nature,* **397,** 417–4120.

Wang, L., Wei, L., Chen, Y., and Jiang, R., 2010, *J. Biotechnol.,* **150,** 57–63.

Wang, Z. G., Wilner, O. I., and Willner, I., 2009, *Nano Lett.,* **9,** 4098–4102.

Wang, L. A., Xu, R., Chen, Y. A., and Jiang, R. R., 2011, *J. Mol. Catal. B Enzymatic,* **69,** 120–126.

Wang, L., Zhang, H., Ching, C. B., Chen, Y., and Jiang, R., 2012, *Appl. Microbiol. Biotechnol.,* **94,** 1233–1241.

Weigele, P. R., Sampson, L., Winn-Stapley, D., and Casjens, S. R., 2005, *J. Mol. Biol.,* **348,** 831–844.

Wen, F., Sun, J., and Zhao, H., 2010, *Appl. Environ. Microbiol.,* **76,** 1251.

Whaley, S. R., English, D. S., Hu, E. L., Barbara, P. F., and Belcher, A. M., 2000, *Nature,* **405,** 665–668.

Wheeldon, I. R., Barton, S. C., and Banta, S., 2007, *Biomacromolecules,* **8,** 2990–2994.

Wheeldon, I. R., Campbell, E., and Banta, S., 2009, *J. Mol. Biol.,* **392,** 129.

Wheeldon, I. R., Gallaway, J. W., Barton, S. C., and Banta, S., 2008, *Proc. Natl. Acad. Sci. U. S. A.,* **105,** 15275–15280.

Wilner, O. I., Weizmann, Y., Gill, R., Lioubashevski, O., Freeman, R., and Willner, I., 2009, *Nat. Nanotechnol.,* **4,** 249–255.

Wilson, D. S., Keefe, A. D., and Szostak, J. W., 2001, *Proc. Natl. Acad. Sci. U. S. A.,* **98,** 3750–3755.

Wong, L. S., Khan, F., and Micklefield, J., 2009, *Chem. Rev.,* **109,** 4025–4053.

Wong, L. S., Okrasa, K., and Micklefield, J., 2010, *Org. Biomol. Chem.,* **8,** 782–787.

Wong, L. S., Thirlway, J., and Micklefield, J., 2008, *J. Am. Chem. Soc.,* **130,** 12456–12464.

Wong, C.-H., and Whitesides, G. M., 1994, *Enzymes in Synthetic Organic Chemistry,* Academic Press.

Wong, N. Y., Zhang, C., Tan, L. H., and Lu, Y., 2011, *Small,* **7,** 1427.

Woolfson, D. N., 2005, *Adv. Protein Chem.* (David, A. D. P., and John, M. S., eds.), Academic Press, vol. 70, pp. 79–112.

Wu, J. C., Hutchings, C. H. Lindsay, M. J., Bundy, B. C., 2014, Conference Paper: 386310;14 AIChE Annual Meeting.

Wu, J. C., Hutchings, C. H., Lindsay, M. J., Werner, C. J., and Bundy, B. C., 2015, *J. Biotechnol.,* **193,** 83–90.

Xie, J., and Schultz, P. G., 2005, *Curr. Opin. Chem. Biol.,* **9,** 548–554.

Xie, J., and Schultz, P. G., 2006, *Nat. Rev. Mol. Cell Biol.,* **7,** 775–782.

Yamamura, Y., Hirakawa, H., Yamaguchi, S., and Nagamune, T., 2011, *Chem. Commun. (Camb.),* **47,** 4742–4744.

Yang, M., Choi, B. G., Park, T. J., Heo, N. S., Hong, W. H., and Lee, S. Y., 2011, *Nanoscale,* **3,** 2950–2956.

Yin, J., Lin, A. J., Golan, D. E., and Walsh, C. T., 2006, *Nat. Protoc.,* **1,** 280–285.

Yin, J., Straight, P. D., McLoughlin, S. M., Zhou, Z., Lin, A. J., Golan, D. E., Kelleher, N. L., Kolter, R., and Walsh, C. T., 2005, *Proc. Natl. Acad. Sci. U. S. A.,* **102,** 15815–15820.

Zeng, Y.-F., Hung, Y.-J., Chen, M.-J., Peng, C.-C., Tzen, J. T. C., and Liu, J.-R., 2009, *J. Chem. Technol. Biotechnol.,* **84,** 1480–1485.

Zhang, K., Diehl, M. R., and Tirrell, D. A., 2005, *J. Am. Chem. Soc.,* **127,** 10136–10137.

Zhang, Y., Yang, Y., Ma, W., Guo, J., Lin, Y., and Wang, C., 2013, *ACS Appl. Mater. Interfaces,* **5,** 2626–2633.

Zheng, J., Constantinou, P. E., Micheel, C., Alivisatos, A. P., Kiehl, R. A., and Seeman, N. C., 2006, *Nano Lett.,* **6,** 1502–1504.

Zhou, N. E., Kay, C. M., and Hodges, R. S., 1993, *Biochemistry,* **32,** 3178–3187.

Chapter 6

Graphene-Based Nanobiocatalytic Systems

Michaela Patila,[a] George Orfanakis,[a] Angeliki C. Polydera,[a] Ioannis V. Pavlidis,[b] and Haralambos Stamatis[a]

[a]*Laboratory of Biotechnology, Department of Biological Applications and Technologies, University of Ioannina, 45510 Ioannina, Greece*
[b]*Group of Biotechnology, Department of Biochemistry, University of Kassel, 34132 Kassel, Germany*

hstamati@uoi.gr

6.1 Introduction

Graphene, a single atomic layer of carbon atoms that are chemically bonded with a hexagonal symmetry, shows a true two-dimensional (2D) crystalline structure and unique thermal, mechanical, and electrical properties (Novoselov et al., 2004; Novoselov et al., 2005). Graphene (G) is the basic building block of carbon-based material of other dimensionalities, i.e. it can be wrapped up into 0D fullerenes, rolled up into 1D carbon nanotubes or stacked into three-dimensional (3D) layered structures (Geim et al., 2007). Depending on the number of layers, composition, surface chemistry, lateral dimensions and defect density, graphene-based nanomaterials (GBNs) can be broadly classified as mono layer graphene, few layer graphene (FLG), ultrathin graphite, graphene quantum dots (GQDs), graphene oxide (GO) and reduced graphene oxide (rGO) (Sanchez et al., 2012).

Biocatalysis and Nanotechnology
Edited by Peter Grunwald

ISBN 978-981-4613-69-9 (Hardcover), 978-1-315-19660-2 (eBook)
www.panstanford.com

GO is the highly oxidized form of graphene with random distributed hydroxyl, epoxy, and carboxyl groups at edges and basal planes, which make GO hydrophilic. The peripheral carboxylate groups provide colloidal stability and pH-dependent negative surface charge, while other uncharged oxygen-containing groups distributed on the basal plane are polar, allowing weak interactions and hydrogen bond formations with other molecules. In addition, the π-conjugated system of the basal plane facilitates π–π interactions and non-covalent functionalization (Kim et al., 2010; Park et al., 2009).

GO under reducing conditions, like thermal, UV and chemical treatment, loses its oxygen containing groups and resembles graphene, due to its excessive exfoliation (Park et al., 2009). Reduced GO regains its conductivity, while the surface charge and hydrophilicity are reduced. Due to excellent properties, such as very good mechanical and thermal stability, chemical inertness, and exceptional electronic properties, these GBNs have attracted great interest in biotechnology and biomedicine such as in gene and drug delivery, bioimaging and biosensing, as well as in the construction of biosensors and biomedical devices (Goenka et al., 2014; Bitounis et al., 2013; Krishna et al., 2013; Du et al., 2012; Wang et al., 2011). Particularly in the field of nanobiotechnology, the use of graphene and its oxidized form, graphene oxide, as immobilization supports for the development of effective nanobiocatalytic systems is being intensively researched. The large surface area (two accessible sides), the abundant oxygen-containing surface functionalities, and the high water solubility of GBNs create an ideal immobilization support for various bioactive molecules (Fig. 6.1) such as genes, drugs, antibodies and other proteins, including enzymes (Pavlidis et al., 2014; Gokhale et al., 2013; Liu et al., 2013; Ge et al., 2012; Jin et al., 2012).

GBNs can be engineered for grafting desirable functional groups (such as epoxide, amine, carboxylic, hydroxyl and thiol groups) onto their surface providing functionalized nanomaterials with tailor-made properties and improved suitability for their application as nanoscaffolds for biomolecule immobilization (Pavlidis et al., 2014). There have been extensive studies on the immobilization of enzymes onto GBNs which have shown that the geometry and surface chemistry of these nanomaterials significantly influence the interactions with biomolecules and thus affect their conformation and biological function (Jin et al., 2012; Pavlidis et al., 2012a,b). Furthermore, the surface modification of GBNs offers the possibility of introduction of functionalities that enhance the catalytic properties and the stability of the immobilized enzymes, such as protection from proteolytic cleavage (Wang et al., 2011), increased thermostability and operational stability (Pavlidis et al., 2012a), increased electron transfer in case of oxidoreductases (Putzbach and Ronkainen, 2013; Walcarius et al., 2013; Kuila et al., 2011), as well as the facilitation of the incorporation of enzymes in nanodevices and microchip bioreactors (Bosio et al., 2015; Liang et al., 2013; Bao et al., 2011a).

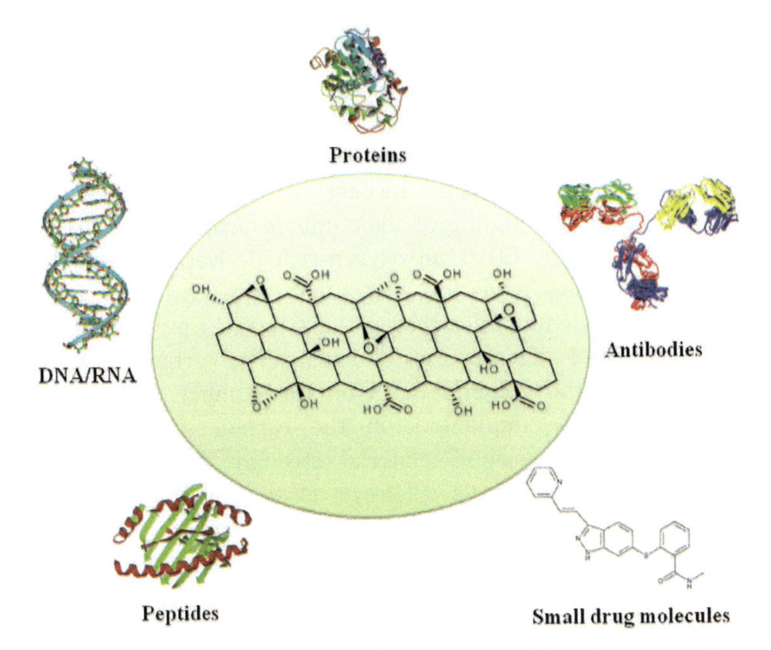

Figure 6.1 Application of graphene oxide (GO) for biomolecule immobilization.

In this chapter, we discuss current advances in the development of nanobiocatalytic systems through the immobilization of different enzymes onto various GBNs. Recent applications of these nanobiocatalysts in various biocatalytic processes of industrial interest as well as in biofuel production, degradation of pollutants, in situ protein digestion, and biosensing have been summarized.

6.2 Graphene-Based Nanomaterials as Enzyme Immobilization Supports

The development of effective nanobiocatalytic systems mainly depends on the use of suitable nanosupports for enzyme immobilization. Among various nanosupports used so far, carbon-based nanomaterials including GBNs such as graphene, GO and rGO, due to their physicochemical and surface properties, have recently emerged as promising and competitive nanosupports for enzyme and protein immobilization (Pavlidis et al., 2014; Zhang et al., 2013). The immobilization of enzymes onto such graphene-based nanosupports was shown to lead to changes in enzyme conformation and catalytic behavior and usually increased enzyme stability and improved specificity, allowing the recovery and multiple usages of the nanobiocatalysts (Pavlidis et al., 2012a).

In order to elaborate the potential of GO and its derivatives as immobilization supports for the development of nanobiocatalytic systems, the understanding of their unique properties is crucial. As indicated before, GO is an oxygen-rich derivative of graphite created by strong oxidation, decorated with hydroxyl, epoxy, and carboxyl functionalities that are distributed randomly on the basal planes and edges of the GO sheets. The carbon to oxygen (C/O) atomic ratio which reflects on the number of functional groups that were implanted on the carbon grid upon the oxidation of the parental material strongly depends on the synthetic method that is used (Dreyer et al., 2010; Gengler et al., 2013). Owing to the existence of such hydrophilic functional groups, GO is highly dispersible in water and polar organic solvents. It must be noted that oxygen content, surface charge, and subsequently water dispersibility of rGO is significantly reduced compared to GO (Bagri et al., 2010).

The high surface area of GO in combination with the abundance of chemically reactive sites, as well as the π-conjugated system, provides a high proportion of catalytically active moieties per weight unit of nanomaterial that facilitate its further functionalization, providing functionalized GO-based nanomaterials with tailor-made properties and improved suitability for their application as nanoscaffolds for enzymes and other biomolecules (Goenka et al., 2014; Pavlidis et al., 2014).

The functionalization of GO-based nanomaterials includes different types of reactions (Dreyer et al., 2010) such as (i) covalent attachment to the carboxylic acid groups, which are located usually at the edges of the graphene sheets, using nucleophiles such as amine or hydroxyl groups; (ii) covalent attachment to the epoxy groups at the basal planes of the sheet, via ring opening reactions with amines; and (iii) noncovalent functionalization that includes van der Waals interactions with polymers, surfactants, and other small molecules, or π–π interactions, with polyaromatic hydrocarbon derivatives. The introduction of such surface functional groups onto GO-based nanomaterials, such as GO and rGO (Fig. 6.2), provides plenty of reaction sites for covalent and non-covalent immobilization of enzymes and other biomacromolecules (Pavlidis et al., 2014). Several studies indicate that the surface chemistry of these nanomaterials significantly influence their interactions with enzymes affecting their conformation and catalytic properties (Zhang et al., 2013).

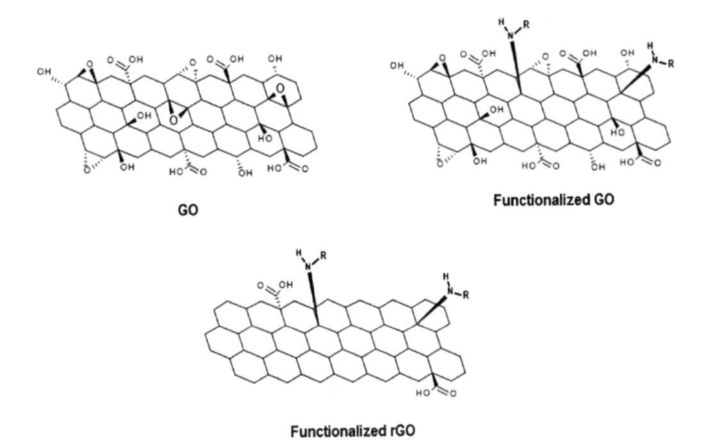

Figure 6.2 Graphic representation of GBNs (R: alkyl chains with different length and terminal functional groups).

6.3 Immobilization Approaches for Enzyme Attachment onto Graphene-Based Nanomaterials

Depending on the biocatalyst and nanomaterials involved, various immobilization approaches have been developed so far for either the physical adsorption or the covalent binding of the biomolecule onto reduced and non-reduced GBNs (Putzbach and Ronkainen, 2013; Verma et al., 2013). These immobilization approaches involve both single and multipoint single-type enzyme attachment onto GBNs (Campbell et al., 2014, Fig. 6.3), while the GBNs-enzyme conjugates are characterized by various spectroscopic and microscopic techniques including circular dichroism (CD) spectroscopy, Fourier transform infrared (FTIR) and Raman spectroscopy, X-ray photoelectron spectroscopy (XPS), energy dispersive X-ray analysis (EDX), scanning electron microscopy (SEM), and atomic force microscopy (AFM) (Pavlidis et al., 2014).

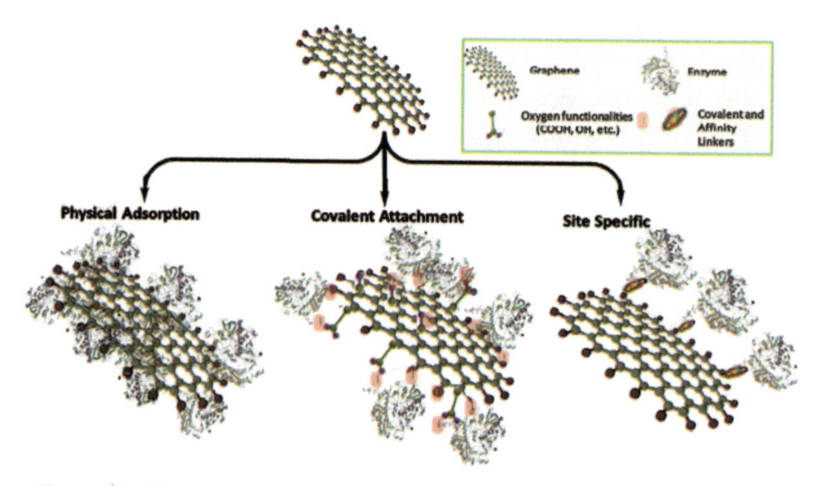

Figure 6.3 Various strategies that can be used to immobilize enzymes onto graphene through physical adsorption, covalent attachment, and site-specific affinity interactions. Reprinted with permission from Karimi et al., 2015.

The non-specific binding via physical adsorption is a simple process which enhances the immobilization efficiency without disrupting the surface of GBNs, unlike chemical functionalization does (Li et al., 2013a; Mesarič et al., 2013; Alwarappan et al., 2012; Hua et al., 2012; Jiang et al., 2012a). During the non-covalent immobilization approach, the nanomaterials are dispersed in the solution (usually by sonication) and then an amount of aqueous solution of the enzyme is added to this dispersion; the mixture is incubated for a specific time interval usually at reduced temperature (4°C). The nanomaterial–enzyme conjugates are recovered by centrifugation or filtration, while rinsing steps are crucial for the removal of the loosely bound enzyme molecules.

The surface chemistry and the geometry of the nanomaterial may affect the interactions developed between GBNs and protein molecules during non-specific binding. The enzymes interact with the surface of GO-derivatives mostly through van der Walls and electrostatic forces, hydrophobic or π–π stacking interactions (Zhang et al., 2012; Zhang et al., 2010b; Gao and Kyratzis, 2008; Jegannathane et al., 2008). Electrostatic interactions can be developed between charged groups of the functionalized GBNs and the protein molecule. Moreover, attractive non-covalent interactions (π–π stacking) can also be developed between the aromatic rings of the surface of GBNs and any aromatic amino acids exposed on the surface of the protein molecule. When the electrostatic interactions are the driving forces of the immobilization, the immobilization yield strongly depends on the charge status of the surface amino acid residues of the biocatalyst and therefore on the pH value, and the ionic strength of the buffer used. Therefore, through electrostatic interaction, different enzymes could exhibit different enzyme loadings and catalytic properties after immobilization on GBNs (Zhang et al., 2013).

On the other hand, hydrophobic interactions between the hydrophobic surface or hydrophobic functional group of the GBNs and the hydrophobic amino acids on the surface of the protein could also play significant role for the non–specific binding of enzymes (Zhang et al., 2012; Gao and Kyratzis, 2008). For instance, the immobilization of horseradish peroxidase (HRP) and oxalate oxidase (OxOx) on chemically reduced graphene oxide is attributed to hydrophobic interactions between the enzymes and the rGO (Zhang et al., 2012). In this case, enzymes can be adsorbed onto rGO directly with a 10-fold higher enzyme loading than onto non-reduced GO, which is richer in chemical groups, indicating that the electrostatic interactions don't play the major role.

To enhance the efficiency of the non-covalent immobilization, various approaches have recently been proposed including the decoration of the GBNs with chemical species and molecules such as calcium ions (Cazorla et al., 2012) or ionic liquids (Jiang et al., 2012b), enhancing the immobilization efficiency without disrupting the nanomaterial's surface.

Recently, an interesting approach based on the pre-adsorption of a highly cationized bovine serum albumin (BSA) to passivate GO, which served as a protein glue for chemically controllable enzyme binding, has been presented (Pattammattel et al., 2013). The cationization of BSA was achieved by reaction of its side chain carboxyl groups with tetraethylenepentamine (TEPA) via carbodiimide coupling. The cationized BSA (cBSA) was bound with high affinity to negatively charged GO, and the cBSA-loaded GO served as a benign host for enzyme binding (Fig. 6.4). It was shown that the immobilization efficiency for hemoglobin and glucose oxidase (GOx) was increased when using this protein glue, compared to parental GO. Furthermore, it was shown that the increased loading of cBSA on GO enhanced the catalytic activity of enzymes tested, as well as improved the retention of their secondary structure, indicating that the sacrificial cBSA on GO provides a protein-compatible surface, which may be used for further manipulation and control of the nanobiointerfaces.

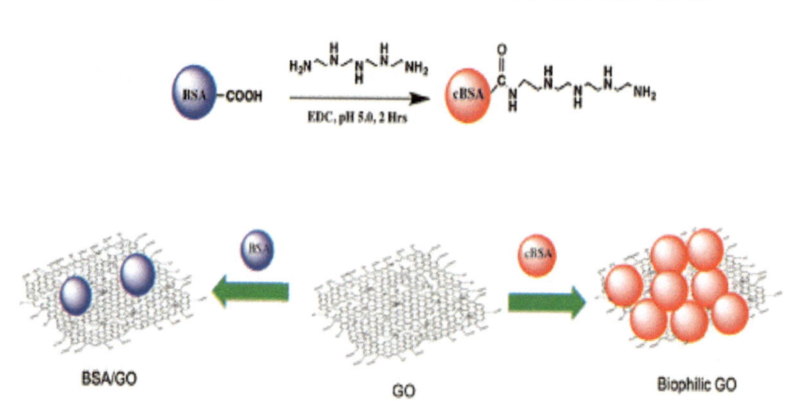

Figure 6.4 Cationization of BSA by reaction of its side chain -COOH groups with tetraethylenepentamine (TEPA) via carbodiimide coupling. Reprinted with permission from Pattammattel et al., 2013.

A major drawback of physical adsorption is the protein leakage from the surface of the nanomaterial (Zhang et al., 2010b; Gao and Kyratzis, 2008). Covalent immobilization can address this drawback by the formation of a covalent bond between the GBNs and the protein molecule which may lead to a more robust nanobiocatalyst, suggesting higher stability and preventing enzyme leakage (Stavyiannoudaki et al., 2009). The development of novel functionalized GBNs with a vast diversity of active terminal functional groups facilitates the immobilization of proteins and other biomacromolecules through various covalent linkage approaches. Such approaches usually require the use of a chemical reagent to act as a cross linker between terminal functional groups of the nanomaterials and the amino acid side chains on the protein surface.

The most common approach applied to GBNs which have plenty of carboxyl groups on their surface, such as GO, is the use of carbodiimide chemistry (Fig. 6.5), usually using 1-ethyl-3-(3-dimethylaminopropyl) carbodiimide (EDC) and *N*-hydroxysuccinimide (NHS) or the more hydrophilic *N*-hydroxysulfosuccinimide (sulfo-NHS). EDC attacks the carboxyl groups of the nanomaterial to form an amine-reactive *O*-acylisouria intermediate which subsequently reacts with amine groups on the surface of the protein to produce stable amine bonds (Gao and Kyratzis, 2008). The addition of NHS stabilizes the intermediate by converting it to a semi-stable amine-reactive NHS ester, thus increasing the coupling efficiency (Sehgal and Vijay, 1994, Staros et al., 1986). This approach was successfully applied for the immobilization of various enzymes such as glucose oxidase (Liu et al., 2010b), lipase (Lau et al., 2014), esterase (Lee et al., 2013a) and trypsin (Xu et al., 2012).

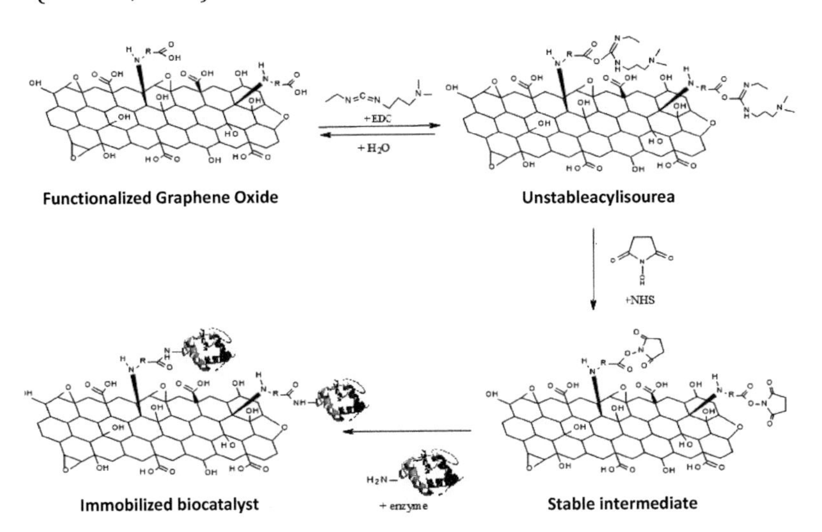

Figure 6.5 Immobilization procedure through EDC/NHS approach.

An alternative approach for enzyme covalent immobilization is the cross-linking of free amine groups located on the surface of the nanomaterials and the lysine residues on the surface of the protein molecule. The most common cross-linker used is glutaraldehyde (Fig. 6.6). To provide amine groups on the graphene surface, graphene is functionalized with several compounds including biomacromolecules. For instance, bovine serum albumin was used to functionalize graphene, to provide the free amine groups and cross-link with cholesterol oxidase and cholesterol esterase (Manjunatha et al., 2012). Hexethylenediamine has recently been used to introduce terminal amine groups onto GO-based nanomaterials for the covalent immobilization of lipases and esterases (Pavlidis et al., 2012a). Su et al. have also used the same cross-linking agent for the immobilization of alkaline protease on GO (Su et al., 2012). In this case, GO was firstly activated with glutaraldehyde and then incubated with alkaline protease to form a stable covalent bond. An interesting process combining both approaches was proposed by Zhu et al. for the covalent immobilization of hemoglobin (Zhu et al., 2012). In this approach, Fe_3O_4 nanoparticles modified with aminopropyltriethoxysilane were covalently bound onto GO nanomaterials through EDC/NHS chemistry and then glutaraldehyde was used as a cross-linker to attach hemoglobin to the silane-derivative of GO.

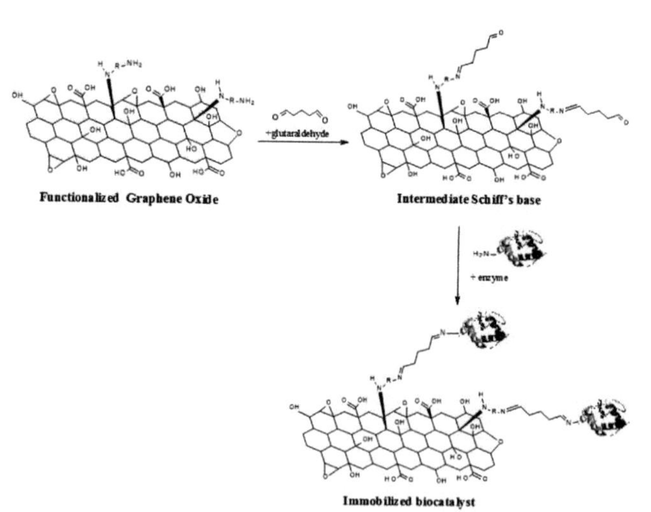

Figure 6.6 Immobilization procedure using glutaraldehyde as cross-linker.

An interesting immobilization technique is based on the modification of carbon-based nanomaterials including graphene and carbon nanotubes by 1-pyrenebutanoic acid succinimidyl ester. The pyrene moiety of this compound interacts with the surface of the nanomaterial by irreversible π–π stacking, while amine groups on a protein react with the anchored succinimidyl ester resulting in the formation of an amide bond for protein immobilization (Chen et al., 2001). Based on this approach, the immobilization of glucose oxidase and glutamic acid dehydrogenase onto graphene for the development of real-time nanoelectronic biosensors was reported (Huang et al., 2010). Lately, a click chemistry approach for the site-specific immobilization of a small laccase has been reported (Guan et al., 2015). Laccase was modified so that it bears the unnatural amino acid 4-azido-L-phenylalanine (AzF) at a defined position on its surface. This approach requires modification of the enzyme at only the tethering site, in contrast to methods of site-specific immobilization that rely on the surface presence of amino acids with a given chemistry (e.g., thiols on cysteines). This way, the site-specific immobilization of the enzyme was feasible onto MWCNTs that were functionalized with a cycloalkyne. The resulting catalyst exhibited high direct electron transfer in a biocathode. This click-chemistry approach is a new approach of rational immobilization that leads to immobilizates of extreme uniformity.

Recently, several affinity immobilization approaches have been developed to produce robust enzyme-GBNs conjugates. In these techniques, the attachment of protein molecules onto nanomaterials is reversible facilitating the regeneration of the nanomaterial and the enzyme. The affinity immobilization approaches are usually based on the use of GBNs modified with antibodies that can specifically bind to the biomacromolecule. For instance, a thrombin detector was developed based on the immobilization of a biotin-modified aptamer onto GO functionalized with avidin (Loo et al., 2013).

The oriented immobilization of enzymes by various specific interactions including enzyme–substrate and lectin–sugar interactions has also been proposed (Takahashi et al., 2011). More specifically, the lectin–sugar biospecific affinity has attracted much attention due to its ability to immobilize glycoenzymes. Zhou and coworkers have proposed a novel selective oriented immobilization of glucose oxidase on GO nanomaterials based on the lectin–sugar biospecific interaction (Zhou et al., 2012). In this approach, concanavalin A (Con A) was covalently attached on pre-activated GO through diimide-activated amidation. The oriented immobilization of glucose oxidase was achieved through strong specific affinity interactions between sugar residues of the enzyme and Con A. The oriented immobilized glucose oxidase exhibited increased stability against pH, temperature, guanidine hydrochloride and higher storage stability compared to the free enzyme.

Other studies indicate that, although affinity-based immobilization produces more stable bioconjugates than physical absorption, the selectivity is reduced, questioning the accuracy of the affinity (Loo et al., 2013).

It is interesting to note that in most studies concerning covalent or affinity immobilization approaches, physical adsorption of protein onto the surface of nanomaterials is not hindered, resulting in bioconjugate systems with two different enzyme populations that probably express different biocatalytic properties (Azamian et al., 2002).

6.4 Effect of GBNs on Structure and Catalytic Behavior of Enzymes

The interactions developed between enzymes and GBNs during various immobilization procedures significantly affect the structure and function of biomolecules and therefore the effectiveness of the nanobiocatalytic systems. GBNs can interact with biomolecules through electrostatic and/or hydrophobic interactions, π–π stacking and van der Waals interactions or hydrogen bonding. In the case of the negatively charged nanomaterials such as GO, electrostatics play a significant role in enzyme–nanomaterial interactions (Yang et al., 2013; Zhang et al., 2013; Zhang et al., 2010b).

These interactions could be very complicated and strongly depend on the nature of the immobilized protein and the physicochemical properties of the nanomaterials used, such as surface chemistry, hydrophobicity and nanomaterial curvature (Patila et al., 2013; Raffaini and Ganazzoli, 2013; Jin et al., 2012;

Pavlidis et al., 2012b). Some of these properties, such as the density of the oxygen-containing groups on the surface of GBNs, varied with the preparation procedure and storage conditions of the nanomaterials complicating the understanding of the interactions between GBNs and enzymes, and thus the rational design of functional GBNs-enzyme conjugates.

The influence of GBNs on the structure and therefore on the catalytic properties of biomolecules, especially enzymes, has been widely reported (Shao et al., 2013; Wei and Ge, 2013; Jiang et al., 2012a; Jin et al., 2012). Several kinetic and spectroscopic studies indicated that the interactions between proteins and GBNs induce conformational changes and affect the catalytic behavior of biocatalysts. In several published works, cytochrome c (cyt c), a small heme protein which is among the best characterized redox proteins able to catalyze peroxidase-like reactions in the presence of an electron acceptor, was used as a model protein to quantitatively assess any structural perturbations resulting from protein–nanoparticles interactions (Shang et al., 2009).

Our group has already presented the influence of functionalized GO on the catalytic and structural characteristics of cyt c in buffer (Patila et al., 2013). UV–Vis and circular dichroism spectroscopy studies suggested that the presence of some functionalized GO nanomaterials resulted in the reorientation of the heme active center to a conformation more accessible to substrate, which led to higher protein peroxidase activity. These changes depended on both hydrophobic interactions between the protein and the surface of the nanomaterial, and electrostatic interactions between cyt c (which is positively charged at pH values lower than its isoelectric point) and the terminal functional groups of the functionalized GO. Similar structural changes of cyt c, which led in good electron transfer capacity, were also observed in the case of its immobilization on different GO nanoplatelets and GBNs (Gupta and Irihamye, 2015). Yang et al. also reported changes in the heme microenvironment of cyt c after its adsorption on GO, accompanied with a loss of the α-helical content of the protein, due to the strong electrostatic interactions developed between the nanomaterial and the biomolecule (Yang et al., 2013). On the other hand, cyt c does not deform significantly after adsorption on graphene, and the secondary structure is preserved to a large extent, as recently was reported (Hu et al., 2015). It was there proposed that the driving forces for the stable conjugation of GBNs and cyt c are the interactions developed between the aromatic amino acids of the protein and graphene, leading to the promotion of electron transfer between graphene and cyt c. It is interesting to note that in the presence of functionalized (Patila et al., 2013) or reduced GO (Yang et al., 2013), the peroxidase activity of cyt c decreased which was correlated to a more rigid protein structure and less accessible heme active site, indicating that the surface chemistry of GO strongly affects the structure and function of cyt c.

Shao et al., using quantitative second-derivative infrared analysis, showed that the adsorption of GOx on GO sheets resulted in the conversion of α-helix to β-sheet structures and therefore led to substantial conformational changes, or even to unfolding of the protein, which caused a significant decrease in the catalytic activity of GOx (Shao et al., 2013). Similar conformational changes induced by GO have also been reported for catalase (Wei and Ge, 2013). In this case, GO leads to unfolding of the protein skeleton, thereby causing change of the framework conformation and reduction of α-helical content in catalase. A strong inhibitory action of GO has also been reported for chymotrypsin (De et al., 2011) suggesting that this inhibition is probably the result of coexistence of anionic, hydrophobic, and π–π stacking interactions.

By contrast, acetylcholinesterase (AChE) retained its native conformation and most of its activity when immobilized onto GO (Mesarič et al., 2013), while the electrostatic interactions developed between HRP or lysozyme with GO led to improved thermal stability and a wide active pH range of enzyme (Zhang et al., 2010a; Zhang et al., 2010b). Experimental and simulation studies on the effect of GO and two other carbon-based nanomaterials such as carbon black (CB) and fullerene (C60) on the catalytic behavior of AchE indicated that nanomaterials interact with AChE affecting in different extent its enzymatic activity but these interactions were not in the active site region. It was proposed that these interactions are primarily influenced by the curvature and surface characteristics of the nanomaterials (Mesarič et al., 2013). Namely, the presence of oxygen rich groups on the surfaces of GO leads to hydrogen bonding and electrostatic interactions between enzyme and GO, while the adsorption of AChE on CB and C60 is mainly driven by hydrophobic interactions.

Moreover, the most efficient AChE inhibitor is CB, while when the enzyme is adsorbed on the GO surface, it retains its native conformation and activity. On the other hand, due to its high surface curvature, C60 is not an efficient adsorbent of AChE.

The conformational changes and the catalytic behavior of enzymes upon immobilization onto GBNs depend on the surface chemistry of the nanomaterials used. The use of biophilic GO, functionalized with BSA, as immobilization matrix for the adsorption of different enzymes, such as horseradish peroxidase, glucose oxidase and catalase, has shown that both the catalytic activity and the secondary structure of the proteins were retained upon immobilization compared to the parental GO nanomaterial (Pattammattel et al., 2013). In a recent study, HRP has been immobilized on GO, graphene and rGO and changes on the secondary structure of the immobilized enzyme have been investigated using CD spectroscopy (Zhang et al., 2015). HRP underwent large structural changes upon immobilization on graphene while lesser extent of conformational changes was observed when the enzyme was immobilized on GO and rGO. It is suggested that when hydrophobic effects are the predominant driving force for adsorption (i.e. in the case of graphene), soluble proteins undergo conformational changes to facilitate adsorption. On the other hand, electrostatic interactions and hydrogen bonding between functionalized nanomaterials (in this case GO and rGO) and enzymes could decrease hydrophobic-induced denaturation. These different conformational changes clearly indicate that the surface chemistry of the nanomaterial affects the interactions between the nanomaterial and the enzymes and thus their conformational state.

The surface chemistry of the nanomaterial seems to play a critical role not only in the conformational state but also in the biocatalytic activity of immobilized enzymes. The catalytic behavior of lipase and cholesterol oxidase was investigated in the presence of different concentrations of graphene (G) and GO (Silva et al., 2015). The catalytic activity of lipase was increased in the presence of GO and especially in the presence of G. On the other hand, the catalytic activity of cholesterol oxidase was increased in the presence of G, while total inactivation was observed in the case of GO. The CD analysis in the far-UV showed that lipase maintained its conformational state in the presence of both G and GO, while cholesterol oxidase showed partial denaturation in the presence of G and total loss of its secondary structure in the presence of GO. This indicates that the surface chemistry of the nanomaterial plays a crucial role in the biocatalytic activity and conformational state of immobilized enzymes, while these effects depend also on the nature of the immobilized enzyme.

The study of the effect of various functionalized GO nanomaterials and multi-wall carbon nanotubes (MWCNTs) on the catalytic behavior of various lipases and esterases, indicated that, the interactions between protein molecules and nanomaterials led, in the case of lipases, to a more rigid structure which enhanced their catalytic efficiency and thermal stability, while in the case of esterases, these interactions led to destabilization and unfolding (Pavlidis et al., 2012a; Pavlidis et al., 2012b). These results indicate that the interactions between GBNs and enzymes depend not only on the surface chemistry of the nanomaterials, but also on the nature of the enzymes involved.

6.5 Applications of Enzymes Immobilized onto GBNs

The engineering of GBNs produces functionalized nanosupports that significantly improve the properties of the immobilized enzymes through introduction of functionalities that enhance the fine-tuned actions of biocatalyst. Namely, the use of functionalized GBNs offers the possibility of manipulating the microenvironment of the immobilized biocatalyst, facilitating the regulation of its activity and/or specificity, the improvement of electron transfer in proteins, the increase of enzymes' thermal and operational stability and the protection from proteolytic cleavage (Pavlidis et al., 2014).

The development of robust conjugates of various biocatalysts with GBNs could be promising for various practical applications including the preparation of enzyme biosensors and enzymes-based biofuel cells, as well as for the development of effective nanobiocatalysts for specific biocatalytic transformations, the degradation of pollutants and agro industrial waste, the wastewater treatment, the proteomic analysis and the development of microchip bioreactors and nanodevices.

6.5.1 Enzyme-Based Biosensors

Graphene is an attractive material for electrode design due to its high surface area, acceptable biocompatibility, unique heterogeneous electron transfer rate, chemical and electrochemical stability and good electrical conductivity. Various graphene derivatives including multilayer graphene nanoplatelets, GO, rGO, and graphene-based hybrids with noble metals (Au, Pt, Pd, Ag), semiconductor quantum dots, organic molecules such as cyclodextrin, ionic liquids and conducting polymers have been developed and used in several types of enzyme-based biosensors (Fang and Wang, 2013; Li et al., 2013b; Walcarius et al., 2013; Li and Han, 2012; Kuila et al., 2011).

Recently, numerous graphene-based biosensors have been developed using several enzymes and proteins such as glucose oxidase, horseradish peroxidase, acetylcholinesterase, cholesterol oxidase, alcohol dehydrogenase, tyrosinase, cyt c and hemoglobin. These biosensors were used for the electrochemical detection of various compounds such as glucose, H_2O_2, and O_2, phenolic pollutants and organophosphates, catechol, ethanol, NADH, nitric oxide, and nitromethane. Table 6.1 presents several selected biosensors designed by combining graphene nanocomposite materials with various enzymes and other proteins.

Several bienzymatic biosensors based on the co-immobilization of various enzymes onto GBNs have been developed for the determination of various biomolecules and metabolites. For instance, amperometric biosensing platforms for the sensing of H_2O_2 and cholesterol using cholesterol oxidase and cholesterol esterase immobilized on hybrid material derived from Pt nanoparticles and graphene (Dey and Raj, 2010, Fig. 6.7) or functionalized graphene modified graphite electrode have been reported.

The possibility of integrating the excellent electrochemical properties of graphene and multienzyme systems in a single multilayer film that could be used for the development of advanced biosensing systems was demonstrated by Zeng et al. (Zeng et al., 2010). Namely, a bienzyme biosensing system for the detection of maltose was prepared by successive layer-by-layer assembly of graphene, glucose oxidase and glucoamylase. Glucoamylase catalyzes the hydrolysis of maltose to glucose, which is then oxidized by glucose oxidase to release H_2O_2. Graphene layer detects the released H_2O_2 to monitor the concentration of maltose.

Table 6.1 Graphene-based biosensors with enzymes

Enzyme	Nanomaterial	Compound detected	Ref.
Glucose oxidase	GO	Glucose	Zhang et al., 2014; Liu et al., 2010b
	rGO	Glucose	Eskandari et al., 2014
	$ZnO/Cu_2O/GO$	Glucose	Elahi et al., 2014
Tyrosinase	ZnO/GO	Hydroxylated polychlorobiphenyls	Rather et al., 2014
	Gr/Au	Bisphenol A	Pan et al., 2015
	GO and rGO	Catechol	Baptista-Pires et al., 2014
Horseradish peroxidase	P-L-His-rGO	Dopamine, H_2O_2	Vilian and Chen, 2014
	AuNPs-GO	H_2O_2	Yu et al., 2015
Hemoglobin	Nafion/GO-IL	Trichloroacetic acid, H_2O_2, $NaNO_2$	Sun et al., 2014
	$GO\text{-}TiO_2$	Trichloacetic acid	Sun et al., 2013
Cholesterol oxidase	CS-G	Cholesterol	Li et al., 2015c
Hemin	GO	Carbofuran	Wong et al., 2014
Alcohol dehydrogenase	IL-graphene/CS	NADH	Shan et al., 2010
Acetyl-cholinesterase	IL-graphene/CS	Organophosphate pesticides	Li and Han, 2012
Cytochrome c	CS/G	Nitric oxide	Wu et al., 2010
	PIL-G	Nitric oxide	Chen and Zhao, 2012

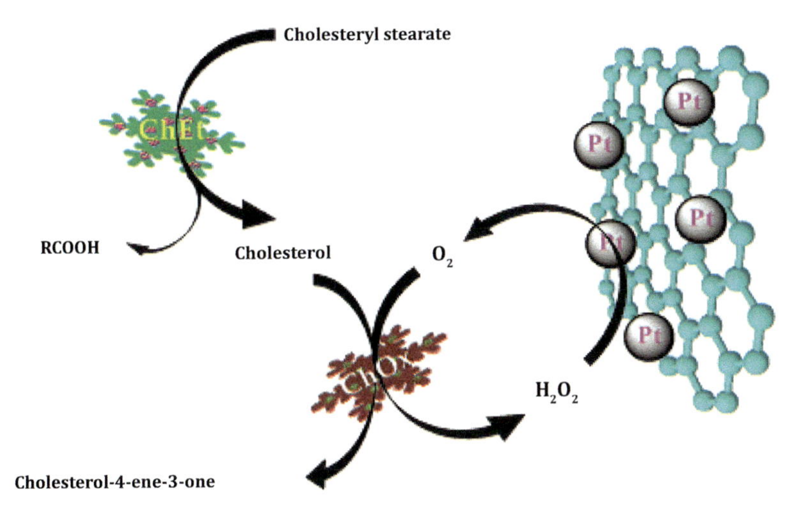

Figure 6.7 The biosensing of cholesterol ester with the GNS-nPt-based biosensor. Reprinted with permission from Dey and Raj, 2010.

Some of the recent works on biosensing have been focused on the development of bioelectrodes for the detection of compounds in living cells or on the design of platforms for mimicking of the natural metabolic pathways in vitro. Gong et al. have recently reported the development of a graphene–cyclodextrin–cytochrome c (GN–CD–Cyt c) assembly with improved electron transfer rate (Gong et al., 2014). This assembly could also mimic the confined environments of the intermembrane space of mitochondria. The assembly showed improved electron transfer rate and high sensitive supramolecular recognition (glycine detection limit ~3.0×10^{-8} mol L^{-1}) to six investigated amino acids. An interesting work has recently been reported by Yu et al. (Yu et al., 2015). This research group developed an effective platform for sensing H_2O_2 inside biological cells. Specifically, a platform of GO decorated with Au nanoparticles was fabricated to immobilize HRP for amperometric detection of H_2O_2. Due to the excellent analytical performance on H_2O_2 detection that this electrode exhibited, the biosensor was further used for real-time monitoring of the flux of H_2O_2 on exposure to ascorbic acid in normal, as well as in cancer cells.

The development of an artificial multi-enzyme complex has been reported by Lu et al., (2015). They designed a cytochrome

P450 (CYP) bienzyme complex on the Au nanoparticle/chitosan/reduced graphene oxide nanocomposite sheets (Au/CS/rGO) to investigate the drug cascade metabolism using an electrochemically-driven approach. The CYP bienzyme-complex exhibited excellent enzymatic activity and synergistic catalytic effect toward the cascade metabolism process of clopidogrel in vitro. The apparent Michaelis constant Kmapp of the CYP bienzyme complex towards clopidogrel was calculated to be 10.82 mM, and catalytic rate constant k_{cat} and catalytic efficiency k_{cat}/K_m were calculated to be 2.42 s^{-1} and 0.2234 $\mu M^{-1}s^{-1}$, respectively, indicating the construction of a powerful tool for sequential enzyme biocatalysis and the potential for investigation and intervention of drug cascade metabolism in vitro.

6.5.2 Biofuel Cells

Enzyme-based biofuel cells (EBFCs) are devices capable of directly transforming chemical to electric energy via electrochemical reactions involving enzymatic catalysis (Willner et al., 2009; Minteer et al., 2007). EBFCs generate power from biomass-derived energy carriers or biofuel substrates, such as glucose, fructose, ethanol and oil under mild operating conditions (e.g. ambient temperature, neutral pH). The use of renewable non-toxic, natural materials as energy sources, makes EBFCs promising candidates as alternative clean energy generators. Moreover, other applications of EFBCs to supply some microscale electronic devices or to use biological fluids as fuel-sources for the activation of implantable biomedical devices are of particular interest (Karimi et al., 2015; Fang and Wang, 2013).

Various oxidoreductases, such as GOD and alcohol dehydrogenase, are used for the oxidation of fuels at the anode of a biofuel cell to generate protons and electrons. At the cathode, oxidases, such as laccase or bilirubin oxidase, are used to catalyze the reaction of an oxidant (usually oxygen) with these electrons and protons, generating water (Fig. 6.8). Over the last decade, major improvements in EBFCs have been made due to the use of carbon-based materials such as graphene or GO to fabricate enzyme-functionalized electrodes (Prasad et al., 2014; Lee et al., 2013b; Devadas et al., 2012; Chang et al., 2011).

Graphene was used for the construction of membraneless EBFCs based on silica sol–gel immobilized graphene sheets/enzyme composite electrodes (Liu et al., 2010a). Glucose oxidase was used as the anode enzyme and bilirubin oxidase as the oxygen reduction catalyst in the cathode. This graphene-based EBFC yields a maximum power density threefold higher than that generated by single wall CNT-based EBFCs.

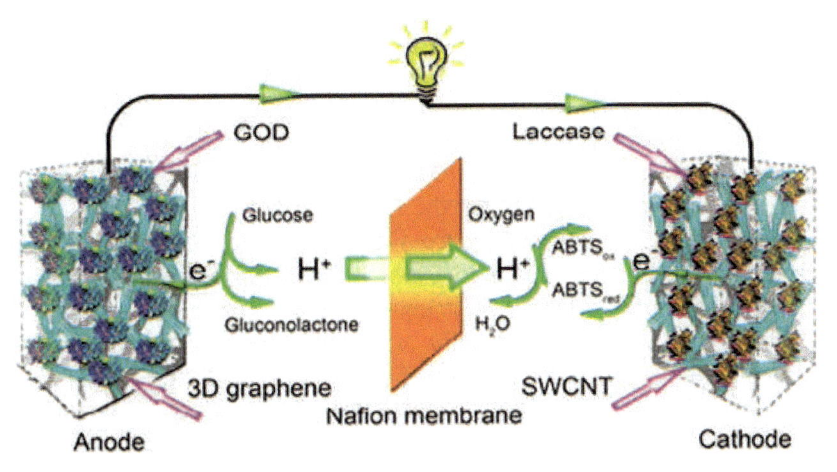

Figure 6.8 The configuration of the GOx/Laccase graphene-SWCNTs hybrid bioelectrodes. Reprinted with permission from Karimi et al., 2015.

Recently, an effective self-power competitive immunosensor driven by biofuel cell was designed for carcinoembryonic antigen (CEA) detection (Li et al., 2015a). Glucose dehydrogenase (GDH)–gold nanoparticles bioconjugate modified with CEA acted as a biocatalyst for enhancing glucose oxidation in the bioanode and nanoporous PtNi/bilirubin oxidase (BOD) acted as a biocatalyst for enhancing O_2 reduction in the biocathode. This BFC-based self-powered immunosensor for CEA possessed largely increased linear detection range from 1 pg mL^{-1} to 0.5 μg mL^{-1} with a detection limit of 0.7 pg mL^{-1}. The proposed BFC-based self-powered immunosensor shows high sensitivity, stability, and reproducibility and can become a promising platform for other protein detection.

In another study, enzymes were bound to the hydrophilic, carboxyl group functionalized graphene–gold nanoparticle hybrid (Chen et al., 2015). In the bioanode compartment, GOx was bound to the graphene–AuNP hybrid, and glucose was oxidized to gluconolactone without a redox mediator under anaerobic conditions; gluconolactone was further oxidized to gluconic acid by the graphene–AuNP hybrid. The biocathode was composed of laccase bound to the graphene–AuNP hybrid as a biocatalyzer, and 2,2-azinobis(3-ethylbenzothiazoline-6-sulfonic acid) diammonium salt (ABTS) as a redox mediator. The two compartments were separated with a nafion membrane. In the EBFC, the E^{ocv} and the P_{max} reached 1.16 ± 0.02 V and 1.96 ± 0.13 mW cm^{-2}, respectively, and E^{ocv} and P_{max} retained 80% and 66% of their optimal values after 70 days.

The performances of GO and graphene–platinum hybrid nanoparticles (Gr-Pt hybrid NPs) were tested for EBFC systems (Tepeli and Anik, 2015). For the fabrication of bioanodes, GO or G–Pt hybrid NPs were incorporated into a composite glassy carbon paste electrode (GCPE), which contained glucose oxidase. For biocathode, a laccase-included GCPE was prepared. The two electrodes were combined in a single cell, and a membraneless EBFC was obtained. Power and current densities of these systems were calculated as 2.40 and 211.90 μA cm^{-2} for GO based BFC and 4.88 μW cm^{-2} and 246.82 μA cm^{-2} for Gr-Pt hybrid NPs based EBFC, indicating that with this type of BFC, it will be possible to obtain better current density and power density values.

Recent works in EBFCs have explored the implementation of more advanced configurations of 3D graphene based nanostructures (Bao et al., 2011b; Kakran et al., 2011; Wang et al., 2010) to increase the number of anchoring sites for the enzyme and facilitate mass transport through the conductive microporous network. Song et al. designed a mediatorless enzymatic micro-biofuel cell based on graphene/enzyme-encrusted 3D micropillar arrays (Song et al., 2015). The fabrication process of this system combines top-down C-MEMS technology to fabricate the 3D micropillar array platform and bottom-up electrophoretic deposition (EPD) to deposit graphene/glucose oxidase and graphene/laccase composites onto the 3D micropillar array-based anode and cathode, respectively (Fig. 6.9). A comparison of the EBFC performances of the graphene/enzyme-encrusted 3D micropillar arrays and the bare 3D micropillar arrays indicates that the graphene-based EBFC generated a maximum power density (136.3 μW cm^{-2} at 0.59 V) almost seven times that of the bare carbon-based EBFC.

Despite the research progress on the fabrication of effective graphene-based EBFCs, several drawbacks, such as short lifetime and low bioactivity and power density, have limited their practical applications. However, functionalized graphene-based nanomaterials, due to their large surface area and excellent electrical conductivity, facilitate the high enzyme loading and enhance the turnover rates between the biocatalysts and the electrode that is expected to provide much more efficient biofuel cells.

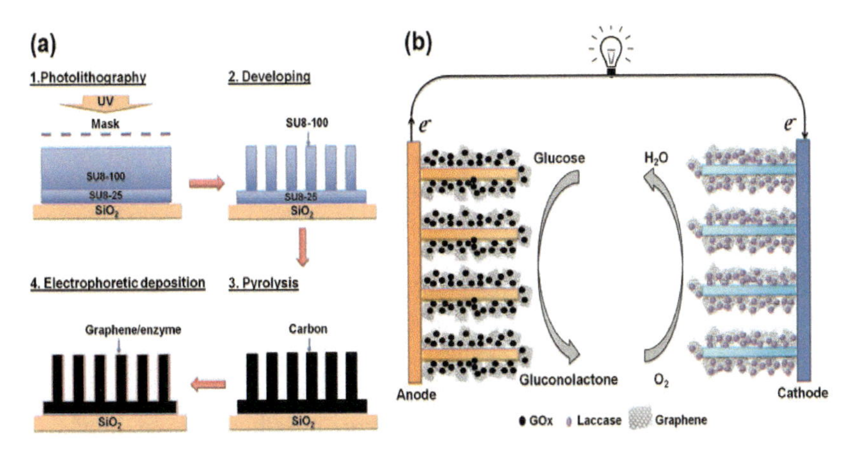

Figure 6.9 (a) Schematic showing of the fabrication of EBFC based on C-MEMS micropillar arrays. (b) Illustration of the EBFC with graphene/enzyme-encrusted 3D carbon micropillar arrays. Reprinted with permission from Song et al., 2015.

6.5.3 Biocatalytic transformations

Different kind of enzymes, mainly hydrolases and oxidoreductases, have been immobilized onto graphene-based nanomaterials and used for various biocatalytic applications in aqueous and non-aqueous media, including waste-water treatment, degradation of pollutants, specific biocatalytic transformations and proteolysis in proteomic analysis. In this section, we discuss some of the biocatalytic applications of various graphene-based nanobiocatalytic systems.

6.5.3.1 Application of GBNs-hydrolases conjugates

Graphene-based nanomaterials have been used as scaffolds for the immobilization of enzymes with lipolytic and esterolytic activity. Our group has reported the use of amine-functionalized GO and MWCNTs as nanoscaffolds for the covalent and non-covalent immobilization of various microbial lipases and esterases of industrial interest and the use of enzymes-nanomaterials conjugates for the esterification reactions in non-aqueous media (Pavlidis et al., 2012a). In all cases studied, the biocatalytic activity of nanoconjugates expressed for the esterification of caprylic acid with 1-butanol in *n*-hexane, was significantly higher when enzymes were immobilized on functionalized MWCNTs compared to that observed when functionalized GO was used as immobilization support. The esterification activity depended on the immobilization approach followed, as well as on the enzyme–nanomaterial weight ratio used. Immobilized enzymes presented high operational stability retaining more than 50% of their esterification activity after five repeated uses (120 h of total operation time) at 60°C.

Lau et al. have reported enhanced esterification activity of *Candida rugosa* lipase upon immobilization on chitosan/GO beads (Lau et al., 2014). The activity of the immobilized lipases for the esterification of lauric acid and oleyl alcohol was increased compared to that of the free enzyme, possibly due to beads' ability to expose the active sites upon immobilization onto the polymer matrix. Patel et al. have recently demonstrated the use of exfoliated graphene oxide (EGO) as nanoscaffold for the immobilization of *Candida rugosa* lipase (Patel et al., 2015). The biocatalyst efficiently catalyzed the synthesis of ethyl caprylate in cyclo-octane retaining 50% of initial activity after 30 reaction cycles.

Similar beneficial effect of functionalized GBNs on enzyme catalytic behavior was also observed for β-galactosidase. Namely, the use of covalently immobilized enzyme enhances the activity of enzyme for hydrolysis of whey and milk lactose (Kishore et al., 2012). The K_m value of the enzyme for the hydrolysis of lactose is reduced from 10 mM for the free enzyme to 5.78 mM for the immobilized one indicating that the affinity of the enzyme for lactose increases after immobilization. The use of carboxylated GO (GO-COOH) as support for the lipase immobilization for the enantioselective resolution of (*R*,*S*)-1-phenylethanol in heptane was also reported (Li et al., 2013a). The immobilized lipase increased its selectivity after immobilization, whereas its catalytic efficiency was up to 1.6-fold higher compared to the free protein.

The development of protein digestion systems based on trypsin immobilized on graphene-based nanomaterials has been widely reported. A trypsin immobilized reactor was designed through the immobilization of the enzyme on GO functionalized with amino-Fe_3O_4 nanoparticles (Jiang et al., 2012a). The enzyme reactor was used to perform digestion of proteins extracted from rat liver. In a similar manner, the design of protein digestion system based on trypsin immobilization onto functionalized GO with poly-L-lysine and polyethylene glycol (PEG)–diglycolic acid was also reported (Xu et al., 2012). The microwave-assisted on-plate proteolysis with the immobilized trypsin followed by MALDI-TOF MS analysis resulted in high efficiency, as hundreds of samples could be digested within a short time (about 15 s), indicating that the enzyme-GO conjugates are a promising strategy for protein digestion and peptide mapping.

Graphene-based nanomaterials were employed for the development of immobilized enzyme systems that could be promising for wastewater treatment, as well as for the conversion of biomass to biofuels. Alkaline protease was immobilized onto GO for the hydrolysis of casein or waste-activated sludge to free amino acids (Su et al., 2012). The thermostability and reusability of the immobilized enzyme was significantly improved, as it retained almost 90% of its initial activity after 20 days of storage, while free alkaline protease retained 75% of its initial activity.

Immobilized cellulases on GBNs have been used for the hydrolysis of cellulose. Cellulose, as an energy source, has been widely used in the form of wood and plant fibers for building materials, paper, textiles, and clothing (Brinchi et al., 2013). Employing cellulases for the hydrolysis of cellulose into sugar has been proved to be of vital importance for producing bioethanol which is a promising clean energy source (Li et al., 2007). The covalent immobilization of cellulase on Fe_3O_4@SiO_2–GO composites has been recently reported (Li et al., 2015b). High immobilization yield and hydrolytic efficiency (above 90%) were observed, while thermal stability of immobilized cellulase was improved compared to that of free enzyme. After heat treatment at 60°C for 2 h, immobilized cellulase retained more than 85% of its original activity while the remaining activity of free enzyme was ~60%, indicating that rigidity and thermostability were both improved by immobilization. Moreover, the immobilized cellulase exhibited great reusability, as the enzyme retained its catalytic activity at 80% after seven cycles of usage.

Gokhale et al. developed a pH tunable, temperature sensitive magnetoresponsive graphene-based nano-bio carrier (MNP-GNP) for cellulase (Gokhale et al., 2013). The maghemite-magnetite decoration and the immobilization of cellulase were carried out by using a combination of quenched and annealed polyelectrolytes. Strong polyelectrolytes are used for efficient electrostatic binding of anionic iron oxide nanoparticles. The use of a pH and heat sensitive polyelectrolyte such as polyacrylic acid (PAA) makes the surface of MNP–GNP biocompatible. The supramolecular assembly of oppositely charged quenched polyelectrolytes and maghemite–magnetite nanoparticles on 2D graphene supports followed by covalent immobilization of cellulase lead to biocatalysts with high *endo*- and *exo*-glucanase activities for soluble, as well as insoluble cellulosic substrates and good operational stability.

6.5.3.2 Application of GBNs-oxidoreductases conjugates for the degradation of pollutants

GBNs have been also used for the immobilization of different oxidoreductases for the degradation of pollutants of environmental concern, such as phenolic compound and chlorinated aromatic compounds. These compounds are widely used in the manufacture of plastics, pharmaceuticals, printing and dyeing materials, pesticides, wood preservatives, and petrochemicals (Zhang et al., 2009) and their degradation is of high importance.

The application of laccase from *Trametes versicolor* immobilized onto GO-based nanomaterials for the removal of bisphenol and catechol has been recently reported (Pang et al., 2015). The immobilized laccase exhibited high degradation efficiency up to 99.9% for bisphenol and catechol after 24 h of incubation. Similarly, immobilized laccase on polyacrylonitrile/montmorillonite/graphene oxide (PAN/MMT/GO) was efficiently used for the degradation of catechol in water (Wang et al., 2014) indicating that these laccase-based nano-conjugates formulated with GO could be promising for the phenolic pollutants in industrial applications.

Immobilized horseradish peroxidase on graphene oxide was also applied in phenolic compound removal (Zhang et al., 2010a). The immobilized enzyme exhibited about 50% remaining activity, while the free enzyme solution exhibited only 20% residual activity after 120 min of incubation at 50°C. The catalytic efficiency for the phenolic compound removal of immobilized enzyme was comparable and, in most cases, higher than that of the free protein indicating that the GO-immobilized enzyme systems should be promising in wastewater treatment, and in other enzyme catalytic protocols.

Recently, the use of covalently immobilized HRP on Fe_3O_4/GO for the degradation of chlorophenols such as 2,4-dichlorophenol, 4-chlorophenol, and 2-chlorophenol has also been reported (Chang et al., 2015). The immobilized enzyme exhibited high storage stability and could be reused up to 4 times with ~70% remaining activity for the degradation of chlorophenols.

Huang et al. prepared and characterized as an artificial enzyme-like catalyst, a composite of graphene oxide and nano-Fe_3O_4 (GO/Fe_3O_4) (Huang et al., 2014). By combining this artificial enzyme-like catalyst and horseradish peroxidase as a binary enzymatic catalyst, the catalytic oxidative removal of 2,4–dichlorophenol (2,4-DCP) was achieved in the presence of H_2O_2 in aqueous media at pH 6.4 and room temperature. The results obtained indicated that there was a synergistic effect between GO/Fe_3O_4 and HRP. The 2,4-DCP was ~50% removed after 120 min. The catalytic mechanism suggested in this work, relied on the donation of lone-pair electrons from the amino groups of HRP to GO which has excellent ability to accommodate electrons and to hold and release electrons, followed by electron transfer from GO to H_2O_2, accelerated through hydrophilic moieties at the interface of GO.

6.6 Conclusions

Because of the remarkable structural, chemical and mechanical properties of graphene and GO-based nanomaterials, their application as nanoscaffolds for the immobilization of enzymes and development of nanobiocatalytic systems has grown rapidly in the past years. Recent studies in the field clearly indicate that the progress in synthesis and surface engineering of GBNs together with the development of effective enzymes immobilization approaches lead to effective GBNs-based nanobiocatalytic systems with great potential in various applications. Although significant and exciting progress has been made in this field, several fundamental issues remain to be resolved for the rational design and the production of nanobiocatalytic systems with superior properties. For instance, the interactions that take place at the nanosupport interface upon enzyme immobilization have yet to be fully understood. The difficulty arises from the fact that these interactions among other factors significantly depend on the distribution of the oxygen containing groups on the GBNs surface which is difficult to be controlled at the moment. The progress in the field of the design, synthesis and characterization of new functionalized GBNs with tailor-made physicochemical properties, as well as the development of techniques and methods to record in situ conformational change of the enzyme during immobilization is expected to facilitate the understanding of the interaction mechanisms and thus the effect of these interactions on the structure and function of immobilized biocatalysts.

Several modern biocatalytic processes do not rely on a single enzymatic step, but require the development of enzymatic cascades. The design and the production of nano-enzyme assemblies based on the oriented immobilization of different enzymes onto nanomaterials, is expected to produce novel tailor-made enzymatic cascades. The remarkable progress in surface engineering of GBNs and the synthesis of graphene-based nanoscaffolds with specific functionalities is expected to facilitate the rational immobilization of different enzymes onto such nanomaterials, producing nano-enzyme assemblies that will combine easily several enzymes in a cascade and thus expanding their application in biocatalytic processes and transformations.

Abbreviations

ABTS	2,2-Azinobis(3-ethylbenzothiazoline-6-sulfonic acid) diammonium salt
AChE	Acetylcholinesterase

AFM	Atomic force microscopy
AzF	4-Azino-L-phenylalanine
BOD	Bilirubin oxidase
BSA	Bovine serum albumin
CB	Carbon black
cBSA	Cationized BSA
CD	Circular dichroism
CEA	Carcinoembryonic antigen
C-MEMS	Carbon microelectromechanical systems
Con A	Concanavalin A
CS	Chitosan
Cyt c	Cytochrome c
CYP	Cytochrome P450
C60	Fullerene
DCP	2,4-Dichlorophenol
EBFCs	Enzyme-based biofuel cells
EDC	1-Ethyl-3-(3-dimethylaminopropyl) carbodiimide
EDX	Energy dispersive X-ray analysis
EPD	Electrophoretic deposition
FLG	Few-layer graphene
FTIR	Fourier transform infrared
G	Graphene
GBNs	Graphene-based nanomaterials
GCPE	Glassy carbon paste electrode
GDH	Glucose dehydrogenase
GN–CD	Graphene–cyclodextrin
GO	Graphene oxide
GOx	Glucose oxidase
GQDs	Graphene quantum dots
HRP	Horseradish peroxidase
IL	Ionic liquid
MALDI-TOF	Matrix-assisted laser desorption-ionization time-of-flight
MMT	Montmorillonite
MNP-GNP	Magnetoresponsive graphene-based nano-bio carrier

MS	Mass spectrometry
MWCNTs	Multi-wall carbon nanotubes
NHS	*N*-hydroxysuccinimide
NPs	Nanoparticles
OxOx	Oxalate oxidase
PAA	Polyacrylic acid
PAN	Polyacrylonitrile
PEG	Polyethylene glycol
PIL	Polymerized ionic liquid
P-L-His	Poly-L-histidine
rGO	Reduced graphene oxide
ROS	Reactive oxygen species
SEM	Scanning electron microscopy
Sulfo-NHS	*N*-hydroxysulfosuccinimide
SWCNTs	Single-wall carbon nanotubes
TEPA	Tetraethylenepentamine
UV–vis	Ultraviolet–visible
XPS	X-ray photoelectron spectroscopy

References

Alwarappan S., Boyapalle S., Kumar A., Li C. Z., Mohapatra S., *J. Phys. Chem. C*, **116** (2012), 6556–6559.

Azamian B. R., Davis J. J., Coleman K. S., Bagshaw C. B., Green M. L. H., *J. Am. Chem. Soc.*, **124** (2002), 12664–12665.

Bagri A., Mattevi C., Acik M., Chabal Y. J., Chhowalla M., Shenoy V. B., *Nat. Chem.*, **2** (2010), 581–587.

Bao H., Chen Q., Zhang L., Chen G., *Analyst*, **136** (2011a), 5190–5196.

Bao H., Pan Y., Ping Y., Sahoo N. G., Wu T., Li L., Li J., Gan L. H., *Small*, **7** (2011b), 1569–1578.

Baptista-Pires L., Pérez-López B., Mayorga-Martinez C. C., Morales-Narváez E., Domingo N., Esplandiu M. J., Alzina F., Sotomayor-Torres C. M., Merkoçi A., *Biosens. Bioelectron.*, **61** (2014) 655–662.

Bitounis D., Ali-Bucetta H., Hong B. H., Min D. H., Kostarelos K., *Adv. Mater.*, **25** (2013), 2258–2268.

Bosio V. E., Islan G. A., Martínez Y. N., Durán N., Castro G. R., *Crit. Rev. Biotechnol.*, **26** (2015), 1–18.

Brinchi L., Cotana F., Fortunati E., Kenny J. M., *Carbohydr. Polym.*, **94** (2013), 154–169.

Campbell A. S., Dong C., Meng F., Hardinger J., Perhinschi G., Wu N., Dinu C. Z., *ACS Appl. Mater. Interfaces*, **6** (2014), 5393–5403.

Cazorla C., Rojas-Cervellera V., Rovira C., *J. Mater. Chem.*, **22** (2012), 19684–19693.

Chang Q., Jiang G., Tang H., Li N., Huang J., Wu L., *Chinese J. Catal.*, **36** (2015), 961–968.

Chang L., Zhongfang C., Chen-Zhong L., *IEEE Trans. Nanotechnol.*, **10** (2011) 59–62.

Chen Y., Gai P., Zhang J., Zhu J. J., *J. Mater. Chem. A*, **3** (2015), 11511–11516.

Chen H., Zhao G., *J. Solid State Electrochem.*, **16** (2012), 3289–3297.

Chen R. J., Zhang Y., Wang D., Dai H., *J. Am. Chem. Soc.*, **123** (2001), 3838–3839.

De M., Chou S. S., Dravid V. P., *J. Am. Chem. Soc.*, **133** (2011), 17524–17527.

Devadas B., Mani V., Chen S. M., *Int. J. Electrochem. Sci.*, **7** (2012) 8064–8075.

Dey R. S., Raj C. R., *J. Phys. Chem. C*, **114** (2010), 21427–21433.

Dreyer D. R., Park S., Bielawski C. W., Ruoff R. S., *Chem. Soc. Rev.*, **39** (2010), 228–240.

Du D., Yang Y., Lin Y., *MRS Bull.*, **37** (2012), 1290–1296.

Elahi M. Y., Khodadadi A. A., Mortazavi Y., *J. Electrochem. Soc.*, **161** (2014), B81–B87.

Eskandari K., Sajjadi S., Keihan A. H., Kamali M., Rashidiani J., Safiri Z., *J. Appl. Biotechnol. Rep.*, **1** (2014), 85–88.

Fang Y., Wang E., *Chem. Commun.*, **49** (2013), 9526–9539.

Gao, Y., Kyratzis, I., *Bioconjug. Chem.*, **19** (2008), 1945–1950.

Ge J., Yag C., Zhu J., Lu D., Liu Z., *Top. Catal.*, **55** (2012), 1070–1080.

Geim A. K., Novoselov K. S., *Nat. Mater.*, **6** (2007), 183–91.

Gengler R. Y. N., Badali D. S., Zhang D., Dimos K., Spyrou K., Gournis D., Miller R. J. D., *Nat. Commun.*, **4** (2013), 2560.

Goenka S., Sant V., Sant S., *J. Control. Release*, **173** (2014), 75–88.

Gokhale A. A., Lu J., Lee I., *J. Mol. Catal. B Enzym.*, **90** (2013), 76–86.

Gong C. B., Guo C. C., Jiang D., Tang Q., Liu C. H., Ma X. B., *Mater. Sci. Eng. C*, **39** (2014), 281–287.

Guan D., Kurra Y., Liu W., Chen Z., *Chem. Commun.*, **51** (2015), 2522–2525.

Gupta S., Irihamye A., *AIP Adv.*, **5** (2015), 037106.

Hu B., Ge Z., Li X., *J. Nanosci. Nanotechnol.*, **15** (2015), 4863–4869.

Hua B. Y., Wang J., Wang K., Li X., Zhu X. J., Xia X. H., *Chem. Commun.*, **48** (2012), 2316–2318.

Huang J., Chang Q., Ding Y., Han X., Tang H., *Chem. Eng. J.*, **254** (2014), 434–442.

Huang Y., Dong X., Shi Y., Li C. M., Li L. J., Chen P., *Nanoscale*, **2** (2010), 1485–1488.

Jegannathan K. R., Abang S., Poncelet D., Chan E. S., Ravindra P., *Crit. Rev. Biotechnol.*, **28** (2008), 253–264.

Jiang B., Yang K., Zhao Q., Wu Q., Liang Z., Zhang L., Peng X., Zhang Y., *J. Chromatogr. A*, **1254** (2012a), 8–13.

Jiang Y., Zhang Q., Li F., Niu L., *Sens. Actuators B*, **161** (2012b), 728–733.

Jin L., Yang K., Yao K., Zhang S., Tao H., Lee S. T., Liu Z., Peng R., *ACS Nano*, **6** (2012), 4864–4875.

Kakran M., Sahoo N. G., Bao H., Pan Y., Li L., *Curr. Med. Chem.*, **18** (2011), 4503–4512.

Karimi A., Othman A., Uzunoglu A., Stanciu L., Andreescu S., *Nanoscale*, **7** (2015), 6909–6923.

Kim J., Cote L. J., Kim F., Yuan W., Shull K. R., Huang J., *J. Am. Chem. Soc.*, **132** (2010), 8180–8186.

Kishore D., Talat M., Srivastava O. N., Kayastha A. M., *PLoS ONE*, **7** (2012), e40708.

Krishna K. V., Ménard-Moyon C., Verma S., Bianco A., *Nanomedicine*, **8** (2013), 1669–1688.

Kuila T., Bose S., Khanra P., Mishra A. K., Kim N. H., Lee J. H., *Biosens. Bioelectron.*, **26** (2011), 4637–4648.

Lau S. C., Lim H. N., Basri M., Masoumi H. R. F., Tajudin A. A., Huang N. M., Pandikumar A., Chia C. H., Andou Y., *PLoS ONE*, **9** (2014), e104695.

Lee H., Jeong H. K., Han J., Chung H. S., Jang S. H., Lee C. W., *Bioresour. Technol.*, **148** (2013a) 620–623.

Lee U. H., Yoo H. Y., Lkhagvasuren T., Song Y. S., Park C., Kim J., Kim S. W., *Biosens. Bioelectron.*, **42** (2013b), 342–348.

Li C., Yoshimoto M., Fukunaga K., Nakao K., *Bioresour. Technol.*, **98** (2007), 1366–1372.

Li Q., Fan F., Wang Y., Feng W., Ji P., *Ind. Eng. Chem. Res.*, **52** (2013a), 6343–6348.

Li S., Wang Y., Ge S., Yu J., Yan M., *Biosens. Bioelectron.*, **71** (2015a), 18–24.

Li Y., Han G., *Analyst*, **137** (2012), 3160–3165.

Li Y., Wang X. Y., Jiang X. P., Ye J. J., Zhang Y. W., Zhang X. Y., *J. Nanopart. Res.*, **17** (2015b), 8–20.

Li Z., He M., Xu D., Liu Z., *J. Photochem. Photobiol. C Photochem. Rev.*, **18** (2013b), 1–17.

Li Z., Xie C., Wang J., Meng A., Zhang F., *Sensor Actuat. B*, **208** (2015c), 505–511.

Li Z. J., Xia Q. F., *Rev. Anal. Chem.*, **31** (2012), 57–81.

Liang R. P., Wang X. N., Liu C. M., Meng X. Y., Qiu J. D., *J. Chromatogr. A*, **1315** (2013), 28–35.

Loo A. H., Bonanni A., Pumera M., *Chem. Asian J.*, **8** (2013), 198–203.

Liu C., Alwarappan S., Chen Z., Kong X., Li C. Z., *Biosens. Bioelectron.*, **25** (2010a), 1829–1833.

Lu J., Cui D., Li H., Zhang Y., Liu S., *Electrochim. Acta*, **165** (2015), 36–44.

Liu J., Cui L., Losic D., *Acta Biomater.*, **9** (2013), 9243–9257.

Liu Y., Yu D., Zeng C., Miao Z., Dai L., *Langmuir*, **26** (2010b), 6158–6160.

Manjunatha R., Shivappa Suresh G., Savio Melo J., D'Souza S. F., Venkatesha T. V., *Talanta*, **99** (2012), 302–309.

Mesarič T., Baweja L., Drašler B., Drobne D., Makovec D., Dušak P., Dhawan A., Sepčić K., *Carbon*, **62** (2013), 222–232.

Minteer S. D., Liaw B. Y., Cooney, M. J., *Curr. Opin. Biotechnol.*, **18** (2007), 228–234.

Novoselov K. S., Geim A. K., Morozov S. V., Jiang D., Katsnelson M. I., Grigorieva I. V., Dubonos S. V., Firsov A. A., *Nature*, **438** (2005) 197–200.

Novoselov K. S., Geim A. K., Morozov S. V., Jiang D., Zhang Y., Dubonos S. V., Grigorieva I. V., Firsov A. A., *Science*, **306** (2004), 666–669.

Pan D., Gu Y., Lan H., Sun Y., Gao H., *Anal. Chim. Acta*, **853** (2015), 297–302.

Pang R., Li M., Zhang C., *Talanta*, **131** (2015), 38–45.

Park S., An J., Jung I., Piner R. D., An S. J., Li X., Velamakanni A., Ruoff R. S., *Nano Lett.*, **9** (2009) 1593–1597.

Patel V., Gajera H., Gupta A., Manocha L., Madamwar D., *Biochem. Eng. J.*, **95** (2015), 62–70.

Patila M., Pavlidis I. V., Diamanti E. K., Katapodis P., Gournis D., Stamatis H., *Process Biochem.*, **48** (2013), 1010–1017.

Pattammattel A., Puglia M., Chakraborty S., Deshapriya I. K., Dutta P. K., Kumar C. V., *Langmuir*, **29** (2013), 15643–15654.

Pavlidis I. V., Patila M., Bornscheuer U. T., Gournis D., Stamatis H., *Trends Biotechnol.*, **32** (2014), 312–320.

Pavlidis I. V., Vorhaben T., Gournis D., Papadopoulos G. K., Bornsheuer U. T., Stamatis H., *J. Nanopart. Res.*, **14** (2012b), 842–852.

Pavlidis I. V., Vorhaben T., Tsoufis T., Rudolf P., Bornscheuer U. T., Gournis D., Stamatis H., *Bioresour. Technol.*, **115** (2012a), 164–171.

Prasad K. P., Chen Y., Chen P., *ACS Appl. Mater. Interfaces*, **6** (2014), 3387–3393.

Putzbach W., Ronkainen N. J., *Sensors*, **13** (2013), 4811–4840.

Raffaini G., Ganazzoli F., *Langmuir*, **29** (2013), 4883–4893.

Rather J. A., Pilehvar S., De Wael K., *Sensor Actuat. B*, **190** (2014), 612–620.

Sanchez V. C., Jachak A., Hurt R. H., Kane A. B., *Chem. Res. Toxicol.*, **25** (2012), 15–34.

Sehgal D., Vijay I. K., *Anal. Biochem.*, **218** (1994), 87–91.

Shan C., Yang H., Han D., Zhang Q., Ivaska A., Niu L., *Biosens. Bioelectron.*, **25** (2010) 1504–1508.

Shang W., Nuffer J. H., Muñiz-Papandrea V. A., Colón W., Siegel R. W., Dordick J. S., *Small*, **5** (2009), 470–476.

Shao Q., Qian Y., Wu P., Zhang H., Cai C., *Colloid. Surf. B: Biointerfaces*, **109** (2013), 115–120.

Silva R. A., Souza M. L., Bloisi G. D., Corio P., Petri D. F. S., *J. Nanopart. Res.*, **17** (2015), 187–199.

Song Y., Chen C., Wang C., *Nanoscale*, **7** (2015), 7084–7090.

Staros J. V., Wright R. W., Swingle D. M., *Anal. Biochem.*, **156** (1986), 220–222.

Stavyiannoudaki V., Vamvakaki V., Chaniotakis N., *Anal. Bioanal. Chem.*, **395** (2009), 429–435.

Su R., Shi P., Zhu M., Hong F., Li D., *Bioresour. Technol.*, **115** (2012), 136–140.

Sun W., Gong S., Shi F., Cao L., Ling L., Zheng W., Wang W., *Mater. Sci. Eng. C*, **40** (2014), 235–241.

Sun W., Guo Y., Ju X., Zhang Y.,Wang X., Sun Z., *Biosens. Bioelectron.*, **42** (2013), 207–213.

Takahashi S., Sato K., Anzai J. I., *Anal. Bioanal. Chem.*, **402** (2011), 1749–1758.

Tepeli Y., Anik U., *Electrochem. Commun.*, **57** (2015), 31–34.

Verma M. L., Barrow C. J., Puri M., *Appl. Microbiol. Biotechnol.*, **97** (2013), 23–39.

Vilian A. T. E., Chen S. M., *RSC Adv.*, **4** (2014), 55867–55876.

Walcarius A., Minteer S. D., Wang J., Lin Y., Merkoçi A., *J. Mater. Chem. B*, **1** (2013), 4878–4908.

Wang Q., Cui J., Li G., Zhang J., Li D., Huang F., Wei Q., *Molecules*, **19** (2014), 3376–3388.

Wang Y., Li Z., Hu D., Lin C. T., Li J., Lin Y., *J. Am. Chem. Soc.*, **132** (2010), 9274–9276.

Wang Y., Li Z., Lin Y., *Trends Biotechnol.*, **29** (2011), 205–212.

Wei X. L., Ge Z. Q., *Carbon*, **60** (2013), 401–409.

Willner I., Yan Y. M., Willner B., Tel-Vered R., *Fuel Cells*, **9** (2009), 7–24.

Wong A., Materon E. M., Sotomayor M. P. T., *Electrochim. Acta*, **146** (2014), 830–837.

Wu J. F., Xu M. Q., Zhao G. C., *Electrochem. Commun.*, **12** (2010), 175–177.

Xu G., Chen X., Hu J., Yang P., Yang D., Wei L., *Analyst*, **137** (2012), 2757–2761.

Yang X., Zhao C., Ju E., Ren J., Qu X., *Chem. Commun.*, **49** (2013), 8611–8613.

Yu C., Wang L., Li W., Zhu C., Bao N., Gu H., *Sensor Actuat. B*, **211** (2015), 17–24.

Zeng G., Xing Y., Gao J., Wang Z., Zhang X., *Langmuir*, **26** (2010), 15022–15026.

Zhang C., Chen S., Alvarez P. J. J., Chen W., *Carbon*, **94** (2015), 531–538.

Zhang X., Liao Q., Chu M., Liu S., Zhang Y., *Biosens. Bioelectron.*, **52** (2014), 281–287.

Zhang Y., Wu C., Guo S., Zhang J., *Nanotechnol. Rev.*, **2** (2013), 27–45.

Zhang J. B., Xu Z. Q., Chen H., Zong Y. R., *Biochem. Eng. J.*, **45** (2009), 54–59.

Zhang Y., Zhang J., Huang X., Zhou X., Wu H., Guo S., *Small*, **8** (2012), 154–159.

Zhang, F., Zheng, B., Zhang, J., Huang, X., Liu, H., Guo, S., Zhang, J., *J. Phys. Chem. C*, **114** (2010a), 8469–8473.

Zhang J., Zhang J., Zhang F., Yang H., Huang X., Liu H., Guo S., *Langmuir*, **26** (2010b), 6083–6085.

Zhou L., Jiang Y., Gao J., Zhao X., Ma L., Zhou Q., *Biochem. Eng. J.*, **69** (2012), 28–31.

Zhu J., Xu M., Meng X., Shang K., Fan H., Ai S., *Proc. Biochem.*, **47** (2012), 2480–2486.

Chapter 7

Immobilization of Biocatalysts onto Nanosupports: Advantages for Green Technologies

Alan S. Campbell,[a,b] Andrew J. Maloney,[a] Chenbo Dong,[a] and Cerasela Z. Dinu[a]

[a]*Department of Chemical Engineering, West Virginia University, Morgantown, West Virginia 26506, USA*
[b]*Current affiliations: Department of Biomedical Engineering, Carnegie Mellon University, Pittsburgh, Pennsylvania 15213, USA*

Cerasela-zoica.dinu@mail.wvu.edu

7.1 Introduction

The growing demand for the development of user- and environmentally friendlyzprocesses and products has resulted in a movement to design new or modify existing systems and technologies around the concept of "green chemistry" and its 12 principles as introduced by Anastas and Warner (Fig. 7.1) (Anastas and Warner, 1998). This demand has stemmed from the large volume of hazardous wastes generated annually across the globe as well as from the environmental disasters caused by human activities (Sanderson, 2011; Clark et al., 2012; Philp et al., 2013). The foundation of this movement is centered on reducing the environmental impact of chemical processes and applied

Biocatalysis and Nanotechnology
Edited by Peter Grunwald

ISBN 978-981-4613-69-9 (Hardcover), 978-1-315-19660-2 (eBook)
www.panstanford.com

technologies through the use of eco-friendly strategies that maximize efficiency while minimizing toxicity (Poliakoff et al., 2002; Noyori, 2005). In industrial syntheses, for instance, these considerations require careful selection of solvents, catalysts, and reaction pathways as well as feasible conditions to meet rising societal expectations (Dunn, 2012). Likewise, for device development, renewable, biocompatible, and biodegradable materials must be designed and implemented to ensure limited harsh byproducts synthesis or lifetime toxic decay.

An ideal approach to meeting such standards is based on green catalysis or biocatalysis. Biocatalysis is defined as the utilization of biological molecules derived from an organic source to catalyze the chemical transformation of other organic molecules or compounds (Bornscheuer et al., 2012). The types of catalysts generally implemented in green processes based on biocatalysis are isolated enzymes, immobilized enzymes or whole-cell catalysts (Lopez-Serrano et al., 2002; Woodley, 2008; Tao and Kazlauskas, 2011; Wenda et al., 2011), with enzyme-based systems being particularly well suited and preferred for lowering the environmental impact of chemical syntheses, again, in adherence to the 12 principles of green chemistry as outlined in Fig. 7.1.

In this chapter, the application of enzyme-based technologies is discussed as it relates to green chemistry practices as well as the development of green technologies principles and devices. For the viability of such methods, immobilization strategies are being discussed in order to ensure enzyme functionality and efficient product recovery. The focus for immobilization is on carbon-based nanomaterials with such focus being based on their unique properties (e.g., surface curvature (Asuri et al., 2006; Campbell et al., 2014), optical adsorptivity (Gorji et al., 2015; Tian et al., 2015), controllable aspect ratios (Liang et al., 2013), etc.), proven and identified modes of interactions with several enzyme candidates, as well as the fundamental understanding of how such carbon-based nanomaterials interfaced with enzymes can be manipulated in order to increase enzyme functionality and stability or conjugate functionality. An overview of immobilization methods and common results is presented followed by discussion of each of the mentioned applications including how their implementation improve adherence to the 12 principles of green chemistry.

12 Principles of Green Chemistry
1. Prevention – plan to prevent rather than clean up waste.
2. Atom Economy – design to maximize materials incorporation into final product.
3. Less Hazardous Chemical Syntheses – design to use and generate less hazardous substances.
4. Designing Safer Chemicals – design products to affect desired function with minimal toxicity.
5. Safer Solvents and Auxiliaries – minimize use of auxiliary substances and their toxicity.
6. Design for Energy Efficiency – minimize energy requirements and strive to operate at ambient conditions.
7. Use of Renewable Feedstocks – utilize renewable feedstocks wherever possible.
8. Reduce Derivatives – minimize unnecessary derivitization.
9. Catalysis – utilize selective catalytic reagents over stoichiometric reagents.
10. Design for Degradation – design for products that break down to innocuous compounds that do not persist.
11. Real-Time Analysis for Pollution Prevention – develop analytical methodologies to monitor and control hazardous materials prior to formation.
12. Inherently Safer Chemistry for Accident Prevention – choose substances used that minimize potential for chemical accidents.

Figure 7.1 The 12 principles of green chemistry presented by Anastas and Warner as a guide for chemists and engineers attempting to design new or modify existing processes to minimize their environmental impact (Anastas and Warner, 1998).

7.2 The Need for Enzyme Immobilization in Enzyme-Based Technologies

Processes built around biocatalysts possess extremely high energy and chemical efficiencies due to the mild operating conditions at which their induced reactions take place (i.e., temperature, pH, pressure and capability to operate with water as a solvent) as well as due to their green nature. Furthermore, the organic source of the catalyst itself allows for renewable raw materials and biodegradable products including the enzymes themselves (Dunn, 2012), while the high selectivity and specificity of enzymes provides for efficient catalysis of stereo- and regio-selective processes thus eliminating costly protection or deprotection sequences, and thus proving advantageous in the pharmaceutical industry for increased production efficiency (Tao and Xu, 2009). These advantages complemented by recent strides in enzyme processing as resulted from the discoveries made in protein expression (Assenberg et al., 2013; Liu et al., 2013; Rosano and Ceccarelli, 2014), directed evolution (Lane and Seelig, 2014; Packer and Liu, 2015), metabolic engineering or large scale DNA sequencing, make enzymatic biocatalysis extremely desirable (Straathof et al., 2002; Tao et al., 2007). Moreover, incorporation of enzymatic reactions into pre-existing production schemes and the design of new syntheses has proven in many instances to reduce both the cost and environmental burden or "E-factor" of the processes, where the E-factor of a chemical process is defined as the weight ratio of waste generated to product formed (Sanderson, 2011) In multiple instances, for instance, the transition to enzymatic catalysis from chemical-based methods has dramatically reduced the E-factor of major chemical syntheses (Sanderson, 2011). For example, Idris and Bukhari thoroughly reviewed the capability of immobilized *Candida Antarctica* lipase B (one of the most widely used enzymes in industrial processing with applications ranging from lipid degradation to detergent and biofuel production (Houde et al., 2004) to reduce the E-factor for the synthesis of biodegradable polymers and outlined important factors to be considered when optimizing immobilization conditions (Idris and Bukhari, 2012)).

The high efficiency and reusability of enzymes can, however, rapidly degrade within a system of interest due to multiple factors encountered during systems' storage or operation. For instance, changes in the system pH or temperature, the presence of chemical inhibitors, and physical interactions can all alter the sensitive secondary and/or tertiary structures of the enzymes, or even sterically hinder them, thus reducing their ability to interact with and turnover the substrate (Asuri et al., 2006; Campbell et al., 2014; Cummings et al., 2014; Murata et al., 2014). For instance, it was found that members of the peroxidase family of enzymes such as horseradish peroxidase (HRP) and soybean peroxidase (SBP) may be affected by reaction conditions. As this family is commonly used in industrial wastewater management (Gray and Montgomery, 1997), because of their ability to degrade a wide spectra of compounds such as aromatic molecules (Chen et al., 2014), anilines (Nakamoto and Machida, 1992), phenols

(Eker et al., 2009), azo dyes (Ali et al., 2013), polyaromatic hydrocarbons (Baborova et al., 2006), and polychlorinated biphenyls (Koller et al., 2000), all known to be relevant products of industrial contamination (Cheng et al., 2006; Husain, 2010; Husain and Ulber, 2011; Pradeep et al., 2015), understanding of this interaction is critical. The wastewater treatment mechanism of these peroxidases practically involves the creation of free radicals containing organic compounds, which then polymerize and eventually become insoluble in water (Manta et al., 2003; Mossavarali et al., 2006). However, the active site of such enzymes must remain clear to accept a new substrate molecule for feasible and efficient biocatalytic reactions. If a free radical containing compound may adsorb to the peroxidase active site, however, it can cause permanent deactivation and a continuous decrease in bulk enzymatic activity (Mossavarali et al., 2006; Libertino et al., 2008) with such losses in activity being common among many various types of industrial enzymes (Sung and Bae, 2003), and being shown to reduce the efficiency of the biocatalyst implementation.

For enzymatic-based processes to be feasible, a common practice is enzyme immobilization.

- In industrial process analyses, for instance, it was shown that immobilization improves the stability and recoverability of the enzyme, which led to increased reusability and cost effectiveness of the overall process implementation. Studies also showed that the increased ease of catalyst separation from product reduced product contamination and downstream processing complexity (Sheldon and van Pelt, 2013).
- In sensing systems relying on enzyme free in solution, sensor stability, accuracy, and applicability are generally far poorer than if the enzyme were confined/immobilized onto the electrode surface (Sassolas et al., 2012). As such, proper immobilization methods have been found to be necessary for the direct use of biosensors in test systems from whole blood to food products (Zhang and Li, 2004).
- In bioelectronics based on the direct electron transfer (DET) with enzymes, which reduces system complexity and increases biocompatibility, selectivity and maximum possible voltage, require close contact of the enzyme active site with a conductive support material (Zhang and Li, 2004).

- In enzyme enzyme-based bioactive coatings, immobilization also yields increased retention and stability, thus reducing the need for reapplication (Dinu et al., 2012).
- For enzymatic biosensors and biofuel cells, immobilization of the electro-active enzymes onto or within a conductive material is required for efficient charge collection and extended operational lifetime with consistent output (Cooney et al., 2008; Sassolas et al., 2012).

Based on the above-identified needs, optimization of the immobilization process and materials used are areas of intense research focus with many diverse approaches currently under study (Mateo et al., 2007; Sheldon, 2007; Garcia-Galan et al., 2011; Sheldon and van Pelt, 2013). Specifically, there are three main categories of immobilization methodologies that will be discussed next, namely intermolecular cross-linking without the use of a support, encapsulation within a carrier and lastly adsorption or binding to a support (Sheldon and van Pelt, 2013).

7.2.1 Enzyme Immobilization by Cross-Linking

The cross-linking of enzymes was implemented in the early 1960s utilizing glutaraldehyde as the cross-linking agent (Quiocho and Richards, 1966; Sheldon, 2007). Cross-linked enzyme crystals (CLECs) for instance, proved to possess significantly improved resiliency to denaturation by high temperature and organic solvents thus greatly increasing such systems applicability in industrial processes. This fact coupled with a greater ease of recovery allowed these systems commercialization in the 1990s, but the need for highly pure enzyme capable of crystallization resulted in production processes not economically feasible, which led to the transition of use to cross-linked enzyme aggregates (CLEAs) (Roy and Abraham, 2004; Sheldon et al., 2005; Sheldon and van Pelt, 2013).

CLEAs are formed through the cross-linking of aggregated enzyme that has precipitated from solution. The elimination of the need to crystallization allows for a more cost-effective production scheme. Further, such cross-linking can be coupled with the use of supports to thus add another level of functionality to the final product (Lee et al., 2005). Applicable CLEAs have been formed from a variety of enzymes, including hydrolases, proteases, oxidoreductases, and lipases (Sheldon, 2011). For instance, CLEAs formed from glucose oxidase (GOX) using glutaraldehyde vapor that were subsequently immobilized onto a carbon nanotube (CNT)-based electrode exhibited increased current generation and operational stability compared to non-cross-linked enzyme. It was also reported that an optimal extent of cross-linking was observed that further allowed for increase in the conducting density of the material and enzyme retention without inhibiting mass or charge transport, thus resulting in improved current generation (MacAodha et al., 2012). In a separate study, the cross-linking of a perhydrolase with aldehyde dextran enzyme followed by the incorporation of the CLEAs into a bioactive coating showed increased anti-sporicidal capability and activity retention upon CNT immobilization and incorporation into the coating compared to free enzyme. These attributes were a direct result of the stabilizing effects imparted to the aggregates through the cross-linking process (Dinu et al., 2012). As for industrial enzyme utilization, the formation of CLEAs of penicillin acylase proved capable of retaining the same activity as free enzyme in the formation of ampicillin while greatly reducing unwanted hydrolysis and allowing for the reaction to be carried out in organic solvent. Again, stabilizing effects were found to be a direct result of aggregate formation (Cao et al., 2000). Such methods avoid the need for an additional carrier particle or molecule and can result in a high catalyst concentration within the formed aggregate; however, activity retention can be fairly low due to the reduced enzyme accessibility for substrate and limited substrate or product mass transport (Sheldon and van Pelt, 2013).

7.2.2 Enzyme Immobilization through Entrapment/Encapsulation within a Carrier

The second heavily studied avenue for enzyme immobilization is encapsulation or entrapment within a support or carrier. Similar to cross-linking, this method can provide increased stability and ease of separation while reducing leaching but it can also lead to hindered mass transport of the substrate molecules. Commonly used materials include sol gels, hydrogels and nanofibers (Pierre, 2004; Ansari and Husain, 2012). As examples, hydrogels are hydrated matrices of polymeric materials with tunable characteristics (e.g., stiffness, density, functionality and pore size) based on composition and processing techniques. (Lee et al., 2015).

In addition to enzymatic entrapment (Nguyen and Minteer, 2015; Sun et al., 2015), and due to their biocompatible nature and multitude of potential functionalities (Wilson and Turner, 1992; de Lathouder et al., 2008; Shen et al., 2011), hydrogels such as alginate (Tran et al., 2014), polyethylene glycol (PEG) (Lee et al., 2015), hyaluronic acid (Han et al., 2002), cyclodextrin (Diez et al., 2012)., and their chemical derivatives (Lee et al., 2015) have been used in a variety of biomedical applications such as tissue engineering (Sundaramoorthy et al., 1998; Kamal and Behere, 2003; Tran et al., 2014; Lee et al., 2015), and drug delivery (Peppas, 1997; Hoare and Kohane, 2008). For example, Milašinović et al. (2014) showed the benefits of *N*-isopropylacrylamide and itaconic acid hydrogels to deliver a lipase through utilization of pH modulation. Specifically, the hydrogel was shown to bind lipase strongly in a low-pH environment (similar to the gastrointestinal tract), as well as the ability to release the protein in more basic conditions similar to those found in the small intestine (Zobnina et al., 2012).

Enzyme entrapment is also quite promising for increased applicability and reuse of industrial enzymes. Since hydrogels are a network of polymers, diffusion of large molecules into the polymer is hindered, which can offer protection from enzymatic deactivation by adsorption to insoluble molecules and thus further allow for greater control over the enzyme micro-environment. The single enzyme encapsulation of HRP by acryloylation for instance, followed by in situ polymerization was shown by Yan et al. to permit for significantly enhanced thermal stability and operational capabilities in organic solvents, all relative to free enzyme used in industrial processing. Further, it was shown that this method of encapsulating individual enzymes can also reduce the mass transfer limitations encountered upon encapsulation of larger quantities of enzyme (Yan et al., 2006), and can find possible applicable to a wide array of other enzymes.

An additional method of controlled enzyme encapsulation that has gained increasing research interest is the modification of individual enzymes using the controlled radical polymerization method or atom transfer radical polymerization (ATRP) (Matyjaszewski and Tsarevsky, 2014). The conjugation of a protein via ATRP using the "grafting-from" technique is a two-step process. First, a halogen containing ATRP initiator complex is attached to the surface of the protein to further serve as the site of monomer attachment during the polymerization procedure. An *N*-hydroxysuccinimide (NHS) functional group on the ATRP initiator reacts with primary amines present in lysine residues on the enzyme surface or the N-terminus.

Analyses showed that stability augmentation can be achieved via ATRP polymer growth from the enzyme surface (Cummings et al., 2013; Murata et al., 2013; Cummings et al., 2014). ATRP also allowed for the highly controlled growth of polymers with precise lengths, especially from initiator groups that were attached to primary amines on the surfaces of enzymes (Matyjaszewski and Xia, 2001; Averick et al., 2012). Further, Cummings et al. showed the capability to tune the pH and temperature stability as well as substrate and inhibitor binding affinity of the industrially relevant enzyme chymotrypsin using a "grafting-from" ATRP approach (Cummings et al., 2013; Murata et al., 2013; Cummings et al., 2014; Murata et al., 2014).

An initial limitation in the "grafting-from" ATRP approach was that the commonly used initiator compounds possessed low solubility in aqueous solution and thus had to be used in mixtures of water and organic solvents such as dimethylformamide (DMF) (De et al., 2008; Zhang et al., 2011) or dimethyl sulfoxide (DMSO) (Nicolas et al., 2006; Magnusson et al., 2010), which are known to lead to enzyme denaturation or inactivation. However, a subsequently developed ATRP initiator was proven to be water-soluble and functionalize the majority of lysine groups available on an enzyme surface (Murata et al., 2013). The second step of the process is the actual growth of the polymer. ATRP has been used to polymerize a variety of monomers. According to Matyjaszewski and Xia, the most typically used monomers included styrenes (Al-Harthi et al., 2007; Tom et al., 2010), meth(acrylates) (Averick et al., 2012; Murata et al., 2013; Silva et al., 2013), and (meth)acrylamides (Cummings et al., 2014) among others that contain substituents that help to stabilize the propagating radicals (Matyjaszewski and Xia, 2001; Matyjaszewski and Tsarevsky, 2014).

Studies showed that ATRP allows for the generation of polymer systems with very low polydispersity indices (PDI) through the control of reactant molar ratios. This high degree of control provides for the ability to generate user-defined systems with unique properties that can use any desired enzyme containing accessible lysine residues. Non-canonical amino acids can even be incorporated into the enzyme structure to direct polymer growth by ATRP from a desired site on the enzyme three-dimensional structure (Averick et al., 2013).

7.2.3 Enzyme Immobilization onto a Prefabricated Carrier/Support

The third category of enzyme immobilization techniques consists of the immobilization of the target enzymes onto a prefabricated carrier or support. The key goal of this process, similar to previously discussed ones, is to retain the maximum enzymatic activity possible within a certain volume, while enhancing stability, retention, and functionality of the immobilized enzyme.

The most widely studied methods of attachment to nanosupports are either through physical adsorption (Campbell et al., 2014, 2014a) or covalent binding (Mateo et al., 2007; Campbell et al., 2013). Physical adsorption of enzymes is achieved through the natural affinity of the material for various enzymes resulting mostly from hydrophobic-hydrophobic interactions. Due to these forces, adsorption occurs spontaneously upon contact between the two structures/surfaces. However, this interaction can also cause partial denaturation of the enzyme evident from the loss of alpha-helical content observed using circular dichroism in various studies resulting in decreased kinetic activity (Karajanagi et al., 2004; Feng and Ji, 2011). Covalent binding of enzymes is generally achieved through the use of 1-ethyl-3-(3-dim ethylaminopropyl)carbodiimide (EDC) and *N*-hydroxysuccinimide (NHS) chemistry. In this reaction scheme, carboxyl functionalities are introduced onto a support scaffold material via strong acid oxidation; such functional groups are subsequently used as hangers to primary amines present in lysine residues on the surface of the target enzyme via the EDC linker. EDC first reacts with a carboxyl group on the support to form a reactive intermediate that is stabilized from hydrolysis by NHS until it can react with a primary amine on the enzyme surface to form a stable amide bond (Min and Yoo, 2014).

The resulting interaction between the enzyme and the nanosupport leads to the formation of a conjugate, possessing specific kinetic and chemical properties (Sheldon and van Pelt, 2013). Such bionano conjugates have shown increased chemical and thermal stability, improved ease of recovery, and enhanced functionality when compared to native enzymes; however, deactivation by chemical inhibitors or aggregation were still concerns for their future implementation (Caza et al., 1999; Klein et al., 2013; Verma et al., 2013). Additional alternative methods aimed to preserve enzymatic structure and function were proposed using chemical compounds like PEG and PEG derivatives to help serve a protective function and bring the enzyme away from the nanosupports (Manta et al., 2003; Libertino et al., 2008). The process of attached PEG chains to proteins known as pegylation is widely used to shield protein-based drugs from proteolytic enzymes and improve pharmacokinetics (Harris and Chess, 2003).

Optimization of the immobilization strategy and the choice of a suitable nanosupport are crucial for optimal performance of enzyme-nanomaterial conjugates. As such, nanoscale materials serving as enzyme supports or nanosupports, have very high specific surface area (SSA) as well as a variety of material properties such as mechanical strength, conductivity, biocompatibility, and magnetism that vary with the type of material chosen (Min and Yoo, 2014). Commonly studied nanosupports have included carbon-based materials (Dinu et al., 2010; Krueger and Lang, 2012; Campbell et al., 2013; Campbell et al., 2014), metal-oxide based materials (Campbell et al., 2014), polymer-based networks (Singh et al., 2006; Li et al., 2011; Wang et al., 2013), gold-based materials (Manso et al., 2008; Kwon et al., 2010; Marx et al., 2011), and silica-based materials (Choi et al., 2011; Schuabb and Czeslik, 2014; Perez-Anguiano et al., 2015), just to name a few. Further, since a common concurrent effect of immobilization is the partial loss of native enzyme activity due to denaturation or inactivation at the enzyme-support interface as a result of enzyme-nanomaterial or adjacent protein-protein interactions (Asuri et al., 2006; Asuri et al., 2007), a bevy of characterization studies to allow knowledgeable design of conjugate systems with improved functional characteristics has been performed.

For instance, Fig. 7.2 shows as an example the utilization of carbon-based nanomaterial immobilized enzymes in four key applications: industrial biocatalysis, greenhouse gas conversion, biosensors and biofuel cells, and bioactive surfaces. The extremely high theoretical SSA (2630 m^2/g for graphene and 1315 m^2/g for single-wall carbon nanotubes (SWCNTs), high thermal and electrical conductivities, ease of functionalization and high mechanical strength of these materials (Walcarius et al., 2013;

Min and Yoo, 2014) make them uniquely suited for the four key applications discussed above (Peigney et al., 2001; Baughman et al., 2002; Zhu et al., 2010). The high aspect ratios of carbon-based nanomaterials provided for enhanced conjugate retention in the system of interest while the capability to easily add functional groups coupled with high SSA allowed for large enzyme loading via strong interactions to reduce enzyme mobility and thus increase stability in solution without enzyme leaching (Sheldon and van Pelt, 2013; Min and Yoo, 2014). However, the breadth of operational possibilities posed by enzyme-based systems utilizing nanosupport immobilization has been thoroughly summarized in excellent reviews (Ran et al., 2008; Feng and Ji, 2011; Sheldon and van Pelt, 2013; Min and Yoo, 2014). From these reviews and other independent studies, a summary of nanosupport immobilized enzyme-based systems designed to improve upon the discussed applications as highlighted in Fig. 7.2 is presented in Table 7.1. The in detail discussion of these green chemistry and green technology applications is the target of subsequent sections of this chapter, with a focus on the utilization of nanosupport-based enzyme immobilization techniques.

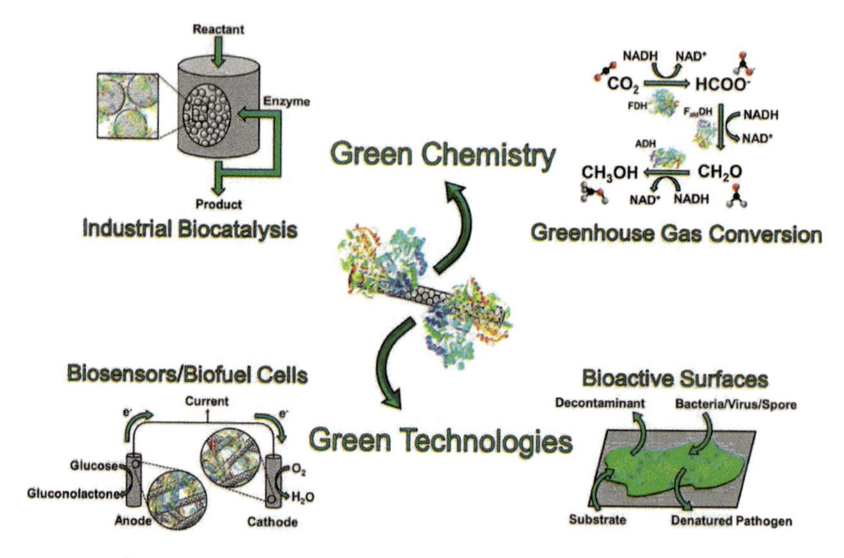

Figure 7.2 Schematic representation of the four key applications of enzyme immobilization onto carbon-based nanomaterials discussed herein. (enzyme abbreviations: FDH, formate dehydrogenase; $F_{ald}DH$, formaldehyde dehydrogenase; ADH, alcohol dehydrogenase).

Table 7.1 Summary of nanosupport immobilized enzyme-based systems and their target applications

Enzyme(s) of Interest	Nanosupport	Target application	Reference
Glucose oxidase/bilirubin oxidase	SWCNT/graphene cogel	Biofuel cell	Campbell et al. (2015)
Glucose oxidase/chloroperoxidase	MWCNT	Active surface decontamination	Campbell et al. (2013)
Glucose oxidase/laccase	Compressed MWCNT disks	Biofuel cell	Zebda et al. (2011)
Lipase	Polyacrylamide nanofiber	Triolein transesterification	Li et al. (2011)
Glucose oxidase	Polyacrylamide/gold nanoparticle composite	Glucose sensing	Jose et al. (2012)
Formate/formaldehyde/alcohol/glutamate dehydrogenases	Polystyrene nanoparticles	CO_2 reduction to methanol	El-Zahab et al. (2008)
Horseradish peroxidase	SWCNT forest	H_2O_2 sensing	Yu et al. (2003)
Gluose oxidase/laccase	SWCNT/graphene foam	Biofuel cell	Prasad et al. (2014)
Alcohol dehydrogenase	Magnetic iron oxide nanoparticles	Enantioselective reduction of 7-methoxy-2-tetralone	Ngo et al. (2012)
Cholesterol esterase, cholesterol oxidase, peroxidase	Polyaniline film	Cholesterol sensing	Singh et al. (2006)

(*Continued*)

Table 7.1 *(Continued)*

Enzyme(s) of Interest	Nanosupport	Target application	Reference
Glucose oxidase/bilirubin oxidase	Polyethylene dioxythiophene coated MWCNT yarns	Biofuel cell	Kwon et al. (2014)
Horseradish peroxidase	MWCNT-titanate nanotube nanocomposite	H_2O_2 sensing	Liu et al. (2015)
Lipase	Smectite nanoclay	α-Pinene epoxidation	Tzialla et al. (2010)
Formate/formaldehyde/alcohol dehydrogenases	Chemically converted graphene	CO_2 reduction to methanol	Yadav et al. (2014)
α-Amylase	Magnetic iron oxide nanoparticles	Starch hydrolysis	Khan et al. (2012)
PQQ-dependent alcohol/aldehyde dehydrogenases	MWCNT/polyamido-amine dendrimers	Biofuel cell	Neto et al. (2013)
Glucose oxidase	Gold nanoparticle/graphene/SWCNT hybrid	Glucose sensing	Yu et al. (2014)
Perhydrolase	MWCNT	Active surface decontamination	Grover et al. (2013)
Perhydrolase	SWCNT	Active surface decontamination	Dinu et al. (2012)
Tyrosinase	SWCNT/polypyrrole composite	Dopamine sensing	Min and Yoo (2009)

Enzyme(s) of Interest	Nanosupport	Target application	Reference
Acetylcholinesterase	Zinc oxide nanoparticles on graphene	Insecticide sensing	Wang et al. (2014)
Fructose dehydrogenase/ bilirubin oxidase	Carbon cryogel	Biofuel cell	So et al. (2014)
Hydrogenase/bilirubin oxidase	Carbon nanofibers	Biofuel cell	de Poulpiquet et al. (2014)
Glucose oxidase	Graphene/gold nanoparticle/ chitosan composite	Glucose sensing	Shan et al. (2010)
Acetylcholinesterase	MWCNT	Pesticide and nerve agent sensing	Lin et al. (2004)
Cellulase	MWCNT	Carboxylmethyl cellulose hydrolysis	Mubarak et al. (2014)
Cellulase	CNT coated polyurethane foam	Lignocellulose hydrolysis	Lu et al. (2013)
Lactate dehydrogenase/ bilirubin oxidase	Buckypaper	Biofuel cell	Reid et al. (2015)
Serine protease	Poly(methyl methacrylate) coated SWCNT	Active antifouling film	Asuri et al. (2007)
Horseradish peroxidase	Poly(dimethyldiallyl-ammonium chloride) coated MWCNT	Cancer biomarker sensing	Bi et al. (2009)

(Continued)

Table 7.1 (*Continued*)

Enzyme(s) of Interest	Nanosupport	Target application	Reference
Lysozyme	DNA coated SWCNT	Antimicrobial coating	Nepal et al. (2008)
Glucose oxidase/bilirubin oxidase	Palladium-based aerogel	Biofuel cell	Wen et al. (2014)
PQQ-dependent glucose dehydrogenase/laccase	Buckypaper	Biofuel cell	Southcott et al. (2013)
Glucose oxidase	Graphene quantum dots	Glucose sensing	Razmi and Mohammad-Rezaei (2013)
Alcohol dehydrogenase	Gold nanoparticle/MWCNT composite	Biofuel cell	Neto et al. (2015)
Formate dehydrogenase	Pyrolytic graphite	CO_2 reduction to formate	Reda et al. (2008)
Glucose oxidase	Hydroxyl fullerenes	Glucose sensing	Gao et al. (2014)
D-amino acid oxidase	MWCNT/copper nanoparticle/polyaniline composite	D-amino acid sensing	Lata et al. (2013)
Organophosphorus hydrolase	Mesoporous carbon	Nerve reagent sensing	Lee et al. (2010)
Laccase	Arylated SWCNT	Biobattery	Stolarczyk et al. (2012)

7.3 Applications of Immobilized Enzymes in Consumer Applications: From the Lab to Industrial Scale

7.3.1 Demand and Benefits of Enzyme Immobilization in Industrial Settings

Industrial chemical syntheses of widely ranging types and scales are ubiquitous throughout the modern world, for producing the necessary goods and chemicals to be used in everyday life. Such processes are predominantly designed based on an economic optimization scheme to produce the target product while meeting all relevant standards and maximizing revenue. In recent decades, an additional design consideration that has been introduced is in regard to the responsible operation of these systems according to the 12 principles of green engineering (Anastas and Warner, 1998).

The implementation of biocatalysis at the industrial scale has been found to be a highly effective means of reducing the environmental impact of a given synthesis when adhering to listed principles. A schematic representation of immobilized enzyme utilization in industrial catalysis is shown in Fig. 7.3, which focuses on stabilization of the working enzyme under operational conditions, while simplifying separations processing to allow enzyme recovery and reuse.

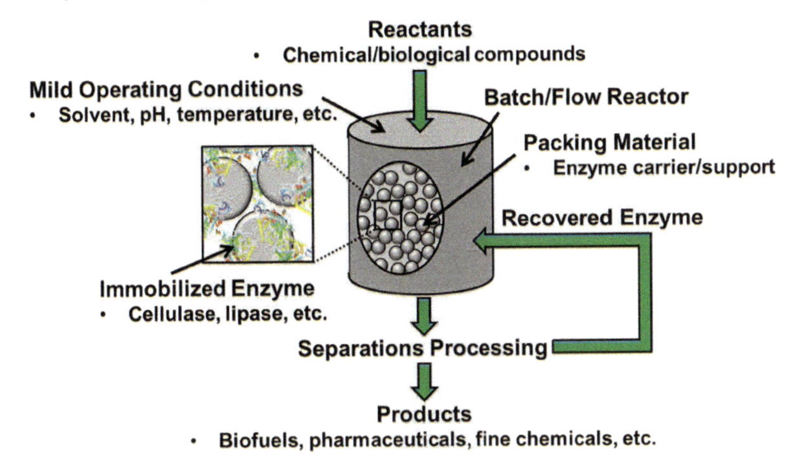

Figure 7.3 Schematic representation of immobilized enzyme utilization in industrial catalysis.

One of the best ways to express the improvements gained upon switching from chemical to enzymatic catalysis is by examining the E-factor (kg of waste produced/kg of product formed) of the overall process before and after the switch. The pharmaceutical industry possesses the highest average E-factor in all of chemical industry at an average value of 25 to more than 100, so there is much demand for processes with increased efficiency that has in many cases been filled using biocatalysis (Sheldon, 2008). For example, the production of pregabalin, an active pharmaceutical ingredient (API) of the anticonvulsant drug Lyrica®, was originally performed using multiple reaction steps that required harsh chemicals and solvents resulting in an overall E-factor of 86 with a mere 20% product yield.

To improve this procedure, modifications were made to the reaction scheme including the incorporation of a reaction step catalyzed by the enzyme lipolase, which improved the product yield to 40%. This change coupled with two other good manufacturing practice steps employed by Pfizer allowed the E-factor of the overall process to drop to 17 thus saving an estimated 10 million gallons of organic solvents and close to 2000 metric tons of raw materials per year (Tao and Xu, 2009). Similarly, a portion of the production process employed by Codexis for an intermediate of Atorvastatin (the API of the Pfizer drug Lipitor®) was transitioned to a three-enzyme reaction using a ketoreductase, a glucose dehydrogenase and a halohydrin dehalogenase, which again eliminated the need for several harsh chemicals. This resulted in a product yield greater than 90% and in 98% product purity with an enantiomeric excess greater than 99.9%. The E-factor of the overall process was, as a result, lowered to 5.8 excluding water or 18 if water is included (Dunn et al., 2010). Both examples exhibit the major impact biocatalysis can have not only on reducing environmental impact but also on improving the overall process efficiency of major chemical syntheses in a green-like fashion.

7.3.2 Enzyme Immobilization in Goods and Chemicals Production

Currently, the enhancement of enzymes used in the production of penicillin via immobilization techniques is one of the major implementations of this approach in the pharmaceutical industry (Es et al., 2015). Specifically, the enzyme penicillin G acylase (PGA), which is used to catalyze the hydrolysis of penicillin G to 6-aminopenicillanic acid in the production pathway of synthetic penicillin, has been the target of significant efforts to improve enzyme stability and activity in conditions relevant to the production process. Immobilization of PGA onto a metal affinity membrane was found to impart enhanced temperature and pH stability as well as reusability to the enzyme allowing for decreased upstream costs in the overall process (Chen et al., 2011). In a separate study, immobilization of PGA onto macro-mesoporous silica spheres was proven to enhance not only operational stability of the enzyme but also the residence time and yield of recovery in a packed bed reactor (Zhao et al., 2011). In terms of the worldwide production of penicillin, which makes up roughly 19% of the estimated global antibiotics market, these increases in applicability were shown to translate to high operational savings (Parmar et al., 2000; Es et al., 2015).

The benefits of similar immobilization practices further extend to other enzymatic systems such as increasing the half-life of intravenously injected streptokinase in blood used to treat deep vein thrombosis and pulmonary embolisms or improving substrate interaction and reusability of chemotherapeutical L-asparaginase used in leukemia treatment (Arenas et al., 2012; Ghosh et al., 2012). Another example is of the enzyme pectinase commonly used in the food and drink industry to improve the clarity of fruit and vegetable juices as well as wines, but such enzyme was shown to be rather sensitive to operational conditions. Immobilization of pectinase onto silica-coated magnetite nanoparticles yielded increased pH and temperature tolerance, storage stability and ease of recovery, which allowed for greater levels of reuse and implementation in a broader window of media conditions (Mosafa et al., 2014). The analysis of consumable products through the use of enzymatic biosensors is also a key use of immobilized enzyme technology, which is discussed in the subsequent section.

As for enzymes in wastewater treatment, laccase-based nanoconjugates have been introduced as candidates for the elimination of recalcitrant pollutants (Corvini and Shahgaldian, 2010). Laccase is a multicopper oxidase capable of effectively inducing polymerization and precipitation of such pollutants as an alternative to peroxidases, if enzyme retention and operational stability can be increased. To that end, Hommes et al. reported the enhanced activity and stability of laccase as a result of sorption onto fumed silica nanoparticles. These conjugates exhibited drastically increased catalytic activity retained over time with roughly 80% residual activity under operationally relevant conditions for one week, whereas free enzyme lost all activity after only 1.5 days (Hommes et al., 2012). Laccase is also a highly relevant chemical in the textile industry, aiding in bleaching processes, and in the treatment of effluent in the pulp and paper industry (Kirk et al. (2002). Similar to laccase utilization in wastewater remediation, immobilization using nanomaterials has been shown to improve recovery and thus allow for economic feasibility when its implementation is considered while also maintaining its operational capabilities (Niladevi and Prema, 2008; Bayramoglu et al., 2010). One of the largest applications of enzymes in industry is as additives in detergents; however, wash conditions can be highly oxidizing and most detergents contain various components detrimental to enzyme performance (Es et al., 2015). Again, immobilization onto prefabricated nanosupports has proven beneficial to the operational performance of target enzymes in this application. In a characteristic study, Soleimani et al. showed that adsorption of the detergent additive α-amylase onto silica nanoparticles not only increased enzyme storage stability, but also improved the starch soil cleaning efficiency of a detergent containing the conjugates whereas addition of free enzyme had little effect (Soleimani et al., 2012).

7.3.3 Enzyme Immobilization in the Production of Biofuels

Additionally, immobilized enzyme technology has found extensive use in the production of biofuels such as methanol and biodiesel (Zhang et al., 2012; Yadav et al., 2014). The production of such fuels has gained increased interest over recent years in an attempt to reduce global dependency on fossil fuels. Biodiesel by definition is a fuel comprised of mono alkyl esters of long chain fatty acids that can be derived from vegetable oils or animal fats and is capable of fuelling diesel engines as an alternative to conventional diesel fuel (Robles-Medina et al., 2009). Biodiesel is most commonly generated through transesterification using methanol as an acyl-acceptor and an alkaline, acidic, enzymatic or inorganic-heterogeneous catalyst (Marchetti et al., 2007). Due to key disadvantages of the other catalysts including high enzyme costs and relatively slower reaction rates when the enzymatic catalyst is implemented, the alkaline catalyzed reaction accounts for nearly all biodiesel production, but yet produces significant waste and requires extensive downstream processing (Robles-Medina et al., 2009).

The most commonly studied enzyme used to catalyze the transesterification reaction is lipase, which offers substantial advantages to the process as a whole, but is not widely implemented due to the high costs mentioned. However, due to strides in enzyme immobilization technology, many immobilized enzyme-based systems have become commercially available to meet the need of more environmentally friendly biodiesel production (Bajaj et al., 2010; Zhang et al., 2012). Currently, lipase systems commercialized under the names *Novozyme* 435 produced by Novozyme and *LS-10* A produced by Beijing CTA New Century Biotechnology Co., Ltd. are two of the most widely utilized systems. Adsorption of lipase onto a textile membrane lowered methanol toxicity to the enzyme and increased recoverability in the case of *LS-10 A*, allowing for an esterification rate greater than 95% with a stability of 210 h. Similarly, adsorption onto an acrylic resin increased the stability of lipase in acidic conditions for *Novozyme 435* allowing for biodiesel yields of 90% and a stability of 500 h, but decreases in the cost of lipase are needed for more widespread use (Bajaj et al., 2010; Zhang et al., 2012). Similar trends have also been reported using supports such as silica nanocomposites, CNTs, gold nanoparticles and magnetic iron oxide nanoparticles, with greater increases in stability and recovery due to increased enzyme loadings, improved mass transfer of substrate and product to the enzymes and enhanced mobility of the conjugates themselves (Wang et al., 2011; Ranjbakhsh et al., 2012; Tran et al., 2012; Verma et al., 2013). For instance, Pavlidis et al. investigated the covalent attachment of lipase onto CNTs and reported increased temperature stability and activity with a 60% improvement of catalytic efficiency (Pavlidis et al., 2012). Incorporation of these nanosupport immobilized enzyme systems has the potential to drastically improve biodiesel production practices.

The biocatalytic generation of methanol via bioconversion of the greenhouse gases carbon dioxide (CO_2) and methane has gained increased research interest due to the possibility of not only reducing the emissions of these gases but also the prospect of generating product, and therefore revenue, from previously inaccessible sources (i.e., remote natural gas deposits, industrial emissions, petroleum drilling flaring, anaerobic digestion in landfills, etc.) (Hwang et al., 2014; Yadav et al., 2014). The naturally occurring oxidation of methane to carbon dioxide is a reversible process that can be utilized with the proper conditions to produce methanol from both sources (Fig. 7.4) (Cazelles et al., 2013; Yadav et al., 2014).

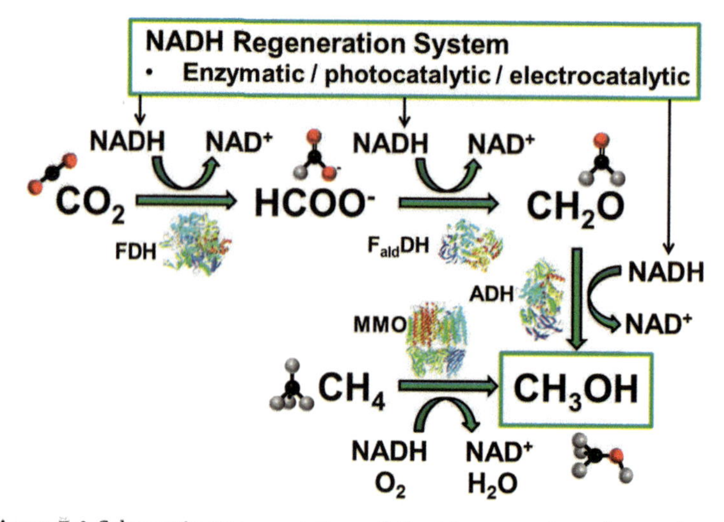

Figure 7.4 Schematic representation of the bioconversion of greenhouse gases into methanol. (enzyme abbreviations: FDH, formate dehydrogenase; $F_{ald}DH$, formaldehyde dehydrogenase; ADH, alcohol dehydrogenase; MMO, methane monooxygenase).

There are four key enzymes involved in this process: methane monooxygenases (MMO), alcohol dehydrogenases (ADH), formaldehyde dehydrogenases ($F_{ald}DH$) and formate dehydrogenases (FDH). The utilization of enzymes is much less energetically intensive and possesses higher conversion and selectivity compared to complex chemical methods, but the rate of conversion is very low (Hwang et al., 2014). Further, oxidation to CO_2 is the energetically favored route of the reaction, thus requiring constant regeneration of the reducing equivalent (NADH) to push the reduction reaction. This regeneration can be accomplished photochemically, electrochemically or enzymatically (Cazelles et al., 2013). Electrochemical regeneration requires an additional source of electrical current while enzymatic regeneration necessitates the use of an additional enzyme and substrate, so photochemical NADH regeneration is truly the greenest and most self-sustainable method of methanol production from CO_2 (Yadav et al., 2014).

Additional complications of these bioconversion reactions are low solubility of CO_2 in water, product inactivation of the enzymes and intermediate accumulation (Cazelles et al., 2013). As discussed, nanosupport immobilization can improve the stability of enzymes against many deactivating factors, but also conductive nanomaterials can enhance electron transfer to improve NADH regeneration. For instance, Yadav et al. reported the performance of a hybrid phocatalytic/biocatalytic system of methanol generation from CO_2 using an isatin-porphyrin chromophore covalently attached to chemically converted graphene as the NADH regeneration mechanism coupled with the reducing enzymatic cascade shown in Fig. 7.4. Specifically, Fig. 7.4 shows the reversible enzymatic pathways capable of forming liquid methanol from either carbon dioxide or methane. This setup was capable of continuously generating methanol at a rate of roughly 0.12 μM/min under visible light from a 0.5 mL/min CO_2 flow, which exhibited an inexpensive and selective method of methanol formation with CO_2 and visible light as the only inputs (Yadav et al., 2014). Continued work in this area will focus on increasing enzyme stability in more favorable solvents and improving the efficiency of product formation.

7.3.4 Enzyme Immobilization in Biosensors

A biosensor is an analytical device that utilizes a biological sensing element in combination with an electrical transducer to sense the presence and concentration of a target analyte via redox reactions (Eggins, 1996). In the case of enzyme-based biosensors, the sensing component is an enzyme that generates the measured current response through product diffusion to the transducer surface (first generation), mediated electron transfer (MET) using specially incorporated redox mediators (second generation) or DET between the enzyme active site and transducer (third generation) (Zhang and Li, 2004). All generations of biosensors greatly benefit from immobilization of the working enzyme on or near the electrode surface. This attachment provides more consistent and efficient current responses by reducing enzyme leaching, increasing operational stability and decreasing the physical distance for electron transfer (Sassolas et al., 2012). First generation biosensors such as those based on the pioneering work of Clark and Lyons on GOX-based electrodes for glucose detection suffer several key limitations (Clark and Lyons, 1962). GOX is the most widely studied anodic enzyme in glucose-based enzymatic biosensors and biofuel cells due to its high intrinsic specificity and selectivity for glucose, but also naturally uses oxygen as a co-substrate (Bankar et al., 2009). Due to the use of molecular oxygen as the natural mediator between the working enzyme and transducer in first generation biosensors, applied voltages must be relatively high leading to interference and the dissolved oxygen concentration tends to fluctuate in solution, which decreases detection accuracy (Zhang and Li, 2004). Thus, second and third generation biosensors have been designed and thoroughly studied using more reliable mediator molecules or no intermediate electron acceptor/donator between the enzyme and electrode, respectively (Sassolas et al., 2012).

Enzyme-based biosensors are highly advantageous due to the natural selectivity and specificity of enzymes toward a target molecule, and their capability to sense such analytes in trace amounts. The lack of harmful by products also allows for the implementation of these sensors in consumable products or in vivo (Mello and Kubota, 2002; Prodromidis and Karayannis, 2002; Vaddiraju et al., 2010). For MET in second-generation enzyme-based biosensors, the introduction of the redox mediators such as ferrocene, osmium, or quinone containing compounds adds additional toxicity and stability concerns to the overall system while also increasing design complexity (Kavanagh and Leech, 2013). In many biosensor designs utilizing MET, the small molecule mediator groups serve to diffusively shuttle electrons between the enzyme active site and electrode surface. However, the diffusive nature of the compounds leads to increased leaching from the system of interest, which rapidly decreases sensor performance and can induce a toxic response in the surrounding environment (Kavanagh and Leech, 2013). These effects can be reduced by incorporating the redox groups into polymer chains where the electrons hop along immobile redox sites in the polymer backbone to the electrode surface (Mao et al., 2003). However, the presence of non-enzymatic redox groups can still decrease sensor selectivity (Zhang and Li, 2004). These complications are eliminated in the operation of third generation biosensors utilizing DET.

To achieve DET with redox-active enzymes, suitable nanosupports capable of achieving close communication with the enzyme active site upon immobilization are required. Commonly used nanomaterials include gold nanoparticle-based composites and graphene-based materials such as graphene/graphene oxide and CNTs (Falk et al., 2012). In some cases, DET can be difficult to achieve due to the conformation of the enzyme. For instance, the flavin adenine dinucleotide (FAD)-based active site of GOX is buried deeply below the protein surface resulting in resistance to electron transfer (Wohlfahrt et al., 1999). The possibility of DET of GOX with graphene-based nanomaterial surfaces has been disputed (Goran et al., 2013; Liang et al., 2015). However, an abundance of studies have exhibited DET characteristics in GOX-based systems such as increases in anodic current responses upon glucose introduction in inert environments suggesting that DET with GOX is possible with careful nanosupport selection (Falk et al., 2012; Martins et al., 2014; Campbell et al., 2015).

Many high performing DET-based biosensors reported have utilized graphene-based electrode systems. For example, Mani et al. studied the glucose sensing capabilities of GOX adsorbed onto a graphene oxide–multi-wall carbon nanotube (MWCNT) composite. This setup achieved a detection limit of 4.7 μM with a linear detection range of 0.01–6.5 mM (Mani et al., 2013). Such a system would be theoretically capable of sensing physiological glucose concentrations at and below the average value of ~5.5 mM. The electron transfer rate constant (k_s) of this system, which is a key measurement of electron transfer efficiency, was found to be 3.02 s^{-1}. The efficiency was comparatively higher to other similar systems, displaying the capability of CNTs to efficiently collect electrons directly from the GOX active site (Mani et al., 2013). Similar results were reported using a MWCNT/chitosan composite in glucose sensing. Namely, Liu et al. observed a glucose detection sensitivity of 0.52 μA/mM^{-1} with a linear detection range up to 7.8 mM and a k_s of 7.73 s^{-1} using GOX adsorbed to MWCNTs within a chitosan matrix (Liu et al., 2005). The sensing capabilities of enzyme-functionalized CNTs extend to other enzyme/analyte combinations as well. Yu et al. reported the H_2O_2 sensing capabilities of HRP covalently immobilized onto SWCNTs yielding a detection limit of 0.05 μM and a linear detection range up to 8 μM and Tominaga et al. showed the fructose detecting characteristics of fructose dehydrogenase adsorbed to MWCNTs with a detection limit of 5 mM and a linear detection range up to 40 mM (Yu et al., 2003; Tominaga et al., 2009). These systems have found extensive application in mainstream applications such as product analysis in the food and beverage industry, in vivo detection of target molecules (most notably glucose detection in diabetes monitoring) and environmental monitoring. There are also multiple other interesting applications that have been presented, including continuous glucose monitoring in diabetes patients via "smart" contact lenses and analysis of sweat composition in wearable biosensors (Falk et al., 2013; Jia et al., 2013).

7.3.5 Enzyme Immobilization in Biofuel Cells

Enzyme-based biofuel cells generate electrical power via oxidation of a fuel at an enzyme-functionalized anode and reduction of a final electron acceptor such as molecular oxygen at an enzyme-functionalized cathode. This process is depicted in Fig. 7.5. Specifically, the target fuel (i.e., glucose, fructose, alcohol, etc.) is oxidized by the corresponding anodic enzyme and electrons are transferred through an external circuit to the cathode, where oxygen is reduced by the cathodic working enzyme. Such systems are ideally suited for the powering of implantable devices due to their biocompatible nature and the biological source of necessary substrates (Cosnier et al., 2014). Again, the operation of the enzyme-based schemes relies on the efficient turnover of substrate and transfer of electrons between the active sites of the enzymes and the external circuit via the electrode surface. The performance of a biofuel cell is generally reported using two main parameters: the maximum power density and open circuit voltage (OCV). The maximum power density is defined as the maximum total power per unit area capable of being produced. The OCV of a given system is governed by the maximum theoretical voltage difference between the oxidation and reduction reactions occurring on the respective electrodes. Losses in the form of overpotentials decrease the observed value in the form of activation, ohmic, and mass transport resistances (Bullen et al., 2006). Kinetic resistances are observed at low currents at which either the anode or cathode is limited by the rate of substrate turnover and electron transfer. At higher currents, the system becomes limited by the transport of substrate and product to and from the enzyme active site. This relationship results in an optimal operating current, and thus voltage, at which the maximum power density is reached (Bullen et al., 2006). The two key limitations currently hindering enzyme-based biofuel cell application are low power output and poor long-term stability. Thus, enzyme immobilization onto nanosupports is one of the leading candidates for improvement, offering increased enzyme stability and the capability to incorporate high enzyme densities within a given volume while promoting efficient charge collection (Walcarius et al., 2013; de Poulpiquet et al., 2014).

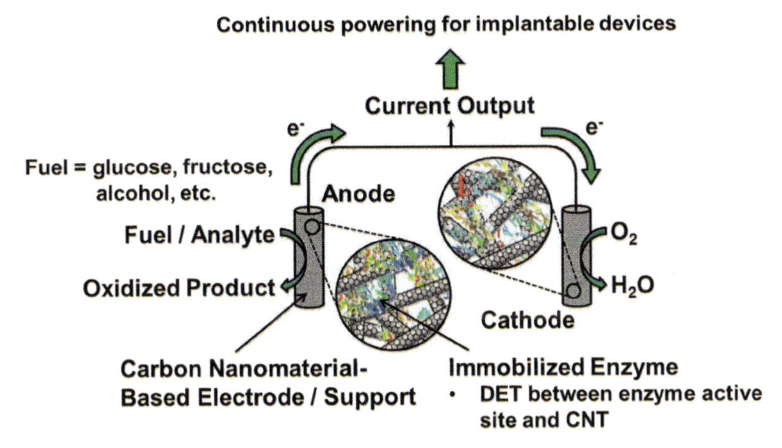

Figure 7.5 Schematic representation of an enzymatic biofuel cell utilizing carbon nanomaterial-based electrodes and DET.

The anodic enzyme of a given system is chosen based on the target fuel source (i.e., glucose, fructose, alcohol, lactose, etc.) with the most commonly studied models being GOX or pyrroloquinoline quinone-dependent glucose dehydrogenase (PQQ-GDH) for glucose, fructose dehydrogenase for fructose and ADH for alcohols (Cooney et al., 2008). Multicopper oxidases such as laccase and bilirubin oxidase (BOD) are the most commonly utilized cathodic enzymes as they have been shown to efficiently accept and transfer electrons to oxygen resulting in the reduction to water as the only by product (Shleev et al., 2005). The distinguishing characteristic that separates the two main categories of enzyme-based biofuel cells is the type of electron transfer utilized: MET or DET. MET-based biofuel cells suffer many of the same drawbacks as MET-based biosensors. Leaching of the mediator groups not only causes increased instability and toxicity, but also can prompt the need for membrane separation of the anodic and cathodic compartments, which greatly hinders miniaturization. Further, in order for the electron transfer via mediator species to be feasible, the redox potentials of the included mediators must lie within those of the enzyme active sites, effectively reducing the theoretical maximum voltage possible to be obtained (de Poulpiquet et al., 2014). However, MET-based fuel cells generally exhibit increased electron transfer efficiency and thus power output due to decreased electron transfer resistances encountered by the small molecule mediators that possess an increased ease of accessibility to both the enzyme active site and electrode surface (Kavanagh and Leech, 2013).

A prime example of the differences between MET- and DET-based biofuel cells is presented in the excellent work performed by Cosnier and coworkers. Specifically, it was shown that a biofuel cell comprised of a GOX-functionalized anode and a laccase-functionalized cathode using compressed MWCNT disks as the support material was able to produce a power density of 1.3 mW/cm^2 with an OCV of 0.95 V making this system one of the highest performing enzymatic biofuel cells to date. Moreover, 96% of the original power output was retained over 1 month of storage (Zebda et al., 2011). In a subsequent study, a naphoquinone mediator was incorporated into the construction of the anodic portion of the fuel cell while the rest of the setup was kept the same. The incorporation of the mediator group resulted in a 23% increase in power output. However, decreases in both OCV and stability were observed with the new OCV reaching 0.76 V and a loss of 40% of the original power after only one week (Reuillard et al., 2013). These trends have resulted in a drive for the development of higher performing enzyme-based biofuel cells capable of utilizing DET.

The previously mentioned study also highlights the benefits of graphene-based nanosupports as electrode materials with many high performing systems relying on graphene or CNTs for electron conduction (Kwon et al., 2014; Prasad et al., 2014; Campbell et al., 2015). These materials provide ample available surface area for high enzyme loadings while their nanoscale dimensions promote DET and can have a stabilizing effect on enzyme conformation (Yang et al., 2012). Further, CNTs and graphene can be used to construct conductive, three-dimensional matrices capable of achieving extremely high enzyme loadings while surrounding the catalysts with conductive material (Tamaki, 2012). Campbell et al. produced graphene/SWCNT cogels with SSA ~800 m^2/g that were shown to load roughly 9×10^{-9} mol/cm^2 of GOX via physical adsorption. Coupled with a similarly functionalized BOD-based cathode, the formed biofuel cell proved capable of producing a power density of 0.19 mW/cm^2 with an OCV of 0.61 V. The observed power density was only about 0.08% of the possible power output relative to native GOX activity at the loading concentration, showing the enormous promise of this electrode material if electron transfer efficiency can be improved (Campbell et al., 2015). In a similar system, Prasad et al. functionalized SWCNT coated graphene foams with covalently attached GOX or laccase yielding a power output of 2.3 mW/cm^2 and an OCV of 1.2 V, which was proven capable of lighting an LED (Prasad et al., 2014). Such systems have already been shown to power devices such as digital displays and pacemakers in vitro as well as operate in implantable models (Halamkova et al., 2012; MacVittie et al., 2013; Southcott et al., 2013; Zebda et al., 2013). With these results, the future is bright for green technologies powered by enzyme-based biofuel cells.

7.3.6 Enzyme Immobilization in Bioactive Coatings

An alternative application of nanosupport-immobilized enzymes is in the development of antifouling and active-surface decontaminating coatings (Olsen et al., 2007; Grover et al., 2013).

In these systems, working enzymes incorporated into a coating matrix serve to produce strong decontaminants such as peraacetic acid (PAA), H_2O_2, or hypochlorous acid (HOCl) in the presence of substrate. In turn, this decontaminant interacts with possible pathogenic or biofilm forming molecules that come into contact with the modified surface causing deactivation of the harmful compounds. Figure 7.6 represents a model surface that has been functionalized with an active-surface decontaminating coating. Specifically, a coating matrix that contains working enzyme immobilized onto a nanosupport is applied to a target surface (i.e., wall, medical device, etc.) and pathogens are continuously denatured by enzyme-produced decontaminant in the presence of substrate. Enzyme immobilization is crucial in these composites to retain enzyme activity during operation and prevent enzyme leaching from the modified surface (Grover et al., 2013).

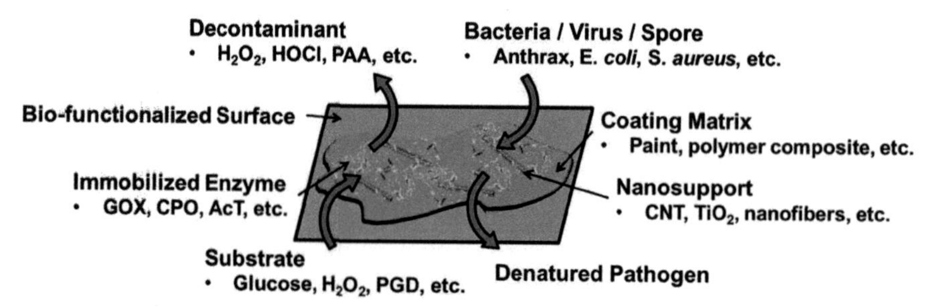

Figure 7.6 Schematic representation of a bioactive self-decontaminating coating comprised of nanosupport immobilized enzyme within a coating matrix. (*Abbreviations*: GOX, glucose oxidase; CPO, chloroperoxidase; AcT, perhydrolase; PAA, peraacetic acid; PGD, propylene glycol diacetate).

Nanomaterials beneficially used for the immobilization of the working enzymes include CNTs, nanofibers and photocatalytic materials such as titanium dioxide (TiO_2) (Grover et al., 2013). Photocatalytic materials capable of generating reactive oxygen species (ROS) in aqueous environments (i.e., TiO_2) are already commonplace in water treatment (Lazar et al., 2012). However, the generated ROS are often not strong enough to fully and efficiently decontaminate the target system. Thus, enzymes such as CPO, which uses the ROS H_2O_2 to generate the much stronger decontaminant HOCl, can be incorporated into the system along with TiO_2 or a second enzyme like GOX that generates H_2O_2 from glucose for increased decontamination efficiency (Campbell et al., 2013). Other enzymes that have been applied to this technology include perhydrolase (AcT) and hexose oxidase, which produce PAA from various acyl donors and H_2O_2 from glucose, respectively (Olsen et al., 2010; Dinu et al., 2012). These types of coatings are important in numerous settings such as pesticide removal from wastewater, rapid warfare agent decontamination on the battlefield, sensitive surface sterilization in hospitals and prevention of biofilm formation in equipment used in marine environments (Olsen et al., 2007; Grover et al., 2013). Due to the operational nature of bioactive coatings, the composite must retain its enzymatic activity for a sufficient time without the need for reapplication to become economically feasible. In other words, the enzymatic component of the coating needs to be stabilized against deactivation from environmental conditions and be prevented from leaching into solution.

Nanosupport immobilization has been shown to greatly benefit the function of these coatings. For example, Dinu et al. reported on the decontamination capabilities of cross-linked AcT covalently immobilized onto MWCNTs and subsequently incorporated into latex paint. The immobilization process was proven to allow for the retention of ~40% of native AcT activity while providing increased thermostability at 75°C and retention within the coating. Specifically, roughly 50% of free AcT incorporated into the latex paint was observed to leach into solution within 30 min whereas MWCNT immobilized AcT exhibited no leaching after 15 days. The AcT-SWCNT conjugates further proved capable of decontaminating 99% of 10^6 CFU/mL B. *cereus* spores in only 1 h with about 60% killing in the first 15 min and complete killing of 10^6 CFU/mL *E. coli* in only 5 min (Dinu et al., 2012). Similar results were observed for MWCNT immobilized AcT challenged with *B. cereus*, *B. anthracis*, and influenza virus (Dinu et al., 2010; Grover et al., 2013). In a separate study, a serine protease was first physically adsorbed onto SWCNTs and then dispersed in poly(methyl methacrylate) (PMMA) and incorporated into films or paints tested as antifouling coatings. It was reported that immobilization onto SWCNTs improved enzyme retention with more than 90% initial activity retained after 30 days compared to complete loss of enzyme activity after washing during preparation for free enzyme. The conjugates also possessed greater thermostability with activity retained up to 70°C. Finally, protease-SWCNT/PMMA conjugate containing coatings were capable of repelling the adhesions of the fouling proteins human serum albumin and fibrinogen with an 85% decrease in protein adsorption after 6 days compared to control (Asuri et al., 2007). These films represent necessary gains in the design and development of protective coatings effectively imparting stable biological function to non-biological surfaces.

7.4 Concluding Remarks

Throughout this chapter, the undeniable benefits of enzyme immobilization in terms of green chemistry and green technologies have been discussed. The many various methods of immobilization offer a wide array of potential characteristics and applications of the resulting conjugates. In particular, enzymes immobilized onto prefabricated carriers, specifically nanosupports, present a unique approach to improving the operational capabilities of various systems while greatly reducing environmental impact.

The applications discussed in this chapter list the enormous potential of these applications, namely the following:

- The capability of industrial biocatalysis-based processes capable of eliminating harsh by products and solvents, reducing energy demands and increasing specific yield is indispensable in moving forward with improvements in green chemistry.
- The increased utilization of biofuels, biofuel production via biocatalysis and the bioconversion of greenhouse gases into biofuels offer an even greater route to reducing industrial environmental effects.
- Continued design and development of enzymatic biosensors and biofuel cells represents growing possibilities not only in the evaluation of industrial product quality and the monitoring biologically relevant compounds for medical treatment, but also in the continuous powering of micro-electronic systems for the next generation of implantable devices.
- Incorporation of enzyme-functionalized bioactive coatings can provide self-sustaining antifouling/decontaminating

surfaces to improve to safety and efficiency of systems in various industries. In short, the implementation of enzymatic catalysis through nanosupport immobilization in all of these applications represents the next steps in designing efficient and effective processes and materials aimed at meeting the goals of green chemistry.

References

Al-Harthi M., Cheng L. S., Soares J. B. P., Simon L. C., *J. Polym. Sci. Pol. Chem.*, **45** (2007), 2212–2224.

Ali L., Algaithi R., Habib H. M., Souka U., Rauf M. A., Ashraf S. S., *BMC Biochem.*, **14** (2013), 1–13.

Anastas P. T., Warner J. C., *Green Chemistry: Theory and Practice*, Oxford University Press, Oxford; New York (1998).

Ansari S. A., Husain Q., *Biotechnol. Adv.*, **30** (2012), 512–523.

Arenas E., Castillon F. F., Farias M. H., *Des. Monomers Polym.*, **15** (2012), 369–378.

Assenberg R., Wan P. T., Geisse S., Mayr L. M., *Curr. Opin. Struc. Biol.*, **23** (2013), 393–402.

Asuri P., Bale S. S., Karajanagi S. S., Kane R. S., *Curr. Opin. Biotechnol.*, **17** (2006), 562–568.

Asuri P., Bale S. S., Pangule R. C., Shah D. A., Kane R. S., Dordick J. S., *Langmuir*, **23** (2007), 12318–12321.

Asuri P., Karajanagi S. S., Kane R. S., Dordick J. S., *Small*, **3** (2007), 50–53.

Asuri P., Karajanagi S. S., Sellitto E., Kim D. Y., Kane R. S., Dordick J. S., *Biotechnol. Bioeng.*, **95** (2006), 804–811.

Asuri P., Karajanagi S. S., Yang H. C., Yim T. J., Kane R. S., Dordick J. S., *Langmuir*, **22** (2006), 5833–5836.

Averick S. E., Bazewicz C. G., Woodman B. F., Simakova A., Mehl R. A., Matyjaszewski K., *Eur. Polym. J.*, **49** (2013), 2919–2924.

Averick S., Simakova A., Park S., Konkolewicz D., Magenau A. J. D., Mehl R. A., Matyjaszewski K., *ACS Macro Lett.*, **1** (2012), 6–10.

Baborova P., Moder M., Baldrian P., Cajthamlova K., Cajthaml T., *Res. Microbiol.*, **157** (2006), 248–253.

Bajaj A., Lohan P., Jha P. N., Mehrotra R., *J. Mol. Catal. B Enzym.*, **62** (2010), 9–14.

Bankar S. B., Bule M. V., Singhal R. S., Ananthanarayan L., *Biotechnol. Adv.*, **27** (2009), 489–501.

Baughman R. H., Zakhidov A. A., de Heer W. A., *Science*, **297** (2002), 787–792.

Bayramoglu G., Yilmaz M., Arica M. Y., *Bioresour. Technol.*, **101** (2010), 6615–6621.

Bi S., Zhou H., Zhang S. S., *Biosens. Bioelectron.*, **24** (2009), 2961–2966.

Bornscheuer U. T., Huisman G. W., Kazlauskas R. J., Lutz S., Moore J. C., Robins K., *Nature*, **485** (2012), 185–194.

Bullen R. A., Arnot T. C., Lakeman J. B., Walsh F. C., *Biosens. Bioelectron.*, **21** (2006), 2015–2045.

Campbell A. S., Dong C., Dordick J. S., Dinu C. Z., *Process Biochem.*, **48** (2013), 1355–1360.

Campbell A. S., Dong C., Maloney A., Hardinger J., Hu X., Meng F., Guiseppe-Elie A., Wu N., Dinu C. Z., *Nano LIFE*, **04** (2014a), 1450005.

Campbell A. S., Dong C. B., Meng F. K., Hardinger J., Perhinschi G., Wu N. Q., Dinu C. Z., *ACS Appl. Mater. Interfaces*, **6** (2014), 5393–5403.

Campbell A. S., Jeong Y. J., Geier S. M., Koepsel R. R., Russell A. J., Islam M. F., *ACS Appl. Mater. Interfaces*, **7** (2015), 4056–4065.

Cao L. Q., van Rantwijk F., Sheldon R. A., *Org. Lett.*, **2** (2000), 1361–1364.

Caza N., Bewtra J. K., Biswas N., Taylor K. E., *Water Res.*, **33** (1999), 3012–3018.

Cazelles R., Drone J., Fajula F., Ersen O., Moldovan S., Galarneau A., *New J. Chem.*, **37** (2013), 3721–3730.

Chen C. I., Ko Y. M., Shieh C. J., Liu Y. C., *J. Membr. Sci.*, **380** (2011), 34–40.

Chen Z. Y., Li H., Peng A. P., Gao Y. Z., *Environ. Sci. Pollut. R*, **21** (2014), 10696–10705.

Cheng J., Yu S. M., Zuo P., *Water Res.*, **40** (2006), 283–290.

Choi O., Kim B. C., An J. H., Min K., Kim Y. H., Um Y., Oh M. K., Sang B. I., *Enzyme Microb. Technol.*, **49** (2011), 441–445.

Clark J. H., Luque R., Matharu A. S., *Annu. Rev. Chem. Biomol.*, **3** (2012), 183–207.

Clark L. C., Lyons C., *Ann. N. Y. Acad. Sci.*, **102** (1962), 29–45.

Cooney M. J., Svoboda V., Lau C., Martin G., Minteer S. D., *Energy Environ. Sci.*, **1** (2008), 320–337.

Corvini P. F. X., Shahgaldian P., *Rev. Environ. Sci. Bio.*, **9** (2010), 23–27.

Cosnier S., Le Goff A., Holzinger M., *Electrochem. Commun.*, **38** (2014), 19–23.

Cummings C., Murata H., Koepsel R., Russell A. J., *Biomaterials,* **34** (2013), 7437–7443.

Cummings C., Murata H., Koepsel R., Russell A. J., *Biomacromolecules,* **15** (2014), 763–771.

de Lathouder K. M., Smeltink M. W., Straathof A. J., Paasman M. A., van de Sandt E. J., Kapteijn F., Moulijn J. A., *J. Ind. Microbiol. Biotechnol.,* **35** (2008), 815–824.

De P., Li M., Gondi S. R., Sumerlin B. S., *J. Am. Chem. Soc.,* **130** (2008), 11288–11289.

de Poulpiquet A., Ciaccafava A., Gadiou R., Gounel S., Giudici-Orticoni M. T., Mano N., Lojou E., *Electrochem. Commun.,* **42** (2014), 72–74.

de Poulpiquet A., Ciaccafava A., Lojou E., *Electrochim. Acta,* **126** (2014), 104–114.

Diez P., Villalonga R., Villalonga M. L., Pingarron J. M., *J. Colloid Interface Sci.,* **386** (2012), 181–188.

Dinu C. Z., Borkar I. V., Bale S. S., Campbell A. S., Kane R. S., Dordick J. S., *J. Mol. Catal. B Enzym.,* **75** (2012), 20–26.

Dinu C. Z., Zhu G., Bale S. S., Anand G., Reeder P. J., Sanford K., Whited G., Kane R. S., Dordick J. S., *Adv. Funct. Mater.,* **20** (2010), 392–398.

Dunn P. J., *Chem. Soc. Rev.,* **41** (2012), 1452–1461.

Dunn P. J., Wells A. S., Williams M. T., *Green Chemistry in the Pharmaceutical Industry,* Wiley-VCH, Weinheim (2010).

Eggins B. R., *Biosensors: An Introduction,* Wiley-Teubner, Chichester; New York (1996).

Eker B., Zagorevski D., Zhu G. Y., Linhardt R. J., Dordick J. S., *J. Mol. Catal. B Enzym.,* **59** (2009), 177–184.

El-Zahab B., Donnelly D., Wang P., *Biotechnol. Bioeng.,* **99** (2008), 508–514.

Es I., Vieira J. D. G., Amaral A. C., *Appl. Microbiol. Biotechnol.,* **99** (2015), 2065–2082.

Falk M., Andoralov V., Silow M., Toscano M. D., Shleev S., *Anal. Chem.,* **85** (2013), 6342–6348.

Falk M., Blum Z., Shleev S., *Electrochim. Acta,* **82** (2012), 191–202.

Feng W., Ji P. J., *Biotechnol. Adv.,* **29** (2011), 889–895.

Gao Y. F., Yang T., Yang X. L., Zhang Y. S., Xiao B. L., Hong J., Sheibani N., Ghourchian H., Hong T., Moosavi-Movahedi A. A., *Biosens. Bioelectron.,* **60** (2014), 30–34.

Garcia-Galan C., Berenguer-Murcia A., Fernandez-Lafuente R., Rodrigues R. C., *Adv. Synth. Catal.,* **353** (2011), 2885–2904.

Ghosh S., Chaganti S. R., Prakasham R. S., *J. Mol. Catal. B Enzym.*, **74** (2012), 132–137.

Goran J. M., Mantilla S. M., Stevenson K. J., *Anal. Chem.*, **85** (2013), 1571–1581.

Gorji T. B., Ranjbar A. A., Mirzababaei S. N., *Sol. Energy*, **119** (2015), 332–342.

Gray J. S. S., Montgomery R., *Glycobiology*, **7** (1997), 679–685.

Grover N., Dinu C. Z., Kane R. S., Dordick J. S., *Appl. Microbiol. Biotechnol.*, **97** (2013), 3293–3300.

Grover N., Douaisi M. P., Borkar I. V., Lee L., Dinu C. Z., Kane R. S., Dordick J. S., *Appl. Microbiol. Biotechnol.*, **97** (2013), 8813–8821.

Halamkova L., Halamek J., Bocharova V., Szczupak A., Alfonta L., Katz E., *J. Am. Chem. Soc.*, **134** (2012), 5040–5043.

Han Y. J., Watson J. T., Stucky G. D., Butler A., *J. Mol. Catal. B Enzym.*, **17** (2002), 1–8.

Harris J. M., Chess R. B., *Nat. Rev. Drug. Discov.*, **2** (2003), 214–221.

Hoare T. R., Kohane D. S., *Polymer*, **49** (2008), 1993–2007.

Hommes G., Gasser C. A., Howald C. B. C., Goers R., Schlosser D., Shahgaldian P., Corvini P. F. X., *Bioresour. Technol.*, **115** (2012), 8–15.

Houde A., Kademi A., Leblanc D., *Appl. Biochem. Biotechnol.*, **118** (2004), 155–170.

Husain Q., *Rev. Environ. Sci. Bio.*, **9** (2010), 117–140.

Husain Q., Ulber R., *Crit. Rev. Environ. Sci. Technol.*, **41** (2011), 770–804.

Hwang I. Y., Lee S. H., Choi Y. S., Park S. J., Na J. G., Chang I. S., Kim C., Kim H. C., Kim Y. H., Lee J. W., Lee E. Y., *J. Microbiol. Biotechnol.*, **24** (2014), 1597–1605.

Idris A., Bukhari A., *Biotechnol. Adv.*, **30** (2012), 550–563.

Jia W. Z., Bandodkar A. J., Valdes-Ramirez G., Windmiller J. R., Yang Z. J., Ramirez J., Chan G., Wang J., *Anal. Chem.*, **85** (2013), 6553–6560.

Jose M. V., Marx S., Murata H., Koepsel R. R., Russell A. J., *Carbon*, **50** (2012), 4010–4020.

Kamal J. K. A., Behere D. V., *J. Inorg. Biochem.*, **94** (2003), 236–242.

Karajanagi S. S., Vertegel A. A., Kane R. S., Dordick J. S., *Langmuir*, **20** (2004), 11594–11599.

Kavanagh P., Leech D., *Phys. Chem. Chem. Phys.*, **15** (2013), 4859–4869.

Khan M. J., Husain Q., Azam A., *Biotechnol. Bioproc. Eng.*, **17** (2012), 377–384.

Kirk O., Borchert T. V., Fuglsang C. C., *Curr. Opin. Biotechnol.*, **13** (2002), 345–351.

Klein M. P., Fallavena L. P., Schoffer Jda N., Ayub M. A., Rodrigues R. C., Ninow J. L., Hertz P. F., *Carbohydr. Polym.*, **95** (2013), 465–470.

Koller G., Moder M., Czihal K., *Chemosphere*, **41** (2000), 1827–1834.

Krueger A., Lang D., *Adv. Funct. Mater.*, **22** (2012), 890–906.

Kwon C. H., Lee S. H., Choi Y. B., Lee J. A., Kim S. H., Kim H. H., Spinks G. M., Wallace G. G., Lima M. D., Kozlov M. E., Baughman R. H., Kim S. J., *Nat. Commun.*, **5** (2014), 3928.

Kwon K. Y., Yang S. B., Kong B. S., Kim J., Jung H. T., *Carbon*, **48** (2010), 4504–4509.

Lane M. D., Seelig B., *Curr. Opin. Chem. Biol.*, **22** (2014), 129–136.

Lata S., Batra B., Kumar P., Pundir C. S., *Anal. Biochem.*, **437** (2013), 1–9.

Lazar M. A., Varghese S., Nair S. S., *Catalysts*, **2** (2012), 572–601.

Lee B. H., Kim M. H., Lee J. H., Seliktar D., Cho N. J., Tan L. P., *PLoS One*, **10** (2015), e0118123.

Lee J., Lee D., Oh E., Kim J., Kim Y. P., Jin S., Kim H. S., Hwang Y., Kwak J. H., Park J. G., Shin C. H., Kim J., Hyeon T., *Angew. Chem. Int. Ed.*, **44** (2005), 7427–7432.

Lee J. H., Park J. Y., Min K., Cha H. J., Choi S. S., Yoo Y. J., *Biosens. Bioelectron.*, **25** (2010), 1566–1570.

Li S. F., Fan Y. H., Hu J. F., Huang Y. S., Wu W. T., *J. Mol. Catal. B Enzym.*, **73** (2011), 98–103.

Liang B., Guo X. S., Fang L., Hu Y. C., Yang G., Zhu Q., Wei J. W., Ye X. S., *Electrochem. Commun.*, **50** (2015), 1–5.

Liang H. W., Liu J. W., Qian H. S., Yu S. H., *Acc. Chem. Res.*, **46** (2013), 1450–1461.

Libertino S., Aiello V., Scandurra A., Renis M., Sinatra F., *Sensors*, **8** (2008), 5637–5648.

Lin Y. H., Lu F., Wang J., *Electroanalytical*, **16** (2004), 145–149.

Liu Y., Wang M. K., Zhao F., Xu Z. A., Dong S. J., *Biosens. Bioelectron.*, **21** (2005), 984–988.

Liu X. Q., Yan R., Zhang J. M., Zhu J., Wong D. K. Y., *Biosens. Bioelectron.*, **66** (2015), 208–215.

Liu L., Yang H. Q., Shin H. D., Li J. H., Du G. C., Chen J., *Appl. Microbiol. Biotechnol.*, **97** (2013), 9597–9608.

Lopez-Serrano P., Cao L., van Rantwijk F., Sheldon R. A., *Biotechnol. Lett.*, **24** (2002), 1379–1383.

Lu J., Weerasiri R. R., Lee I., *Biotechnol. Lett.*, **35** (2013), 181–188.

MacAodha D., Ferrer M. L., Conghaile P. O., Kavanagh P., Leech D., *Phys. Chem. Chem. Phys.*, **14** (2012), 14667–14672.

MacVittie K., Halamek J., Halamkova L., Southcott M., Jemison W. D., Lobeld R., Katz E., *Energy Environ. Sci.*, **6** (2013), 81–86.

Magnusson J. P., Bersani S., Salmaso S., Alexander C., Caliceti P., *Bioconjug. Chem.*, **21** (2010), 671–678.

Mani V., Devadas B., Chen S. M., *Biosens. Bioelectron.*, **41** (2013), 309–315.

Manso J., Mena M. L., Yanez-Sedeno P., Pingarron J. M., *Anal. Biochem.*, **375** (2008), 345–353.

Manta C., Ferraz N., Betancor L., Antunes G., Batista-Viera F., Carlsson J., Caldwell K., *Enzyme Microb. Technol.*, **33** (2003), 890–898.

Mao F., Mano N., Heller A., *J. Am. Chem. Soc.*, **125** (2003), 4951–4957.

Marchetti J. M., Miguel V. U., Errazu A. F., *Renew. Sust. Energy Rev.*, **11** (2007), 1300–1311.

Martins M. V. A., Pereira A. R., Luz R. A. S., Lost R. M., Crespitho F. N., *Phys. Chem. Chem. Phys.*, **16** (2014), 17426–17436.

Marx S., Jose M. V., Andersen J. D., Russell A. J., *Biosens. Bioelectron.*, **26** (2011), 2981–2986.

Mateo C., Palomo J. M., Fernandez-Lorente G., Guisan J. M., Fernandez-Lafuente R., *Enzyme Microb. Technol.*, **40** (2007), 1451–1463.

Matyjaszewski K., Tsarevsky N. V., *J. Am. Chem. Soc.*, **136** (2014), 6513–6533.

Matyjaszewski K., Xia J. H., *Chem. Rev.*, **101** (2001), 2921–2990.

Mello L. D., Kubota L. T., *Food Chem.*, **77** (2002), 237–256.

Milašinović N., Knežević-Jugović Z., Milosavljević N., Škorić M. L., Filipović J., Kalagasidis Krušić M., *Biomed. Res. Int.*, **2014** (2014), 364930.

Min K., Yoo Y. J., *Talanta*, **80** (2009), 1007–1011.

Min K., Yoo Y. J., *Biotechnol. Bioproc. Eng.*, **19** (2014), 553–567.

Mosafa L., Shahedi M., Moghadam M., *J. Chin. Chem. Soc.*, **61** (2014), 329–336.

Mossavarali S., Hosseinkhani S., Ranjbar B., Mirohaei M., *Int. J. Biol. Macromol.*, **39** (2006), 192–196.

Mubarak N. M., Wong J. R., Tan K. W., Sahu J. N., Abdullah E. C., Jayakumar N. S., Ganesan P., *J. Mol. Catal. B Enzym.*, **107** (2014), 124–131.

Murata H., Cummings C. S., Koepsel R. R., Russell A. J., *Biomacromolecules*, **14** (2013), 1919–1926.

Murata H., Cummings C. S., Koepsel R. R., Russell A. J., *Biomacromolecules*, **15** (2014), 2817–2823.

Nakamoto S., Machida N., *Water Res.*, **26** (1992), 49–54.

Nepal D., Balasubramanian S., Simonian A. L., Davis V. A., *Nano Lett.*, **8** (2008), 1896–1901.

Neto S. A., Almeida T. S., Belnap D. M., Minteer S. D., De Andrade A. R., *J. Power Sources*, **273** (2015), 1065–1072.

Neto S. A., Suda E. L., Xu S., Meredith M. T., De Andrade A. R., Minteer S. D., *Electrochim. Acta*, **87** (2013), 323–329.

Ngo T. P. N., Zhang W., Wang W., Li Z., *Chem. Commun.*, **48** (2012), 4585–4587.

Nguyen K. V., Minteer S. D., *Chem. Commun.*, **51** (2015), 13071–13073.

Nicolas J., San Miguel V., Mantovani G., Haddleton D. M., *Chem. Commun.*, **45** (2006), 4697–4699.

Niladevi K. N., Prema P., *World J. Microbiol. Biotechnol.*, **24** (2008), 1215–1222.

Noyori R., *Chem. Commun.*, **14** (2005), 1807–1811.

Olsen S. M., Kristensen J. B., Laursen B. S., Pedersen L. T., Dam-Johansen K., Kiil S., *Prog. Org. Coat.*, **68** (2010), 248–257.

Olsen S. M., Pedersen L. T., Laursen M. H., Kiil S., Dam-Johansen K., *Biofouling*, **23** (2007), 369–383.

Packer M. S., Liu D. R., *Nat. Rev. Genet.*, **16** (2015), 379–394.

Parmar A., Kumar H., Marwaha S. S., Kennedy J. F., *Biotechnol. Adv.*, **18** (2000), 289–301.

Pavlidis I. V., Vorhaben T., Gournis D., Papadopoulos G. K., Bornscheuer U. T., Stamatis H., *J. Nanopart. Res.*, **14** (2012), 1–10.

Peigney A., Laurent C., Flahaut E., Bacsa R. R., Rousset A., *Carbon*, **39** (2001), 507–514.

Peppas N. A., *Curr. Opin. Colloid Interface Sci.*, **2** (1997), 531–537.

Perez-Anguiano O., Wenger B., Pugin R., Hofmann H., Scolan E., *ACS Appl. Mater. Interfaces*, **7** (2015), 2960–2971.

Philp J. C., Ritchie R. J., Allan J. E. M., *Trends Biotechnol.*, **31** (2013), 219–222.

Pierre A. C., *Biocatal. Biotransform.*, **22** (2004), 145–170.

Poliakoff M., Fitzpatrick J. M., Farren T. R., Anastas P. T., *Science*, **297** (2002), 807–810.

Pradeep N. V., Anupama S., Navya K., Shalini H. N., Idris M., Hampannavar U. S., *Appl. Water Sci.*, **5** (2015), 105–112.

Prasad K. P., Chen Y., Chen P., *ACS Appl. Mater. Interfaces*, **6** (2014), 3387–3393.

Prodromidis M. I., Karayannis M. I., *Electroanalytical*, **14** (2002), 241–261.

Quiocho F. A., Richards F. M., *Biochemistry*, **5** (1966), 4062–4076.

Ran N. Q., Zhao L. S., Chen Z. M., Tao J. H., *Green Chem.*, **10** (2008), 361–372.

Ranjbakhsh E., Bordbar A. K., Abbasi M., Khosropour A. R., Shams E., *Chem. Eng. J.*, **179** (2012), 272–276.

Razmi H., Mohammad-Rezaei R., *Biosens. Bioelectron.*, **41** (2013), 498–504.

Reda T., Plugge C. M., Abram N. J., Hirst J., *Proc. Natl. Acad. Sci. U. S. A.*, **105** (2008), 10654–10658.

Reid R. C., Minteer S. D., Gale B. K., *Biosens. Bioelectron.*, **68** (2015), 142–148.

Reuillard B., Le Goff A., Agnes C., Holzinger M., Zebda A., Gondran C., Elouarzaki K., Cosnier S., *Phys. Chem. Chem. Phys.*, **15** (2013), 4892–4896.

Robles-Medina A., Gonzalez-Moreno P. A., Esteban-Cerdan L., Molina-Grima E., *Biotechnol. Adv.*, **27** (2009), 398–408.

Rosano G. L., Ceccarelli E. A., *Front. Microbiol.*, **5** (2014), 1–2.

Roy J. J., Abraham T. E., *Chem. Rev.*, **104** (2004), 3705–3721.

Sanderson K., *Nature*, **469** (2011), 18–20.

Sassolas A., Blum L. J., Leca-Bouvier B. D., *Biotechnol. Adv.*, **30** (2012), 489–511.

Schuabb V., Czeslik C., *Langmuir*, **30** (2014), 15496–15503.

Shan C. S., Yang H. F., Han D. X., Zhang Q. X., Ivaska A., Niu L., *Biosens. Bioelectron.*, **25** (2010), 1070–1074.

Sheldon R. A., *Adv. Synth. Catal.*, **349** (2007), 1289–1307.

Sheldon R. A., *Chem. Commun.* (2008), 3352–3365.

Sheldon R. A., *Appl. Microbiol. Biotechnol.*, **92** (2011), 467–477.

Sheldon R. A., Schoevaart R., Van Langen L. M., *Biocatal. Biotransform.*, **23** (2005), 141–147.

Sheldon R. A., van Pelt S., *Chem. Soc. Rev.*, **42** (2013), 6223–6235.

Shen Q., Yang R., Hua X., Ye F., Zhang W., Zhao W., *Process Biochem.*, **46** (2011), 1565–1571.

Shleev S., Tkac J., Christenson A., Ruzgas T., Yaropolov A. I., Whittaker J. W., Gorton L., *Biosens. Bioelectron.*, **20** (2005), 2517–2554.

Silva T. B., Spulber M., Kocik M. K., Seidi F., Charan H., Rother M., Sigg S. J., Renggli K., Kali G., Bruns N., *Biomacromolecules*, **14** (2013), 2703–2712.

Singh S., Solanki P. R., Pandey M. K., Malhotra B. D., *Sens. Actuators B*, **115** (2006), 534–541.

So K., Kawai S., Hamano Y., Kitazumi Y., Shirai O., Hibi M., Ogawa J., Kano K., *Phys. Chem. Chem. Phys.*, **16** (2014), 4823–4829.

Soleimani M., Khani A., Najafzadeh K., *J. Mol. Catal. B Enzym.*, **74** (2012), 1–5.

Southcott M., MacVittie K., Halamek J., Halamkova L., Jemison W. D., Lobel R., Katz E., *Phys. Chem. Chem. Phys.*, **15** (2013), 6278–6283.

Stolarczyk K., Sepelowska M., Lyp D., Zelechowska K., Biernat J. F., Rogalski J., Farmer K. D., Roberts K. N., Bilewicz R., *Bioelectrochemistry*, **87** (2012), 154–163.

Straathof A. J. J., Panke S., Schmid A., *Curr. Opin. Biotechnol.*, **13** (2002), 548–556.

Sun H. F., Yang H., Huang W. G., Zhang S. J., *J. Colloid Interface Sci.*, **450** (2015), 353–360.

Sundaramoorthy M., Terner J., Poulos T. L., *Chem. Biol.*, **5** (1998), 461–473.

Sung W. J., Bae Y. H., *Biosens. Bioelectron.*, **18** (2003), 1231–1239.

Tamaki T., *Top. Catal.*, **55** (2012), 1162–1180.

Tao J., Kazlauskas R. J., *Biocatalysis for Green Chemistry and Chemical Process Development*, John Wiley & Sons, Hoboken N. J. (2011).

Tao J. H., Xu J. H., *Curr. Opin. Chem. Biol.*, **13** (2009), 43–50.

Tao J. H., Zhao L. S., Ran N. Q., *Org. Process Res. Dev.*, **11** (2007), 259–267.

Tian Y., Jiang H., Anoshkin I. V., Kauppinen L. J. I., Mustonen K., Nasibulin A. G., Kauppinen E. I., *RSC Adv.*, **5** (2015), 102974–102980.

Tom J., Hornby B., West A., Harrisson S., Perrier S., *Polym. Chem. UK*, **1** (2010), 420–422.

Tominaga M., Nomura S., Taniguchi I., *Biosens. Bioelectron.*, **24** (2009), 1184–1188.

Tran D. T., Chen C. L., Chang J. S., *J. Biotechnol.*, **158** (2012), 112–119.

Tran N. M., Dufresne M., Helle F., Hoffmann T. W., Francois C., Brochot E., Paullier P., Legallais C., Duverlie G., Castelain S., *PLoS One*, **9** (2014), e109969.

Tzialla A. A., Pavlidis I. V., Felicissimo M. P., Rudolf P., Gournis D., Stamatis H., *Bioresour. Technol.*, **101** (2010), 1587–1594.

Vaddiraju S., Tomazos I., Burgess D. J., Jain F. C., Papadimitrakopoulos F., *Biosens. Bioelectron.*, **25** (2010), 1553–1565.

Verma M. L., Barrow C. J., Puri M., *Appl. Microbiol. Biotechnol.*, **97** (2013), 23–39.

Verma M. L., Naebe M., Barrow C. J., Puri M., *PLoS One*, **8** (2013), e73642.

Walcarius A., Minteer S. D., Wang J., Lin Y. H., Merkoci A., *J. Mater. Chem. B*, **1** (2013), 4878–4908.

Wang X., Liu X. Y., Zhao C. M., Ding Y., Xu P., *Bioresour. Technol.*, **102** (2011), 6352–6355.

Wang G. C., Tan X. C., Zhou Q., Liu Y. J., Wang M., Yang L., *Sens. Actuators B*, **190** (2014), 730–736.

Wang R., Zhang Y. F., Huang J. H., Lu D. N., Ge J., Liu Z., *Green Chem.*, **15** (2013), 1155–1158.

Wen D., Liu W., Herrmann A. K., Eychmuller A., *Chem. Eur. J.*, **20** (2014), 4380–4385.

Wenda S., Illner S., Mell A., Kragl U., *Green Chem.*, **13** (2011), 3007–3047.

Wilson R., Turner A. P. F., *Biosens. Bioelectron.*, **7** (1992), 165–185.

Wohlfahrt G., Witt S., Hendle J., Schomburg D., Kalisz H. M., Hecht H. J., *Acta Crystallogr. Sect. D Biol. Crystallogr.*, **55** (1999), 969–977.

Woodley J. M., *Trends Biotechnol.*, **26** (2008), 321–327.

Yadav R. K., Oh G. H., Park N. J., Kumar A., Kong K. J., Baeg J. O., *J. Am. Chem. Soc.*, **136** (2014), 16728–16731.

Yan M., Ge J., Liu Z., Ouyang P. K., *J. Am. Chem. Soc.*, **128** (2006), 11008–11009.

Yang X. Y., Tian G., Jiang N., Su B. L., *Energy Environ. Sci.*, **5** (2012), 5540–5563.

Yu X., Chattopadhyay D., Galeska I., Papadimitrakopoulos F., Rusling J. F., *Electrochem. Commun.*, **5** (2003), 408–411.

Yu Y. Y., Chen Z. G., He S. J., Zhang B. B., Li X. C., Yao M. C., *Biosens. Bioelectron.*, **52** (2014), 147–152.

Zebda A., Cosnier S., Alcaraz J. P., Holzinger M., Le Goff A., Gondran C., Boucher F., Giroud F., Gorgy K., Lamraoui H., Cinquin P., *Sci. Rep.*, **3** (2013), 1516.

Zebda A., Gondran C., Le Goff A., Holzinger M., Cinquin P., Cosnier S., *Nat. Commun.*, **2** (2011), 370.

Zhang J. J., Du J. J., Yan M., Dhaliwal A., Wen J., Liu F. Q., Segura T., Lu Y. F., *Nano Res.*, **4** (2011), 425–433.

Zhang W. J., Li G. X., *Anal. Sci.*, **20** (2004), 603–609.

Zhang B. H., Weng Y. Q., Xu H., Mao Z. P., *Appl. Microbiol. Biotechnol.*, **93** (2012), 61–70.

Zhao J. Q., Wang Y. J., Luo G. S., Zhu S. L., *Bioresour. Technol.*, **102** (2011), 529–535.

Zhu Y. W., Murali S., Cai W. W., Li X. S., Suk J. W., Potts J. R., Ruoff R. S., *Adv. Mater.*, **22** (2010), 3906–3924.

Zobnina V. G., Kosevich M. V., Chagovets V. V., Boryak O. A., Vekey K., Gomory A., Kulyk A. N., *Rapid Commun. Mass Spectrom.*, **26** (2012), 532–540.

Chapter 8

Enzyme Immobilization on Membrane and Its Application in Bioreactors

Peng-Cheng Chen,[b] Xue-Yan Zhu,[a] Jin Li,[a] and Xiao-Jun Huang[a,b]

[a]*MOE Key Laboratory of Macromolecular Synthesis and Functionalization, Department of Polymer Science and Engineering, Zhejiang University, Hangzhou 310027, China*
[b]*The Key Laboratory of Industrial Biotechnology, Ministry of Education, School of Biotechnology, Jiangnan University, Wuxi 214122, China*

hxjzxh@zju.edu.cn

8.1 Introduction

Enzymes are Nature's sustainable catalysts that possess a high degree of specificity. They are biocompatible, biodegradable and are often derived from renewable resources. Due to their catalytic mildness and specificity, the interest in them continues to grow. Furthermore, in the field of chemical production, the adoption of enzymes generally obviates the need for functional group protection and/or activation, thus affording synthetic routes for more step economic, more energy efficient, and more environmental friendly than conventional organic synthesis. Consequently, in the last two decades, enzymatic catalysis has appeared as an important methodology for meeting the growing demand for green and sustainable chemicals manufacture, for example, in the synthesis of flavor, fragrances, pharmaceuticals, and vitamins.

Biocatalysis and Nanotechnology
Edited by Peter Grunwald

ISBN 978-981-4613-69-9 (Hardcover), 978-1-315-19660-2 (eBook)
www.panstanford.com

Notwithstanding all these advantages, free enzymes, after being optimized via natural evolution and always catalyzing in complex metabolic pathways, usually perform unsatisfactorily in industrial application where operational conditions are far from those in the natural biological environment. To overcome the fragile nature limitation associated with free enzymes and realize enzyme recovery and recyclability, the methodology of enzyme immobilization is utilized, making them industrially and commercially viable. In this methodology, enzymes are incorporated or attached into or onto an inert, insoluble support. In addition to more convenient handling of enzymes as a solid rather than a liquid formulation, this methodology provides continuous operation and simplified product purification. Moreover, this methodology offers better control of the catalytic process; some enzymes, mostly lipases, even show an increased activity when immobilized on a suitable support with an appropriate immobilization technique.

In general immobilization cases, the multipoint attachment to support unavoidably hampers the free conformation of enzymes and sometimes non-biospecific interactions of enzyme-support result in the denaturation of enzyme protein, making the activity retention of the immobilized enzymes lower than 100%. It is thus important that the properties of supports and immobilization processes should be well understood in order to improve the activity retention of immobilized enzymes. The approaches to enzyme immobilization can be categorized into three types: non-covalent adsorption, covalent attachment, and entrapment, each with their own benefits and drawbacks, and these immobilization techniques are closely related to support properties. When a support is used to immobilize enzymes, considerations on three levels of structure should be in this order: macroscopic level, microscopic level, and further submicroscopic level (Balcao et al., 1996). At the macroscopic level, there may be one dominant dimension (e.g., the length of hollow fiber), two dominant dimensions (e.g., the surface of flat-sheet membrane), or three dominant dimensions (e.g., the volume of bead and micelle). The dominant dimension is determined by manufacture processes. In terms of microscopic level, the thickness and the porous structure of solid supports are of vital importance. Decreasing thicknesses and increasing porosity can usually minimize diffusional limitations upon substrates and maximize available area for enzyme attachments. The submicroscopic character is a direct result of molecular characteristics of supports and can range from hydrophilic to hydrophobic, which are influenced by both the material property and the manufacture process.

This chapter focuses on the use of membranes for enzyme immobilizations. A membrane, besides its porous structure which is beneficial for enzyme immobilizations, can be advantageous for its product separation capabilities along with the potential for biocatalytic conversions. Also, the continuous removal of products assisted by membranes can shift the reaction equilibrium toward the product side thus increasing the reaction productivity, which is a notable advantage of enzyme-immobilized membrane bioreactors. The decline in by-product formation and the low energy requirements make enzyme-immobilized membrane bioreactors to be environmentally friendly produced.

In recent decades, one-dimensional nanostructured materials have attracted continuous attention. They are normally in forms of fibers, wires, rods, belts, tubes, spirals, and rings. Among them, nanofibers are exceptionally long, uniform in diameter, and diversified in composition. These unique features ensure them in different aspects, such as templates, reinforcement, catalysis, biomedicine, and optical devices. Notably, when nanofibers are piled together, a nanofiber membrane can be achieved, which shows distinctive characteristics in combining the benefits of membranes and nanostructured materials. Thus, we divide the membranes discussed in this chapter into two categories: (1) general mesoporous membranes appearing as a flat sheet and hollow fiber and (2) nanofiber membranes piled by nanofibers.

This chapter analyzes the research in enzyme immobilizations on membranes from the perspectives of (a) the enzyme immobilizations on mesoporous and nanofiber membranes and (b) the structure and application of enzyme-immobilized membrane bioreactors. Although pristine membranes can be directly used for enzyme immobilization in most cases, they are still often modified to meet specific immobilization requirements. Moreover, depending on the solubility of the substrates and reaction products, the configuration of enzyme-immobilized membrane bioreactors can be monophasic, or biphasic regarding to the flowing fluid phases, which will also be discussed in this chapter.

8.2 Enzyme Immobilization on Mesoporous Membranes

As an important unit operation, membrane separations present unique advantages of high selectivity, high surface-area-per-unit-volume, and potential for controlling the level of contact and/or mixing between two phases. Therefore, different membranes, possessing microfiltration, ultrafiltration, or nanofiltration functions, have been widely reported in enzyme immobilization endeavors.

8.2.1 Non-Covalent Adsorption on Mesoporous Membranes

Non-covalent adsorption is the easiest immobilization technique. It can be performed by physical (assisted by hydrophobic and van der Waals interactions), and ionic interactions between the enzyme and the support. Physical interaction is generally known to be too weak to keep the enzyme stably fixed under industrial conditions, usually suffering from enzyme leakage during use. Comparatively, ionic binding is stronger, with the strength depending on surface charges between the membrane and the enzyme. Non-covalent adsorption is superior with respect to convenience and mildness. In this technique, enzymes are dissolved in an aqueous solution, and then the solid support is placed in contact with the enzyme solution for a period time to fix. Some membranes that have been used in this method were polysulfone (PSf), polysulfone (PS), polyethylenimine (PEI), polyacrylonitrile (PAN), PS-ZrO_2, alumina and polypropylene (PP) with corresponding enzymes lipase, glucose oxidase, and β-galactosidase (Cheng and Richard, 2010; Ebrahimi et al., 2010; Pedersen et al., 1985).

Sometimes, modifications are performed to mediate membrane properties for enzyme immobilizations. Tailoring the surface chemistry toward biocompatibility is commonly used for promoting the activity of immobilized enzymes, which is stimulated by biomimetic methodology. Mimicking the natural mode in living cells where enzymes exist can stabilize the structure of enzymes and thus retain their activities. Deng et al. tethered poly(γ-ethyl-L-glutamate) (PELG), poly(γ-stearyl-

L-glutamate) (PSLG), and poly(α-allyl glucoside) (PAG) onto the microfiltration PP flat-sheet membrane functionalized by nitrogen and ammonia plasma for lipase immobilization (Deng et al., 2004a, 2004b). It was found that the thermal stability of the immobilized enzyme was obviously improved on the PAG modified membranes. Activity retention of lipase increased from 57.5 ± 2.8% to 62.8 ± 3.3% and 72.4 ± 3.9% after PELG and PSLG modification, respectively. Moreover, they also modified phospholipid analogous polymers (PAPs) containing hydrophobic octyloxy, dodecyloxy, and octadecyloxy groups (abbreviated as 8-PAP, 12-PAP, and 18-PAP, respectively) onto the surface of hollow fiber PP membranes to create a natural microenvironment for lipase immobilization (Deng et al., 2004c). The activity retention of immobilized lipase on the modified membranes increased to 74.1 ± 3.2%, 77.5 ± 3.7%, and 83.2 ± 3.3%, respectively.

Surface modification is also applied to increase the enzyme loading amount on support. A polymer brush possessing aminoethanol (AE) functional groups was grafted onto a polyethylene hollow fiber membrane by radiation-induced graft polymerization. Figure 8.1 shows that the polymer brushes unfold through positive charge repulsion between the AE groups and enable multi-layer immobilization of lipases (Okobira et al., 2016). Moreover, the hydroxyl groups in AE can retain water molecules around the hydrophilic part of the lipase, thus achieving efficient utilization of enzymes in organic media.

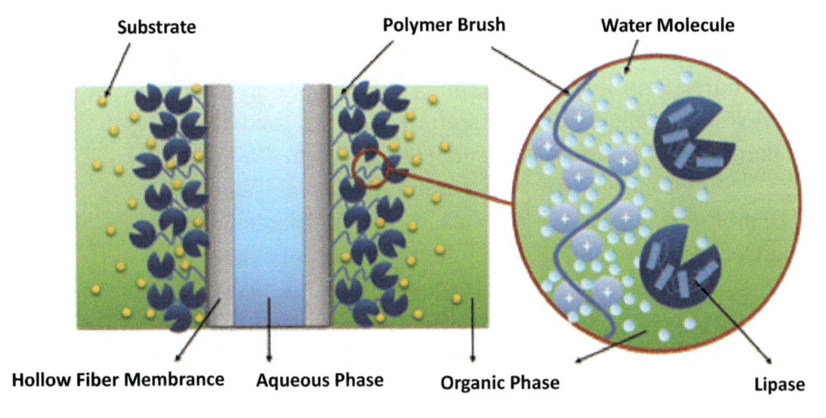

Figure 8.1 Polymer brushes containing hydroxyl groups and amino groups for binding lipase and water molecules.

8.2.2 Covalent Attachment on Mesoporous Membranes

Covalent attachment can stabilize the enzymes and efficiently prevent the enzymes leaching from the membrane surface and thus is widely used in practical applications. However, there can also be one major drawback: If the enzyme is irreversibly deactivated, both the enzyme and the support become unusable. A wide range of functional groups can form covalent bonds with enzymes, such as –OH, –COOH, $-NH_2$, and –CHO. For example, Ye et al. fabricated a poly(acrylonitrile-*co*-maleic acid) (PANCMA) ultrafiltration hollow fiber membrane and used 1-ethyl-3-(dimethyl-aminopropyl) carbodiimide hydrochloride/*N*-hydroxyl succinimide (EDC/NHS) as activation reagent for covalent immobilization of enzymes (Ye et al., 2005). Also, the –CN group on PAN ultrafiltration membrane was hydrolyzed for binding horseradish peroxidase (Wang et al., 2016).

Biomimetic modification is also applied in the covalent immobilization protocols. Tyrosinase was reported be to immobilized on a flat-sheet membrane composed of poly (acrylonitrile-*co*-vinyl acetate) (PAN-co-PVAc) and polyvinyl alcohol (PVA) with glutaraldehyde used as linker (Sandu et al., 2015). Besides, the previously mentioned PANCMA flat-sheet membranes were biomimetically modified by the natural macromolecules chitosan, leading to the formation of a dual-layer biomimetic support. Results showed a 30.3% increase in activity retention when lipase was immobilized on the chitosan-tethered membrane compared with that on the pristine membrane (Ye et al., 2005). Similarly, another biomacromolecule, gelatin, was tethered on PANCMA hollow fiber membranes; the activity retention of immobilized lipases increased from 33.9% to 49.2%.

Sometimes, different immobilization methods can be combined together. For example, β-galactosidase was first adsorbed on a PSF hollow-fiber membrane, then PEI was used as a polyelectrolyte intermediate layer to provide a positively charged character and afterwards, the β-galactosidase adsorbed on the PEI layer was cross-linked with glutaraldehyde (Gonawan et al., 2015). The stabilized β-galactosidase was used to convert lactose to galactooligosaccharides.

8.2.3 Entrapment in Mesoporous Membranes

The entrapment method of immobilization is based on the localization of enzymes embedded in membranes. The facility of this method lies in that enzymes not only can be immobilized on the outer surface of membrane (sometimes even forming an enzyme layer) but also can penetrate into the membrane (that is, enzymes can be immobilized in the membrane pores). Immobilization of enzyme entrapment into membrane can be divided into two types. One is entrapment after membrane fabrication process; in this type, enzymes can be entrapped by dead-end or cross-flow filtration (Giorno et al., 2009). The other is entrapment process coinciding with membrane fabrication. Up to now, PVA and its derivatives, cellulose acetate (CA), polytetrafuoroethylene (PTFE), PSf, PAN and its derivatives, and poly(methyl methacrylate) (PMMA) membranes have all be used to entrap enzymes.

Generally, after entrapment process, the flux of modified membranes decreases due to cake layer formation or pore blocking. Hydrolases were filtrated into flat-sheet polyvinylidene fluoride (PVDF) membranes with a nominal pore size of 0.45 mm with the aim of enhancing membrane performances in anaerobic membrane bioreactors (Wong et al., 2015). One distinct character of the entrapment method is that enzyme, mediators, and additives can be simultaneously deposited in the same layer. A thin alginate layer induced on the surface of an ultrafiltration PSF membrane was used as a matrix for immobilization of alcohol dehydrogenases (Marpani et al., 2015), as shown in Fig. 8.2. Despite the expected decrease of flux across the membrane resulting from the coating, such a system allowed an enzyme loading as high as 44.8 $\mu g/cm^2$, which resulted in a 40% conversion of formaldehyde to methanol as compared to the control setup (without alginate) where only 6.9 $\mu g/cm^2$ enzyme was loaded, with less than 5% conversion. Such conversion even increased to 60% when polyethylene glycol (PEG) was added during the construction of the gel layer, as a strategy to increase flux.

Asymmetric hollow fiber can also be performed as an interesting support for enzyme immobilizations using this method. Lipase was entrapped into PSF membrane by dead-end filtration with glutaraldehyde used as cross-linker to better stabilize the entrapped lipases (Zhu et al., 2016).

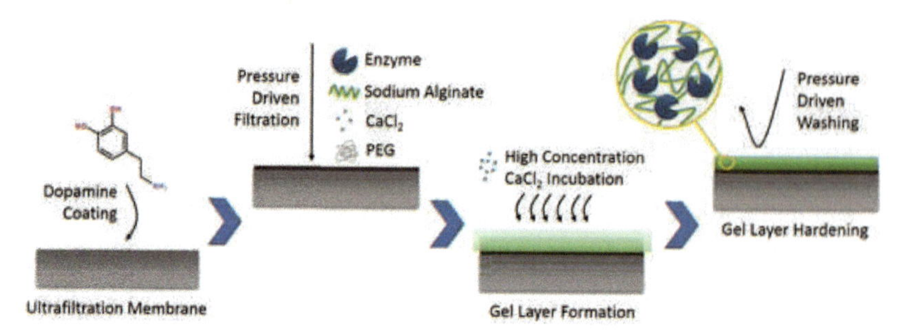

Figure 8.2 In situ formation of a biocatalytic alginate membrane by enhanced concentration polarization for immobilization of alcohol dehydrogenases.

8.3 Enzyme Immobilization on Nanofiber Membranes

Nanofiber membranes, which can be simply prepared by an electrospinning process, present a promising choice with a large specific area for high enzyme loading and fine porous structure allowing ready mass accessibility. Generally, the qualification of nanofibers as excellent supports can be attributed to the following: (i) A variety of polymers can be electrospun and meet different requirements as supports. (ii) The high porosity and interconnectivity of electrospun supports permit a low hindrance for mass transfer. (iii) The nanofiber surfaces can be modified to benefit enzyme activity.

8.3.1 Non-Covalent Adsorption on Nanofiber Membranes

Similar to mesoporous membranes, enzymes can be attached onto pristine and modified nanofiber membranes by non-covalent methods.

Wang et al. modified the PS nanofiber membrane using biocompatible polymers poly (*N*-vinyl-2-pyrrolidone) (PVP) and poly(ethylene glycol) (PEG) (Wang et al., 2006). Increasing the levels of PVP and PEG decreases the amount of protein that can be immobilized. That is because the PVP and PEG can increase the viscosity of the PS so that the resulting membrane surface area decreases and the surface could occupy by smaller enzymes. The modified membranes were positive for retaining lipase activity, and the immobilized lipase showed less sensitivity for pH and heat. In another example, air-dielectric barrier discharge (DBD) plasma at atmospheric pressure was used to modify the surface of electrospun chitosan/polyethylene oxide nanofibers with NH^{3+} group enabling negatively charged acetylcholinesterase to adsorb on the modified nanofiber membrane with higher activity retention and enzyme loading (Dorraki et al., 2015).

8.3.2 Covalent Attachment on Nanofiber Membranes

Enzymes can be covalently immobilized onto nanofibers with functional groups on the surface. In our group, nanofibers electrospun from poly(acrylonitrile-comaleic acid) were used for lipase immobilization via the activation of carboxyl groups in the presence of EDC/NHS (Ye et al., 2006a). It was found that compared with the PANCMA hollow fiber membrane, the enzyme loading and the activity retention of the immobilized lipase on the nanofiber membrane increased from 2.36 ± 0.06 to 21.2 ± 0.7 mg/g and from 33.9% to 37.6%, respectively. Combined with the kinetic parameters, it can be concluded that nanofiber membrane presents a low mass transfer limitation. A similar immobilization method was applied to immobilize anti-staphylococcus enterotoxin B on polyethersulfone (PES) nanofiber membrane (Mahmoudifard et al., 2016). In another study, a poly(vinyl alcohol-co-ethylene) nanofiber membrane was first activated by cyanuric chloride with triazinyl groups, and then reacted with 1,3-propanediamine and biotin to immobilize firefly luciferase (Wang et al., 2015). The bioluminescent reaction catalyzed by the immobilized firefly luciferase was used for the assay of adenosine triphosphate with high detection sensitivity.

Some reactive groups on the nanofibers have potential for further modification of the nanofiber surface. In the study by Huang et al., poly(acrylonitrile-*co*-2-hydroxy-ethyl methacrylate) was electrospun into nanofiber membrane for lipase immobilization. The hydroxyl groups on the surface were activated with three chemicals: epichlorohydrin, cyanuric chloride and p-benzoquinone (Huang et al., 2007). It was shown that enzyme loading and the stabilities of the immobilized lipase were obviously improved.

To increase the activity and stability of the immobilized enzymes, much attention has also been paid to surface modification of nanofibers toward biocompatibility. Biomacromolecules such as chitosan or gelatin were tethered on PANCMA nanofiber membranes to build dual-layer biomimetic surface for enzyme immobilization (Ye et al., 2006b), as is shown in Fig. 8.3. Similarly, collagen, gelatin, protein hydrolysate from egg skin, and bovine serum albumin were tethered on the PANCAA nanofibrous membranes for lipase immobilization, showing an obvious increase in the stabilities of the immobilized lipases (Huang et al., 2009; Wang et al., 2009). Moreover, Huang et al. prepared nanofiber membranes from the copolymer of acrylonitrile and 2-methacryloyloxyethyl phosphorylcholine for lipase immobilization (Huang et al., 2006). It was found that the introduction of phospholipid moieties, principal components of natural biomembranes, can obviously enhance the activity of lipase, and retained the enzyme loading through electrostatic interaction between phospholipid moieties and enzyme molecules.

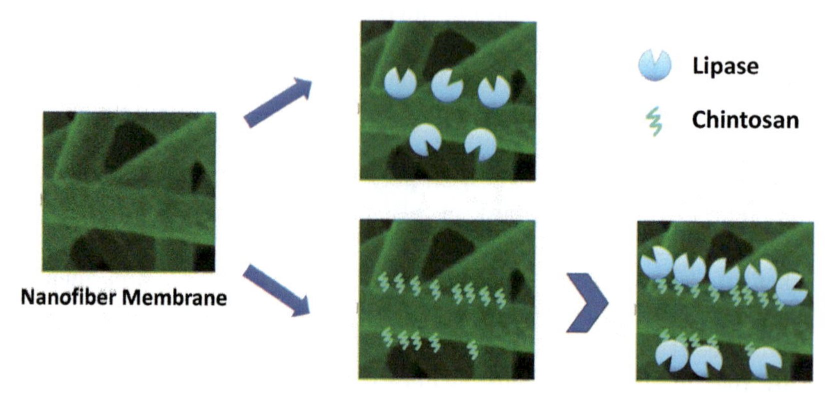

Figure 8.3 PANCMA nanofiber membranes modified with chitosan or gelatin for enzyme immobilization.

Many natural polymers such as cellulose and chitosan, which possess excellent membrane forming properties, good thermal, mechanical stability and biocompatibility, have been successfully electrospun into nanofiber membrane and used for enzyme immobilization (Huang et al., 2007; Huang et al., 2011). The stabilities of the immobilized enzyme were obviously improved on these natural polymeric membranes.

Another way to improve the specific activity and stabilities of immobilized enzymes is to introduce a flexible bifunctional spacer onto the supports to offer the immobilized enzyme high freedom of movement as well as to minimize unfavorable steric hindrance posed by solid supports. Chen et al. introduced pentaethylenehexamine as a spacer arm onto cellulose nanofiber membranes for covalent immobilization of lipase (Chen et al., 2011). The result showed that a residual activity of 52.1 U/g was observed with an activity retention of 54.3%, which was a 76.3% increase in residual activity compared with membranes without pentaethylenehexamine modification.

8.3.3 Entrapment in Nanofiber Membranes

The entrapment of enzymes in nanofiber membranes can be achieved by direct co-electrospinning of enzymes and bulk fiber-forming material. Most proteins can only be dissolved in aqueous media. Therefore, in many cases, the bulk material is required to be water-soluble so that it can form homogeneous solution with the enzymes. In one study, lipase was co-electrospun with PVA or polylactic acid (PLA). Fine dispersion of the immobilized enzymes in the polymer matrix resulted in an increase in the activity compared to the non-immobilized crude powder forms of the lipases (Soti et al., 2016). Good stability of the enzyme-immobilized membrane was observed in ten repeated cycles. In Monier's study, lipase was entrapped in modified photo-cross-linkable chitosan membranes, presenting better thermal stability than the free one (Monier et al., 2010). Sometimes, organic solvent was used to suspense the enzymes provided that they can tolerate the organic solvent used. Sakai et al. electrospun PS nanofiber membranes from a suspension of crude lipase powder in an *N,N*-dimethylformamide solution (Sakai et al., 2010). The encapsulated and moistened lipase showed 77% of residual activity after 10 cycles of use.

8.4 Enzyme-Immobilized Membrane Bioreactors

When it comes to the industrial applications of the enzyme-immobilized membranes, bioreactors are one of the most promising research areas (Fang et al., 2011). Compared with the common advantages of traditional bioreactors, such as continuous stirred tank reactors (CSTR) (Nunes et al., 2014), packed bed reactors (Nguyen et al., 2016; Ma et al., 2016), fluidized bed reactors (Lorenzoni et al., 2015), etc., the enzymatic membrane bioreactor (EMBR) is prominent as a unique mode for the combination of the product separation process with enzymatic catalysis in continuous operation (Eş et al., 2015). In recent years, EMBRs exhibited huge applicative prospects in fine chemicals synthesis, food, textile, detergent industries, waste water treatment and biomedical application and so on (Rehman et al., 2013; Ma et al., 2015; Soares et al., 2011; Nguyen et al., 2015).

There are many classification criteria of the EMBR. According to the form of the support membrane, the enzymatic membrane can be simply divided into two categories, nanofiber membrane and mesoporous membrane, such as hollow fiber membrane, or flat-sheet membrane. Both mesoporous and nanofiber membranes can be integrated into a membrane bioreactor. Depending on the solubility of the substrates and reaction products, the bioreactor can be monophasic or biphasic with respect to the flowing fluid phases.

8.4.1 Monophasic Membrane Bioreactors

The most common method to achieve a catalytic process in monophasic EMBR is filtration. The substrate solution filtrates through the enzymatic membrane by pressure; the filtrate may contain part of the substrate and the product, as shown in Fig. 8.4A. Besides, the membrane filtration also may be presented in cross-flow filtration (Fig. 8.4B), which can increase the disturbance between the substrate and the immobilized enzyme. Hence, not only the membrane fouling can be reduced, but also the mass transfer efficiency can be improved.

In order to better understand the fouling process affected by the membrane and threshold flux operation, an EMBR with casein glycomacropeptide (CGMP) as the substrate was fabricated by dead-end filtration (Luo et al., 2014). The results indicated that higher hydrophilicity of the membrane, elevated pH and agitation, and lower CGMP concentration were found to increase the threshold flux and decrease membrane fouling.

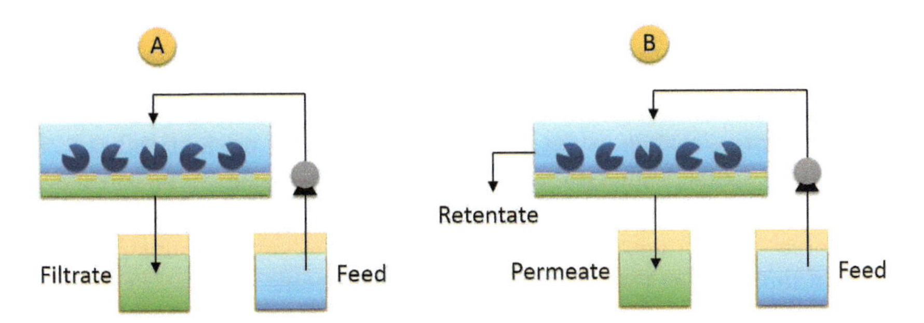

Figure 8.4 Monophasic membrane bioreactor: (A) Dead-end filtration process; (B) Cross flow filtration process.

Hollow fiber PSf microfiltration membranes with a perfect radial gradient of pores were selected as substrates and the subsequent enzyme-immobilization process was achieved in a facile way by dead-end filtration and cross-linking, to finally construct an enhanced EMBR system (Zhu et al., 2016). Lipase from *Candida rugosa* was introduced as the functional enzyme and was cross-linked by glutaraldehyde, and the hydrolysis of glycerol triacetate was used as the model reaction. The complete EMBR system with lipase from *Candida rugosa* as the functional enzyme cross-linked by glutaraldehyde and the hydrolysis of glycerol triacetate used as the model reaction showed an excellent performance of around 0.178 $mmol \cdot min^{-1} \cdot g^{-1}$ under optimum operating conditions and operational stability about 90% of the original membrane activity remaining after six reuses.

Sawada et al. (2015) investigated the recycling of a gold nanofibrous membrane as an enzyme-immobilizing carrier. The enzyme-immobilized gold nanofibrous membrane was prepared via electrospinning of a PAN/gold salt solution, electroless gold plating, and laccase immobilization. The results showed that electrochemical treatment reduced the enzymatic activity on the membrane by 98.5%, and re-immobilization restored the activity to 90%, which means a gold nanofibrous membrane can be successfully recycled by electrochemical desorption of the immobilized enzyme. In addition, a flow-through reactor was used to test the stability of the immobilized enzyme. The activities of laccase immobilized and re-immobilized on the membrane were 62.7% and 58.5% of the activity on the first day for each run after 7 days.

The monophasic membrane bioreactor can have a simple fabrication process and high catalytic performance with more intensive flow. However, the separation of the substrate and the product to a certain extent is hard to achieve. Also, the immobilized enzyme leakage due to transmembrane pressure and fluid shear force is a potential risk. In contrast, the biphasic membrane bioreactor is a better choice to catalyze the oil-soluble substrates and separate the water-soluble products simultaneously.

8.4.2 Biphasic Membrane Bioreactors

In most cases, the biphasic EMBR is formed by two immiscible fluids, a continuous organic phase and a continuous aqueous phase, separated by the enzyme–membrane interfaces, as shown in Fig. 8.5. Immobilization of enzymes at organic–aqueous interfaces is necessary for EMR implementation (Agustian et al., 2011). A lot of enzymes are active at these interfaces, and lipase is a typical example for catalyzing a series of diverse reactions, such as hydrolysis, alcoholysis, aminolysis, and transesterication, with various organic substrates (Sheldon et al., 2007).

Chen et al. (2014) fabricated a biphasic lipase-immobilized cellulose membrane bioreactor with high enzyme loading of 28.9 mg/g and activity retention of 44.3%. This bioreactor was assembled with electrospun cellulose nanofiber membranes that were fixed in a spiral form and wound to increase their specific surface area. To improve the catalytic efficiency of the immobilized enzymes, the supports went through a series of surface modification before covalently binding the lipase. The performance of this bioreactor were investigated with continuous hydrolysis of olive oil, and the result showed that under optimum operational conditions, 100% hydrolysis conversion of olive oil was achieved after 9 organic phase circulations.

Surface modification of native PVDF flat-sheet membrane was performed by wet chemical strategy (Vitola et al., 2015). The lipase was attached to the GA-activated modified membrane by covalent bond. Enzymatic activity of the immobilized lipase was tested in a biphasic system using the hydrolysis of triglycerides as model reaction. The observed specific activity of lipase immobilized on PVDF was about 40% of the activity of free lipase, which is quite good considering that this observed value is strongly influenced by mass transfer properties and therefore significantly underestimated.

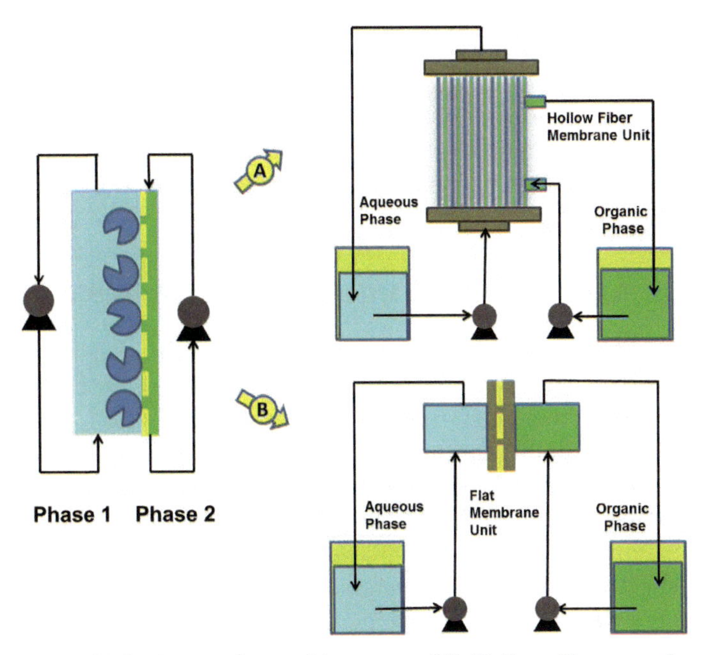

Figure 8.5 Biphasic membrane bioreactor: (A) Hollow fiber membrane bioreactor. (B) Flat membrane bioreactor.

Generally, the Michaelis-Menten kinetics is often used to estimate the effect of the reaction parameters, but this kinetic model does not involve the diffusive and convective transport of the reactant inside of the membrane. Nagy et al. (2015), systematically investigated the diffusive-convective mass transport in a hollow-fiber biphasic biocatalytic membrane reactor, based on the immobilization of the β-glucosidase enzyme in the sponge layer of the asymmetric, hydrophilic membrane layer. The results showed that a series of parameters, such as the lumen radius, membrane thickness, location of the inlet of the substrate, the inlet concentration and Peclet number, as well as the effect of the external mass transfer resistances have a strong effect on the membrane performance, which can enable the user to select the right choice between operating conditions, providing a high efficiency membrane reactor.

Compared with the monophasic reaction, the biphasic membrane reactor is often applied when the substrate is an oil-soluble compound and the product is soluble in the aqueous phase (Nagy et al., 2015). During the catalysis process, the water-soluble product can pass through the membrane for the purpose of separation. However, the fabrication of biphasic membrane bioreactors is more complex and requires higher equipment cost. Whether the monophasic membrane bioreactor or the biphasic membrane reactor, the polymer membrane plays the role as a barrier or a support.

8.4.3 Membrane as a Barrier or Support in EMBR

In an EMBR, the selective membrane aims to separate the enzyme from the reaction products and as well as to control to some extent the mass transfer rate during the whole catalysis process. The EMBR can be divided into two categories according to the location and function of the membrane as shown in Fig. 8.6. First, the enzyme can be rejected at one side of the EMBR using the membrane as a barrier, allowing the enzyme to contact with the substrate in a more favorable environment (Jochems et al., 2011). This can be achieved by size exclusion effect or charge repulsion effect. When the size of the enzyme molecules or cross-linked enzyme aggregates (CLEAs) is larger than the membrane pore size, the biocatalyst can be rejected by the membrane pores (Nguyen et al., 2015).

Enzymes with molecular sizes 10–500 kDa can be retained by ultrafiltration membranes, with a typical molecular weight cut-off (MWCO) of 10 kDa. A hollow fiber PES membrane with MWCO of 10-kDa was used to fabricate a EMBR as a barrier (Gracie Cid et al., 2014) to investigate the prevention of β-glucosidase inhibition by high molecular mass compounds during enzymatic wine aroma enhancement. Only the low-molecular-mass fraction (<10 kDa) of the Tenant wine could pass through the membrane pores and contact the β-glucosidase enzyme. The interesting result showed that enzymatic activity and stability in the membrane reactor were significantly higher than when the enzyme is present in a free state since the catalyst is only in contact with the low-molecular-mass components of this beverage.

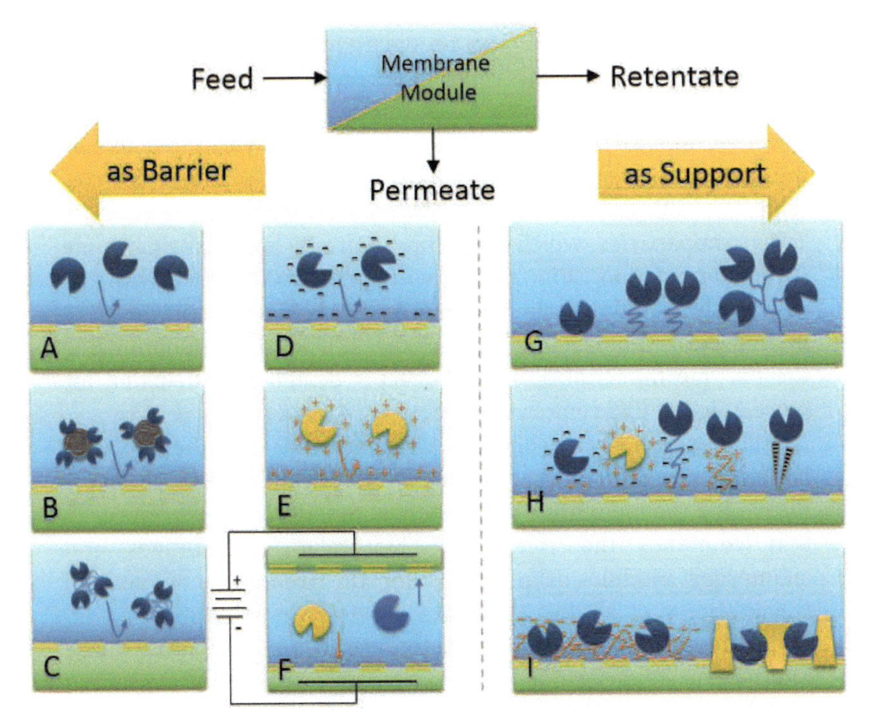

Figure 8.6 Enzyme membrane bioreactor: (A–F) Membrane as selective barrier. (A–C) Membrane retains enzyme by size exclusion. (D–F) Membrane retains enzyme by electrostatic repulsion. (G–H) Membrane as support. (G) Enzyme immobilized on membrane by adsorption, covalent binding. (H) Enzyme immobilized on membrane by electrostatic interactions. (I) Enzyme immobilized on membrane by entrapment.

Retention also could be achieved by immobilizing the enzyme into a hydrogel, so that the immobilized enzyme could not be passed through the membrane barrier. Zhang et al. (2010) developed a novel hollow fiber membrane reactor containing immobilized enzyme for selective separation of low concentration CO_2 from mixed gas streams. In the reactor, two bundles of PVDF hollow fiber membranes were aligned staggered parallel in a tube module, and lab-made nanocomposite hydrogel was filled between fibers, in which carbonic anhydrase (CA enzyme) was immobilized. The results showed that the hydrogel layer accounts for the major transport resistance which decreased greatly with immobilization of CA enzyme in hydrogel. Moreover, immobilized CA could retain over 76% enzymatic activity and thermal stability was also improved. The enzyme could be enlarged by immobilizing on a nanoparticles so that it can be retained at one side of the membrane.

Ji et al. (2016) fabricated a membrane hybrid reactor by immobilizing *Trametes versicolor* laccase on TiO_2 nanoparticles with the effluent recirculating through suspended biocatalytic TiO_2 nanoparticles, which can achieve an efficient carbamazepine (CBZ) degradation ratio of 71% within 96 h. Compared with the previously studied whole cell fungi systems, further improvement in their current biocatalytic system specially in terms of optimizing the enzyme immobilization and reactor design is anticipated for more efficient CBZ removal.

The second category is characterized by the enzyme immobilized on/in the membrane by physical/chemical methods furnishing the membrane with catalytic function. A novel approach that combines the concept of biocatalytic membranes and submerged modules is proposed for the treatment of biomass (Chakraborty et al., 2012). Lipase has been immobilized in PES hollow fiber membranes in order to develop a two separate phase biocatalytic submerged membrane reactor in which the membrane works as both catalytic and separation unit. After the optimization of parameters by response surface methodology (RSM), the immobilized enzyme-loaded membrane showed high performance and worked efficiently producing fatty acids from waste oils and separating the products of glyceride hydrolysis reaction successfully.

When the membrane acts as a barrier, the enzyme remains at one side of the membrane in the free form, which may retain its specific activity as much as possible. On the other hand, the free enzyme is easily inactivated in the reaction system and hence unable to meet the requirement of long-term use in industrial application. The solution to this problem is to exchange the enzyme regularly, resulting in a higher cost. Another problem of such EMBR is the unavoidable membrane fouling due to blocking and adsorption, aggravating the equipment's energy consumption.

Compared with this, the membrane as a support is more attractive in industrial application due to the advantages such as the stability of long-term and repeated use, the more effective mass transfer process, smaller cover area of the whole equipment and others. Enzyme immobilization process is indispensable to this type of EMBR, and the immobilized enzyme will lose part of the specific activity in most cases.

8.4.4 The Performance of the Enzymatic Membrane Bioreactor

In industry, EMBRs have a large application potential. Three important parameters of the EMBR are particularly paid attention, e.g., productivity, lifetime, and stability. Only the enzyme and membrane performances optimized can make EMBR meet the standard demands of industrial (Jochems et al., 2011).

First, the optimization of enzyme performance should be taken into consideration, since different enzymes respond differently to an immobilization method. Therefore, there is no universal immobilization method applicable to all the enzymes so far. Moreover, no matter which immobilization method is applied, the property of the immobilized enzyme will be quite different from the free enzyme, because of the interaction between the enzyme and the support membrane or the changed microenvironment making the transfer to a best conformation. Besides, the performance of the enzyme-immobilized membrane will change accordingly. The most important parameters of the immobilized enzyme are enzyme activity, and operational stability. In many articles, there is a negative effect on the activity of the immobilized enzyme reported. However, when the immobilization method is mild and the support membrane is suitable, the result turn out to be positive. A catalytic performance of a biocatalytic membrane reactor using β-glucosidase immobilized in PSf capillary membranes was investigated by Mazzei et al. (2009). Results evidenced that intrinsic kinetics of the enzyme immobilized in the PSf membrane reactor was not negatively affected (K_m = 3.80 ± 0.14 mM) compared to the free enzyme (K_m = 4.80 (±0.11) mM). On the contrary, it seemed slightly improved, because the lower K_m value indicates higher enzyme-substrate affinity. This showed that lower catalytic activity of immobilized enzyme, commonly reported in the literature, is not an intrinsic technological drawback and that appropriate control of biohybrid microstructured systems, microenvironment conditions and transport properties can even improve their catalytic efficiency. Additionally, the stability of immobilized enzyme is strengthened in most cases of the EMBR which ensures the reuse of the EMBR and lowers the operating cost (Yurekli et al., 2011).

Second, two crucial parameters of the membrane are flux and selectivity. Almost in all membrane processes, the membrane fouling is inevitable, which is responsible for a decline of the membrane permeability while the mass transfer resistance increases (Yurekli et al., 2011). Since enzymes in EMR will inevitably foul the membrane, Luo et al. (2014) hypothesized that controlled fouling could be used deliberately to immobilize enzymes in an efficient manner. At the same time, the activity and stability of the enzymes could be improved by manipulating filtration variables according to fouling formation mechanisms. Fouling-induced enzyme immobilization was applied to study the role of selected variables as, e.g., applied pressure, enzyme concentration, pH and membrane properties by dead-end filtration process. Interestingly, the results showed that pore blocking as a fouling mechanism permitted a higher enzyme loading but generated more permeability loss, while cake layer formation increased enzyme stability but resulted in low loading rate.

Third, it is true that the enzyme-immobilized membrane bioreactor is an integrated system, in addition to the appropriate immobilization method and support membrane, the operating parameters (temperature, pH, flow rate, substrate concentration, etc.) are also very important for the efficiency and long-term use of the EMBR. Chen et al. (2014) reported a fabrication of a biphasic lipase-immobilized nanofiber membrane bioreactor by covalent binding. The influences of membrane diameters and operation variables on bioreactor efficiency were studied using the catalytic hydrolysis of olive oil as the model reaction. In various aqueous-phase systems, the PBS buffer (0.05 M, pH7.0) showed best catalytic efficiency, because the immobilized lipase required some ionic strength as well as a relatively stable pH to maintain its activity. Response surface methodology is a statistical and mathematical technique used for improving and optimizing processes in which a response of interest is influenced by several variables, and the objective is to optimize this response (Bas et al., 2007). The graphical presentation of RSM function is called response surface, as shown in Fig. 8.7. RSM has become popular for optimization studies in recent years and has been applied in biochemical and biotechnological processes (Chang et al., 2008). Zhu et al. (2016) (see also Section 8.4.1) applied RSM to optimize operating variables that significantly affect the activity of the EMBR. The result showed that the membrane activity increased rapidly with increasing concentration of substrate; when the membrane flux exceeds 197.7 $L \cdot m^{-2} \cdot h^{-1}$ the membrane activity decreases gradually. The response surface plot showed an increase up to a maximum value of 0.178 $mmol \cdot min^{-1}$ g^{-1} while optimizing the reaction temperature and pH.

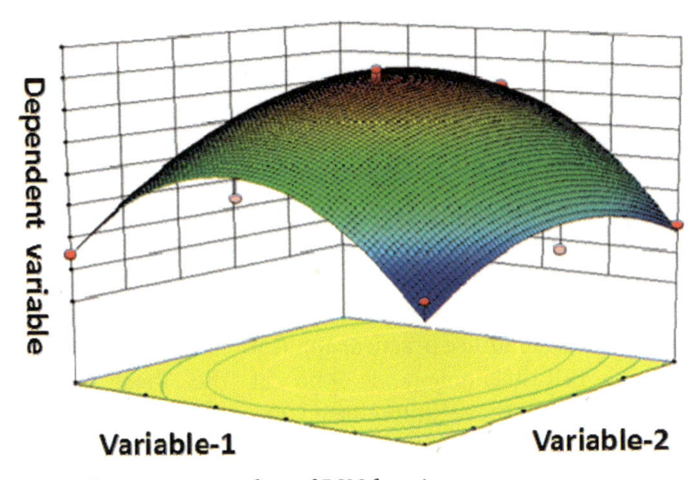

Figure 8.7 The response surface of RSM function.

8.5 Conclusion and Outlook

Enzyme immobilization continues to be of immense interest in both academic field and industrial applications. It is admitted that the commercial viability of industrial biotransformations develops or falls with the contributions of enzymes. In addition to providing an active and stable biocatalyst, the technology of enzyme immobilization is a relatively simple operation, not requiring a highly pure enzyme preparation or an expensive support that may not be commercially available. It is also clear that every enzyme is different, and thus there is no "one size fits all" solution to enzyme immobilizations. Moreover, the variables affecting the performances of immobilized enzymes are of large quantity, leading to a difficult comparison between different working cases. Thanks to new developments in the analysis of structure–function relationships of enzymes along with statistical methodologies, a more rational approach toward the optimization of immobilization strategies is feasible, accompanied by an adequate development in materials applicable as supports for enzymes. The parallel progress of these two research fields will enable the full exploitation of the catalytic potential of enzymes in industry.

Nanofiber membranes have been proven to be beneficial for enzyme immobilization because of their large surface area-to-volume ratios, functionalized surfaces, and low mass-transfer limitation. However, the studies of this issue are still limited into a small extent, as there are still problems in their large-scale applications. First of all, technical problems still exist in fabricating nanofiber membranes at a large scale; second, very few tools can be used to evaluate the effect of the nanofiber surfaces on the behaviors of the immobilized enzymes. Nevertheless, based on their unique advantages, researchers may still anticipate that the resultant biocatalytic materials will enable new and expanded uses of enzymes in future practical applications.

The EMBR technology is one of the most important applications with those immobilized enzymes, and it has been widely used in the field of synthesis, hydrolysis and kinetic resolution of compounds. However, most of the EMBR are only tools from laboratory rather than systematically used in broad industrials. Driven by the industrial and societal needs for sustainable chemical products and processes, as well as the attractive features of biocatalysis, it is expected that interest in improving the operational performances of enzymes in bioreactors and stabilizing long-term stabilities of EMBRs will continue to be unabated in the future.

References

Agustian J., Kamaruddin A. H., Bhatia S. J., *Chem. Technol. Biotechnol.*, **86** (2011), 1032–1048.

Balcao V. M., Paiva A. L., Malcata F. X., *Enzyme Microb. Technol.*, **18** (1996), 392–416.

Bas D., Boyaci I. H. J., *Food Biochem.*, **78** (2007), 836–845.

Chakraborty S., Drioli E., Giorno L., *Biomass Bioenerg.*, **46** (2012), 574–583.

Chang S. W., Shaw J. F., Yang K. H., Chang S. F., Shieh J. C., *Bioresour. Technol.*, **99** (2008), 2800–2805.

Chen P. C., Huang X. J., Huang F., Ou Y., Chen M. R., Xu Z. K., *Cellulose*, **18** (2011), 1563–1571.

Chen P. C., Huang X. J., Xu Z. K., *RSC Adv.*, **4** (2014), 6151–6158.

Chen P. C., Huang X. J., Xu Z. K., *Cellulose*, **21** (2014), 407–416.

Cheng H. N., Richard A. G., *Green Polymer Chemistry*: Biocatalysis and Biomaterials (ACS Symposium Series #1043). Northamptonshire, Great Britain: Oxford University Press (2010).

Cid A. G., Daz M., Ellenrieder G., *Food Technol. Biotech.*, **52** (2014), 334–341.

Deng H. T., Xu Z. K., Huang X. J., Wu J., Seta P., *Langmuir*, **20** (2004c), 10168–10173.

Deng H. T., Xu Z. K., Liu Z. M., Wu, J., and Ye, P. *Enzyme Microb. Technol.*, **35**(2004a), 437–443.

Deng, H. T., Xu Z. K., Wu J., Ye P., Liu Z. M., Seta P. J., *Mol. Catal. B Enzym.*, **28** (2004b), 95–100.

Dorraki N., Safa N. N., Jahanfar M., Ghomi H., Ranaei-Siadat S., *Appl. Surf. Sci.*, **349** (2015), 940–947.

Ebrahimi M., Placido L., Engel L., Ashaghi K. S., Czermak P., *Desalination*, **250** (2010), 1105–1108.

Eş I., Vieira J. D. G., Amaral A. C., *Appl. Microbiol. Biotechnol.*, **99** (2015), 2065–20858.

Fang Y., Huang X. J., Chen P. C., Xu Z. K., *BMB Rep.*, **44** (2011), 87–95.

Giorno L., Mazzei R., Drioli E., Biochemical membrane reactors in industrial processes in *Membrane Operations: Innovative Separations and Transformations*. Weinheim, Germany: Wiley-VCH Verlag GmbH & Co. KGaA (2009).

Gonawan F. N., Kamaruddin A. H., Bakar M. Z. A., Karim K. A., *Ind. Eng. Chem. Res.*, **55** (2016), 21–29.

Huang X. J., Chen P. C., Huang F., Ou Y., Chen M. R., Xu Z. K., *J. Mol. Catal. B Enzym.*, **70** (2011). 95–100.

Huang X. J., Ge D., Xu Z. K., *Eur. Polym. J.*, **43** (2007), 3710–3718.

Huang X. J., Xu Z. K., Wan L. S., Innocent C., Seta P., *Rapid Commun.*, **27** (2006), 1341–1345.

Huang X. J., Yu A. G., Jian, J., Pan C., Qian J. W., Xu Z. K., *J. Mol. Catal. B Enzym.*, **57** (2009), 250–256.

Ji C., Hou J., Wang K., Zhang Y., Chen V. J., *Membrane. Sci.*, **502** (2016), 11–20.

Jochems P., Satyawali Y., Diels L., Dejonghe W., *Green Chem.*, **13** (2011), 1609.

Lorenzoni A. S. G., Aydos L. F., Klein M. P., Ayub M. A. Z., Rodrigues R. C., Hertz P. F., *J. Mol. Catal. B Enzym.*, **111** (2015), 51–55.

Luo J., Marpani F., Brites R., Frederiksen L., Meyer A. S., Jonsson G., Pinelo M., *J. Membrane. Sci.*, **459** (2014), 1–11.

Luo J., Morthensen S. T., Meyer A. S., Pinelo M. J., *Membrane Sci.*, **469** (2014), 127–139.

Ma X., Deng S., Su E., Wei D., *Biochem. Eng. J.*, **95** (2015), 1–8.

Ma B. D., Yu H. L., Pan J., Xu J. H., *Biochem. Eng. J.*, **107** (2016), 45–51.

Mahmoudifard M., Soudi S., Soleimani M., Hosseinzadeh S., Esmaeili E., Vossoughi M., *Mater. Sci. Eng. C*, **58** (2016), 586–594.

Marpani F., Luo J. Q., Mateiu R. V., Meyer A. S., Pinelo M., *ACS Appl. Mater. Interfaces*, **7** (2015), 17682–17691.

Mazzei R., Giorno L., Piacentinia E., Mazzuca S., Drioli E., *J. Membrane. Sci.*, **339** (2009), 215–223.

Monier M., Wei Y., Sarhan A. A., *J. Mol. Catal. B Enzym.*, **63** (2010), 93–101.

Nagy E., Dudás J., Mazzei R., Drioli E., Giorno L. J., *Membrane. Sci.*, **482** (2015), 144–157.

Nguyen L. N., Hai F. I., Dosseto A., Richardson C., Price W. E., Nghiem L. D., *Bioresource Technol.*, **210** (2016), 108–116.

Nguyen L. N., Hai F. I., Kang J., Leusch F. D. L., Roddick F., Price W. E., McAdam E., Magram S. F., Nghiem L. D., Magram S. F., Nghiem L. D., *Int. Biodeter. Biodegr.*, **99** (2015), 115–122.

Nguyen L. T., Neo K. R. S., Yang K. L., *Enzyme. Microb. Tech.*, **78** (2015), 34–39.

Nunes M. A. P., Rosa M. E., Fernandes P. C. B., Ribeiro M. H. L., *Bioresource Technol.*, **164** (2014), 362–370.

Okobira T., Matsuo A., Matsumoto H., Tanaka T., Kai K., Minari C., Goto M., Kawakita H., Uezu K. J., *Biosci. Bioeng.*, **120** (2015), 257–262.

Pedersen H., Furler L., Venkatasubramanian K., Prenosil J., Stuker E., *Biotechnol. Bioeng.*, **27** (1985), 961–971.

Rehman H. U., Aman A., Silipo A., Qader S. A. U., Molinaro A., Ansari A., *Food Chem.*, **139** (2013), 1081–1086.

Sakai S., Yamaguchi T., Watanabe R., Kawabe M., Kawakami K., *Catal. Commun.*, **11** (2010), 576–580.

Sandu T., Sarbu A., Damian C. M., Patroi D., Iordache T. V., Budinova T., Tsyntsarski B., Yardim M. F., Sirkecioglu A., *React. Funct. Polym.*, **96** (2015), 5–13.

Sawada K., Sakai S., Taya M., *Chem. Eng. J.*, **208** (2015), 558–563.

Sheldon R. A., *Adv. Synth. Catal.*, **349** (2007), 1289–1307.

Soares J. C., Moreira R. P., Queiroga. C. A., Morgado J., Malcata X. F., Pintado M. E., *Biocatal. Biotransform.*, **29** (2011), 223–237.

Soti P. L., Weiser D., Vigh T., Nagy Z. K., Poppe L., Marosi, G., *Bioprocess Biosyst. Eng.*, **39** (2016), 449–459.

Vitola G., Mazzei R., Fontananova E., Giorno L. J., *Membrane. Sci.*, **476** (2015), 483–489.

Wang S., Liu W., Zheng J. W., Xu X. P., *Can. J. Chem. Eng.*, **94** (2016), 865–871.

Wang Z. G., Wan L. S., Xu Z. K., *Soft Matter*, **9** (2009), 4161–4168.

Wang Z. G., Wang J. Q., Xu Z. K., *J. Mol. Catal. B Enzym.*, **42** (2006), 45–51.

Wang W. W., Zhao Q. H., Luo M. Y., Li M. F., Wang D., Wang Y. D., Liu Q. Z., *ACS Appl. Mater. Interfaces*, **7** (2015), 20046–20052.

Wong P. C. Y., Lee J. Y., Teo C. W., *J. Membr. Sci.*, **491** (2015), 99–109.

Ye P., Xu Z. K., Che A. F., Wu J., Seta P., *Biomaterials*, **26** (2005), 6394–6403.

Ye P., Xu Z. K., Wang Z. G., Wu J., Deng H. T., Seta P. J., *Mol. Catal. B Enzym.*, **28** (2005), 115–121.

Ye P., Xu Z. K., Wu J., Innocent C., Seta P., *Biomaterials*, **27** (2006b), 4169–4176.

Ye P., Xu Z. K., Wu J., Innocent C., Seta P., *Macromolecules*, **39** (2006a), 1041–1045.

Yurekli Y., Altinkaya S. A., *J. Mol. Catal. B Enzym.*, **71** (2011), 36–44.

Zhang Y. T., Zhan. L., Chen H. L., Zhang H. M., *Chem. Eng. Sci.*, **65** (2010), 3199–3207.

Zhu X. Y., Chen C., Chen P. C., Gao Q. L., Fang F., Li J., Huang X. J., *RSC Adv.*, **6** (2016), 30804–30812.

Chapter 9

Potential Applications of Nanobiocatalysis for Industrial Biodiesel Production

Avinesh R. Byreddy and Munish Puri

Centre for Chemistry and Biotechnology, School of Life and Environmental Sciences, Waurn Ponds Campus, Deakin University, Victoria 3216, Geelong, Australia

Munish.puri@deakin.edu.au

9.1 Introduction

Biodiesel refers to long-chain alkyl fatty acid esters and is emerging as an alternative fuel for diesel engines (Byreddy et al., 2016). The increase in the petroleum prices and environmental benefits resulted in the increased production of biodiesel. Biodiesel occupies 10% of total biofuel production and the production is about 6 billion liters/year globally (Nogueira, 2011). The basic requirements for biodiesel production are feedstock (oil), an alcohol and catalyst (e.g. base, acid and enzyme). The reaction of biodiesel production occurs in the following steps: production of free fatty acids from triacylglycerols and transesterification of free fatty acids to methanol, resulting in the formation of new

Biocatalysis and Nanotechnology
Edited by Peter Grunwald

ISBN 978-981-4613-69-9 (Hardcover), 978-1-315-19660-2 (eBook)
www.panstanford.com

chemical compounds called methyl esters. The important process variables during the production of biodiesel are reaction temperature, ratio of alcohol to vegetable oil, amount of catalyst, mixing intensity (RPM), raw oils use and catalyst (Avhad and Marchetti, 2015; Marchetti et al., 2007). In the alkali-based production process generally sodium hydroxide (NaOH) or potassium hydroxide (KOH) are used as the catalyst. During the initial phase of the reaction, the catalyst reacts with the alcohol and forms an alkoxy moiety which reacts with the TAG to produce biodiesel and glycerol. Glycerol and biodiesel can be separated easily based on gravity. There may be a chance of soap formation by free acid or water contamination that makes the separation process difficult (Gerpen, 2005; Meher et al., 2006). In the acid-based method, sulfuric acid is the most preferable catalyst for converting triacylglycerol (Tran et al., 2013). The acid catalyst gives a high yield of esters, but the reaction for conversion requires a long time (Su and Guo, 2014). Although the production cost of biodiesel produced by the conventional methods such as acid and alkali catalysis is cheap, however there are challenges with respect to the recovery of glycerol and repeated washing requirement to purify methyl esters (Gog et al., 2012).

Vegetable oils (edible oils) as a renewable source, are also used as a feedstock for the production of biodiesel by transesterification process. The conventional way of biodiesel production was to blend the vegetable oils with the diesel fuel in a suitable ratio, but this direct mixing is technically not possible due to the high viscosity, low stability against oxidation and low volatility (Robles-Medina et al., 2009). This resulted in an increased price of edible oils and also increased the production cost of biodiesel that ultimately restricted their use, although it has advantages over conventional oils (Hoekman et al., 2012). There is a pressing need to use sustainable sources for biodiesel production keeping in view escalating oil supply demand. In recent times, microalgae have drawn attention as feedstock for the biodiesel production due to more advantages than plants and other microorganisms (Maity et al., 2014; Topf et al., 2014). Although the industrial production of biodiesel from microalgae is still in the developmental stage, scientists are investigating the right policies and strategies to implement alternative feedstocks for industrial biodiesel production.

Transesterification process involves the use of fatty acids (oils), short-chain alcohols and a catalyst. Methanol is the most commonly used solvent in the transesterification process because of its low price compared to other solvents; so biodiesel is chemically referred to as fatty acid methyl esters (FAMEs) (Tan et al., 2010). The synthesis of biodiesel is classified as chemical or enzymatic based on the catalyst used in the reaction. Chemical transesterification is more rapid and results in high product yields when compared to the enzymatic reaction. However, disadvantages are associated with the chemical process (acid or alkali as a catalyst), such as high energy requirements, difficulties in the recovery of the catalyst and glycerol yields, and potential pollution to the environment (Lara Pizarro and Park, 2003).

Biodiesel synthesis using enzymes is attracting many researchers because of its green approach to producing renewable fuels. Moreover, the many advantages associated with the enzymatic transesterification process include high product purity, less wastewater generation, and mild reaction conditions (Guldhe et al., 2015). Lipase is the most commonly used enzyme in biodiesel production and many researchers have reported its use for treating a variety of feedstocks for biodiesel production (Zhao et al., 2015). The limitations in the production of biodiesel using lipases are (i) the high cost of the enzyme, (ii) longer reaction time, and (iii) organic solvents and water requirement in the reaction mixture. Recently many researchers have focused on overcoming the limitation of lipase-based biodiesel production. This chapter details the enzymatic conversion of various feedstocks into biodiesel, different immobilization techniques, lipase immobilization, and factors that affect enzymatic conversion.

9.2 Transesterification Process

Transesterification is the most common process for biodiesel production which can be done in many ways using a various catalyst(s) such as an alkali, acid, enzymes, and heterogeneous systems or using alcohols in their supercritical state. This process is a three-step reaction in which (i) triglycerides are converted to diglycerides, (ii) diglycerides to monoglycerides, and (iii) monoglycerides to glycerol. A monoalkyl ester of fatty acid is produced in each of the three steps (Fig. 9.1) (Fukuda et al.,

2001; Guldhe et al., 2015). The conventional way of producing biodiesel includes the use of acid or alkali as the catalyst along with methanol or ethanol. Stoichiometrically, 3 moles of methanol are added to 1 mole of oil to synthesize biodiesel, but usually, a little more methanol is added to drive the reaction in the forward direction (Leca et al., 2010). Enzymatic transesterification has been found to be an efficient alternative using lipase as biocatalyst by immobilizing it on a suitable support. Several advantages are associated with immobilizing lipases, such as the enzyme can be reused without separation and the operating temperature of the process is low (50°C) compared to other techniques (Aguieiras et al., 2015).

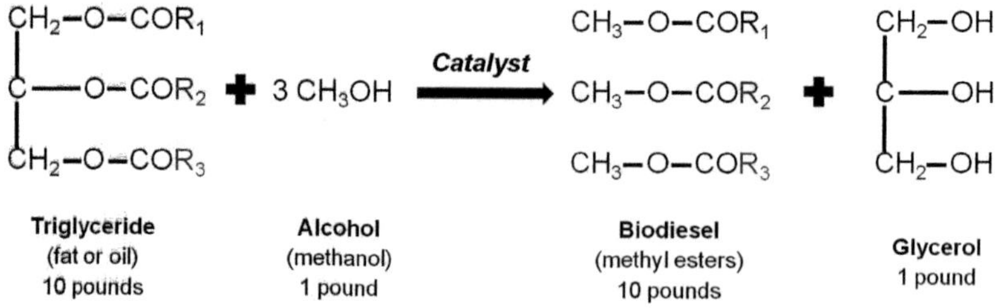

$$\begin{array}{l} CH_2-O-COR_1 \\ | \\ C-O-COR_2 \\ | \\ CH_2-O-COR_3 \end{array} + 3\,CH_3OH \xrightarrow{\textit{Catalyst}} \begin{array}{l} CH_3-O-COR_1 \\ CH_3-O-COR_2 \\ CH_3-O-COR_3 \end{array} + \begin{array}{l} CH_2-OH \\ | \\ C-OH \\ | \\ CH_2-OH \end{array}$$

Triglyceride (fat or oil) 10 pounds — **Alcohol** (methanol) 1 pound — **Biodiesel** (methyl esters) 10 pounds — **Glycerol** 1 pound

Figure 9.1 Transesterification process, where R^1, R^2, and R^3 represent alkyl chains.

9.3 Lipase Immobilization

Lipases (EC 3.1.1.3) are widely found in animals, plants, and microorganisms and play a major role in the metabolism of oils and fats. Lipases are extensively employed in various chemical reactions such as hydrolysis, alcoholysis, esterification, and transesterification of carboxylic esters. Excellent catalytic activity and stability of lipases in non-aqueous media make them a suitable catalyst for esterification and transesterification process during biodiesel production (Feng et al., 2013; Villeneuve et al., 2000). There are several limitations in commercial biodiesel production via enzymatic transesterification such as enzyme inactivation by solvents (ethanol, methanol, and glycerol), high enzyme costs, and barriers to scale-up. The high enzyme cost can be reduced by immobilizing lipase to suitable supports/matrixes.

Lipase immobilization is defined as confinement of lipase to a matrix/support with retention of their catalytic activities allowing preparations to be used repeatedly and continuously (Jegannathan et al., 2008). Immobilized lipase (IL) shows many advantages compared to free lipases such as easy recovery and reuse, higher adaptability for continuous operation, less effluent problems, greater pH and thermal stability, and higher tolerance to reactants and products. However, ILs have some disadvantages such as (1) loss of enzymatic activity during immobilization, (2) high cost of the carriers, (3) low stability in oil–water systems, and (4) requirement of novel reactors for well mixing and maximizing oil-to-biodiesel conversion (Zhao et al., 2015).

9.3.1 Immobilization Methods

Several lipase immobilization methods have been developed as shown in Fig. 9.2. These methods are further classified into irreversible and reversible depending on the enzyme and carrier interaction (Brena and Batista-Viera, 2006). In irreversible immobilization, enzymes are attached to support material only single time; they cannot be detached without damaging either the enzyme or the support material. In reversible immobilization, enzymes can be recovered by gentle treatment conditions. Covalent bonding, entrapment, and cross-linking are the well-known techniques for irreversible lipases immobilization. Physical adsorption and various non-covalent bonding, such as affinity and chelation binding, are reversible in nature and commonly used for immobilizing enzymes (Zhao et al., 2015). Every immobilization technique has its own merits and demerits; some are discussed in this chapter with respect to lipase immobilization.

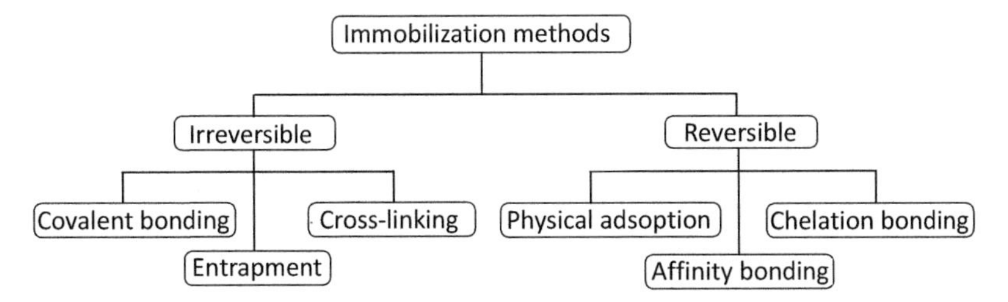

Figure 9.2 Various techniques used for lipase immobilization.

9.3.1.1 Lipase immobilization by physical adsorption

Physical adsorption is the most commonly used technique for lipase immobilization. This immobilization technique involves several non-covalent interactions such as non-specific physical adsorption, bio-specific adsorption, affinity adsorption, electrostatic interaction (also ionic binding), and hydrophobic interaction (Cao, 2006). Adsorption technique is advantageous compared to other immobilization techniques in the following aspects: (i) mild conditions for easier operation, (ii) relatively low cost of carrier materials for immobilization, (iii) no requirement of chemical additives during adsorption, (iv) easy regeneration of carriers thus promoting recycling, and (v) high lipase activity recovery (Zhang et al., 2012). In physical adsorption, enzymes are attached to support materials by non-specific interactions such as van der Waals forces, hydrogen bonds, and hydrophobic interactions. Different cost-effective carrier materials such as polymer resins (polystyrene, polypropylene, polyacrylate) have been tested to adsorb lipases. Novozym® 435 is the popular commercial lipase, which was immobilized on acrylic resin by adsorption (Laszlo et al., 2011).

The immobilization efficiency of the enzyme depends on the carrier properties and immobilization conditions. The important chemical and physical properties of the carrier material such as polarity, the molar ratio of hydrophilic to hydrophobic groups, particle size and surface area, porosity and pore size influence the amount of lipase bound and activity of lipase after immobilization (Verma et al., 2013a). The immobilization conditions such as enzyme concentration, carrier-to-enzyme ratio, pH, ionic strength also affect the immobilization efficiency. A recent study demonstrated the ratio of polystyrene resin beads to lipase significantly affected the immobilization efficiency. Increasing of resin-to-lipase ratio enhanced the adsorption rate continuously but decreased the adsorption capacity (Zhao et al., 2013). The pH of the buffer, where the immobilization conducted is a key parameter since ionic interactions are crucial for immobilization (Sun et al., 2010). Lipase immobilization by physical adsorption is commercially advantageous because of easy process and high immobilization efficiency in biodiesel production, but the IL stability needs to be further improved (Jegannathan et al., 2008).

9.3.1.2 Other immobilization techniques

There are several other techniques such as covalent binding, entrapment, encapsulation and cross-linking that can be used for immobilizing lipases. The advantages and disadvantages of different immobilized techniques are summarized in Table 9.1.

Table 9.1 Comparison of different lipase immobilized methods (Tan et al., 2010)

Methods	Advantages	Disadvantages
Adsorption	Preparation conditions are mild and easy with low cost. The carrier can be regenerated for repeated use.	The interaction between the lipase and the carrier is weak, so the immobilized lipase is sensitive to pH, ionic strength, temperature, etc. The adsorption capacity is small and the protein might be stripped off from the carrier.
Covalent binding	Stable because of the strong forces between the protein and the carrier.	The preparation conditions are rigorous, so the lipase might lose its activity during the immobilized process. Some coupling reagents are toxic.
Cross-linking	The interaction between the lipase and the carrier is strong and the immobilized lipase is stable.	The cross-linking conditions are intense and the mechanical strength of the immobilized lipase is low.
Entrapment	The entrapment conditions are moderate, and the immobilization method is applicable to a wide range of carriers and lipases.	Immobilized lipase always suffers from mass transfer restriction during the catalytic process, so the lipase is only effective for low molecular weight substrates.

Immobilization process by covalent binding refers to the chemical interaction between the amino acid residues outside the active site of the enzyme and functional groups of the support material. The functional groups of enzymes involved in covalent binding are thiol and amine groups (Trevan, 1988). Different carriers have been employed for covalent immobilization of lipase including polymers, silica-gel, chitosan, and magnetic particles, etc. IL obtained through covalent binding shows strong stability during transesterification process since binding is strong between lipase and the carrier. This IL is also stable under extreme conditions such as variable pH and temperature with almost no lipase leaching (Dizge et al., 2009; Tan et al., 2010). A recent study compared the lipase immobilization efficiency by different strategies and found that the covalent immobilization on epoxy-SiO_2-PVA in organic medium gave the highest hydrolytic activity (Mendes et al., 2011). However, Zhang et al, summarized that the covalent immobilization conditions are extreme which may lead to denaturation of the enzyme during immobilization process (Zhang et al., 2012).

Entrapment is defined as the encapsulation of lipase within the polymeric network, where the substrate and product can pass through (Jegannathan et al., 2008). In entrapment, lipases are not attached to the polymer matrix or support, but their diffusion is constrained. Entrapped lipases are more stable than physically adsorbed lipases. Entrapment of lipase is comparatively simple to perform and it is fast, cheap, and usually involves mild conditions, while the activity of lipases is maintained (Sassolas et al., 2013). A study demonstrated immobilization of lipase by entrapping in a hydrophobic sol–gel, where IL was used in biodiesel production with soybean oil as feedstock. The authors achieved final conversion of around 67% (Noureddini et al., 2005). The entrapped lipases showed relatively low activity and stability during biodiesel production. However, the encapsulation of *Burkholderia cepacia* lipase into κ-carrageenan with an encapsulation efficiency of 42.6% demonstrated better activity. The IL retained 72% of its activity after 6 cycles of hydrolysis of *p*-NPP (*p*-nitrophenylpalmitate). Surprisingly, when κ-carrageenan-encapsulated lipase was used for biodiesel production from palm oil, the yield was decreased to 40% only after 10 cycles (Jegannathan et al., 2010). The relatively low conversion rate of entrapped lipase is due to the poor diffusion and erosion of lipase from the surface of the support during the processing procedures (Jegannathan et al., 2008). The mass transfer is the major challenge in entrapped lipases in biodiesel production. Therefore, enhancing the mass transfer is a critical step in the industrial production of biodiesel using entrapped lipases.

9.4 Nanomaterials as Immobilizing Supports

Several immobilization support materials have been developed for enzyme immobilization in order to improve enzyme stability and reusability to reduce the enzyme cost. However, there is no universal material that is suitable for all enzymes (de Lathouder et al., 2008). The selection of the suitable material for enzyme immobilization is a complex process and depends on many parameters such as affinity of protein, availability of reactive functional groups on support, mechanical stability, rigidity, feasibility of regeneration, non-toxicity and biodegradability (Foresti and Ferreira, 2007). Moreover, it also depends on the type of enzyme, reaction medium, safety policy in the field of hydrodynamic conditions and reaction conditions (Khan and Alzohairy, 2010). Different support materials possess different physical and chemical properties and offer variability in their pore size, hydrophilic/hydrophobic balance and surface chemistry for enzyme binding because support morphology can influence the immobilization efficiency and catalytic property (Cao et al., 2003).

Immobilization supports can be classified into natural and synthetic polymers based on their origin and further subdivided into organic and inorganic types according to their physical and chemical characteristics. The most widely used supports are carboxymethyl-cellulose, starch, collagen, modified sepharose, ion exchange resins, active charcoal, silica, clay, aluminum oxide, titanium, diatomaceous earth, hydroxyapatite, ceramic, celite, agarose, or treated porous glass which is an inorganic material and certain polymers (Datta et al., 2013; Mohamad et al., 2015; Wu et al., 2012). Recently nano-sized (size in nm) supports such as nanoparticles, nanofibers, nanotubes, multi-walled carbon nanotubes and nanocomposites are considered for enzyme immobilization. Nanomaterials offer excellent characteristics for enzyme immobilization which improve the efficiency of biocatalysts, because of large surface area and high mechanical properties that allow effective enzyme binding with minimum diffusion limitation (Kim et al., 2006; Verma et al., 2013b; Puri et al., 2013).

Nanoparticles possess several important properties over bulk material in enzyme immobilization process. Nanoparticles provide a large surface to volume ratio and high enzyme loading which makes nanoparticles a suitable candidate for lipase immobilization (Verma et al., 2013a). Immobilizing lipases onto nanoparticles showed encouraging results in terms of its reusability and stability. It was observed that the nanoparticles are too small and require an energy intensive process to recovery from the reaction mixture. This issue was overcome by introducing the magnetic nanoparticles in enzyme immobilization and it was found very easy in separation by using an external magnetic field (Verma et al., 2013a). It indicates that the immobilization of lipase on magnetic nanoparticles enhances the reusability of the biocatalyst. Figure 9.3 illustrates the immobilization of lipase on different nanoparticles for biodiesel production.

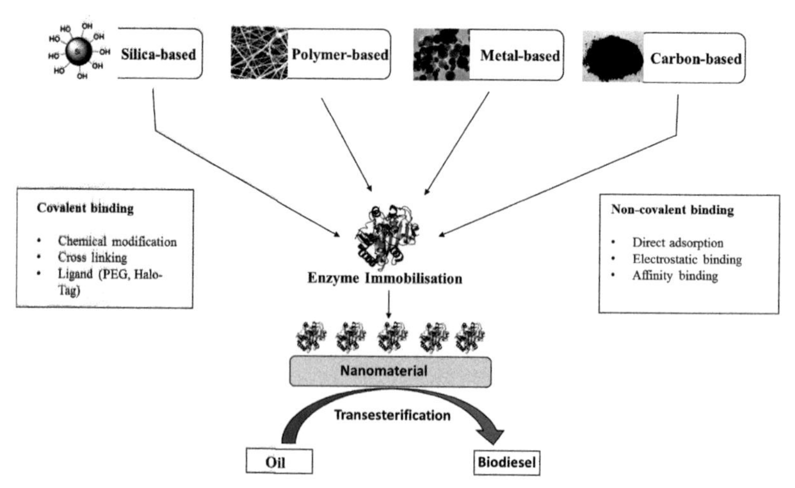

Figure 9.3 Flow diagram of enzyme immobilization to various nanomaterials. Modified from Misson et al., 2015.

9.5 Use of Immobilized Lipase for Biodiesel Production

Lipase-catalyzed transesterification for biodiesel production has gained a lot of attention due to mild reaction conditions, environmental friendliness, and wide adaptability for feedstocks. Lipase immobilization on various nanoparticles improves the reusability and stability of the enzyme and reduces the cost of the process relative to free enzyme. Immobilizing lipases onto nanoparticles showed encouraging results in terms of its reusability and stability. Royo et al. (2007) observed a 97% of

Table 9.2 Summary of the list of lipase immobilization methods used for biodiesel production using various feedstocks

Raw material/oil	Name of enzyme	Lipase origin	Immobilized method	Carrier particle	Solvent used	Yield/ conversion	References
Sunflower	Novozyme 435	*Mucor miehei, Candida antarctica*	Adsorption	Macroporous resin	—	—	(Ranganathan et al., 2008)
Cotton seed	Novozyme 435	*Candida antarctica*	Adsorption	Macroporous resin	*t*-Butanol	97%	(Royon et al., 2007)
Canola	Lipozyme TL	*Thermomyces lanuginosus*	Covalent binding	Polyurethane foams	Methanol	90%	(Dizge and Keskinler, 2008)
Soybean	Lipase	*Candida rugosa*	Covalent binding	Magnetic chitosan	Methanol	87%	(Xie and Wang, 2012)
Soybean	Lipase	*Pseudomonas cepacia*	Covalent binding	Activated magnetic silica nanocomposite Particles	Methanol	54%	(Kalantari et al., 2013)
Jatropha	Lipase	*Burkholderia cepacia*	Cross-linking	Modified attapulgite (ATP)	Methanol	94%	(You et al., 2013)
Simulated waste cooking	Lipase	*Candida antarctica, Pseudomonas cepacia*	Adsorption	Ceramic beads	Methanol, *n*-hexane	40% without *n*-hexane	(Al-Zuhair et al., 2009)

(*Continued*)

Table 9.2 (*Continued*)

Raw material/oil	Name of enzyme	Lipase origin	Immobilized method	Carrier particle	Solvent used	Yield/ conversion	References
Sunflower and soybean waste cooking	Lipase	*Thermomyces lanuginosus*	Covalent binding	Microporous polymeric biocatalyst (Bead, powder, monolithic)	Methanol	63.8% (sunflower), 55.5% (soybean), 50.9% (waste cooking)	(Dizge et al., 2009)
Sapium sebiferum	Lipase	*Pseudomonas cepacia G63*	—	—	Methanol	96.22%	(Li and Yan, 2010)
Waste grease	Lipase	*Thermomyces lanuginosus Lipase* (*TLL*), *Candida antarctica Lipase B*	Covalent binding	Iron oxide (magnetic nanobiocatalyst aggregates (MNA)	Methanol	>97%	(Ngo et al., 2013)
Palm	Novozyme 435, Lipozyme TL IM	*Candida antarctica B*, *Thermomyces lanuginosus*	Adsorption, covalent binding	Macroporous acrylic resin, silica gel	Dimethyl carbonate (DMC)	90.5%, 11.6%	(Zhang et al., 2010)
Waste, corn	Lipase	*Penicillium expansum*	Adsorption	Resin D4020	Methanol	92.8%,	(Li et al., 2009)

Raw material/oil	Name of enzyme	Lipase origin	Immobilized method	Carrier particle	Solvent used	Yield/ conversion	References
Soybean	Lipase	*Candida rugosa* (Amano AY-30)	Adsorption	Polyvinylidene fluoride (PVDF) membrane	Methanol, *n*-hexane	97.2%	(Kuo et al., 2013)
Soybean	Lipozyme TL	*Thermomyces lanuginosa*	Covalent binding	Magnetic nanoparticles	Methanol	>90	(Xie and Ma, 2010)
Chlorella protothecoides	Lipase	*Candida* sp. 99–125	Adsorption	Macroporous resin	Methanol	98%	(Xiong et al., 2008)
Chlorella vulgaris ESP-31	Lipase	*Burkholderia* sp. C20	Covalent binding	Alkyl-grafted Fe_3O_4- SiO_2	Methanol	97.3 wt% oil	(Tran et al., 2012)
Chlorella sp. KR-1	Novozyme 435	*Candida antarctica*	Adsorption	Macroporous resin	DMC mixture	>90%	(Lee et al., 2013)
Chlorella pyrenoidosa, Chlorella vulgaris, Botryococcus braunii (BB763, BB764)	Novozyme 435	*Candida antarctica,*[a] *Penicillium expansum*	Adsorption	Acrylic resins	Ionic liquid [BMIm] [PF_6], *tert*-butanol	90.7% and 86.2% in ionic liquid; 48.6% and 44.4% in *tert*-butanol	(Lai et al., 2012)
Chlorella protothecoides	Lipase	*Candida* sp. 99–125	Adsorption	—	Methanol	98.15%	(Li et al., 2007)
Tetraselmis sp.	Lipase	*Candida* sp.		Sodium alginate	Methanol	7 folds increased	(Teo et al., 2014)

biodiesel yield using *Candida antarctica* lipase immobilized on macroporous resin. Dizge et al. (2008) developed a lipase immobilization method from hydrophilic polyurethane foams using polyglutaraldehyde. The immobilized lipase at optimized reaction condition showed 90% efficiency methyl esters conversion and remained 100% stable after 10 batches. Kyu et al. (2013) immobilized lipase sourced from *Candida rugosa* (Amano AY-30) onto Polyvinylidene fluoride (PVDF) membrane via adsorption. The transesterification conditions with the immobilized enzyme were also optimized by response surface methodology (RSM) for biodiesel production. A transesterification efficiency of more than 95% was achieved. A functionalized magnetic nanoparticle was used for covalent immobilization of lipase for 30 min and applied in biodiesel production using soybean oil in continuous reaction (Wang et al., 2011). The immobilized lipase remained active after 5 cycles of reaction and the product yield was higher than obtained with a free enzyme (Macario et al., 2013). The high content of free fatty acids present in waste grease was used as a source for enzymatic biodiesel production. Lipases from *Thermomyces lanuginosus* and *Candida antarctica* Lipase B were covalently immobilized onto magnetic nanoparticle and achieved 97% product conversion in 12 h (Ngo et al., 2013). Several attempts have been made in immobilizing lipases onto nanoparticles for reducing production cost thus improving biodiesel production. Some of the nanoparticle-immobilized lipases for biodiesel production are presented in Table 9.2.

Several lipase immobilizations methods have been developed at the lab scale, however, only a few methods have been commercialized so far. The major drawback in the technical transfer is the high cost of immobilization steps. For instance, Novozym® 435 market price is ~$1000/kg (Stoytcheva et al., 2011). To minimize the immobilization cost, the support must be available at the low price or easy to synthesize. The immobilization techniques should be highly efficient in recovering enzymes with high stability to avoid enzyme leaching or activity loss and retaining their enzymatic activities. Among the commercially available ILs Novozym® 435 is the most widely-investigated in literatures for biodiesel production. Lipozyme® TL IM and Novozym® 435 both were used for enzymatic biodiesel production in first commercial biodiesel production plant built in 2006, with a capacity of 20,000 t/y (Du et al., 2008). Another biodiesel production plant was built in 2001 with the capacity of 10,000 t/y using immobilized *Candida* sp. 99–125. Candida sp. 99–125 was immobilized on textile membranes and the estimated lipase cost was only ~$32/t biodiesel (Tan et al., 2010). Some of the commercially available immobilized lipases are listed in Table 9.3.

Table 9.3 Some of the commercially available immobilized lipases

Commercial name	Enzyme origin	Support	Hydrophobicity/ philicity	Producer or inventor
Novozym® 435	*Candida antarctica* form B	Lewatit VP OC 1600	Medium hydrophobic	Novozymes (Denmark)
Lipozyme® RM IM	*Rhizomucor miehei*	Duolite A568	Hydrophilic	Novozymes (Denmark)
Lipozyme® TL IM	*Thermomyces lanuginosa*	Silica granules	Hydrophilic	Novozymes (Denmark)
Lipase PS Amano IM	*Burkholderia cepacia*	Diatomaceous earth	Hydrophilic	Amano (Japan)
—	*Candida* sp. 99–125	Textile membrane	Hydrophobic	Beijing University of Chemical Technology (China)

Source: Information reproduced from Tan et al. (2010) and Zhao et al. (2015).

9.6 Conclusion

Biodiesel production from chemical conversion of oil to FAMEs is widely accepted. This involves some additional steps to recover highly purified products and wastewater generation, which are the major constraints. Enzymatic methods are increasingly preferred compared to conventional chemical methods in industrial production. Lipase catalyzed biodiesel production offers benefits such as lower energy requirement, pure quality of biodiesel and glycerol, high product yield, easy recovery of products and no wastewater generation. The cost of lipase is the major concern in further scaling-up of this process to industrial production level. The production cost can be reduced by reusing the enzyme by immobilizing them on suitable supports for which immobilization of lipase on nanostructured material should be considered.

References

Aguieiras, E. C. G., Cavalcanti-Oliveira, E. D., Freire, D. M. G., *Fuel*, **159** (2015), 52–67.

Al-Zuhair, S., Dowaidar, A., Kamal, H., *Biochem. Eng. J.*, **44** (2009), 256–262.

Avhad, M. R., Marchetti, J. M., *Renew. Sust. Energ. Rev.*, **50** (2015), 696–718.

Brena, B. M., Batista-Viera, F., Immobilization of enzymes. In: *Immobilization of Enzymes and Cells* (Guisan, J. M., ed.), Humana Press. Totowa, NJ (2006), pp. 15–30.

Byreddy, A. R., Puri, M., Advances in nanobiotechnology in enhancing microalgal lipid production, harvesting and biodiesel production. In: *Marine OMICS: Principles and Applications* (Kim, S.-K., ed.), CRC Press Boca Raton, Florida, USA (2016), pp. 463–477.

Cao, L., Covalent enzyme immobilization. In: *Carrier-Bound Immobilized Enzymes*, Wiley-VCH Verlag GmbH & Co. KGaA (2006), pp. 169–316.

Cao, L., Langen, L. V., Sheldon, R. A., *Curr. Opin. Biotechnol.*, **14** (2003), 387–394.

Datta, S., Christena, L. R., Rajaram, Y. R. S., 3 *Biotech.*, **3** (2013), 1–9.

de Lathouder, K. M., van Benthem, D. T. J., Wallin, S. A., Mateo, C., Lafuente, R. F., Guisan, J. M., Kapteijn, F., Moulijn, J. A., *J. Mol. Catal. B Enzym.*, **50** (2008), 20–27.

Dizge, N., Aydiner, C., Imer, D. Y., Bayramoglu, M., Tanriseven, A., Keskinler, B., *Bioresour. Technol.*, **100** (2009), 1983–1991.

Dizge, N., Keskinler, B., *Biomass. Bioenerg.*, **32** (2008), 1274–1278.

Du, W., Li, W., Sun, T., Chen, X., Liu, D., *Appl. Microbiol. Biotechnol.*, **79** (2008), 331–337.

Feng, X., Patterson, D. A., Balaban, M., Emanuelsson, E. A. C., *Chem. Eng. Res. Des.*, **91** (2013), 1684–1692.

Foresti, M. L., Ferreira, M. L., *Enzyme Microb. Technol.*, **40** (2007), 769–777.

Fukuda, H., Kondo, A., Noda, H., *J. Biosci. Bioeng.*, **92** (2001), 405–416.

Gerpen, J. V., *Fuel. Process. Technol.*, **86** (2005), 1097–1107.

Gog, A., Roman, M., Toşa, M., Paizs, C., Irimie, F. D., *Renew. Energ.*, **39** (2012), 10–16.

Guldhe, A., Singh, B., Mutanda, T., Permaul, K., Bux, F., *Renew. Sust. Energy Rev.*, **41** (2015), 1447–1464.

Hoekman, S. K., Broch, A., Robbins, C., Ceniceros, E., Natarajan, M., *Renew. Sust. Energ. Rev.*, **16** (2012), 143–169.

Jegannathan, K. R., Abang, S., Poncelet, D., Chan, E. S., Ravindra, P., *Crit. Rev. Biotec.*, **28** (2008), 253–264.

Jegannathan, K. R., Jun-Yee, L., Chan, E. S., Ravindra, P., *Fuel*, **89** (2010), 2272–2277.

Kalantari, M., Kazemeini, M., Arpanaei, A., *Biochem. Eng. J.*, **79** (2013), 267–273.

Khan, A. A., Alzohairy, M. A., *Res. J. Biol. Sci.*, **5** (2010), 565–575.

Kim, J., Jia, H., Wang, P., *Biotechnol. Adv.*, **24** (2006), 296–308.

Kuo, C. H., Peng, L. T., Kan, S. C., Liu, Y. C., Shieh, C. J., *Bioresour. Technol.*, **145** (2013), 229–232.

Lai, J.-Q., Hu, Z.-L., Wang, P.-W., Yang, Z., *Fuel*, **95** (2012), 329–333.

Lara Pizarro, A. V., Park, E. Y., *Process. Biochem.*, **38** (2003), 1077–1082.

Laszlo, J. A., Jackson, M., Blanco, R. M., *J. Mol. Catal. B Enzym.*, **69** (2011), 60–65.

Leca, M., Tcacenco, L., Micutz, M., Staicu, T., *Rom. Biotech. Lett.*, **15** (2010), 5618–5630.

Lee, O. K., Kim, Y. H., Na, J. G., Oh, Y. K., Lee, E. Y., *Bioresour. Technol.*, **147** (2013), 240–245.

Li, X., Xu, H., Wu, Q., *Biotechnol. Bioeng.*, **98** (2007), 764–771.

Li, Q., Yan, Y., *Appl. Energy*, **87** (2010), 3148–3154.

Li, N. W., Zong, M. H., Wu, H., *Process. Biochem.*, **44** (2009), 685–688.

Macario, A., Verri, F., Diaz, U., Corma, A., Giordano, G., *Catal. Today*, **204** (2013), 148–155.

Maity, J. P., Bundschuh, J., Chen, C.-Y., Bhattacharya, P., *Energy.*, **78** (2014), 104–113.

Marchetti, J. M., Miguel, V. U., Errazu, A. F., *Renew. Sust. Energ. Rev.*, **11** (2007), 1300–1311.

Meher, L. C., Vidya Sagar, D., Naik, S. N., *Renew. Sust. Energ. Rev.*, **10** (2006), 248–268.

Mendes, A. A., Freitas, L., de Carvalho, A. K. F., de Oliveira, P. C., de Castro, H. F., *Enzyme Res.*, **2011** (2011), 8.

Misson, M., Zhang, H., Jin, B. J. R., *Soc. Interface.*, **12** (2015), 0891.

Mohamad, N. R., Marzuki, N. H. C., Buang, N. A., Huyop, F., Wahab, R. A., *Biotechnol. Biotec. Eq.*, **29** (2015), 205–220.

Ngo, T. P. N., Li, A., Tiew, K. W., Li, Z., *Bioresour. Technol.*, **145** (2013), 233–239.

Nogueira, L. A. H., *Energy*, **36** (2011), 3659–3666.

Noureddini, H., Gao, X., Philkana, R. S., *Bioresour. Technol.*, **96** (2005), 769–777.

Puri, M., Barrow, C. J., Verma, M. L., *Trends Biotechnol.*, **31** (2013), 215–216.

Ranganathan, S. V., Narasimhan, S. L., Muthukumar, K., *Bioresour. Technol.*, **99** (2008), 3975–3981.

Robles-Medina, A., González-Moreno, P. A., Esteban-Cerdán, L., Molina-Grima, E., *Biotechnol. Adv.*, **27** (2009), 398–408.

Royon, D., Daz, M., Ellenrieder, G., Locatelli, S., *Bioresour. Technol.*, **98** (2007), 648–653.

Sassolas, A., Hayat, A., Marty, J.-L., Enzyme immobilization by entrapment within a gel network. In: *Immobilization of Enzymesand Cells*, 3rd Edition (Guisan, J. M., ed.), Humana Press. Totowa, NJ (2013), pp. 229–239.

Stoytcheva, M., Montero, G., Toscano, L., Gochev, V., Valdez, B., *Biodiesel: Feedstocks and Processing Technologies.* (2011), 397–410.

Su, F., Guo, Y., *Green. Chem.*, **16** (2014), 2934–2957.

Sun, J., Jiang, Y., Zhou, L., Gao, J., *New. Biotech.*, **27** (2010), 53–58.

Tan, T., Lu, J., Nie, K., Deng, L., Wang, F., *Biotechnol. Adv.*, **28** (2010), 628–634.

Teo, C. L., Jamaluddin, H., Zain, N. A. M., Idris, A., *Renew. Energy*, **68** (2014), 1–5.

Topf, M., Koberg, M., Kinel-Tahan, Y., Gedanken, A., Dubinsky, Z., Yehoshua, Y., *Biofuels*, **5** (2014), 405–413.

Tran, H.-L., Ryu, Y.-J., Seong, D., Lim, S.-M., Lee, C.-G., *Biotechnol. Bioprocess Eng.*, **18** (2013), 242–247.

Tran, D. T., Yeh, K. L., Chen, C. L., Chang, J. S., *Bioresour. Technol.*, **108** (2012), 119–127.

Trevan, M. D., Enzyme immobilization by covalent bonding. In: *New Protein Techniques* (Walker, J. M., ed.), Humana Press. Totowa, NJ (1988), pp. 495–510.

Verma, M. L., Barrow, C. J., Puri, M., *Appl. Microbiol. Biotechnol.*, **97** (2013a), 23–39.

Verma, M. L., Naebe, M., Barrow, C. J., Puri, M., *PloS One*, **8** (2013b), e73642.

Verma, M. L., Puri, M., Barrow, C. J., *Crit. Rev. Biotechnol.*, **36** (2016), 108–119.

Villeneuve, P., Muderhwa, J. M., Graille, J., Haas, M. J., *J. Mol. Catal. B Enzyme*, **9** (2000), 113–148.

Wang X., Liu X., Zhao C., Ding Y., Xu P., *Bioresour. Technol.*, **102** (2011), 6352–6355.

Wu, C., Zhou, G., Jiang, X., Ma, J., Zhang, H., Song, H., *Process. Biochem.*, **47** (2012), 953–959.

Xie, W., Ma, N., *Biomass. Bioenerg.*, **34** (2010), 890–896.

Xie, W., Wang, J., *Biomass. Bioenerg.*, **36** (2012), 373–380.

Xiong, W., Li, X., Xiang, J., Wu, Q., *Appl. Microbiol. Biotechnol.*, **78** (2008), 29–36.

You, Q., Yin, X., Zhao, Y., Zhang, Y., *Bioresour. Technol.*, **148** (2013), 202–207.

Zhang, L., Sun, S., Xin, Z., Sheng, B., Liu, Q., *Fuel*, **89** (2010), 3960–3965.

Zhang, B., Weng, Y., Xu, H., Mao, Z., *Appl. Microbiol. Biotechnol.*, **93** (2012), 61–70.

Zhao, X., Fan, M., Zeng, J., Du, W., Liu, C., Liu, D., *Enzyme Microb. Technol.*, **52** (2013), 226–233.

Zhao, X., Qi, F., Yuan, C., Du, W., Liu, D., *Renew. Sust. Energ. Rev.*, **44** (2015), 182–197.

Chapter 10

Enzymogel Nanoparticles Chemistry for Highly Efficient Phase Boundary Biocatalysis

Ananiy Kohut,[a,b] Scott W. Pryor,[c] Andriy Voronov,[a] and Sergiy Minko[d]

[a]*Department of Coatings and Polymeric Materials, North Dakota State University, Fargo, ND 58108, USA*
[b]*Department of Organic Chemistry, Lviv Polytechnic National University, Lviv 79013, Ukraine*
[c]*Department of Agricultural and Biosystems Engineering, North Dakota State University, Fargo, ND 58108, USA*
[d]*Nanostructured Materials Laboratory, University of Georgia, Athens, GA 30602, USA*

sminko@uga.edu.

10.1 Introduction

Enzymatic catalysis is extremely important for the control of chemical reactions in living cells. Many examples of compartmentalized enzymatic reactions can be found in nature. These reactions are often confined in different cellular compartments where the organelle's internal environment is optimized to maintain the highest level of enzyme activity and selective transport of reactants and metabolites. Although enzymes may be localized in compartments, they can move freely within the compartment,

Biocatalysis and Nanotechnology
Edited by Peter Grunwald

ISBN 978-981-4613-69-9 (Hardcover), 978-1-315-19660-2 (eBook)
www.panstanford.com

which is very important for their biological functions (Henzler-Wildman et al., 2007). Living systems also demonstrate numerous examples of enzymatic catalysis when enzymes are secreted and released by cells to their environment. The environmental conditions have a significant impact on extracellular biocatalytic activity of the released enzymes. The secreted exoenzymes survive for a limited period of time while securing the supply of degraded extracellular substrates to the cell. Generally, enzymes secreted by cells are lost after a certain period of time due to degradation, diffusion, or nonproductive binding. Using exoenzymes in industrial technologies is quite efficient, but enzyme costs are indeed very high for singular use. Recycling of enzymes could be a possible solution, however typically enzyme recovery is too expensive (Stewart, 2001).

Solid-phase biocatalysis is a very efficient approach for industrial synthesis which utilizes and combines high selectivity of biocatalytic reactions and simple separation of reactants and the catalysts for recycling. At the same time, solid-phase biocatalysis has a number of limitations. Dynamics of catalytic centers is considered to be the main difference between manmade phase-boundary catalysis and enzymatic biocatalysis in living organisms (Henzler-Wildman et al., 2007). High diffusivity is a typical property of exoenzymes. Low mobility of the catalytic centers limits manmade solid-phase catalysis because of reduced biocatalytic activity in comparison with soluble enzymes. Low mobility of enzymes turns to be even more critical in the case of insoluble substrates. Soluble enzymes are able to reach the solid substrate surface and fit their spatial orientation at the interface (Bornscheuer, 2003), while immobilized biocatalysts are less capable of such spatial adjustments.

Several approaches to solve the problems of delivery and recycling of enzymes have been developed: (i) enzyme immobilization on solid nonporous supports and (ii) enzyme encapsulation using semipermeable capsules. Enzymes immobilized on solid supports are easily recovered and reused thus reducing costs of many industrial biotechnological processes (Bornscheuer, 2003). Enzyme immobilization can be conducted by either covalent or physical adsorption on a carrier, or by using carrier-free cross-linking methods; both techniques have been reported to have advantages and disadvantages (Hwang et al., 2013). Enzyme immobilization can lead to altering their structure and changing activity, selectivity, and specificity (Rodrigues et al., 2013). Strong covalent binding to the carrier can result in irreversible losses of the catalytic performance or deactivation of enzymes (Veum et al., 2006). Physical adsorption of enzymes on solid supports may also result in a lowering of their activity because of considerable changes in the enzyme structure (Czeslik et al., 2001; Norde et al., 1978; Czeslik, 2004a). Various methods of modification of substrate surfaces are being applied in order to avoid a direct contact between the solid substrate and the immobilized enzyme (Lu et al., 2009). One of recently reported approaches for retaining enzyme activity of physically or covalently attached enzymes explores a soft polymer gel (Coradin et al., 2003; Kobayashi et al., 2006). The approach involves smart polymer networks that undergo conformational changes in response to external stimuli such as pH, temperature, and the ionic strength of the environment (Ivanov et al., 2003; Lutz et al., 2006; Chiu et al., 2011). Enzyme immobilization in gels can preserve them and minimize structural changes. However, a low enzyme loading and low diffusivity in the host matrix are main limitations of the process (Pessela et al., 2002).

Attaching enzyme molecules to soft polymer-based carriers such as flexible polyelectrolyte brushes is one of the most efficient approaches to ensure higher loading and mobility of the engulfed enzymes. The brushes consist of polymer macromolecules end-grafted to solid colloidal particles at high grafting density (Haupt et al., 2005; Henzler et al., 2008; Becker et al., 2011). The ionic strength in the system has been shown to be the most important parameter for adsorption of enzymes by polyelectrolyte brushes (Wittemann et al., 2006; Czeslik et al., 2004b). The activity (Pessela et al., 2002) and the native secondary structure (Wittemann et al., 2004; Wittemann et al., 2005) of the adsorbed enzyme molecules are preserved in the brushes. Thus, colloidal particles with densely grafted polyelectrolytes brushes are promising candidates for enzyme recovery and reuse (Wittemann et al., 2007). Protein molecules adsorbed on polyelectrolyte brushes retain their mobility within the brush (Hollmann et al., 2008).

Although many methods for enzyme immobilization and encapsulation by polymer networks have been developed and tested, it is still challenging to design efficient, stable, and economically reliable hybrid biocatalytic systems for biotechnological and other applications when the enzymes are recovered and recycled with no losses of their activity. Here, we describe a method of encapsulation of exoenzymes by nanoparticles with grafted polymer brushes. The brush-decorated nanoparticles are based on a superparamagnetic particle core to facilitate recycling. We term the particles as enzymogel in this publication. The enzymogel has revealed unique properties making it a novel type of phase-boundary biocatalyst. The enzymogel combines (i) biocatalysis involving enzymes in the particle, (ii) stimuli-triggered release of enzymes to the nanoparticle's surrounding and extracapsular biocatalysis, (iii) biocatalytic conversion of substrates contacting the enzymogel when the enzymes act to bridge the particle and the substrate and engulf the substrate, and (iv) stimuli-triggered reattachment of released enzymes for their reuse. The enzymogel nanoparticle imitates catabolism of either intracapsular or external substrates. Owing to this versatility, the nanoparticle is turned into a universal biocatalytic "supercapsule" for a broad range of applications, including synthesis of biofuels as well as the storage, delivery, and reuse of enzymes in biotechnology. The enzymogel nanoparticles could be also used for the development of multifunctional biomaterials which could be remotely directed to the surface of temporal implants (absorbable sutures) or containers loaded with drugs and decompose them during contact.

10.2 Development of Enzymogel Nanoparticles

The approach to enzymogel nanoparticles discussed here is based on the engineering of core–shell nanoparticles with an inorganic core and a polymer brush shell—a shell made of end-tethered polymer chains that are stretched radially in the crowded polymer brush (Fig. 10.1a). Typically, the structure is composed of a 100 ± 10 nm silica core with a 30 ± 5 nm (in the dry state) poly(acrylic acid) (PAA) brush. The core is prepared from one or several 15-nm γ-Fe_2O_3 superparamagnetic particles embedded into the silica matrix. The iron oxide nanoparticles possess superparamagnetic properties when they experience magnetization in an external magnetic field but have zero residual magnetization outside of external magnetic fields. The iron oxide nanoparticles are introduced in order to enable magnetic separation of the enzymogel particles.

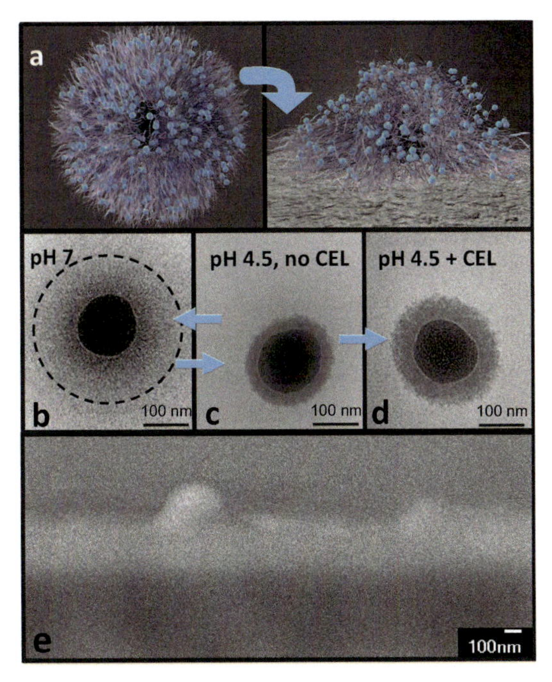

Figure 10.1 An enzymogel nanoparticle: schematic (a) of the particle that consists of an γ-iron oxide core, a silica shell, and a PAA brush loaded with enzymes: the particle with spherical symmetry in solution and interacting with a solid substrate; cryo-TEM images of the particle with a 120 nm in diameter silica core and (b) swollen and (c) shrunken PAA brush at pH 7 and pH 4.5, respectively. (d) The brush is uniformly loaded with CEL enzymes at pH 4.5. (e) The spreading of the brush over the substrate surface is shown in the SEM image. With permission from Kudina et al. (2014).

The PAA brush is formed by polymerization of *tert*-butyl acrylate monomer initiated from the surface of the particles followed by the hydrolysis of poly(*tert*-butyl acrylate) to synthesize the PAA brush. For the carboxylic acid monomer units of PAA, the pK_a is 4.75. Thus, at pH > 5, PAA chains are negatively charged. As evidenced by cryo-TEM images (Fig. 10.1b), the enzymogel particle in an aqueous medium at pH of 5–7 has a swollen homogeneous PAA brush which is negatively charged. At a pH value of 4.5, the PAA shell shrinks but it still maintains some net negative charge (Fig. 10.1c). PAA is a weak acid and poly(arylic acid) chains are negatively charged at pH of 4–8. Most enzymes possess the highest enzymatic activity in this pH range. Conversely, enzymes with an isoelectric point above 4.5 carry an overall positive charge and can be loaded onto negatively charged enzymogel particles. The PAA brush is highly swollen in water and can accommodate quite substantial amount of enzyme cargo. Thus, the enzymogel can be easily loaded with positively charged enzymes driven by electrostatic interactions.

In this chapter, we discuss the application of cellulase (CEL) enzymes to cleave cellulose chains and transform them to glucose which could be used for the production of biofuels or biochemicals (Lynd et al., 2005). Many cellulases have an isoelectric point near 4.9 and, hence, are moderately positively charged at pH 4.5. The loading of CEL into the enzymogel particles was conducted at pH 4.5 (Fig. 10.1d). This process is reversible and the enzyme can be released at pH 7 and in a high ionic strength environment. The kinetics of CEL adsorption and desorption (Zhang et al., 2012) was monitored using in situ ellipsometry, the Bradford protein quantification assay, particle size analysis methods, and thermogravimetric analysis. It was found that the amount of the immobilized CEL can be as high as 300% of the PAA by weight. The adsorption isotherm at pH 4 (Fig. 10.2) reveals the saturation of protein layer with the enzyme at its concentration of above 1%.

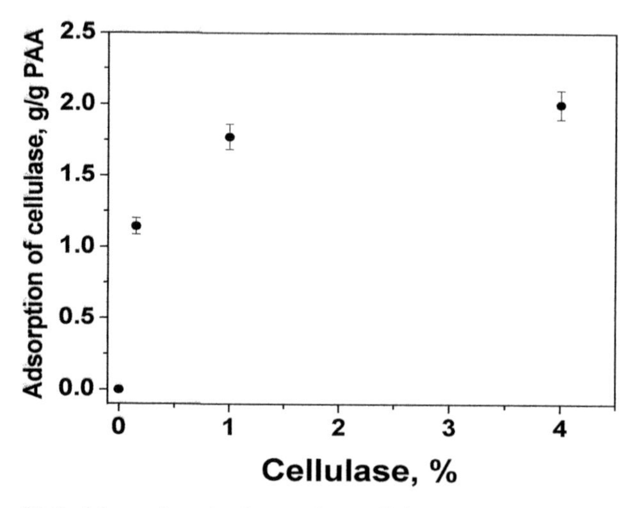

Figure 10.2 Adsorption isotherm for cellulose enzymes on the PAA brush (a Si-wafer grafted brush) plotted as an increase of the film thickness *vs.* concentration of cellulase at pH 4, 20°C. With permission from Kudina et al. (2014).

Obviously, adsorption of CEL at pH 4 was found to increase with thickness of the grafted PAA brush (Fig. 10.3a). The normalized thickness of the CEL layer, determined as a ratio of the adsorbed CEL to the PAA brush, decreases with thickness of the poly(acrylic acid) brush (Fig. 10.3b). Thicker brushes are characterized with a lower normalized enzyme load. A plausible explanation for this effect is entropic penalty for loading the densely grafted PAA brushes.

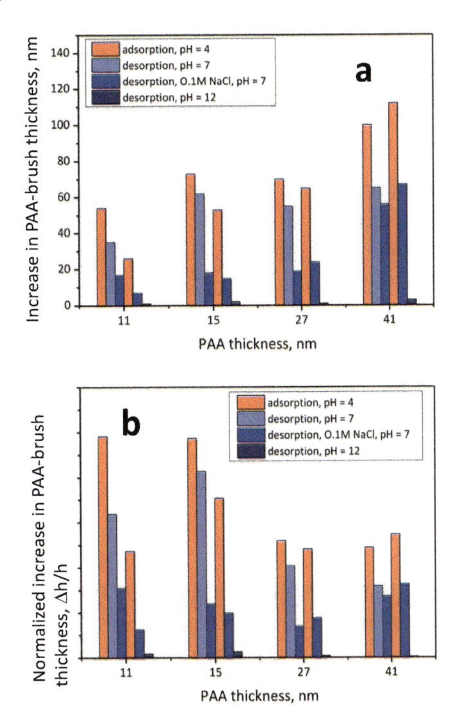

Figure 10.3 Adsorption-desorption equilibrium at different pH and salt concentration plotted as an increase in the PAA-brush thickness, Δh (a) and a normalized increase in the brush thickness, $\Delta h/h$ *vs.* PAA-brush thickness, h. With permission from Kudina et al. (2014).

10.3 Mobility of Cellulase in the Enzymogel

The incubation of the enzymogel in aqueous buffer at pH 7 leads to desorption of a small amount of CEL from the polymer brushes. An essential part of the protein was observed to desorb at pH 7 in 0.1 M NaCl aqueous solutions (Fig. 10.3). CEL can be repeatedly adsorbed and desorbed at pH 4 and pH 7, respectively. The most CEL can be desorbed from the PAA brush at pH 12. Thus, enzyme extraction is determined by both pH and the ionic strength and can reach about 80% of the adsorbed amount of CEL. The enzyme uptake/release triggered by changes in pH of the medium is well reproducible.

Exchange of the enzyme between the enzymogel nanoparticles and solution was studied using a dye-labeled CEL with UV-Vis spectroscopy. In order to test the CEL exchange between adsorbed molecules and solution, two experiments were carried out. In the first experiment, Test A, a non-labeled CEL was loaded into polymer brushes and dispersed in solution of labeled CEL (Fig. 10.4A). Then, in the second test, Test B, a labeled CEL was loaded onto the enzymogel and dispersed in solution of a non-labeled CEL (Fig. 10.4B). For both tests, we observed a small change of UV-Vis absorbance (Table 10.1) indicating a very slow exchange between the brush and solution enzymes. Thus, the CEL-loaded enzymogel reveals a high affinity of the enzyme to the PAA brush. CEL is localized in the enzymogel nanoparticles even after multiple rinsing in an aqueous medium at pH 4.5. Small CEL leakage (approximately 5%) was recorded after washing of the fully loaded enzymogel. The similar amount of CEL was exchanged in the tests with the labeled CEL. This indicates a high affinity and quasi-irreversible adsorption of cellulase by the PAA brushes. We attribute this behavior to the strong steric hindrance of CEL-loaded PAA brush, when only some fraction of CEL molecules at the brush interface is available for the exchange.

Table 10.1 Concentration of labeled cellulase in liquid dispersion phase for tests A and B

	Cellulase concentration, mg/mL	
Time	**Test A**	**Test B**
20 min	0.05	0.66
2 h	0.05	0.84
6 h	0.02	0.83
24 h	0.01	0.75
48 h	0.02	0.77

Source: With permission from Kudina et al. (2014).

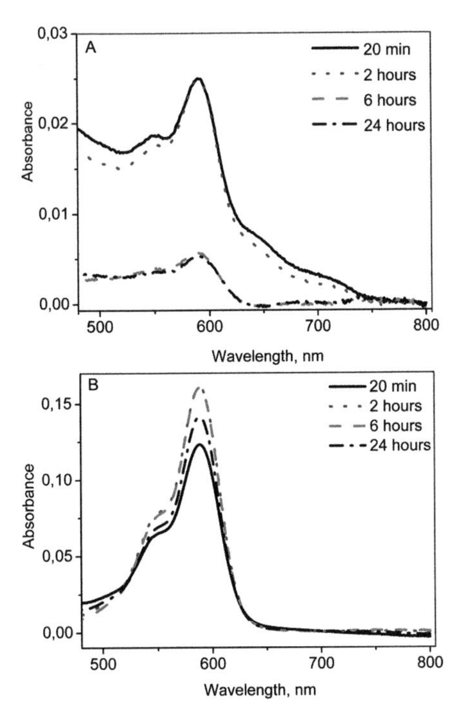

Figure 10.4 UV-Vis spectra of dye-labeled cellulose recorded for supernatant after different period of time in two different experiments: (A) the enzymogels (0.56 mg/ml) loaded with non-labeled CEL were dispersed in aqueous solution of the dye-labeled labeled CEL (0.56 mg/ml); (B) the enzymogels (0.87 mg/ml) loaded with dye-labeled CEL were dispersed in aqueous solution of the non-labeled CEL (0.87 mg/ml). With permission from Kudina et al. (2014).

An important feature of CEL dynamics in the enzymogel nanoparticle was discovered from tests with cellulase labeled with a fluorescent dye (FL-CEL). Fluorescence recovery was used after photobleaching experiments in order to study CEL mobility in the grafted PAA brush. To this end, a reference poly(acrylic acid) brush grafted to the Si-wafer was applied as a model scaffold to study the CEL diffusivity in the enzymogel using a fluorescence recovery after photobleaching (FRAP) technique (Fig. 10.5). The experiment is based on photobleaching of FL-CEL molecules with an intense laser source. The bleached area then was monitored to estimate kinetics of the fluorescent signal recovery affected by diffusivity of FL-CEL in the brush.

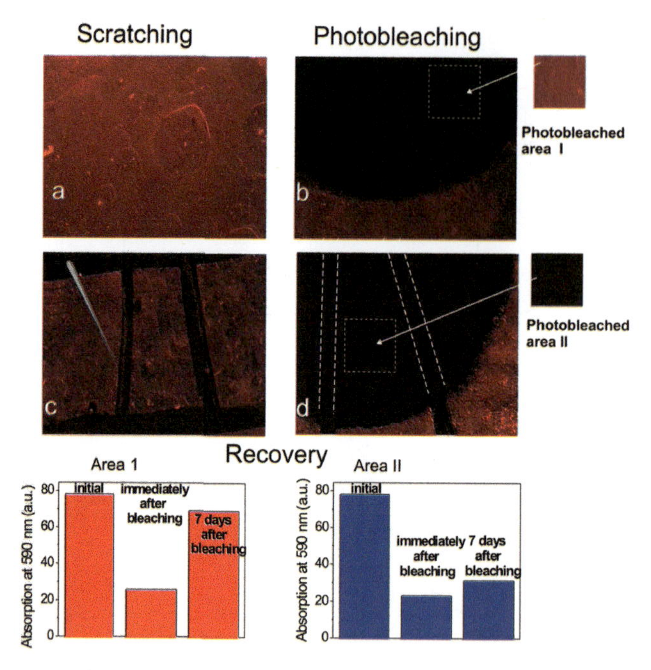

Figure 10.5 Fluorescence recovery after photobleaching experiments (fluorescent microscope images and schematics) with the PAA brush on the Si-wafer loaded with FL-CEL enzymes (a) and scratched to form a rectangular area (b). Both samples were photobleached and recovery of the fluorescent emission from the photobleached areas marked with dashed squares was monitored for 7 days. The images immediately after photobleaching are shown in panels b and c. Changes of intensity over 7 days are shown in the rectangular inserts. The fluorescent intensities at 590 nm are shown in the insert diagrams for the original samples, immediately after photobleaching and 7 days after photobleaching. The recovery in the scratched sample (d) is much slower than for the sample in the panel B providing evidence for a dominating recovery mechanism by transport in the brush. With permission from Kudina et al. (2014).

Two possible mechanisms should be considered for transport of the FL-CEL to the bleached region: (1) diffusion of CEL through the PAA brush and (2) desorption of enzyme molecules to the bulk solution, diffusion in the solution, and re-adsorption on the bleached region. Two parallel experiments were conducted in order to distinguish and estimate contributions of the mechanisms using FL-CEL-loaded brushes prepared in a pH 4.5 aqueous solution. In the first experiment, a region (about 1 cm^2)

was bleached with a laser light and the fluorescence recovery was monitored with a fluorescent microscope (Fig. 10.5a,b). In the second experiment, the bleached region was encircled with a metal needle to scratch the PAA brush around the designated area. In such a way, a track of the brush was removed from the wafer's surface to form a rectangular scratch (Fig. 10.5c,d). In the latter test, recovery of the photobleached region is only possible through the above-mentioned desorption-adsorption process. Two orders of magnitude slower fluorescence recovery (inserts of the Fig. 10.5) for the test with the scratch was revealed from the comparison of the results. Thus, it could be concluded that the major recovery mechanism consists in CEL transport through the PAA brush. The diffusion coefficient of CEL in the PAA brush at pH 4.5, estimated in this experiment, is 1.3×10^{-10} cm^2s^{-1}, which is close to the diffusion coefficient of the CEL adsorbed on cellulose (Jervis et al., 1997). The result is consistent with the previously described CEL exchange experiments. Hence, the enzymogel nanoparticles uniquely combine very high affinity to the cellulase with mobility of the loaded enzymes.

This combination results in many benefits for the application of the enzymogel with a high loading capacity along with unaltered mobility and biocatalytic activity of CEL. This behavior is explained by the polyelectrolyte properties of the PAA brushes and the ampholytic nature of the enzymes. The high affinity of CEL to the brush is caused by release of counterions due to the development of the polyelectrolyte complex. This fact provides a substantial entropic gain and higher activation barrier for the exchange of the enzyme molecules in the enzymogel and in the solution. Simultaneously, the activation energy for diffusion of CEL through PAA segments decreases owing to the ampholytic character of the CEL carrying both positive and negative charges.

10.4 Biocatalytic Activity of the Enzymogel

Two different mechanisms of biocatalytic activity of the enzymogel were discovered in the research. The enzymogel nanoparticles consisting of the superparamagnetic core, the silica shell, and the polymer brush were applied to cleave cellulose macromolecules by loading CEL from the solution at pH 4.5 and releasing it into the bioreactor at pH 7 which utilized the stimuli-responsive properties of the enzymogel. This mechanism of enzymogel biocatalytic activity is based on the released CEL. The CEL loaded into the polymer brush and the CEL released from the brush retain their initial biocatalytic activity.

The second revealed mechanism is based on the fact that even localized in the PAA brush, the enzyme remains active that is the most remarkable feature. To study and compare biocatalytic activity of CEL, eight experimental systems were used (Fig. 10.6a). Depending on a cellulose substrate applied for the study, two groups of the systems could be distinguished: (I–V) cleavage of filter paper standard (insoluble cellulose) and (VI–VIII) cleavage of semi-soluble colloidal dispersion of α-cellulose with a molecular weight of 9000 g mol^{-1}.

For both types of substrates, several different methods to introduce CEL were used: I and VI—CEL was chemically bound to silica particles with no PAA brush using the EDC-coupling protocol; II and VII—free CEL was added to the substrate solution; III and VIII—CEL was loaded into enzymogel particles; and IV—CEL was released from enzymogel particles by adding NaCl, increasing pH and extracting unloaded enzymogel prior to loading the released enzyme in the bioreactor. In system V, CEL-loaded enzymogel particles were recycled in four consecutive bioconversion cycles. In systems I-V, a filter paper standard was used as a substrate, while in systems VI-VIII colloidal cellulose was a substrate. In all the experiments, concentrations of the enzyme and cellulose in the studied solutions at pH 4.8 are the same. In experiments II and VII (reference tests), free CEL shows a reference level of biocatalytic activity that is lower for filter paper (II) than for colloidal cellulose (VII). The enzyme released from the nanoparticles retains initial biocatalytic activity (compare IV and II). In both systems (I and VI), much lower activity was observed for the covalently attached CEL. Evidently, the difference between systems VII and VI is because of the loss of enzyme activity due to grafting, whereas the difference between systems I and VI demonstrates the efficiency loss caused by a poor interfacial contact between the attached CEL and the filter paper. While loaded in the polymer brush, CEL shows almost unaltered biocatalytic activity (systems III and VIII) for both cellulose substrates. Furthermore, hydrolysis of cellulose can be reproduces in several cycles if the nanoparticles of enzymogel are reused (V).

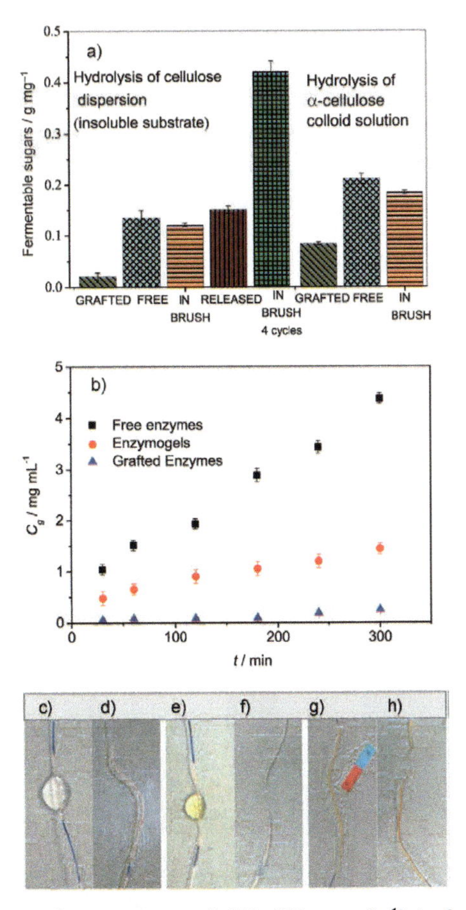

Figure 10.6 Biocatalytic activity of CEL (40 μg mL^{-1}) in buffer solutions at pH 5.0: (a) for insoluble (I-V) and semisoluble colloidal (VI-VIII) cellulose (5 mg mL^{-1}) presented as amount of synthesized fermentable sugars per CEL (g mg^{-1}) for CEL dissolved in the buffer (II,VII), CEL grafted to 200-nm silica particles (I,VI), CEL in the PAA brush in the enzymogel nanoparticles (III, V and VIII), CEL released from the PAA brush (IV), and after 4 cycles of reuse of the same enzymogel particles with a magnetic core (V). (b) Kinetics of glucose production shown as glucose concentration (C_g) versus time (t) using free enzymes in solution (squares), grafted enzymes (triangles), and the enzymogel (circles). Degradation of a 500-μm thick cotton floss by deposition of CEL using (c) a free CEL enzyme solution, (e) the enzymogel, (g) and magnetic field-directed deposition (d) when no rupture of the floss was observed for the free enzyme in 14 h, (f) rupture was documented for the enzymogel-treated floss in 2.5 h and (h) for the magnetic field-localized enzymogel in 1 h. With permission from Kudina et al. (2014).

A multicycle hydrolysis of cellulose could be realized using both mechanisms of enzymogel bioactivity: multiple release and uptake of CEL by recyclable enzymogel or multiple use of the enzymogel with no release of CEL. These two mechanisms are compared in Fig. 10.7. The enzyme was released into the medium at pH 7.0 followed by the fermentation for 72 h and recovery at pH 4.0. The initial amount of the enzyme in the brush, fractions of released and recovered CEL are shown in Fig. 10.7. The amount of the recovered CEL decreases in each cycle. The sugar amount formed in each cycle is given in Fig. 10.8. The cellulose fermentation with multiple release/recovery cycles of CEL in and from the medium is clearly seen to be less efficient obviously due to some loss of CEL after each load/release step.

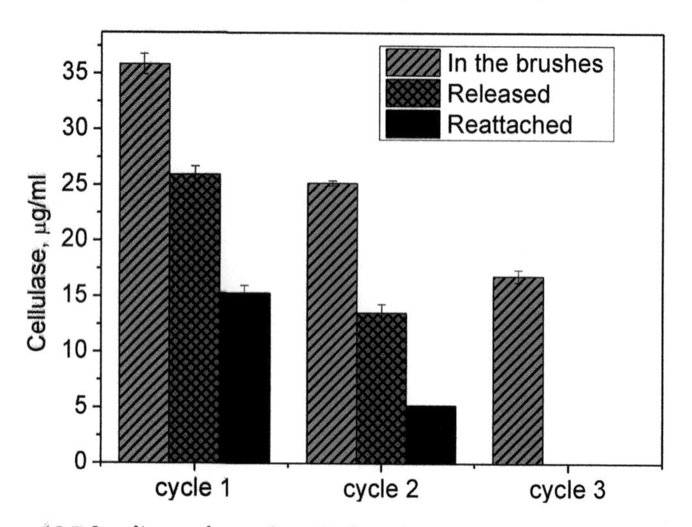

Figure 10.7 Loading, release (at pH 4) and capture (at pH 7) of cellulase enzymes in fermentation cycles. Orange bars show the initial amount of CEL in the brush and the total amounts of CEL in the enzymogel after reattachment (capture) of CEL from the biomass, in the first, second and third cycles, respectively. Red bars show the amount of CEL released from the brush into biomass at pH 7 in the first and the second cycles (no release was conducted after the third cycle). Blue bars show the amounts of CEL captured at pH 7 adjusted after fermentation in the first and the second cycles. With permission from Kudina et al. (2014).

These data demonstrate that after two cycles, the biocatalytic activity is mainly due to the adsorbed enzymes. Thus, the most efficient way of enzymogel application is using and recycling enzymogel as is with no release/uptake of CEL cycle. According to this mechanism, the enzymogel particles are immobilized on the substrate surface (i.e., cellulose fibers) and hydrolyze the cellulose macromolecules by CEL shuttling between the interior of the polymer brush and nanoparticle–cellulose interface. This shuttling is the reason for the high efficacy of cellulose hydrolysis. The nearly unaltered biocatalytic activity of the enzyme provides evidence that the most CEL molecules in the enzymogel nanoparticles take part in the biocatalytic hydrolysis. It seems probable that only a small part of the enzyme is entrapped and deactivated by the brush thus slightly decreasing a sugar yield in comparison with free CEL.

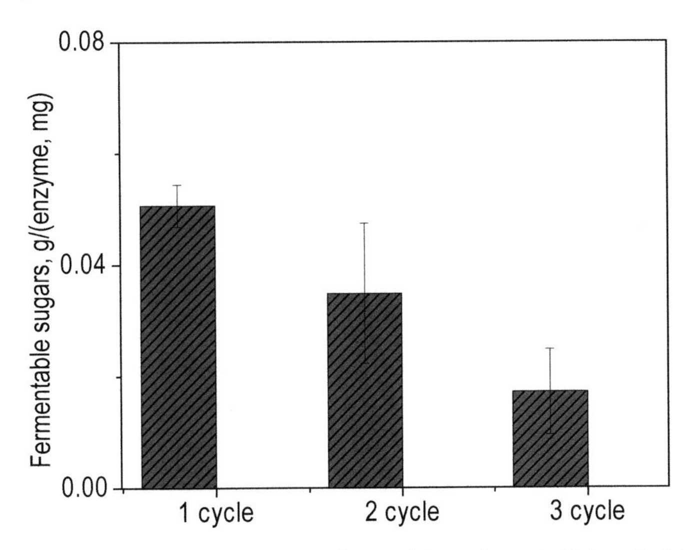

Figure 10.8 Amount of sugars synthesized in each cycle (20 μg/ml CEL) that corresponds to cycles in Fig. 10.7. With permission from Kudina et al. (2014).

Although, the biocatalytric activity of enzymogel versus free CEL evaluated by the cumulative release of sugars is only slightly different, the kinetics of the bioconversion process is slower for enzymogel, however enzymogel based bioconversion is much faster than that for covalently bound CEL (Fig. 10.6b). The slower enzymogel kinetics versus free enzyme process is explained by the difference in size between free CEL molecules and enzymogel particles. Enzymogel particles cannot occupy and cover the entire surface of cellulose substrates due to the limitations of the surface packing of the particles. The density of the surface packing is limited by steric and electrostatic repulsion between enzymogel particles adsorbed on the surface of cellulose substrates. Assuming the experimentally proven fact (Fig. 10.6a) that free CEL and CEL in enzymogel have the same biocatalytic activity and that all enzymogel is adsorbed on the cellulose substrate, the ratio of the rates of the glucose synthesis catalyzed by free CEL and enzymogel was used to estimate the surface area of the cellulose-enzymogel contact and an average diameter per enzymogel particle adsorbed on the cellulose surface. The estimated contact diameter is about 140 nm for 180 nm in diameter enzymogel particles. That implies that the PAA brush loaded with CEL spreads over the cellulose surface (for a rigid particle the contact diameter would be much smaller). On the other side, the contact diameter is smaller than the enzymogel particle diameter. The latter explains the difference in kinetics between free CEL and enzymogel.

In situ AFM study of single enzymogel nanoparticles adsorbed on the cellophane surface (a model of cellulose substrate) confirmed the estimated value of the surface contact area (Fig. 10.9a). The spreading of the enzymogel nanoparticles is clearly evidenced by Figs. 10.9b and 10.9c (3D topography images and topographical cross-sections) and Fig. 10.1e (SEM image). The estimated contact area of the reaction is about an order of magnitude greater for enzymogel nanoparticles than for a solid particle of the same size.

Hence, industrial cellulase recovery could be realized using enzymogel nanoparticles with unique properties. Due to the high level of CEL activity in the enzymogel there is no necessity to release and extract enzymes. The enzymogel nanoparticles can be applied as an entire system for cellulose hydrolysis. The enzymogel can be extracted after cleavage of the cellulose and used again for conversion of a freshly loaded cellulosic biomass. The results show that this approach enables a four-time growth of glucose yield per CEL in comparison with the common application of enzymes for cellulose hydrolysis (Fig. 10.6a, V).

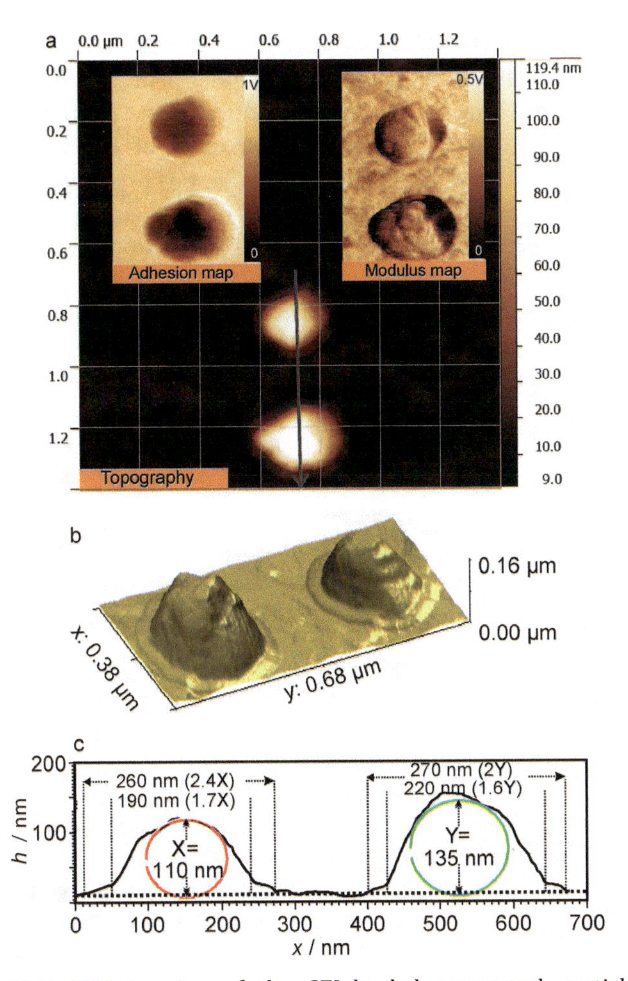

Figure 10.9 AFM imaging of the CEL-loaded enzymogel particles in 10 mm citrate buffer on the surface of a cellophane film using (a) topography, modulus, and adhesion contrasts; 3D topographical images of the particles and their topographical cross-sectional profiles (b and c). With permission from reference Kudina et al. (2014).

10.5 Enzymogel in Localized Bioconversion

One of unique and interesting aspect of enzymogel particles is related to magnetic field directed bioconversion process. The magnetic core of enzymogel could be used to direct enzymogel particles to specific location using nonuniform magnetic fields. This mechanism is less relevant to biofuel technologies, however, could be interesting for specific applications, for example in devices when localized and on-command synthesis of biofuel could be achieved. The latter concept for the application of enzymogel nanoparticles is illustrated with Figs. 10.6c–h using experiments with a cotton floss. A droplet of a free CEL solution (Fig. 10.6c) and a dispersion of enzymogel nanoparticles with no magnetic field (Fig. 10.6e) and in the presence of the magnet (Fig. 10.6g) were deposited on the flosses and allowed to dry. In all experiments, the CEL content was the same. Afterword, both flosses were immersed in an aqueous medium at pH 5.5. The enzymogel-treated floss was broken in 2.5 h. The floss treated with the free CEL was unharmed after 14 h. In the experiment with the magnet, the floss failure was much faster in 1 h. Therefore, directing enzymogel nanoparticles with magnetic field allows for high efficacy of localized bioconversion.

10.6 Impact of Loading on the Efficacy and Recovery of Enzymes Loaded in Enzymogel Nanoparticles

Attachment of CEL to the enzymogel can be carried out at different initial concentrations of CEL charged to a dispersion of nanoparticles at the loading process that allows for controlling the loaded amount (Bayramoglu et al., 2010). High enzyme loadings make the use of nanoparticles more efficacious as long as a biocatalytic activity of the enzyme does not decrease significantly at increased loadings. Enzymes could be adsorbed into the polymer brush as multilayers at higher loadings, which could decrease interaction of the enzyme with the substrate (Srere et al., 1990). The best ratio of an enzyme to nanoparticles is the highest loading of the enzyme on the enzymogel that does not diminish enzyme efficiency or ability to be recovered. Here, the influence of enzyme-to-nanoparticle ratio on efficiency and recoverability of cellulase and β-glucosidase enzymogels is considered.

Enzyme loading on the enzymogel, plotted as a function of CEL protein concentration in the solution, is given in Fig. 10.10. Enzymogels with 33, 72, and 104 μg cellulase mg^{-1} nanoparticles

(A, B, and C, respectively, Fig. 10.10) were utilized for biocatalytic hydrolysis of cellulose. The impact of enzyme loading on the rate of hydrolysis was determined for these samples. In all experiments (free enzymes and enzymogels) the enzyme-to-substrate ratio was same, but enzyme loadings into enzymogel were different. Fig. 10.11 shows the concentrations of low-molecular-weight soluble sugars formed after hydrolysis of filter paper (filter paper units, FPU) (Ghose, 1987) and Solka-Floc cellulose.

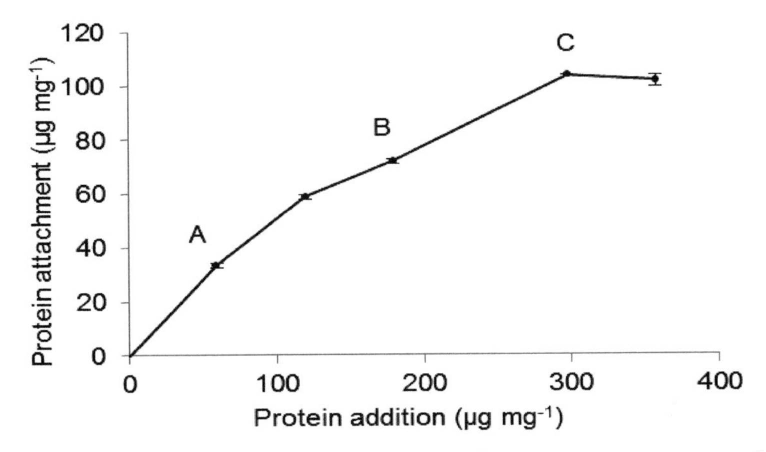

Figure 10.10 Cellulase enzymogel loading for increasing amounts of initial enzyme added for a fixed amount of nanoparticles. Samples from points A, B, and C were used for later experiments (error bars represent standard deviations in protein testing). With permission from Samaratunga et al. (2015a).

For filter paper and Solka-Floc, initial (24-h) rates of hydrolysis with cellulase enzymogels were respectively 45% and 53% of those with the free CEL. Applying free CEL and cellulase enzymogels, close bioconversion yields were shown after 72 h (Fig. 10.6a). It implies that similar conversions can be reached using enzymogel nanoparticles although rates of the cellulose cleavage are reduced at initial stages. Longer reaction times or higher enzyme-to-substrate ratio may be needed to produce the same results as for free CEL. Lower initial rates could result if a part of loaded enzymes were physically hindered from reaching the substrate surface due to the large diameter of enzymogel particles as discussed above. When an enzymogel nanoparticle contacts the substrate, enzyme molecules localized on the opposite side of the nanoparticle obviously are not in contact with the cellulose substrate. CEL mobility within polymer brushes and comparable efficiency for CEL released from enzymogel nanoparticles were shown above to suggest limited impacts of the adsorbed CEL.

In order to compare the total concentrations of soluble carbohydrates (sum of glucose and glucose equivalents of cellobiose) formed during cleavage of each cellulose substrate with free CEL and enzymogel nanoparticles, one-way analysis of variance (ANOVA) was carried out (Table 10.2). CEL content in the enzymogel nanoparticles did not have a significant impact on the concentrations of soluble sugars in the case of filter paper or Solka-Floc ($p > 0.05$). Cellulases retain mobility within the polymer brush to enhance the effective enzyme concentration at the surface of cellulose substrate. Applying enzymogel nanoparticles with a higher CEL loading, one can reduce the amount of enzymogel without lowering hydrolysis rates.

Table 10.2 ANOVA for the comparison of the total soluble sugars obtained during hydrolysis of filter paper and Solka-Floc with three enzymogel cellulase concentrations

	Filter paper				**Solka-Floc**			
Source of variation	**df**	**MS**	***F***	***p* value**	**df**	**MS**	***F***	***p* value**
Between groups	2	0.150	1.84	0.239	2	0.016	1.45	0.336
Within groups	6	0.082			4	0.011		
Total	8				6			

Source: With permission from Samaratunga et al. (2015a).
Abbreviations: df, degrees of freedom; MS, mean square.

Free CEL and enzymogels afforded different ratios between cellobiose and glucose in hydrolyzates (Fig. 10.11). For both filter paper and Solka-Floc hydrolyzates, cellobiose-to-glucose ratios were 80–90% higher in hydrolyzates from enzymogels than in those from free CEL. Multiple possible reasons for this include a lower affinity of β-glucosidase to poly(acrylic acid) compared to endoglucanases and exoglucanases, higher affinity of exoglucanases, and lower accessibility to cellobiose (a soluble substrate) for enzymes immobilized within the enzymogels, or reduced activity of β-glucosidase when it is adsorbed within the enzymogel.

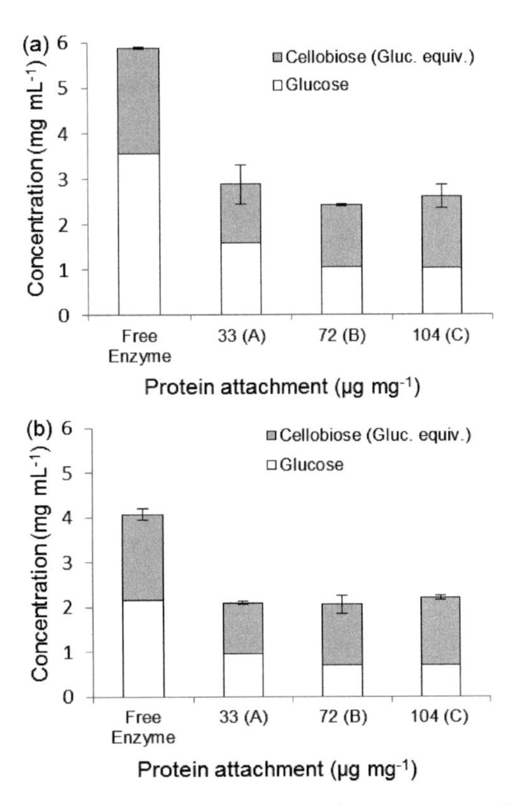

Figure 10.11 Twenty-four-hour soluble sugar concentrations for hydrolysis of (a) filter paper (10 FPU g^{-1}) and (b) Solka-Floc (5 FPU g^{-1}) with free cellulases and cellulase enzymogels (with increasing enzymogel protein loading). Points *A*, *B*, and *C* are from Fig. 10.10 (*error bars* represent standard deviations of the sum of glucose and glucose equivalents of cellobiose). With permission from Samaratunga et al. (2015a).

Adsorption of proteins in poly(acrylic acid) is a selective process depending on the charges of both the polymer brush and the protein in certain medium (Gautrot et al., 2009; Minko, 2006). The isoelectric point of cellulases from *T. reesei* is 4.7–5.1 for exoglucanases, 4.2–7.6 for endoglucanases, and 5.3–6.4 for β-glucosidases (Cantarel et al., 2009). Although the ranges overlap, differences in the isoelectric points of these individual enzymes at the immobilization pH of 4.5 would impact affinity. Isoelectric point of CEL also varies depending on the organism from which CEL was produced. Endoglucanases from *Trichoderma viride* and *Trichoderma harzianum* have the isoelectric points of 4.32

and 5.0, respectively, and a mixture of cellulases from *T. harzianum* has the isoelectric point within the range from 6.4 to 7.6 (Huang et al., 2010; Thrane et al., 1997).

It is predicted that adsorbed β-glucosidase have a lower accessibility to a cellulose surface due localization of enzymes in the enzymogels. For endoglucanases and exoglucanases, substrates are cellulose macromolecules which are insoluble in water. Endoglucanases and exoglucanases adsorbed in the polymer brush could repeatedly catalyze multiple hydrolysis reactions at the same time owing to the localization of the enzymes at the cellulose surface. On the contrary, soluble cellobiose is the substrate for β-glucosidase. Hence, adsorbed β-glucosidase is expected to be less effective because the enzyme is localized within the polymer brush while cellobiose is dispersed throughout the environment.

The recovery of grafted enzymes after cellulose cleavage is shown in Fig. 10.12. Regardless of the enzyme loading, CEL recovery was over 95% after hydrolysis of both filter paper and Solka-Floc. Unhydrolyzed residue could be pelleted with the attached enzymes during centrifugation (Gregg et al., 1996). Preliminary tests with enzymogels having a magnetic Fe_3O_4 cores with silica shells and poly(acrylic acid) brush layer grafted from a silica surface demonstrated a similar CEL recovery applying magnetic separation after cellulose cleavage.

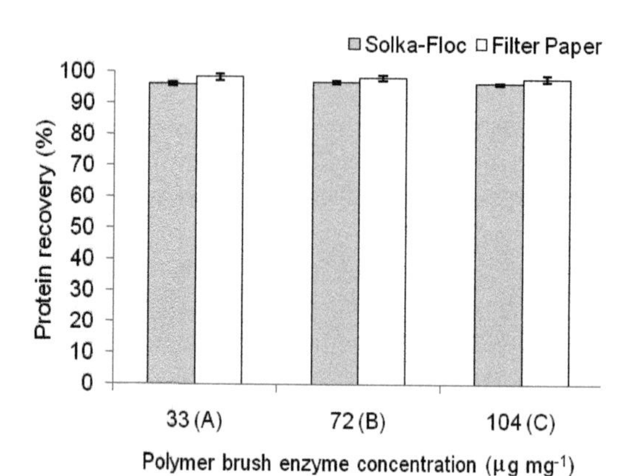

Figure 10.12 Protein recovery after 72 h of hydrolysis with cellulase enzymogels for (a) filter paper (10 FPU g^{-1}) and (b) Solka-Floc (5 FPU g^{-1}). Points *A*, *B*, and *C* are from Fig. 10.10 (*error bars* represent standard deviations in protein recovery). With permission from Samaratunga et al. (2015a).

The β-glucosidase loading into the enzymogel nanoparticles as a function of enzyme concentration is given in Fig. 10.13. It shows a gradual increase in enzymogel loading with a saturation enzymogel loading of 113 μg mg^{-1} (point C). For tests on biocatalytic cleavage of cellulose, enzymogel nanoparticles with 31, 66, and 113 μg mg^{-1} (A, B, and C, respectively, Fig. 10.13) were used. Fig. 10.14 shows the concentrations of glucose after initial hydrolysis with these three enzymogels.

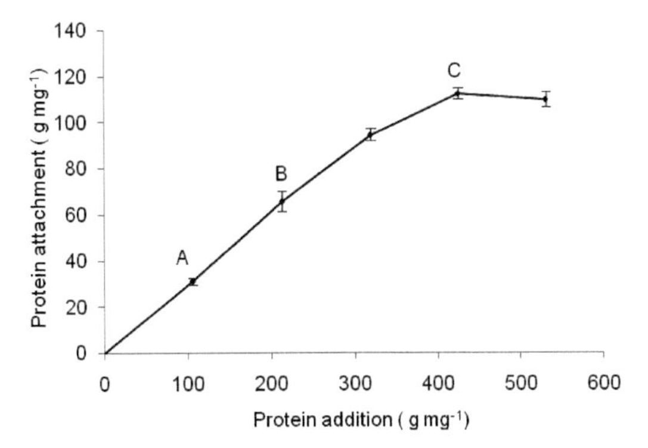

Figure 10.13 β-Glucosidase enzymogel loading for increasing amounts of initial enzyme added for a fixed amount of nanoparticles (*error bars* represent standard deviations in protein testing). With permission from Samaratunga et al. (2015a).

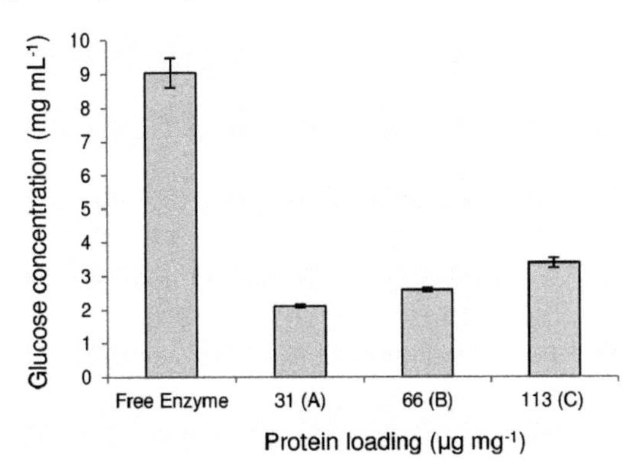

Figure 10.14 Six-hour glucose concentrations from hydrolysis using free β-glucosidase and β-glucosidase enzymogels (with increasing enzymogel protein loading) at a protein equivalent of 10 CBU g^{-1} cellobiose. Points *A*, *B*, and *C* are from Fig. 10.13 (*error bars* represent standard deviations in glucose concentrations). With permission from Samaratunga et al. (2015a).

Free β-glucosidase showed higher rates of hydrolysis as compared with β-glucosidase enzymogels, implying lowering of net activity of the attached enzymes. It could be explained by increased heterogeneity of the system consisting of the concentrated and localized enzyme and a soluble, dispersed substrate. After biocatalytic hydrolysis with β-glucosidase enzymogels, glucose concentrations were 23–38% of the concentration using free β-glucosidase (Fig. 10.14). On the other hand, CEL enzymogels formed 49% of the soluble sugar using the free enzyme (Fig. 10.11). This fact supports the idea of a heterogeneity limitation for cellobiose hydrolysis with enzymogel nanoparticles.

Unlike attached CEL, enzymogel nanoparticles with higher β-glucosidase loading revealed more glucose in hydrolyzates. Higher protein loading in the enzymogels leads to increased content of enzymes since the same amount of the enzyme is localized on a smaller nanoparticle. Also, enzymes may be buried deeper into the brush at lower loadings, hence limiting interaction between the enzyme and substrates. The recovery of loaded β-glucosidase at increasing loadings is shown in Fig. 10.15. For the three β-glucosidase loadings, recovery of protein was significantly different ($p = 0.004$) as determined with ANOVA. Recovery of protein for loaded β-glucosidase at 66 and 113 μg protein mg^{-1} brush (B and C) were not very different as shown by Tukey's multiple comparison test, but nevertheless higher than that for loading of 31 μg mg^{-1} (A). Despite the statistical difference between these values, differences have little significance from a practical standpoint.

Hence, it could be concluded that for both β-glucosidase and cellulase, the initial hydrolysis rates of enzymes adsorbed within the PAA brush of the enzymogel nanoparticles were lower than those for free enzymes. Higher CEL loading on the enzymogel nanoparticles did not change rates of biocatalytic hydrolysis; therefore, increased loadings can be efficiently used. More cellobiose in cellulose hydrolyzates, formed by using grafted cellulases, implies a potentially decreased affinity of β-glucosidase or endoglucanase to poly(acrylic acid), a lower substrate accessibility to the β-glucosidase, or a lower activity of loaded β-glucosidase. Glucose concentrations in hydrolyzates increased at higher β-glucosidase loadings in the enzymogel. For either

β-glucosidase or cellulase, enzyme recovery did not depend significantly on enzyme loading. Strong potential for reuse was demonstrated for attached cellulases, recovery of which exceeded 95%. However, the enzymogel nanoparticles are less appropriate for attaching β-glucosidase since its recovery (below 68%) and hydrolysis rates are lower.

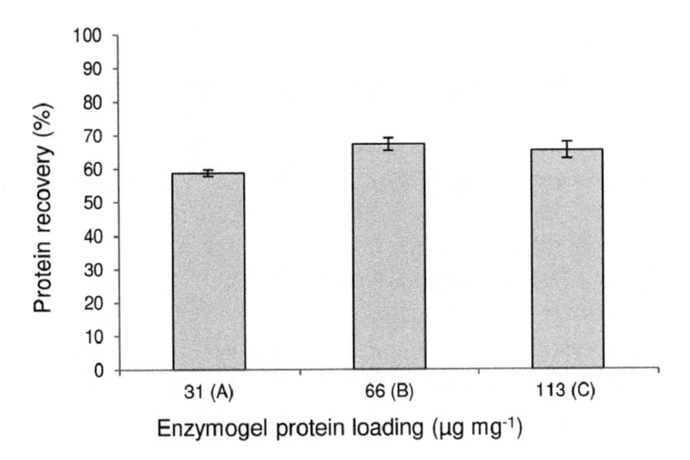

Figure 10.15 Protein recovery after 6 h of cellobiose hydrolysis using free β-glucosidase and β-glucosidase enzymogels at a protein equivalent of 10 CBU g^{-1} cellobiose. Points *A*, *B*, and *C* are from Fig. 10.13 (*error bars* represent standard deviations in protein recovery). With permission from Samaratunga et al. (2015a).

10.7 Effect of pH and Temperature for Cellulases Loaded in Enzymogel Nanoparticles

Glucose is the final product of cellulose hydrolysis; nevertheless, the total amount of all reducing sugars is a true criterion for evaluating the efficiency of endoglucanase, exoglucanase, and β-glucosidase mixtures. Thus, concentrations of glucose and total reducing carbohydrates after partial biocatalytic cleavage of cellulose with free enzymes and enzymogel nanoparticles were monitored to reveal the impact of pH and temperature.

Free CEL and enzymogel nanoparticles perform best at 50°C (the highest tested temperature), but the optimal pH for the free and loaded CEL is different. The free enzymes performed best at pH 4.4. Enzymogels nanoparticles, however, show an optimal pH of 5.0, close to the most commonly used pH value (4.8) for commercial cellulases (Tu et al., 2007; Tu et al., 2009; Jeya et al., 2009). CEL enzymogel nanoparticles were generally more sensitive than free enzymes to changes in pH of the medium. At the least effective pH, the concentration of reducing carbohydrates at 50°C (optimal temperature) was 86% of that at the optimal pH for free CEL and only 54% for the enzymogel nanoparticles. Both systems, however, exhibited similar temperature sensitivity; reducing carbohydrate concentrations after low-temperature hydrolysis were about 75% of those at the higher optimal temperature for the free CEL and enzymogel nanoparticles. Carbohydrate concentrations for enzymogel nanoparticles were normally 60–75% of those for free CEL.

Depending on a specific enzyme, substrate, and loading technique, responses to temperature and pH after loading differ, but improved temperature and pH tolerance is generally recognized (Miletić et al., 2012). Enzymogel nanoparticles showed reduced hydrolysis rates at low pH values, which are not used normally. It could be explained by the fact that the physical structure of poly(acrylic acid) brushes alters with changing solution pH. These brushes tend to shrink at low pH, thus diminishing the effective surface area for each nanoparticle and limiting access of enzyme to the substrate since the enzymes evidently will be buried in the PAA brush (Minko, 2006). In a similar way, the CEL loaded into the enzymogel nanoparticles readily spread out on the brush nanoparticle surface at the higher pH, enhancing accessibility of the insoluble substrates to the enzymes. The reduced rates of a biocatalytic cleavage could be defeated by higher enzyme loadings as CEL can be recovered for multiple cycles.

CEL loading into the brushes is not covalent; mobility of the loaded enzymes allows them for diffusing throughout the PAA brush. The strongest CEL attachment to PAA brushes was observed at pH values of about 4.5; at higher pH, the attachment strength decreases. This could improve diffusion of the endoglucanases, exoglucanases, and β-glucosidase within the PAA brush and hydrolysis of cellulose macromolecules at a higher pH.

The CEL recovery was more sensitive to pH of the environment than to temperature. Over 87% of loaded enzyme was recovered, and recovery was predicted to be greater than 89% under the expected range of operating conditions (between pH 4.4 and 5.0). Some loss of proteins attached on polymeric substrates is common owing to the non-covalent nature of attachment (Miletić et al., 2012).

The highest enzymogel recovery was observed at the lowest studied pH value of 4.0. Nevertheless, CEL efficiency was also reduced at low pH and about 30% of substrate remained as unhydrolyzed residue. CEL recovery was measured by determining enzyme concentrations of supernatant after centrifugation. Our preliminary tests on enzymogel nanoparticles with magnetic (Fe_3O_4) cores showed a mean recovery of 87% CEL after biocatalytic hydrolysis of filter paper substrate and recovery with magnetic separation. Almost complete biocatalytic hydrolysis was achieved, and thus, the impact of unhydrolyzed residue as sedimentation was considered minimal. This supports that the CEL recovery was primarily favored by CEL attachment on the enzymogel nanoparticles.

The experimental data on concentration of glucose after hydrolysis using free β-glucosidase and enzymogels show that free β-glucosidase is most efficient at the higher temperatures (>48°C) and lower pH values (<4.5). Glucose concentrations in this region were close to 100% of theoretical values. Enzymogel nanoparticles loaded with β-glucosidase were most efficient in the same region of temperature and pH, however rates of hydrolysis were reduced. Predicted concentrations of glucose in hydrolyzate using enzymogel nanoparticles were generally between 30 and 60% of those when using the free β-glucosidase and 53% under optimum conditions (50°C, pH 4.4). It implies lower feasibility of this approach for loading β-glucosidase as compared with CEL.

The application of CEL enzymogel nanoparticles supplemented with free β-glucosidase may be beneficial for producing industrial carbohydrates. Activity of immobilized β-glucosidases vary significantly (Figueira et al., 2011; Singh et al., 2011; Yan et al., 2010; Yang et al., 2011). Some other studies on β-glucosidase attachment have focused on its stability but have not reported on activity comparisons between the free and immobilized enzyme (Chen et al., 2014; Tan et al., 2014; Zhou et al., 2013).

Figure 10.16 shows protein recovery after a biocatalytic hydrolysis with β-glucosidase enzymogel nanoparticles as a function of the medium pH. β-Glucosidase recovery decreased as pH increases. The reduced affinity between poly(acrylic acid) brushes and β-glucosidase at higher pH is consistent with the low recovery with increasing pH.

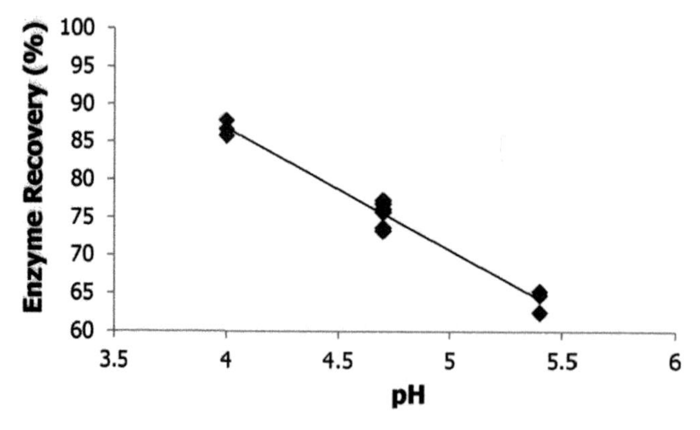

Figure 10.16 Enzyme recovery after 6-h cellobiose hydrolysis using β-glucosidase enzymogels. With permission from Samaratunga et al. (2015b).

Therefore, free cellulase enzymes and enzymogel nanoparticles respond to changes in pH and temperature similarly despite the fact that biocatalytic hydrolysis with free cellulase suggested a lower optimal pH of 4.4. Highest hydrolysis rates were observed for both systems at 50°C. Under the typical operating conditions, cellulase recovery exceeded 89% and pH and temperature had a relatively small impact on it. For cellulase enzymogel nanoparticles, operating pH had a larger effect on efficiency as compared with free enzymes. For cellulase enzymogels, pH values below 4.2 have a negative impact on β-glucosidase activity in comparison with exoglucanases and endoglucanases. Overall activity of cellulase enzymogel nanoparticles was generally 60–70% of the free cellulase. β-Glucosidase enzymogel nanoparticles were less efficient than free enzyme and their recovery rate was lower; these enzymogel are not recommended. Recovery reduced at higher pH for both enzymes tested.

10.8 Conclusion

In summary, the novel phase boundary biocatalysis utilizes enzymes engulfed by polymer brush nanoparticles—enzymogel. The enzymes retain their mobility in polymer brushes and are able to cleave insoluble cellulose due to their mobility in the brush and high affinity to both the enzymogel brush and to the substrate surface. The enzyme attached to the polymer brush catalyzes the cleavage of the insoluble substrates. Biocatalytic activity of the enzymogel in terms of kinetics is lower versus native cellulases; however, recycling of the enzymes results in the increase of the overall efficacy by at least fourfold.

References

Bayramoglu G., Arica M. Y., *J. Mol. Catal. B: Enzym.*, **62** (2010), 297–304.

Becker A. L., Welsch N., Schneider C., Ballauff M., *Biomacromolecules*, **12** (2011), 3936–3944.

Bornscheuer U. T., *Angew. Chem. Int. Ed.*, **42** (2003), 3336–3337.

Cantarel B. L., Coutinho P. M., Rancurel C., Bernard T., Lombard V., Henrissat B., *Nucleic Acids Research*, **37** (2009), D233–D238.

Chen T., Yang W., Guo Y., Yuan R., Xu L., Yan Y., *Enzyme Microb. Technol.*, **63** (2014), 50–57.

Chiu Y.-C., Cheng M.-H., Engel H., Kao S.-W., Larson J. C., Gupta S., Brey E. M., *Biomaterials*, **32** (2011), 6045–6051.

Coradin T., Nassif N., Livage J., *J. Appl. Microbiol. Biotechnol.*, **61** (2003), 429–434.

Czeslik C., Winter R., *Phys. Chem. Chem. Phys.*, **3** (2001), 235–239.

Czeslik C., *Z. Phys. Chem.*, **218** (2004a), 771–801.

Czeslik C., Jansen R., Ballauff M., Wittemann A., Royer C. A., Gratton E., Hazlett T., *Phys. Rev. E.*, **69** (2004b), 021401.

Figueira J. D. A., Dias F. F. G., Sato H. H., Fernandes P., *Enzyme Res.*, **2011** (2011), 8.

Gautrot J. E., Huck W. T. S., Welch M., Ramstedt M., *ACS Appl. Mater. Interfaces*, **2** (2009), 193–202.

Ghose T. K., *Pure Appl. Chem.*, **59** (1987), 257–268.

Gregg D. J., Saddler J. N., *Biotechnol. Bioeng.*, **51** (1996), 375–383.

Haupt B., Neumann T., Wittemann A., Ballauff M., *Biomacromolecules,* **6** (2005), 948–955.

Henzler K., Haupt B., Ballauff M., *Anal. Biochem.,* **378** (2008), 184–189.

Henzler-Wildman K., Kern D., *Nature,* **450** (2007), 964–972.

Hollmann O., Steitz R., Czeslik C., *Phys. Chem. Chem. Phys.,* **10** (2008), 1448–1456.

Huang X. M., Yang Q., Liu Z. H., Fan J. X., Chen X. L., Song J. Z., Wang Y., *Appl. Biochem. Biotechnol.,* **162** (2010), 103–115.

Hwang E. T., Gu M. B., *Eng. Life Sci.,* **13** (2013), 49–61.

Ivanov A. E., Edink E., Kumar A., Galaev I. Y., Arendsen A. F., Bruggink A., Mattiasson B., *Biotechnol. Prog.,* **19** (2003), 1167–1175.

Jervis E. J., Haynes C. A., Kilburn D. G., *J. Biol. Chem.,* **272** (1997), 24016–24023.

Jeya M., Zhang Y.-W., Kim I.-W., Lee J.-K., *Bioresour. Technol.,* **100** (2009), 5155–5161.

Kobayashi J., Mori Y., Kobayashi S., *Chem. Commun.*, (2006), 4227–4229.

Kudina O., Zakharchenko A., Trotsenko O., Tokarev A., Ionov L., Stoychev G., Puretskiy N., Pryor S. W., Voronov A., Minko S., *Angew. Chem. Int. Ed.,* **53** (2014), 483–487.

Lu Y., Wittemann A., Ballauff M., *Macromol. Rapid Commun.,* **30** (2009), 806–815.

Lutz J.-F., Akdemir Ö., Hoth A., *J. Am. Chem. Soc.,* **128** (2006), 13046–13047.

Lynd L. R., van Zyl W. H., McBride J. E., Laser M., *Curr. Opin. Biotechnol.,* **16** (2005), 577–583.

Miletić N., Nastasović A., Loos K., *Bioresour. Technol.,* **115** (2012), 126–135.

Minko S., *J. Macromol. Sci., Polym. Rev.,* **46** (2006), 397–420.

Norde W., Lyklema J., *J. Colloid Interface Sci.,* **66** (1978), 266–276.

Pessela B. C. C., Mateo C., Carrascosa A. V., Vian A., García J. L., Rivas G., Alfonso C., Guisan J. M., Fernández-Lafuente R., *Biomacromolecules,* **3** (2002), 107–113.

Rodrigues R. C., Ortiz C., Berenguer-Murcia A., Torres R., Fernandez-Lafuente R., *Chem. Soc. Rev.,* **42** (2013), 6290–6307.

Samaratunga A., Kudina O., Nahar N., Zakharchenko A., Minko S., Voronov A., Pryor S. W., *Appl. Biochem. Biotechnol.,* **175** (2015a), 2872–2882.

Samaratunga A., Kudina O., Nahar N., Zakharchenko A., Minko S., Voronov A., Pryor S. W., *Appl. Biochem. Biotechnol.,* **176** (2015b), 1114–1130.

Singh R., Zhang Y.-W., Nguyen N.-P.-T., Jeya M., Lee J.-K., *Appl. Biochem. Biotechnol.*, **89** (2011), 337–344.

Srere P. A., Ovadi J., *FEBS Lett.*, **268** (1990), 360–364.

Stewart J. D., *Curr. Opin. Chem. Biol.*, **5** (2001), 120–129.

Tan I. S., Lee K. T., *Bioresour. Technol.*, **184** (2014), 386–394.

Thrane C., Tronsmo A., Jensen D. F., *Eur. J. Plant Pathol.*, **103** (1997), 331–344.

Tu M., Chandra R. P., Saddler J. N., *Biotechnol. Progr.*, **23** (2007), 398–406.

Tu M., Zhang X., Paice M., MacFarlane P., Saddler J. N., *Bioresour. Technol.*, **100** (2009), 6407–6415.

Veum L., Hanefeld U., *Chem. Commun.*, (2006), 825–831.

Wittemann A., Ballauff M., *Anal. Chem.*, **76** (2004), 2813–2819.

Wittemann A., Ballauff M., *Macromol. Biosci.*, **5** (2005), 13–20.

Wittemann A., Ballauff M., *Phys. Chem. Chem. Phys.*, **8** (2006), 5269–5275.

Wittemann A., Haupt B., Ballauff M., *Z. Phys. Chem.*, **221** (2007), 113–126.

Yan J., Pan G., Li L., Quan G., Ding C., Luo A., *J. Colloid Interface Sci.*, **348** (2010), 565–570.

Yang Y.-S., Zhang T., Yu S.-C., Ding Y., Zhang L.-Y., Qiu C., Jin D., *Molecules*, **16** (2011), 4295–4304.

Zhang Y., Liu Y. Y., Xu J. L., Yuan Z. H., Qi W., Zhuang X. S., He M. C., *Bioresources*, **7** (2012), 345–353.

Zhou Y., Pan S., Wu T., Tang X., Wang L., *Electron. J. Biotechnol.*, **16** (2013), 1–13.

Chapter 11

Recent Advances in Nanostructured Enzyme Catalysis for Chemical Synthesis in Organic Solvents

Zheng Liu,[a] Jun Ge,[a] Diannan Lu,[a] Guoqiang Jiang,[a] and Jianzhong Wu[b]

[a]*Key Laboratory of Industrial Biocatalysis, Ministry of Education, Department of Chemical Engineering, Tsinghua University, Beijing 100084, China*

[b]*Department of Chemical and Environmental Engineering, University of California, Riverside, California 92521, USA*

liuzheng@mail.tsinghua.edu.cn

11.1 Introduction

Enzymes have unique capabilities to catalyze chemical reactions with high chemo-, regio-, and stereo-selectivity, high turnover rate and mild solution conditions. Utilizing enzymes for chemical synthesis would enable high-yield production of complex molecules and circumvention of intermediate separation, leading to a shortened synthetic route with reduced energy consumption and minimized waste discharge (Liese et al., 2006; Pollard and Kosjek, 2008; Patel, 2011; Torrelo et al., 2015). Such efficiency is illustrated with the industrial synthesis of Sitagliptin, an antidiabetic drug that inhibits dipeptidyl peptidase and therefore

Biocatalysis and Nanotechnology
Edited by Peter Grunwald

ISBN 978-981-4613-69-9 (Hardcover), 978-1-315-19660-2 (eBook)
www.panstanford.com

decreases the blood glucose level in human body. The conventional route for chemical synthesis of Sitagliptin involves a rhodium-catalyzed asymmetric amine hydrogenation at high pressure. A transaminase is able to convert prositagliptin ketone to sitagliptin with enzyme efficiency over 99.95% and a yield of 92%. In comparison with the rhodium-catalyzed process, the enzymatic process increases the overall yield by 10–13%, reduces the total waste by 19%, and excludes the use of hazardous heavy metals (Savile et al., 2010).

Organic solvents are extensively used in the synthesis of fine chemicals. The roles of organic solvents in chemical synthesis include solubilizing substrates, inhibiting reverse reactions, and eliminating adverse effects of water (Zaks and Klibanov, 1998). Despite recent progress in enzymatic catalysis in organic media (Adlercreutz, 2013; Ge et al., 2012), its application to industrial productions remains at early stage. One major huddle resides in the low stability of enzymes in an organic solvent, particularly in a polar solvent such as DMSO, DMF or THF. Such solvents extract water molecules from the enzyme surface essential for its functionality and dehydration often induces protein denaturation. Another major problem is the low activity of enzyme in an organic medium; enzyme activity is further reduced if it exists as aggregates. Aggregation limits enzyme accessibility to its substrate, which could reduce the apparent activity up to 4 to 5 orders of magnitude in comparison to that under the native condition. Besides, an organic solvent may also undermine the enzyme selectivity due to partial denaturation of the protein.

Recent advances in nanotechnology offer immense opportunities to address above-mentioned problems hindering the widespread application of enzymes for chemical synthesis. A number of novel strategies have been explored in recent years to incorporate enzymes into nanostructured materials such as mesoporous materials (Luckarift et al., 2004) or nanoparticles (Kim and Grate, 2003; Dyal et al., 2003). Excellent reviews are available on fabrication of nanostructured enzyme catalysts as well as interpretation of the mechanisms underpinning the improved catalytic performance (Johnsona et al., 2014; Zhang et al., 2015). In this chapter, we review progresses in our laboratory over the past decade on the design and synthesis of nanostructured enzyme catalysts that exhibit high stability and activity in organic solvents at high temperature. Our discussion focuses on two kinds of aqueous synthetic methods in terms of in situ polymerization and conjugation. Special attention is given to the fundamentals of nanostructured enzyme catalyst gained upon using complementary inputs from molecular simulation, structural characterization and reaction kinetics. We will demonstrate the successful applications of these nanostructured enzyme catalysts for chemical synthesis in organic solvents. This chapter also discusses existing problems and future prospects in the fundamental study and application of nanostructured enzyme catalysts.

11.2 Enzyme Nanogel from in situ Polymerization

Our experimental efforts for preparation of enzyme nanogels were motivated by an earlier simulation study on the conformation transition of a confined protein, which indicates that an enhanced thermal stability can be achieved once an enzyme is encapsulated in a hydrophilic confinement (Lu et al., 2006). An aqueous-based, two-step in situ polymerization procedure was developed to synthesize enzyme nanogels (Yan et al., 2006; Fig. 11.1). The first step introduces vinyl groups on the protein surface by acryloylation (Wang et al., 1997; Yang et al., 1995), and the second step involves in situ polymerization that encapsulates the acryloylated protein. We have demonstrated that the enzyme nanogels are able to reproduce the catalytic performance of their native counterpart in free form. While the nanogels yield similar K_m and k_{cat} parameters in the Michaelis–Menten equation, they greatly enhance protein stability at high temperature or in the presence of polar organic solvents. The nanogel approach can be similarly applied to other enzymes. Its robustness in catalytic performance has been validated for horseradish peroxidase (Yan et al., 2006), carbonic anhydrase (Yan et al., 2007); lipase (Ge et al., 2008), and urate oxidase (Liu et al., 2009).

A combination of molecular dynamics (MD) simulations and diverse experimental characterization methods allows us to gain further insights on the fundamentals of enzyme nanogel synthesis and catalysis. Using lipase as an example, we find that acrylamide assembly around the enzyme is driven by hydrogen bonding with functional groups at the enzyme surface. Acrylamide assembly is essential to the fabrication of enzyme nanogels. As shown in Fig. 11.2, both MD simulation and the fluorescence resonance energy transfer (FRET) spectrum indicate that the enhanced enzyme stability can be attributed to the multipoint linkage with the tailor-made hydrophilic network of polymers. An increase in the number of intramolecular hydrogen bonds of the encapsulated protein was evident from the blue shift of the fluorescence spectrum of the encapsulated lipase. These hydrogen bonds contribute to a greatly enhanced the stability of encapsulated enzyme at high temperature (Ge et al., 2008), as confirmed by the unchanged fluoresce emission spectrum as well as the circular dichroism (CD) spectrum. Moreover, the porous gel offers a hydrophilic microenvironment for the encapsulated protein, preventing the extraction of water molecules by the polar solvent and thus rendering the encapsulated protein an enhanced tolerance to the organic solvent (Ge et al., 2009).

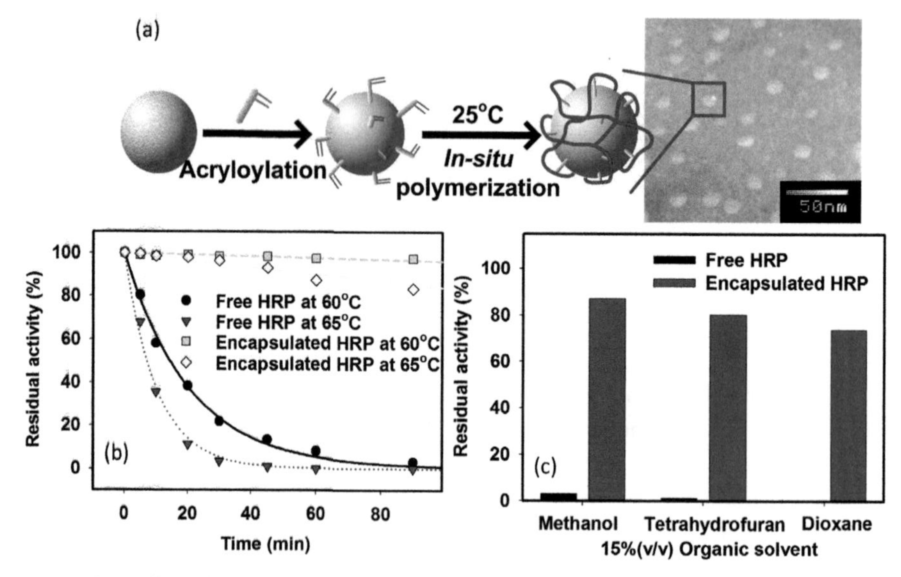

Figure 11.1 Aqueous synthesis of enzyme nanogel: (a) synthetic scheme; (b) thermal stability; (c) tolerance to polar solvents. Reprinted with permission from Yan et al. (2006), Copyright (2006), American Chemical Society.

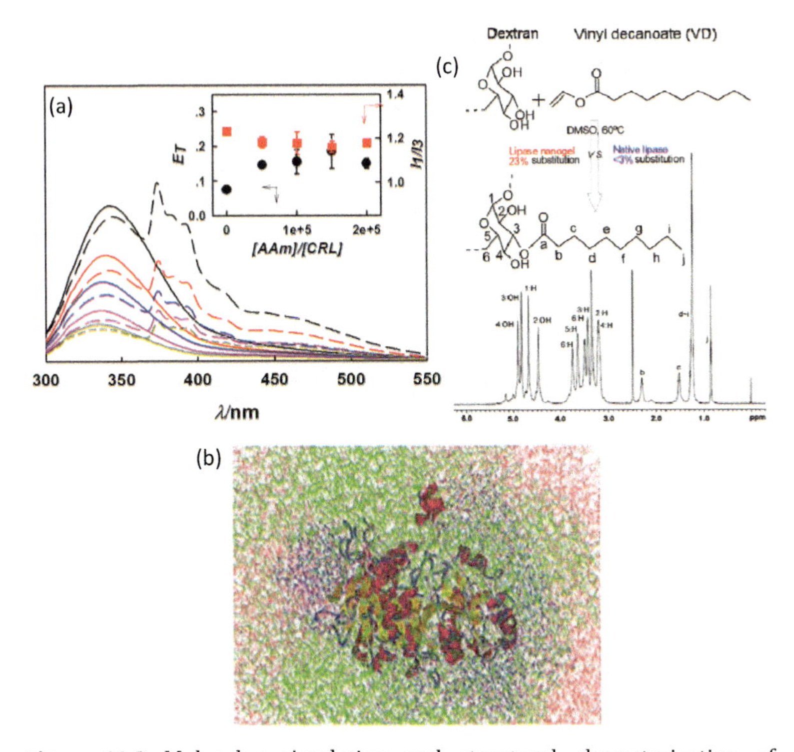

Figure 11.2 Molecular simulation and structural characterization of enzyme nanogel (a) FRET from lipase to pyrene in absence of AM and in presence of 0.5% AM in solution; (b) Structure of lipase nanogel at 60°C in DMSO (red, DMSO; blue, water; green, polyacrylamide shell; (c) Conversion of transesterification between dextran and VD catalyzed by lipase and lipase nanogel. Reprinted with permission from Ge et al. (2008, 2009), Copyright (2008, 2009), American Chemical Society.

Encapsulation of enzymes in nanogels may lead to an activation effect. In the case of lipase nanogel, in which glycidyl methacrylate (GMA) was used as the acryloylation reagent (Xu et al., 2012), the nanogel has a lower K_m and a higher k_{cat} in comparison to its native counterpart, as shown in Table 11.1. Both MD simulation and experiment suggest that the activation effect can be partially attributed to lipase interaction with GMA, which reinforces the protein "open" configuration and thus facilitates more efficient mass transport to and from the modified lipase in both free and encapsulated forms. Such an activation effect shows the potential of tailored design of nanostructured enzyme catalyst.

Table 11.1 Enzymatic kinetic parameters of native CRL and its modifiers

	Native CRL	CRL-NAS	NANOGEL(N)	CRL-GMA	NANOGEL(G)
K_m(mM)	0.263±0.051	0.216±0.083	0.236±0.048	0.441±0.097	0.594±0.045
k_{cat} (s^{-1})	3.55±0.083	3.16±0.474	1.54±0.432	8.86±1.17	7.79±1.15
k_{cat}/K_m (mM^{-1} s^{-1})	13.8±1.39	16.2±4.02	6.4±0.52	20.5±1.83	13.0±0.94

To enhance the uptake of hydrophobic substrate, we synthesized a polyacrylamide (PA) and poly(*N*-isopropylacrylamide) (PNIPAM) interpenetrating network encapsulating lipase. The modified acryloylation-polymerization scheme involves two monomers (i.e., acrylamide and isopropylacrylamide) that are loaded sequentially during synthesis. Compared to the lipase polyacrylamide nanogel, this novel lipase nanogel facilitates the uptake of hydrophobic substrates and thus gives an enhanced activity in non-polar solvents. Besides, the modified polymer network enhances the thermal stability. It was shown by an all-atom MD simulation that above interpenetrating polymer matrix formed a more hydrophobic environment and this was confirmed by the fluorescence spectrum. Such a hydrophobic network favors the uptake of hydrophobic substrates and thus increases the overall rate of the enzymatic reaction. We demonstrated the enhanced stability and catalytic performance of this novel lipase nanogel in aqueous and non-polar organic solvent using hydrolysis reaction of *p*-NPP in aqueous and esterification reaction of ibuprofen in isooctane (Du et al., 2013).

To further enhance the catalytic performance of enzyme nanogels in non-polar solvents, Wang et al. (2013) introduced a substrate-imprinting procedure (Fig. 11.3). Using lipase as the model enzyme, we demonstrated that the apparent activity of the enzyme nanogels in toluene is 200% of that corresponding to its native counterpart. The adsorption experiments suggest that the improved activity is mainly attributed to the facilitated transport of substrates through the imprinted channel. More recently, Wang et al. (2014) proposed a polyethyleneglycol (PEG)-substrate joint imprinting method, which activates the encapsulated lipase and facilitates the substrate uptake, leading to an enhanced apparent catalytic activity in organic media. Compared to native lipase, the substrate-PEG imprinted lipase nanogel displays an increased apparent activity in organic solvents by 2.5–4.7 folds. It enables a one-step synthesis of chloramphenicol palmitate with a yield of 99% and purity of 99%.

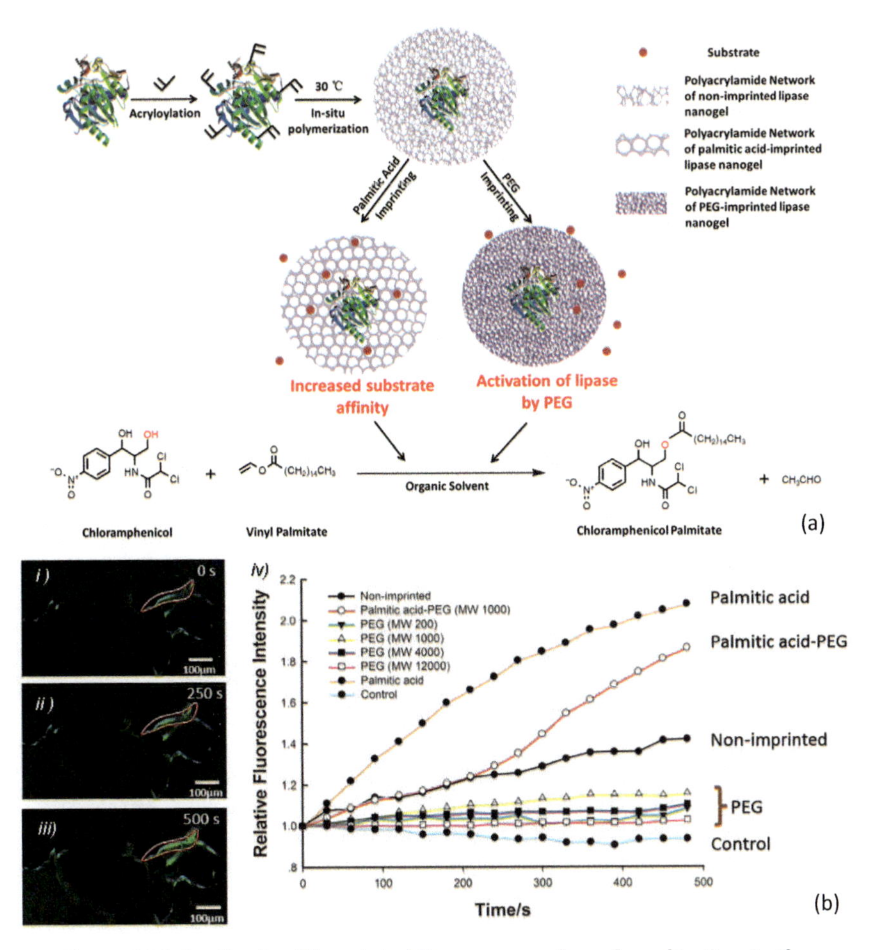

Figure 11.3 Synthesis of imprinted lipase nanogels and application in the enzymatic synthesis of chloramphenicol palmitate in organic solvents (a) Synthetic scheme; (b) Laser confocal scanning microscopy (LSCM) images of palmitic acid-imprinted lipase nanogel incubatedwith FITC-labeled palmitic acid in acetonitrile for (i) 0 s, (ii) 250s, (iii) 500 s. (Red circles indicate the selected region for analysis). (iv): Increase in relative fluorescence intensity within different lipase nanogels (all data were obtained from three parallel experiments). Reprinted with permission from Wang et al. (2013, 2014), Copyright (2013, 2014), The Royal Society of Chemistry.

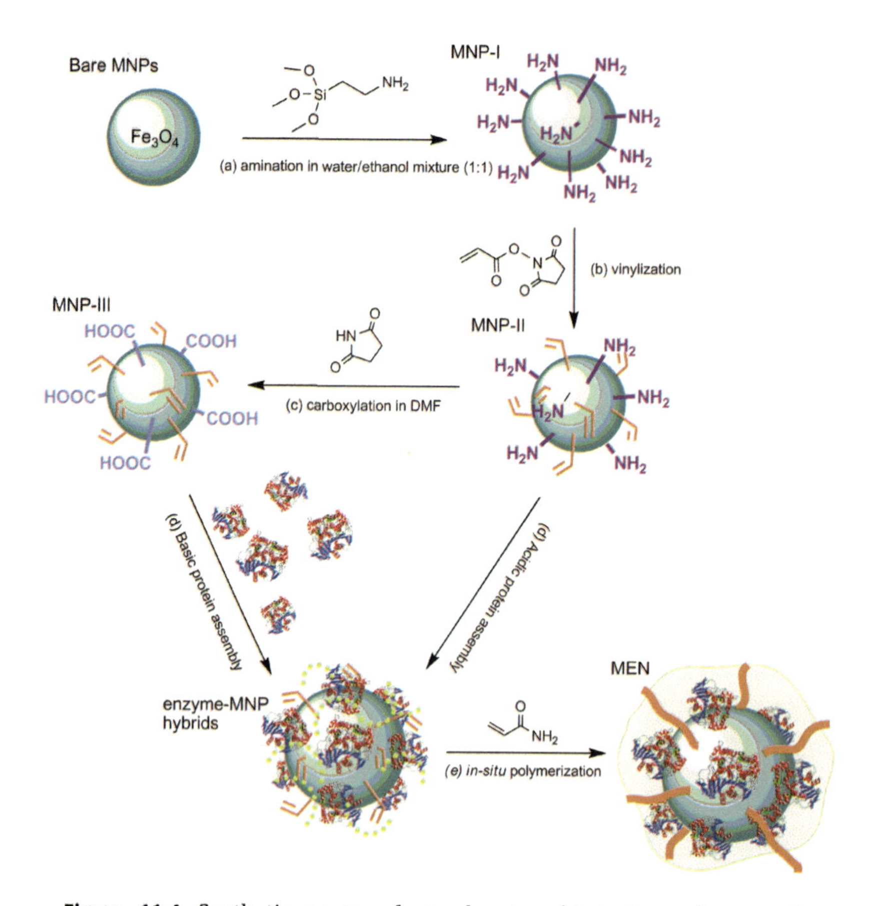

Figure 11.4 Synthetic route of step-by-step fabrication of magnetic enzyme nanogels (MENGs): (a) MNP-I were modified with APTES to generate surface amine functionalities (MNP-II); (b) MNP-II were reacted with NAS to generate vinyl groups; (c) the remaining amine groups on MNP-II were converted to carboxylate groups by succinimide (MNP-III); (d) enzyme deposition onto the oppositely charged MNPs (II or III depending on pI of the target proteins) via electrostatic interaction; (e) in situ polymerization taken place on the surface vinyl groups of modified MNPs, acrylamide (AM) as monomers, bis-methyleneacrylamide (MBA) as cross-linkers, and ammonium persulfate as initiators. Reprinted with permission from Lin et al. (2012), Copyright (2012), The Royal Society of Chemistry.

To facilitate the recovery and operation of enzyme nanogel, Lin et al. (2012) proposed a novel strategy to fabricate magnetic enzyme nanoparticles (MENGs). In this case, enzyme encapsulation is achieved through the electrostatically driven assembly of the enzyme around the magnetic nanoparticles, followed by in situ polymerization from the surface of magnetic nanoparticles, forming a polyacrylamide network (Fig. 11.4). We demonstrated the effectiveness of this method for enhancing enzyme stability without a significant compromise of enzyme activity. The versatility of this method was validated using lipase, horseradish peroxidase, trypsin, and cytochrome *C* as enzyme samples. Thc electrostatically driven assembly method is generally applicable to virtually all enzymes because of the ubiquity of electrostatic interactions. This method circumvents chemical modification of enzyme thus simplifies the experimental procedure. Besides, formation of a polyacrylamide network through aqueous polymerization prevents deterioration of the enzyme conformation. It is a combination of high catalytic activity, enhanced thermal stability, and ease of recovery and reuse that ensures a wide spectrum of utility for enzymatic catalysis.

11.3 Aqueous Synthesis of Enzyme-Polymer Conjugate for Chemical Synthesis in Non-Polar Solvents

To improve enzyme performance in common organic solvents, we synthesized a new class of enzyme-polymer nanoconjugates (Zhu et al., 2013). The nano-conjugates consist of enzymes and Pluronic, a copolymer with a hydrophobic backbone (poly (propylene oxide)) flanked by two hydrophilic side chains (poly(ethylene oxide)). The first step is to oxidize Pluronic F-127 (POH) with Dess–Martin periodinane to convert its hydroxyl end-groups into aldehyde functionalities (PCHO). The PCHO is then conjugated to the lysines of protein by forming a Schiff base followed by a reduction with $NaCNBH_3$. The resulted nanoconjugates are readily dissolvable in organic solvents at high temperature and show greatly enhanced catalytic activities compared with their native counterparts. For example, the enzyme activity gains a 67-fold increase for lipase and a 670-fold increase for cytochrome c (Fig. 11.5).

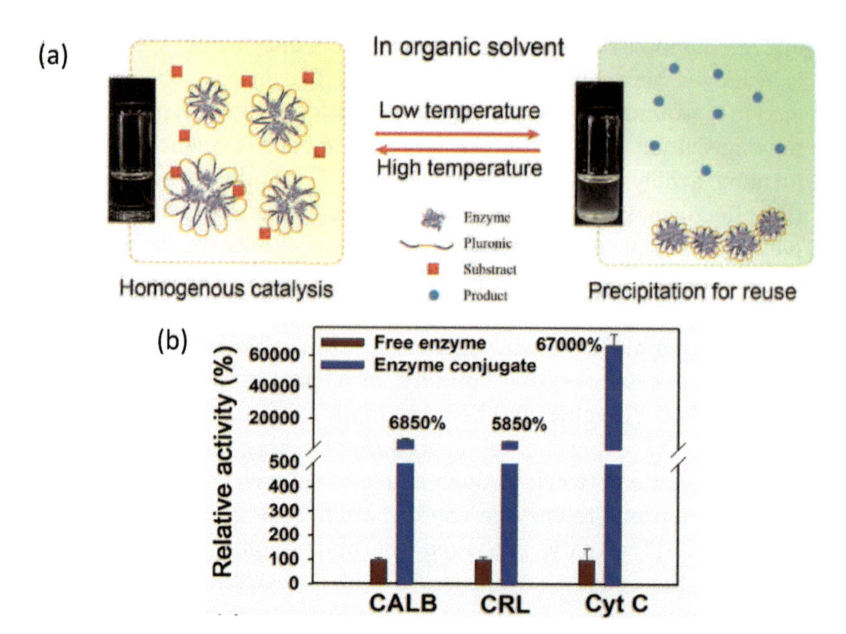

Figure 11.5 Temperature responsiveness of Pluronic conjugated enzymes. (a) Solubilization of the conjugates in toluene for homogenous biocatalysis at 40°C, precipitation of the conjugates in toluene for recycling at 4°C. Photograph insets show (left) solubilization and (right) precipitation of the nanoconjugates in toluene at 40°C and 4°C; (b) The activities of enzyme–Pluronic nanoconjugates compared to native enzymes in organic solvents at 40°C. Reprinted with permission Zhu et al. (2013), Copyright (2013), The Royal Society of Chemistry.

To gain insight into enzyme activity in nanocongjuates, Chen et al. (2015) performed a coarse-grained MD simulation for cytochrome c (Cyt c) conjugated with a polypropylene oxide (PPO) block connected to two poly(ethylene oxide) (PEO) block copolymer (ABA) in toluene (Fig. 11.6). It was shown that at low temperature, the PEO tails tend to form a hairpin structure outside the conjugated protein, whereas the Cyt c–ABA conjugates tend to form larger aggregates. At high temperature, however, the PEO tails adsorb onto the protein surface, thus improving the suspension of the Cyt c–ABA conjugates and, consequently, the contact with the substrate. Moreover, the temperature increase drives conformational transition of Cyt c–ABA from an "inactive state" to an "activated state" and as experimentally observed, results in an enhanced activity.

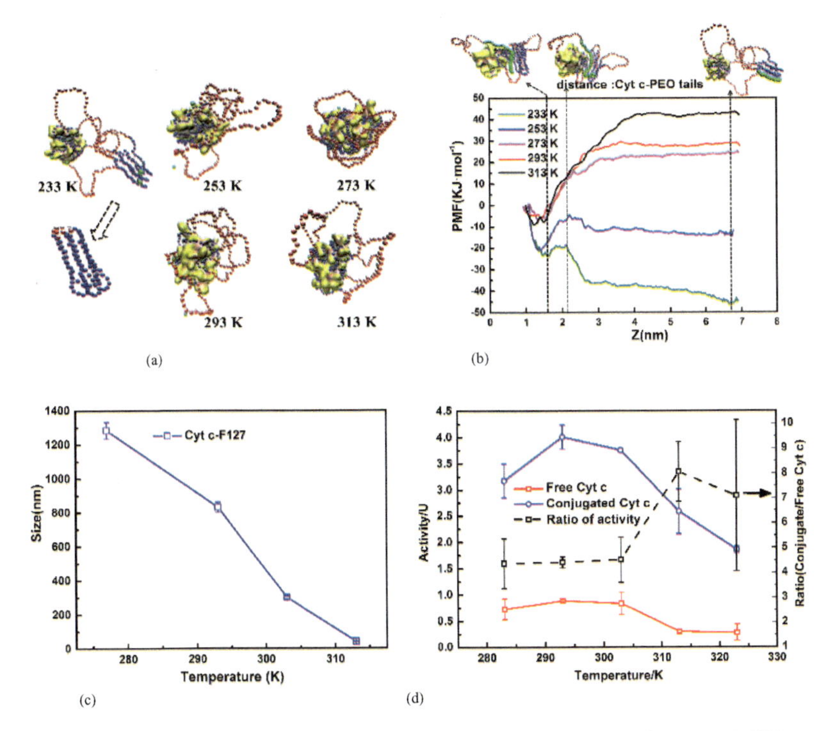

Figure 11.6 Molecular simulation of Cytc conjugated with a PEO-PPO-PEO block copolymer. (a) The geometry of the conjugates; (b) PMF curves for pulling PEO tails towards protein at different temperatures; (c) average size of Cyt c–F127 conjugates detected by DLS as a function of temperature; (d) specific activity of free and F127-conjugated Cyt c as a function of temperature). Reprinted with permission from Chen et al. (2015), Copyright (2015), The Royal Society of Chemistry.

The conjugation efficiency of enzyme–polymer aggregates can be improved by adding surfactants or protein cross-linkers. Wu et al. (2015) carried out the conjugation experiment in reverse emulsions using Pluronic as a reactive surfactant (Fig. 11.7). Here the aldehyde-functionalized Pluronic serves as both the surfactant required in the formation of a reverse emulsion and the polymer modifier for the synthesis of enzyme–polymer conjugates. Micelle formation resulted in concentrated aldehyde groups of Pluronic at the oil–water interface and thus favors conjugation between Pluronic and enzymes in water droplets. Bovine serum albumin (BSA) can also be used as cross-linker for the formation of enzyme–Pluronic conjugates. Addition of BSA increases the conjugation efficiency and leads to a more uniformed size distribution in comparison with the direct conjugation procedure in aqueous solution. We validated the versatility of this method using horseradish peroxidase (HRP), *Candida rugosa* lipase (CRL) and *Candida antarctica* lipase B (CALB). The resulting enzyme–Pluronic conjugates showed greatly enhanced apparent activity compared to free enzymes in organic media.

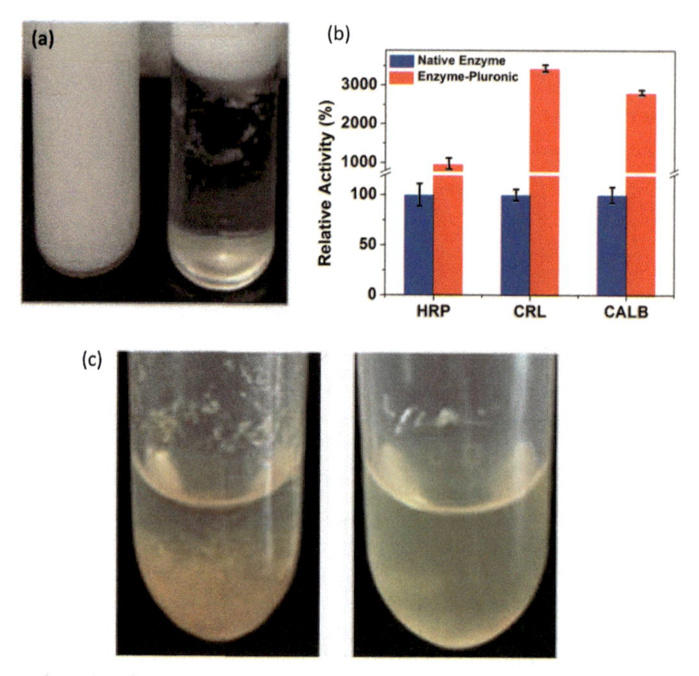

Figure 11.7 Synthesizing enzyme–polymer conjugates in reverse emulsions using Pluronic as a reactive surfactant. (a) Photos showing reverse emulsion of aldehyde-functionalized Pluronic F-127 (left) and instable emulsion of Pluronic F-127 (right); (b) The activities of enzyme–Pluronic conjugates in organic media; (c) photographs showing enzymes in toluene at room temperature: free HRP as precipitates (left) and dissolved HRP–Pluronic conjugate (right). Reprinted with permission from Wu et al. (2015), Copyright (2015), The Royal Society of Chemistry.

Our enzyme nanoconjugates have been tested for chemical synthesis in organic media. For example, Zhang et al. (2013) applied the lipase-Pluronic conjugate for the synthesis of valrubicin (*N*-triuoroacetyladriamycin-14-valerate), a FDA approved anti-cancer drug for the treatment of refractory carcinoma in the bladder.

Valrubicin can be produced from a three-step chemical synthesis that involves trifluoroacetylation, iodination and nucleophilic displacement with sodium valerate. The low reactivity of the second and third steps is responsible for a low overall yield of 24% (Israel et al., 1975). The multistep procedure results in a heavy burden of downstream processing, including product purification and waste treatment. Our chemo-enzymatic route starts with trifluoroacetylation of doxorubicin (Dox), followed by the lipase–Pluronic nanoconjugate catalyzed transesterication of the *N*-triuoroacetyl. The enzyme nanoconjugate results in a 10-fold increase of the reaction rate for the synthesis of valrubicin in organic media in comparison to that of the native CALB. Besides, the reaction occurs at a relatively low temperature, which is more convenient for practical operations. Under optimized reaction conditions, the overall yield of valrubicin synthesis was 82% with a purity of 98%. Importantly, the temperature responsiveness of the lipase conjugate offers advantages for the recovery and operation of the lipase nanoconjugates.

We have further demonstrated the practical applicability of Pluronic-lipase conjugates for fine chemical synthesis in organic solvents (Wu et al., 2014). For the synthesis of lutein palmitate, the enzymatic process is able to convert 85% of lutein with a product purity of 96% at relatively mild conditions, whereas the free lipase catalyzes the same process with a lutein conversion efficiency below 10%. Lutein laurate, lutein myristate and lutein stearate were also synthesized with the Pluronic-lipase conjugate, achieving conversions of 73.8%, 88.5% and 84.4%, respectively (Hou et al., 2015). Most recently, we applied *Candida antarctica* lipase B (CALB)-Pluronic F127 to chemo-enzymatic epoxidation of fatty acids in toluene. At optimized conditions, we obtained a yield of 97%, 75% and 67% for epoxidized oleic acid, linoleic acid and linolenic acid, respectively (Oda et al., 2016). All these examples indicate the high potential of the Pluronic-conjugated enzymes for fine chemical synthesis in organic solvents.

11.4 Summary and Outlook

We have engineered non-aqueous enzyme complexes with enhanced activity, selectivity and stability by incorporating tailor-made nanostructures. The enzyme performance can be further improved by tuning the molecular-level interactions among enzymes, substrates, and products in appropriate solution environments. We demonstrate that material synthesis and process design benefit tremendously from complementary inputs from theoretical analysis, simulation and process chemistry. An integration of molecular modeling with experimentation is essential for establishing the engineering fundamentals for the design, synthesis, and application of nanostructured enzyme catalysts.

The man-made nanomaterials, together with today's biotechnological tools, offer unprecedented opportunities to reengineer enzyme catalysts for industrial applications. Enzymatic catalysis for chemical synthesis in organic solvent is just one of them. While successful examples remain limited, we expect more exciting progress in the future with the rapid advances in molecular engineering fundamentals. By combining theory and experiments, we will be able to elucidate the microscopic details underlying interaction of nanostructure with enzyme and the consequential effects on the structure and reaction kinetics. The fundamental studies will allow us to develop novel enzyme-catalyzed processes and associated apparatus for large-scale production of fine chemicals with good performance. For industrial applications, significant hurdles remain in terms of enzyme activity, reducibility, and ease-of-operation and new strategies to overcome these huddles are of great and urgent need. Promising solutions may arise, for example, from recent advancements in microfluidic techniques. These new techniques have proven effectiveness in controlled fluid flow, high-performance mass and heat transport at low shearing force, and thus may offer novel approaches to both production of nanostructured enzyme catalysts with well-defined structure and properties and implementation of the enzymatic reactions in organic media.

Acknowledgment

The authors would like to thank the National Natural Science Foundation of China for the financial support (Grand no. 21036003 and 21520102008). The authors are grateful to all graduate and undergraduate students who contributed to the results discussed in this work.

References

Adlercreutz P., *Chem. Soc. Rev.*, **42** (2013), 6406–6436.

Chen G., Kong X., Zhu J. Y., Lu D. N., and Liu Z. *Phys. Chem. Chem. Phys.*, **17** (2015), 10708–10714.

Du M. L., Lu D. N., Liu Z., *J. Mol. Catal. B Enzymatic*, **80** (2013), 60–68.

Dyal A., Loos K., Noto M., Chang S. W., *J. Am. Chem. Soc.*, **125** (2003), 1684–1685.

Ge J., Lu D. N., Wang J., and Liu Z., *Biomacromolecules*, **10** (2009), 1612–1618.

Ge J., Lu D. N., Wang J., Yan M., Lu Y. F., and Liu Z., *J. Phys. Chem. B*, **112** (2008), 14319–14324.

Ge J., Yang C., Zhu J., Lu D., and Liu Z., *Top. Catal.*, **55** (2012), 1070–1080.

Hou M., Wang R., Wu X. L., Zhang Y. F., Ge J., and Liu Z., *Catal. Lett.*, **145** (2015), 1825–1829.

Israel M., Modest E. J., and Frei E., *Cancer Res.*, **35**(1975), 1365–1368.

Johnsona B. J., Algarc W. R., Malanoskia A. P., Anconab M. G., and Medintza I. L., *Nano Today*, **9** (2014) , 102–131.

Kim J. B. and Grate J. W., *Nano Lett.*, **3** (2003), 1219–1222.

Liese A., Seelbach K., and Wandrey C., *Ind. Biotransformations.* 2nd, Wiley-VCH, Weinheim (2006).

Lin M. M., Lu D. N., Yang C., Zhu J. Y., Zhang Y. F., and Liu Z., *Chem. Commun.*, **48** (2012), 3315–3317.

Liu Z. X., Lu D. N., Li J. M., Chen W., and Liu Z., *Phys. Chem. Chem. Phys.*, **11** (2009), 333–340.

Lu D. N., Liu Z., and Wu J. Z., *Biophys. J.*, **90** (2006), 3224–3238.

Luckarift H. R., Spain J. C., Naik R. R., and Stone M. O., *Nat. Biotechnol.*, **22** (2004), 211–213.

Oda Y., Zhang Y. F., Ge J., and Liu Z., *Catal. Lett.*, 2016, DOI 10.1007/s10562-016-1726-5.

Patel R. N., *ACS Catal.*, **1** (2011), 1056–1074.

Pollard D. and Kosjek B., *Organic Synthesis with Enzymes in Non-Aqueous Media.* Wiley-VCH Verlag GmbH & Co. KGaA, Weinheim (2008).

Savile C. K., Janey J. M., Mundorff E. C., Moore J. C., Tam S., Jarvis W. R., Colbeck J. C., Krebber A., Fleitz F. J., Brands J., Devine P. N., Huisman G. W., Hughes G. J., *Science*, **329** (2010), 305–309.

Torrelo G., Hanefeld U., and Hollmann F., *Catal. Lett.*, **145** (2015), 309–345.

Wang P., Sergeeva M. V., Lim L., and Dordick J. S., *Nat. Biotechnol.*, **15** (1997), 789–793.

Wang R., Zhang Y. F., Ge J., and Liu Z., *RSC Adv.*, **4**(2014), 40301–40304.

Wang R., Zhang Y. F., Huang J. H., Ge J., and Liu Z., *Green Chem.*, **15** (2013), 1155–1158.

Wu X. L., Ge J., Zhu J. Y., Zhang Y. F., Yong Y., and Liu Z., *Chem Commun.*, **51** (2015), 9674–9677.

Wu X. L., Wang R., Zhang Y. F., Ge J., and Liu Z., *Catal. Lett.*, **144** (2014), 1407–1410.

Xu D. D., Tonggu G., Bao X. P., Lu D. N., and Liu Z., *Soft Matter,* **8** (2012), 2036–2042.

Yan M., Ge J., Liu Z., and Ouyang P. K., *J. Am. Chem. Soc.,* **128** (2006), 11008–11009.

Yan M., Liu Z. X., Lu D. N., and Liu Z., *Biomacromolecules,* **8**(2007), 560–565.

Yang Z., Mesiano A. J., Venkatasubramanian S., Gross S. H., Harris J. M., and Russell A. J., *J. Am. Chem. Soc.*, **117** (1995), 4843–4850.

Zaks A. and Klibanov A. M., *J. Biol. Chem.*, **263** (1998), 3194–3201.

Zhang Y. F., Dai Y., Hou M., Li T., Ge J., and Liu Z., *RSC Adv.*, **3** (2013), 22963–22966.

Zhang Y. F., Ge J., and Liu Z., *ACS Catal.*, **5** (2015), 4503–4513.

Zhu J. Y., Zhang Y. F., Lu D. N., Zare R. N., Ge J., and Liu Z., *Chem. Commun.*, **49** (2013), 6090–6092.

Chapter 12

Probing Enzymatic Activity by Combining Luminescent Quantum Dots, Gold Nanoparticles, and Energy Transfer

Naiqian Zhan, Goutam Palui, Wentao Wang, and Hedi Mattoussi
Florida State University, Department of Chemistry and Biochemistry, Tallahassee, Florida 32306, USA
mattoussi@chem.fsu.edu

12.1 Introduction

Enzymes are proteins that are involved in catalyzing an array of biological processes. They maintain the physiological homeostasis by modulating reaction kinetics in cells, or in organisms (White et al., 1959; Neurath and Walsh, 1976). A variety of genetic and non-genetic disruptions can activate or silence the intrinsic activities of enzymes, which often lead to diseases. For example, proteases are known to play an important role in hormone activation and maturation, protein digestion, apoptosis, growth, differentiation and cell signaling (Manning et al., 2002; Puente et al., 2003; Turk, 2006; López-Otín and Matrisian, 2007; Drag and Salvesen, 2010; Barrett et al., 2012). Very often a change in the optimal levels of proteases initiates an irregular hydrolysis of peptide bonds (in a polypeptide chain or in proteins) at post-

Biocatalysis and Nanotechnology
Edited by Peter Grunwald

ISBN 978-981-4613-69-9 (Hardcover), 978-1-315-19660-2 (eBook)
www.panstanford.com

translational level and underlies various diseases such as cancer, AIDS, inflammation and various neurodegenerative disorders (Puente et al., 2003; Turk, 2006). In addition, many infectious microorganisms such as bacteria, viruses, and parasites use protease as virulence factors (Kohl et al., 1988; Wu et al., 2003; Liu et al., 2009).

Over past two decades, several analytical biological and chemical techniques have been developed to probe and understand enzymatic activities and functions. Among them, spectroscopic detection method relying on fluorescence resonance energy transfer (FRET) has been utilized for probing the abundance and activity of various enzymes (Rodems et al., 2002; Giepmans et al., 2006; Medintz et al., 2006). However, for designing a FRET-based fluorescent probe and achieving optimal results, there are many factors to be considered, including optical requirements, substrate structure, and stability of the fluorophores in testing environments. It is also challenging to design a single and universal platform that can probe a whole range of enzymes. Recently, several studies have incorporated quantum dots (QDs) and gold nanoparticles (AuNPs) for designing ET-based sensing platforms to probe the enzyme activity with high sensitivity (Selvin, 2000; Medintz et al., 2005; Medintz and Mattoussi, 2009; Mattoussi et al., 2012; Saha et al., 2012; Geißler et al., 2013). This has been motivated by the fact that these nanomaterials possess unique optical and electronic properties that can be tuned via size- and shape-selective manner. They can act as energy donor or energy acceptor/quencher. These properties in combination yield enhanced energy transfer efficiency and increased detection sensitivity (Clapp et al., 2005; Medintz et al., 2005; Yun et al., 2005; Somers et al., 2007; Charbonnière and Hildebrandt, 2008; Singh and Strouse, 2010; Silvi and Credi, 2015).

In this chapter, we summarize a few representative sensor designs based on the combination of energy transfer interaction with the use of luminescent quantum dots as donor, or acceptor, with metal nanoparticles, fluorescent dyes and proteins to measure the enzymatic activity in solution and in cell cultures. We start with a brief discussion of the basic principles of energy transfer process, including FRET, bioluminescence resonance energy transfer (BRET), and chemiluminescence resonance energy transfer (CRET). We outline the specific features associated with these processes when applied to QDs and AuNPs. We then provide a few representative examples where conjugates of QD–peptide, QD–protein, AuNP–peptide have been utilized to design specific sensing assemblies for targeting enzymatic activity in vitro (solution) or in vivo (cell culture).

12.2 Fluorescence Resonance Energy Transfer: Background

Fluorescence resonance energy transfer is an electrodynamic process that involves the nonradiative transfer of excitation energy from a photo-excited state donor molecule to a proximal ground-state acceptor molecule, as depicted in Fig. 12.1 (Lakowicz, 2006; Valeur and Berberan-Santos, 2012). It is driven by dipole–dipole interactions, and as such achieving efficient FRET interactions requires that two important criteria be obeyed: (a) close proximity between donor (D) and acceptor (A) due to the nature of the dipole–dipole interaction and its strong dependence on the separation distance; (b) finite spectral overlap between the donor emission and acceptor excitation profiles. The rate of energy transfer, $k_{\mathrm{D\text{-}A}}$, between a donor and an acceptor separated by a distance, r, as formulated by *T.* Förster can be expressed as (Förster, 1948; Lakowicz, 2006)

$$k_{\mathrm{D\text{-}A}} = \frac{B \times Q_{\mathrm{D}} I}{\tau_{\mathrm{D}} r^6} = \left(\frac{1}{\tau_{\mathrm{D}}}\right) \times \left(\frac{R_0}{r}\right)^6, \tag{12.1}$$

where Q_{D} and τ_{D} designate the photoluminescence (PL) quantum yield and the exciton radiative lifetime of the donor, respectively; I is the integral of the spectral overlap function defined as

$$I = \int PL_{\mathrm{D\text{-}coor}}(\lambda)\varepsilon_{\mathrm{A}}(\lambda)\lambda^4 d\lambda, \tag{12.2}$$

where $PL_{\mathrm{D\text{-}corr}}(\lambda)$ is the wavelength-dependent normalized fluorescence intensity spectrum of the donor, and ε_{A} is the molar extinction coefficient spectrum of the acceptor (Lakowicz, 2006). The constant B is a function of the refractive index of the medium, n_{D}, Avogadro's number, N_{A}, and the parameter accounting for the relative orientation of the donor and acceptor dipoles, κ_{p}^2:

$$B = \frac{[9000 \times (\ln 10)]\kappa_p^2}{128\pi^5 n_D^4 N_A} \tag{12.3}$$

The FRET efficiency defined as the ratio between the nonradiative decay rate and the total decay rate, including the radiative rate given by τ_D^{-1}, can be expressed as

$$E_{FRET} = \frac{k_{D\text{-}A}}{k_{D\text{-}A} + \frac{1}{\tau_D}} = \frac{R_0^6}{R_0^6 + r^6}. \tag{12.4}$$

R_0 designates the separation distance for which $E_{FRET} = 0.5$ and is given by

$$R_0^6 = BQ_D I \tag{12.5}$$

The term E_{FRET} accounts for the fraction of excitons that are transferred from the donor to the acceptor via non-radiative pathways. Equation (12.4) shows that there is a sixth power dependence of the FRET efficiency on the D-A separation distance (r). FRET interactions are accounted for experimentally by measuring the energy transfer efficiency, E, often referred to as the PL quenching efficiency, using either steady-state or time-resolved fluorescence data via the following expressions (Lakowicz, 2006):

$$E = 1 - \frac{F_{DA}}{F_D}, \text{ for steady-state fluorescence} \tag{12.6a}$$

and

$$E = 1 - \frac{\tau_{DA}}{\tau_D}, \text{ for time-resolved fluorescence} \tag{12.6b}$$

Here, F_{DA} designates the ensemble PL intensity measured for the sample containing the D–A pair while F_D is the intensity measured from a sample made of donors without any acceptors (control/reference). τ_{DA} is the lifetime of the donor in the presence of the acceptor.

The strong dependence of the FRET efficiency on the D-A distance (r) makes this process very sensitive and ideal for probing nanoscale separation distances in the range of 10–100 Å. In biology, this range is sufficient for measuring intracellular protein folding, observing binding interactions between biomolecules, DNA sequencing analysis, etc. FRET as an analytical technique has been widely employed in biology to probe interactions between separate molecules or changes in conformation within large size biomolecules (Weiss, 2000; Sustarsic and Kapanidis, 2015).

For a few decades the use of FRET has relied on employing fluorescent dyes as donors and acceptors (Lakowicz, 2006). Use of organic dyes in FRET though effective, has encountered a few limitations, such as low brightness, substantial overlap between donor and acceptor emission/absorption spectra. The latter produces significant direct excitation contribution to the acceptor emission which complicates experimental conditions and analysis of FRET data (Resch-Genger et al., 2008). FRET-based characterization using fluorescent proteins as donors and/or acceptors has also gained a tremendous interest, following the remarkable development and optimization of fluorescent proteins over the past four decades (Kohl et al., 2002; Lippincott-Schwartz and Patterson, 2003; Rizzo et al., 2004; Chudakov et al., 2005; Shaner et al., 2005). More recently, advances made in designing fluorescent semiconductor nanocrystals (with their high brightness) have generated intense interest for use as stable probes for in vitro and in vivo imaging, sensing and in various energy transfer-transfer modalities (Zrazhevskiy et al., 2010; Cassette et al., 2013; Palui et al., 2015). Indeed, use of QDs in developing FRET-based assays (where they serve as donors combined with dye acceptors, or as acceptors combined with bio/chemiluminescent donors) offers a few clear advantages derived from the unique photophysical properties exhibited by these materials. Similarly, interest in AuNPs and Au nanostructures has grown over the past two decades due to their large extinction coefficients and strong size- and shape-dependent surface plasmon absorption (SPR) features (Link and El-Sayed, 2000; Jana et al., 2001; Liz-Marzán, 2006; Saha et al., 2012; Ye et al., 2013). Although Au nanostructures do not usually exhibit any emission properties (except for cluster materials), they provide highly efficient quenchers via a process often referred to as nanometal energy quenching of dyes and QDs alike (Dulkeith et al., 2005;

Yun et al., 2005; Pons et al., 2007; Breshike et al., 2013). In this report we will briefly summarize the use of QD-based FRET, CRET/BRET and AuNP-induced PL quenching of QDs as means of designing assays targeting enzymatic activity.

12.2.1 A Few Unique Properties of QD Donors

Luminescent QDs exhibit a few optical and spectroscopic properties that make them highly effective FRET donors. They can potentially overcome some of the limitations encountered by dyes and fluorescent proteins (Medintz et al., 2005; Medintz and Mattoussi, 2009; Mattoussi et al., 2012; Cassette et al., 2013). These include high quantum yield, strong resistance to both photo- and chemical-degradation, broad absorption profile, size-tunable narrow PL, and large Stokes shift (Murray et al., 1993; Dabbousi et al., 1997; Resch-Genger et al., 2008; Talapin et al., 2010). Below we focus on two critical factors that make QDs excellent donor in a FRET configuration (Medintz and Mattoussi, 2009).

12.2.1.1 Tuning the spectral overlap by changing the QD emission

Equation 12.1 above indicates that the FRET rate $k_{D\text{-}A}$ varies linearly with the spectral overlap integral, *I*. Because of the ability to size- and/or composition-tune the QD emission, it is possible to control the degree of spectral overlap with any target dye acceptor, by selecting the emission location of QDs to be used. Figure 12.1B shows the PL spectra of a few sets of CdSe nanocrystals with peak location varying from 510 to 610 nm. Figure 12.1C show a plot of the normalized emission of three distinct sets of CdSe-ZnS QD dispersions (λ_{em} = 510 nm, 530 nm, 555 m) together with the absorption profile of Cy3, along with the corresponding spectral overlap function $J(\lambda)$ for each QD-Cy3 pair (Clapp, Medintz et al., 2004). Data show that the overlap function varies depending on the location of the PL peak, with the largest overlap measured for the 555-nm-QD-Cy3 pair. Such tuning of the spectral overlap proves one's capacity to optimize the FRET efficiency for any dye acceptor, by selecting the QD set with maximal spectral overlap with that particular dye.

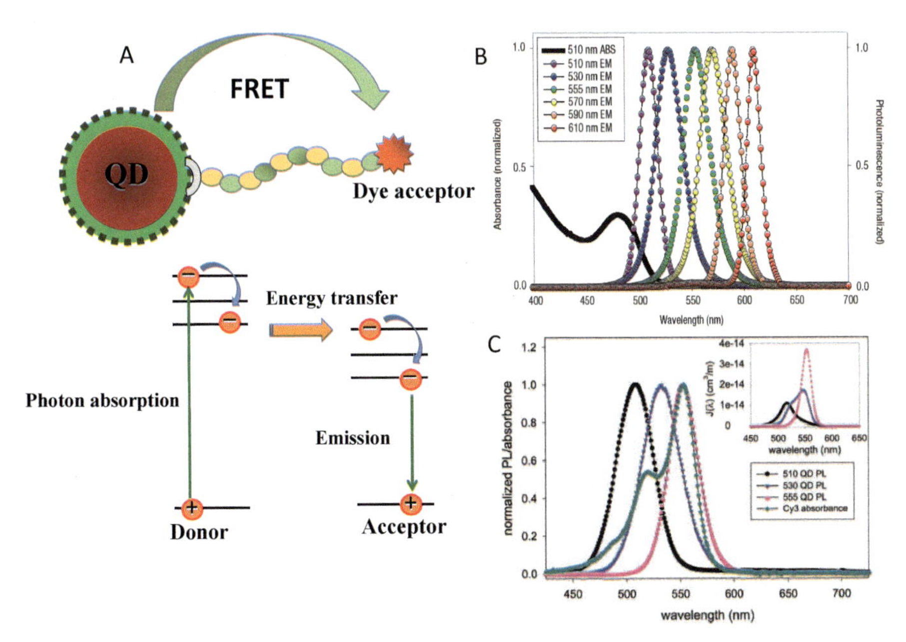

Figure 12.1 (A) Schematic representation of FRET interactions between a QD donor and one acceptor dye molecule. (B) Absorption and emission profiles of six different size CdSe-core QDs. (C) Plot of the normalized absorption profile of Cy3 dye together with the PL spectra of three sets of QDs (with emission wavelength at 510, 530, and 555 nm), the spectra highlight the ease of tuning the overlap integral with a given acceptor dye (here Cy3) by selecting the QDs with appropriate emission profile. Reproduced partially from Clapp et al. (2004) and Medintz et al. (2005), with permissions from the American Chemical Society and NPG.

12.2.1.2 Enhancing the FRET interactions by varying the QD-acceptor conjugate valence

Due to their rather large surface-to-volume ratio, a single QD can be coupled to several dyes, which yields a configuration where the QD acts as a platform for immobilizing multiple molecules and as a donor simultaneously interacting with several energy transfer acceptors. This donor–acceptor configuration produces a substantial increase in the "cross-sectional" interactions of a single QD with the surrounding dyes. These multichannel interactions between a QD and the surrounding dyes can produce a large increase in the rate of FRET and thus the resulting quenching efficiency, compared to one-donor–one-acceptor pair. The total rate is expressed as (Clapp et al., 2004; Medintz and Mattoussi, 2009)

$$k_{\text{FRET}/n} = \sum_{i=1}^{n} k_{\text{D-A},i}. \tag{12.7}$$

This leads to an energy transfer efficiency that can be expressed as

$$E_n = \frac{k_{\text{FRET}/n}}{k_{\text{FRET}/n} + \tau_{\text{D}}^{-1}}. \tag{12.8}$$

Here, each individual energy transfer rate $k_{\text{D-A},i}$ within the QD-dye assembly depends on the spectral overlap integral (shown in Eq. 12.2), n is the total number of acceptors interacting with the central QD donor and τ_{D} is the PL lifetime of the donor alone. For a construct where n identical acceptors are arrayed at the same separation distance (i.e., r is fixed) around the central QD, the FRET efficiency can be further simplified to (Medintz and Mattoussi, 2009):

$$E = \frac{nk_{\text{D-A}}}{nk_{\text{D-A}} + \tau_{\text{D}}^{-1}} = \frac{nR_0^6}{nR_0^6 + r^6} \tag{12.9}$$

Indeed enhancement in the measured FRET efficiency with increasing dye-to-QD ratio has been demonstrated in several instances using conjugates made of QD-dye and QD-fluorescent protein pairs (Medintz et al., 2003; Clapp et al., 2004; Clapp et al., 2005b; Lu et al., 2008; Snee et al., 2011). This enhancement in the measured FRET efficiency has been exploited by several groups to design specific sensing assemblies. Some of these examples will be discussed below.

12.2.2 QDs as Acceptors

While there have been numerous reports on developing FRET-based sensors using QDs as donors, use of QDs as FRET acceptors has been much less exploited (Achermann et al., 2004; Clapp et al., 2005; Ast et al., 2014). A few early studies have found no measurable FRET interactions between dye donors and QD acceptors (Clapp et al., 2005a; Ast et al., 2014). Indeed, the use of QDs as energy transfer acceptors, in particular with dye donors, is very challenging for two practical reasons (Clapp et al., 2005a): (i) Most if not all organic dyes and fluorescent proteins have short (~a few nanosecond) lifetimes, usually shorter than the PL decay of the QDs (10–30 ns). This makes using time-resolved FRET to isolate shortening of the dye lifetime along with lengthening of the acceptor lifetimes, often recorded in conventional FRET pairs, very difficult. (ii) The broad absorption profile of QDs combined with high extinction coefficients always result in high direct excitation of these nanocrystals regardless of the dye used in any donor–acceptor configuration. This also makes isolating the potential FRET sensitization in dye-QD (D-A) pair using steady state-fluorescence measurements very difficult.

Table 12.1 Representative bioluminescent donors combined with fluorescent protein or QD acceptors, along with their respective absorption and emission peak locations

Method	Donor	Substrate	Donor emission (nm)	Acceptor	Acceptor emission (nm)
BRET	RLuc	Coelenterazine	480	eYFP	530
BRET	RLuc	Deep Blue C™	395	GFP	510
eBRET	RLuc8	Deep Blue C™	395	GFP	510
BRET	Firefly	Luciferin	565	DsRed	583
QD-BRET	RLuc/RLuc8	Coelenterazine	480	QDot	605

Bioluminescent enzyme substrates (commonly found in certain beetles, bacteria, and marine species) can provide great energy donors to QDs, bypassing these obstacles via a process called BRET. This process involves the non-radiative transfer of excitation energy (induced by a natural photon-generating chemical or a biological reaction) from the donor to a ground state QD acceptor (Meighen, 1991; So et al., 2006; Haddock et al., 2010). When excitation of the chromophore is promoted by a chemical reaction, this process is called chemiluminescence resonance energy transfer (CRET) (Dodeigne et al., 2000; Sapsford et al., 2006). The principle advantage in BRET or CRET processes is that excitation of the biomolecule (donor) is not promoted optically, and thus it avoids several problems such as light scattering, high background noise, and in particular direct excitation of the QD acceptors. Both BRET and CRET are efficient for D-A separation distances of ~1–10 nm, similar to what has been demonstrated for FRET (Meighen, 1991; Branchini et al., 2011; Branchini et al., 2015; Dubinnyi et al., 2015). Several BRET configurations consisting of donor enzymes and acceptor proteins, along with examples using QD acceptors have been discussed in the literature (see Table 12.1) (Pfleger et al., 2006; Bacart et al., 2008).

12.2.3 AuNPs as Quenchers

Metallic nanostructures, such as those made of gold nanoparticles and gold nanorods have been employed as highly effective quenchers of the photoluminescence of QDs and dyes alike, though fluorescence enhancement of dye emission has been reported for well-defined separation distances and for poorly emitting fluorophores (Toshihiro and Shinji, 2005; Tovmachenko et al., 2006). This has been attributed to a nonradiative process that is more effective than the Förster dipole–dipole interaction mechanism. Instead, interactions with metal surfaces and nanostructures have been discussed within the framework of surface metal quenching; this is shown to extend over much larger separation distances beyond those allowed by FRET (reaching 20–30 nm) (Pons et al., 2007; Singh and Strouse, 2010; Breshike et al., 2013).

12.3 Sensing of Enzymatic Activity

A common nanosensor configuration employs a QD and/or AuNP decorated with several copies of a peptide substrate that can be specifically recognized by an enzyme of interest. The peptide is bound onto the nanocrystal surface via either metal-coordination or chemical coupling at one end, while the other lateral end is labeled with a dye. When assembled the QD–peptide–dye, or dye–peptide–AuNP conjugates exhibit strong fluorescence quenching of the QD or dye emission. Addition of the enzyme cleaves the substrate and induces changes in the fluorescence properties of the system that are dependent on the conjugate and enzyme concentrations (see schematics in Fig. 12.2). These changes in turn provide an indicator for the enzymatic activity. The transduction mechanisms utilized for reporting on the enzymatic activity via changes in the fluorescence signatures of the assemblies

(QD–peptide–dye and QD–peptide–AuNP and dye–peptide–AuNP conjugates) involve FRET and charge transfer interactions. Below, we will focus on a few examples using these configurations, and highlight their sensing mechanisms and advantages.

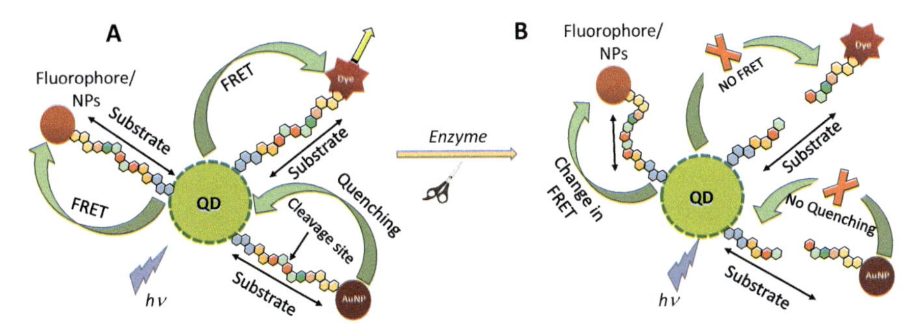

Figure 12.2 (A) Schematic depiction of a self-assembled nanosensor made of either a QD-peptide-dye or a QD-peptide-AuNP conjugate. Acceptor-labeled modular peptides containing a cleavage/recognition site are used. This configuration produces quenching of the QD emission. (B) Subsequent addition of a specific enzyme either can cleave the peptide or drastically alters its conformation, resulting in changes in the QD-to-quencher separation distance; this alters the FRET signature of the system.

12.3.1 QD-Based FRET Sensing

In this configuration, one can immobilize multiple dye-labeled substrates on a single nanocrystal, and thus maximize the energy transfer efficiency (see the above section). Once the QD–peptide–dye assemblies are formed, a quenching of the QD emission coupled with an enhancement of dye signal is measured, promoted by FRET interactions between the QD donor and proximal dye acceptors. Subsequent addition of a specific enzyme to the conjugate sample cleaves the peptide and displaces the dye away from the QD surfaces. This alters the donor–acceptor separation distance, leading to reduction in the FRET efficiency accompanied with a recovery of the QD emission (Medintz et al., 2006; Algar et al., 2012a; Algar et al., 2012b). As chemical modification of the terminal amino acid on the peptides is required for introducing a dye molecule, great care should be used when peptide modification is implemented, as such modification may reduce the substrate affinity to the target enzyme. Use of QDs as FRET platforms for detecting enzymatic activity has been reported by a few groups including ours (Medintz et al., 2006; Shi et al., 2006; Huang et al., 2008). We hereby discuss a few representative examples.

In an earlier report, our group has detailed the design of a few modular peptides each containing a hexahistidine (His_6) tag, a helix-linker spacer to provide rigidity, an exposed cleavage sequence for recognition, and a conjugated terminal dye acceptor (Medintz et al., 2006). These modular peptides can be easily self-assembled on the QDs via metal-His coordination, bringing the QD and dyes in close proximity and promoting efficient FRET interactions. One key advantage of this approach is that the specificity of the substrate toward a certain enzyme can be readily altered by modifying the sequence of the central recognition site, while maintaining the overall design strategy. The cleavage of the peptide upon the addition of a specific enzyme changes the FRET signature, which can then be used to gain information about the enzyme activity. Using this approach, we were able to investigate the activity of different proteases, including caspase-1 (associated with apoptosis), thrombin (associated with blood clotting), collagenase (related cancer metastasis), and chymotrypsin (a digestive enzyme) (Medintz et al., 2006).

In a follow-up study, a fluorescent protein (mCherry) was engineered to express a peptide substrate linker, and the new QD–peptide–mCherry construct was applied to detect the activities of caspase 3 in solution; dihydrolipoic acid (DHLA)-stabilized QDs were used for assembling the peptide–mCherry conjugates (Boeneman et al., 2009). Here, the mCherry plasmid was modified to express a caspase 3 cleavage site (DEVD sequence) along with the terminal polyhistidine tag for metal-histidine coordination on the DHLA-QDs. Figure 12.3 shows that mixing of the peptide–mCherry with DHLA-QDs brought the QDs and mCherry in close proximity, resulting in efficient loss in QD PL. We found that introducing higher numbers of peptide–mCherry improved the FRET efficiency, manifesting in valence-dependent PL loss with increasing mCherry concentration (Boeneman et al., 2009). To study the proteolytic activity of the enzyme, we chose one ratio of peptide–mCherry-to-QD and monitored changes in the PL intensity as a function of changes in the enzyme concentration. We measured large and progressive change in the FRET efficiency upon the addition of caspase 3 enzyme, proving that effective cleavage of the substrate peptide has in fact taken place.

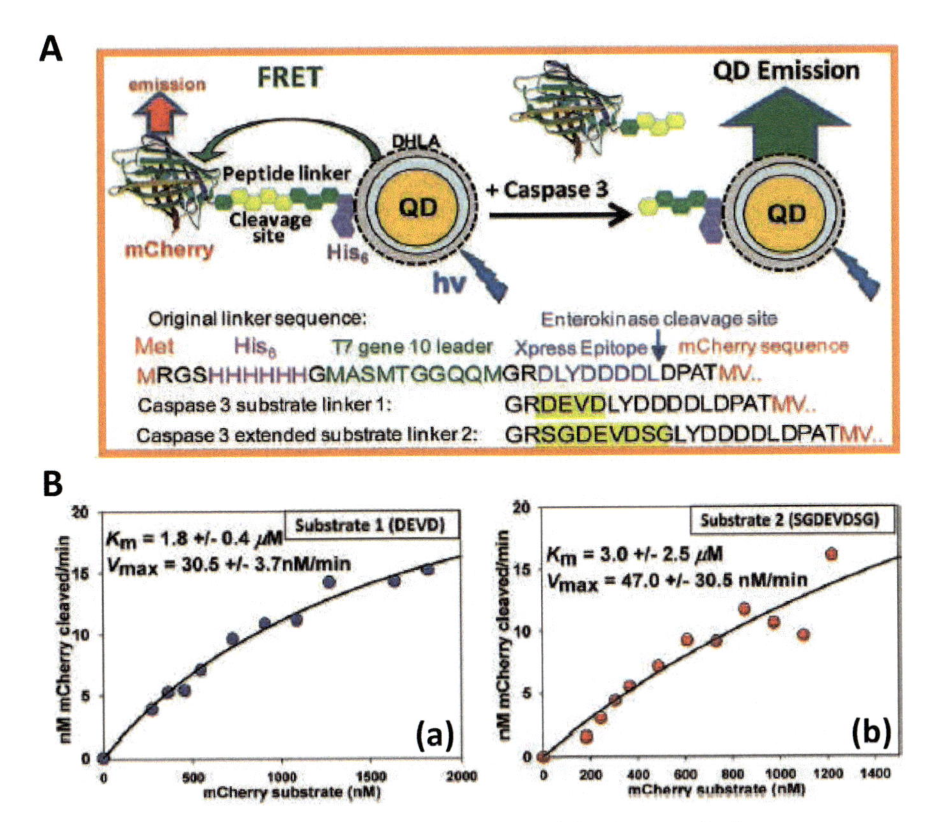

Figure 12.3 (Top) Schematic representation of the QD-peptide-fluorescent-protein sensor: (A) mCherry protein fused onto a caspase 3 cleavage sequence and an N-terminal His_n sequence was self-assembled onto the surface of QDs, resulting in FRET-induced quenching of the QD emission. Addition of the enzyme cleaves the peptide and alters the FRET quenching. (B) Shown are the peptide sequences with the enzyme substrate region highlighted in yellow. (Bottom) Plot of the proteolytic velocity vs. substrate concentration: (A) for a substrate with short linker 1 and (B) for a substrate with an extended linker 2 (using mCherry/QD ratios of 3.5 and 1.5, respectively). The estimated K_M and V_{max} values of the two configurations are also shown. Reproduced from Boenemann et al. (2009), with permission from the American Chemical Society.

Furthermore, we measured comparable changes in the FRET signature when either substrate 1 (without linker) or substrate 2 (with a linker) were used, which confirmed that the substrate sequence was accessible to the enzymes. To probe the kinetics, we fixed the concentration of enzyme at 400 pM but varied the substrate peptide concentration from 200 nM to 2 μM. Changes in the emission ratio of mCherry-to-QD were measured, and then converted to enzymatic velocity Fig. 12.3, bottom panels. The Michaelis constants K_M and maximal velocities V_{max} of the two substrates were then derived using Michaelis–Menten kinetic analysis. In control experiments using peptide–mCherry without the DEVD sequence we measured no activity toward the enzyme. Additionally, this sensor exhibited high sensitivity thus requiring only low concentration of caspase 3 (~20 pM) to be used.

In another example, Maysinger and coworkers reported the use of QD–peptide–rhodamine conjugate assemblies to measure the activity of caspase-1, an enzyme associated with the manifestation of inflammation (Moquin et al., 2013). A rhodamine-labeled peptide containing a caspase 1 cleavable sequence was attached to the QD surface via the thiol group present on the terminal amino acid. This nanosensor was successfully applied to detect the activity of caspase 1 both in vitro and in vivo. For in vitro test, microglia cells were first treated with lipopolysaccharide (LPS). The cell lysates were then exposed to the QD–peptide–rhodamine nanosensor and the ratio of the QD and dye emissions ($I_{QD}/I_{QD\text{-}RCP}$) was measured and used to extract an estimate for the rate of proteolytic cleavage. They found that the cleavage rate consistently increased with incubation time when the cells were pre-treated with LPS. This observation indicates the presence of sufficient amount of caspase 1 that cleaved the bound dye–peptide on the QDs, releasing rhodamine away from the nanoparticle surfaces. In comparison, control experiments using untreated cells or cells treated with DHLA-capped QDs (a negative control for caspase-1 activation) showed much lower enzymatic activity. The authors also demonstrated that this QD-based FRET pair provided higher sensitivity compared to the colorimetric assay (micromolar versus nanomolar concentration of the substrates). For the in vivo test, similar experiments were done by injecting LPS into the brain parenchyma in hairless non-immunosuppressed mice SKH-1; a non-treated animal was used as a control. The effectiveness of the nanosensor (QD-RCP) was measured by the real-time fluorescence imaging of the animals and the emission intensity of the QDs at 530 nm was tracked over time. They found that the QD emission intensity increased rapidly after the injection of the nanosensor, indicating efficient enzymatic activity that cleaved the substrate peptide and released the dye from QD surfaces. The intensity reached a plateau at 2 h for LPS treated animal followed by continuous reduction of emission intensity until it returned to the baseline values after 4 h. The authors attributed the decrease in emission signal to the clearance of QDs from the system. These findings prove the promise of potential use of the QD-nanosensors to assess the enzymatic activity in vivo.

In a recent study, Wang and coworkers detailed the design of a QD-based FRET sensor targeting the enzyme matrix metalloproteinase (MT1-MMP) in solution and on the membrane of live cells (Chung et al., 2015). One of the key features of this sensor is the careful design of the peptide substrate. They developed a modular and multifunctional peptide sequence containing an AHLR sequence, cleavable by MT1-MMP, inserted between a cationic sequence and an anionic sequence. They also inserted a C-terminus Cy3 dye and an N-terminus His6 tag for self-assembly on CdSe-ZnS QDs, along with an RGD sequence for promoting membrane binding and intracellular uptake. The peptide naturally forms a bent structure that is weakly stabilized by electrostatic attractions between the cationic and anionic sequences (Fig. 12.4A). Following self-assembly on the QD surface via metal-histidine binding, the bent sequence imposes a configuration where the Cy3 is brought close to the QD surface, resulting in high FRET quenching of the QDs, referred to as weak FRET ratio and defined by QD-PL/Cy3-PL. This construct was then tested for its ability to sense the activity of the active catalytic domain of MT1-MMP (i.e., MT1-CAT). They found that incubating the QD–peptide–Cy3 conjugates with MT1-CAT indeed resulted in large changes in the FRET ratio (by a factor as high as 5). This change was attributed to cleavage of the AHLR sequence, which separated the two charged domains, displacing the Cy3 away from the QD surface. The measured changes were found to strongly track the enzyme concentration and the incubation time. They also tested this sensor in MDA-MB-231 cells (from a human cancer cell line) that express high levels of MT1-MMP on their membranes.

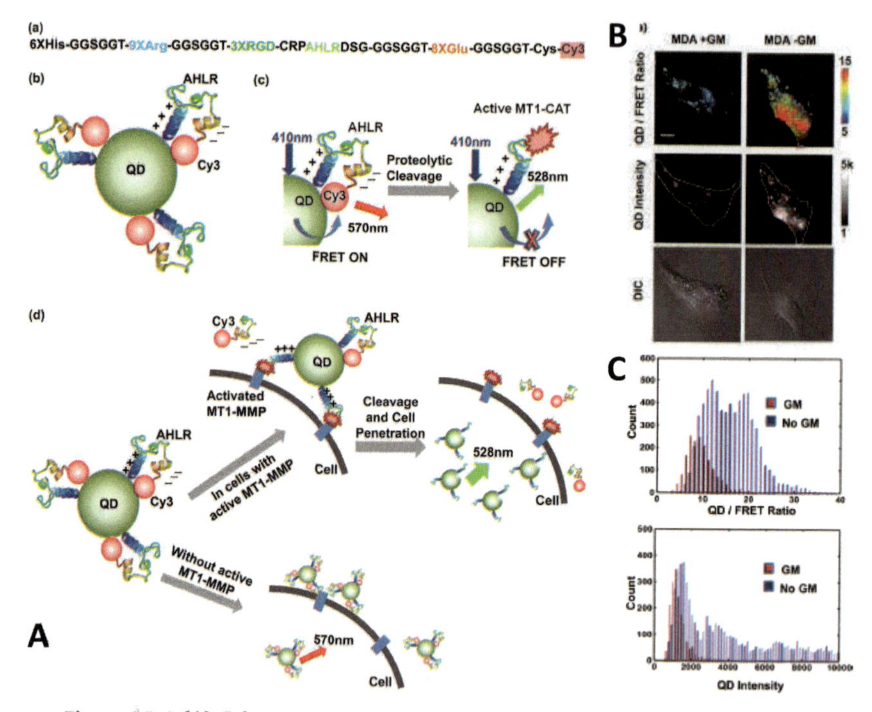

Figure 12.4 (A) Schematic representation of the QD-based nanosensor for the detection of MT1-MMP activity using FRET. (a) The structure of Cy3-dye-labeled peptide substrate. (b) Formation of the hairpin conformation once attached onto the QD surfaces, due to electrostatic attraction between the cation and anionic segments, leading to FRET between QDs and dye. (c) Nanosensor exposed to MT1-CAT in solution; (d) Nanosensor incubated with cancer cells that express active MT1-MMP on the membrane. When incubated with the cell culture, the sensor first bound to the cell membrane, followed by cleavage of the peptide by MT1-MMP, triggering cellular entry of the QD-conjugate via the exposed positively charged arginine-rich sequence on the nanocrystal. (B) Representative fluorescence (QD/FRET ratio and QD intensity) and DIC images of MDA cells following incubation with the nanosensor with (left) and without (right) GM (MT1-MMP inhibitor) pretreatment. (C) The quantification of the distribution of QD/FRET ratio (top) and QD intensity (bottom) within the detected nanosensor clusters of breast cancer cell at the level of individual pixels. Reproduced from Chung et al. (2015), with permission from the American Chemical Society.

The cells were incubated with the QD-conjugates for 3 h. Unbound QDs were removed by washing, and then the cells were imaged using fluorescence microscopy. They measured an increase in the FRET ratio combined with a more pronounced endocytosis of the QDs, a process attributed to the specific binding of the QD–peptide sensor to the membrane-bound integrin. This association promoted cleavage of the peptide bound on the QD by the active MT1-MMP. This cleavage exposed the arginine-rich domain, promoting intracellular uptake of the QD-conjugates with high QD emission (dye-free system). In control experiments they showed that when an MT1-MMP-specific inhibitor was added to the culture, only marginal changes in the FRET ratio combined with minimal intracellular uptake could be measured. These results combined confirm that the sensor specifically interacts with the membrane-bound enzyme (Chung et al., 2015).

Willner and coworkers have reported on the sensing of casein kinase (CK2) and alkaline phosphatase (ALP), using QDs combined with either FRET or CT interactions for optical transduction (Fig. 12.5) (Freeman et al., 2010). In the scheme for detecting CK2, the hydrophilic QDs were decorated with serine-containing peptide. In the presence of ATP, the casein kinase CK2 induced the catalysis of the serine unit and triggered phosphorylation of the bound peptide. The resulting phosphorylated peptides in the QD-conjugates were then mixed with a dye-labeled ATP-antibody, which promoted the binding to the QD-peptide conjugates. The proximity of the dyes with QDs resulted in FRET interactions, thereby providing the optical readout for time- and concentration-dependent phosphorylation via CK2 catalysis. Furthermore, the authors showed that co-incubation of the QD–peptide conjugates with CK2 and dye-labeled ATP, instead of separating the catalysis from the dye-labeling, produced similar changes in the PL composite spectra and yielding similar detection sensitivity. Conversely, using a slightly different rationale and replacing the serine-containing peptide with a phosphorylated peptide, the QD-bioconjugates can be applied for sensing the activity of ALP. When exposed to ALP and then to tyrosinase, the phosphorylated peptide underwent two transformations: (1) hydrolytical cleavage by the ALP to yield tyrosine, and (2) catalytic oxidation of the resulting tyrosine by tyrosinase to produce oxidized dopamine. The latter can strongly quench the QD PL via a complex charge transfer interactions between photoexcited QDs and dopamine. Tracking the time- and ALP-concentration-dependent changes in the QD PL provided a means for analyzing the ALP activity.

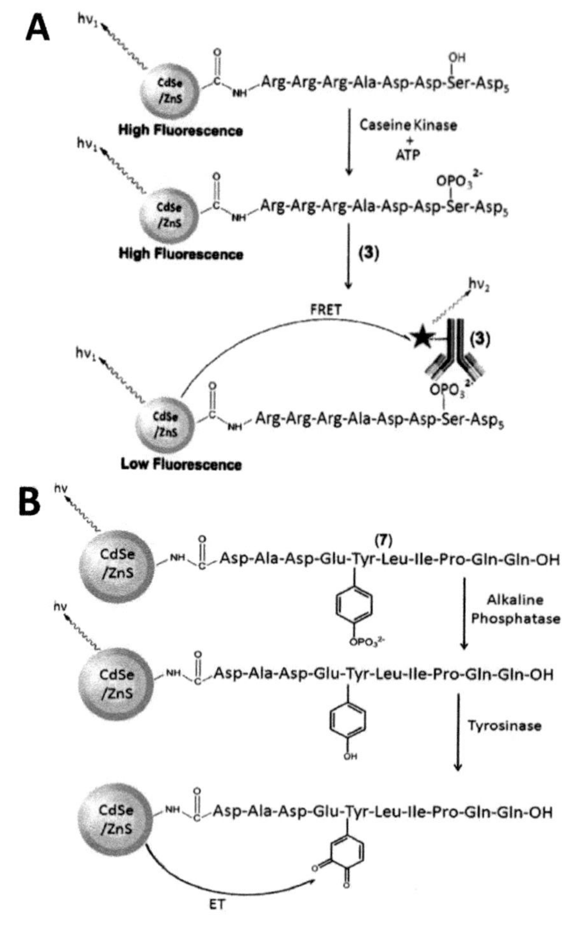

Figure 12.5 Schematic illustration of the sensor design: (A) Detection of the presence of CK2; CK2 catalyzes the phosphorylation of serine which facilitates the assembly of Atto-590-modified anti-phosphoserine-antibody onto the QDs. Successful binding brings the QD donor and Atto-590 dye acceptor in close proximity and induces efficient FRET interactions; the latter produces time- and CK2-dependent decrease in QD emission and enhancement in Atto-590 signal. (B) Sensing of the hydrolytic activity of ALP using QD conjugated to a phosphorylated peptide. The presence of ALP hydrolyzes the phosphoester in the peptide, yielding a tyrosine functional group, which can be further oxidized to dopaquinone by co-added tyrosinase. The resulting quinone present on the QD surfaces quenches the luminescence of QDs via electron-transfer/charge-transfer interactions, manifesting in a decrease in the QD PL intensity. Reproduced from Freeman et al. (2010), with permission from the American Chemical Society.

12.3.2 Sensing of Enzymatic Activity Using QD-Based BRET

QDs serve as ideal BRET acceptor compared to dye and fluorescent proteins. QDs have broad absorption cross-section and they can be excited by almost all bioluminescent protein donors. In addition, the large Stokes shift exhibited by QDs allows for a good spectral separation of donor and acceptor emissions, simplifying the data analysis. A few QD-BRET systems have been explored as effective sensors to detect the enzymatic activities in several reports.

Rao and coworkers were first to report the development of a BRET system using QDs as acceptors (So et al., 2006; Zhang et al., 2006; Yao et al., 2007). They then applied this approach to probe the proteolytic activity of various matrix metalloproteinase enzymes. In one example, they genetically fused a short peptide containing a substrate for the matrix metalloproteinases 2 (MMP-2) and a histidine tag (His_6) to a mutant of Renilla luciferase (Yao et al., 2007). They then applied Ni^{2+} to mediate the assembly of luciferase on the QDs driven by the cooperative coordination between Ni^{2+} and His_6 tag on the peptide with the carboxyl group on the QD surfaces. When exposed to coelenterazine BRET interactions took place, resulting in the pronounced emission from the QD acceptor. Then, cleavage of the peptide sequence by MMP-2 disrupted the conjugates, releasing QDs away from the luciferase and resulting in reduced sensitization of the QDs. The ratio between the QD PL (at 655 nm) to luciferase emission (at 480 nm) decreased from 1.62 (initial value at t = 0 min.) to 0.62 (after 30 min). This BRET-based nanosensor was able to sense the enzyme with concentration of a few nanograms per milliliter of buffer solvent.

The same group expanded the above QD-based BRET design and constructed a QD nanosensor specific for the detection of MMP-7 activity (Xia et al., 2008). They first functionalized the carboxyl QDs with adipic dihydrazine (via EDC coupling), then the nanocrystals were further coupled to a recombinant protein presenting MMP-7 substrate sequence and luciferase 8 (Luc 8) to provide the MMP-7 sensor as shown in Fig. 12.6. They then demonstrated the ability of this sensor to detect the activity of the protease both in a buffer medium and in serum by tracking the BRET ratio before and after the addition of MMP-7 (see

Fig. 12.6B). By varying the peptide sequence, they showed that such QD–peptide conjugate can be used to measure the proteolytic activity of other enzymes, including MMP-2 and urokinase-type plasminogen activator (uPA) with detection limit of 1.0 ng/ml. They also extended this approach to implement the multiplexed detection of a few distinct enzymes in the same sample. For example, they used two sets of QDs coupled to two different peptide substrates to prepare conjugates with distinct emissions, QD705-uPA-Luc8 and QD655-MMP-2-Luc8. Following mixing of the two conjugates in the same dispersion, they found that addition of the two proteases (MMP-2 and uPA) produced a decrease in the emission of both QDs, indicating that the proteases specifically recognized and cleaved both conjugate substrates.

In addition to this one-step process (one-donor–one-acceptor interactions), researchers have recently explored the combination of BRET and FRET (multistep) interactions with the same construct/system. Although no enzymatic activity was involved in this study the use of QDs or QRs (quantum rods) in a relay combining bioluminescence energy transfer (BRET) and FRET is interesting, nonetheless (Carriba et al., 2008). In one particular example, Maye and coworkers used QDs or QRs as inorganic linkers which simultaneously function as energy acceptors (sensitized via BRET interactions) and FRET donors to sensitize bound red fluorescent protein (RFP) acceptors; no photoexcitation was involved in this design (Alam et al., 2013). A firefly protein variant of *Photinus pyralis* (*Ppy*) and RFP emitting at 590 nm were self-assembled on the QD (or the QR) via metal-His conjugation. *Photinus pyralis* can emit light either at 520 nm (when it is combined with a benzothio-phene analogue to the firefly luciferin, $BtLH_2$, as well as ATP and O_2) or at 547 nm (when it is combined with a standard luciferin, LH_2). Using one variant or the other, the authors were able to vary the overlap integral between luciferin donor and nanocrystal acceptor. Varying degrees of spectral overlap between the QD/QR emission and RFP excitation further induced controlled rates of energy transfer from the nanocrystals to the red fluorescent protein. Minimum-to-no energy transfer from luciferase to QDs (λ_{em} = 530 nm) was observed when luciferase was treated with standard luciferin, LH_2, a result attributed to the drastically reduced spectral overlap between the protein emission and absorption of the QDs. However, the enzyme combined with standard luciferin (LH_2), with its red shifted emission, did not promote BRET with green QDs, but direct sensitization (via BRET) of the tagRFP assembled on the QD surface. Here, the weak overlap between Ppy resulted in weak interactions with the QDs, while larger spectral overlap with the PL of the tagRFP allowed direct BRET to the protein acceptor.

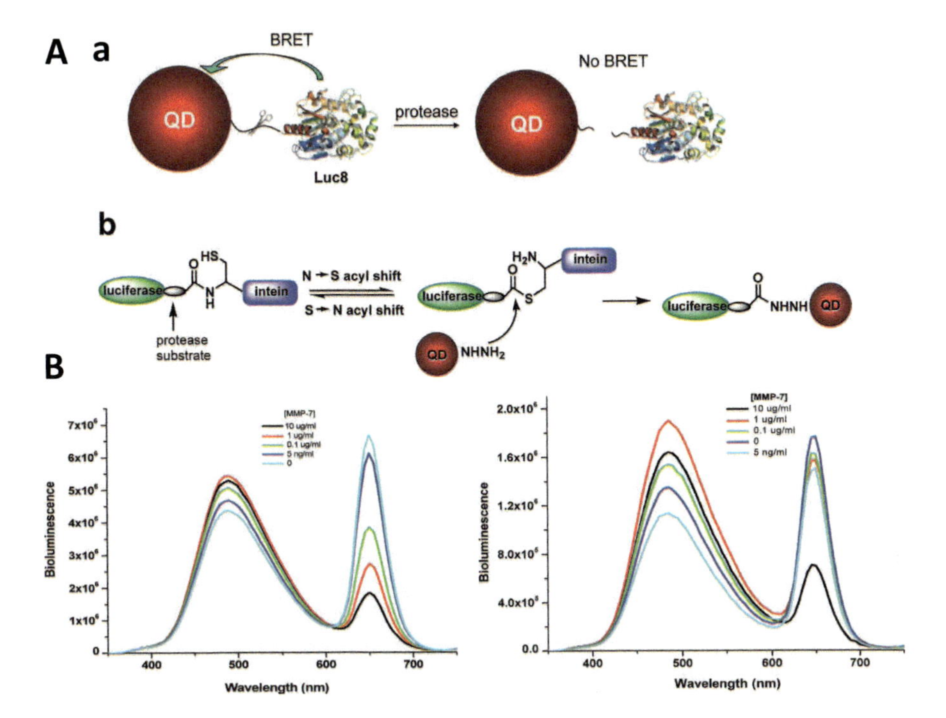

Figure 12.6 (A) (a) Schematic representation of a BRET-based biosensor designed for the detection of the proteolytic activity of MMPs. A BRET donor, luciferase proteins (Luc8), is conjugated onto the BRET acceptor, QDs, through a peptide linker that can be modified to be either MMP-2 or MMP-7 substrate; (b) the site-specific bioconjugation strategy used to couple Luc8 fusion proteins with QDs mediated by intein. (B) Shows the emission spectra of several dispersions of biosensor QD655-MMP-7-Luc8 in the presence of MMP-7 with varying concentrations. The assay was carried out both in buffer (left) and in mouse serum (right). Reproduced from Xia et al. (2008), with permission from the American Chemical Society.

12.3.3 Sensing of Enzymatic Activity Using AuNP Quenchers

In this format the AuNPs act as strong fluorescence quenchers coupled to QD (or dye) donors (Saha et al., 2012). The specificity of the nanosensor is provided by the use of a substrate peptide self-assembled on the AuNPs surface. We will describe two representative examples where the quenching ability of AuNPs has been utilized to probe the enzymatic activity in vitro and in cells.

In the first example, Park and coworkers designed an AuNP-conjugate nanoprobe to sense the heparinase protease, an indicative of metastatic stage of cancer cells (Lee et al., 2010). This design was accomplished by conjugating a dye-labeled heparin to the AuNP surfaces via the thiol group on the peptide terminal. The formation of Au-heparin-dye brought the AuNPs and dye in close proximity resulting in quenching of dye emission. Following assembly the sensor was tested in cell culture by incubating the conjugates with different types of metastatic cancer cell lines such as Hela cells and MCF-7 cells; these are cells that are known to over-express heparinase, as well as noncancerous cells (NIH3T3). They found that after 4 h of incubation, recovery of the dye emission (at 651 nm) tracked the activity of heparinase secreted from cancerous cells. The relative recovery of the dye fluorescence was dependent on the extent of heparinase expression. They also reported a high dye PL recovery using Hela cells (5.5 fold higher) than noncancerous NIH3T3 cells. This behavior is consistent with the fact that HeLa cells have the highest metastatic activity among those three cell lines.

In the second example, Stevens and coworkers reported the assembly of QD–peptide conjugates by substituting the organic dye with QDs (as donors), prepared QD–peptide–AuNP bioconjugates, and used them to sense the activity of the uPA protease (Lowe et al., 2012). They assembled AuNPs (1.4 nm size) and 525 nm QDs via a selective streptavidin-biotin interaction mediated by an uPA substrate inserted between the two nanoparticles. This platform was used in combination with another QD655-dye FRET sensor with high specificity towards Her2 kinase to simultaneously detect the activity of uPA protease and Her2 kinase in the same sample (Fig. 12.7).

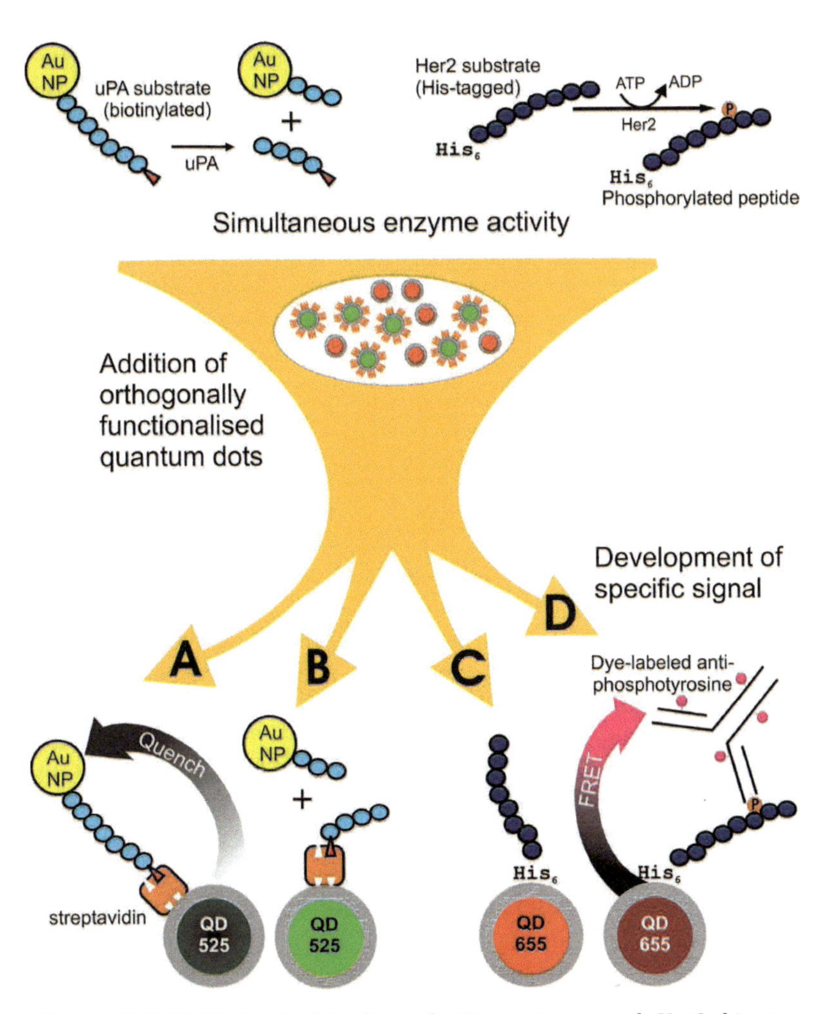

Figure 12.7 Multiplexed detection of uPA protease and Her2 kinase using two distinct sensors containing two different peptide substrates simultaneously incubated with a mixture of uPA and Her2 enzymes. (A) Quenching of the QD emission (λ_{em} = 525 nm) promoted by AuNPs due to self-assembly of the uPA substrates on the QD via biotin-streptavidin interaction. (B) Recovery of QD emission (λ_{em} = 525 nm) upon cleavage of the substrate in the presence of uPA protease. (C) Representation of QDs (λ_{em} = 655 nm) conjugated with Her2 substrate via metal-histidine coordination. (D) FRET interaction between the QD (λ_{em} = 655 nm) and AF660 dye, promoted by binding of AF660-labeled anti-phosphotyrosine antibody on the nanocrystal following phosphorylation of the substrate peptide. Reproduced from Lowe et al. (2012), with permission from the American Chemical Society.

The emission wavelength of QDs and dye were carefully chosen so that the fluorescence signals were well separated. They found that the two sensors functioned independently to detect the uPA protease, by monitoring the emission increase at 525 nm, and the Her2 kinase by measuring the dye-to-QD emission ratio. A detection limit of 50 ng/ml for uPA and 7.5 nM for Her2 kinase were achieved using these conjugates. The use of AuNPs simplified the data analysis and provided well-resolved fluorescence spectra, because the AuNPs function only as quenchers of QD fluorescence.

12.4 Conclusion

We have summarized a few recent representative studies to probe the activity of a few specific enzymes using luminescent QDs (or QRs) and AuNPs combined with energy transfer interactions. We discussed modalities where the transduction mechanism driving the sensor function involves (1) fluorescence resonance energy transfer (promoted by the Förster dipole–dipole interactions) between QD donors and dye or fluorescent protein acceptors (FRET), (2) bioluminescent resonance energy transfer between luminescent biological and chemical substrates and QDs acceptors, and (3) energy transfer quenching of QDs and dye emission by AuNPs. The sensing constructs share a common feature where a peptide containing an amino acid sequence that is recognized and cleaved by the target enzyme is sandwiched between the NP platform and a dye or fluorescent protein; the peptide acts as a bridge between the donor and acceptor/quencher to promote energy transfer interactions. The enzyme activity is subsequently deduced from monitoring changes in the energy transfer efficiency upon the addition of a specific enzyme to the sample and digestion of the peptide substrate. QDs offer a few promising advantages for such sensor configurations, including the ability to enhance the FRET interactions between QDs and bound dye–peptide by either adjusting the valence of the QD–peptide–dye conjugates, or optimizing the spectral overlap. QDs are also highly suitable BRET acceptors due to their high brightness and broad absorption profiles. They can further act as center relay to combine BRET and FRET interactions in a single system. Conversely, AuNPs provide high quenching capacity of dye and QDs alike. The use of the QD- and AuNP-based sensing of enzymatic activity has been shown to offer high sensitivity (pico- to nano-molar concentration) and low background signal. Additionally, real-time fluorescence imaging of animals for in vivo enzyme detection has been achieved.

In view of the significant progress that has been made in the biosensor design, the future direction of nanosensor development targeting enzyme activity using QDs could be centered on a few important aspects: (i) the fabrication of biocompatible nanocrystals with multifunctional surface for immobilizing several peptides substrates; (ii) develop effective means for delivering QD-conjugates into cellular compartments; and (iii) implement the use of QD sensors in vivo where they can be combined with improved imaging resolution to allow spatial localization and sensing of enzyme activity in real time. These developments could certainly exploit the high brightness and photo-stability of QDs.

References

Achermann M., Petruska M. A., Kos S., Smith D. L., Koleske D. D., Klimov V. I., *Nature,* **429** (2004), 642–646.

Alam R., Zylstra J., Fontaine D. M., Branchini B. R., Maye M. M., *Nanoscale,* **5**(12) (2013), 5303–5306.

Algar W. R., Ancona M. G., Malanoski A. P., Susumu K., Medintz I. L., *Nano Lett.,* **12**(7) (2012a), 3793–3802.

Algar W. R., Malonoski A., Deschamps J. R., Blanco-Canosa J. B., Susumu K., Stewart M. H., Johnson B. J., Dawson P. E., Medintz I. L., *ACS Nano,* **6**(12), (2012b), 11044–11058.

Ast S., Rutledge P. J., Todd M. H., *Phys. Chem. Chem. Phys.,* **16**(46) (2014), 25255–25257.

Bacart J., Corbel C., Jockers R., Bach S., Couturier C., *Biotechnol. J.,* **3**(3) (2008), 311–324.

Barrett A. J., Woessner J. F., Rawlings N. D., *Handbook of Proteolytic Enzymes.* Elsevier (2012), vol. 1.

Boeneman K., Mei B. C., Dennis A. M., Bao G., Deschamps J. R., Mattoussi H., Medintz I. L., *J. Am. Chem. Soc.,* **131**(11) (2009), 3828–3829.

Branchini B. R., Behney C. E., Southworth T. L., Fontaine D. M., Gulick A. M., Vinyard D. J., Brudvig G. W., *J. Am. Chem. Soc.,* **137**(24) (2015), 7592–7595.

Branchini B. R., Rosenberg J. C., Fontaine D. M., Southworth T. L., Behney C. E., Uzasci L., *J. Am. Chem. Soc.*, **133**(29) (2011), 11088–11091.

Breshike C. J., Riskowski R. A., Strouse G. F., *J. Phys. Chem. C*, **117**(45) (2013), 23942–23949.

Carriba P., Navarro G., Ciruela F., Ferre S., Casado V., Agnati L., Cortes A., Mallol J., Fuxe K., Canela E. I., Lluis C., Franco R., *Nat. Methods*, **5**(8) (2008), 727–733.

Cassette E., Helle M., Bezdetnaya L., Marchal F., Dubertret B., Pons T., *Adv. Drug Deliv. Rev.*, **65**(5) (2013), 719–731.

Charbonnière L. J., Hildebrandt N., *Eur. J. Inorg. Chem.*, **2008**(21) (2008), 3231–3231.

Chudakov D. M., Lukyanov S., Lukyanov K. A., *Trends Biotechnol.*, **23**(12) (2005), 605–613.

Chung E. Y., Ochs C. J., Wang Y., Lei L., Qin Q., Smith A. M., Strongin A. Y., Kamm R., Qi Y.-X., Lu S., Wang Y., *Nano Lett.*, **15**(8) (2015), 5025–5032.

Clapp A. R., Medintz I. L., Fisher B. R., Anderson G. P., Mattoussi H., *J. Am. Chem. Soc.*, **127**(4) (2005), 1242–1250.

Clapp A. R., Medintz I. L., Mauro J. M., Fisher B. R., Bawendi M. G., Mattoussi H., *J. Am. Chem. Soc.*, **126**(1) (2004), 301–310.

Clapp A. R., Medintz I. L., Uyeda H. T., Fisher B. R., Goldman E. R., Bawendi M. G., Mattoussi H., *J. Am. Chem. Soc.*, **127**(51) (2005b), 18212–18221.

Dabbousi B. O., Rodriguez Viejo J., Mikulec F. V., Heine J. R., Mattoussi H., Ober R., Jensen K. F., Bawendi M. G., *J. Phys. Chem. B*, **101**(46) (1997), 9463–9475.

Dodeigne C., Thunus L., Lejeune R., *Talanta*, **51**(3) (2000), 415–439.

Drag M., Salvesen G. S., *Nat. Rev. Drug Discov.*, **9**(9) (2010), 690–701.

Dubinnyi M. A., Kaskova Z. M., Rodionova N. S., Baranov M. S., Gorokhovatsky A. Y., Kotlobay A., Solntsev K. M., Tsarkova A. S., Petushkov V. N., Yampolsky I. V., *Angew. Chem. Int. Ed.*, **127**(24) (2015), 7171–7173.

Dulkeith E., Ringler M., Klar T. A., Feldmann J., Muñoz Javier A., Parak W. J., *Nano Lett.*, **5**(4) (2005), 585–589.

Förster T., *Ann. Phys.*, **437**(1–2) (1948), 55–75.

Freeman R., Finder T., Gill R., Willner I., *Nano Lett.*, **10**(6) (2010), 2192–2196.

Geißler D., Linden S., Liermann K., Wegner K. D., Charbonniere L. J., Hildebrandt N., *Inorg Chem.*, **53**(4) (2013), 1824–1838.

Giepmans B. N. G., Adams S. R., Ellisman M. H., Tsien R. Y., *Science,* **312**(5771) (2006), 217–224.

Haddock S. H. D., Moline M. A., Case J. F., *Annu. Rev. Mater. Sci.,* **2** (2010), 443–493.

Huang S., Xiao Q., He Z. K., Liu Y., Tinnefeld P., Su X. R., Peng X. N., *Chem. Commun.,* **45** (2008), 5990–5992.

Jana N. R., Gearheart L., Murphy C. J., *J. Phys. Chem. B,* **105**(19) (2001), 4065–4067.

Kohl N. E., Emini E. A., Schleif W. A., Davis L. J., Heimbach J. C., Dixon R. A., Scolnick E. M., Sigal I. S., *Proc. Natl. Acad. Sci.,* **85**(13) (1988), 4686–4690.

Kohl T., Heinze K. G., Kuhlemann R., Koltermann A., Schwille P., *Proc. Natl. Acad. Sci.,* **99**(19) (2002), 12161–12166.

Lakowicz J. R., *Principles of Fluorescence Spectroscopy* (2006), New York, Springer.

Lee K., Lee H., Bae K. H., Park T. G., *Biomaterials,* **31**(25) (2010), 6530–6536.

Link S., El-Sayed M. A., *Int. Rev. Phys. Chem.,* **19**(3) (2000) 409–453.

Lippincott-Schwartz J., Patterson G. H., *Science,* **300**(5616) (2003) 87–91.

Liu Z., Yang F., Robotham J. M., Tang H., *J. Virol.,* **83**(13) (2009), 6554–6565.

Liz-Marzán L. M., *Langmuir,* **22**(1) (2006), 32–41.

López-Otín C., Matrisian L. M., *Nat. Rev. Cancer,* **7**(10) (2007), 800–808.

Lowe S. B., Dick J. A. G., Cohen B. E., Stevens M., *Acs Nano,* **6**(1) (2012), 851–857.

Lu H., Schöps O., Woggon U., Niemeyer C. M., *J. Am. Chem. Soc.,* **130**(14) (2008), 4815–4827.

Manning G., Whyte D. B., Martinez R., Hunter T., Sudarsanam S., *Science,* **298**(5600) (2002), 1912–1934.

Mattoussi H., Palui G., Na H. B., *Adv. Drug Deliv. Rev.,* **64**(2) (2012), 138–166.

Medintz I., Clapp A., Brunel F., Tiefenbrunn T., Tetsuo Uyeda H., Chang E., Deschamps J., Dawson P., Mattoussi H., *Nat. Mater.,* **5**(7) (2006), 581–589.

Medintz I. L., Clapp A. R., Mattoussi H., Goldman E. R., Fisher B., Mauro J. M., *Nat. Mater.,* **2**(9) (2003), 630–638.

Medintz I. L., Mattoussi H., *Phys. Chem. Chem. Phys.,* **11**(1) (2009), 17–45.

Medintz I. L., Uyeda H. T., Goldman E. R., Mattoussi H., *Nat. Mater.,* **4** (2005), 435–446.

Meighen E. A., *Microbiol. Rev.,* **55**(1) (1991), 123–142.

Moquin A., Hutter E., Choi A. O., Khatchadourian A., Castonguay A., Winnik F. M., Maysinger D., *ACS Nano,* **7**(11) (2013), 9585–9598.

Murray C. B., Norris D. J., Bawendi M. G., *J. Am. Chem. Soc.,* **115**(19) (1993), 8706–8715.

Neurath H., Walsh K. A. *Proc. Natl. Acad. Sci.,* **73**(11) (1976), 3825–3832.

Palui G., Aldeek F., Wang W. T., Mattoussi H., *Chem. Soc. Rev.,* **44**(1) (2015), 193–227.

Pfleger K. D. G., Seeber R. M., Eidne K. A., *Nat. Protocols,* **1**(1) (2006), 337–345.

Pons T., Medintz I. L., Sapsford K. E., Higashiya S., Grimes A. F., English D. S., Mattoussi H., *Nano Lett.,* **7**(10) (2007), 3157–3164.

Puente X. S., Sanchez L. M., Overall C. M., Lopez-Otin C., *Nat. Rev. Genet.,* **4**(7) (2003), 544–558.

Resch-Genger U., Grabolle M., Cavaliere-Jaricot S., Nitschke R., Nann T., *Nat. Methods,* **5**(9) (2008), 763–775.

Rizzo M. A., Springer G. H., Granada B., Piston D. W., *Nat. Biotechol.,* **22**(4) (2004), 445–449.

Rodems S. M., Hamman B. D., Lin C., Zhao J., Shah S., Heidary D., Makings L., Stack J. H., Pollok B. A., *Assay Drug Dev. Technol.,* **1**(1) (2002), 9–19.

Saha K., Agasti S. S., Kim C., Li X. N., Rotello V. M., *Chem. Rev.,* **112**(5) (2012), 2739–2779.

Sapsford K. E., Berti L., Medintz I. L., *Angew. Chem. Int. Ed.,* **45**(28) (2006), 4562–4589.

Selvin P. R., *Nat. Struct. Mol. Biol.,* **7**(9) (2000), 730–734.

Shaner N. C., Steinbach P. A., Tsien R. Y., *Nat. Methods,* **2**(12) (2005), 905–909.

Shi L., De Paoli V., Rosenzweig N., Rosenzweig Z., *J. Am. Chem. Soc.,* **128**(32) (2006), 10378–10379.

Silvi S., Credi A., *Chem. Soc. Rev.,* **44**(13) (2015), 4275–4289.

Singh M. P., Strouse G. F., *J. Am. Chem. Soc.,* **132**(27) (2010), 9383–9391.

Snee P. T., Tyrakowski C. M., Page L. E., Isovic A., Jawaid A. M., *J. Phys. Chem. C,* **115**(40) (2011), 19578–19582.

So M.-K., Xu C., Loening A. M., Gambhir S. S., Rao J., *Nat. Biotechnol.,* **24**(3) (2006), 339–343.

Somers R. C., Bawendi M. G., Nocera D. G., *Chem. Soc. Rev.,* **36**(4) (2007), 579–591.

Sustarsic M., Kapanidis A. N., *Curr. Opin. Struct. Biol.,* **34** (2015), 52–59.

Talapin D. V., LeeJ. S., Kovalenko M. V., Shevchenko E. V., *Chem. Rev.,* **110**(1) (2010), 389–458.

Toshihiro N., Shinji H., *Jpn. J. Appl. Phys.,* **44**(9R) (2005), 6833.

Tovmachenko O. G., Graf C., van den Heuvel D. J., van Blaaderen A., Gerritsen H. C., *Adv. Mater.,* **18**(1) (2006), 91–95.

Turk B., *Nat. Rev. Drug. Discov.,* **5**(9) (2006), 785–799.

Valeur B., Berberan-Santos M. N., *Molecular Fluorescence: Principles and Applications* (2012), John Wiley & Sons.

Weiss S., *Nat. Struc. Mol. Biol.,* **7**(9) (2000), 724–729.

White A., Handler P., Smith E., Stetten Jr D., *Principles of Biochemistry* (1959), 2nd ed.

Wu Y., Wang X., Liu X., Wang Y., *Genome Res.,* **13**(4) (2003), 601–616.

Xia Z., Xing Y., So M.-K., Koh A. L., Sinclair R., Rao J., *Anal. Chem.,* **80**(22) (2008), 8649–8655.

Yao H., Zhang Y., Xiao F., Xia Z., Rao J., *Angew. Chem. Int. Ed.,* **46**(23) (2007), 4346–4349.

Ye X., Zheng C., Chen J., Gao Y., Murray C. B., *Nano Lett.,* **13**(2) (2013), 765–771.

Yun C. S., Javier A., Jennings T., Fisher M., Hira S., Peterson S., Hopkins B., Reich N. O., Strouse G. F., *J. Am. Chem. Soc.,* **127**(9) (2005), 3115–3119.

Zhang Y., So M.-K., Loening A. M., Yao H., Gambhir S. S., Rao J., *Angew. Chem. Int. Ed.,* **45**(30) (2006), 4936–4940.

Zrazhevskiy P., Sena M., Gao X. H., *Chem. Soc. Rev.,* **39**(11) (2010), 4326–4354.

Chapter 13

FRET Reporter Molecules for Identification of Enzyme Functions

Jing Mu,[a] Hao Lun Cheong,[a] and Bengang Xing[a,b]

[a]*Division of Chemistry & Biological Chemistry,*
School of Physical & Mathematical Sciences,
Nanyang Technological University, Singapore 637371, Singapore
[b]*Institute of Materials Research and Engineering (IMRE), Agency for Science,*
*Technology and Research (A*STAR), Singapore 117602, Singapore*

bengang@ntu.edu.sg

13.1 Introduction

Protein structures and functions are always the major concerns in medical and biological areas, because proteins serve important roles in intrinsic cellular functions and diseases theranostics (Saghatelian et al., 2005). So far, most drug active or diagnostic targets have been known to focus on various types of biomolecules, of which enzyme is one intriguing and important class and has received considerable attention from many scientists and researchers. Generally, enzymes play fundamental and essential roles in numerous biological processes. For example, some enzymes such as kinases and phosphatases are indispensable in the signal transduction and regulation of intracellular processes. Others including various types of proteases etc. are known to break the large proteins into shorter

Biocatalysis and Nanotechnology
Edited by Peter Grunwald

ISBN 978-981-4613-69-9 (Hardcover), 978-1-315-19660-2 (eBook)
www.panstanford.com

fragments, and thus provide a wide variety of functions in the protein digestion, intracellular protein turnover and damaged protein removal, etc. However, as the biological catalysts, malfunction of enzymes will give rise to a series of biological disorders and subsequent unexpected diseases such as inflammation, osteoarthritis, neurondegeneration and cancer (Hopkins et al., 2002). Therefore, detailed investigations of enzyme functions and development of tools that can, in a direct and systematic manner, real-time identify specific enzyme activities, are highly crucial for the better understanding of biological basis of multistage of diseases, and thus greatly benefiting the rational design towards early diagnosis, drug discovery, and monitoring of treatment efficacy.

As a powerful technique, molecular imaging has been widely applied to real-time monitor cellular functions and biological processes of interests in intact living system. Currently, an array of sophisticated molecular imaging technologies has been well established for understanding of integrative biology, early detection and characterization of disease, and evaluation of treatment efficacy, which includes positron emission tomography (PET), single photon emission computed tomography (SPECT), magnetic resonance imaging (MRI), ultrasound, and optical imaging (Rudin and Weissleder, 2003). Of the different modalities, optical imaging has been extensively applied to real-time visualize the enzyme activitiesnon-invasively monitor cellular events in vitro and in vivo owing to their virtue of high sensitivity, high-throughput capabilities, low costs and safety profile (Baruch et al., 2004). In order to analyze detailed and authentic enzyme processing in a dynamic way, it is highly necessary to generate more specific and quantitative signals in the course of optical imaging. Strategies based on the principle of Förster resonance energy transfer (FRET) stand out owing to its amplified signal generation, the accuracy in fluorescent intensity measurement, better contrast and high detection sensitivity (Sapsford, et al., 2006; Yuan, et al., 2013; Razgulin et al., 2011). So far, various kinds of FRET-based imaging probes have been extensively developed for revealing different enzyme expressions and functions under the disease state. In this chapter, we will mainly focus on recent progress in the design of FRET-based reporters for visualization and sensing of enzymatic activities in living systems. Also, we will provide detailed discussions on the important functions of some typical enzymes in both physiological and pathological conditions.

13.2 Principle of Förster Resonance Energy Transfer

Generally, FRET is a process of energy transfer between two chromophores, where the excited state donor (D) transfers energy to a ground state acceptor (A) through non-radiative dipole–dipole coupling (Jares-Erijman and Jovin, 2003; Sapsford et al., 2006). The acceptor must be able to absorb the energy at the emission wavelength(s) of the donor; however, it is not necessary for the acceptor to remit the energy. This rate of energy transfer is dependent on several factors, such as the spectral overlap (Fig. 13.1) of the chromophores, relative orientation of the transition dipole moment and more importantly, distance between the donor and acceptor chromophores (Eq. 13.1).

$$K_T = (1/\tau_D) \times [R_0/r]^6$$

$$R_0 = 9.78 \times 10^3 \times [\kappa^2 \times J(\lambda) \times \eta^{-4} \times Q_D]^{1/6} \quad (13.1)$$

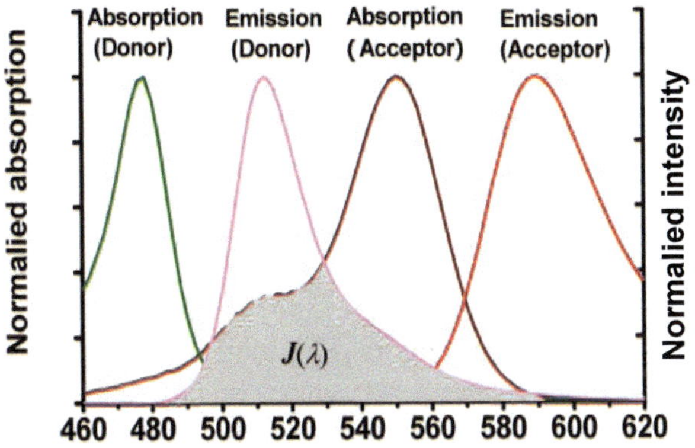

Figure 13.1 Illustration of the spectra overlaps between a FRET pair (Yuan, et al., 2013). Reprinted by copyright permission of the American Chemical Society.

The rate of energy transfer (K_T) of a D/A pair is dependent on the distance (r) between D and A, which can be expressed in terms of Förster distance (R_0). Förster distance is defined as the distance between D and A at which 50% of the excited D decay via energy transfer to A. Therefore, R_0 is dependent on the spectra properties of D and A, which can be expressed by Eq. 13.1. K^2 is a factor that describes the orientation of the transition dipoles of D and A, which range from 0 (perpendicular) to 4 (collinear/parallel). $J(\lambda)$ is the spectra overlap integral of D and A, while η is the refractive index of the medium. Lastly, Q_D describe the quantum yield of the D in the absent of A. This equation explains the high sensitivity nature of a FRET pair system due the D/A separation distance (r). Generally, effective FRET occurs over distances about 10 Å to 100 Å.

13.3 Design of FRET-Based Probes for Various Types of Enzymes

In the process of catalytic reaction, the enzyme grabs on to the substrate at the active site to form the enzyme-substrate complex. Subsequent catalysis will break or build chemical bond to form new molecules, during which the enzyme will return to its original state and get ready to work on another molecule of substrate. By taking the advantages of similar principle in the bond-breaking or bond-formation, the distance between the donor and acceptor is easily adjusted, which results in the enzyme-triggered fluorescence change. So far, various FRET-based probes have been well designed to identify the processes of different enzyme activities.

13.3.1 β-Lactamase

One most well-characterized enzyme applied in the design of FRET-based imaging systems is β-Lactamase (Bla). β-Lactamases (Bla) are a family of bacterial enzymes that efficiently cleave penicillins and cephalosporins and result in the bacterial resistance to these antibiotics. β-Lactamases have four subclasses from class A to class D based on the difference of the nucleotide and amino acid sequences in these enzymes. Due to the great clinic importance for verifying these enzymes, extensive efforts have been made to develop new antibacterial agents for circumventing the resistance caused by β-Lactamases. Except its major impact on the bacterial resistance, as a biological molecule, the

β-Lactamase itself has proven to be an attractive biosensor for detecting biological process and protein-protein interactions due to its great merits, such as being small size (29 kD) to be easily expressed in eukaryotic cells with minimum toxicity and with no interference from mammalian enzymes (Philippon et al., 1998; Campbell, 2004). Based on the reactions with their substrate to generate the fluorescent signals, β-Lactamase has been utilized for a wide variety of biological assays. One important case was that Zlokarnik et al. (1998) developed a β-Lactamase substrate probe called CCF2/AM based on the fluorescence resonance energy transfer (FRET). The donor fluorophore in CCF2 is 7-hydroxycoumarin and acceptor is the fluorescein attached at the 3′ position by a stable thioether linkage. This CCF2/AM (Fig. 13.2) could efficiently cross the cell membrane and diffuse into cytosol, where ester groups are hydrolyzed by intracellular esterases.

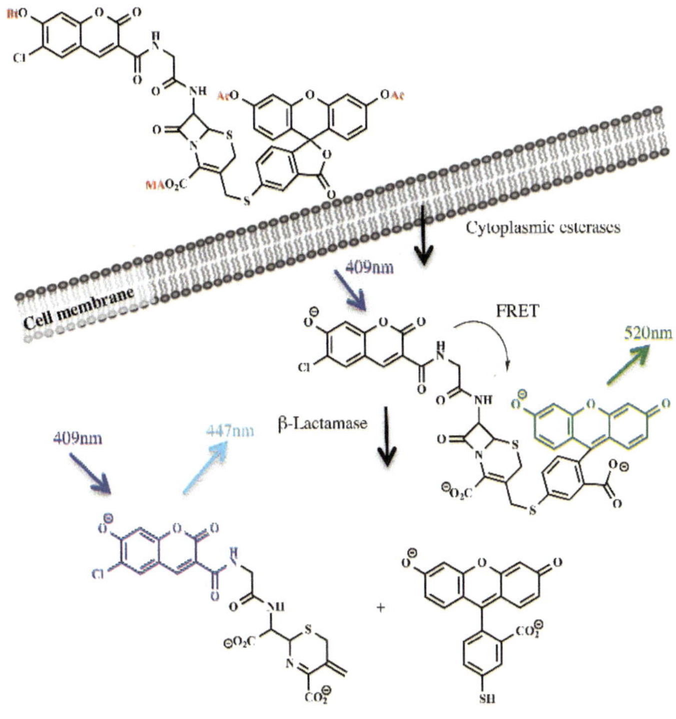

Figure 13.2 Schematic presentation of CCF2/AM for Bla detection in living cells (Zlokarnik et al., 1998). Reprinted by copyright permission from the American Association for the Advancement of Science.

The generated structure can undergo hydrolysis by β-Lactamases, disrupting the FRET between fluorescein and coumarin and leading to the emission shift from 520 to 477 nm. Such FRET-based membrane-permeant, fluorogenic probe with high sensitivity, real-time response and amplified signal output was used for the quantification of enzymatic activities, antibiotic screening and other mechanism studies in living systems.

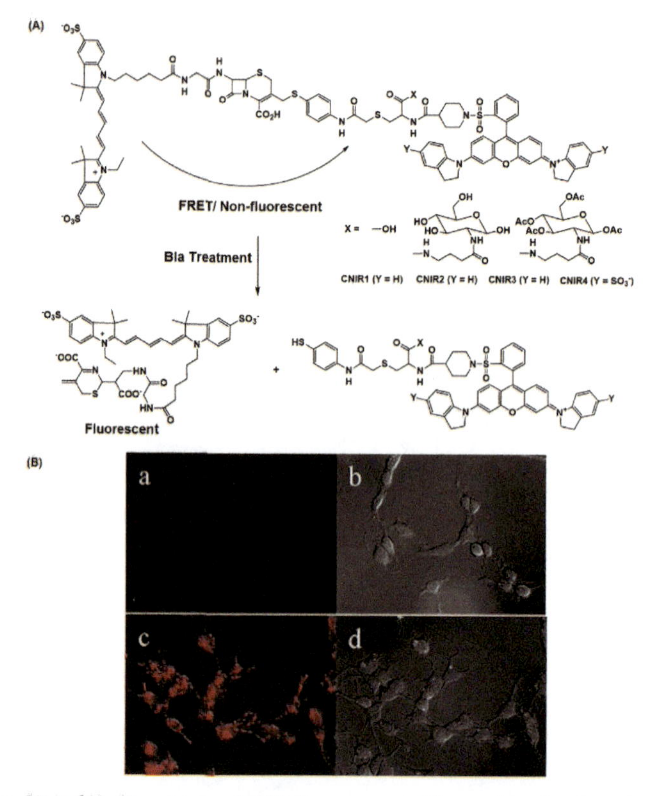

Figure 13.3 (A) Structure of NIR fluorescent Bla probes for imaging Bla activity. (B) Fluorescence (a and c) and differential interference contrast (b and d) images of wild-type (a and b) and Bla stably transfected (c and d) C6 glioma cells loaded with CNIR4 (Xing et al., 2005). Reprinted by copyright permission from the American Chemical Society.

Recently, Xing et al. developed the fluorogenic Bla probes with a class of novel cell-permeable near-infrared (NIR) β-lactamase substrates (Xing et al., 2005). In their design, cyanine dye Cy5 with maximum emission of 670 nm was tethered to 7′-amino of a Bla substrate cephalosporin and a quenching group QSY21 with strong absorbance at 660 nm was connected to the 3′-position through a linker of amino thiophenol and cysteine residue. The whole molecule was not fluorescent due to the FRET quenching mechanism. One fully acetylated D-glucosamine group was also introduced into Cy5-QSY21 cephalosporin probe to assist the molecular penetration across cell membrane. The activation of the NIR fluorochrome cephalosporin probe by Bla enzyme would cleave the connection between the cyanine dye Cy5 and QSY21 quencher, and thus inducing the release of strong fluorescence signal (Fig. 13.3). This probe was found to efficiently image Bla stably transfected C6 glioma cells and it was also modified for the successful imaging of Bla expression in a C6 glioma tumor in living animals (Kong et al., 2010).

13.3.2 Protein Kinase and Alkaline Phosphatase

Apart from β-Lactamase used in the FRET-based imaging, other types of enzymes have also been widely investigated to build up FRET systems for bioanalysis and medical imaging studies, such as protein kinase and alkaline phosphate. Protein kinases modulate the activities of proteins though the phosphorylation process, a very critical step in the intercellular communication and functioning of the nervous and immune systems. Mutation and dysregulation of protein kinase will cause several human diseases like Alzheimer's disease and cancer as well. Phosphatases, the other class of enzymes that is opposite to the action of protein kinases, can remove a phosphate group from its substrate, called dephosphorylation. A representative phosphatase in many organisms is alkaline phosphatase, ALP. In humans, alkaline phosphatases are present in all tissues, but particularly concentrate on liver, kidney and bone. Either elevated levels or lowered levels of ALP in some specific tissues will be responsible for many human liver or bone diseases. Different methods have been developed to monitor the activities of protein kinases or phosphatases. Recently, Freeman et al. (Freeman et al., 2010) developed two FRET-based sensing configurations for analyzing casein kinase (CK2), a serine/threonine selective protein kinase (Fig. 13.4). One approach was semiconductor quantum dots (QDs) was linked with serine-containing peptide sequence that can be recognized by CK2. Subsequently, the CK2 catalyzed reaction would result in the phosphorylation of the serine unit with the fluorophore-labeled γ-phosphate, γ-ATP-Atto-590, leading to the formation the FRET between the QDs and Atto-590 acceptor. The second approach is serine-containing peptide functionalized-QDs first react with CK2 and ATP to yield the phosphorylated peptide. Then interaction with Atto-590-modified antiphosphoserine-antibody would form FRET from QDs to the fluorophore acceptor. Such formed FRET process would provide direct and sensitive readout for the phosphorylation process. Meanwhile, these sensitive methods also provide great possibility to allow the multiplexed analysis of several kinases or several phosphatases in vitro and in vivo.

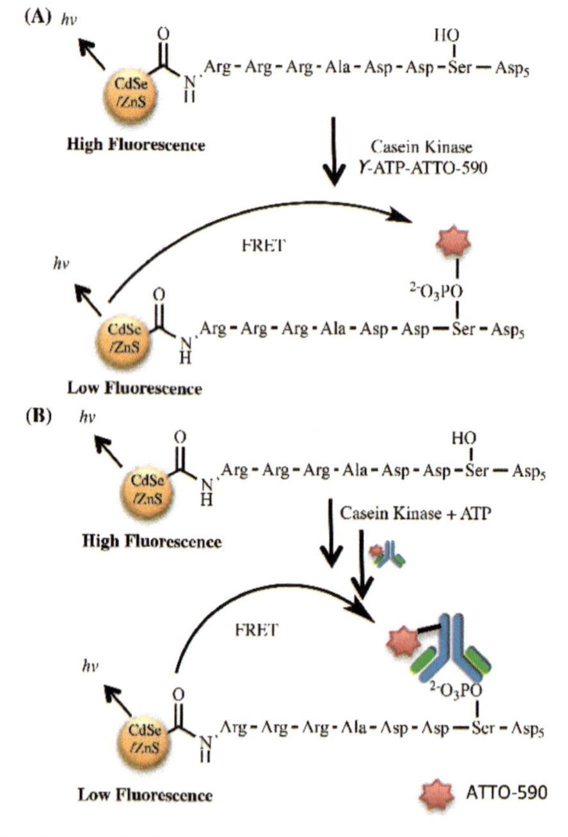

Figure 13.4 Optical Analysis of Casein Kinase, CK2, via FRET (Freeman et al., 2010). Reprinted by copyright permission from the American Chemical Society.

13.3.3 Phosphodiesterase

Another mostly applied enzyme reporters are Phosphodiesterase, which can break a phosphodiester bond and serve as regulators of cyclic nucleotide signaling with diverse physiological functions, like ion channel function, proinflammatory mediator production, and glycogenolysis, etc. Given their functional properties and cellular roles, they are often drug targets for the treatment of various diseases, such as heart failure, inflammation, asthma and depression, etc. Nagano and his coworkers developed an enzyme-cleavable sensor for phosphodiesterase activity based on FRET (Takakusa et al., 2002). They introduced two phenyl linkers between the donor and acceptor moieties to prevent quenching caused by dye-to-dye close contact (Fig. 13.5). After the hydrolysis of the phosphodiester enzyme, the probe CPF4 exhibited a large shift in the emission maximum from 515 to 460 nm. Such developed method can also be easily applicable to other hydrolytic enzymes.

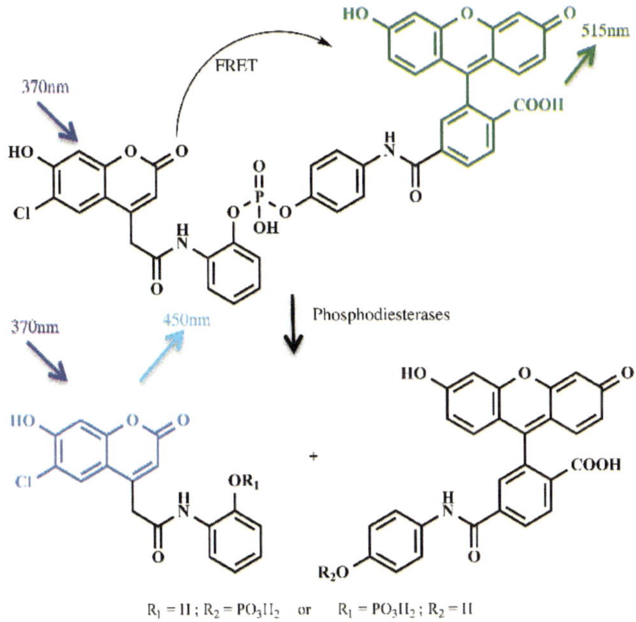

Figure 13.5 The Structure of CPF4 and the Sensor Mechanism (Takakusa et al., 2002). Reprinted by copyright permission from American Chemical Society.

13.4 Design of FRET-Based Probes for Protease Enzymes

Proteases, also known as proteolytic enzymes, are one of the most abundant enzyme families encoded by the human genome with more than 500 active members. Proteases catalyze the breakdown of proteins by hydrolysis of peptide bonds, an enzymatic reaction influential to many physiological and pathological processes such as cell proliferation, tissue remodeling, blood coagulation, wound repair, protein catabolism, inflammation, infection, and cancer progression, etc. (Turk et al., 2006; Drag et al., 2010). Generally, the mammalian proteases are classified as serine, cysteine, threonine, aspartic, and metalloproteases based on different catalytic mechanisms for substrate hydrolysis. Proteases interact with their peptide substrate through hydrogen binding, hydrophobic, and electrostatic reactions between the substrate side chain and active pockets, leading to hydrolysis of peptide bonds (Deu et al., 2012; Edgington et al., 2011). In the N-terminal serine, cysteine and threonine proteases, the acyl-enzyme intermediate was formed by nucleophilic attack of the catalytic side chain residue, whereas in the metalloproteinases and aspartic proteases the nucleophile is an activated water molecule, which results in acid–base catalysis (Fig. 13.6).

Since the enzyme substrate cleavage can be well regulated by the targeting protease, the simple and conventional approach in the design of FRET-based probes is to use the optimal peptide sequence containing the fluorescence donor (fluorophore) and acceptor (another fluorophore or quencher) at the opposite site. Once the active protease interacts with the recognized peptide sequence, the cleavage product will dissociate from each other, leading to the generation of fluorescence signals (Lemke et al., 2011; Edgington et al., 2011). Moreover, another recently proposed approach is based on the "reverse design" (Watzke et al., 2008), in which optimized inhibitors are transformed to cleavable and selective substrates and where new FRET probes can be designed accordingly.

An alternative approach to the design of FRET-based probes for protease enzymes is based on the quenched ABP (qABP) (Paulick et al., 2008; Edgington et al., 2011). A typical qABP consists of three basic components: a reactive functional group, a specific recognition linker and a fluorophore/quencher pair. The recognition linker provides selective affinity to the enzyme of interest, followed by the covalent modification though reacting with reactive group. Then covalent modification will liberate the quencher and the probe-labeled enzyme will become fluorescent.

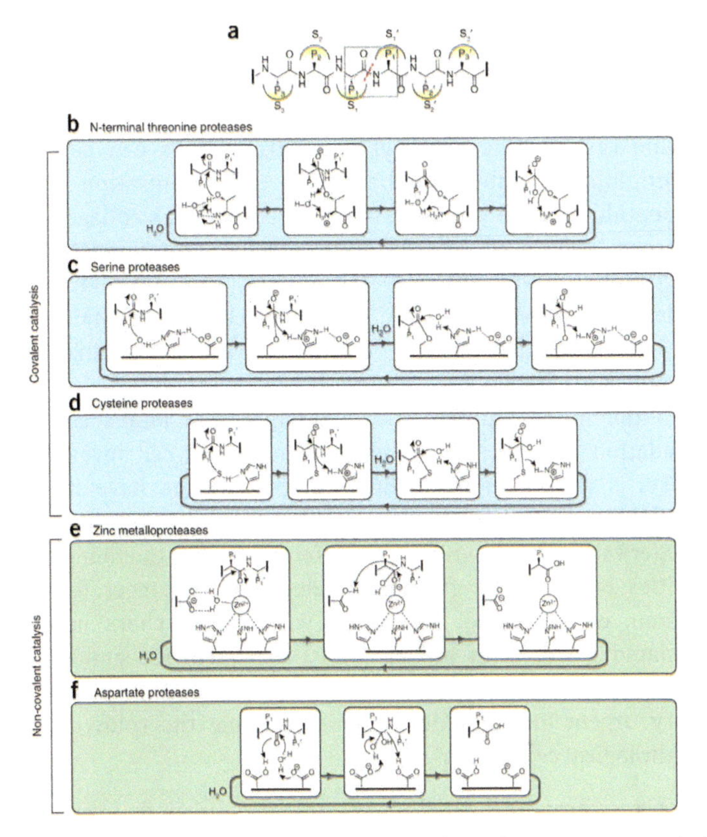

Figure 13.6 (a) Protease substrates bind through interactions of the side chain residues (P and P′ residues) with the substrate pockets of the protease (S and S′ pockets). The red dashed line indicates a scissile bond. (b–f) Catalytic mechanisms of five different types of proteases. Reprinted by copyright permission of Nature Publishing Group.

13.4.1 Matrix Metalloproteinases

Matrix metalloproteinases (MMPs) are a family of 23 human zinc endopeptidases, which are largely responsible for the degradation of extracellular matrix (ECM) and the process of tissue remodeling (Verma et al., 2007). Generically, MMPs structure contains three domains, the pro-peptide, the catalytic domain and the hemopexin-like C-terminal domain. Like the other proteolytic enzyme, MMPs are initially expressed as zymogens and the interaction between cysteine's thiol group in the pro-peptide domain and the zinc ion of the catalytic site will keep them in the dormant state. Subsequently, the active forms can be generated once the interaction is disrupted though proteolytic cleavage of pro-peptide domain or chemical modification of the cysteine residue (Sternlicht et al., 2001). MMPs can be further divided into six sub-group based on the organization of their pro-peptide domain and substrate specificity: (1) collagenases—consist of MMP–1, –8, –13, (2) gelatinases—consist of MMP-2, –9, (3) stromelysins–consist of MMP–3, –10, –11, (4) matrilysins–consist of MMP–7, –26, (5) MT-MMP—consist of MMP–14 to –17, –24, –25, (6) others—consist of the remaining MMPs (Murphy et al., 2008).

In the initial studies, it was thought that MMPs cause the degradation of ECM, which resulted in cancer cell invasion and metastasis (Liotta et al., 1980). Further studies have indicated that MMPs assist in tumor proliferation as well as tumor angiogenesis. This signifies the requirement to monitor the activities of MMPs for further development in cancer treatment (Vihinen, et al., 2005). For many years, researchers have also associated MMPs with inflammatory, osteoarthritis and vascular diseases (Hu, et al., 2007; Fingleton et al., 2007), which make it very urgent and significant for unraveling the roles of MMPs in pathological conditions.

13.4.1.1 FRET probes by using organic molecules

As a common class of proteolytic enzymes to process the peptide-based substrates, MMP was the most widely used enzymes for the design of FRET-based systems. So far, numerous FRET-based probes have been developed for unraveling the functions of MMPs in various pathological conditions. For example, Meyer and Rademann (2012) reported a FRET substrate as a tool to monitor the presence and activities of MMP-11 in cellular levels. In their studies, they evaluated the peptide sequence "Ac-GRRRK(Dabcyl)-GGAANC(MeOBn)RMGG-fluorescein" as the most active substrate for MMP-11 and the results indicated a 25-fold increase of affinity when compared with previously published substrates. In addition, they added a "cell-penetrating unit" consisting of with three arginine residues at the N-terminal to enhance the substrate solubility as shown in Fig. 13.7. Using this designed probe, they conducted imaging studies of various MMP-11 positive and negative cells. It is shown that the probe was able to effectively stain the MMP-11 overexpressed cell. The presences of the fluorescence also enable the localization of MMP-11 in various types of cell lines. Moreover, they were able to disregard non-specific cleavage of their probe through negative control imaging of MMP-11 non-expressed cells (Jurkat cell) and non-MMP-11 cleavable probe sequence.

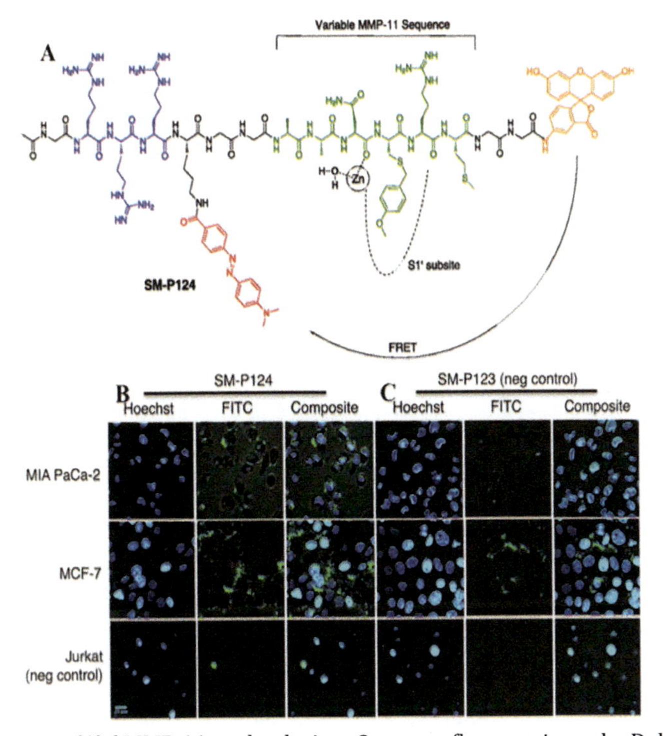

Figure 13.7 (A) *h*MMP-11 probe design: Orange—fluorescein; red—Dabcyl; green—MMP-11 recognition sequence; blue—cell-penetrating unit. (B) Confocal images showing staining of cell membrane with MMP-11 cleavable probe. (C) Confocal imaging using MMP-11 non-cleavable probe (Meyer and Rademann, 2012). Reprinted by copyright permission from the American Society for Biochemistry and Molecular Biology.

Substrate-based FRET probes were also applied to measure MMP-12 activities in the macrophage cells in the mouse pulmonary inflammation model (Cobos-Correa et al., 2009). Their study focused on the development and application of a ratiometric FRET reporter, LaRee1. They demonstrated the reporter's amino acid scaffold (PLGLEEA) is selectively cleaved by MMP-12 over other MMPs. For a more holistic detection of MMP-12 activities, they synthesized soluble FRET reporters (LaRee2-5) that are used for extracelluar activity detection. During their studies, they observed that the probe internalized into the cells only after cleavage activities by MMP-12. The activated fluorescence revealed that MMP-12 activity at the plasma membrane after macrophage cells were stimulated by lipopolysaccharide.

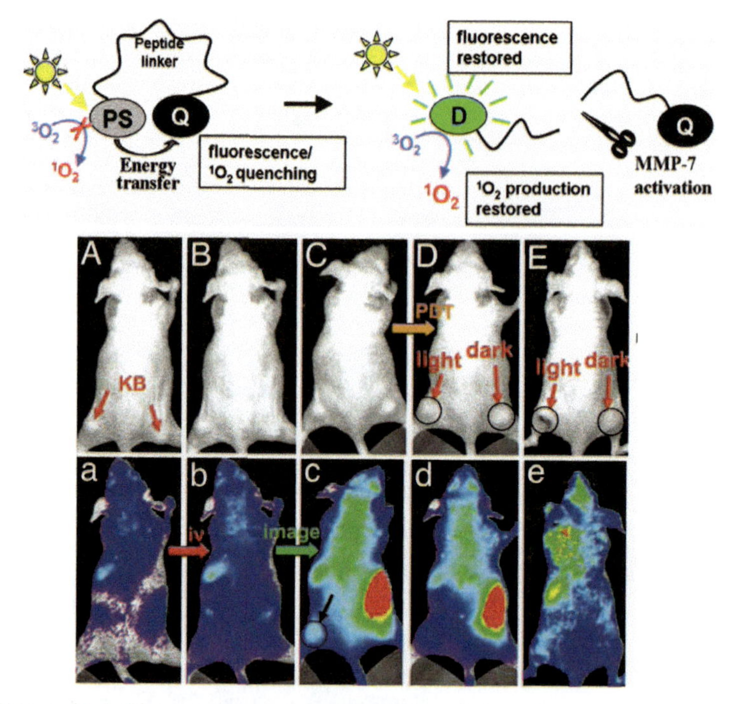

Figure 13.8 The concept of PPMMP7B and in vivo images of mice showing bright field (A-E) and fluorescence (a-e). (A–E) PPMMP7B-administered mouse (A, prescan; B, 10 min after i.v. injection; C, 3 h after i.v. injection; D, 5 h after i.v. injection and 1 h after PDT; E, 3 d after PDT) Light-treated tumors are marked as "light" and non-light-treated tumors as "dark" (Zheng et al., 2007). Reprinted by copyright permission from the National Academy of Sciences.

The detection of MMP-7 activities was also achieved by Mclntyre et al. (2004) though the use of the substrate fluorescein-Ahx-Arg-Pro-Leu-Ala~Leu-Trp-Arg-Ser-Ahx-Cys (Fl-M7). Additional MMP-7 activated photodynamic molecular beacon (PMB) was created by Zheng et al. through the use of photosensitizer pyropheophorbide (Pyro) and black hole quencher 3 (BHQ3) (Zheng et al., 2007). The photoactivity of Pyro is inactive only until target enzyme processes the linker and liberates the quencher. Further in vivo studies also revealed the MMP-7-activated PDT efficacy of this PMB. By combining the two principles of FRET and PDT (photodynamic therapy), the selectivity of PDT-induced cell death can be controlled only in the target enzyme-overexpressed tumor area (Fig. 13.8).

13.4.1.2 FRET probes by using nanomaterials

Apart from organic fluorophore-based FRET used for enzyme imaging, similar concept has been also applied by using nanomaterials for the detection systems (Li et al., 2015a). Due to the attractive enhanced permeability and retention (EPR) effect by which molecules with certain size tend to accumulate in tumor areas more than in normal tissues, nanoparticle-assisted systems are able to greatly improve the uptake and accumulation of probes within cancer areas. Currently, a variety of nanomaterials, like quantum dots (QDs), polymer nanoparticles, gold, silica, and iron oxide nanoparticles etc. have proven useful for monitoring various proteases functions in vitro and in vivo (Lee et al., 2008; Li et al., 2012).

Among the various MMPs, MT1-MMP is known to degrade and disrupt structural barriers composed of type I collagen within the ECM (Poincloux et al., 2009). The critical role of MT1-MMP in cancer cell invasion makes it crucial to detect this enzyme activity. Chung et al. (2015) had recently developed a FRET QDs nanoprobe to visualize the activity of MT1-MMP in cells. In their design, they utilized a bent peptide sequence, which is stabilized by electrostatic attraction of nine cationic arginines and eight anionic glutamates. With the bending of the peptide, the acceptor (Cy3) is brought closer to the donor; thereby enhance the energy transfer efficiency of the FRET pair. In addition, the bent peptide consists of MT1-MMP cleavable sequences (AHLR) that will detach Cy3 in the presence of MT1-MMP. Using this system, they were able to detect and conduct imaging of MT1-MMP by using QD/FRET emission ratio (shown in Fig. 13.9). Single cell imaging results demonstrated that the probe was also able to profile the cancer cell's potential in tissue degradation, migration, and invasion.

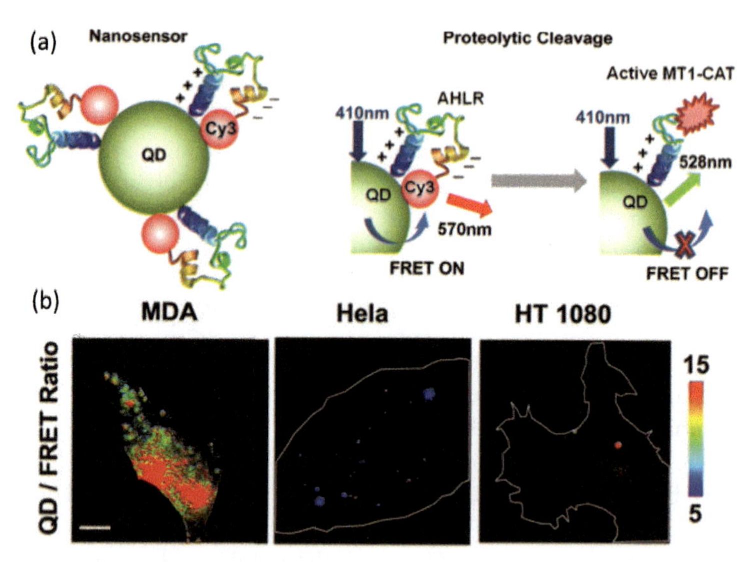

Figure 13.9 (a) Illustration of the design and mechanism of the QD probe (b) QD/FRET ratio imaging of different cell lines (Chung et al., 2015). Reprinted by copyright permission from the American Chemical Society.

MMP-2 is often overexpressed in solid tumors such as colon, breast, and prostate cancers. Unraveling the cancer progression renders the importance to detect MMP-2 (Klein et al., 2004). In a recent publication, Wang et al. (2015) reported a new MMP-2 targeting nanoprobe used for tumor detection and imaging. Their nanoprobe design comprises of two main units, (1) low molecular weighted heparin modified quantum dots (QDs) and (2) QSY21 quencher dye with MMP-2 cleavable peptide sequence (ALMWP). These two units provided the FRET effect, thus making the QDs inactive. Once inside, the QDs are localized in the MMP-2 expressed tumor, and QSY21 will be cleaved from the QD. These will "activate" the QDs, thus allowing it to fluoresce. In addition, with the use of the low molecular weighted heparin, non-specific uptake of QDs by reticuloendothelial system was significantly reduced. This allows the nanoprobes to accumulate at the tumor due to the enhanced permeability and retention effect. For further application, Wang's group used the nanoprobe for brain tumor imaging. It was noted that MMP-2 is overexpressed in gliomas, which is accounted for 80% of all malignant brain tumor. To overcome the poor penetration of the blood-brain barrier (BBB) of the probe, they modified their initial probe design. By incorporating the brain-targeting T7 sequence into their cleavable peptide sequence, they were able to improve the accumulation of the QDs in the brain tumor with significant fluorescence effect. Another type of FRET-based sensor to determine and quantify the MMP-2 activities is based on the use of graphene oxide (GO) as a quencher (Feng et al., 2011). Due to its attractive electronic and optical properties of such two-dimensional material, GO has been applied in a great variety of research areas, including the biomedical studies. Based on this design (Fig. 13.10), it is possible to develop more GO-based FRET probes for detecting various types of proteases and other enzymes.

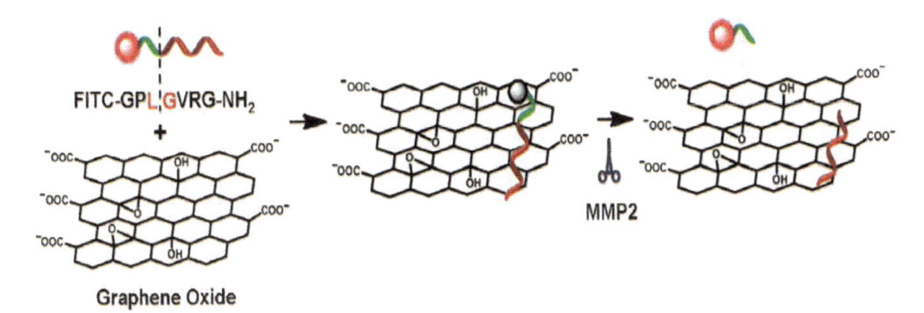

Figure 13.10 GO–peptide fluorescence sensor for the assay of MMP2 (Feng et al., 2011). Reprinted by copyright permission from Royal Society of Chemistry.

13.4.2 Cysteine Proteases

Cysteine proteases, which are characterized by the cysteine residue in the active site, are present in intracellular compartments like lysosome or cytosol and extracellular environment in certain types of pathological conditions. Cysteine proteases involve in the intracellular protein catabolism and extracellular protein degradation (Gora et al., 2015). Mostly notably, numerous researches have shown that caspases and cathepsins are highly related to the tumor progression. Caspases mediate the programmed cell-death process–apoptosis and play important roles in the processing of cytokines (Mcllwain et al., 2013). Cathepsins, on the other hand, can assist the turnover of divergent intracellular proteins and involve in the angiogenesis, apoptosis, and tumor cell invasion (Mohamed et al., 2006; Palermo et al., 2008).

13.4.2.1 Cysteine cathepsin

The cysteine cathepsins, which are predominantly located in endo/lysosomal vesicles, belong to the papain subfamily of cysteine proteases. This group consists of cysteine cathepsins: B, C, F, H, K, L, O, S, V, W, and X, which share a conserved active site that is formed by cysteine, histidine, and asparagine residues (Palermo et al., 2008). A variety of fluorescent probes have been developed for studying the functional roles of different types of cathepsins. Among cysteine cathepsins, the cathepsins B, L, S, K, and X are most well studied due to their implication in the progression, invasion and metastasis of tumors. For the substrate-based FRET probes, various types of reporters have been designed for monitoring cathepsins B and S activities based on the specific and optimal peptide sequences (Hu et al., 2014).

Another recently proposed approach is based on the "reverse design" (Watzke et al., 2008), in which specific inhibitors are transferred to cleavable and selective substrates and new FRET probes can be designed accordingly. Several studies have engaged in this strategy to develop new probes for cathepsin investigations in proteomic lysates and in vivo molecular imaging. Caglič et al. (2011) introduced a near-infrared quenched fluorescent probe, which enabled the monitoring of cathepsin S activity in vitro and in a mouse model of inflammatory paw edema. More recently, Na et al. (2012) identified two potent inhibitors of cathepsin L by using a small molecule microarray as a high-throughput screening platform. Subsequently, these two inhibitors were converted into cell-permeable and two-photon small molecule imaging probes, which were further applied to detect endogenous cysteine cathepsin activities from cell lysates or live mammalian cells of HepG2 cancer cells (Fig. 13.11).

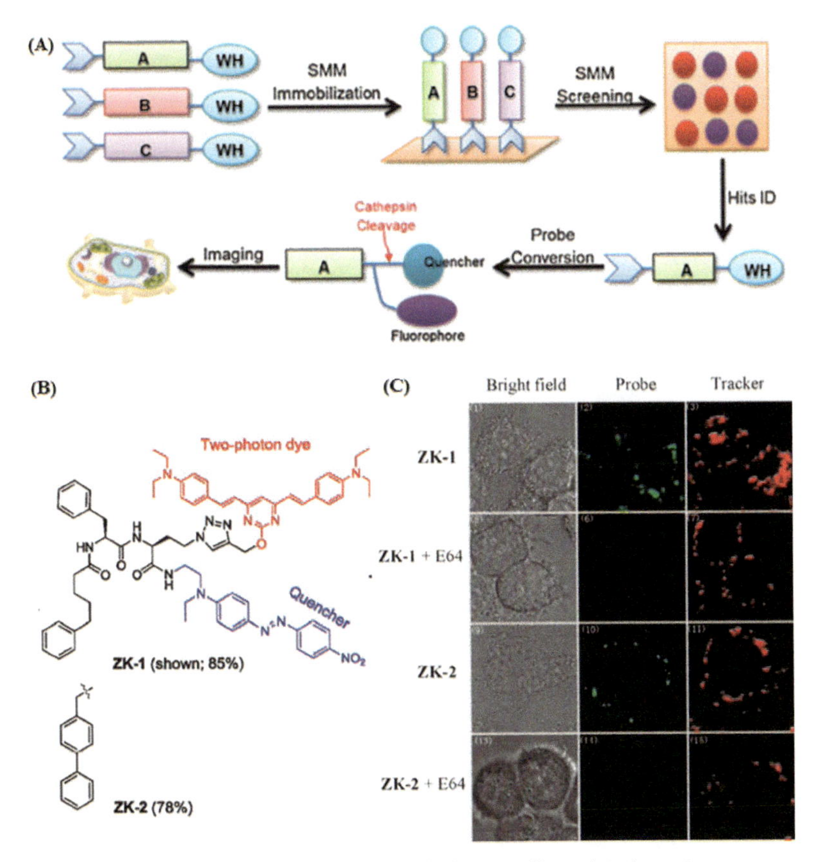

Figure 13.11 (A) Overall strategy of the small molecule microarrays (SMM)-guided, high-throughput discovery of cell-permeable, two photon probes for imaging of endogenous cathepsins; (B) structure of imaging probes (ZK-1 and ZK-2); (C) Two-photon confocal images of live HepG2 cells upon treatment with ZK-1/-2 (2 mM) for 2 h (Na et al., 2012). Reprinted by copyright permission from the Royal Society of Chemistry.

Blum et al. (2005) developed a series of quenched activity-based probes (qABP) (Paulick et al., 2008), based on the epoxy-succinyl- and acyloxymethyl ketone reactive groups (Fig. 13.12). These pre-quenched probes become fluorescent upon activity-dependent covalent modification. These reagents can be used to dynamically image protease activities in living cells. Subsequent near-infrared versions of fluorescent probes were able to provide whole-body non-invasive imaging (Blum et al., 2007).

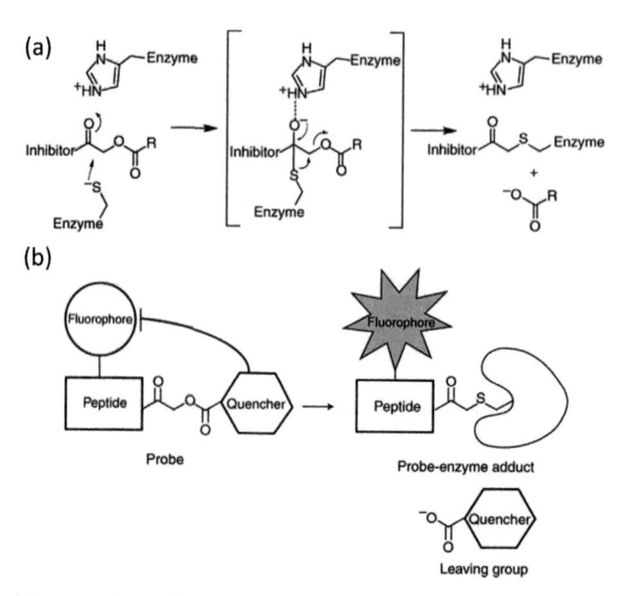

Figure 13.12 Design of a qABP. (a) Mechanism of covalent inhibition of a cysteine protease by an acyloxymethyl ketone. (b) Activity-dependent labeling of a cyseine protease target by a qABP (Blum et al., 2005). Reprinted by copyright permission of Nature Publishing Group.

13.4.2.2 Caspases

Caspases belong to a family of endoproteases that play crucial rules in inflammatory signaling and cell death. According to the structural and functional features, caspases are usually classified by their involvement in apoptosis (caspase-3, -6, -7, -8, and -9 in mammals), and in inflammation (caspase-1, -4, -5, and -12 in humans). A detailed overview on the substrate-based probes for the respective caspases can be found in the recent review (McIlwain et al., 2013). Although the specific binding regions have been optimized based on the positional scanning studies, the developed probes still suffer from the lack of selectivity for the specific caspase (Fig. 13.13), mostly due to the similarity of catalytic mechanism (Poręba et al., 2013).

Bullok et al. (2005) first synthesized a novel, membrane-permeant probe, TcapQ647, which comprises an effector caspase recognition sequence, DEVD, and a flanking optically activatable pair Alexa Fluor 647/QSY 21 to realize the imaging of caspase activity during apoptosis (Fig. 13.14). In the development of caspase activity-based probes, the most prominent case was the new probe

AB50 (Cy5-EPD-AMOC) designed by Edgington et al. (2009). This developed type of imaging agent with NIRF fluorophores enabled direct visualization and quantification of apoptosis in vivo. These probes were able to covalently label active caspase and monitor apoptosis in the thymi of mice treated with dexamethasone and in tumor-bearing mice treated with the apoptosis-inducing monoclonal antibody Apomab. Until now, there are not sufficient reports about the quenched ABPs for the detection of caspases.

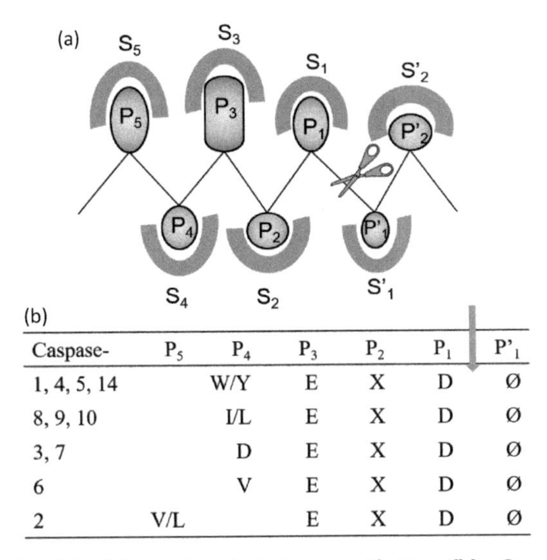

Caspase-	P_5	P_4	P_3	P_2	P_1	P'_1
1, 4, 5, 14		W/Y	E	X	D	Ø
8, 9, 10		I/L	E	X	D	Ø
3, 7		D	E	X	D	Ø
6		V	E	X	D	Ø
2	V/L		E	X	D	Ø

Figure 13.13 (a) Inherent substrate specificity; (b) Caspases bind substrates through interactions of the active-site cleft with amino acids in the cleavage site. Residues like Gly, Ala, Thr, Ser, and Asn are preferred at P_1' (φ symbol). Adapted from Poręba et al. (2013).

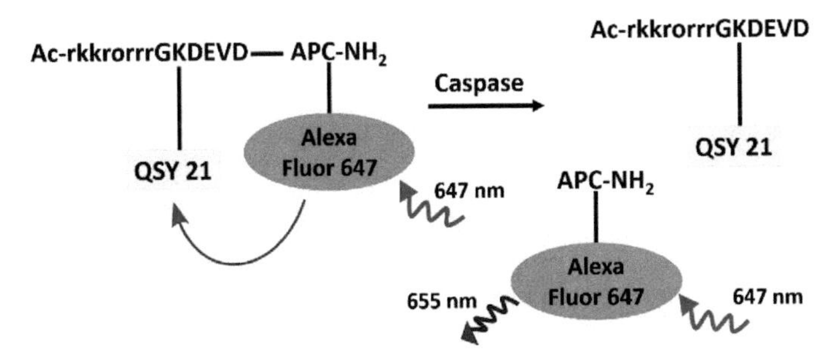

Figure 13.14 Model of TcapQ647 cleavage by executioner caspases (Bullok et al., 2005). Reprinted by copyright permission from the American Chemical Society.

13.4.3 Serine Proteases

Serine (Ser) proteases are a family of endopeptidase, which is identified by the nucleophilic serine residue located at its enzymatic site. They are one of the largest proteases families, where they contributed to more than one third of all known proteolytic enzymes. Serine protease can be classified into 13 clans and 40 families (Page et al., 2008). Generically, they are divided into the various clans based on their catalytic mechanism and common ancestry shown in Table 13.1. Typically, the catalytic site of the serine proteases is dependent on the catalytic triad of aspartate (asp), histidine (his) and serine.

Table 13.1 General Properties of serine protease

Clan	Families	Representative member	Catalytic residues
PA	12	Trypsin	His, Asp, Ser
SB	2	Subtilisin	Asp, His, Ser
SC	2	Prolyl oligopeptidase	Ser, Asp, His
SE	6	D-A, D-A carboxypeptidase	Ser, Lys
SF	3	LexA peptidase	Ser, Lys/His
SH	2	Cytomegalovirus assembling	His, Ser, His
SJ	1	Lon peptidase	Ser, Lys
SK	2	Clp peptidase	Ser, His, Asp
SP	3	Nucleoporin	His, Ser
SQ	1	Aminopeptidase DmpA	Ser
SR	1	Lactoferrin	Lys, Ser
SS	1	L,D-carboxypeptidase	Ser, Glu, Hi

Source: Reprinted by copyright permission from Springer.

Using this catalytic triad, the proteases are able to catalyze hydrolysis of peptide bonds (Hedstrom et al., 2002). Further studies on serine protease had also led to the discovery novel and simpler catalytic residues.

Serine proteases hold pivotal roles in many physiological processes in our body, which include digestion, apoptosis and immune response. However, dysfunction of the proteases can lead to inflammation, tumor proliferation, invasion and even metastasis (Sloane et al., 2013). Being closely associated with tumor progression, the ability to detect and monitor serine proteases will help have better understanding of its roles thereby develop better cancer treatment.

13.4.3.1 Proprotein convertases

Proprotein convertases are the subfamily of subtilases, which belongs to the family of subtilisin-like serine proteases. These convertases associate with the activation of cellular and pathogenic precursor proteins like polypeptide hormones, growth factors and bacterial pathogens, etc. They consist of nine family members: convertase 1 (PC1), PC2, furin, PC4, PC5, paired basic amino acid cleaving enzyme 4 (PACE4), PC7, subtilisin kexin isozyme 1 (SKI-1) and proprotein convertase subtilisin kexin 9 (PCSK9) (Seidah et al., 2012). Currently, proprotein convertases are considered to be attractive targets due to their important functions in the progression of various diseases such as cancer, inflammation and pathogenic infections (Seidah et al., 2012). Recently, Mu et al. (2014) utilized a small molecule based FRET reporter to real-time monitor furin activities (Fig. 13.15). They designed the probe by using furin recognizable cleaving peptide sequence (RVRRSVK) as the framework for their probe. Using a FRET pair fluorescein (FITC) and quencher 4-(dimethylaminoazo)benzene-4-carboxamide (Dabcyl) in the probe, they were able to observe a "turn-on" fluorescence effect upon furin cleavage. In addition, the lipid membrane anchor was designed to effectively immobilize the unique reporter onto the cell surface. Using this system, they tested it on various furin positive and negative cells lines. It was observed that the probe was selective to furin expressed cell lines. Furthermore, the probe was capable of staining the cell membrane for relatively quite long time (up to 2 h). In addition, they used the probe in two-photon imaging to prove the utility of the probe in monitoring living cells or tissues. The successful mouse's ear tissues imaging proved the potential of the probe in monitoring furin activities in vivo and clinic trials.

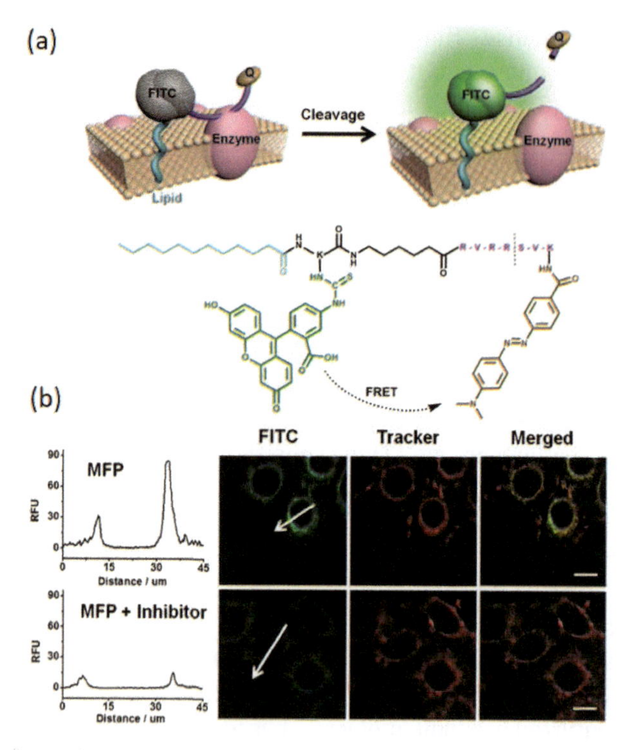

Figure 13.15 (a) Illustration of the structure of membrane-anchored and furin-responsive probe (MFP) and surface associated furin cleavage on the cell membranes. (b) Fluorescence imaging of U251 cells by MFP. Cells were incubated with MFP alone, MFP and inhibitor. CellMask Deep Red was used as a standard to image the membrane (Mu et al., 2014). Reprinted by copyright permission from WILEY-VCH Verlag GmbH & Co. KGaA, Weinheim.

13.4.3.2 Neutrophil elastase (NE)

Neutrophil elastase (NE) is a PA clan, serine proteases that is secreted by neutrophils or macrophage during inflammation. Even with its protective roles in extracellular matrix remodeling and host defense against bacterial infection, dysfunction, NE is also associated with proteolytic damage of airways and lung in chronic neutrophilic inflammation. Such damage often leads to respiratory failure and even death (Demkow et al., 2010). Therefore, it's crucial to understand the activity distribution of NE in order to protect the lung from uncontrolled proteolytic damage. Gehrig et al. (2012) reported a new NE reporter capable of detecting both soluble and cell-surface NE. They used a long peptide sequence, QPMAVVQSVPQ to improve the specificity and sensitivity of the probe. In addition, the lengthen peptide substrate allows the uses of this probe in both mouse and human NE. Thereafter, Gehrig and his coworkers added the lipidated variant of the probe (NEmo-2) shown in Fig. 13.16A as reporter for cell-surface NEs. In their design, they utilized two fluorophores instead of the conventional FRET pair to allow ratiometric readout of the cleavage activities. This allows the detection of NE activities independent of the probe concentration. Using their design, they used NEmo-2 on wild type and NE deficient mouse neutrophils. They were able to observe significant D/A ratio changed on the surface of wild type cell caused by NE cleavage. Combining the information with the NE reporter (NEmo-1), they concluded that NE is predominantly active on the surface of neutrophil and not in the extracellular fluids in lung inflammation.

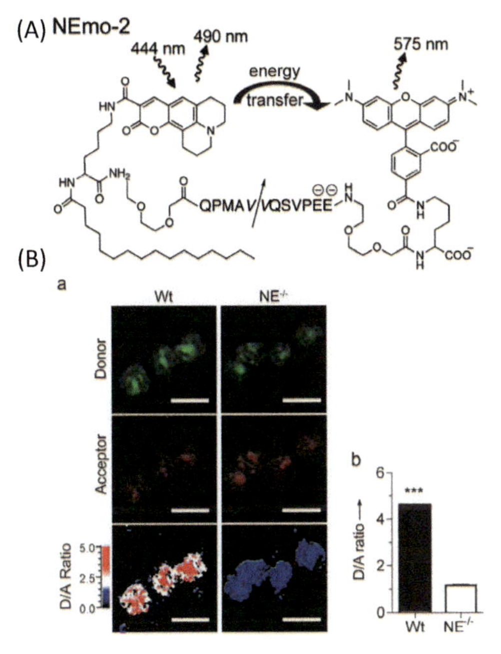

Figure 13.16 (A) Membrane-targeting neutrophil elastase probe (NEmo-2). (B) NE activity imaging using NEmo-2; blue—low D/A ratio indicating intact NEmo-2; red—high D/A ratio indicate cleavage of NEmo-2 (Gehrig et al., 2012). Reprinted by copyright permission from WILEY-VCH Verlag GmbH & Co. KGaA, Weinheim.

13.4.3.3 Fibroblast activation protein-alpha

Fibroblast activation protein-α (FAPα) is membrane-bound gelatinase belonging to the serine protease family. It has been found to be highly and selectively expressed in over 90% of human epithelial cancers, but not in normal adult tissues (Scanlan et al., 1994). FAPα play important roles in tumor growth and invasion though dipeptidyl peptidase and collagenase proteolytic activities. As a biomarker and potential target of the cancer-associated fibroblasts, FAPα attracted a lot of attention and the design of sensitive FAPα-based probes become of great importance. Li et al. (2012) developed a near-infrared (NIR) fluorescent probe by linking the peptide sequence short peptide sequence (KGPGPNQC) with Cy5.5/QSY21 FRET pairs for in vivo optical imaging of FAPα (Fig. 13.17). The in vivo imaging results demonstrated that in the FAPα-expressed U87MG tumor models, the probes presented faster tumor uptake and higher fluorescence signals when compared to FAPα-negative C6 tumor models. Results indicated that the designed probe could be used for early detection of FAPα-expressing tumors in future studies.

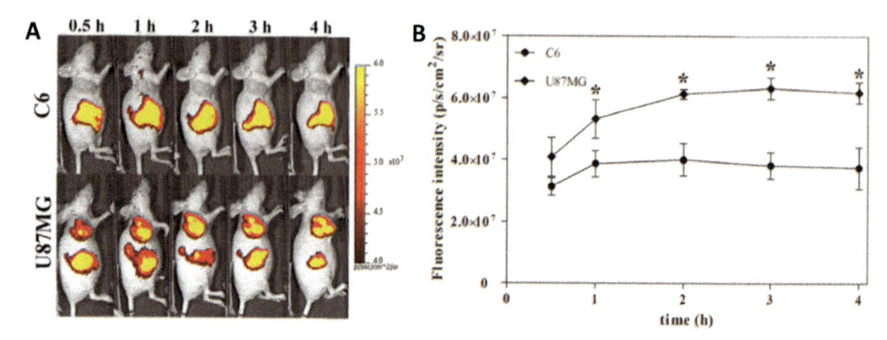

Figure 13.17 (A) In vivo fluorescent imaging of subcutaneous C6 and U87MG tumor-bearing mice at 0.5, 1, 2, 3, and 4 h post tail vein injection of ANPFAP (1 nmol). (B) Fluorescence ROI analysis of tumor and muscle in mice bearing C6 and U87MG tumor from 0.5 to 4 h p.i. (Li et al., 2012). Reprinted by copyright permission from the American Chemical Society.

Another case reported by Ji et al. (2013) recently was based on the peptide conjugated-ferritin nanoparticle due to its good biocompatibility and sufficient stability. They incorporated carboxyfluorescein-tagged (FAM-tagged) peptide (FAM-DRGETGPAC) and black hole quencher (BHQ-1) into the ferritin respectively and further reassembled to the optimal hybrid ferritin nanoparticles. After intravenously injecting this nanoprobe into mice, the fluorescence imaging was realized with high specificity and stimuli-responsive sensitivity in tumor microenvironment.

13.4.4 Aspartic Proteases

Although aspartic protease has been found as the smallest class in the human genome, many of them have been treated as validated and potential drug targets in the pharmaceutical industry. Generally human aspartic proteases are divided into two clans based on their different tertiary structures: clan AA and clan AD (Eder et al., 2007). Clan AA contains the classical aspartic proteases (renin, pepsin A, pepsin C, cathepsin D, cathepsin E, BACE1, etc. Clan AD contains intra-membrane cleaving proteases like the presenilins and signal peptide peptidase.

13.4.4.1 β-Secretase

β-Secretase (BACE) is a membrane bound aspartic protease ubiquitously expressed in the brain and pancreas tissues. BACE was identified as the proteolytic enzyme that cleaves amyloid precursor protein (APP). Proteolytic cleavages of APP by BACE, creates a C99 membrane-bound C-terminal fragments, which are further processed by secretase to release toxic amyloid-β peptides. Such insoluble peptides aggregate to induce neurodegeneration, which are known as Alzheimer's disease (AD). Therefore, effective tools to elucidate the secretase functions, especially in cellular levels and living animal are quite necessary for better understanding the pathological feature of AD (Hardy et al., 2002; Evin et al., 2010).

Typical assay for evaluating BACE activity is the sandwich enzyme linked immunosorbent assay (ELISA) for detection of Aβ production, which is unfortunately costly and time-consuming. Alternative FRET-based assay has been developed based on dye labeled synthetic peptides or recombinant APP-reporter fusion protein constructs to address those issues. Pietrak et al. (2005) established an efficient BACE-1 inhibition assay in which HEK293T cells were stably transfected with APP693 containing the optimized NFEV cleavage sequence to assess the level of secreted amyloid EV40_NF. Another peptide-based FRET probe (β-MAP), which use the peptide sequence EVNLDAHFWADR flanked with DMACA–DABCYL pairs, were applied to monitor real-time BACE activity and screen libraries of potential BACE inhibitors in living cells (Folk et al., 2012).

Another novel fluorogenic nanoprobe developed by Choi et al. (2012) was from the assembly of CdSe/ZnS quantum dot (QD) and gold (Au) nanoparticle via Ni-nitrilotriacetate (Ni-NTA)–histidine (His) interaction (Fig. 13.18). The specific binding of Ni-NTA-AuNP with QD conjugated β-secretase substrate peptide via Ni-His interaction contributes to highly efficient quenching of QD fluorescence. The robust QD fluorescence signal was recovered upon BACE1 enzymatic activity, allowing visualizing of BACE1 activity in living cells.

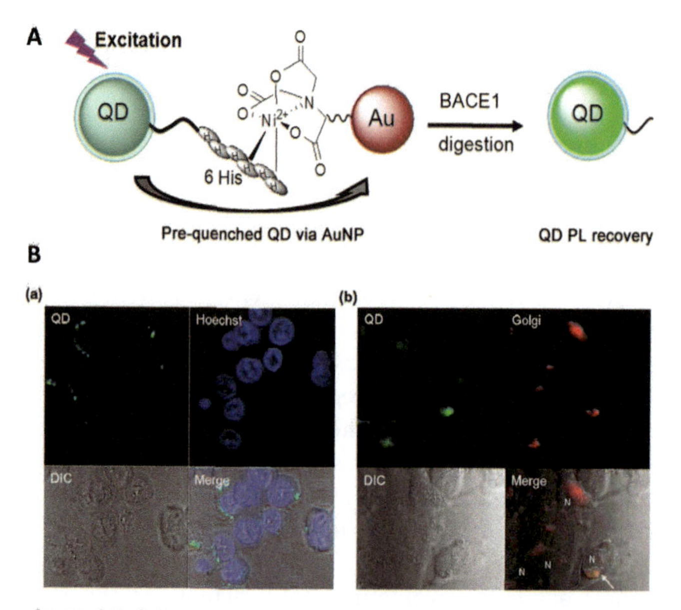

Figure 13.18 (A) Schematic representation of BACE1 enzyme assay using pre-quenched polymer-coated QD and Ni-NTA-modified Au nanoparticle assembly. (B) (a) Live cell images of HEK293 cells transfected with BACE1 plasmid after 1 h of incubation with 20 nM of QD-Au assembly; (b) colocalization of recovered fluorescence of the QD-Au assembly mainly with the Golgi apparatus and endosomal compartments as stained with the NBD-ceramide-BSA complex reagent (Choi et al., 2012). Reprinted by copyright permission from the American Chemical Society.

13.4.4.2 Cathepsin D

Cathepsin D is a lysosomal aspartyl protease, which has shown 2–50 folds increased expression levels in epithelial breast cancer cells. Clinic studies have associated its overexpression with higher metastatic potential, invasion, angiogenesis and degradation of basement membrane, etc. (Eder et al., 2007). Early studies reported a near-infrared fluorescence (NIRF) probe for in vivo imaging of enzyme activity (Tung et al., 1999). They attached cathepsin D-sensitive peptide sequences to the free poly-L-lysine residues and further linked to synthetic graft copolymer for efficient tumor delivery. The close spatial proximity of the multiple fluorochromes induced self-quenching of fluorescence. Once the peptide sequence was cleaved by enzyme, the amplified fluorescence signals would be observed in living cells (Tung et al., 1999). So far, limited FRER-based reporters have been reported for the detection of cathepsin D activities, mostly due to lack of specificity.

13.4.4.3 Cathepsin E

Cathepsin E, an endosomal/endoplasmic reticulum protease, shares many common characteristics with cathepsin D, like substrate preferences, molecular weight, and inhibition susceptibility, etc. However, the difference of localization and distribution in cells and tissues between these cathepsins leads to their different physiological roles (Eder et al., 2007). Although until now the whole functions of cathepsin E has not been fully understood, it may serve important immunological roles in the host defense process. Recently, Abd-Elgaliel and Tung (2010) developed a fluorogenic peptide substrates Mca-Ala-Gly-Phe-Ser-Leu-Pro-Ala-Lys(Dnp)-*D*Arg-CONH2, which could selectively distinguish cathepsin E activity from cathepsin D. After the proteolytic cleavage of peptide substrate, the fluorescent donor, 7-methoxycoumarin-4-acetic acid (Mca) dissociated with the energy acceptor, dinitrophenyl (Dnp) and 265 folds' fluorescence difference was observed between cathepsin E and cathepsin D activity. This study provided promising possibilities of development of selective probes for monitoring in vivo cathepsin E activities.

13.5 FRET-Based Probes for Multiple Enzyme Imaging

In the complex microenvironment of our body, these proteases work inter-dependently. When using conventional individual proteases fluorescent probe as biomarker, it may give "false positives" results during clinical diagnosis due to the complex natural of biological function (Sidransky et al., 1997). Therefore, studies on multiple proteases imaging allow us to have better clarity in the working relationship and the biological processes.

13.5.1 MMP-2 and MMP-7

Wang et al. (2012) developed a dual-luminophore labeled gold nanoparticles (AuNP) for multienzyme detection with completely resolved emission peaks under single-wavelength excitation. They target on two matrix metalloproteinases, MMP-2 and MMP-7, which involves in tumor metastasis and both are expressed in higher levels in tumor cells. In their design, they attached a cleavable peptide spacer comprising of MMP-2 recognizable substrate (GPLGVRG) as well as MMP-7 substrate (VPLSLTMG) on the surface of AuNP. At each terminus of the peptide spacer, a different fluorophores was attached. Then a well-quenched nanoprobe is formed though the FRET effect between two labeled dyes as donors and Au NPs as quenchers. Utilizing the different emission wavelength of the fluorophores, the presence of MMP-2 and MMP-7 will result in the increase fluorescent intensity at different wavelength.

13.5.2 MMP-2 and Caspase-3

Li et al. (2015b) developed a dual-FRET based fluorescence probe for the detection of MMP-2 and Caspase-3. By using these two enzymes, they hope to achieve early cancer diagnosis due to MMP-2 involvement in cancer progression and quick optimization of drug dosage during clinical trials through caspase-3 apoptosis detection. In their design, the fluorescent probe was quenched by dual FERT process to the quencher (Dabcyl). Each of the quencher was attached with an enzyme specific cleavable peptide sequence, comprises of PLGVR sequence for MMP-2 and DEVD sequence for caspase-3 as shown in Fig. 13.19. In general, MMP-2 mainly in the extracellular matrix, while caspase-3 is typical intracellular enzyme. By utilizing the difference of spatial distribution, the probe can sequentially detect these two enzymes. Furthermore, they successfully applied this probe to monitor the Dox-induced early cell apoptosis and UV-induced real-time apoptosis signals.

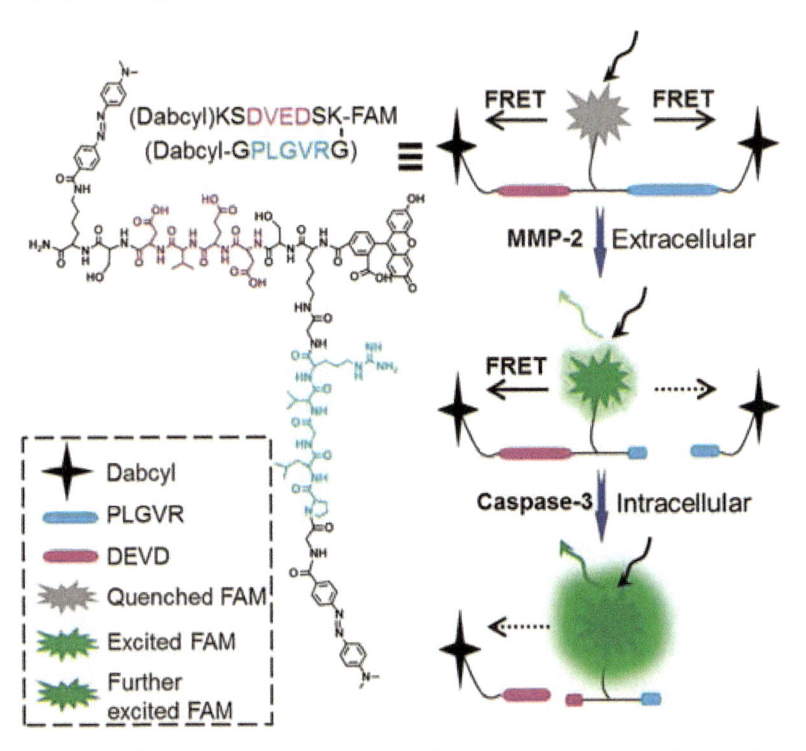

Figure 13.19 The chemical structure, and the proposed mechanism of the probe for the sequential MMP-2 and caspase-3 detection (Li et al., 2015b). Reprinted by copyright permission from Royal Society of Chemistry.

13.6 Conclusions

Optical imaging has been considered as a powerful tool for widespread applications ranging from clinical diagnosis to molecular biology. This sensitive, fast and non-invasive technology is expected to have a substantial impact on the prevention and treatment of a variety of human diseases. In order to provide more sensitive and quantitative results in the process of optical imaging, FRET has been well recognized as one of the most common strategies applied for real-time monitoring of the specific molecular events and biological pathways of interests. In this chapter, we have presented an overview of recent progress on the design of FRET-based reporters for visualizing numerous enzyme activities and non-invasively monitoring cellular events in vitro and in vivo. A wide variety of platforms such as small molecules, peptides, polymers and nanomaterials, etc. have been presented in the probe design for sensing and imaging of enzyme activities. The strategy based on FRET has shown its superior advantages in the amplified signal generation, the accuracy in fluorescent intensity measurement, better contrast, and high detection sensitivity. In the FRET system containing fluorophores and quenchers, reporters can be activated to "fluorescence on" state by the targets under the "dark background," allowing better contrast and more sensitive analysis of enzyme activities. Besides, ratiometric FRET probes are able to offer well-resolved emission wavelength shift before and after interaction with the enzyme of interest. Unlike with those of single signal probes, the ratiometric FRET probes eliminate the interference of probe concentration, excitation source fluctuations, which can provide more accurate and quantitative analysis.

Despite great success has been obtained in the FRET-based reporters, many of these reporters are verified effectively only in cellular levels while few have been practically tested in the animal models. The far more dynamic and complex in vivo environment demands improved performances of these reporters. Currently, in order to make the imaging of enzyme activities feasible in living animals, NIR fluorophores are used as they process superior properties for the good tissue penetration and low autofluorescence. However, the fluorescence intensity of these organic molecules are highly susceptible to external environment, like the temperature pH and water solubility, making it quite difficult for accurate measurement of fluorescence change. Besides, the bulky structures of organic fluorophores and quenchers result in the difficulties of conjugation with enzyme substrates. Thus, the continued efforts in the development of novel fluorophores and reporters with improved photophysical and chemical properties are highly required. Moreover, another solution is to develop biocompatible and NIR light-activated nanomaterials and polymers to complement the organic fluorophores. Moreover, the strong signal amplification generated from the continuous product conversion catalyzed by enzymes indeed facilitates the detection of enzymatic activity. Nevertheless, owing to the similar substrate preferences for the same enzyme family, it is difficult to identify the specific enzyme that is responsible in generating the signal. Hence, creative chemistry and approaches in resolving the specificity will be needed to ensure the correct signal measurement during the catalytic studies. Additionally, quantitative imaging of enzyme activities in the living systems is still challenging and complicated. The administration and circulation process will vary the distribution of reporters at different tissue locations, thus making it difficult for comparison at different sites. Therefore, such uneven distribution of reporters may result in inaccurate quantitative measurement of the enzyme activities at target site. To resolve such issue, solution such as multimodality of imaging should be considered for future design of enzyme-activated reporters to get more accurate and independent measurement of enzyme activities.

In summary, the development of these FRET-based reporters provides the opportunities for better understanding of enzyme functions and their roles in the disease progression. Continued development of new strategies for design and biological applications of more FRET-based reporters is expected to gain further insight into the molecular information of enzyme activities. More importantly, unraveling the enzyme functions in the pathological conditions will greatly facilitate the disease diagnosis and efficient therapy in the future.

Acknowledgment

This work was partially supported by Start-Up Grant (SUG), Tier 1 (RG 64/10), (RG 11/13) and (RG 35/15) awarded in Nanyang Technological University, Singapore.

References

Abd-Elgaliel W. R., Tung C.-H., *BBA Gen. Sub.,* **1800** (2010), 1002–1008.

Baruch A., Jeffery D. A., Bogyo M., *Trends Cell Biol.,* **14** (2004), 29–35.

Blum G., Mullins S. R., Keren K., Fonovič M., Jedeszko C., Rice M. J., Sloane B. F., Bogyo M., *Nat. Chem. Biol.,* **1** (2005), 203–209.

Blum G., Von Degenfeld G., Merchant M. J., Blau H. M., Bogyo M., *Nat. Chem. Biol.,* **3** (2007), 668–677.

Bullok K., Piwnica-Worms D., *J. Med. Chem.,* **48** (2005), 5404–5407.

Caglič D., Globisch A., Kindermann M., Lim N.-H., Jeske V., Juretschke H.-P., Bartnik E., Weithmann K. U., Nagase H., Turk B., *Bioorgan. Med. Chem.,* **19** (2011), 1055–1061.

Campbell R. E., *Trends Biotechnol.,* **22** (2004), 208–211.

Choi Y., Cho Y., Kim M., Grailhe R., Song R., *Anal. Chem.,* **84** (2012), 8595–8601.

Chung E. Y., Ochs C. J., Wang Y., Lei L., Qin Q., Smith A. M., Strongin A. Y., Kamm R., Qi Y.-X., Lu S., *Nano Lett.,* **15** (2015), 5025–5032.

Cobos-Correa A., Trojanek J. B., Diemer S., Mall M. A., Schultz C., *Nat. Chem. Biol.,* **5** (2009), 628–630.

Demkow U., van Overveld F., *Eur. J. Med. Res.,* **15** (2010), 1–9.

Deu E., Verdoes M., Bogyo M., *Nat. Struct. Mol. Biol.,* **19** (2012), 9–16.

Drag M., Salvesen G. S., *Nat. Rev. Drug Discov.,* **9** (2010), 690–701.

Eder J., Hommel U., Cumin F., Martoglio B., Gerhartz B., *Curr. Pharm. Des.,* **13** (2007), 271–285.

Edgington L. E., Berger A. B., Blum G., Albrow V. E., Paulick M. G., Lineberry N., Bogyo M., *Nat. Med.,* **15** (2009), 967–973.

Edgington L. E., Verdoes M., Bogyo M., *Curr. Opin. Chem. Biol.,* **15** (2011), 798–805.

Evin G., Barakat A., Masters C. L., *Int. J. Biochem. Cell Biol.,* **42** (2010), 1923–1926.

Feng D., Zhang Y., Feng T., Shi W., Li X., Ma H., *Chem. Commun.,* **47** (2011), 10680–10682.

Fingleton B., *Curr. Pharm. Des.,* **13** (2007), 333–346.

Folk D. S., Torosian J. C., Hwang S., McCafferty D. G., Franz K. J., *Angew. Chem. Int. Ed.,* **51** (2012), 10795–10799.

Freeman R., Finder T., Gill R., Willner I., *Nano. Lett.,* **10** (2010), 2192–2196.

Gehrig S., Mall M. A., Schultz C., *Angew. Chem. Int. Ed.,* **51** (2012), 6258–6261.

Gora J., Latajka R., *Curr. Med. Chem.,* **22** (2015), 944–957.

Hardy J., Selkoe D. J., *Science,* **297** (2002), 353–356.

Hedstrom L., *Chem. Rev.,* **102** (2002), 4501–4524.

Hopkins A. L., Groom C. R., *Nat. Rev. Drug Discov.,* **1** (2002), 727–730.

Hu H. Y., Gehrig S., Reither G., Subramanian D., Mall M. A., Plettenburg O., Schultz C., *Biotechnol. J.,* **9** (2014), 266–281.

Hu J., Van den Steen P. E., Sang Q.-X. A., Opdenakker G., *Nat. Rev. Drug Discov.,* **6** (2007), 480–498.

Jares-Erijman E. A., Jovin T. M., *Nat. Biotechnol.,* **21** (2003), 1387–1395.

Ji T., Zhao Y., Wang J., Zheng X., Tian Y., Zhao Y., Nie G., *Small,* **9** (2013), 2427–2431.

Klein G., Vellenga E., Fraaije M., Kamps W., De Bont E., *Crit. Rev. Oncol. Hematol.,* **50** (2004), 87–100.

Kong Y., Yao H., Ren H., Subbian S., Cirillo S. L., Sacchettini J. C., Rao J., Cirillo J. D., *Proc. Natl. Acad. Sci. U. S. A.,* **107** (2010), 12239–12244.

Lee S., Park K., Kim K., Choi K., Kwon I. C., *Chem. Commun.,* (2008), 4250–4260.

Lemke E. A., Schultz C., *Nat. Chem. Biol.,* **7** (2011), 480–483.

Li J., Chen K., Liu H., Cheng K., Yang M., Zhang J., Cheng J. D., Zhang Y., Cheng Z., *Bioconjug. Chem.,* **23** (2012), 1704–1711.

Li J., Cheng F., Huang H., Li L., Zhu J.-J., *Chem. Soc. Rev.,* **44** (2015a), 7855–7880.

Li S. Y., Liu L. H., Cheng H., Li B., Qiu W. X., Zhang X. Z., *Chem. Commun.,* **51** (2015b), 14520–14523.

Liotta L., Tryggvason K., Garbisa S., Hart I., Foltz C., Shafie S., *Nature,* **284** (1980), 67–68.

Mcllwain D. R., Berger T., Mak T. W., *CSH Perspect. Biol.,* **5** (2013), a008656.

Mclntyre J. O., Fingleton B., Wells K. S., David W., Lynch C. C., Gautam S., Matrisian L. M., *Biochem. J.,* **377** (2004), 617–628.

Meyer B. S., Rademann J., *J. Biol. Chem.,* **287** (2012), 37857–37867.

Mohamed M. M., Sloane B. F., *Nat. Rev. Cancer,* **6** (2006), 764–775.

Mu J., Liu F., Rajab M. S., Shi M., Li S., Goh C., Lu L., Xu Q. H., Liu B., Ng L. G., *Angew. Chem. Int. Ed.,* **126** (2014), 14585–14590.

Murphy G., Nagase H., *Mol. Aspects Med.,* **29** (2008), 290–308.

Na Z., Li L., Uttamchandani M., Yao S. Q., *Chem. Commun.*, **48** (2012), 7304–7306.

Page M., Di Cera E., *Cell. Mol. Life Sci.*, **65** (2008), 1220–1236.

Palermo C., Joyce J. A., *Trends Pharmacol. Sci.*, **29** (2008), 22–28.

Paulick M. G., Bogyo M., *Curr. Opin. Genet. Dev.*, **18** (2008), 97–106.

Philippon A., Dusart J., Joris B., Frere J.-M., *Cell. Mol. Life Sci.*, **54** (1998), 341–346.

Pietrak B. L., Crouthamel M.-C., Tugusheva K., Lineberger J. E., Xu M., DiMuzio J. M., Steele T., Espeseth A. S., Stachel S. J., Coburn C. A., *Anal. Biochem.*, **342** (2005), 144–151.

Poincloux R., Lizárraga F., Chavrier P., *J. Cell Sci.*, **122** (2009), 3015–3024.

Poręba M., Stróżyk A., Salvesen G. S., Drąg M., *CSH Perspect. Biol.*, **5** (2013), a008680.

Razgulin A., Ma N., Rao J., *Chem. Soc. Rev.*, **40** (2011), 4186–4216.

Rudin M., Weissleder R., *Nat. Rev. Drug Discov.*, **2** (2003), 123–131.

Saghatelian A., Cravatt B. F., *Nat. Chem. Biol.*, **1** (2005), 130–142.

Sapsford K., E. Berti L., Medintz I. L., *Angew. Chem. Int. Ed.*, **45** (2006), 4562–4589.

Scanlan M. J., Raj B., Calvo B., Garin-Chesa P., Sanz-Moncasi M. P., Healey J. H., Old L. J., Rettig W. J., *Proc. Natl. Acad. Sci. U. S. A.*, **91** (1994), 5657–5661.

Seidah N. G., Prat A., *Nat. Rev. Drug Discov.*, **11** (2012), 367–383.

Sidransky D., *Science*, **278** (1997), 1054–1058.

Sloane B. F., List K., Fingleton B., Matrisian L., Proteases in cancer: Significance for invasion and metastasis. In: *Proteases: Structure and Function.* Springer: 2013; pp. 491–550.

Sternlicht M. D., Werb Z., *Annu. Rev. Cell Dev. Biol.*, **17** (2001), 463.

Takakusa H., Kikuchi K., Urano Y., Sakamoto S., Yamaguchi K., Nagano T., *J. Am. Chem. Soc.*, **124** (2002), 1653–1657.

Tung C.-H., Bredow S., Mahmood U., Weissleder R., *Bioconjug. Chem.*, **10** (1999), 892–896.

Turk B., *Nat. Rev. Drug Discov.*, **5** (2006), 785–799.

Verma R. P., Hansch C., *Bioorgan. Med. Chem.*, **15** (2007), 2223–2268.

Vihinen P., Ala-aho R., Kahari V.-M., *Curr. Cancer Drug Targets*, **5** (2005), 203–220.

Wang X., Xia Y., Liu Y., Qi W., Sun Q., Zhao Q., Tang B., *Chem. Eur. J.*, **18** (2012), 7189–7195.

Wang Y., Lin T., Zhang W., Jiang Y., Jin H., He H., Yang V. C., Chen Y., Huang Y., *Theranostics,* **5** (2015), 787.

Watzke A., Kosec G., Kindermann M., Jeske V., Nestler H. P., Turk V., Turk B., Wendt K. U., *Angew. Chem. Int. Ed.,* **47** (2008), 406–409.

Xing B., Khanamiryan A., Rao J., *J. Am. Chem. Soc.,* **127** (2005), 4158–4159.

Yuan L., Lin W., Zheng K., Zhu S., *Acc. Chem. Res.,* **46** (2013), 1462–1473.

Zheng G., Chen J., Stefflova K., Jarvi M., Li H., Wilson B. C., *Proc. Natl. Acad. Sci. U. S. A.,* **104** (2007), 8989–8994.

Zlokarnik G., Negulescu P. A., Knapp T. E., Mere L., Burres N., Feng L., Whitney M., Roemer K., Tsien R. Y., *Science,* **279** (1998), 84–88.

Chapter 14

Quantum Dot Architectures on Electrodes for Photoelectrochemical Analyte Detection

Mark Riedel, Daniel Schäfer, and Fred Lisdat

Biosystems Technology, Institute of Applied Life Sciences, Technical University of Wildau, Hochschlring 1, D-15745 Wildau, Germany

fred.lisdat@th-wildau.de

14.1 Introduction

Continuous progress in the construction and synthesis of new nanostructures has revolutionized various research fields. Numerous reviews show that in particular quantum dots (QDs)—following the breakthrough in optical applications—have increasingly drawn researchers' attention for combination with electrochemical methods (Devadoss et al., 2015; Lisdat et al., 2013; Yue et al., 2013; Zhao et al., 2014, 2015).

QDs are colloidal nanoparticles consisting of a semiconductor material and are defined by the occurrence of quantum confinement. As typical for semiconductors, the possible electronic states of electrons are described by a valence band (VB) and a conduction band (CB), which are separated by a bandgap. In bulk

Biocatalysis and Nanotechnology
Edited by Peter Grunwald

ISBN 978-981-4613-69-9 (Hardcover), 978-1-315-19660-2 (eBook)
www.panstanford.com

semiconductors, the energetic states can be described as a band-like model, since the energetic level of many individual atoms overlaps and merge to one VB and one CB, respectively. In contrast, due to the small size of QDs (few nanometers), a comparatively small amount of atoms defines the energetic structure and therefore resulting in a more pronounced formation of discrete energetic levels.

As a result of light–semiconductor interaction in form of photon absorption the bandgap can be overcome, creating free electrons in the CB and holes (or defect electrons) in the VB. This excited state is unstable and possesses limited lifetime in the range of nanoseconds. In the simplest case, the excited electrons relax in the initial state and light, which corresponds to the energy of the bandgap, is emitted (fluorescence). The "exciton Bohr radius" indicates the most probable distance between the electron and the hole. In QDs this distance is restricted in at least one dimension by the size of the particle. The resulting quantum confinement influences the energetic structure and leads to a functional correlation between the bandgap and the size of the nanoparticle. More specifically the bandgap increases with decreasing particle size.

This has some consequences for the physicochemical behavior of QDs. As illustrated in Fig. 14.1, most impressively and obviously this can be followed by the optical properties of the nanoparticles. While the absorption of light by QDs occurs over a relatively wide range with at least one maximum peak, the emission spectra show a rather sharp peak if the size distribution of the QD solution is narrow disperse. Both, the absorbance and the photoluminescence maxima depend on the size of the QDs and shift to smaller wavelength with decreasing particle diameter. The size-tunable fluorescence spectra of QDs make them very interesting as a label for bioanalytical approaches. This means by coupling of various molecules to different sized QDs, the fluorescence properties can be used for the optical discrimination of different biomolecules and the read-out of binding assays (Medintz et al., 2005). Thus, a multiplexed analysis becomes feasible. In comparison to organic fluorophores, QDs possess broad and intense absorption, high fluorescence quantum yields and better resistance against photobleaching, enabling long-term investigations (Resch-Genger et al., 2008).

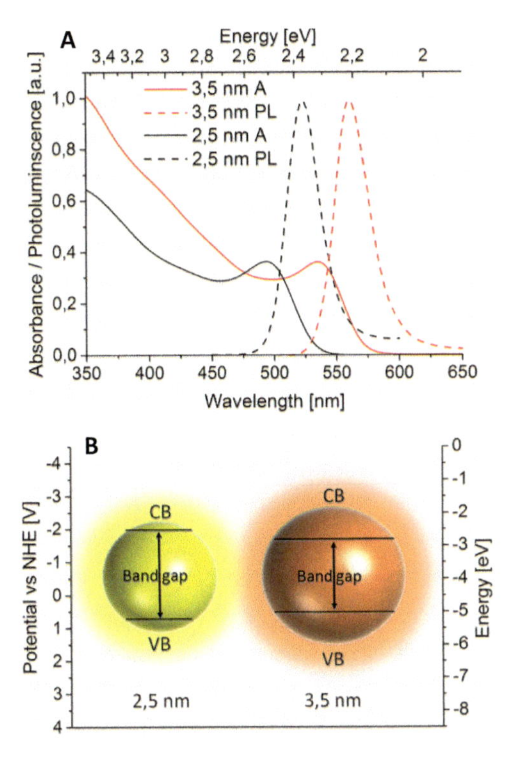

Figure 14.1 (A) Absorption and photoluminescence spectra of 2.5 and 3.5 nm CdSe/ZnS quantum dots, respectively. (B) Schematic illustration of the energetic level (CB—conduction band; VB—valence band) of CdSe/ZnS quantum dots with a diameter of 2.5 and 3.5 nm according to Jasieniak et al. (2011).

In addition to the use of QDs in optical assays, also the optoelectronic features of QDs have been considered to design analytical methods for the detection of biomolecules and chemicals. In principle, one can distinguish between an electrical-to-optical or optical-to-electrical transduction. The first transduction method is based on electrochemiluminescence (ECL) mechanisms. Here, QDs are excited by potential-controlled electrochemical reactions, resulting in the generation of light. However, this approach will be not considered further in this chapter. The second transduction method mainly represents the photoelectrochemical (PEC) approach.

By coupling QDs to electrodes electron transfer reactions can be controlled by switching on/off the light source (see Fig. 14.2).

In the absence of light, QDs can act as an insulating layer and therefore prevent reactions with the electrode and the analyte in solution. By illuminating the QDs, charge carriers are generated inside the particles and the electrode is "switched on," resulting in the generation of a current, which depends on the analyte concentration.

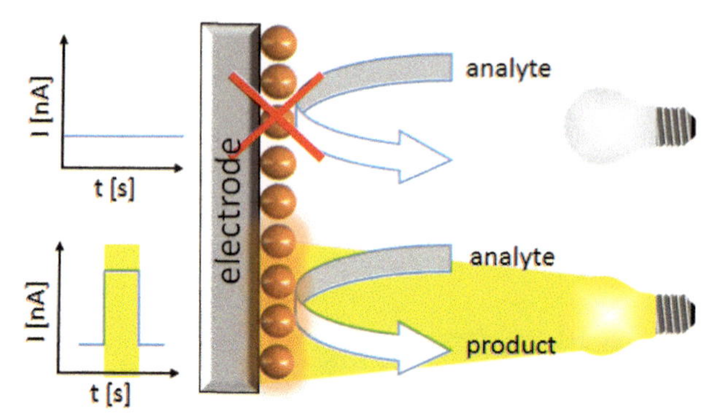

Figure 14.2 Schematic illustration of a QD-based photoelectrochemical sensor. On the left side of the scheme the current-output is demonstrated with and without illumination. The right side shows the light-controlled behavior of the QD-layer immobilized at an electrode. While in the absence of light, the QDs act as a rather insulating layer and prevent electron transfer reactions, by switching-on the light source charge carriers are generated inside the particles, resulting in a photocurrent, which depends on the analyte concentration in solution.

The PEC technique combines the advantages of electrochemical methods, e.g., low cost, simple instrumentations, high sensitivity, and portability, with the light-triggered read-out without the necessity of expensive optical equipment. This means the reaction can be controlled by the light source as well as the applied potential, which offers a higher flexibility during the experimental procedure.

While for the excitation process a simple white light source is sufficient, also monochromatic light can be utilized for kinetic investigations or for triggering complex reaction schemes. Furthermore, due to the light-induced read-out, the background signal can be reduced, leading to potentially higher sensitivities of PEC systems compared to conventional electrochemical methods. Therefore, significant efforts have been invested during the past decade to establish QD-based PEC processes in sensing and biosensing. New detection schemes have been developed by using new semiconductor materials as well as improving the efficiency of current generation, sensitivity, and selectivity of the analysis.

14.2 Functional Principles of QD-Based Electrodes

14.2.1 Electron Transfer Reactions

PEC QD-based systems rely on sequential charge transfer processes inside the nanoparticles, between the electrode and the QDs and between the QDs and the analyte in solution. In the absence of light, only very limited free charge carriers are available within the QDs. By illumination of the QD electrode electron–hole pairs are created inside the particles. Subsequently, three reactions can take place: electrode-QD electron transfer, redox molecule-QD electron transfer, and also unwanted charge carrier recombination (Curri et al., 2002; Katz and Shipway, 2005; Stoll et al., 2006). In Fig. 14.3, possible light-induced electron flow pathways are schematically illustrated. As a result of excitation and application of a negative bias a cathodic photocurrent is induced. In this case electrons from the electrode can fill up the holes in the valence band of the QDs and subsequently reduce an acceptor in solution via electron transfer from the conduction band. In the opposite direction, by applying a positive potential, electrons can flow from the QDs to the electrode and generate an anodic photocurrent. This is accompanied by the oxidation of a donor compound in solution, reacting with the holes of the QDs.

For the design of highly efficient photocatalytic applications, all electron transfer steps have to be adapted to each other to be limited mainly by one desired process only. In detail, the photocatalytic properties of the semiconductor material have to be adjusted to the analyte of interest, moreover efficient charge transfer to the electrode has to be ensured. The electron transfer between the electrode and the QDs is mainly influenced by the immobilization strategy (see Section 14.3), which defines the electron tunneling distance as well as the electronic properties in the gap between QDs and electrode (Khalid et al., 2011a; Nevins et al., 2011). Furthermore, the difference between the energetic level inside the QDs (CB and VB) and the Fermi level of the electrode material under a certain applied potential plays a crucial role in facilitating efficient electron transfer. Similarly, the reaction rates of redox molecules with photo-excited QDs are strongly influenced by the electronic states of the reaction partners. Consequently, the redox potential of the molecule in relation to the energetic structure of the QDs defines whether oxidation or reduction of the molecule is preferred under a certain potential. In addition, also surface properties of the nanoparticle have to be controlled to ensure the accessibility for the analyte, but also to favor stronger QD–analyte interactions. This can be achieved by surface modification strategies, e.g., by the introduction of surface charges (Stoll et al., 2006, 2008).

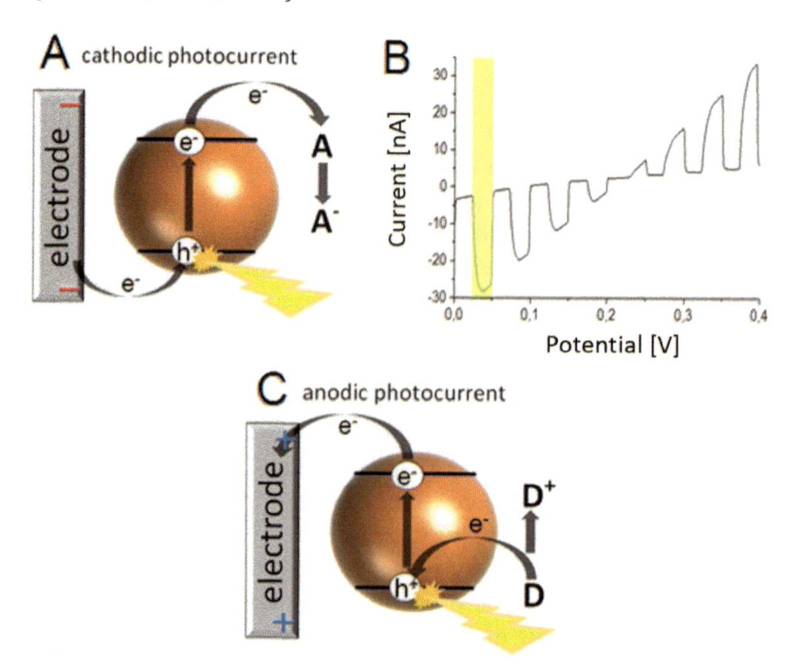

Figure 14.3 Light-initiated electron transfer steps at a QD electrode as cascade of reduction reactions (A) and oxidation reactions (C), respectively. A represents an acceptor and D a donor in solution. (B) Potential dependence of the photocurrent of a CdSe/ZnS QD-modified gold electrode (0.1 M HEPES pH 7, scan rate 5 mV/s, illumination duration 5 s).

By illumination a rather complex situation is created at QD electrodes where several electron transfer processes can compete and significantly influence the amplitude, shape, direction, and stability of the photocurrent. Also, surface trap states have shown to play a major role in the charge carrier behavior inside the QDs (see Section 14.2.2). Accordingly, the photocurrent represents a steady-state situation. If the electron–hole pair recombination extent is small and constant, and the electron transfer kinetics between electrode and QDs is fast and under constant conditions (constant temperature, pH, potential, etc.), the photocurrent only depends on the electron transfer rate between the QDs and the acceptor/donor. Thus, the magnitude of the photocurrent enhances with increasing concentration of the acceptor or donor in solution, enabling the utilization of QD electrodes for light-controlled analysis. At this point it has also to be noted that the application of QD electrodes for analysis is a relative new research area which benefits from other directions of science, also based on (photo) excitation of charge carriers in QDs:

- Solar cell developments based on the conversion of light to electrical energy
- Photocatalysis (utilization of light in combination with a photocatalyst to favor a reaction)
- ECL (electrochemiluminescence)

14.2.2 The Role of Surface Trap States

Surface trap states are energetic states, mostly in the bandgap of the semiconductor and can usually be found at the particle surface. The appearance of trap states significantly influences the energetic structure and, thus, the semiconductor properties. Since nanoparticles consist of a large part of surface atoms, also the susceptibility for such traps increases and therefore influence the behavior stronger compared to the bulk. In the presence of such traps, excited electrons (or holes) can be trapped in these electronic states, resulting in a decreasing fluorescence due to a diminished electron–hole pair recombination rate. However, also strong effects on the photoelectrochemical behavior have been observed and controversially discussed in the literature. While in the absence of surface trap states the direction of the photocurrent is controllable by the applied potential, the presence of traps can result in a unidirectionality of the photocurrent over a wide bias range and thus, limit the applicability to the detection of acceptor or donor compounds, respectively. For example CdS quantum dots have shown to tend to unidirectional photocurrents as a result of hole states (Hickey et al., 2000). Other reports demonstrated that a bias control of photocurrent direction is feasible with CdSe (Stoll et al., 2006, 2008) and CdTe QDs (Xu et al., 2014). Probably the different observations are not only attributable to the difference in semiconductor material, but much more to the different QD qualities used in the respective reports. Accordingly, the continuous progress in improved synthesis protocols may decrease the influence of surface trap states on the design of photoelectrochemical systems. Furthermore, other approaches utilize a coating of the QD core with a thin film of a second semiconductor (shell) with a higher bandgap to passivate the surface and reduce trap states (e.g., CdSe/ZnS QDs). Such core–shell particles result in improved fluorescence properties. Beyond that, potential-controlled photocurrents have also been demonstrated for these particles.

14.2.3 Influence of Charge Carrier Separation

The recombination of photoexcited charge carriers inside the QDs limits the performance of photoelectrochemical systems. Therefore, an improved charge carrier separation of the photogenerated electron–hole pairs is crucial to enhance the photocurrent response and sensitivity for the detection of redox species in solution.

The lifetime of electron–hole pairs is in the range of only a few nanoseconds (Gao et al., 2004), which makes the recombination process very fast and restricts the efficiency of charge carrier transfer reactions. Consequently, different approaches have been studied to improve the charge carrier separation and therefore increasing the photocurrent. Here, either the extension of the electron–hole pair lifetime inside the QDs or the utilization of nanomaterials and redox substances, which enable a faster electron transfer with external partners, have been considered.

For instance, by doping QDs with metal ions the photophysical properties of the QDs can be tuned. Recent reports have shown that the lifetime of electron hole pairs in QDs doped with manganese can be increased significantly (Beaulac et al., 2008; Vlaskin et al., 2010). Such long-lived charge carrier can be advantageous to boost the efficiency of electron transfer reactions (Santra and Kamat, 2012).

Also donor compounds such as ascorbic acid and sulfite or acceptor compounds like methylene blue have shown to favor the separation of charge carriers, which results in enhanced and stabilized photocurrents. Consequently, such molecules are interesting analytes for the photoelectrochemical detection. Since methylene blue is often used as a mediator for enzymatic reactions or as DNA intercalator this becomes also interesting for the combination with bioanalytical systems.

Furthermore, nanomaterials have shown to allow rapid electron transfer with the excited QDs and are thus intensively studied. Here, particularly gold nanoparticles, carbon nanotubes, titanium dioxide, and tin dioxide are widely used to improve the charge carrier separation rate. The construction of such hybrid nanostructures can be realized by two approaches, which influences the efficiency of the system. The most frequently used method is based on the assembly of the QDs together with nanostructures of different materials. Even if the assembly conditions must be strictly controlled this method is easy to apply. The other method uses advanced synthesis protocols to create dimeric nanostructures, which consist of at least one semiconductor part and another material. Due to the tight connection between the semiconductor nanostructure and the other nanomaterial the probability of charge carrier separation can be further increased, resulting in improved photoelectrochemical properties.

14.3 Construction of QD-Modified Electrodes

The modification and immobilization of QDs represents the essential part for the construction of light triggered PEC sensors. QDs typically consist of semiconductors from the II/VI family, such as CdS, CdS/ZnS, CdSe, CdSe/ZnS, CdTe, ZnO, and PbS. The synthesis of QDs has been shown to be feasible in organic as well as aqueous solution by bottom-up approaches. However, the synthesis in the organic phase is most often used, since it is easier, more controllable and favors the synthesis of monodisperse solutions (Gaponik et al., 2002; Mussa Farkhani and Valizadeh, 2014). The synthesis in aqueous solution utilizes thiols such as mercaptopropanoic acid as capping ligands. This is advantageous since the as-prepared QDs can directly be coupled by the carboxy group to an amine-modified electrode without further reaction steps. In contrast, for the stabilization of QDs which are prepared in the organic phase hydrophobic capping ligands like trioctylphosphine oxide, trioctylamine, and dodecylamine are used. Since these ligands act rather insulating and possess no functional groups for the attachment of QDs on electrodes, different modification and immobilization procedures have been established to overcome this problem.

Mostly a displacement of the organic ligands via ligand exchange is performed in the presence of other bifunctional ligands. These molecules act as linker between the electrode and the QDs. Therefore, the linker has to have at least two functional groups, which allow the interaction with the nanoparticles as well as with the electrode. Here, mainly two strategies have been applied for the fixation of QDs. One option is to modify the electrode with a linker molecule, followed by attachment of QDs to the modified electrode surface by partial ligand exchange and adsorption (Katz et al., 2006; Khalid et al., 2011a; Sabir et al., 2015; Stoll et al., 2006). Another option is to perform a ligand exchange in solution and immobilize the modified QDs afterwards to the electrode (Göbel et al., 2011; Nevins et al., 2011; Riedel et al., 2013; Tanne et al., 2011). The last variant offers the advantage that the organic ligand can nearly completely be displaced by several ligand exchange steps and do not further influence the interaction with the QD electrode and the analyte in solution.

Particularly thiol containing compounds are widely used for the modification of QDs. In Fig. 14.4 some often utilized modification compounds are compiled. The thiol–QD interaction is based on the high attraction between the semiconductor surface of the QDs and the thiol group, which results in a chemisorptive binding and a replacement of the organic ligand. Since the modified QDs have to be fixed to the electrode afterwards, the choice of the second functional group of the linker molecule depends on the electrode material or the modifier attached to the electrode.

For gold surfaces the utilization of dithiol or disulfide compounds (e.g., 1,6-hexandithiol, 1,4-dithiane, 1,4-benzenedithiol, stilbenedithiol) have been shown to be beneficial for the controlled attachment, since thiol groups bind both QDs and gold rather strongly. Here, the quality of the formed thiol layer and the electronic properties strongly determine the photocurrent behavior. In particular, the tunneling distance which depends on the length of the linker molecule affects the electron transfer rate. Large distances between the QDs and the electrode have shown to prevent photocurrents when aliphatic molecules are used. However, improved electron transfer rates can be achieved by means of linker molecules with conjugated π systems. Due to the conjugated π system a more effective electronic coupling is promoted, compared to linker molecules composed of aliphatic chains. Here, for example, stilbenedithiol and tetrahydro-4H-thiopyranylidene have shown to form well-ordered and rigid monolayers, resulting in functional surfaces for the attachment of the QDs with defined photocurrents (Bakkers et al., 2000; Khalid et al., 2011a).

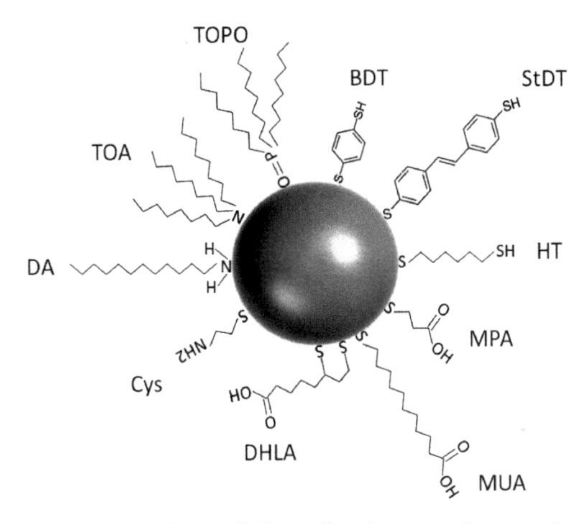

Figure 14.4 Overview of used ligands during the synthesis and for the modification of QDs. TOPO—trioctylphosphine oxide; TOA—trioctylamine; DA—dodecylamine; BDT—1,4-benzenedithiol; StDT—stilbenedithiol; HT—1,6-hexandithiol; MPA—mercaptopropionic acid; MUA—mercaptoundecanoic acid; DHLA—dihydroxylipoic acid; Cys—cysteamine.

However, it has to be mentioned that dithiols possess the tendency for the formation of disulfide bonds, the attachment of both thiol groups to the surface and are sensitive to sulfur oxidation. Therefore, the quality of the resulting dithiol layer has to be controlled to ensure efficient and stable PEC sensors. Different measurement techniques such as impedance spectroscopy, cyclic voltammetry, atomic force microscopy, scanning tunneling microscopy, X-ray photon spectroscopy, surface plasmon resonance, and quartz crystal microbalance are used to optimize the layer properties and the nanoparticle fixation. For instance, the amount of immobilized QDs at BDT modified gold surfaces have been investigated via quartz crystal microbalance, verifying a coverage in the range of a monolayer (Schubert et al., 2010).

Besides dithiols, also carboxy group containing thiols (e.g., mercaptopropionic acid, mercaptoundecanoic acid, dihydroxylipoic acid) are studied for the construction of PEC systems. In this case, the thiol binds to the QD surface and the carboxylic acid group can be used for the covalent or electrostatic fixation to amino group containing surfaces. For the covalent coupling carboxylic acid activation reagents like carbodiimides are often utilized. This approach is used for the QD immobilization on ITO by means of (3-aminopropyl)triethyoxy silane and coupling the carboxylic acid group to the amine of the silane (Xu et al., 2014). Furthermore, the strong interaction between the carboxy group and TiO_2 has been studied for the attachment of QDs (Nevins et al., 2011).

An alternative procedure for the construction of nanoparticle layers takes advantage of the biospecific interaction between avidin and biotin-conjugated QDs (Baş and Boyacý, 2009). Since several enzymes such as glucose oxidase have already been integrated in avidin architectures, the combination with biotin conjugated QDs can be also interesting for the construction of photoelectrochemical sensors.

Also the electrochemical deposition by means of electropolymerization has been shown to be beneficial for the immobilization of nanoparticles at electrodes. Therefore, CdS QDs capped with a mixture of *p*-aminothiophenol and mercaptoethane sulfonic acid are electrochemically assembled at a *p*-aminothiophenol modified gold electrode which results in a covalent binding of the nanoparticle via dianiline bridges (Granot et al., 2004). The amount of immobilized QDs can be controlled by the number of electrochemical deposition steps and followed by an increasing photocurrent response. Furthermore, it is found that the dianiline linker between the QDs participates on the charge transfer process at certain potentials and acts as tunneling medium from the nanoparticle to the electrode (Granot et al., 2004). Although many efforts are invested in the optimization of QD monolayers at electrodes by using various bifunctional molecules, the photocurrent output is limited by the amount of material within a QD monolayer. Therefore, the construction and performance of QD multilayers at electrodes has been studied by different assembly methods such as chemical linkage, bioaffinity binding, spin coating, and electrostatic adsorption (layer-by-layer assembly). Generally, for all approaches an enhancement and an improved stability of the photocurrent can be observed. Some QD multilayer architectures discussed below are illustrated in Fig. 14.5.

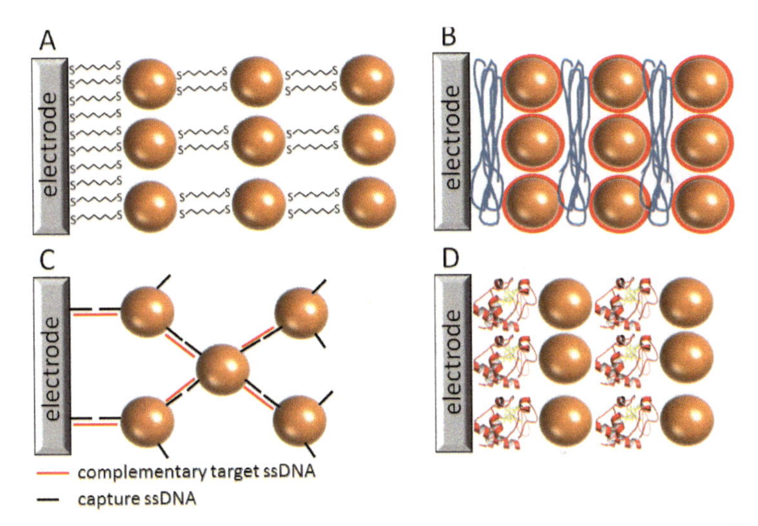

Figure 14.5 Multilayer architectures based on the alternating assembly of QDs and the different building blocks. (A) Layer-by-layer assembly of QDs by the use of chemical linkers such as alkanedithiols (chemisorption) according to Nakanishi et al. (1998). (B) Electrostatic layer-by-layer assembly of negatively charged QDs with positively charged polyelectrolytes according to Göbel et al. (2011). (C) Formation of QD/DNA structures by DNA-modified QDs and hybridization with complementary analyte ssDNA according to Willner et al. (2001). (D) Electrostatic layer-by-layer assembly of negatively charged QDs with the positively charged redox protein cytochrome c according to Göbel et al. (2011).

For example mono- and multilayer of CdS QDs have been build up on gold electrodes with alkanedithiols (see Fig. 14.5A) (Nakanishi et al., 1998). Therefore, alternating incubation steps in the alkanedithiol solution and the QD dispersion are performed, resulting in a layer-by-layer structure with enhanced photocurrent output. Wavelength dependent measurements show that the current spectrum corresponds to the absorbance of the nanoparticles and confirm that the QDs keep their photoelectrical properties even after immobilization.

Willner et al. (2001) have utilized the biospecific binding between complementary single-stranded DNA for the construction of QD multilayers (schematically shown in Fig. 14.5C). Therefore, the CdS QDs have been modified with short, single-stranded probe DNA and are cross-linked via hybridization with complementary DNA at the electrode. This results in an increasing number of attached QD layers and consequently enhances the photocurrent output. As discussed in Section 14.5.4, this approach has also been used for the analysis of hybridization events.

For the multilayer formation of nanoparticles electrostatic interactions have also been studied. This method is called electrostatic layer-by-layer assembly and is based on the construction of multiple films by deposition of oppositely charged species (see Fig. 14.5B). Interestingly, the properties of such films (roughness, thickness, porosity) can be controlled by the applied experimental conditions such as temperature, pH and ionic strength. Very often cationic polyelectrolytes such as polyallylamine hydrochloride (PAA) or polydiallyldimethylammonium chloride are used in combination with negatively charged QDs to achieve a layered deposition. The polyelectrolyte does not only facilitate the multilayer construction, but also provides advantageous conditions for interparticle electron transfer and enables the insertion of additional molecules or particles. Besides polyelectrolytes, the multilayer construction with charged proteins have been comparatively investigated by Göbel et al. (2011). In detail, one QD multilayer system has been constructed with the redox protein cytochrome c (cyt c) as building block (see Fig. 14.5D) and another with the polyelectrolyte PAA. While the construction of cyt c/QD multilayers results only in a minor photocurrent enhancement, the PAA-assisted multilayer assembly shows a proportional increase of the photocurrent with the number of deposited layers. Accordingly, cyt c cannot facilitate the electron transfer between the QD layers. In contrast PAA does not disturb the electron transfer between the QDs in the layered format. This comparison demonstrates the crucial influence of the used building block in a multilayer system on the interparticle electron transfer and subsequently on the photocurrent magnitude.

14.4 Concepts of Photochemical Systems for Analytical Detection

The various approaches of photoelectrochemical analysis can be classified in two main categories (I/II)—mainly based on the function of the QDs in the detection scheme. While one approach utilizes QDs as light switchable layer on electrodes (I), the other uses QDs as label for the detection of recognition events (II).

For the first category, the nanoparticles have to be immobilized on electrodes as described in Section 14.3. Due to the semiconductor properties of QDs, this results in the formation of a rather insulating layer in the absence of light. Accordingly, electrochemical conversions at the electrode are hindered to occur. However, if the QD electrode is illuminated the formation of a photocurrent is facilitated.

(Ia) As described in Chapter 14.2.1, under fixed conditions (e.g., potential, pH, etc.) the photocurrent is only influenced by the presence of redox active molecules which can act as donor or acceptor compounds. Thus, the determination of analyte concentrations by following the photocurrent magnitude becomes feasible.

(Ib) Furthermore, a light switchable QD layer attached to an electrode can be used to investigate binding events, if a recognition element such as DNA or antibodies is immobilized on the QDs. Upon binding of the target antigen, changes of the QD surface or a restricted access of redox molecules to the QD electrode is induced, which influences the photocurrent response. This allows label-free binding analysis. But since such a detection scheme is rather unspecific, the use of QDs as a label can be an alternative.

(II) In this case the photocurrent is used as a transduction principle for the detection of the binding event. This approach is focused on the construction of affinity sensors. Therefore, the recognition element is typically fixed to the electrode and can bind the analyte. For the detection step, one binding partner has to be labeled with QDs to introduce a photoelectrochemical activity. Consequently, as a result of the binding event a photocurrent can be generated, provided that the QD label is in close contact to the electrode. It is also possible to add mediators to shuttle electrons between the QDs and the electrode. This approach is beneficial to overcome weak electron transfer between the label and the electrode, because of larger distances.

For both concepts the use of light for switching on/off the sensor provides an additional tool to control the analysis and thus extends the possible application range of electrochemical methods compared to light-insensitive approaches. Since only the illuminated spot of the QD electrode is addressed, multiplex analysis become feasible when different biochemical systems are immobilized spatially separated on the electrode. As schematically illustrated in Fig. 14.6, the analytes can be individually read out by photoexcitation of the respective area with a spatially focused light beam. This means after read-out of the analyte of spot A the light is focused to spot B for the detection of another analyte. The spot area can be miniaturized and thus enables in principle the parallel detection of a large number of different analytes at one electrode. This approach is mainly limited by the resolution of the used light system and the spatially resolved immobilization of the biocomponent.

In all the approaches described before, current measurements (amperometric mode) have been applied for the light-triggered detection. However, signal transduction for light addressable sensors can also be performed in the potentiometric mode. Such potentiometric approaches are well known as light-addressable potentiometric sensors (LAPS) and are based on the potential formation at an ion-selective layer (Yoshinobu et al., 2015). In comparison to LAPS, an amperometric sensor system provides a better lateral resolution, since the photoexcited charge carriers diffuse less within the QD layer than in a bulk semiconductor. Additionally, current measurements reach a higher sensitivity but have a smaller dynamic measurement range compared to the potentiometric read-out. Furthermore, the amperometric mode needs no potential forming step, which is required for a potentiometric sensor and thus increases the applicability of such photoelectrochemical systems.

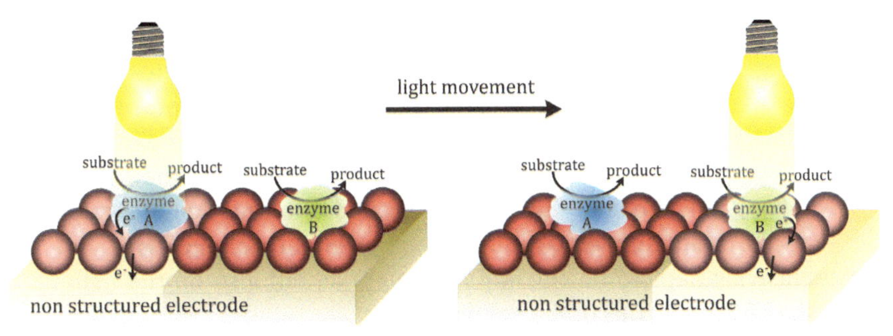

Figure 14.6 Schematically illustration of the spatially resolved sensor read-out of a QD electrode with different immobilized biochemical systems by illumination of the respective area.

14.5 Analyte Detection with QD Electrodes

The interest in using of QD electrodes for photoelectrochemical analysis and bioanalysis has led to different sensor formats for the sensitive and specific detection of a number of important analytes. Therefore, QDs act as an analyte-converting interface or nanoparticles have been combined with biological recognition elements or catalytically active compounds to address analytical purposes. In the following section, we introduce some photoelectrochemical approaches for analytical application. This will be subdivided into the direct analysis of redox active molecules, enzymatic sensing, DNA analysis, and aptamer/immunodetection.

14.5.1 Direct Analysis of Small Redox-Active Molecules

The electronic states of the QDs (conduction and valence band) and of the molecule to be detected (redox potential) strongly determine whether a molecule is able to be reduced or oxidized at the QD electrode under certain experimental conditions (schematically illustrated in Fig. 14.3). Since such light-triggered detection schemes can become rather complicated, first attempts have concentrated on the detection of simple redox systems like hexacyanoferrate (II/III). When hexacyanoferrate (III) ions are present in the solution, increasing cathodic photocurrents can be observed at CdSe/ZnS QD electrodes at negative potentials (Stoll et al., 2006). Consequently, electrons are transferred from the electrode via the illuminated QDs to hexacyanoferrate (III). As illustrated in Fig. 14.7, suitable conditions for the oxidation of hexacyanoferrate (II) ions at the QDs can also be found, which result in enhanced anodic photocurrents. First anodic photocurrent responses have been observed at electrode potentials below the redox potential of hexacyanoferrate (II/III), indicating the strong oxidizing power of holes in the photoexcited QDs. This demonstrates that the oxidation and reduction of hexacyanoferrate does not exclusively depend on the applied potential, but also on the light-triggered properties of the nanoparticles.

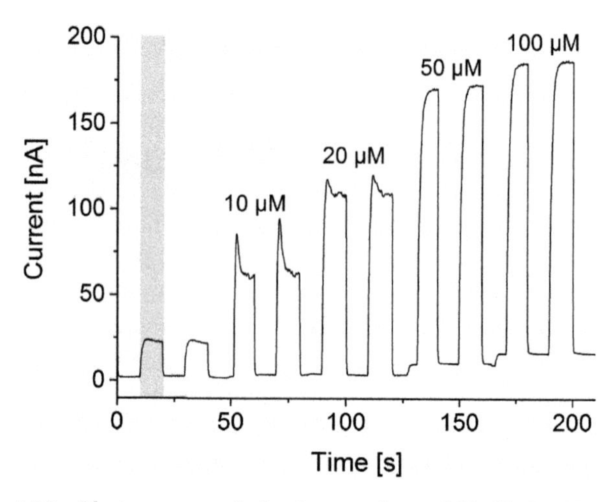

Figure 14.7 Photocurrent behaviour of a CdSe/ZnS QD-modified gold electrode after successive addition of the electron donor hexacyanoferrate (II) in solution. (+300 mV, illumination duration 10 s, 0.1 M HEPES pH 7).

Furthermore, various thiols have been analyzed at QD electrodes in a light-triggered format. Pardo-Yissar et al. (2003) used immobilized CdS QDs for the detection of thiocholine. Here, only in the presence of the donor thiocholine anodic photocurrents can be observed, which enhance with increasing analyte concentration.

Another approach has used methyl viologen modified CdS QDs immobilized on ITO for the detection of the amino acid cysteine (Long et al., 2011). Methyl viologen acts as electron acceptor, which transfers electrons from the conduction band of the excited QDs to the electrode. In the presence of cysteine, an enhanced anodic photocurrent can be observed, indicating that the oxidation takes place at the illuminated QDs by refilling the valence band. Therefore, the photoelectrochemical sensor shows a good sensitivity and selectivity against other interfering molecules.

In another study, glutathione has been analyzed with graphene–CdS nanocomposites immobilized on ITO, enabling the PEC detection of the analyte by direct oxidation in the range between 10 μM and 1.5 mM (Zhao et al., 2012d).

Various groups have also studied QD-based photoelectrochemcial systems for the detection of different metal cations such as copper and silver (Wang et al., 2010; Shen et al., 2011; Wang et al., 2012). However, the principle of all approaches is rather similar. The metal ions interact with the QDs and form metal containing compounds (e.g., Cu_xS) on the QD surface, which leads to the insertion of electron–hole pair trapping sites. This means, that the electron transfer between the electrode and the QDs is disrupted by the formed metal compound. Consequently, a decreasing photocurrent can be observed with increasing metal ion concentration, starting in the lower nanomolar range. However, the reusability of such detection schemes can be an issue due to the irreversible binding of the metal ions on the QD surface.

Furthermore, the photoelectrochemical response of modified QD-electrodes for phenolic compounds such as *p*-aminophenol (Khalid et al., 2011b) and dopamine (Wang et al., 2014a; Yan et al., 2015b) has been analyzed. Khalid et al. (2011b) have demonstrated the direct oxidation of *p*-aminophenol at CdS-modified gold electrodes under illumination, resulting in an increase of the anodic photocurrent. Signal changes are obtained in the range form 25 μM to 1.5 mM *p*-aminophenol (Khalid et al., 2011b).

In another approach, dopamine has been directly detected at graphene QD/TiO_2-nanocomposites (Yan et al., 2015b). Increasing anodic photocurrents are obtained with raising dopamine concentrations during illumination, showing the direct photooxidation of the analyte at the nanocomposites in the range between 20 nM and 105 μM (Yan et al., 2015b).

One special system for the detection of dopamine has been demonstrated by Wang et al. (2014a). In a first step poly-dopamine is prepared by oxidation of dopamine in weakly alkaline solution. Operation relies on the photocatalytic activity of CdS QDs for the reduction of benzoquinone groups of the formed poly-dopamine, leading to decreased anodic photocurrents. Consequently, the PEC detection is based on the competitive situation between the poly-dopamine and the electrode for the photoexcited electrons of the QDs. Since the amount of benzoquinone groups in the formed poly-dopamine correlates to the initial dopamine concentration, the signal response reflects the amount of analyte in solution. Here, the photoelectrochemical system provides high sensitivity and selectivity, but longer analysis time (Wang et al., 2014a).

Schubert et al. (2010) have studied the photoelectrocatalytic oxidation of the enzymatic co-factor nicotinamide adenine dinucleotide (NADH) at CdSe/ZnS QD-modified gold electrodes. Such an electrode design allows the detection of NADH in the range of 20 μM to 2 mM in a rather wide potential window. At a potential where anodic photocurrents are generated the addition of NADH results in an enhancement of the anodic photocurrent (see Fig. 14.8). In contrast the addition of the oxidized species NAD^+ shows no photocurrent response. This confirms that NADH can be selectively oxidized at the illuminated QD electrodes, resulting in an electron transfer from NADH over the QDs to the electrode. At potentials (below +100 mV vs. Ag/AgCl) where an initial cathodic photocurrent is generated the photocurrent decreases or even reverses to an anodic photocurrent in the presence of NADH. This can be attributed to the competitive situation between the gold electrode and NADH for the transfer of electrons to the excited QD layer. This also demonstrates that the direction of the photocurrent can be reversed by a chemical reaction.

In some cases catalytic activity towards an analyte cannot be provided by the QDs itself, but has to be incorporated by a modification of the nanoparticles. For example, a large number of different porphyrin complexes have been combined with the light-triggered read-out of QD electrodes to add or improve the catalytic activity (Tu et al., 2010; Golub et al., 2012; Wang et al., 2015; Yan et al., 2015a). Tu et al. (2010) studied porphyrin-functionalized TiO_2 nanoparticles as photoelectrochemcial platform for the detection of glutathione. The modified nanoparticles are prepared by the binding of the sulfonic acid groups of [meso-tetrakis(4-sulfonatophenyl) porphyrin] iron(III) monochloride to TiO_2. The authors report that glutathione is oxidized at the photo-excited porphyrin and subsequently the electrons are transferred to the TiO_2 nanostructures. Enhanced anodic photocurrents are detected in the range from 0.05 to 2.4 mM with a detection limit of 0.03 mM glutathione.

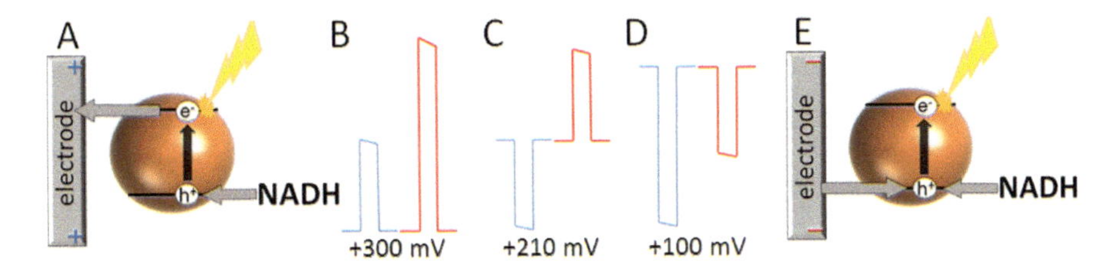

Figure 14.8 NADH oxidation at a CdSe/ZnS QD electrode: (I) Anodic light-initiated electron pathways (A) and photocurrent change (B) at a CdSe/ZnS QD-modified gold electrode in the absence (blue) and presence (red) of NADH by applying a potential of +300 mV vs. Ag/AgCl. The addition of NADH results in an enhancement of the anodic photocurrent. (II) Cathodic light-initiated electron pathways (E) and photocurrent change (C/D) at a CdSe/ZnS QD-modified gold electrodes in the absence (blue) and presence (red) of NADH by applying a potential of +210 mV (C) or +100 mV (D) vs. Ag/AgCl. The addition of NADH results in a decrease of the cathodic photocurrent (D) or a complete inversion to an anodic photocurrent (C) due to the competitive situation for filling up the light-initiated holes of the excited QDs. The figure is according to Schubert et al. (2010).

In another approach, the detection of the biologically interesting analyte hydrogen peroxide has been investigated at CdSe/ZnS and CdS QD-electrodes. However, no photocurrent response is found in the presence of the molecule. To overcome these limitations various approaches have been followed to introduce a catalytic activity towards H_2O_2 in a light-triggered system. Khalid et al. (2011a) have studied the combination of CdS QDs with FePt nanoparticles for the design of a light-switchable H_2O_2 sensor in two approaches (see Fig. 14.9).

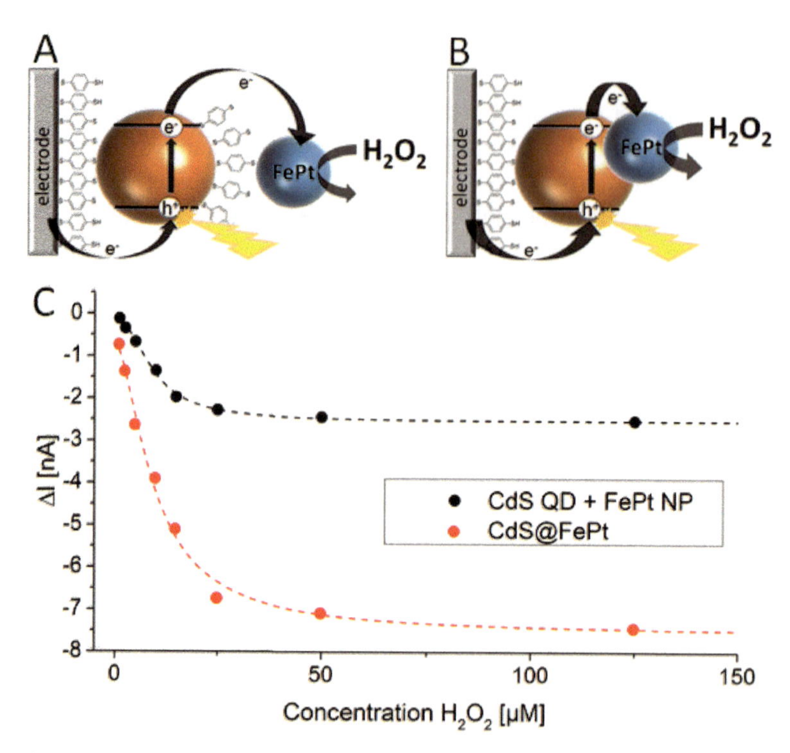

Figure 14.9 Detection of H_2O_2 with different QD/FePt nanoparticle electrode assemblies. (A) Schematic illustration of a CdS QD-modified gold electrode with FePt nanoparticles fixed on top of the QDs via a dithiol. (B) Schematic illustration of a gold electrode modified with hybrid CdS/FePt nanoparticles. (C) Photocurrent response of the respective QD/FePt nanoparticle electrode assemblies in dependence on hydrogen peroxide concentration in solution according to Khalid et al. (2011a). The black and the red points correspond to electrode assembly (A) and (B), respectively (E = −200 mV vs. Ag/AgCl).

In one system the QDs and the FePt nanoparticles have been co-immobilized on gold electrodes by the use of dithiols. Here, an increasing cathodic photocurrent can be observed in the presence of H_2O_2, which demonstrates the reduction of the molecule at the co-assembled nanostructure. Consequently, electrons are transferred from the electrode via the QDs to the FePt nanoparticles and subsequently to the analyte. Since the reduction of H_2O_2 involves H^+ and e^- the reaction is preferred in acidic region and can be facilitated with a decreasing potential. In a more advanced approach, hybrid nanoparticles, which consist of a CdS- and a FePt-part, have been synthesized and immobilized on gold electrodes (Khalid et al., 2011a). As described in Section 2.3, such nanohybrid materials can result in improved photoelectrochemical performance due to the tight connection. In the presence of H_2O_2 a 2–3-fold higher response can be obtained for electrodes modified with the dimeric nanoparticles compared to electrodes, which are modified by the co-assembly of CdS QDs and FePt nanoparticles. Enhanced signal responses can be obtained in the range from 1 to 30 μM H_2O_2.

14.5.2 Combination of Redox Molecules with Enzymatic Reactions

As described in the previous section, some analytes can directly be detected on various modified QD-electrodes via a light-triggered read-out. However, the limited number of detectable molecules as well as the specificity of such systems can be an issue. In order to improve this situation particularly biocatalysts are often used due to their good substrate specificity. Here, several approaches have been followed to combine enzymatic reactions with QD electrodes. This is summarized in Fig. 14.10. Mainly, the detection of enzymatically formed products or consumed co-substrates have been studied at the illuminated QDs (Fig. 14.10A–D). Not much work has been devoted so far to mediator-based approaches (Riedel et al., 2017) (Fig. 14.10E) or the direct protein electrochemistry at QD electrodes (Fig. 14.10F) (see Section 14.5.3).

In one approach, acetylcholine esterase (AChE) has been covalently bound to a CdS-modified gold electrode for the investigation of enzyme inhibitors (Pardo-Yissar et al., 2003). While in the absence of the substrate acetylthiocholine no photocurrent is observed at the QD electrode, after addition of the substrate the enzyme catalyzes the conversion to acetate and thiocholine, resulting in a generation of a photocurrent during illumination (schematically illustrated in Fig. 14.10A). The product thiocholine acts as electron donor and therefore is the driving force for the formation of a photocurrent. Consequently, the magnitude of the photocurrent can be controlled by the concentration of acetylcholine and depends on the activity of the enzyme. Since acetylcholine esterase is inhibited by various insecticides as well as toxins, the signal response for the selective inhibitor 1,5-bis(4-allyldimethylammoniumphenyl)pentane-3-one dibromide has been evaluated. In the presence of the inhibitor (and at constant substrate concentration), a decreasing photocurrent can be observed, which is attributed to the lower yields during the enzymatic production of thiocholine (Pardo-Yissar et al., 2003). In a similar approach, a QD/AChE hybrid system has been utilized for the detection of organophosphorus pesticides with a detection limit of 2.5 pM for dichlorvos and 0.06 pM for paraoxon (Li et al., 2015). The PEC sensor shows a good sensitivity, stability, reproducibility and has been successful applied for the detection of organophosphorus pesticides in real fruit samples (Li et al., 2015).

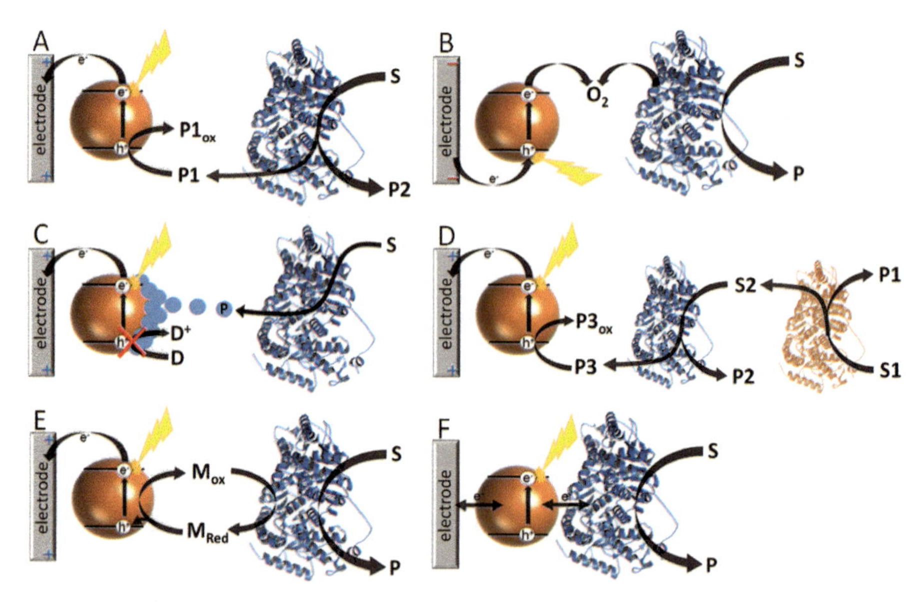

Figure 14.10 Different PEC detection schemes based on enzymatic reactions. Detection of enzymatically formed products at the illuminated QDs by the use of a single enzyme (A) or a sequential enzymatic reaction cascade (D). Detection of the consumed co-substrate results in a competitive reaction between the QDs and the enzyme (B). Production of an insoluble product, which precipitates at the QD electrode and thus disturbs the access of the donor to the particle surface (C). Mediator-based (E) and direct electron transfer (F) between the QDs and the enzyme.

In two other PEC approaches the enzyme alkaline phosphatase (ALP) has been combined with semiconductor nanoparticles to establish sensorial detection schemes for the analysis of *p*-aminophenylphosphate (pAPP) (Khalid et al., 2011b), ascorbic acid 2-phosphate and enzyme inhibitors (Zhao et al., 2013). The first system is based on the oxidation of the phenolic compound *p*-aminophenol (pAP) at excited CdS nanoparticles. Since the ALP catalyzes the reaction of pAPP to pAP the substrate can be detected by the light-controlled oxidation of the produced pAP at the QD electrode (schematically illustrated in Fig. 14.10A). In order to construct a sensor the enzyme has been immobilized in a layered format by using positively charged polyelectrolytes. Increasing anodic photocurrents have been found in the range between 0.025 and 1 mM pAPP (Khalid et al., 2011b).

In the second system, alkaline phosphatase has been attached to TiO_2 nanoparticles. The enzymes catalyzes the dephosphorylation of the substrate ascorbic acid 2-phosphate to ascorbic acid (Zhao et al., 2013). Ascorbic acid contains an enediol structure, which has a high affinity for TiO_2 surfaces. The authors report on a self-coordination of the enzymatic product to the nanoparticles, which influences the photoelectrochemical properties of the system. In detail, for the bare TiO_2 nanoparticles the absorption is shifted from the ultra violet region (<400 nm) to the visible range in the presence of ascorbic acid. Consequently, after the reaction with ascorbic acid the TiO_2 nanoparticles can also be excited by illuminating with wavelengths above 400 nm. By performing photoelectrochemical measurements at 410 nm illumination an increasing photocurrent response can be observed after addition of ascorbic acid 2-phosphate. The applicability of this approach has also been tested for the investigation of the ALP inhibition by using the model inhibitor 2,4-dichlorophenoxyacetic acid (Zhao et al., 2013). In the presence of the inhibitor, diminished photocurrents have been observed compared to the measurements without inhibitor. This shows that the inhibition of ALP by 2,4-dichlorophenoxyacetic acid can be effectively followed by the TiO_2 nanoparticle-based PEC detection scheme.

Schubert et al. (2010) have applied the NADH sensitivity of CdSe/ZnS QD-modified electrodes for the combination with NADH-producing enzymes for the light-triggered analysis. Here, the amount of enzymatically produced NADH corresponds to the concentration of the substrate of the biocatalyst. In one example, glucose dehydrogenase has been combined with QD electrodes (Schubert et al., 2010). The enzyme catalyzes the oxidation of glucose to gluconolactone in the presence of its cofactor NAD^+. The cofactor is reduced to NADH and can be detected at the QD electrode (schematically illustrated in Fig. 14.10A). Since the glucose signal is converted to NADH by enzymatic electron transfer the NADH-dependent signal increase directly correlates to the substrate concentration. In such a detection scheme, the sensitivity for glucose is similar to that for NADH and provides a signal response up to 1 mM glucose.

In another example, the NADH sensitivity of QD electrodes has been utilized for the light-triggered detection of guanosine monophosphate (GMP) in a enzymatic reaction assay (Sabir et al., 2015). The general principle of such an assay is schematically illustrated in Fig. 14.10D. First, GMP is phosphorylated in the presence of ATP by the guanylate kinase (GMPK) to GDP and ADP. Second, the pyruvate kinase (PK) catalyzes the conversion of ADP (and GDP with lower reaction rate) after addition of phosphoenolpyruvate to ATP (GTP) and pyruvate. In a last step, NADH is consumed by lactate dehydrogenase (LDH) and forms lactate and NAD^+. Consequently, NADH is no longer available for the oxidation at the illuminated QD electrode and the NADH-dependent part of the photocurrent decreases. Since the concentration of the enzymes and co-substrates is constant, the enzymatic NADH oxidation and thus the photocurrent change exclusively depends on the GMP concentration in solution. The enzymatic reaction assay is given by the following equations:

$$\text{GMP} + \text{ATP} \xrightarrow{\text{GMPK}} \text{GDP} + \text{ADP}$$

$$\text{ADP(GDP)} + \text{PEP} \xrightarrow{\text{PK}} \text{ATP(GTP)} + \text{pyruvate}$$

$$\text{pyruvate} + \text{NADH} + \text{H}^+ \xrightarrow{\text{LDH}} \text{lactate} + \text{NAD}^+$$

The feasibility of this concept has been first evaluated with all enzymes in solution, resulting in a well-defined response in the concentration range from 0.05 to 1 mM GMP (Sabir et al., 2015). Subsequently, a mixture of all enzymes has been immobilized on top of the QD electrode to build up a photoelectrochemical GMP sensor. For the stabilization of the enzyme layer the QD/enzyme electrode is covered by two bilayers of the polyelectrolytes poly(sodium-4-styrene sulfonate) and poly(allylamine) hydrochloride. The sensor shows a quiet similar signal response compared to that detected with the enzymes in solution, showing the functionality of the system. The combination of QD electrodes with enzymatic reaction cascades also provides the potential for the detection of other nucleotides if the nucleotide-consuming enzyme is simply exchanged.

Various studies which have investigated the optical properties of QDs in solution reveal that the photoluminescence is influenced by the oxygen concentration (van Sark et al., 2001). Inspired by these findings, Tanne et al. (2011) have studied whether oxygen can act as an electron acceptor in a QD-based photoelectrochemical system. Therefore, photocurrent measurements have been performed in air-saturated and oxygen-free (argon-purged) buffer with CdSe/ZnS QD electrodes. When the electrode is negatively polarized a smaller cathodic photocurrent can be observed in oxygen-free solution compared to the air-saturated solution. This shows that oxygen accepts electrons from the QDs and participates in the cathodic photocurrent generation. Consequently, the electron transfer from the electrode to the valence band of the excited QDs is facilitated by the reduction of oxygen. Another study, which utilizes CdTe QDs immobilized on ITO as photoelectrochemical platform, has indicated that the reaction product of oxygen reduction is superoxide (Xu et al., 2014).

As illustrated in Fig. 14.10B, the oxygen sensitivity of QDs immobilized on electrodes can be coupled with an oxidase in order to detect the substrate of the respective enzyme. Here, a competitive situation between the QDs and the oxidase for oxygen is created in the presence of substrate, which results in a decrease of the cathodic photocurrent. First, this principle has been evaluated with glucose oxidase (GOx) for the light-controlled glucose detection (Tanne et al., 2011). In order to reach a sufficient oxygen depletion in front of the QD electrode a rather high enzyme density has to be provided. Therefore different coupling approaches like the cross-linking of GOx with glutaraldehyde or a layer-by-layer assembly with the oppositely charged polyelectrolyte poly(allylamine) hydrochloride can be used. The latter one provides the advantage that the amount of enzyme can easily be controlled by the number of attached polyelectrolyte/enzyme-layers and that—compared to the approach with glutaraldehyde—no reactive groups can influence the photocurrent behavior. The deposition of four enzyme layers allows to use the full range of the oxygen-dependent photocurrent and therefore provides optimal conditions for the photoelectrochemical glucose sensing. In such a layered format, the sensor achieves short response times and a dynamic range between 0.1 and 5 mM glucose (Tanne et al., 2011).

In fact, this PEC concept is also applicable to other oxygen-consuming enzymes. This has already been shown in a similar approach by immobilization of sarcosine oxidase on QD-modified electrodes for the light-triggered sarcosine detection (Riedel et al., 2013).

One simple PEC detection principle is based on the enzymatic production of an insoluble product in the presence of the analyte (schematically shown in Fig. 14.10C). Subsequently, the insoluble product precipitates at the transducer, which disturbs the electron transfer between the QD electrode and donor or acceptor substances in solution and thus, decreases the photocurrent. Such a PEC detection scheme has been presented by Zhao et al. (2011a) for the detection of H_2O_2. Therefore, horseradish peroxidase (HRP) has been immobilized on a TiO_2/CdS QD matrix modified ITO electrode. The enzyme catalyzes the oxidation of 4-chloro-1-naphthol in the presence of H_2O_2 to insoluble benzo-4-chloronaphthol, which precipitates at the QD electrode and forms an insulting layer, resulting in decreasing photocurrents. In this case, a detection limit of 0.5 nM H_2O_2 has been achieved. However, since the binding of the insoluble product on the QDs is irreversible, such a detection scheme is only suitable for single use application.

Zheng et al. (2011) have demonstrated the construction of a PEC glucose sensor by using a mediator-based approach to transfer the electrons from the enzyme glucose oxidase to the excited QDs (schematically illustrated in Fig. 14.10E). Therefore, the enzyme has been attached in combination with the tris-1, 10-phenantroline complex of cobalt $[Co(Phen)_3]Cl_2$ to the electrode modified with TiO_2 and CdSe/CdS QDs. Given that a rather high oxidation potential (+400 mV vs. Ag/AgCl) has been used for the analysis, the mediated electron transfer can be influenced by the direct sugar oxidation.

14.5.3 Direct Protein Electrochemistry

Many efforts have been invested during the last decade to establish direct protein electrochemistry on electrodes to overcome the needs of expensive redox mediators or co-substrates and to avoid interfering side reactions. Such approaches are based on the direct heterogeneous electron transfer (DET) between the redox center of the protein and the electrode and have been demonstrated for redox proteins and enzymes. Nanostructures have shown to be beneficial for electrode modification in order to obtain DET with biomolecules. Besides gold nanoparticle and carbon nanotubes also QDs have been studied for direct electron transfer reactions of proteins (schematically illustrated in Fig. 14.10F).

In one approach, the redox protein cytochrome c (cyt c) has been considered for the establishment of a light-triggered direct electron transfer with CdSe/ZnS QDs immobilized on gold electrodes (Stoll et al., 2006). Here, particularly the surface properties of the QDs have to be controlled to enable an efficient electron transfer between cyt c and the illuminated nanoparticles. While hydrophobic organic capping agents on the QD surface prevent the electron transfer with cyt c, the exchange of the ligand by a mercaptopropionic acid or mercaptosuccinic acid promotes the QD–protein interaction. An increasing cathodic photocurrent can be observed by applying a negative bias and adding oxidized cyt c to the solution. The photocurrent change depends on the concentration of cyt c in solution and thus demonstrates that the redox protein can be reduced at the excited QDs. Conversely, reduced cytochrome c can be oxidized by the illuminated nanoparticle under positive polarization, resulting in enhanced anodic photocurrents. Thus, a DET between cyt c and the illuminated QDs has been observed, the direction of which depends on the potential and the redox state of the protein. This provides the basis for the design of analytical signal chains.

In one cyt c-based system the superoxide radical sensitivity of the redox protein has been utilized to construct a light-switchable radical sensor (Stoll et al., 2008). Since superoxide is generated in biological systems under several physiological and pathophysiological conditions the analysis of this reactive oxygen species has been intensively followed. In the presence of superoxide radicals, the heme center of cytochrome c is reduced. When the electrode is positively polarized the reduced protein can be re-oxidized at the illuminated QDs (see Fig. 14.11A). Consequently, the electron transfer from the excited QDs to the electrode is amplified and an increasing anodic photocurrent can be observed. Thereby the signal enhancement correlates with the concentration of the radical and enables the detection in the nanomolar range (Stoll et al., 2008).

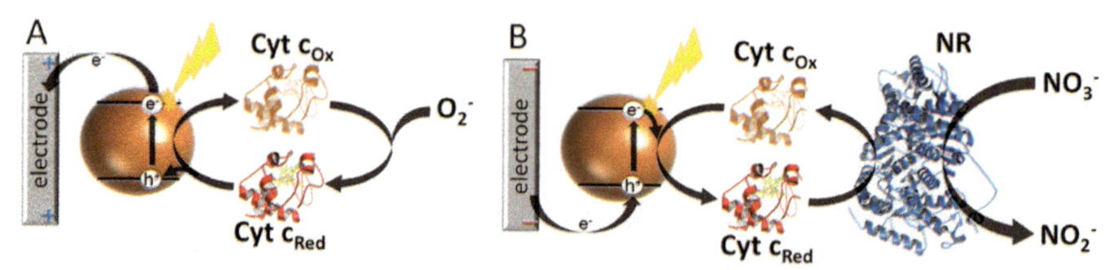

Figure 14.11 Photoelectrochemical signal chains on QD electrodes with cytochrome c based on DET. (A) Superoxide detection via cyt c according to Stoll et al. (2008). (B) Photoreduction of cyt c by the QD electrode can be amplified in the presence of nitrate (NO_3^-) by nitrate reductase (NR). This allows the photoelectrochemical light-triggered determination of nitrate according to Katz et al. (2006).

Furthermore, the interaction of cyt c with enzymes has been evaluated for the construction of light controlled signal chains on QD-modified electrodes. In such a way the enzyme catalyzes the redox reaction of cyt c in the presence of substrate and amplifies the photoelectrochemical signal compared to the situation without the enzyme and substrate. The feasibility of this concept has been shown by Katz et al. (2006) by coupling of QD electrodes with cyt c and the enzyme nitrate reductase (NR) for the nitrate detection (see Fig. 14.11B). The cathodic photocurrent increases in the presence of NR and nitrate up to a substrate concentration of 80 mM. This corresponds to an 8-fold signal amplification compared to the signal obtained for the measurement with oxidized cytochrome c only. Consequently, the enzyme regenerates oxidized cyt c, which is now available again for the reduction at the conduction band of the excited QD and amplifies the photocurrent.

Similarly, the cyt c/QD-system has been coupled to the enzyme lactate dehydrogenase (LDH) for lactate detection (Katz et al., 2006). LDH catalyzes the reduction of cyt c in the presence of lactate, which results in an up to 9-fold enhancement of the anodic photocurrent. In this case, reduced cyt c is regenerated by the enzyme and is once again ready for the oxidation at the illuminated QD electrode.

Based on the work of Ipe and Niemeyer (2006), the direct electron transfer between QDs and the enzyme cytochrome P450 2C9 (CYP2C9) has been applied to tune a light-driven drug metabolism. Cyt P450s are a family of heme-containing proteins which are involved in the metabolism of drugs and the bioactivation of xenobiotic substances into reactive compounds. Xu et al. (2014) coupled CYP2C9 covalently to functionalized CdTe QDs and assembled both to ITO electrodes. As model substrate tolbutamide has been chosen, which is metabolized by CYP2C9 to 4-hydroxytolbutamide in the presence of oxygen (hydroxylation reaction). Upon illumination cathodic photocurrents are generated under negative polarization and enhance after successive addition of tolbutamide. In a similar approach, cytochrome P450 2D6 has been combined with QDs for the light driven metabolism of tramadol, showing the applicability of this concept to other enzymes from the Cyt P450 family.

14.5.4 Analysis of Biospecific Interactions

The PEC analysis of binding events exploiting biospecific interactions has been followed with growing interest. The applications can be divided on the basis of the used recognition elements such as DNA, aptamers, antibodies and molecular imprinted polymers (MIP). All approaches use the specificity of the recognition element and the close proximity between the QDs and the electrode for the detection of the bound analyte. Here, different detection principles have been developed, which are presented in the following section:

- Label-free analysis of binding events based on changes in accessibility for donor/acceptor compounds at the QD electrode after formation of biospecific complexes.

- Direct oxidation/reduction of the bound analyte at the illuminated QDs.
- Redox-active label (metal nanoparticles and hemin compounds), which influence the photocurrent behavior of the QD electrode after the biospecific interaction due to the close proximity between the QDs and the label.
- Production of electroactive or insulting products by enzymatic labels near the QDs
- Application of QDs as photoelectrochemical label in binding assays.
- Chemiluminescence resonance energy transfer (CRET). Generation of light by a chemical reaction, which triggers the excitation of the QDs and overcomes the needs for an external light source.

One of the most extensively investigated PEC biosensing area represents the DNA analysis. Much efforts have been put into the study of DNA-hybridization, DNA-protein interactions and aptamer sensing. Hybridization describes the specific association of two single DNA strands to one double-stranded DNA. The precondition is that the single strands are complementary to each other. Consequently, if one single DNA strand is immobilized on the electrode it can act as recognition element (probe) for the detection of complementary target DNA.

In a first approach, Willner et al. (2001) have utilized CdS QDs modified with short single-stranded probe DNA for the sensing of target DNA. In the presence of target DNA the QDs are cross-linked via hybridization at the electrode, resulting in an increasing number of attached QDs (see Fig. 14.5C). Consequently, the generated photocurrent enhances with increasing concentration of target DNA. Additionally, redox complexes such as ruthenium hexamine ($[Ru(NH_3)_6]^{3+}$), which binds electrostatically to duplex DNA, are added in order to improve the conductivity of the DNA linker between the QDs (Willner et al., 2001).

Zhao et al. (2012a) have used QD-electrodes modified with single-stranded probe DNA for the sensing of DNA (schematically shown in Fig. 14.12A). Stable anodic photocurrents are observed in the presence of the electron donor ascorbic acid. After hybridization with silver nanoparticle-labeled target DNA the signal response decreases. The authors explain this by the interaction between the light-induced electron–hole pairs and the plasmon of the silver nanoparticle, resulting in a decrease of the photocurrent after binding of the labeled target DNA. The signal response depends on the distance between the QDs and the silver nanoparticle and is therefore determined by the length of the hybridized DNA strand. For 36 nucleotide-long DNA, decreasing anodic photocurrents have been found in the range between 2 fM and 20 pM labeled target DNA (Zhao et al., 2012a). In contrast, control experiments with labeled noncomplementary DNA show only a negligible photocurrent change of 2%, indicating a high selectivity (Zhao et al., 2012a). Similar results have been obtained by using probe DNA-modified QD-electrodes for the detection of gold nanoparticle-labeled target DNA (Zhao et al., 2011).

In a rather unique approach a PEC detection scheme has been designed, which is based on CRET and needs no external light irradiation for the excitation of the QDs (Golub et al., 2012). The principle is based on the generation of light by a chemical reaction, which is crucial for the photoactivation of the QDs (see Fig. 14.12B). In detail, guanine-rich DNA (aptamer), which can specifically bind hemin and result in the formation of hemin/G-quadruplex complexes, is used as a recognition element. Hemin/G-quadruplex complexes act as an electrocatalyst for the reduction of H_2O_2 and are able to oxidize simultaneously luminol, generating light by the chemiluminescence reaction. This light switches the excitation of the QDs. Consequently, a photocurrent can be observed, the magnitude of which depends on the number of hemin/G-quadruplexes if the luminol and H_2O_2 concentration are constant. For DNA detection probe DNA has been immobilized on CdS QD-electrodes. Afterwards, a mixture which contains the target DNA, the aptamer and hemin is added. The target DNA possesses two sections which bind to the attached probe and at the same time to the aptamer. The non-hybridized part of the aptamer forms the hemin/quadruplex complex and catalyzes the chemiluminescence reaction. The signal response depends on the amount of the bound hemin/quadruplex complexes and accordingly represents the concentration of target DNA bound to the probe. This system is highly selective for a specific analyte sequence and provides a photoelectrochemical response up to a detection limit of 2 nM (Golub et al., 2012).

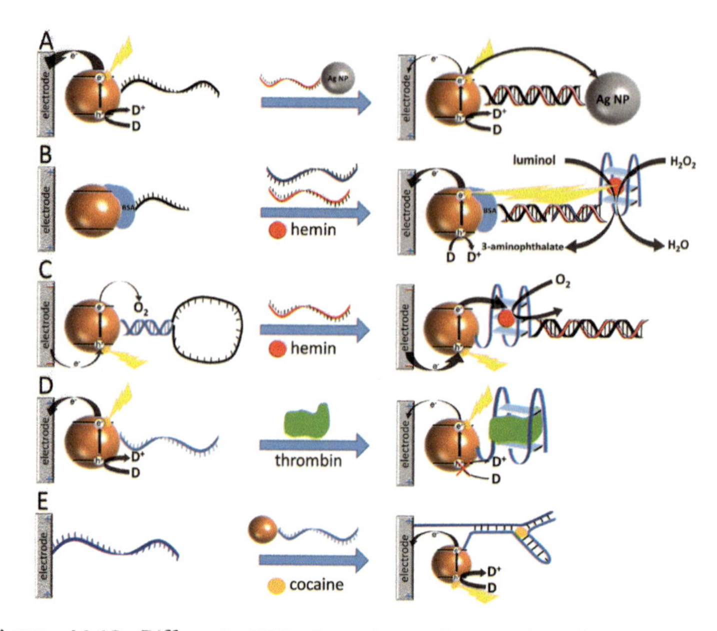

Figure 14.12 Different PEC detection schemes based on DNA as recognition element. Probe DNA, aptamer and target DNA are illustrated in black, blue and red, respectively. (A) Probe DNA-modified QD electrode before and after hybridization with silver nanoparticle-labeled target DNA according to Zhao et al. (2012a). The interaction between the QDs and the silver nanoparticles results in a decrease of the photocurrent. (B) Probe DNA-modified QD electrode before and after hybridization with target DNA and a second probe according to Golub et al. (2012). The nonhybridized section of the second probe is an aptamer and forms a hemin/G-quadruplex complex in the presence of hemin. Luminol is oxidized in the presence of H_2O_2 by the formed hemin/G-quadruplex, generating light by the chemiluminescence reaction (CRET). This light triggers-on the excitation of the QDs and generates a photocurrent. (C) Hairpin DNA-modified QD electrode before and after hybridization with target DNA according to Wang et al. (2015). Due to the hybridization of target DNA the hairpin opens and allows the formation of hemin/G-quadruplex complexes, resulting in enhanced cathodic photocurrents compared to the situation without the complexes. (D) Aptamer-modified QD electrode before and after aptamer/thrombin-complex formation (G-quadruplex) according to Zhang et al. (2011). The formed complex changes the access for the donor compound at the QD electrode, resulting in decreased signals. (E) Aptamer-modified electrode before and after addition of QD-labeled DNA, resulting in the formation of a supramolecular aptamer after addition of cocaine according to Golub et al. (2009). Due to the close proximity between the electrode and the QD-label, light-induced electron transfer reactions can occur.

In another study, the formation of hemin/G-quadruplex complexes on top of PbS QD modified ITO electrodes has also been utilized to detect hybridization events in a different way (Wang et al., 2015). As illustrated in Fig. 14.12C, the hybridization of target DNA results in an opening of an immobilized self-hybridized hairpin probe DNA. After opening, the nonhybridized overhang of the probe acts as an aptamer for hemin and forms a hemin/G-quadruplex complex. By illumination the electrons are transferred from the electrode over the excited QDs to the hemin/G-quadruplex complex, resulting in enhanced cathodic photocurrents compared to the situation without the complexes. The enhancement effect has been explained by the continuous regeneration of the hemin/G-quadruplex through oxidation with ambient oxygen. The detectable concentrations of target DNA ranging from 0.8 fM to 10 pM (Wang et al., 2015).

In the same study, the applicability of this detection principle has been demonstrated for the analysis of the blood coagulation enzyme thrombin (Wang et al., 2015). In this case a thrombin aptamer which forms a G-quadruplex and can specifically bind thrombin is used. Additionally, hemin can intercalate into the thrombin/aptamer-complex to form the hemin/G-quadruplex structure. Also for this approach an increase of the photocurrent in the range of 0.1 pM to 10 nM thrombin is found when a constant concentration of hemin is added. However, the background photocurrent (without analyte) is higher compared to that for the DNA detection, indicating that hemin is also loosely bound to the aptamer in the absence of thrombin (Wang et al., 2015). The aptasensor enables the discrimination of thrombin concentrations in solution and provide a good selectivity against BSA, IgG, and L-Cystein.

Zhang et al. (2011) have also used thrombin, as model analyte to demonstrate another photoelectrochemical detection principle based on a thrombin specific aptamer. Therefore, the aptamer has been immobilized onto a multiple QD-layer electrode. Stable anodic photocurrents are generated upon illumination in the presence of the electron donor ascorbic acid. The addition of thrombin results in the formation of an aptamer/thrombin complex direct in front of the QDs. The formed complex blocks the QD surface and prevents the electron transfer between ascorbic acid and the excited QDs (see Fig. 14.12D). This results in a decline of the initial photocurrent, which depends on the surface density of aptamer/thrombin complexes. Signal changes have been observed between 1 pM and 100 pM (Zhang et al., 2011). A similar detection principle has been used to the investigate DNA-protein interactions by means of TATA-binding proteins (TATA: thymine and adenine-rich DNA sequence for the gene regulation) (Ma et al., 2015).

Various PEC aptasensorial detection principles have also been designed for small molecules such as tetracycline (Liu et al., 2015) and cocaine (Golub et al., 2009) or for the metal atom Pb^{2+} (Zang et al., 2014). For the detection of tetracycline graphite carbon nitride/CdS hybrid nanoparticles are combined with tetracycline specific aptamers acting as a recognition element. After incubation of the hybrid nanoparticle/ aptamer electrode in the analyte containing solution and a thorough washing step, only the bound tetracycline is in close proximity to the nanoparticle. Due to the high sensitivity of the nanohybrids for the oxidation of tetracycline an enhancement of the anodic photocurrent can be observed if the analyte is bound to the aptamer. The PEC sensor show a concentration-dependent signal change between 10 nM and 250 nM tetracycline (Liu et al., 2015).

Golub et al. (2009) have developed a PEC aptasensor for the detection of cocaine by using a supramolecular aptamer complex and CdS QDs as a label. The supramolecular aptamer consists of two DNA sequences, which only forms a complex in the presence of the analyte cocaine (as illustrated in Fig. 14.12E). For the sensorial construction, one DNA sequence has been immobilized on an electrode. The other DNA sequence is labeled with a QD and is added to the solution to form the supramolecular aptamer complex if cocaine present. Due to the complex formation the QD-label is in close proximity to the electrode and can exchange electrons with the electrode upon illumination. The resulting photocurrent corresponds to the number of complexes and therefore correlates with the cocaine concentration and achieves sensitivities up to 1 μM (Golub et al., 2009).

In another approach, the previously described detection principle based on silver nanoparticle-labeled DNA has been adapted in a competitive assay for the detection of Pb^{2+} (Zang et al., 2014). Therefore, probe DNA, which acts as an aptamer for the metal atom but can also hybridize to the complementary nanoparticle-labeled DNA, is immobilized on the QDs. If both, Pb^{2+} and nanoparticle-labeled DNA are present, a competitive situation for the binding at the probe is created. Since the nanoparticle-labeled DNA concentration is kept constant, the signal change only depends on the Pb^{2+} concentration. This competitive PEC assay format provides a signal response in the range of 0.1 to 50 nM Pb^{2+} and has been successfully applied to environmental water analysis (Zang et al., 2014).

Besides the use of aptamers, antibodies have also come into the focus for the light-triggered detection of biospecific binding events. Wang et al. (2009a) have developed the first label-free PEC sensor for the detection of the model analyte mouse IgG. Therefore, CdS multilayer films have been build up on ITO electrodes by means of the polyelectrolyte poly(dimethyldiallyl ammonium chloride), resulting in defined anodic photocurrents in the presence of the electron donor ascorbic acid. After immobilization of the antibodies at the QD electrode and blocking the unmodified electrode surface with bovine serum albumin (BSA) in order to avoid unspecific interactions a decreasing anodic photocurrent is observed. The authors explain this by steric hindrance caused by the attached proteins, interfering the photooxidation of ascorbic acid at the illuminated QDs. In the presence of mouse IgG the initial signal further declines, which is attributed to a further blocking of the electrode surface and restricting the accessibility for the electron donor. The PEC immunosensor shows a linear concentration range within 10 pg/ml and 100 ng/ml mouse IgG, and provides an acceptable selectivity against goat IgG or rabbit IgG (Wang et al., 2009a).

This detection principle, based on changes in the accessibility for donor/acceptor compounds by the formed immune complex has been further developed by using more advanced QD electrodes and applied to other analytically relevant molecules like the tumor marker α-fetoprotein (Wang et al., 2009b). Surprisingly, it has also been reported that small molecules such as the pesticide pentachlorophenol (Kang et al., 2010), the plant hormone indole-3-acetic acid (Sun et al., 2014) and the toxin ochratoxin A (Yang et al., 2015) can be analyzed in a photoelectrochemcial detection scheme, which relies on access changes in front of the electrode. The principle is exemplified illustrated in Fig. 14.13A.

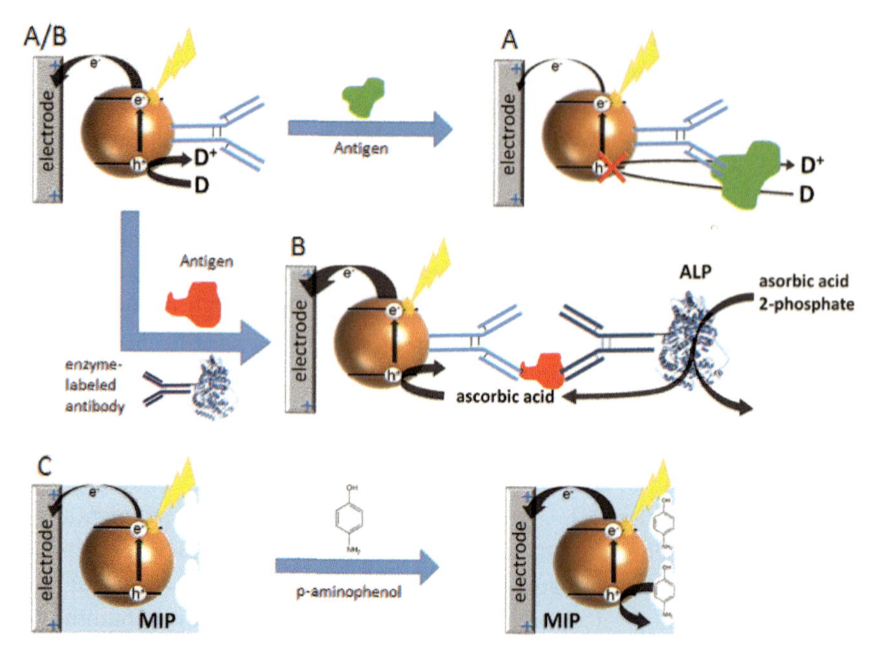

Figure 14.13 Different PEC detection schemes based on antibodies and molecular imprinted polymers as recognition element. (A) Antibody-modified QD electrode before and after binding of an antigen. The formed immune complex influences the access of the donor to the QD electrode and thus decreases the photocurrent. (B) Antibody-modified QD electrode before and after binding of an antigen followed by a second incubation step with an alkaline phosphatase (ALP)-labeled antibody according to Zhao et al. (2012b). The enzyme generates the donor ascorbic acid and thus enhances the anodic photocurrent. (C) Molecular imprinted polymer (MIP)-modified QD electrode before and after addition of the analyte *p*-aminophenol according to Wang et al. (2014b). Upon binding of pAP enhanced anodic photocurrents can be observed due to the light-induced oxidation of the analyte.

Even if label-free immunoassays are preferred due to low cost and simplicity, for the detection of small trace amounts an amplification of the binding signal is often necessary. Therefore, enzyme-based PEC immunoassays have been investigated. In this case, after formation of the antibody/antigen immune complex on top of the electrode a second incubation step with an enzyme-labeled antibody is carried out, resulting in a sandwich-like arrangement of antibody/antigen/antibody-enzyme complexes. Consequently, after a washing step and addition of the substrate the biocatalytic turnover induces the PEC signal change. Zhao et al. (2012c) have used horseradish peroxidase as a label to produce insoluble precipitating compounds direct in front of the QD electrode. Due to the insulating effect of the precipitate the barrier for the electron donor is increased and consequently the signal is diminished. For the model analyte mouse IgG a high sensitivity with a lower limit of detection of 0.5 pg/ml IgG has been found (Zhao et al., 2012b). This amplification principle is based on the detection of enzyme reactions as exemplified in Fig. 14.10C. It has also been applied in another study for the PEC detection of the tumor marker carcinoembryonic antigen (Zeng et al., 2014).

In a different approach the enzyme-based PEC immunoassay is performed by product analysis as illustrated in Fig. 14.13B. The biocatalytic conversion of ascorbic acid 2-phosphate by ALP to ascorbic acid is used for the PEC immunosensing of the tumor marker prostate-specific antigen (PSA) (Zhao et al., 2012c). After binding of PSA to the antibody-modified QD/TiO_2 nanotube electrode gold nanoparticles with immobilized antibody and ALP bind in a sandwich-type assay to the attached antigen. Consequently, the enzymatic production of ascorbic acid results in an increase of the anodic photocurrent, since ascorbic acid is an excellent electron donor for the excited QDs. The PEC sensor shows a signal response in the range between 10 and 0.5 ng/ml PSA (Zhao et al., 2012c).

Another principle is followed in a PEC immunoassay which is constructed by using TiO_2 electrodes modified with antibodies against α-fetoprotein and bioconjugates (AFP-CdTe-GOx) consisting of CdTe QDs modified with glucose oxidase and α-fetoprotein (Li et al., 2012). The detection principle based on a dual signal amplification strategy by means of the bioconjugate. On the one hand, due to the binding of the bioconjugate to the antibodies photoexcited electrons from the conduction band of the CdTe QDs inject fast to the conduction band of the TiO_2 upon illumination and increase the photoelectrochemical signal. On the other hand, the GOx produce H_2O_2 in the presence of glucose, which can act as an electron donor at the valence band of the excited QDs, causing a further amplification of the photocurrent signal. The assay is performed in a competitive assay with a constant concentration of the bioconjugate AFP-CdTe-GOx and the analyte α-fetoprotein in the sample solution. Both, the bioconjugate and the analyte competing for the binding sites at the antibody-modified TiO_2 electrode and thus the analyte concentration determines the photocurrent output. This amplification strategy enables the detection of α-fetoprotein in the linear range between 0.5 pg/ml and 10 μg/ml with a detection limit of 0.13 pg/ml (Li et al., 2012).

In addition to natural recognition elements a first approach shows the combination of an artificial recognition element with a QD-based PEC sensor system. Therefore, Wang et al. (2014b) have attached a molecularly imprinted polymer (MIP) to a CdS-graphene nanocomposite-modified FTO electrode. MIPs are made by polymerization of functional monomers and cross-linkers around a template molecule (the later analyte). After the dissolution of the template cavities remain, which mimic the binding function of an antibody and allow a specific recognition of the analyte. Wang et al. (2014b) have polymerized polypyrrole MIPs for the binding of *p*-aminophenol (see Fig. 14.13C). The bound analyte is oxidized at the valence band of the QDs upon illumination, resulting in an enhanced anodic photocurrent with increasing concentration of bound *p*-aminophenol. The sensor shows a linear response in the range from 50 nM to 3.5 μM *p*-aminophenol with a good reproducibility and selectivity against similar compounds such as phenol, 2-aminophenol, 3-aminophenol, 4-aminobenzene sulfanilic acid, 4-acetamidophenol, 4-nitrophenol, 4-chlorophenol, and 4-chloroaniline (Wang et al., 2014b).

14.6 Summary

Photoelectrochemical sensors represent a rather new area of research, which has developed fast due to the progress in nanoparticle synthesis, light-directed read-out methods, and high sensitivities. As a rather simple and cost-effective technique, the PEC sensing based on quantum dots in combination with electrodes provides an interesting approach for bioanalysis. Upon illumination of the QDs with light of sufficient energy electron–hole pairs can be generated inside the particles. The separated charge carriers can be transferred to the electrode, resulting in a generation of a photocurrent the magnitude and direction of which depend on the applied potential. Furthermore, light-induced electron transfer reactions between QDs and redox-active molecules in solution are feasible, resulting in concentration-dependent photocurrents. This enables the construction of analytical signal chains with light as additional tool for controlling the sensorial system. Thus, multiplexed analysis by using a defined light beam and spatially resolved immobilization of the respective analytical system becomes feasible.

Various PEC detection principles have been developed to apply QD electrodes for the detection of redox active molecules, for enzymatic sensing, DNA analysis as well as for the detection of binding events with aptamers and antibodies. While enzymatic PEC sensors are mostly based on the detection of enzymatic products, co-substrates or the direct electron transfer with a protein, approaches based on biospecific interactions rely on direct oxidation/reduction of the bound analyte, changes in access of the e^- donor/acceptor and the production of substances by an enzymatic label. Also the QDs itself can act as a label. Alternatively, light can be generated by chemical reactions, which overcomes the need for an external light source.

The chapter illustrates that the PEC sensing is considered to be an important approach in bioanalysis and remains a hot topic in research. Even if much efforts have been put into new improved PEC systems further studies of the photoelectrochemical properties are required to meet the practical analytical demands.

Abbreviations

AA: Ascorbic Acid
AChE: Acetylcholine esterase
ADP: Adenosine diphosphate
Ag: Silver
AgCl: Silver chloride
ALP: Alkaline phosphatase
ATP: Adenosine triphosphate
BDT: 1,4-Benzenedithiol
BSA: Bovine serum albumin
CB: Conduction band
CdSe: Cadmium selenide
CdS: Cadmium sulfide
CdTe: Cadmium telluride
CRET: Chemiluminescence resonance energy transfer

CuS: Copper sulfide
Cys: Cysteamine
Cyt c: Cytochrome c
DA: Dodecylamine
DET: Direct electron transfer
DHLA: Dihydroxylipoic acid
DNA: Desoxyribonucleic acid
e^-: Electron
ECL: Electrochemiluiminescence
FePt: Iron-platinum
FTO: Fluorine doped tin oxide
GDP: Guanosine diphosphate
GMP: Guanosine monophosphate
GMPK: Guanylate kinase
GOx: Glucose oxidase
GTP: Guanosine triphosphate
HT: 1,6-Hexandithiol
H_2O_2: Hydrogen peroxide
IgG: Immunoglobulin G
ITO: Indium tin oxide
LAPS: Light-addressable potentiometric sensor
LDH: Lactate dehydrogenase
MIP: Molecular imprinted polymer
MPA: 3-Mercaptopropionic acid
MUA: 11-Mercaptoundecanoic acid
NADH: Nicotinamide adenine dinucleotide (reduced)
NAD^+: Nicotinamide adenine dinucleotide (oxidized)
NR: Nitrate reductase
PAA: Polyallylamine(hydrochloride)
pAP: *p*-Aminophenol
pAPP: *p*-Aminophenylphosphate
PbS: Lead sulfide
Pb^{2+}: Lead ion
PEC: Photoelectrochemical
PK: Pyruvate kinase
PSA: Prostate-specific antigen
QDs: Quantum dots
StDT: Stilbenedithiol
TiO_2: Titanium dioxide
TOA: Trioctylamine

TOPO: Trioctylphosphine oxide
VB: Valence band
ZnO: Zinc oxide
ZnS: Zinc sulfide

References

Bakkers E. P. A. M., Roest A. L., Marsman A. W., Jenneskens L. W., de Jong-van Steensel L. I., Kelly J. J., Vanmaekelbergh D., *J. Phys. Chem. B*, **104** (2000), 7266–7272.

Baş D., Boyacý, Ý. H., *Electroanalysis*, **21** (2009), 1829–1834.

Beaulac R., Archer P. I., Liu X., Lee S., Salley G. M., Dobrowolska M., Furdyna J. K., Gamelin D. R., *Nano Lett.*, **8**, (2008), 197–1201.

Curri M. L., Agostiano A., Leo G., Mallardi A., Cosma P., Della Monica M., *Mater. Sci. Eng. C*, **22** (2002), 449–452.

Devadoss A., Sudhagar P., Terashima C., Nakata K., Fujishima A., *J. Photochem. Photobiol. C Photochem. Rev.*, **24** (2015), 43–63.

Gao X., Cui Y., Levenson R. M., Chung L. W. K., Nie S., *Nat. Biotechnol.*, **22** (2004), 969–976.

Gaponik N., Talapin, D. V., Rogach A. L., Hoppe K., Shevchenko E. V., Kornowski A., Eychmüller A., Weller H., *J. Phys. Chem. B*, **106** (2002), 7177–7185.

Göbel G., Schubert K., Schubart I. W., Khalid W., Parak W. J., Lisdat F., *Electrochim. Acta*, **56** (2011), 6397–6400.

Golub E., Pelossof G., Freeman R., Zhang H., Willner I., *Anal. Chem.*, **81** (2009), 9291–9298.

Golub E., Niazov A., Freeman R., Zatsepin M., Willner I., *J. Phys. Chem. C*, **116** (2012), 3827–13834.

Granot E., Patolsky F., Willner I., *J. Phys. Chem. B*, **108** (2004), 5875–5881.

Hickey S. G., Riley D. J., Tull E. J. J., *Phys. Chem. B*, **104** (2000), 7623–7626.

Ipe B. I., Niemeyer C. M., *Angew. Chem. Int. Ed.*, **45** (2006), 504–507.

Jasieniak J., Califano M., Watkins S. E., *ACS Nano*, **5**, (2011), 5888–5902.

Kang Q., Yang L., Chen Y., Luo S., Wen L., Cai Q., Yao S., *Anal. Chem.*, **82** (2010), 9749–9754.

Katz E., Shipway A. N. (2005). Molecular optobioelectronics. In *Bioelectronics*, Willner I., Katz E., eds. (Wiley-VCH Verlag GmbH & Co. KGaA), pp. 309–338.

Katz E., Zayats M., Willner I., Lisdat F., *Chem. Commun.*, **13** (2006), 1395–1397.

Khalid W., El Helou M., Murböck T., Yue Z., Montenegro J.-M., Schubert K., Göbel G., Lisdat F., Witte G., Parak W. J., *ACS Nano*, **5** (2011a), 9870–9876.

Khalid W., Göbel G., Hühn D., Montenegro J.-M., Rivera-Gil P., Lisdat F., Parak W. J., *J. Nanobiotechnology*, **9** (2011b), 46.

Li X., Zheng Z., Liu X., Zhao S., Liu S., *Biosens. Bioelectron.*, **64** (2015), 1–5.

Li Y.-J., Ma M.-J., Zhu J.-J., *Anal. Chem.*, **84** (2012), 10492–10499.

Lisdat F., Schäfer D., Kapp A., *Anal. Bioanal. Chem.*, **405** (2013), 3739–3752.

Liu Y., Yan K., Zhang J., *ACS Appl. Mater. Interfaces*, **8** (2015), 28255–28264.

Long Y.-T., Kong C., Li D.-W., Li Y., Chowdhury S., Tian H., *Small*, **7**, 1 (2011), 624–1628.

Ma Z.-Y., Ruan Y.-F., Zhang N., Zhao W.-W., Xu J.-J., Chen H.-Y., *Chem. Commun.*, **51** (2015), 8381–8384.

Medintz I. L., Uyeda H. T., Goldman E. R., Mattoussi H., *Nat. Mater.*, **4** (2005), 435–446.

Mussa Farkhani S., Valizadeh A., Review: *IET Nanobiotechnol.*, **8** (2014), 59–76.

Nakanishi T., Ohtani B., Uosaki K., *J. Phys. Chem. B*, **102** (1998), 1571–1577.

Nevins J. S., Coughlin K. M., Watson D. F., *ACS Appl. Mater. Interfaces*, **3**, **4** (2011), 242–4253.

Pardo-Yissar V., Katz E., Wasserman J., Willner I., *J. Am. Chem. Soc.*, **125** (2003), 622–623.

Resch-Genger U., Grabolle M., Cavaliere-Jaricot S., Nitschke R., Nann T., *Nat. Methods*, **5** (2008), 763–775.

Riedel M., Göbel G., Abdelmonem A. M., Parak W. J., Lisdat F., *ChemPhysChem*, **14** (2013), 2338–2342.

Riedel M., Sabir N., Scheller F. W., Parak W. J., Lisdat F., *Nanoscale* (2017), DOI: 10.1039/C7NR00091J.

Sabir N., Khan N., Völkner J., Widdascheck F., del Pino P., Witte G., Riedel M., Lisdat F., Konrad M., Parak W. J., *Small*, **11** (2015), 5844–5850.

Santra P. K., Kamat P. V., *J. Am. Chem. Soc.*, **134** (2012), 2508–2511.

Schubert K., Khalid W., Yue Z., Parak W. J., Lisdat F., *Langmuir*, **26** (2010), 1395–1400.

Shen Q., Zhao X., Zhou S., Hou W., Zhu J.-J., *J. Phys. Chem. C*, **115** (2011), 17958–17964.

Stoll C., Kudera S., Parak W. J., Lisdat F., *Small*, **2** (2006), 741–743.

Stoll C., Gehring C., Schubert K., Zanella M., Parak, W. J., Lisdat F., *Biosens. Bioelectron.*, **24** (2008), 260–265.

Sun B., Chen L., Xu Y., Liu M., Yin H., Ai S., *Biosens. Bioelectron.*, **51** (2014), 164–169.

Tanne J., Schäfer D., Khalid W., Parak W. J., Lisdat F., *Anal. Chem.*, **83** (2011), 7778–7785.

Tu W., Dong Y., Lei J., Ju H., *Anal. Chem.*, **82** (2010), 8711–8716.

van Sark W. G. J. H. M., Frederix P. L. T. M., Van den Heuvel D. J., Gerritsen H. C., Bol A. A., van Lingen J. N. J., de Mello Donegá C., Meijerink A., *J. Phys. Chem. B*, **105** (2001), 8281–8284.

Vlaskin V. A., Janssen N., van Rijssel J., Beaulac R., Gamelin D. R., *Nano Lett.*, **10** (2010), 3670–3674.

Wang G.-L., Yu P.-P., Xu J.-J., Chen H.-Y., *J. Phys. Chem. C*, **113** (2009a), 11142–11148.

Wang G.-L., Xu J.-J., Chen H.-Y., Fu S.-Z., *Biosens. Bioelectron.*, **25** (2009b), 791–796.

Wang G.-L., Xu J.-J., Chen H.-Y., *Nanoscale*, **2** (2010), 1112–1114.

Wang G.-L., Jiao H.-J., Liu K.-L., Wu X.-M., Dong Y.-M., Li Z.-J., Zhang C., *Electrochem. Commun.*, **41** (2014a), 47–50.

Wang G.-L., Shu J.-X., Dong Y.-M., Wu X.-M., Zhao W.-W., Xu J.-J., Chen H.-Y., *Anal. Chem.*, **87** (2015), 2892–2900.

Wang P., Ma X., Su M., Hao Q., Lei J., Ju H., *Chem. Commun.*, **48** (2012), 10216.

Wang R., Yan K., Wang F., Zhang J., *Electrochim. Acta*, **121** (2014b), 102–108.

Willner I., Patolsky F., Wasserman J., *Angew. Chem. Int. Ed.*, **40** (2001), 1861–1864.

Xu X., Qian J., Yu J., Zhang Y., Liu S., *Chem. Commun.*, **50** (2014), 7607–7610.

Yan X., Wang K., Xie D., Xu L., Han Q., Qi H., Pei R., *Anal. Methods*, **7** (2015a), 3697–3700.

Yan Y., Liu Q., Du X., Qian J., Mao H., Wan K., *Anal. Chim. Acta*, **853** (2015b), 258–264.

Yang J., Gao P., Liu Y., Li R., Ma H., Du B., Wei Q., *Biosens. Bioelectron.*, **64** (2015), 13–18.

Yoshinobu T., Miyamoto K., Wagner T., Schöning M. J., *Sens. Actuators B Chem.*, **207**, Part B, (2015), 926–932.

Yue Z., Lisdat F., Parak W. J., Hickey S. G., Tu L., Sabir N., Dorfs D., Bigall N. C., *ACS Appl. Mater. Interfaces*, **5** (2013), 2800–2814.

Zang Y., Lei J., Hao Q., Ju H., *ACS Appl. Mater. Interfaces*, **6** (2014), 15991–15997.

Zeng X., Bao J., Han M., Tu W., Dai Z., *Biosens. Bioelectron.*, **54** (2014), 331–338.

Zhang X., Li S., Jin X., Zhang S., *Chem. Commun.*, **47** (2011), 4929–4931.

Zhao W.-W., Wang J., Xu J.-J., Chen H.-Y., *Chem. Commun.*, **47** (2011), 10990–10992.

Zhao W.-W., Yu P.-P., Shan Y., Wang J., Xu J.-J., Chen H.-Y. (2012a), *Anal. Chem.*, **84**, 5892–5897.

Zhao W.-W., Ma Z.-Y., Yu P.-P., Dong X.-Y., Xu J.-J., Chen H.-Y., *Anal. Chem.*, **84** (2012b), 917–923.

Zhao W.-W., Ma Z.-Y., Yan D.-Y., Xu J.-J., Chen H.-Y., *Anal. Chem.*, **84** (2012c), 10518–10521.

Zhao W.-W., Ma Z.-Y., Xu J.-J., Chen H.-Y., *Anal. Chem.*, **85** (2013), 8503–8506.

Zhao W.-W., Xu J.-J., Chen H.-Y., *Chem. Rev.*, **114** (2014), 7421–7441.

Zhao W.-W., Xu J.-J., Chen H.-Y., *Chem. Soc. Rev.*, **44** (2015), 729–741.

Zhao X., Zhou S., Shen Q., Jiang L.-P., Zhu J.-J., *Analyst*, **137** (2012d), 3697.

Zheng M., Cui Y., Li X., Liu S., Tang Z. J., *Electroanal. Chem.*, **656** (2011), 167–173.

Chapter 15

Inorganic Nanoparticles as Enzyme Mimics

Ruben Ragg,[a] Karsten Korschelt,[a] Karoline Herget,[a] Filipe Natalio,[b] Muhammad Nawaz Tahir,[a] and Wolfgang Tremel[a]

[a]*Institut für Anorganische und Analytische Chemie, Johannes-Gutenberg-Universität, Duesbergweg 10–14, 55099 Mainz, Germany*
[b]*Institut für Chemie, Kurt Mothes-Str. 2, Naturwissenschaftliche Fakultät II, Martin-Luther-Universität Halle-Wittenberg, 06120 Halle, Germany*

tremel@uni-mainz.de

15.1 Introduction

The ambitious goal of biomimetic chemistry to mimic the structural and functional aspects of natural enzymes is at the center of interest by contemporary scientists. In the past decades, intensive efforts have been made to synthesize inorganic nanomaterials capable of mimicking natural enzymes termed as "artificial enzymes," that are more stable and cost efficient compared to their natural counterpart (Wei and Wang, 2013; Breslow, 2006; Breslow and Overman, 1970). Based on previous studies of catalytically active model compounds, including metal complexes

Biocatalysis and Nanotechnology
Edited by Peter Grunwald

ISBN 978-981-4613-69-9 (Hardcover), 978-1-315-19660-2 (eBook)
www.panstanford.com

(Kirby and Hollfelder, 2009), polymers (Kirkorian et al., 2012; Klotz, 1984; Kofoed and Reymond, 2005; Wang et al., 2014; Wulff, 2001), supramolecules (Dong et al., 2011; Raynal et al., 2014) and biomolecules (Aiba et al., 2011; Breaker and Joyce, 1994; Mader and Bartlett, 1997; Pollack et al., 1986; Tramontano et al., 1986), new materials have been identified to imitate the biological functions of natural enzymes. Still, mimicking enzymatic reactions inside living organisms—especially in the presence of other competing reactions—remains a great challenge. Recently, several biocompatible inorganic nanomaterials were found to exhibit enzyme-like activities. Therefore, applications inside living cells or organisms could be possible in the future (Fan et al., 2012; Kim et al., 2012; Ragg et al., 2014). However, it is still an open question whether and in which form inorganic nanoparticles (NPs) can mimic the high efficiency and exceptional specificity of their natural counterparts (André et al., 2013). At the same time, enzyme-catalyzed reactions are greatly dependent on specific reaction conditions like temperature, pH or chemical structure of the substrates. For example, enzymes generally suffer from low stability in body fluids or organic solvents, short shelf life and high production costs. In contrast, inorganic NPs provide significant advantages compared to their natural counterpart, e.g., cost-efficient synthesis up to industrial scale and tolerance of major changes in reaction conditions, like temperature, pH or solvent. Additionally, NPs constitute the essential feature of enhanced chemical activity due to their large surface area leading to an increase in catalytic activity. Furthermore, the surface of NPs can be modified by post-synthetic steps. Suitable stabilizing ligands are based on chelating agents that bind tightly to the NP surface, which makes them versatile tools in various applications. Due to the functionalization process, NPs offer the possibility of specific cellular targeting in combination with drug delivery, enhancement of the solubility in different media or increase of their physiological compatibility as well as active site-substrate interactions (Breslow and Overman, 1970; André et al., 2013; D'Souza et al., 1987). Recently, great efforts have been made in identifying new materials with enzyme-like properties that are equally or even more efficient than their natural counterparts. An exceeding number of peroxidase mimics are among the reported materials, whereas only a few other enzymatic systems (e.g., superoxide dismutases, catalases, oxidases, haloperoxidases) have been explored so far.

15.2 Iron Oxide Nanomaterials as Peroxidase Mimics

Peroxidases are the first group of enzymes that could efficiently be mimicked by inorganic nanomaterials. They play a very important role in many organisms by reducing cellularly generated toxic hydrogen peroxide (Wei and Wang, 2013; Dunford and Stillman, 1976). They are a widely distributed and structurally diverse class of enzymes, which catalyze the oxidation of several substrates using hydrogen peroxide and other peroxides (Fig. 15.1). Additionally, catalase, a special form of peroxidase, uses only hydrogen peroxide (H_2O_2) instead of a second substrate finally forming oxygen and water.

$$H_2O_2 + AH_2 \longrightarrow A + 2\,H_2O$$

Figure 15.1 Peroxidase activity: Reduced substrates and their oxidation products are described by AH_2 and A, respectively, with A usually being organic aliphatic and aromatic compounds. In case of catalase a second H_2O_2 is used as substrate.

In most peroxidases heme prosthetic groups are located in the active site of the enzyme including iron, usually in the oxidation state +III. Apart from that, manganese or vanadium containing peroxidases have been described (Dunford, 2010), while the best-studied example, horseradish peroxidase (HRP), also contains an iron-heme co-factor (Dawson, 1988). Peroxidases, including HRP, are used in analytical and clinical chemistry for the enzyme-catalyzed conversion of colorimetric substrates in signaling and imaging applications (Breslow and Overman, 1970). Since the production of natural enzymes is relatively costly, it was a long-standing goal to find materials that are able to mimic the activity of HRPs. The Fenton reagent, a solution containing Fe^{2+}/Fe^{3+} ions, is the oldest known peroxidase mimic that catalyzes the breakdown of peroxide (Fenton, 1894). It is indeed surprising to see that it took more than a century to show that aqueous solutions of other transition metal ions (e.g., Fe^{2+}, Cu^{+}, Co^{2+}, and Mn^{3+}) also exhibit a peroxidase-like activity (Sawyer, 1997). Furthermore, peroxidase mimics based on iron-porphyrin complexes, which are structurally related to the HRP prosthetic group, have already been reported (Fruk and Niemeyer, 2005; Huang et al., 2003, Wang et al., 2007). In recent years, different inorganic nanomaterials were identified that mimic the enzymatic activity of peroxidases (Table 15.1).

In a first study by Yan and co-workers, magnetite NPs (Fe_3O_4 NPs) with different sizes (Fig. 15.2; 30, 50 and 300 nm) were used to oxidize the peroxidase substrate 3,3′,5,5′-tetramethylbenzidine (TMB) (Fig. 15.2). It was found that Fe_3O_4 NPs with the smallest size exhibited the highest catalytic activity, suggesting that the surface area of the nanostructured material is an important factor for an efficient enzyme mimic (Gao et al., 2007).

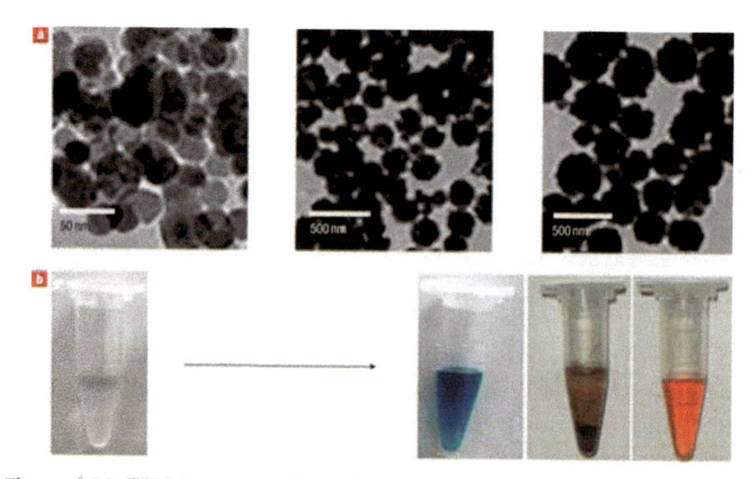

Figure 15.2 TEM images and catalytic peroxidase activity of Fe_3O_4 NPs. (a) Transmission electron microscopy (TEM) images of Fe_3O_4 NPs showing sizes of 30, 150 and 300 nm. (b) Conversion of various peroxidase substrates catalyzed by Fe_3O_4 NPs producing different color reactions. Reprinted with permission from Gao et al., 2007; copyright (2007) Nature Publishing Group.

Compared to the native enzyme Fe_3O_4 mimics are much more robust as they retain their catalytic efficiency over a wide range of temperature (4–90°C) and pH (0–12) (Fig. 15.3). Additionally, the peroxidase mimic reacts faster than native HRP with both substrates H_2O_2 and TMB as indicated by an increased catalytic turnover number k_{cat} (8.58×10^4 s^{-1} and 3.02×10^4 s^{-1} for Fe_3O_4; 4.00×10^3 s^{-1} and 4.48×10^3 s^{-1} for native HRP). Michaelis–Menten substrate binding constants K_m for Fe_3O_4 are higher for H_2O_2 (154 mM Fe_3O_4 and 3.70 mM for natural HRP) and slightly lower for TMB (0.098 mM for Fe_3O_4 and 0.434 mM for HRP) (Gao et al., 2007). Fe_3O_4 peroxidase mimics were further used as HRP substitute in a classical enzyme-linked immunosorbent assay (ELISA) by Wang and coworkers for the detection of H_2O_2 and glucose (Wei and Wang, 2008).

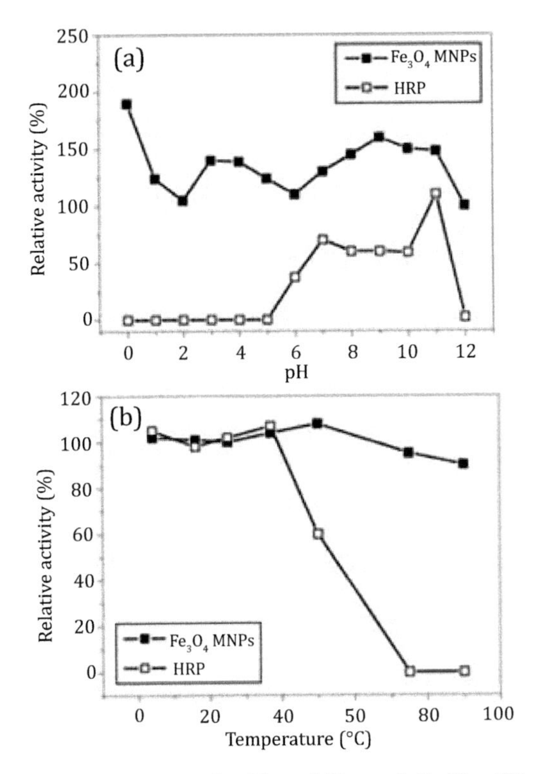

Figure 15.3 Temperature and pH stability of Fe_3O_4 NPs and HRP. Peroxidase activity of Fe_3O_4 NPs and HRP at (a) pH values ranging from 0 to 12 showing an increased stability of the nanomaterial especially in very acidic and basic environment and (b) at temperatures ranging from 4 to 90°C showing an increased stability at temperatures above 40°C. Reprinted with permission from Gao et al., 2007; copyright (2007) Nature Publishing Group.

In a similar approach, Yan's group also used Fe_3O_4 NPs for tumor targeting and visualization (Fan et al., 2012). The NPs were functionalized with a recombinant human heavy-chain ferritin (HFn) shell which binds to tumor cells overexpressing the transferrin receptor 1 (TfR 1). The Fe_3O_4core catalyzes the oxidation of peroxidase substrates in the presence of H_2O_2, which is accompanied by a color change that can be used to visualize the tumor (Fan et al., 2012).

Besides the Fe_3O_4 NPs mentioned above, a variety of different inorganic nanoparticles have been reported to exhibit intrinsic peroxidase like activities, e.g., γ-Fe_2O_3 (Chen et al., 2012), γ- FeOOH (Peng et al., 2011), $FePO_4$ (Wang et al., 2012b) FeS (Dai et al., 2009), FeTe (Roy et al., 2012), FePt (Fan et al., 2011), CuO (Liu et al., 2014), $Cu_3(PO_4)_2$ (Huang et al., 2015), CuS (He et al., 2012), Cu (Hu et al., 2013), $CuInS_2$ (Liu et al., 2015) Co_3O_4 (Mu et al., 2012) CoFe LDHs (Zhang et al., 2012), $CoFe_2O_4$ (He et al., 2010a), CeO_2 (Jiao et al., 2012), $CePO_4$:Tb, Gd (Wang et al., 2012a), MnO_2 (Liu et al., 2012), MnSe (Qiao et al., 2014), ZnO (Biparva et al., 2014), $ZnFe_2O_4$ (Zhao et al., 2013), $BiFeO_3$ (Luo et al., 2010), MoS_2 (Lin et al., 2014), WC (Li et al., 2014), VO_2 (Zhang et al., 2015), V_2O_3 (Han et al., 2015), IrO (Su et al., 2015), Pd-Ir (Xia et al., 2015), Au (Jv et al., 2010), Pt (Ma et al., 2011), Au/CuS (Cai et al., 2014), Bi/Au (Lien et al., 2012), Ag/Pd (He et al., 2010b), Ag/Pt (He et al., 2010b; Zheng et al., 2014), Ag/Au (He et al., 2010b), Au/Pt (He et al., 2011), Au/Pd (Nangia et al., 2012), V_2O_5 (André et al., 2011) and silicon dots (Chen et al., 2014) (Table 15.1). The described peroxidase mimics have been used in various immunoassays (Gao et al., 2007; Liu et al., 2012, Xia et al., 2015; He et al., 2011; Chen et al., 2014), for the removal of environmental pollutants (Liu et al., 2014; Huang et al., 2015; Cai et al., 2014), but usually as sensors for H_2O_2 (Wei and Wang, 2008; Wang et al., 2012b; Dai et al., 2009; Hu et al., 2013; Liu et al., 2015; Mu et al., 2012; Zhang et al., 2012; Jiao et al., 2012; Qiao et al., 2014; Luo et al., 2010; Jv et al., 2010; Cai et al., 2014), glucose (Kirkorian et al., 2012; Wei and Wang, 2008; Roy et al., 2012; Hu et al., 2013; Mu et al., 2012; Zhang et al., 2012; Qiao et al., 2014; Zhao et al., 2013; Luo et al., 2010; Lin et al., 2014; Jv et al., 2010; Chen et al., 2014) and other biomolecules (Zheng et al., 2014) (Table 15.1).

Table 15.1 Inorganic nanomaterials with enzyme like activities[a,b]

Nanomaterial	Enzyme-like activity	Application	Ref.
Fe_3O_4 NPs	Peroxidase	Tumor targeting and visualization	Fan et al., 2012
		Immunoassays	Gao et al., 2007
		H_2O_2 and glucose detection	Wei and Wang, 2008
γ-Fe_2O_3 NPs	Peroxidase	—	Chen et al., 2012
γ-FeOOH graphene	Peroxidase	Phenole degradation	Peng et al., 2011
$FePO_4$ microflowers	Peroxidase	H_2O_2 biosensor	Wang et al., 2012b
FeS nanosheets	Peroxidase	H_2O_2 detection	Dai et al., 2009
FeTe nanorods	Peroxidase	Glucose detection	Roy et al., 2012
FePt NPs	Peroxidase	—	Fan et al., 2011
CuO nanostructures	Peroxidase	Pollutant removal in wastewater	Liu et al., 2014
$Cu_3(PO_4)_2$ NC	Peroxidase	Pollutant removal	Huang et al., 2015
CuS polyhedral structures	Peroxidase	—	He et al., 2012
Cu nanoclusters	Peroxidase	H_2O_2 and glucose detection	Hu et al., 2013
$CuInS_2$ NC	Peroxidase	H_2O_2 detection	Liu et al., 2015

(*Continued*)

Table 15.1 *(Continued)*

Nanomaterial	Enzyme-like activity	Application	Ref.
Co_3O_4 NPs	Peroxidase	H_2O_2 detection	Mu et al., 2012
CoFe LDHs	Peroxidase	H_2O_2 and glucose detection	Zhang et al., 2012
$CoFe_2O_4$/cyclodex.	Peroxidase	H_2O_2 detection	He et al., 2010a
CeO_2 NPs	Peroxidase	H_2O_2 and glucose detection	Jiao et al., 2012
$CePO_4$:Tb, Gd NPs	Peroxidase	Magnetic-fluorescent imaging	Wang et al., 2012a
MnO_2 NPs	Peroxidase	Immunoassays	Liu et al., 2012
MnSe NPs	Peroxidase	H_2O_2 and glucose detection	Qiao et al., 2014
ZnO NPs	Peroxidase	Carvedilol determination	Biparva et al., 2014
$ZnFe_2O_4$/ZnO NC	Peroxidase	Glucose detection	Zhao et al., 2013
$BiFeO_3$ NPs	Peroxidase	H_2O_2 and glucose detection	Luo et al., 2010
MoS_2 nanosheets	Peroxidase	Glucose detection in blood	Lin et al., 2014
WC nanorods	Peroxidase	Catalysis in organic solvents	Li et al., 2014
VO_2 nanoplates	Peroxidase	—	Zhang et al., 2015
V_2O_3 NC	Peroxidase	Glucose detection	Han et al., 2015

Nanomaterial	Enzyme-like activity	Application	Ref.
Ir NPs	Peroxidase	Cell protect. against H_2O_2	Su et al., 2015
Pd-Ir core shell nanocubes	Peroxidase	Immunoassays	Xia et al., 2015
Au NPs	Peroxidase	H_2O_2 and glucose detection	Jv et al., 2010
Pt nanocrystals	Peroxidase		Ma et al., 2011
Au/CuS NC	Peroxidase	Pollutant removal in wastewater	Cai et al., 2014
Bi/Au NPs	Peroxidase	Thrombin determination	Lien et al., 2012
Ag/Pt, Ag/Pd, Ag/Au NC	Peroxidase	—	He et al., 2010b
Ag/Pt NC	Peroxidase	Thrombin determination	Zheng et al., 2014
Au/Pt NC	Peroxidase	Immunoassays	He et al., 2011
Au/Pd NC	Peroxidase	Immunoassays	Nangia et al., 2012
Si-dots	Peroxidase	Glucose detection	Chen et al., 2014
V_2O_5 nanowires	Peroxidase	Catalysis in organic solvents	André et al., 2011
V_2O_5 nanowires	Haloperoxidase	Antifouling agent	Natalio et al., 2012

(*Continued*)

Table 15.1 (*Continued*)

Nanomaterial	Enzyme-like activity	Application	Ref.
V_2O_5 nanowires	Glutathione peroxidase	ROS reduction	Vernekar et al., 2014
Fe_3O_4 NPs	Catalase		Chen et al., 2012
γ-Fe_2O_3 NPs	Catalase		Chen et al., 2012
FePt NPs	Catalase		Fan et al., 2011
Ir NPs	Catalase	Cell protect. against H_2O_2	Su et al., 2015
Co_3O_4 NPs	Catalase	Calcium detection in milk	Mu et al., 2012; 2014
Au/Pt NC	Catalase		He et al., 2011
CeO_2 NPs	Catalase		Pirmohamed et al., 2010; Celardo et al., 2011b
$LaNiO_3$ nanofibers	Catalase	H_2O_2 and glucose detection	Ligtenbarg et al., 2003
CeO_2 NPs	SOD	Stroke treatment Cell protect. in radiation therapy Antioxidant properties Anti-inflammation Neuroprotection	Kim et al., 2012; Korsvik et al., 2007 Tarnuzzer et al., 2005 Celardo et al., 2011a Hirst et al., 2009 Chen et al., 2006

Nanomaterial	Enzyme-like activity	Application	Ref.
$FePO_4$ microflowers	SOD		Wang et al., 2012b
Pt NPs	SOD	ROS reduction	Zhang et al., 2009
CeO_2 NPs	Oxidase	Immunoassay for tumor detection	Asati et al., 2009
$CoFe_2O_4$ NPs	Oxidase	Luminol oxidation	Zhang et al., 2013
$MnFe_2O_4$ NC	Oxidase		Vernekar et al., 2015
$LaNiO_3$ nanofibers	Oxidase	H_2O_2 and glucose detection	Wang et al., 2013
MnO_2 NPs/nanowires	Oxidase	Immunoassays	Liu et al., 2012
Mn_3O_4 octahedra	Oxidase	Sensor for phenols and tannic acid	Zhang et al., 2015
Au NPs	Oxidase	Glucose oxidation and detection	Comoti et al., 2004
Pt NPs	Oxidase		Yu et al., 2014
Au/Pt NC	Oxidase	Immunoassays	He et al., 2011
MoO_3 NPs	Sulfite Oxidase	SuOx deficiency treatment	Ragg et al., 2014

[a]NP, nanoparticle; NC, nanocomposite; LDH, layered-double-hydroxide.
[b]ROS, reactive oxygen species; SOD, superoxide dismutase; SuOx, sulfite oxidase.

Besides their peroxidase-like activity, Fe_3O_4 and γ-Fe_2O_3 NPs were also described to exhibit an intrinsic catalase activity (Chen et al., 2012). Catalases are found in nearly all living organisms and catalyze the decomposition of hydrogen peroxide into water and oxygen (Fig. 15.1) (Murthy et al., 1981). Besides iron oxide, catalase-like activities have also been found for some other nanomaterials like CeO_2 (nanoceria) (Pirmohamed et al., 2010; Celardo et al., 2011b), $LaNiO_3$ (Wang et al., 2013), Co_3O_4 (Mu et al., 2012; 2014), FePt (Fan et al., 2011), Ir (Su et al., 2015), and Au/Pt nanostructures (He et al., 2011) (Table 15.1). So far, the only application reported with nanomaterial-based catalase mimics are calcium ion biosensors, where Co_3O_4 nanoparticles were used. They showed a high selectivity against other metal ions and were successfully used to determine the calcium concentration in milk (Mu et al., 2014).

15.3 Vanadium Pentoxide Nanowires as Haloperoxidase Mimics

Haloperoxidases represent a class of enzymes with the capability of catalyzing the oxidation of halides (Cl^-, Br^-, I^-) using hydrogen peroxide (H_2O_2) to form hypohalous acids. These acids can lead to the halogenation of suitable nucleophilic acceptors (Fig. 15.4) (Wever and Hemrika, 2006).

$$H_2O_2 + H^+ + X^- \longrightarrow HOX + H_2O$$

$$HOX + AH \longrightarrow AX + H_2O$$

Figure 15.4 Reaction scheme for halogenation activity. Halides are represented by X, nucleophilic substrates and halogenated products with AH and AX, where A are usually organic aliphatic and aromatic compounds.

Several classes of haloperoxidases have been identified, where the so-called vanadium-dependent haloperoxidases (V-HPOs) are the best-studied examples (Messerschmidt et al., 1997; Neumann et al., 2008). V-HPOs are widely used for halogenations (Martinez et al., 2001; Coughlin et al., 1993) and sulfoxidations (ten Brink et al., 1998) while reacting in a highly stereo- and regioselective manner (Butler and Sandy, 2009). Based on these findings the research in the field of inorganic and organometallic haloperoxidase mimics with vanadium and other inorganic elements was encouraged. A wide variety of V-HPO mimetic compounds, mostly organometallic vanadium or vanadium peroxo complexes, have been identified so far (Ligtenbarg et al., 2003). These compounds exhibit good efficiencies in different oxidation reactions like halogenation (Clague et al., 1993), sulfoxidation (Smith and Pecoraro, 2002), epoxidation (Mimoun et al., 1983), and oxidation of alcohols (Velusamy and Punniyamurthy, 2003). In addition to vanadium, mimics based on other metal species like molybdenum (VI), tungsten (VI) and rhenium (VII) have been reported (Espenson et al., 1994; Hansen and Espenson, 1995; Meister and Butler, 1994). Besides chemical modifications haloperoxidases were used as additives in antifouling paints (Wever et al., 1995). Since the generated hypohalous acids are highly toxic to a wide variety of microorganisms, these compounds control the biofilm formation through simple biocidal activity or halogenate signaling molecules that are involved in intracellular communication (Borchardt et al., 2001; Rosenhahn et al., 2010). An application of V-HPOs as component of an underwater paint close to the water surface would be possible, but it is limited by high production costs, long-term stability of the enzymes and the reaction conditions for proper activity (Soedjak et al., 1995).

Recently, we described the intrinsic catalytic haloperoxidase activity of vanadium pentoxide (V_2O_5) nanowires that allows the formation of hypobromous acid (HOBr) under seawater conditions and builds the basis for the formulation in antifouling agents (Natalio et al., 2012). In general V_2O_5 can form activated peroxo-complexes with H_2O_2, which is formed in sea water in small concentration through UV activation by sunlight. The activated complex leads to the formation of the respective hypohalous acids (HOCl, HOBr, HOI) (Zampella et al., 2014; Vernekar et al., 2014). V_2O_5 is one of the most widely used catalysts, due to its stability, high availability, and cost efficiency (Biette et al., 2005; Hu et al., 2009). The haloperoxidase activity of the V_2O_5 nanowires was analyzed spectrophotometrically by varying the concentration of the nanomaterial in buffered solution (Tris-SO_4, pH 8.3). When keeping the concentrations of the haloperoxidase substrates 2-chlorodimedone (MCD), Br^- and H_2O_2 constant a linear dependence of the bromination activity on the V_2O_5 nanowire concentration was observed (Fig. 15.5b, blue diamonds). Replacing the V_2O_5 nanowires with commercially available bulk V_2O_5 powder under otherwise identical experimental conditions led to a significantly smaller bromination activity, suggesting that the intrinsic haloperoxidase activity is closely related to the surface area of the nanomaterial (Fig. 15.5b, green triangles).

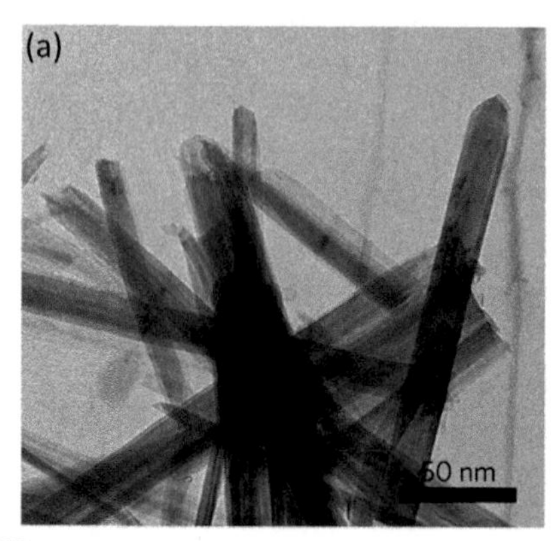

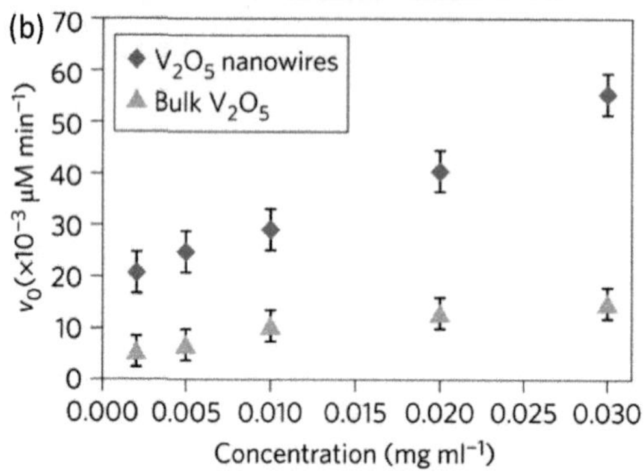

Figure 15.5 TEM image and catalytic haloperoxidase activity of V_2O_5 nanowires. (a) TEM image of V_2O_5 nanowires showing an average length of 300 nm and a width of 20 nm. (b) Haloperoxidase activity of the V_2O_5 nanowires (blue diamonds) compared to bulk V_2O_5 (green triangles), determined by measuring the reaction rates of the monochlorodimedone bromination from absorption at 290 nm. Reprinted with permission from Natalio et al. (2012); copyright (2012) Nature Publishing Group.

A typical Michaelis–Menten behavior of the V_2O_5 nanowires towards both, H_2O_2 and Br^-, was observed by varying the concentrations of H_2O_2 and Br^- while keeping the concentrations of the remaining components constant. K_m values of 7.3 μM for H_2O_2 and 0.24 mM for Br^- and a V_{max} of 4.3×10^{-2} Ms^{-1} were determined, from which a k_{cat} of 8 s^{-1} was derived (Yu et al., 2014). These values are of the same order of magnitude as those determined for the *C. inaequalis* V-chloroperoxidase mutant ($K_m(H_2O_2)$ = 16 μM, $K_m(Br^-)$ = 3.1 mM) (Hasan et al., 2006) and *A. nodosum* V-bromoperoxidase ($K_m(H_2O_2)$ = 22 μM, $K_m(H_2O_2)$ = 18.1 mM) (de Boer and Wever, 1988). Inspired by previous descriptions of the use of natural haloperoxidases as paint additives (Wever and Dekker, 1995), the possibility of using V_2O_5 nanowires as antifouling agent in underwater paints was tested. The nanowires were applied onto stainless steel plates (Fig. 15.6a, +V_2O_5 NW), placed on a boat hull together with a control sample without V_2O_5 (Fig. 15.6b, –V_2O_5 NW) and incubated for 60 days in the Atlantic Ocean. Severe biofouling was found on the untreated control sample (Fig. 15.6c), whereas the V_2O_5 containing plate did not show any signs of biofouling (Fig. 15.6d).

Figure 15.6 Antifouling activity of V_2O_5 nanowires. Digital images of stainless steel plates (2 × 2 cm) coated (a) without V_2O_5 nanowires (–V_2O_5 nw) and (b) with V_2O_5 nanowires. The plates were placed on the outside of a boat hull. The boat remained in seawater for 60 days (lagoon in Atlantic Ocean). (c) Pronounced natural biofouling was seen on the stainless steel plates without V_2O_5. (d) Plates with V_2O_5 nanowires did not show any signs of biofouling. Reprinted with permission from Natalio et al. (2012); copyright (2012) Nature Publishing Group.

Based on the reported intrinsic peroxidase and haloperoxidase activity, V_2O_5 nanowires were recently used to mimic the enzymatic activity of glutathione peroxidase (GPx) (Vernekar et al., 2014). Glutathione (GSH), a tripeptide consisting of glutamate, glycine and cysteine, contains a free sulfhydryl residue in its reduced form and is therefore able to act as an intracellular antioxidant by reducing potentially toxic reactive oxygen species (ROS), while being oxidized to glutathione disulfide dimer (GSSG) (Masella et al., 2005). Native glutathione peroxidase and V_2O_5 nanowires use H_2O_2 as co-substrate to oxidize GSH to GSSG leading to reduced intracellular H_2O_2 levels. V_2O_5 nanowires readily enter into mammalian cells, which are therefore protected when challenged with intrinsic or extrinsic oxidative stress by scavenging ROS (Vernekar et al., 2014).

15.4 Cerium Oxide Nanoparticles as Superoxide Dismutase and Oxidase Mimics

Apart from the reports on peroxidase mimics, superoxide dismutases (SOD) attracted attention due to their ability to reduce the concentration of intracellular reactive oxygen species. SOD catalyzes the dismutation of superoxide radicals into elemental oxygen and hydrogen peroxide (Fig. 15.7) and play a major role in aerobic organisms in their defense mechanisms against oxidative stress (McCord et al., 1971).

$$O_2^{\bullet -} + M^{n+1} \longrightarrow O_2 + M^n$$

$$O_2^{\bullet -} + M^n + 2H^+ \longrightarrow H_2O_2 + M^{n+1}$$

Figure 15.7 Reaction scheme for superoxide dismutase activity. The oxidized and reduced metal centers are summarized by M^{n+1} and M^n, respectively (where M = Fe, Mn, Cu).

It has been shown more than 30 years ago, that manganese ions (Mn^{2+}) protect against oxy-radical-mediated damage (Archibald and Fridovich, 1981), but the chemical mechanism became controversial and is still under discussion. Recently, Barnese et al. (2008) reported that Mn^{2+} reacts with superoxide to form the short-lived MnO_2^+, which disproportionates rapidly to give Mn^{2+}, O_2, and H_2O_2. In the same study only manganese phosphate, among different manganese salts, was found to remove superoxide from solution catalytically at physiologically relevant concentrations (Korsvik et al., 2007). Additionally manganese porphyrins were described to exhibit SOD-like activities (Raynal et al., 2014). Based on the knowledge about the active site present in most SODs and the reported activity of manganese compounds it can be resumed, that the presence of an oxygen-affine metal, which can easily change its oxidation state by one unit, is necessary to efficiently mimic SOD activity (André et al., 2013).

Cerium oxide (ceria) qualifies as a suitable candidate to the above-mentioned criteria, because it can easily change between oxidation states Ce3+/Ce4+ (Celardo et al., 2011a). Even before proposing the SOD-mimetic activity using nanoceria (Korsvik et al., 2007). Seal and co-workers reported a cell-protective mechanism against radiation induced cell damage (Tarnuzzer et al., 2005), which later led to the discovery of their antioxidant capacity (Kim et al., 2012). Seal et al. reported that the SOD activity of cerium oxide NPs with an average diameter of 3–5 nm (3.6×10^9 $M^{-1}s^{-1}$) is even more efficient than that of native CuZn SOD ($1.3–2.8 \times 10^9$ $M^{-1}s^{-1}$), where the presence of higher ratios of Ce^{3+}/Ce^{4+} led to enhanced activities (Fig. 15.8) (Korsvik et al., 2007). By reducing the Ce^{3+}/Ce^{4+} ratio, a catalase-like activity of the nanoceria could be obtained, leading to the consumption of the produced H_2O_2 (Fig. 15.7) (Pirmohamed et al., 2010). Besides CeO_2 NPs, $FePO_4$ microflowers and platinum NPs show SOD-like activities as well (Table 15.1).

Nanoceria is widely used in vitro and in vivo for different biomedical applications due to its ability to reduce ROS (Kim et al., 2012; Tarnuzzer et al., 2005; Celardo et al., 2011a; Xu and Qu, 2014). A very appealing example is the use of nanoceria to treat ischemic stroke, where they significantly reduced the infarct volumes in rat brain tissues (Fig. 15.9) (Kim et al., 2012), but other examples are equally impressive including the use of nanoceria-based materials that exhibit anti-inflammatory (Hirst et al., 2009), antioxidant (Celardo et al., 2011b) and neuro-protective-effects (Chen et al., 2006) as well as protection of normal cells during radiation therapy (Tarnuzzer et al., 2005).

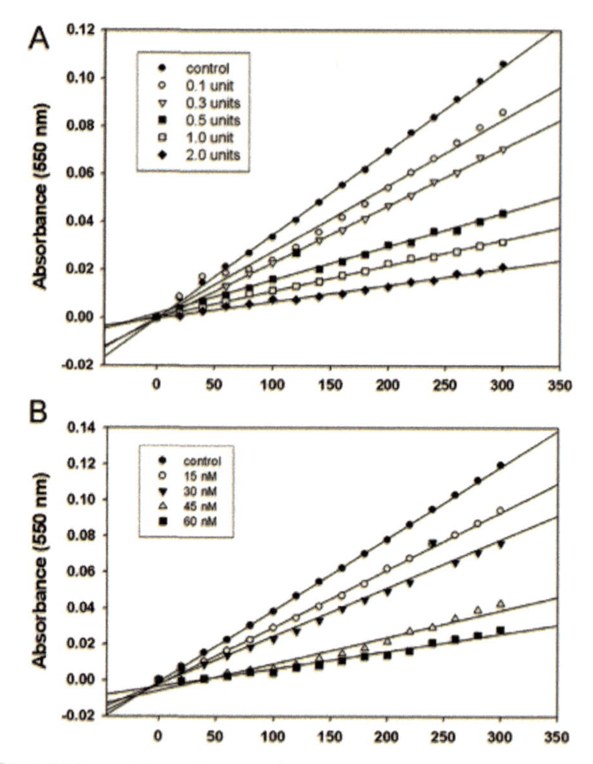

Figure 15.8 SOD activity of natural SOD compared to CeO_2. SOD activity of the native CuZn SOD enzyme (A) and ceria nanoparticles with high Ce^{3+}/Ce^{4+} ratio (B) were determined spectrophotometrically by observing the ferricytochrome c reduction at 550 nm. Reprinted with permission from Korsvik et al. (2007); copyright (2007) Royal Society of Chemistry.

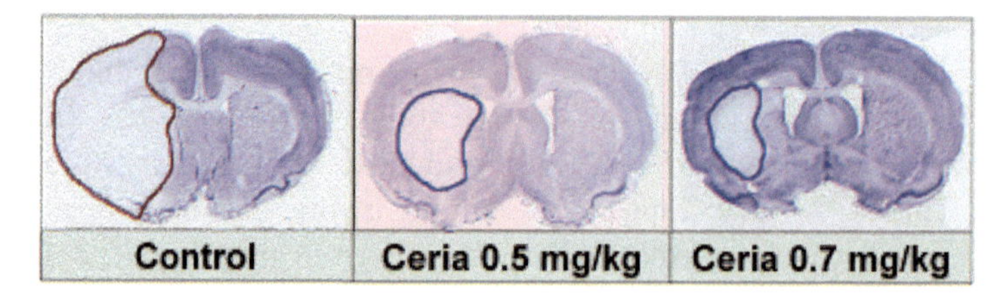

Figure 15.9 Effect of nanoceria treatment on infarct volume after ischemic stroke. Direct comparison of infarct volumes in rat brain tissues after ischemic stroke and treatment with 0.5 and 0.7 $mg\,kg^{-1}$ ceria nanoparticles clearly shows the efficiency of the treatment in reducing ROS induced brain damage. Reprinted with permission from Kim et al. (2012); copyright (2012) John Wiley and Sons.

Besides their SOD mimetic activity cerium oxide NPs were also shown to exhibit an intrinsic oxidase activity (Asati et al., 2009). Oxidases are a large family of enzymes that catalyze the oxidation of a substrate by molecular oxygen, which is converted into water, hydrogen peroxide, or superoxide (Bolwell and Wojtaszek, 1997). The nanoceria oxidase mimics have been used in an immunoassay similar to traditional ELISAs, where they could selectively detect tumor cells (Asati et al., 2009). The polymer-coated cerium oxide NPs were modified with folic acid, which can be specifically recognized by a folate receptor that is present in various tumors but absent in normal tissue except choroid plexus, lung, thyroid and kidney (Sudimack and Lee, 2000). Besides nanoceria, oxidase-like activities have also been reported for several other nanomaterials like MnO_2 (Liu et al. 2012), Mn_3O_4 (Zhang and Huang, 2015), $CoFe_2O_4$ (Zhang et al., 2013), $MnFe_2O_4$ (Vernekar et al., 2015), $LaNiO_3$ (Wang et al., 2013), Au (Comotti et al., 2004), Pt (Yu et al., 2014), Au/Pt (He et al., 2011), and MoO_3 (Ragg et al., 2014) (Table 15.1).

15.5 Molybdenum Oxide Nanoparticles as Sulfite Oxidase Mimics

Molybdenum trioxide (MoO_3) is a well-known and widely used material for selective oxidation catalysis (Haber et al., 1997), but its biocatalytic behavior is virtually unknown. In nature, molybdenum is utilized as part of the molybdenum cofactor in enzymes, where it is responsible for catalyzing different redox and oxygen transfer reactions (Hille, 2013). In mammals, the molybdenum cofactor is incorporated in several important enzymes like sulfite oxidase (SuOx), a 104 kDa dimeric enzyme that is located in the mitochondrial intermembrane space of liver and kidney cells (Schwarz et al., 2009). SuOx is a special kind of oxidase, that catalyzes the oxidation of sulfite to sulfate as the final step in the catabolism of sulfur containing amino acids (Fig. 15.10), but also exogenously supplied sulfite and sulfur dioxide (e.g., from pollution, preservatives). A deficiency in SuOx leads to severe neurological damage resulting in early childhood death (Arnold et al., 1993; Garrett et al., 1998).

$$SO_3^{2-} + H_2O \longrightarrow SO_4^{2-} + 2H^+ + 2e^-$$

Figure 15.10 Reaction equation of sulfite oxidase activity. The occurring electrons are transferred to cytochrome c as the physiological electron acceptor (Garrett et al., 1998).

The fundamental chemistry behind SuOx as well as the oxidation of sulfite has been studied using several model compounds, but none of those biomimetic compounds has shown any cellular activity (Mader and Bartlett, 1997; Peng et al., 2011; Das et al., 1994; Groysman and Holm, 2009; Xiao et al., 1992). Recently we demonstrated that molybdenum oxide (MoO_3) NPs exhibit an intrinsic SuOx-like activity (Ragg et al., 2014). The MoO_3 NPs with an approximate diameter of 2 nm were functionalized with a mitochondria targeting ligand containing a triphenylphosphonium ion (TPP, Fig. 15.11a), after which MoO_3-TPP NPs could cross the cellular membrane and accumulate specifically at the mitochondria, where SuOx is located.

The SuOx activity of the MoO_3-TPP NPs was analyzed spectrophotometrically by measuring the reduction of potassium hexacyanoferrate(III) (ferricyanide) at 420 nm. By varying the concentration of MoO_3-TPP, while keeping the concentrations of sulfite and ferricyanide constant, a linear dependence for the SuOx activity was observed (Fig. 15.11b, blue diamonds). The activity of MoO_3-TPP was significantly higher than commercially available bulk (Fig. 15.11b, orange diamonds). In contrast to other kinetic experiments for nanomaterials used as artificial enzyme mimics, where a typical Michaelis–Menten behavior was reported when the concentration of the corresponding substrate was varied, a sigmoidal curve was observed in case of the sulfite oxidase mimic, which is typically found for cooperative binding of substrates to the active site. Based on the Hill-equation for cooperative binding, the experimental K_m values of 0.59 ± 0.02 mM for sulfite and a Hill coefficient n (cooperativity constant) of 2.35±0.15 indicate a positive cooperative behavior. The MoO_3-TPP NPs mediate the sulfite oxidation with a V_{max} of 35.23 ± 1.13 μM/min, through which a turnover frequency (k_{cat}) of 2.78 ± 0.09 s^{-1} was determined. These kinetic values have the same order of magnitude as those found for the human SuOx mutant R160Q ($K_m(SO_3^{2-})$ = 1.7 mM, k_{cat} = 16 s^{-1}) and native human SuOx ($K_m(SO_3^{2-})$ = 0.017 mM, k_{cat} = 2.4 s^{-1}) (Garrett et al., 1998; Karakas et al., 2005).

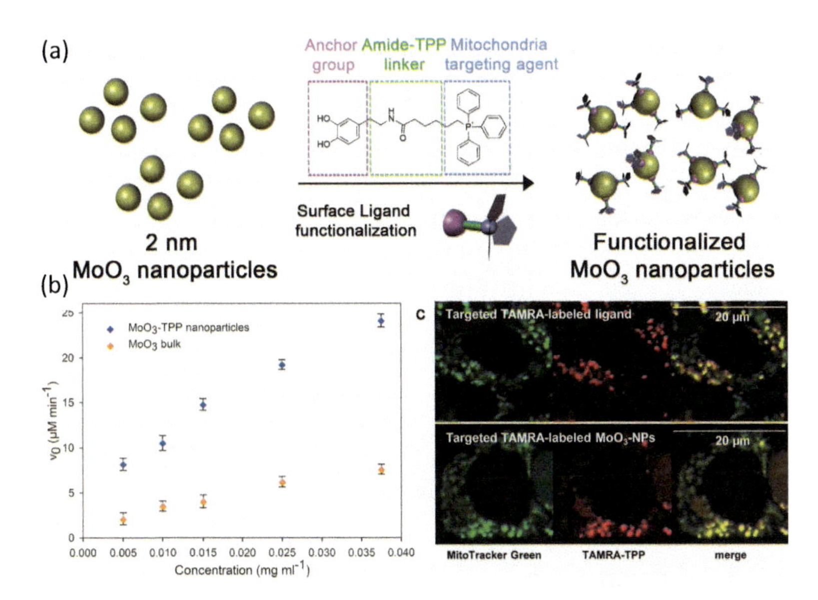

Figure 15.11 Sulfite oxidase activity and mitochondria targeting of functionalized MoO_3-TPP nanoparticles. (a) Mitochondria-specific surface functionalization of 2 nm MoO_3 nanoparticles with ligand containing dopamine as an anchor group and triphenylphosphonium ion (TPP) as mitochondria targeting agent. (b) Sulfite oxidase activity of functionalized MoO_3-TPP nanoparticles (blue diamonds) and bulk MoO_3 (orange diamonds) with constant concentrations of sodium sulfite and potassium ferricyanide determined by measuring spectrophotometrically the ferricyanide reduction at 420 nm. (c) Confocal laser scanning microscopy localization studies in liver cells using the commercially available mitochondria targeting dye MitoTracker Green (λ_{ex} = 488 nm) with fluorescently-labeled (TAMRA) MoO_3-TPP nanoparticles (MoO_3-TAMRA-TPP, λ_{ex} = 543 nm) and TAMRA-labeled ligand (Dopa-TAMRA-TPP, λ_{ex} = 543 nm) showing selective targeting of the mitochondria by co-localization with MitoTracker Green (merge). Reprinted with permission from Ragg et al. (2014); copyright (2014) American Chemical Society.

The selective mitochondria targeting of the functionalized MoO_3-TPP NPs (Fig. 15.11c, merge) was demonstrated by co-localization of the MoO_3-TAMRA-TPP NPs (Fig. 15.11c, red channel), additionally labeled with the fluorescent marker 5-carboxytetramethylrhodamine (TAMRA), with the commercially available mitochondria targeting dye MitoTracker Green (Fig. 15.11c, green channel). Pearson's correlation coefficient (PCC) is used to quantitatively describe the co-localization of the two dyes.

The fluorescent ligand Dopa-TAMRA-TPP was co-localized with MitoTracker Green (Fig. 15.11c) showing a PCC of 0.84, whereas MoO_3-TAMRA-TPP NPs exhibit a PCC of 0.91, which lies in the same range compared to other mitochondria targeting compounds (Chyan et al., 2014; Wisnovsky et al., 2013). Finally the MoO_3-TPP NPs were tested as a possible SuOx replacement in vitro, on chemically modified cells with reduced levels of functional sulfite oxidase. The SuOx activity of the deficient cells could be fully recovered when treated with functionalized MoO_3-TPP NPs, whereby the potential of the NPs as a possible SuOx replacement in vitro was shown (Ragg et al., 2014). Up to now no efficient treatment for SuOx deficiency is known (Veldman et al., 2010; Schwarz et al., 2004), so MoO_3 NPs could be a promising option for the treatment of this severe disease.

15.6 Conclusion

We have given a general overview about enzyme mimics based on inorganic nanoparticles and highlighted the progress that has been made in recent years. The research in this field is highly active, indicated by the rapidly growing number of publications. Besides the overwhelming mass of reports on peroxidase mimics and their applications, only a few other enzymatic systems have been explored so far which leaves much space for future developments.

Acknowledgment

RR and KK are thankful to the Max Planck Graduate Center Mainz (MPGC) for financial support.

References

Aiba, Y., J. Sumaoka, and M. Komiyama, *Chem. Soc. Rev.*, **40** (2011), 5657–5668.

André, R., F. Natálio, M. Humanes, J. Leppin, K. Heinze, R. Wever, H. C. Schröder, W. E. G. Müller and W. Tremel, *Adv. Funct. Mater.*, **21** (2011), 501–509.

André, R., F. Natálio, and W. Tremel, Nanoparticles as enzyme mimics, in *New and Future Developments in Catalysis*, L. S. Steven (ed); chapter 6, Elsevier, Amsterdam, 2013.

Archibald, F. S., and I. Fridovich, *J. Bacteriol.*, **146** (1981), 928–936.

Arnold, G. L., C. L. Greene, J. Patrick Stout, and S. I. Goodman, *J. Pediatr.*, **123** (1993), 595–598.

Asati, A., S. Santra, C. Kaittanis, S. Nath, and J. M. Perez, *Angew. Chem. Int. Ed.*, **48** (2009), 2308–2312.

Barnese, K., E. B. Gralla, D. E. Cabelli, and J. Selverstone Valentine, *J. Am. Chem. Soc.*, **130** (2008), 4604–4606.

Biette, L., F. Carn, M. Maugey, M. F. Achard, J. Maquet, N. Steunou, J. Livage, H. Serier, and R. Backov, *Adv. Mater.*, **17** (2005), 2970–2974.

Biparva, P., S. M. Abedirad, and S. Y. Kazemi, *Talanta*, **130** (2014), 116–121.

Bolwell, G. P., and P. Wojtaszek, *Physiol. Mol. Plant Pathol.*, **51** (1997), 347–366.

Borchardt, S. A., E. J. Allain, J. J. Michels, G. W. Stearns, R. F. Kelly, and W. F. McCoy, *Appl. Environ. Microbiol.*, **67** (2001), 3174–3179.

Breaker, R. R., and G. F. Joyce, *Chem. Biol.*, **1** (1994), 223–229.

Breslow, R., *Artificial Enzymes*; Wiley-VCH, Weinheim (2006).

Breslow, R., and L. E. Overman, *J. Am. Chem. Soc.*, **92** (1970), 1075–1077.

Butler, A., and M. Sandy, *Nature*, **460** (2009), 848–854.

Cai, Q., S. Lu, F. Liao, Y. Li, S. Ma, and M. Shao, *Nanoscale*, **6** (2014), 8117–8123.

Celardo, I., M. De Nicola, C. Mandoli, J. Z. Pedersen, E. Traversa, and L. Ghibelli, *ACS Nano*, **5** (2011a), 4537–4549.

Celardo, I., J. Z. Pedersen, E. Traversa, and L. Ghibelli, *Nanoscale*, **3** (2011b), 1411–1420.

Chen, J., S. Patil, S. Seal, and J. F. McGinnis, *Nat Nano*, **1** (2006), 142–150.

Chen, Q., M. Liu, J. Zhao, X. Peng, X. Chen, N. Mi, B. Yin, H. Li, Y. Zhang, and S. Yao, *Chem. Commun.*, **50** (2014), 6771–6774.

Chen, Z., J.-J. Yin, Y.-T. Zhou, Y. Zhang, L. Song, M. Song, S. Hu, and N. Gu, *ACS Nano*, **6** (2012), 4001–4012.

Chyan, W., D. Y. Zhang, S. J. Lippard, and R. J. Radford, *Proc. Natl. Acad. Sci. U. S. A.*, **111** (2014), 143–148.

Clague, M. J., N. L. Keder, and A. Butler, *Inorg. Chem.*, **32** (1993), 4754–4761.

Comotti, M., C. Della Pina, R. Matarrese, and M. Rossi, *Angew. Chem. Int. Ed.*, **43** (2004), 5812–5815.

Coughlin, P., S. Roberts, C. Rush, and A. Willetts, *Biotechnol. Lett.*, **15** (1993), 907–912.

D'Souza, V. T., X. L. Lu, R. D. Ginger, and M. L. Bender, *Proc. Natl. Acad. Sci. U. S. A.*, **84** (1987), 673–674.

Dai, Z., S. Liu, J. Bao, and H. Ju, *Chem. Eur. J.*, **15** (2009), 4321–4326.

Das, S. K., P. K. Chaudhury, D. Biswas, and S. Sarkar, *J. Am. Chem. Soc.*, **116** (1994), 9061–9070.

Dawson, J. H., *Science*, **240** (1988), 433–439.

de Boer, E., and R. Wever, *J. Biol. Chem.*, **263** (1988), 12326–12332.

Dong, Z., W. Yongguo, Y. Yin, and J. Liu, *Curr. Opin. Colloid Interface Sci.*, **16** (2011), 451–458.

Dunford, H. B., *Peroxidases and Catalases: Biochemistry, Biophysics, Biotechnology and Physiology*; John Wiley & Sons, Hoboken (2010).

Dunford, H. B., and J. S. Stillman, *Coord. Chem. Rev.*, **19** (1976), 187–251.

Espenson, J. H., O. Pestovsky, P. Huston, and S. Staudt, *J. Am. Chem. Soc.*, **116** (1994), 2869–2877.

Fan, K., C. Cao, Y. Pan, D. Lu, D. Yang, J. Feng, L. Song, M. Liang, and X. Yan, *Nat. Nano*, **7** (2012), 765–765.

Fan, J., J.-J. Yin, B. Ning, X. Wu, Y. Hu, M. Ferrari, G. J. Anderson, J. Wei, Y. Zhao, and G. Nie, *Biomaterials*, **32** (2011), 1611–1618.

Fenton, H. J. H., *J. Chem. Soc. Trans.*, **65** (1894), 899–910.

Fruk, L., and C. M. Niemeyer, *Angew. Chem.*, **117** (2005), 2659–2662.

Gao, L., J. Zhuang, L. Nie, J. Zhang, Y. Zhang, N. Gu, T. Wang, J. Feng, D. Yang, S. Perrett, and X. Yan, *Nat. Nano*, **2** (2007), 577–583.

Garrett, R. M., J. L. Johnson, T. N. Graf, A. Feigenbaum, and K. V. Rajagopalan, *Proc. Natl. Acad. Sci. U. S. A.*, **95** (1998), 6394–6398.

Grasselli, R., *Top. Catal.*, **21** (2002), 79–88.

Groysman, S., and R. H. Holm, *Biochemistry*, **48** (2009), 2310–2320.

Haber, J., and E. Lalik, *Catal. Today*, **33** (1997), 119–137.

Han, L., L. Zeng, M. Wei, C. M. Li, and A. Liu, *Nanoscale*, **7** (2015), 11678–11685.

Hansen, P. J., and J. H. Espenson, *Inorg. Chem.*, **34** (1995), 5839–5844.

Hasan, Z., R. Renirie, R. Kerkman, H. J. Ruijssenaars, A. F. Hartog, and R. Wever, *J. Biol. Chem.*, **281** (2006), 9738–9744.

He, W., H. Jia, X. Li, Y. Lei, J. Li, H. Zhao, L. Mi, L. Zhang, and Z. Zheng, *Nanoscale*, **4** (2012), 3501–3506.

He, W., Y. Liu, J. Yuan, J.-J. Yin, X. Wu, X. Hu, K. Zhang, J. Liu, C. Chen, Y. Ji, and Y. Guo, *Biomaterials*, **32** (2011), 1139–1147.

He, S., W. Shi, X. Zhang, J. Li, and Y. Huang, *Talanta*, **82** (2010a), 377–383.

He, W., X. Wu, J. Liu, X. Hu, K. Zhang, S. Hou, W. Zhou, and S. Xie, *Chem. Mater.*, **22** (2010b), 2988–2994.

Hille, R., *Dalton Trans.*, **42** (2013), 3029–3042.

Hirst, S. M., A. S. Karakoti, R. D. Tyler, N. Sriranganathan, S. Seal, and C. M. Reilly, *Small*, **5** (2009), 2848–2856.

Hu, Y.-S., X. Liu, J.-O. Müller, R. Schlögl, J. Maier, and D. S. Su, *Angew. Chem. Int. Ed.*, **48** (2009), 210–214.

Hu, L., Y. Yuan, L. Zhang, J. Zhao, S. Majeed, and G. Xu, *Anal. Chim. Acta*, **762** (2013), 83–86.

Huang, Y., W. Ma, J. Li, M. Cheng, J. Zhao, L. Wan, and J. C. Yu, *J. Phys. Chem. B*, **107** (2003), 9409–9414.

Huang, Y., X. Ran, Y. Lin, J. Ren, and X. Qu, *Chem. Commun.*, **51** (2015), 4386–4389.

Jiao, X., H. Song, H. Zhao, W. Bai, L. Zhang, and Y. Lv, *Anal. Methods*, **4** (2012), 3261–3267.

Jv, Y., B. Li, and R. Cao, *Chem. Commun.*, **46** (2010), 8017–8019.

Karakas, E., H. L. Wilson, T. N. Graf, S. Xiang, S. Jaramillo-Busquets, K. V. Rajagopalan, and C. Kisker, *J. Biol. Chem.*, **280** (2005), 33506–33515.

Kim, C. K., T. Kim, I.-Y. Choi, M. Soh, D. Kim, Y.-J. Kim, H. Jang, H.-S. Yang, J. Y. Kim, H.-K. Park, S. P. Park, S. Park, T. Yu, B.-W. Yoon, S.-H. Lee, and T. Hyeon, *Angew. Chem. Int. Ed.*, **51** (2012), 11039–11043.

Kirby, A. J., and F. Hollfelder, *From Enzyme Models to Model Enzymes*; The Royal Society of Chemistry (2009).

Kirkorian, K., A. Ellis, and L. J. Twyman, *Chem. Soc. Rev.*, **41** (2012), 6138–6159.

Klotz, I. M., *Ann. N. Y. Acad. Sci.*, **434** (1984), 302–320.

Kofoed, J., and J.-L. Reymond, *Curr. Opin. Chem. Biol.*, **9** (2005), 656–664.

Korsvik, C., S. Patil, S. Seal, and W. T. Self, *Chem. Commun.*, **0** (2007), 1056–1058.

Li, N., Y. Yan, B.-Y. Xia, J.-Y. Wang, and X. Wang, *Biosens. Bioelectron.*, **54** (2014), 521–527.

Lien, C.-W., C.-C. Huang, and H.-T. Chang, *Chem. Commun.*, **48** (2012), 7952–7954.

Ligtenbarg, A. G. J., R. Hage, and B. L. Feringa, *Coord. Chem. Rev.*, **237** (2003), 89–101.

Lin, T., L. Zhong, L. Guo, F. Fu, and G. Chen, *Nanoscale*, **6** (2014), 11856–1862.

Liu, H., C. Gu, W. Xiong, and M. Zhang, *Sens. Actuators B Chem.*, **209** (2015), 670–676.

Liu, X., Q. Wang, H. Zhao, L. Zhang, Y. Su, and Y. Lv, *Analyst*, **137** (2012), 4552–4558.

Liu, Y., G. Zhu, C. Bao, A. Yuan, and X. Shen, *Chin. J. Chem.*, **32** (2014), 151–156.

Luo, W., Y.-S. Li, J. Yuan, L. Zhu, Z. Liu, H. Tang, and S. Liu, *Talanta*, **81** (2010), 901–907.

Ma, M., Y. Zhang, and N. Gu, *Colloids Surf. Physicochem. Eng. Aspects*, **373** (2011), 6–10.

Mader, M. M., and P. A. Bartlett, *Chem. Rev.*, **97** (1997), 1281–1302.

Martinez, J. S., G. L. Carroll, R. A. Tschirret-Guth, G. Altenhoff, R. D. Little, and A. Butler, *J. Am. Chem. Soc.*, **123** (2001), 3289–3294.

Masella, R., R. Di Benedetto, R. Varì, C. Filesi, and C. Giovannini, *J. Nutr. Biochem.*, **16** (2005), 577–586.

McCord, J. M., B. B. Keele, and I. Fridovich, *Proc. Natl. Acad. Sci. U. S. A.*, **68** (1971), 1024–1027.

Meister, G. E., and A. Butler, *Inorg. Chem.*, **33** (1994), 3269–3275.

Messerschmidt, A., L. Prade, and R. Wever, *Biol. Chem.*, **378** (1997), 309.

Mimoun, H., L. Saussine, E. Daire, M. Postel, J. Fischer, and R. Weiss, *J. Am. Chem. Soc.*, **105** (1983), 3101–3110.

Mu, J., Y. Wang, M. Zhao, and L. Zhang, *Chem. Commun.*, **48** (2012), 2540–2542.

Mu, J., L. Zhang, M. Zhao, and Y. Wang, *ACS Appl. Mater. Interfaces*, **6** (2014), 7090–7098.

Murthy, M. R. N., T. J. Reid Iii, A. Sicignano, N. Tanaka, and M. G. Rossmann, *J. Mol. Biol.*, **152** (1981), 465–499.

Nangia, Y., B. Kumar, J. Kaushal, and C. Raman Suri, *Anal. Chim. Acta*, **751** (2012), 140–145.

Natalio, F., R. Andre, A. F. Hartog, B. Stoll, K. P. Jochum, R. Wever, and W. Tremel, *Nat. Nano*, **7** (2012), 530–535.

Neumann, C. S., D. G. Fujimori, and C. T. Walsh, *Chem. Biol.*, **15** (2008), 99–109.

Peng, C., B. Jiang, Q. Liu, Z. Guo, Z. Xu, Q. Huang, H. Xu, R. Tai, and C. Fan, *Energy Environ. Sci.*, **4** (2011), 2035–2040.

Pirmohamed, T., J. M. Dowding, S. Singh, B. Wasserman, E. Heckert, A. S. Karakoti, J. E. S. King, S. Seal, and W. T. Self, *Chem. Commun.*, **46** (2010), 2736–2738.

Pollack, S. J., J. W. Jacobs, and P. G. Schultz, *Science*, **234** (1986), 1570–1573.

Qiao, F., L. Chen, X. Li, L. Li, and S. Ai, *Sens. Actuators B Chem.*, **193** (2014), 255–262.

Ragg, R., F. Natalio, M. N. Tahir, H. Janssen, A. Kashyap, D. Strand, S. Strand, and W. Tremel, *ACS Nano*, **8** (2014), 5182–5189.

Raynal, M., P. Ballester, A. Vidal-Ferran, and P. W. N. M. van Leeuwen, *Chem. Soc. Rev.*, **43** (2014), 1734–1787.

Reedijk, J., and E. Bouwman, *Bioinorganic Catalysis*; Marcel Dekker, Inc., New York (1999).

Rosenhahn, A., S. Schilp, H. J. Kreuzer, and M. Grunze, *PCCP*, **12** (2010), 4275–4286.

Roy, P., Z.-H. Lin, C.-T. Liang, and H.-T. Chang, *Chem. Commun.*, **48** (2012), 4079–4081.

Sawyer, D. T., *Coord. Chem. Rev.*, **165** (1997), 297–313.

Schwarz, G., R. R. Mendel, and M. W. Ribbe, *Nature*, **460** (2009), 839–847.

Schwarz, G., J. A. Santamaria-Araujo, S. Wolf, H.-J. Lee, I. M. Adham, H.-J. Gröne, H. Schwegler, J. O. Sass, T. Otte, P. Hänzelmann, R. R. Mendel, W. Engel, and J. Reiss, *Hum. Mol. Genet.*, **13** (2004), 1249–1255.

Smith, T. S., and V. L. Pecoraro, *Inorg. Chem.*, **41** (2002), 6754–6760.

Soedjak, H. S., J. V. Walker, and A. Butler, *Biochemistry*, **34** (1995), 12689–12696.

Su, H., D.-D. Liu, M. Zhao, W.-L. Hu, S.-S. Xue, Q. Cao, X.-Y. Le, L.-N. Ji, and Z.-W. Mao, *ACS Appl. Mater. Interfaces*, **7** (2015), 8233–8242.

Sudimack, J., and R. J. Lee, *Adv. Drug Del. Rev.*, **41** (2000), 147–162.

Tarnuzzer, R. W., J. Colon, S. Patil, and S. Seal, *Nano Lett.*, **5** (2005), 2573–2577.

ten Brink, H. B., A. Tuynman, H. L. Dekker, W. Hemrika, Y. Izumi, T. Oshiro, H. E. Schoemaker, and R. Wever, *Inorg. Chem.*, **37** (1998), 6780–6784.

Tramontano, A., K. D. Janda, and R. A. Lerner, *Science*, **234** (1986), 1566–1570.

Veldman, A., J. A. Santamaria-Araujo, S. Sollazzo, J. Pitt, R. Gianello, J. Yaplito-Lee, F. Wong, C. A. Ramsden, J. Reiss, I. Cook, J. Fairweather, and G. Schwarz, *Pediatrics*, **125** (2010), e1249–e1254.

Velusamy, S., and T. Punniyamurthy, *Org. Lett.*, **6** (2003), 217–219.

Vernekar, A. A., T. Das, S. Ghosh, and G. Mugesh, *Chem. Asian J.* (2015), 10.1002/asia.201500942.

Vernekar, A. A., D. Sinha, S. Srivastava, P. U. Paramasivam, P. D'Silva, and G. Mugesh, *Nat. Commun*, **5** (2014), 5301.

Wang, B., S. Gu, Y. Ding, Y. Chu, Z. Zhang, X. Ba, Q. Zhang, and X. Li, *Analyst*, **138** (2013), 362–367.

Wang, W., X. Jiang, and K. Chen, *Chem. Commun.*, **48** (2012a), 6839–6841.

Wang, W., X. Jiang, and K. Chen, *Chem. Commun.*, **48** (2012b), 7289–7291.

Wang, Y., L. Salmon, J. Ruiz, and D. Astruc, *Nat. Commun.*, **5** (2014), 3489.

Wang, Q., Z. Yang, X. Zhang, X. Xiao, C. K. Chang, and B. Xu, *Angew. Chem. Int. Ed.*, **46** (2007), 4285–4289.

Wei, H., and E. Wang, *Anal. Chem.*, **80** (2008), 2250–2254.

Wei, H., and E. Wang, *Chem. Soc. Rev.*, **42** (2013), 6060–6093.

Wever, R., H. L. Dekker, J. W. P. M. Van Schijndel, and E. G. M. Vollenbroek, Antifouling Paint Containing Haloperoxidase and Method to Determine Halide (1995).

Wever, R., and W. Hemrika, *Vanadium Haloperoxidases in Handbook of Metalloproteins*; John Wiley & Sons, Ltd, Hoboken (2006).

Wisnovsky, S. P., J. J. Wilson, R. J. Radford, M. P. Pereira, M. R. Chan, R. R. Laposa, S. J. Lippard, and S. O. Kelley, *Chem. Biol.*, **20** (2013), 1323–1328.

Wulff, G., *Chem. Rev.*, **102** (2001), 1–28.

Xia, X., J. Zhang, N. Lu, M. J. Kim, K. Ghale, Y. Xu, E. McKenzie, J. Liu, and H. Ye, *ACS Nano*, **9** (2015), 9994–10004.

Xiao, Z., C. G. Young, J. H. Enemark, and A. G. Wedd, *J. Am. Chem. Soc.*, **114** (1992), 9194–9195.

Xu, C., and X. Qu, *NPG Asia Mater*, **6** (2014), e90.

Yu, C.-J., T.-H. Chen, J.-Y. Jiang, and W.-L. Tseng, *Nanoscale*, **6** (2014), 9618–9624.

Zampella, G., L. Bertini, and L. De Gioia, *Chem. Commun.*, **50** (2014), 304–307.

Zhang, X., S. He, Z. Chen, and Y. Huang, *J. Agric. Food Chem.*, **61** (2013), 840–847.

Zhang, X., and Y. Huang, *Anal. Methods*, **7** (2015), 8640–8646.

Zhang, L., L. Laug, W. Münchgesang, E. Pippel, U. Gösele, M. Brandsch, and M. Knez, *Nano Lett.*, **10** (2009), 219–223.

Zhang, Y., J. Tian, S. Liu, L. Wang, X. Qin, W. Lu, G. Chang, Y. Luo, A. M. Asiri, A. O. Al-Youbi, and X. Sun, *Analyst*, **137** (2012), 1325–1328.

Zhang, L., F. Xia, Z. Song, N. A. S. Webster, H. Luo, and Y. Gao, *RSC Adv.*, **5** (2015), 61371–61379.

Zhao, M., J. Huang, Y. Zhou, X. Pan, H. He, Z. Ye, and X. Pan, *Chem. Commun.*, **49** (2013), 7656–7658.

Zheng, C., A. Zheng, B. Liu, X.-L. Zhang, Y. He, J. Li, H. Yang, and G. Chen, *Chem. Commun.*, **50** (2014), 13103–13106.

Chapter 16

Enzyme Nanocapsules for Glucose Sensing and Insulin Delivery

Wanyi Tai and Zhen Gu

Joint Department of Biomedical Engineering,
University of North Carolina at Chapel Hill and North Carolina State University,
Raleigh, North Carolina 27695, USA

wtai@ncsu.edu, zgu@email.unc.edu

16.1 Introduction

Diabetes mellitus (DM), also known as diabetes, is a type of common metabolic disease in which glucose is build up in the blood, caused either by pancreas's failure to produce insulin (Type 1 DM) or by insulin resistance from the body tissue (Type 2 DM) (Kasuga, 2006). Diabetes currently affects an estimate 387 million people, which is equal to 8.3% of the adult population worldwide. It is also a rapid growing problem, with an expectation to rise to 592 million by 2035 (Whiting, Guariguata et al., 2011). Diabetes may lead to serious complications including cardiovascular disease, retinopathy, chronic kidney disease, and even cancer (Tai et al., 2014). Moreover, diabetes is a chronic disease and currently there is no cure with exception only in the early stage of Type 2 DM. Despite this, the diabetes-associated complications can be minimized through the tight control of

Biocatalysis and Nanotechnology
Edited by Peter Grunwald

ISBN 978-981-4613-69-9 (Hardcover), 978-1-315-19660-2 (eBook)
www.panstanford.com

blood glucose level (BGL), which requires frequent blood glucose monitoring and careful administration of insulin. Therefore, it is urgent to develop the sensitive glucose biosensor and advanced insulin delivery systems.

Nanomaterials have very small featured sizes in the range of 1–100 nm, which offer significant advantages over the macroscale materials. The extremely large surface area, tunable chemical structure, rapid movement of reactant and enhanced optical properties make the nanomaterials especially valuable for developing advanced biosensor and drug delivery system (Cash et al., 2010). Glucose oxidase (GOx) catalyzes the oxidation of glucose to hydrogen peroxide and D-glucono-δ-lactone. Due to its responsiveness to physiological concentration of glucose and stability, GOx has been widely used as the sensing element in glucose sensors and advanced insulin delivery (Mo et al., 2014). The combination of GOx and nanomaterials, more specifically the enzyme-anchored or -encapsulated nanomaterials, has dramatically increased the sensitivity and reliability of these systems, making them nearing the stage of commercial and clinical implementation (Cash et al., 2010). In this chapter, we summarize the recent development in the field of enzyme nanomaterials for diabetes care. We focus our discussion on GOx anchored nanomaterials that are used for glucose detection and advanced insulin delivery. Additionally, we slightly expand our discussion to other nanomaterials that can be potentially an alternative to GOx nanosystems.

16.2 Glucose, Insulin, and Diabetes

16.2.1 The Control of Glucose Homeostasis by Insulin

Glucose is the basic unit of fuel for our body and cells. Insulin is a peptide hormone produced by beta cells in pancreas for regulation of glucose metabolism (Fig. 16.1a). In order to maintain the proper levels of blood glucose, insulin circulates through the body to generate signals by binding with insulin receptors. Acting as a messenger, insulin promotes the absorption of glucose from blood to skeletal muscles and fatty tissues, effecting a reduced BGL (Saltiel et al., 2001). Here we list the fate of insulin from its secretion in beta cells to its action on the body.

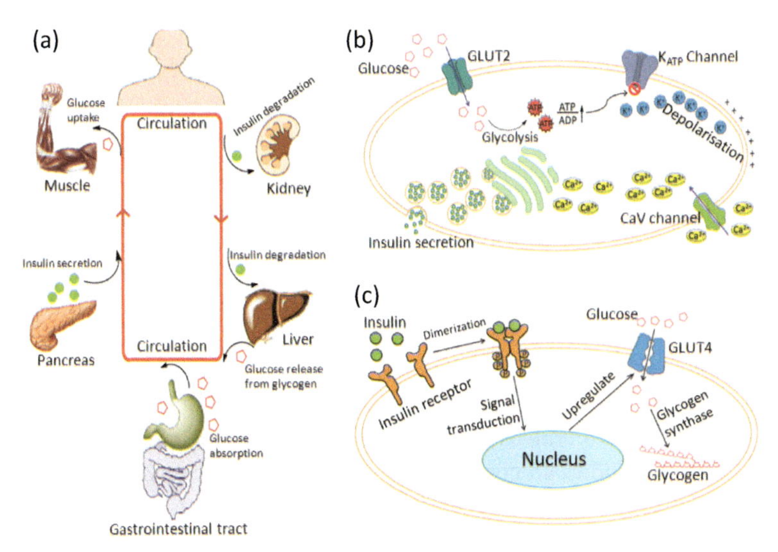

Figure 16.1 Regulation of glucose level by insulin. (a) After nutrients digestion, glucose is absorbed into blood circulation, resulting in BGL rising. Beta cells in pancreas receive the signal and start to secrete the insulin, which circulates to muscle/fat cells and activates the glucose uptake from the circulation. The remaining insulin is quickly degraded by kidney and liver to modulate the insulin action. When BGL falls below the normal level, liver generates free glucose from glycogen to maintain homeostasis. (b) Insulin secretion in beta cells is triggered by rising BGL. The uptake of glucose by GLUT2 facilitates the glycolysis, which generates lots of ATP molecules and causes a rise in ATP/ADP ratio. The rise blocks the K_{ATP} channel and leads to the buildup of potassium ion that depolarizes the membrane. The depolarization opens the CaV channel and calcium ions flow inward, leading to the exocytotic release of insulin from granules. (c) Insulin mediates the uptake of glucose in muscle and fat cells. Insulin binds to insulin receptors and triggers a series of signal transduction which upregulates the GLUT4 expression. Glucoses are transported into cytoplasm and converted into glycogen by glycogen synthase.

Insulin is synthesized by pancreatic beta cells, which are known to cluster as the islets of Langerhans in pancreas (Ashby et al., 1975). Although the islets account for only 1–2% of the total mass of pancreas, they are the only tissues responsible for insulin secretion in the entire body throughout the whole life. Insulin is a small protein that is constituted by two short polypeptide chains, the A- and B-chains, linked together by disulfide bonds (Katsoyannis, 1964). During the production of insulin, it is, however, first synthesized as an inactive, single chain polypeptide called preproinsulin. Preproinsulin contains a 24-residue signal peptide which directs the translocation of the nascent protein to the endoplasmic reticulum (ER) for post-translational processing. Upon entering the ER, the signal peptide is proteolytically removed to form proinsulin, which then folds into the correct conformation and forms three vital disulfide bonds in the ER lumen. Proinsulin undergoes maturation into active insulin through digestion by specific peptidases in the secretary vesicles (Yoi et al., 1979). Finally, the resulting mature insulin is packaged and stored in secretary granules waiting for secretion (Lindall et al., 1963).

The insulin release from beta cells is triggered by the increased blood glucose concentration (Fig. 16.1b). Glucose enters the beta cells through the glucose transporter 2 (GLUT2). As a substrate for glycolysis, glucose is quickly metabolized to generate high-energy ATP molecules, leading to a rise of the ATP/ADP ratio in the cell (Sweet et al., 2004). The increased intracellular ATP/ADP ratio causes the shutdown of ATP-gated potassium (K_{ATP}) channels in the cellular membrane, which prevents potassium ions (K^+) from leaving the cell, leading to a buildup of potassium ions. The ensuring rise of positive charges inside cells leads to the depolarization of the cell surface membrane, which triggers the activation of voltage-gated calcium (CaV) channels. The brisk influx of calcium ion leads to the exocytotic release of insulin from their storage granule (Rajan et al., 1990).

Insulin circulates throughout the blood stream until it binds to an insulin receptor. Binding to insulin causes a conformational change of the insulin receptor that activates its kinase domain and induces the autophosphorylation of tyrosine residues on the C-terminus of the receptor, leading to internal signal cascade that allows the glucose transporter 4 (GLUT4) to transport glucose into cells (Fig. 16.1c) (Yang, 2010). The key consequence of intracellular signal transduction is the increased expression of GLUT4 in the plasma membrane and the immediate activation of glycogen synthase. By the facilitative transport of glucose into the cells, the glucose transporters effectively remove glucose from the blood stream. The glycogen synthase converts the glucose into glycogen and stores it in cells as the glucose reservoir (Halse et al., 2001). The insulin receptors promote the uptake of glucose into various tissues but mainly muscle cells (myocytes) and fat cells (adipocytes) (Czech et al., 1978). When glucose concentration comes down to normal level, the insulin secretion from beta cells slows and stops. The insulin action will be terminated by endocytosis and degradation of GLUT4, which leads to a decrease and finally to an abolished glucose uptake in myoctes and adipocytes (Yang, 2010).

The primary organs for insulin clearance are the liver and the kidney (Rabkin et al., 1986). The liver clears most insulin during first pass metabolism, whereas the kidney clears most of the insulin in systemic circulation. The blood-circulating insulin is broken down with a half-life of around 6 min, which ensures the insulin action to be modulated fast and BGL does not go dangerously low.

16.2.2 Pathology of Diabetes

The blood glucose level is closely regulated by insulin secretion in beta cells and its action in the targeted cells. Any failure of the system involving insulin secretion, action, and clearance might lead to the imbalance of glucose homeostasis. Type 1 DM is characterized by the body's failure to produce enough insulin due to the autoimmune destruction of the beta cells in the pancreas. The loss of the beta cells, the only insulin source of the body, will inevitably lead to the build-up of glucose in the blood. Insulin resistance itself (Type 2 DM), in which cells fail to respond to the normal actions of insulin, also leads to hyperglycemia. The most severe form of diabetes is Type 1 DM. It might be contributed by various factors, including genetics, environment factors, and the exposure to certain viruses. Certain genotypes such as human leukocyte antigen (HLA) appear to confer the strong risk of Type 1 DM (Mychaleckyj et al., 2010). There are also some evidences suggesting a possible link between Coxsackie B4 virus and Type 1 DM (Yoon et al., 1978; Horwitz et al., 1998). The most common forms of diabetes are Type 2 DM, which accounts for approximately 90–95% of DM cases. Unlike Type 1 DM, which usually appears during childhood, Type 2 DM affects mostly adults. The primary causes are lifestyle factors, including excessive body weight, lack of enough exercise, and dietary factors (Howard, 2002).

Insulin treatment is indispensable for Type 1 and advanced Type 2 DM. Despite the effect, the use of insulin is always associated with risk and side effect. Improper dose of insulin might cause deviation of BGL from normal level or even hypoglycemia, which can cause unconsciousness, brain damage, and death (Cryer, 2007). For better management of BGL, frequent glucose monitoring and smart insulin administration are essential, which requires the development of sensitive glucose biosensors and advanced insulin delivery systems.

16.3 Glucose Biosensor

Frequent testing of the BGL by glucose sensor is crucial for attaining effective treatment and avoiding diabetic emergencies. The current method requires the patients to place a drop of blood that is usually obtained by piercing the skin of a finger, onto a chemically active disposable sensor test strip and the BGL is reported by a handheld electronic reader. However, there are obvious limitations in this diagnostic method including painful sampling, glucose level cannot be monitored on a continuous basis and occasional fluctuations between reading (Cash et al., 2010). Significant efforts have been made to improve the usability, accuracy, and lifetime of the sensor for the treatment of diabetes. Nanomaterials and nanosensors have significant impact on the effort owing to the small size, high surface area, better catalytic ability, and rapid diffusion of glucose through sensors. Based on the detection technologies, glucose nanosensors are classified into two categories: electrochemical glucose sensors and fluorescence-based sensors.

16.3.1 Electrochemical Glucose Biosensors

16.3.1.1 GOx-mediated electrochemical nanosensor

Electrochemical detection of glucose by GOx is widely utilized in glucose sensing because these sensors have low detection limits, fast response times and much lower costs compared to the sensors based on other detection mechanisms. The historical commencement of electrochemical glucose sensors dates back to 1962 with the development of the first device by Clark and

Lyons of the Cincinnati Children's Hospital (Clark et al., 1962). The sensor relied on GOx immobilized on an oxygen electrode, to oxidize the glucose to gluconic acid. The detection was made by measuring the oxygen consumed by the enzyme-catalyzed reaction.

$$\text{D-Glucose} + O_2 \xrightarrow{\text{Glucose oxidase}} \text{Gluconic acid} + H_2O_2$$

The accuracy of the detection was obviously affected by the oxygen background variation in samples. In 1973, Guilbault and Lubrano improved the electrochemical biosensor for the measurement of blood glucose based on amperometric (anodic) monitoring of the hydrogen peroxide product H_2O_2, which offered good accuracy and precision in connection with blood samples (Guilbault et al., 1973).

$$H_2O_2 \longrightarrow O_2 + 2H^+ + 2e^-$$

Since then, significant effort has been focused on developing improved electrodes to measure the hydrogen peroxide. The electrochemical glucose biosensors are divided into several generations based on the electrode materials ranging from the bulk metal electrode to nanocomposite sensors (Fig. 16.2).

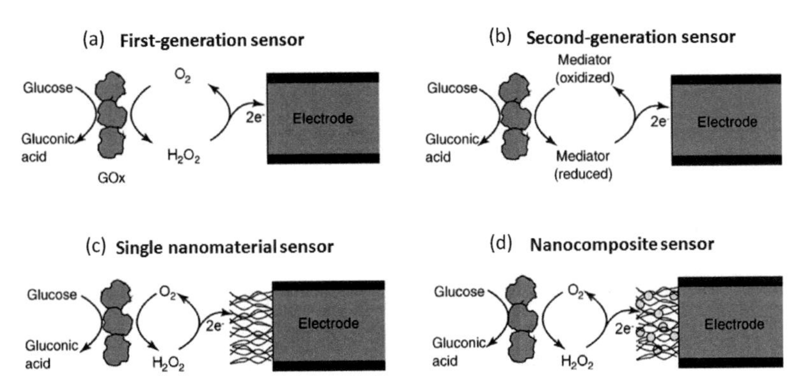

Figure 16.2 GOx based biosensors for glucose detection. Standard electrochemical biosensor use GOx layer to recognize glucose. The signal is transferred from GOx through O_2 reduction to H_2O_2 (a) or the reduction of a mediator (b). Nanomaterials (c) or nanocomposite (d) can be incorporated into the electrode to improve the electron transfer and enhance the signal. Adapted with permission from Cash et al. (2010). Copyright 2010 Elsevier.

The most featured application of nanomaterials for glucose sensor is the use of nanoscaled electrodes to assist the electrochemical detection of the enzymatic reaction of glucose. The nanoelectrodes have at least one structural dimension of the order of 100 nm or less. The integration of nanomaterials into the sensor offers some significant advantages including higher enzyme immobilization capability, more efficient electron transfer from enzyme to electrode and better catalysis, owing to the bulk surface area of nanomaterials.

The single nanomaterial biosensor is the simplest nanomaterial-based sensor. Owing to the nanoscaled dimension, nanomaterial can be used for effective electoral wiring of GOx enzyme, allowing more fast and effective transfer of electrons from enzyme redox center to electrode. Various nanomaterials have been used as electrical connectors; however, gold nanoparticles and carbon nanotubes have demonstrated to be the most effective connectors for glucose sensors, partly because of their excellent electron transfer capabilities and large surface areas. Xiao and his colleagues have constructed such a bioelectrocatalytic nanosensor by nanowiring the GOx and the electrode with gold nanoparticle (Fig. 16.3a). The gold nanoparticle acts as an electron relay to align the GOx enzyme on the conductive electrode support. The gold nanoparticle, chemically conjugated with GOx, is anchored onto the gold electrode by means of a dithiol linker, thus acting as an "electrical nanoplug" to tether the relay units in an optimal positions and alignment with the electrode. The improved surface-reconstitution and electrical contacting between GOx redox centers and electrode leads to a high electron-transfer turnover rate of ~5000 per second, compared with the rate at which molecular oxygen, the natural cosubstrate of the enzyme, accepts electrons (~700 per second) (Xiao et al., 2003). Carbon nanotubes represent another common nanomaterial that is often used as an electrical connector and can be coupled with GOx to provide an optimal surface-mounting relay on the electrode (Fig. 16.3b). The sensor surface is immobilized with the nanofibers, onto which the GOx is incorporated (Gooding et al., 2003; Patolsky et al., 2004). The highly porous structure has a much bigger surface area than conventional bulk metal electrodes, and therefore is capable of binding more GOx to generate stronger electronic signals. Similar to gold nanoparticle connector, the vertically aligned carbon nanotube can act as a conductive nanowire to transfer electron signal quickly from GOx redox center to the macroelectrode. Willner's group demonstrated that such reconstitution represents an extremely efficient approach for "plugging" GOx enzyme on the electrode surface (Patolsky et al., 2004). An additional advantage is that the rate of electron transport can be precisely controlled by the length of the carbon nanotube, as the electron is transported along the distance slightly higher than the length of nanowire. For example, an interfacial electron-transfer rate constant of 42 s^{-1} was estimated as a distance of 50 nm. Another approach to improve the electrical communication is to modify the nanotube with electrochemical mediators such as ferrocene to improve the electron transfer; however, this is often to the detriment of selectivity (Wang, 2008).

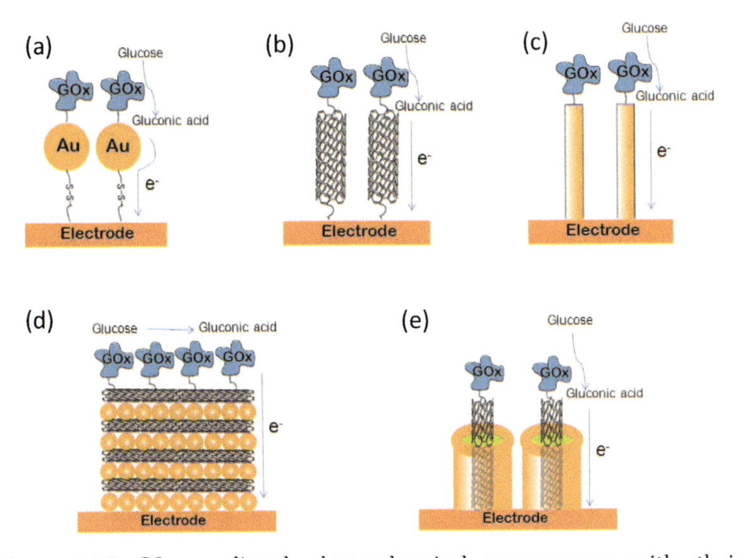

Figure 16.3 GOx mediated electrochemical nanosensors with their electron-transfer relay. (a) Assembly of Au nanoparticle-constituted GOx electrode by absorption of Au nanoparticle to the dithiol monolayer associated with an Au electrode. (b) Electrochemical nanosensor consisted by immobilized carbon nanotube. (c) Nanosensor based on metal nanowire array which is grown directly on the electrode surface. (d) Nanocomposite biosensor prepared by layer-by-layer assembly of carbon nanotube, nanoparticle and polymers. (e) Nanocomposite biosensor constructed by vapor-growing carbon nanotube in TiO_2 nanotube arrays.

A variety of other nanofiber-structured materials can be used for glucose detection as well. Zinc oxide deployed nanowire arrays have an increased surface area and electrochemical activity comparable to carbon nanotube array electrode sensors (Usman Ali et al., 2010). Inspired by this approach, Yang et al. upgraded the ZnO nanowire into highly oriented single-crystal ZnO nanotube arrays which have a larger surface-to-bulk ratio due to the hollow structures of the ZnO nanotubes (Yang et al., 2009). The resulting biosensor enhanced the sensitivity for analytes as demonstrated by the detection of glucose concentrations as low as 1 μM without any electron mediator. Nanowire arrays fabricated from ruthenium and gold have also been used to improve electron transfer ability and increase the surface area of the glucose sensors (Chi et al., 2009; Liu et al., 2009). Another unique advantage of metal nanofibers is that the metal nanowires and nanotubes can be grown directly on a conductive membrane/surface to ensure best electrical contact between nanostructure and the electrode, which is different from carbon nanotube biosensors assembled on the surface of the electrode resulting in a poor contact (Fig. 16.3c) (Kong et al., 2009; Yang et al., 2009).

Carbon nanotubes can work together with other nanomaterials to form nanocomposites for glucose detection, denoted as nanocomposite sensors. Nanoparticle-coupled carbon nanotubes possess many properties such as better electrical conductivity, improved catalysis, enhancement of mass transport, high surface area and tight control over electrode microenvironment (Welch et al., 2006). The most common nanocomposite sensors are fabricated by depositing carbon nanotubes, nanoparticles, and polymers onto the electrode through layer-by-layer assembly (Fig. 16.3d). The assembling process of multilayer films is simple to operate. The response current to H_2O_2 is changed regularly with the increase of the layers, while the maximal signal strength usually can be achieved by optimizing the number of the layers. This approach has been applied by coupling carbon nanotubes with metal nanoparticles of gold, silver, platinum and some alloys, as well as silica nanoparticles and silica/metal hybrid nanoparticles (Kang et al., 2007; Gopalan et al., 2009; Lin et al., 2009; Wang et al. 2009b; Wen et al., 2009; Wu et al., 2009; Baby et al., 2010). The special structure of enzyme/electrode surfaces and excellent electrocatalytic activity results in good characteristics, good reproducibility and stability, fast response time (as less as 3 s), and a detection limit of ~3 μM with a signal/noise ratio of 3 (Wu et al., 2009). Instead of layer-by-layer assemblies, nanocomposite biosensors can be constructed by vapor-growing carbon nanotubes in the titanium oxide (TiO_2) nanotube arrays, which results in uniform decoration of nanocomposites on the electrode (Fig. 16.3e) (Pang et al., 2009). Graphene nanosheets, another form of nanostructured carbon, can replace carbon nanotubes as a connector to couple with platinum, gold or hybrid metal nanoparticles (Baby et al., 2010).

16.3.1.2 Glucose biosensors based on nanostructured metal oxides that directly oxidize glucose

GOx is the most popular glucose recognition element with electrochemical properties for glucose detection. It has a relatively good thermostability, high selectivity, and sensitivity to glucose, explaining both the popularity and commercial success of this enzyme on glucose detection. Compared to nonbiological recognition elements, GOx, however, exhibits several intrinsic drawbacks, including poorer stability and unreliable enzymatic activity that is affected by the temperature, pH, humidity, and toxic chemicals (Schügerl et al., 1996).

To solve the problem, significant efforts have been made to develop enzyme-free biosensors allowing for direct nonenzymatic electrooxidation of glucose on electrodes. Recent progress in this field is mainly categorized on the nanoscale transition metal oxides such as copper oxide that has fast reaction kinetics and good selectivity. Nonenzyme nanosensors constructed from metal oxides are cost-effective and highly sensitive in glucose detection affording minimum fabrication costs and longer storage lifetimes. Moreover, the sensor performance can be further improved by the inclusion of metal nanoparticles, carbon nanofibers, or smaller biorecognition molecules which improve the oxidation characteristics of metal oxides. Rahman provided an overview of progress on nanostructured metal oxides biosensors (Rahman et al., 2010); another excellent source is the review by Tian and colleagues concerning the most recent advances in nonenzymatic glucose sensors (Tian et al., 2014).

16.3.2 Fluorescence-Based Glucose Sensors

Electrochemical glucose biosensors represent the most reliable technology of glucose detection and dominate the field of commercially available glucose sensors. Although providing satisfactory data to the control of blood glucose, the electrochemical glucose monitoring approach implies the risk of missing BGL excursion that occurs between sampling points. Improved technology in this field is rapidly changing the standards of glucose monitoring regime into in vivo continuous glucose monitoring. The most promising one among them is the fluorescence-based sensor, which is often called "smart tattoo." Fluorescence-based glucose sensors represent a large proportion of research into continuous glucose monitoring mainly due to their ability to optically interrogate through the skin rather than have an electrode sensor implanted. Similar to regular tattoo, the sensor probes are intended to inject into the dermis where they are exposed to the interstitial fluid and measure local glucose changes that are correlated with BGL. The glucose concentration is read out by measuring the fluorescence change of the "smart tattoo" after interrogation with simple optical instrumentation. Based on the glucose recognition elements, the optical glucose biosensors can be classified into GOx fluorescent nanosensors and other fluorescence sensors.

16.3.2.1 GOx fluorescent nanosensor

Although extensively used in electrochemical glucose biosensors, GOx can be used as glucose recognition element in fluorescence biosensor as well. One simple optical sensor works on the principle of fluorescence quenching of chromophores, which is caused by hydrogen peroxide (H_2O_2) produced from the GOx-catalyzed oxidation of glucose. The rate of H_2O_2 generation by GOx is well correlated with the glucose concentration. Tethering the enzyme to chromophores allows accurate detection of glucose levels by measuring the fluorescence decay of the sensors in response to H_2O_2 generation. Semiconductor quantum dots (QD), owing to their narrow fluorescence peaks and minimal photobleaching, have been used as preferable chromophores in GOx-based fluorescence sensors. Cao et al. have fabricated a glucose-sensing system by facile covalent conjugation of cadmium telluride (CdTe) QDs with GOx (Cao et al., 2008). The enzyme catalyzes the glucose oxidization and generates H_2O_2. When the peroxide reaches the surface of CdTe QDs, the electron transfer reaction occurs and the H_2O_2 is reduced to O_2, thus forming a fluorescence quenching (Fig. 16.4a). The produced O_2 can further be consumed by GOx, which is favorable for the whole reaction. This sensing system can sensitively determine glucose concentrations as low as 0.1 mM. The wide range of glucose detection also makes it suitable for the direct detection of glucose level in biological samples. Similarly, the direct coupling of GOx with manganese-doped zinc sulfide (Mn-doped ZnS) QDs and TiO_2/SiO_2 nanohybrid has also yielded biosensors capable to detect BGL (Zenkl et al., 2009; Wu et al., 2010).

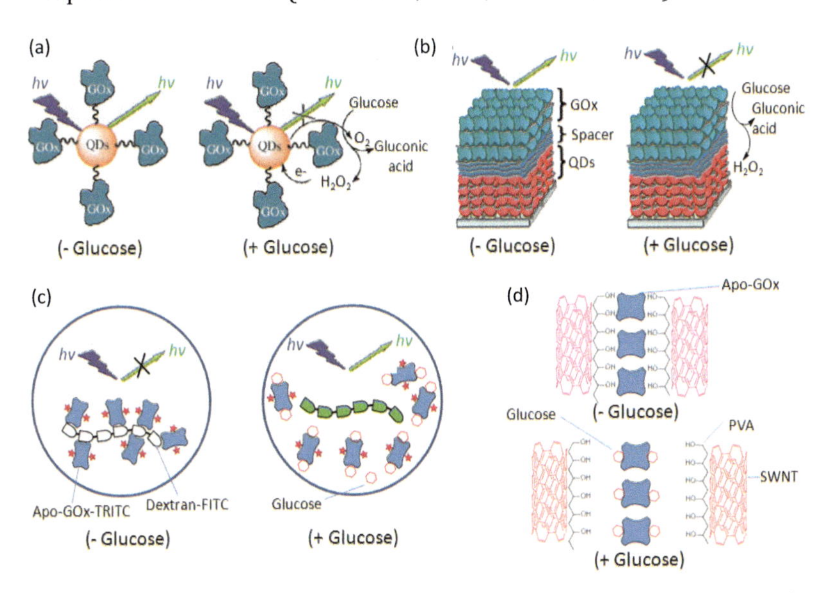

Figure 16.4 GOx-based fluorescent biosensors. (a) The structure of the assembled CdTe QDs–GOx complex and a schematic illustration of its glucose sensing principles. Adapted with permission from Cao et al. (2008). Copyright 2008 Wiley. (b) Glucose biosensor based on multi-layer films of CdTe QD and GOx. Adapted with permission from Wang et al. (2009b). Copyright 2009 American Chemical Society. (c) Apo-GOx mediated fluorescent glucose biosensor based on detection of the FRET signal. (d) Hydrogel swelling as a mechanism to detect glucose through the shift of near-infrared emission of SWNT in a biocompatible hydrogel.

This approach can be also applied on multilayer films of QDs and GOx using the layer-by-layer assembly technique. The nanocomposite film was fabricated by depositing multilayers of CdTe QDs, followed by multilayers of GOx on the glass substrate (Pang et al., 2009). When the upper layers of GOx are contacted with glucose, the enzyme-catalyzed reaction produces H_2O_2 which then quickly diffuse to the bottom layers of CdTe QDs. H_2O_2 chemically etches the QDs to generate many surface defects, leading to quenching of QD fluorescence (Fig. 16.4b). Since the quenching rate is a function of the glucose level, one can calculate the glucose concentration in the samples by monitoring the fluorescence change. By adding one film of QDs layer as a stable color background, the multilayer film sensor can be upgraded to a colorimetric optical glucose sensor that allows direct readout of glucose concentrations using the "glucose ruler" (Wang et al., 2009a). During glucose determination, GOx oxidizes the glucose, which induces the fluorescence change of the sensing layer whereas the signal from the bottom layer remains constant. The sensor displayed distinguishable fluorescent colors from green to red, which can easily be identified with the naked eyes or a camera. By comparing the apparent colors of the sensor strip with the "glucose ruler," glucose determination becomes rapid, easy, and convenient.

Catalytically inactive GOx, namely apo-GOx, has been used to construct fluorescence resonance energy transfer–based competitive binding assay sensors (FRET competition assay sensors). Apo-GOx can still bind glucose but cannot catalyze the glucose oxidization due to the lack of cofactors (FAD). Apo-GOx is labeled with a fluorescent dye, and dextran, a glucose analogue, is conjugated to a quencher. Owing to the affinity between apo-GOx and dextran, the dye and quencher are closely contacted and the energy can be transferred from the donor (fluorescent dye) to the acceptor (quencher). When glucose is introduced into the system, it displaces dextran from Apo-GOx, resulting in an increase of fluorescence, which can be used to measure the glucose concentration. Upon the removal of glucose, this process can be reversed if both dye and quencher are contained in a capsule (Fig. 16.4c). The concept was first demonstrated by encapsulating apo-GOx-TRITC and dextran-FITC in polyelectrolyte microcapsules (Chinnayelka et al., 2005). Later work expanded the system to near-infrared fluorescent dyes which can minimize the background signals from tissue and biological fluids, thereby making the sensor more amenable to in vivo use (Chinnayelka et al., 2008; Chaudhary et al., 2009). Barone and colleagues used hydrogel swelling as mechanism to generate a fluorescent sensor (Barone et al., 2009). They fabricated a poly(vinyl alcohol) hydrogel containing dispensed single wall carbon nanotubes (SWNTs) cross-linked with apo-GOx. In the presence of glucose competition, GOx loses the binding to nanotube and the cross-linking density decreases, allowing hydrogel swelling. Due to declining cross-linking density, photoluminescence emission maxima of the nanotube shift, yielding excellent signal changes in response to glucose concentrations (Fig. 16.4d).

16.3.2.2 Other fluorescent glucose nanosensors

Apart from apo-GOx, a variety of other glucose recognition elements have been developed for fluorescent glucose nanosensors. Such key elements include boronic acid derivatives, glucose/galactose-binding proteins and the lectin Concanavalin A (Conc A) which can sense glucose through interactions between glucose and these elements.

Takeuchi's group has engineered a fluorescent glucose sensor system based on polymeric microbeads incorporating boronic acid derivatives to recognize glucose (Fig. 16.5) (Shibata et al., 2010). A fluorescent dye anthracene is conjugated with two molecules of phenylboronic acid that act as the specific glucose recognition site. In the absence of the sugar, the fluorescence of anthracene is quenched by the photo-induced electron transfer (PET) that occurs from the unshared electron pair of the nitrogen atom to the anthracene (Fig. 16.5a,b). Upon sugar binding to the boronic acid, the strong reaction between the nitrogen atom and the boron atom inhibits PET and the fluorescence of anthracene increases. To apply to fully implantable glucose sensors, the glucose-responsive fluorescent dye was later polymerized with acrylamide and fabricated into porous hydrophilic microbeads. The biosensor shows sensitive responsiveness to glucose in vitro within a range from 0–1000 mg/dL (Fig. 16.5c). At a glucose concentration of 500 mg/dL, the fluorescence intensity was approximately 3 times higher than that of background. Moreover, association and dissociation between glucose and dye occur reversibly, making the biosensor suitable for continuous glucose monitoring. Importantly, this biosensor has been used to continuously monitor the blood glucose of mice in vivo (Fig. 16.5d). The fluorescence response of the biosensor well traces blood glucose concentrations throughout the physiological range although the result has 11 ~ 5 min lag behind the change in blood glucose (Fig. 16.5e). Fluorescent nanosensors have also been developed based on the swelling kinetics which can be detected by distance-dependent FRET. The polymeric nanoparticles containing phenylboronic acid and fluorophore pairs were prepared with an average size of 380 nm (Zenkl et al., 2008). In the absence of glucose molecules, the nanoparticles shrink and hold the fluorophore pair together, allowing efficient FRET. In the presence of glucose, boronic acid polymer binds to glucose and particle swells, which increases the distance between fluorophores, leading to the increase of the donor fluorescent intensity. The sensitivity of this approach was later improved by optimizing the structures of boronic acid derivatives and altering the concentration of fluorophore pairs (Zenkl et al., 2009).

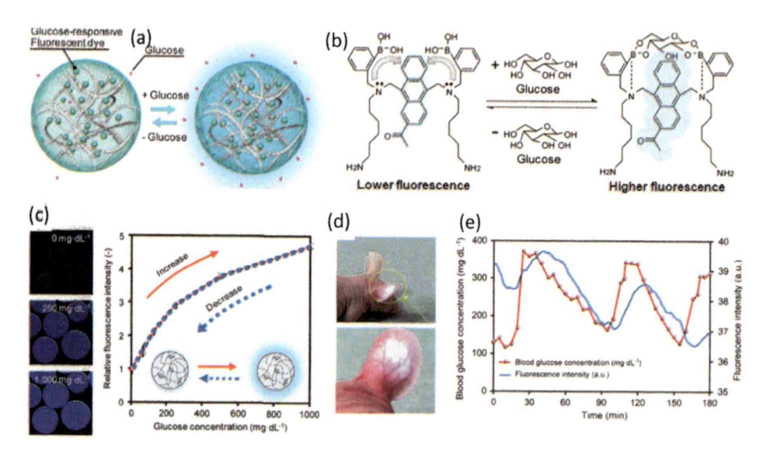

Figure 16.5 Injectable hydrogel beads for fluorescence-based in vivo continuous glucose monitoring. Adapted with permission from Shibata et al. (2010). Copyright 2010 National Academy of Sciences. (a) Schematic illustration of the injectable fluorescent beads. The fluorescence intensity of the beads increases as glucose concentration increases. (b) Fluorescence intensity changes depending on the existence of a glucose molecule. (c) Glucose responsiveness of the fluorescence based hydrogel bead. (d) Fluorescent beads under the dermis of a mouse ear. (e) The fluorescence intensity traces blood glucose concentrations.

Glucose-binding protein (GBP) and Con A also recognize glucose and can be used for glucose detection. Veetil et al. developed a GBP fusion protein GFP-GBP-mCherry, an optical sensor assembly, which undergoes conformational switch upon glucose binding (Veetil et al., 2010). The conformational switch leads to the distance change between GFP and mCherry, thereby generating measureable FRET signals. Although the FRET responses to glucose in a concentration lower than physiological range, the sensor is devoid of interference from other sugars like galactose, fructose, lactose, and mannose, which is an obvious advantage over boronic acid sensors. A similar approach also can be applied to glucose-binding protein Con A. For example, Con A-labeled QDs closely attach to gold nanoparticles modified with β-cyclodextrin (β-CD) due to strong interactions between Con A and β-CD. The short distance between two fluorophores induces the fluorescence quench by FRET. Glucose displaces β-CD, which increases the fluorescence, enabling the concentration to be measured correspondingly (Tang et al., 2008).

16.4 Closed-Loop Insulin Delivery

The major method of treatment of diabetes in the form of a subcutaneous insulin injection has remained the same over 80 years. The endless injection suffering, diet control and the risk of poor glucose control are the central problems that are faced by every diabetic patient. All of these have provoked the desire of alternative approaches which can avoid the injection and improve the quality of patients' lives. Improvement has occurred as a result of several developments including the introduction of long-acting insulin analogues that only require injection once or twice daily, the exploring of noninvasive administration routes and the use of glucose-responsive (closed-loop) insulin delivery system. The prospect of non-invasive insulin delivery is appealing to patients with type 1 diabetes who need frequent insulin injection. This concept is currently explored to deliver insulin using oral, pulmonary, nasal, ocular, and rectal routes. However, to achieve efficient absorption of insulin via oral, pulmonary, or other routes, several barriers have to be overcome: the poor permeability of insulin through epithelial membrane, the chemical stability at harsh, acidic luminal conditions and local enzymatic digestion. The noninvasive administration of insulin has been extensively developed by combining the most advanced nanotechnology and formulation techniques. The corresponding insulin formulations such as Oral-Lyn™, IN-105, AERx®, etc., have been evaluated in preclinical and clinical trials (Mo et al., 2014).

The best approach for insulin delivery is the closed-loop system which mimics pancreatic function of healthy people and releases insulin in response to BGLs. The closed-loop system, functioned by automatically feedback to blood glucose fluctuation, offers obvious advantages over the open loop system. Unlike the latter, the delivery rate and dose of insulin in the closed-loop system is real-timely controlled by BGL in an automatic manner, which dramatically minimizes the blood glucose fluctuation and hypoglycemia risk (Ganeshkumar et al., 2013). Currently, most of the closed-loop insulin delivery systems are constructed on nanometer-scale materials that can speed up the response time and maximize the stability of the system. Based on the closed-loop theory, the system is composed by a sensing element that detects the BGL changes and a controller which regulates the insulin release in responsive to the sensing signal. Currently, glucose-sensing elements are classified into three types of materials: GOx, phenylboronic acid (PBA) or GBP. Upon hyperglycemia, all the three materials can generate somewhat a stimulus and trigger the insulin release by the matrix structure change, polymer hydrolysis, or glucose-binding competition. The progress of chemically controlled closed-loop insulin delivery systems has been well summarized by V. Varaine et al. in 2008 (Ravaine et al., 2008). Herein we focus on the most recent progress of GOx controlled nanosystems.

16.4.1 GOx-Mediated Insulin Delivery

Glucose oxidase catalyzes the conversion of D-glucose into D-gluconic acid by consuming oxygen (O_2) with simultaneously producing hydrogen peroxide (H_2O_2) (Bankar et al., 2009). The generation rate of gluconic acid by GOx is directly correlated with the external glucose concentration, so GOx is widely used as a glucose-sensing element in glucose sensors and glucose-responsive insulin delivery systems (Ravaine et al., 2008; Steiner et al., 2011; Wu et al., 2011). By integrating with GOx, pH-sensitive polymeric matrices can present a glucose-triggered response in terms of volume change or hydrolysis, leading to corresponding insulin release.

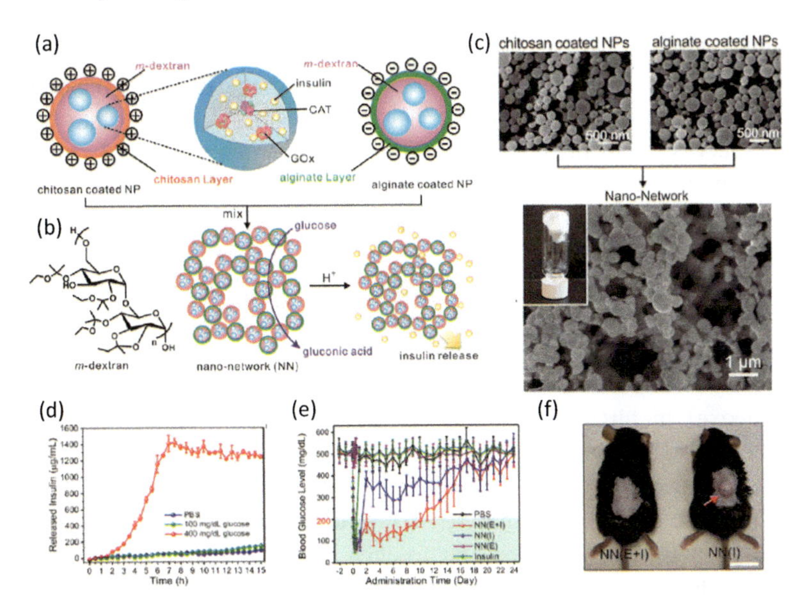

Figure 16.6 (a) Schematic representation of injectable nano-network for glucose-mediated insulin delivery by Gu and coworkers. (b) The chemical structure of acetal-modified dextran (m-dextran). (c) SEM images of nanoparticles (NPs) coated with chitosan and alginate, and formed nano-network. Inset is a gel-like nano-network adhered to the bottom of an inverted vial. (d) In vitro accumulated insulin release of the nano-network in different glucose concentrations at 37°C. (e) In vivo study of the nano-network on blood glucose regulation in STZ-induced C57B6 diabetic mice. NN(E+I): nano-network encapsulated insulin and enzymes; NN(I): nano-network encapsulated insulin only; NN(E): nano-network encapsulated with enzymes only. (f) The dorsum pictures of mice treated with subcutaneous injection of NN(E+I) or NN(I) after 4 weeks. Scale bars: 1 cm. Reprinted with permission from Gu et al. (2013a). Copyright 2013 American Chemical Society.

Gu et al. have extensively developed glucose oxidase (GOx)-based closed-loop insulin delivery systems through acid-degradable materials (Gu et al., 2013a; Gu et al., 2013b; Tai et al., 2014). Appling emulsion-based encapsulation, insulin and GOx solution were fabricated with acetal modified dextran (m-dextran) to form an acid sensitive double emulsion nanoparticle

(Fig. 16.6a,b) (Gu et al., 2013a). GOx catalyzes the conversion of glucose into gluconic acid which further hydrolyzes m-dextran into water-soluble dextran, acetone, and ethanol (Kauffman et al., 2012). M-dextran degradation triggers nanoparticle collapse and the subsequent insulin release in a glucose-responsive manner. To make the nanoparticle injectable, Gu et al. coated the double emulsion nanoparticles with chitosan and alginate, respectively. An injectable nano-network formulation was formed by mixing the oppositely charged nanoparticles together. The charge-charge interaction of nanoparticles allows the nano-network to form a cohesive and porous 3-D structure with microchannels which are essential for glucose and insulin diffusion (Fig. 16.6c). The resulting nano-network demonstrates a fast release at hyperglycemic state (400 mg/mL glucose) and nearly no release at normoglycemic condition (100 mg/mL glucose) in vitro (Fig. 16.6d). Single injection of the developed nano-network in STZ-induced diabetes mice can maintain the BGL in normoglycemic state for up to 10 days (Fig. 16.6e). Moreover, m-dextran-based nano-networks show excellent biocompatibility in vitro and in vivo, making it well suited for multiple injections and long-term therapies (Fig. 16.6f).

Encouraged by these promising results, Gu et al. developed a chitosan-based sponge-like matrix as an insulin reservoir (Gu et al., 2013b). Chitosan is a linear polysaccharide which can be degraded by ubiquitous lysozymes and glycosidases in the body (Kumar et al., 2004). Cross-linked by tripolyphosphate (TPP), chitosan forms a sponge-like matrix to entrap the insulin (Fig. 16.7a). The sponge smart nanocapsules, made of porous polymer, contains GOx and CAT scattered throughout the sponge matrix. Protons (H^+) generated from glucose oxidation, protonate chitosan amine groups and subsequently expand the sponge-like microparticle by more than 5-fold volume (Fig. 16.7b). Structure swelling of microparticle facilitates insulin release from the smart sponge, thereby achieving BGL self-regulation during hyperglycemic condition (Fig. 16.7c–e).

Our group has constructed a glucose-responsive nanocapsule by packing insulin, GOx and CAT into pH-sensitive polymersome (Tai et al., 2014; Fig. 16.8). A polymersome is a self-assembled polymeric capsule, in which an aqueous core is surrounded by a well-organized amphiphilic polymeric bilayer. Composed by high molecular weight polymer, polymersomes have robust mechanical stability, which can prevent premature loss of its cargo (Discher et al., 2006). Glucose can be passively transported across the bilayer membrane of the nanocapsule and oxidized into gluconic acid by GOx, thereby causing a decrease in local pH. The acidic microenvironment causes the hydrolysis of the pH sensitive nanocapsule that in turn triggers the release of insulin in a glucose-responsive fashion. To achieve the best biocompatibility, a new acid-sensitive polymer, PEG-poly(Ser-Ketal), was synthesized for polymersome assembly. In the presence of glucose, this

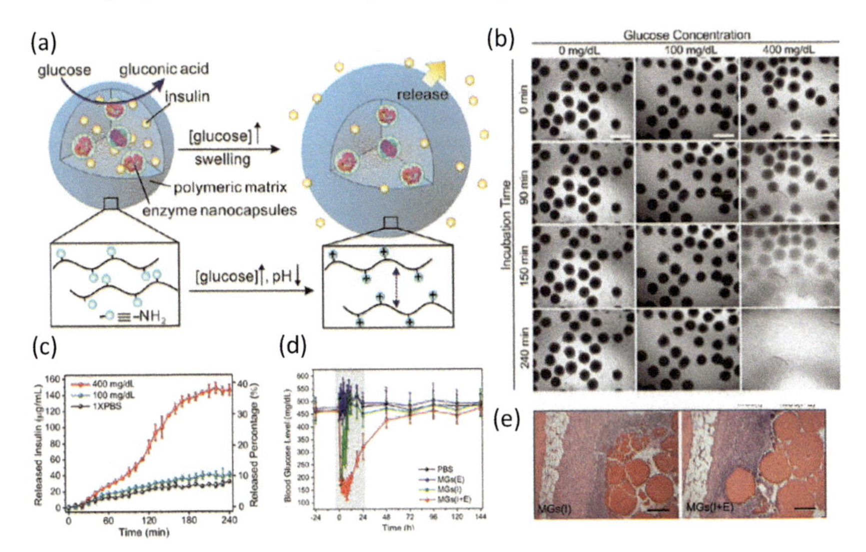

Figure 16.7 Glucose-responsive microgels integrated with enzyme nanocapsules for closed-loop insulin delivery by Z. Gu et al. (a) Schematic of microgels encapsulating insulin and enzyme nanocapsules. The encapsulated glucose-specific enzyme catalyzes glucose into gluconic acid. The subsequent protonation of polymer chains rich in amine groups increases the charge in the gel matrix, leading to swelling of the microgels and release of insulin. (b) Optical microscope images of microgels incubated with 1× PBS solutions at different glucose concentrations (0, 100, and 400 mg/dL) over time at 37°C. (c) In vitro release kinetics of insulin from the microgels in 1× PBS solutions with different glucose concentrations: 0, 100, and 400 mg/dL at 37°C. (d) BGL in STZ-induced C57B6 diabetic mice after subcutaneous injection with 1× PBS, microgels encapsulating insulin and enzymes (MGs(E+I)), microgels encapsulating insulin only (MGs(I)) and microgels encapsulating enzymes only (MGs(E)). (e) Representative H&E staining of mice dermal tissues containing MGs(EþI) and MGs(I). Scale bar in (e) represents 200 μm. Reprinted with permission from Gu et al. (2013b). Copyright 2013 American Chemical Society.

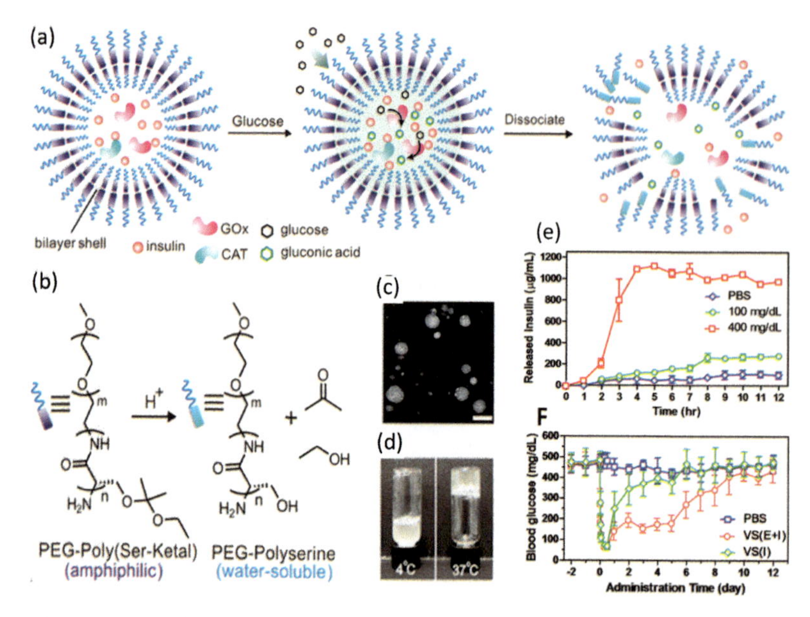

Figure 16.8 GOx-encapsulated polymersome for closed-loop insulin delivery. (a) GOx converts glucose into gluconic acid and acidifies the aqueous core of the polymersome nanocapsule, leading to hydrolysis of the polymeric bilayer shell and subsequent dissociation of the nanocapsule, resulting in glucose-responsive insulin release. (b) The chemical structure of the pH-sensitive diblock copolymer PEG-poly(Ser-Ketal), which can be hydrolyzed into water-soluble PEG-polyserine and acetone/ethanol in an acidic environment. (c) TEM image of a polymersome. Scale bar: 500 nm. (d) The polymersome was mixed with PF127 to form a thermoresponsive suspension. (e) In vitro accumulated insulin release from the vesicles incubated in the solutions with different glucose concentrations. (f) The BGL of STZ-induced diabetic mice after treatment with PBS solution, polymersome encapsulating both enzyme and insulin (VS(E+I)) and polymersome encapsulating insulin only (VS(I)). Adapted with permission from Tai et al. (2014). Copyright 2014 American Chemical Society.

amphiphilic diblock copolymer can be hydrolyzed into water-soluble PEG-polyserine and acetone/ethanol, all of which are biodegradable and biocompatible. In order to (1) hold the formulation underneath the skin for the long-term release intention and (2) generate a catalysis center for synergistically enhanced conversion efficiency, 30% Pluronic-127 (PF127), a thermoresponsive and biodegradable polymer, was mixed with the polymersome to form a suspension. Once subcutaneously injected, the suspension quickly formed a stable hydrogel, in which nanocapsule were evenly dispersed. In vitro studies validated that the release of insulin from such nanocapsule effectively correlated with the external glucose concentration. In vivo experiments, in which diabetic mice were subcutaneously administered with the nanocapsules, demonstrate that a single injection of the developed nanocapsule facilitated stabilization of the BGL in the normoglycemic state (<200 mg/dL) for up to 5 days.

Although GOx can control the insulin release in response to the glucose level by degrading the pH-sensitive materials via a local decrease of pH value, such pH-dependent methods are often compromised by slow responsiveness, sometime even hindered by the hypoxia because of the quick oxygen consumption by GOx. Instead of using enzymatically induced pH changes, our group for the first time utilized the local generation of hypoxia as a trigger for rapid insulin release in response to hyperglycemia (Yu et al., 2015). To achieve hypoxia-responsive transduction, 2-nitroimidazole, a hydrophobic component sensitive to hypoxia, was conjugated to high-molecular-weight hyaluronic acid (HA), which is well known for its excellent biocompatibility and biodegradability. The amphiphilic polymer can readily encapsulate recombinant human insulin and GOx to form nanoscale vesicles. In the presence of a high BGL, the dissolved oxygen can be rapidly consumed by the glucose oxidation catalyzed by GOx, producing a local hypoxic environment. Nitroimidazole groups on the HS-HA were then reduced to hydrophilic 2-aminoimidazoles under this hypoxic condition, which resulted in the dissociation of nanovesicles and subsequent release of insulin (Fig. 16.9a,b). To realize ease of administration, the nanovesicles were loaded into a microneedle-array patch for painless insulin delivery (Fig 16.9c–f). With transcutaneous administration, the microneedles disassembled when exposed to high interstitial fluid glucose in vascular and lymph capillary networks, thereby promoting the release of insulin, which was then taken up quickly through the regional lymph and capillary vessels. It has demonstrated that this "smart insulin patch" with a novel glucose-responsive mechanism displayed rapid responsiveness for glucose regulation and reliable avoidance of hypoglycemia in a mouse model of type 1 diabetes.

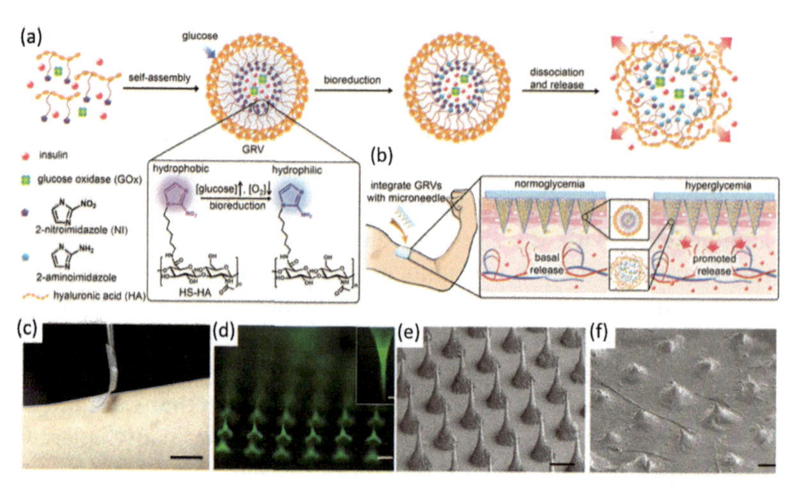

Figure 16.9 The glucose-responsive insulin delivery system using hypoxia-sensitive vesicle-loading microneedle-array patches. Adapted with permission from Shibata et al. (2010). Copyright 2015 National Academy of Sciences. (a) Schematic of the formation and release mechanism of glucose-responsive vesicle (GRV). (b) Schematic of the GRV-containing microneedle-array patch (smart insulin patch) for in vivo insulin delivery triggered by a hyperglycemic state to release more insulin. (c) Photograph of the smart insulin patch with a microneedle array. (Scale bar: 1 cm). (d) Fluorescence microscopy image of microneedle-loaded GRVs with FITC labeled insulin. (Inset) Zoomed-in image of microneedle. (Scale ba'rs: 200 μm). (e) SEM image of microneedle array. (Scale bar: 200 μm). (f) SEM image of GRV-loaded MN after insertion into mouse skin for 4 h. (Scale bar: 200 μm).

C. Gordigo and X. Y. Wu et al. developed bioorganic nanohybrid membranes for closed-loop insulin delivery (Gordijo et al., 2010; Gordijo et al., 2011). The glucose-responsive nanohybrid membrane was prepared by cross-linking MnO_2 nanoparticles, poly(*N*-isopropylacrylamide-co-methacrylic acid) (poly(NIPAM-MAA)) nanoparticles with biomacromolecules GOx, CAT and BSA by glutaraldehyde. Herein GOx and CAT served as glucose-sensing moieties which metabolize Glucose and causes local pH value decreases in response to the glucose concentration. Poly(NIPAM-MAA) is a thermo- and pH-sensitive hydrogel. The acidic condition under hyperglycemic state induces the hydrogel nanoparticles to shrink which increases the porosity of the membrane and leads into the accelerated insulin release (Fig 16.10a,b). The MnO_2 nanoparticles are believed to play an essential role in this system. These inorganic nanoparticles not only reinforce the mechanical strength but also enhance the enzyme stability of GOx and CAT. MnO_2 nanoparticles, which can react with H_2O_2 and produce Mn^{2+} and oxygen, can assist CAT to scavenge H_2O_2 and replenish O_2 (Dong le et al., 2004; Bai et al., 2007), therefore stabilizing CAT and resulting in a MnO_2–CAT nanohybrid with relatively higher catalytic efficiency (Fig 16.10c). The nanohybrid membrane responds quickly to the change of glucose concentration. The permeability of the membrane increased with increasing glucose concentration from 100 to 400 mg/dL, which is repeatable for more than 5 cycles (Fig 16.10d,e). In vitro experiments demonstrated that the nanohybrid membrane can regulate insulin release in response to glucose concentration (Fig. 16.10f and g). In vivo performance of the membrane-containing device in STZ-induced diabetic rats showed that the nanohybrid membrane can maintain the normoglycemic level for up to 4 days (Fig. 16.10h).

GOx capsules can also be fabricated by the layer-by-layer assembly method. Qi et al. utilized insulin particles as template to fabricate a core-shell capsule for glucose-responsive insulin delivery (Qi et al., 2009). Poly(ethylenimine) (PEI) was first absorbed onto the surface of insulin microparticles and then cross-linked with protein (GOx/CAT) by glutaraldehyde. When glucose is introduced into the system, GOx/CAT catalyzes the glucose oxidation and produces protons, thus dropping the pH value of the local environment (Fig. 16.11a,b). The Schiff base, formed by glutaraldehyde cross-linking, will partially break down, which increases the permeability of the capsule shell and lead to insulin release. The permeability of the shell was tested in buffer of pH ranging from 7.8 to 4.8, in which the protein capsule was incubated with dextran-FITC and imaged by confocal microscopy (Fig 16.11c). The capsule shell shows much higher permeability under lower pH condition. In vitro release studies proved that the protein capsule has much higher insulin release rate in glucose solution than the regular buffer (Fig 16.11d). Similarly, insulin can also be encapsulated into mesoporous silica particles and then coated with multi-layers of enzymes cross-linked with glutaraldehyde, which acts as a valve to control the release of insulin in response to the external glucose concentration (Zhao et al., 2011).

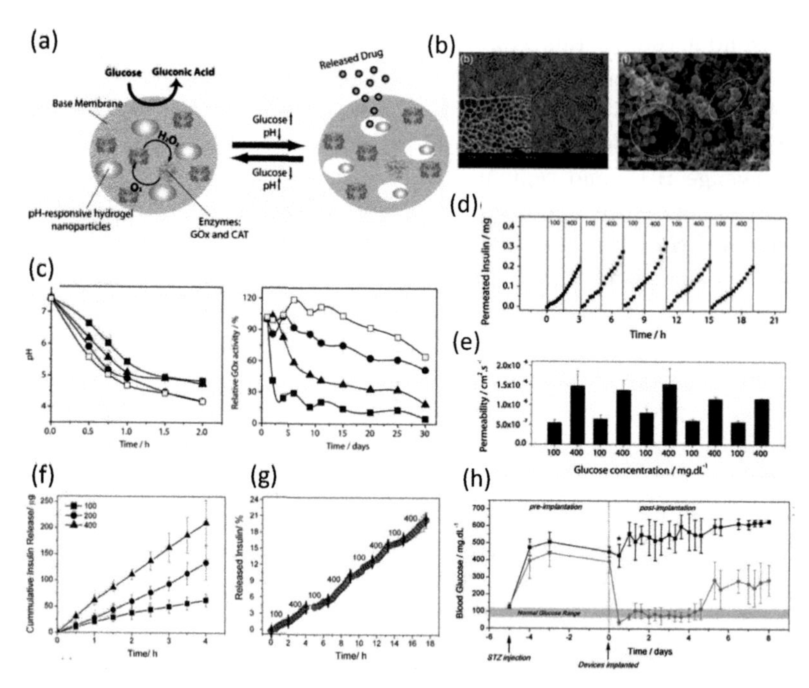

Figure 16.10 (a) Schematic illustration of the mechanism of glucose-responsive membrane. GOx catalyzes the oxidation of glucose and generates lots of protons, which causes local pH value decrease. The polymeric nanoparticle responses to the acidic condition and shrinks to the more compact nanoparticle, which increases the porosity of the membrane and leads to the accelerated insulin release. (b) ESEM images of the surface of the membrane (left) and the fractured cross section of the membrane (right). (c) Profile of pH changes curves of immobilized GOx (left) and stability of immobilized GOx in the different membrane formulations (right). Solid square: control GOx; solid circle: GOx + CAT; solid triangle: GOx + MnO_2 nanoparticle; open square: GOX + CAT + MnO_2 nanoparticle. (d) Profile of insulin permeated across membrane in response to abrupt changes in glucose concentration in five alternated cycles. (e) Regulated profile of membrane showing permeability of insulin (cm^2 s^{-1}) as a function of glucose concentration. (f) The in vitro release of insulin over time in different glucose concentrations: 100,200, and 400 mg dL^{-1}. (g) Profile of Insulin released from the insulin delivery membrane device in response to changes in glucose concentration. (h) Change in blood glucose over the course of 8 days after intraperitoneal implantation with saline-filled membrane device (Sham) or insulin-filled membrane devices in STZ-induced diabetic rats. Adapted with permission from Gordijo et al. (2010) and Gordijo et al. (2011). Copyright 2010, 2011 Wiley-VCH.

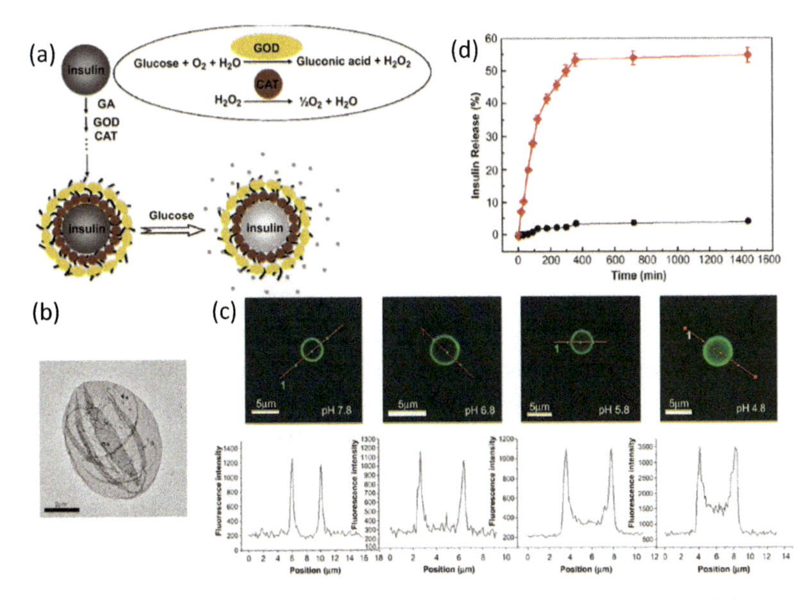

Figure 16.11 GOx microcapsule for glucose-responsive insulin delivery. (a) Schematic representation of coupled reactions of GOx and CAT on the capsule wall and the enhanced permeability of the capsule for release of insulin. (b) TEM image of microcapsule. Insert bar: 2 μm. (c) Permeability of the capsule shell increases as the pH of local environment gets lower as evidenced by the more dextran-FITC penetrates into the core. (d) Release profiles of coated insulin particles in PBS (black) and in glucose solution (red), showing glucose responsive insulin release. Adapted with permission from Qi et al. (2009). Copyright 2009 Elsevier.

16.4.2 Closed-Loop System with Phenylboronic Acid Moieties

The reversible interaction between PBA and sugar was first discovered by Lorand and Edwards (Lorand et al., 1959). Since then, many PBA-based materials have been developed for carbohydrate detection, purification and drug delivery (Jin et al., 2009; Preinerstorfer et al., 2009). The specific and reversible affinity of PBA to glucose also makes it a good alternative to lectins known for their toxicity and immunogenesis (Vilarem et al., 1978; Miyake et al., 2007; Vaz et al., 2013).

Kim and coworkers exploited a monosaccharide-responsive polymersome for closed-loop insulin delivery (Kim et al.,

2012). The amphiphilic copolymer poly(ethylene glycol)-b-Poly(styreneboroxole) (PEG_{45}-b-PBOx) demonstrated a diversity in self-assembly to form micelles, cylindrical micelles, and polymersomes vesicle, well controlled by the PBO block length. The resulting polymersome vesicles, capable of encapsulating insulin, had an average size of 340 nm. Once exposed to monosaccharides, the PBO block interacts with glucose and forms the charged phenylboronate which inverses this hydrophobic block to water soluble. The insulin-encapsulated polymersomes showed a self-regulated release profile in response to the concentrations of monosaccharides. The drawback of this Wulff-type PBA-based polymersome is the low selectivity for glucose and high threshold to elicit a response. However, its ability to control insulin release in aqueous solution at physiological pH condition makes it closer to application than other PBA-based insulin delivery systems.

16.4.3 Glucose-Binding Proteins

Glucose-binding proteins, mainly lectins, are a group of natural carbohydrate-binding proteins known to interact with glycosylated receptors or lipids for cell signal transduction (Geijtenbeek et al., 2009; Brudner et al., 2013). The most commonly used lectin for insulin delivery is Con A, a widely studied lectin member showing affinity to D-glucose and D-glucosyl substance (Sharon et al., 1972). Con A has four glucose-binding sites. By glycosylation, G-insulin can form complexes with a Con A matrix. Upon exposure to external glucose, insulin can be released via the binding competition between glucose and glycosylated insulin with Con A (Brownlee et al., 1979; Seminoff et al., 1989). Moreover, affinity between Con A and natural polysaccharide polymers makes it possible to fabricate glucose-responsive hydrogel for insulin encapsulation (Nakamae et al., 1994; Taylor et al., 1995). When the glucose concentration increases, the hydrogel swells or hydrolyzes, leading to insulin release in a glucose-responsive manner (Yin et al., 2011).

Wu et al. engineered Con A-gated carbohydrate-functionalized mesoporous silica nanoparticles (MSN) for glucose-responsive drug delivery (Wu et al., 2013). The surface of MSN was immobilized with mannose at optimal density. After encapsulating cargo within the pores, multivalent Con A was introduced to cap the pores. Con A-gated pores can be re-opened by either introducing an acidic environment or a competitive binding with glucose. In the presence of glucose, the Con A-gated MSNs gradually release its cargo at physiological pH condition. The release rate correlates well with the concentration of glucose. Although a model drug instead of insulin was used in this paper, it demonstrated a new strategy for Con A-mediated closed-loop insulin delivery.

16.5 Conclusions

The development of enzyme nanomaterials for glucose sensing and insulin delivery is an exciting and important research area. Nanomaterial-based biosensors have demonstrated its accuracy and reliability in testing clinic samples. Several polymer-based nanosystems have also been approved by the FDA for noninvasive insulin delivery. In spite of the rapid progress in this field, the smart nanomaterials that can mimic the pancreas for glucose sensing and insulin release have not yet developed. To have an impact on the current sensor market, the next-generation biosensors must provide not only improvement in accuracy but also the ability to continuously monitor the BGL. Other questions concerning the costs, biocompatibility and sensor lifetime should also be considered. Current diabetes treatment involves fingerprick of blood samples, measuring BGL with glucose sensor, the calculation of insulin dose and injection of drug. The long-term, continuous glucose monitoring by new biosensors as discussed here, eliminate the need of frequent fingerprick, minimizing patient inconvenience and pain. The quality of life can be further improved by the development of closed-loop insulin delivery nanosystems, an artificial pancreas that has been achieved using GOx-encapsulated nanocapsules in our group. Overall, enzyme nanomaterials have shown the capabilities to improve the life quality of patients with diabetes.

References

Ashby J. P., Speake R. N., *Biochem. J.*, **150** (1975), 89–96.

Baby T. T., Aravind S. S. J., Arockiadoss T., Rakhi R. B., et al., *Sens. Actuat. B Chem.*, **145** (2010), 71–77.

Baby T. T., Ramaprabhu S., *Talanta*, **80** (2010), 2016–2022.

Bai Y.-H., Du Y., Xu J.-J., Chen H.-Y., *Electrochem. Commun.*, **9** (2007), 2611–2616.

Bankar S. B., Bule M. V., Singhal R. S., Ananthanarayan L., *Biotechnol. Adv.*, **27** (2009), 489–501.

Barone P. W., Yoon H., Ortiz-Garcia R., Zhang J., et al., *ACS Nano*, **3** (2009), 3869–3877.

Brownlee M., Cerami A., *Science*, **206** (1979), 1190–1191.

Brudner M., Karpel M., Lear C., Chen L., et al., *PLoS One*, **8** (2013), e60838.

Cao L., Ye J., Tong L., Tang B., *Chemistry*, **14** (2008), 9633–9640.

Cash K. J., Clark H. A., *Trends Mol. Med.*, **16** (2010), 584–593.

Chaudhary A., Raina M., Harma H., Hanninen P., et al., *Biotechnol. Bioeng.*, **104** (2009), 1075–1085.

Chi B.-Z., Zeng Q., Jiang J.-H., Shen G.-L., et al., *Sens. Actuat. B Chem.*, **140** (2009), 591–596.

Chinnayelka S., McShane M. J., *Anal. Chem.*, **77** (2005), 5501–5511.

Chinnayelka S., Zhu H., McShane M., *J. Sens.*, **2008** (2008), 11.

Clark L. C., Jr., Lyons C., *Ann. N. Y. Acad. Sci.*, **102** (1962), 29–45.

Cryer P. E., *J. Clin. Invest.*, **117** (2007), 868–870.

Czech M. P., Richardson D. K., Becker S. G., Walters C. G., et al., *Metabolism*, **27** (1978), 1967–1981.

Discher D. E., Ahmed F., *Ann. Rev. Biomed. Eng.*, **8** (2006), 323–341.

Dong le V., Eng K. H., Quyen le K., Gopalakrishnakone P., *Biosens. Bioelectron.*, **19** (2004), 1285–1294.

Ganeshkumar M., Ponrasu T., Sathishkumar M., Suguna L., *Colloids Surf. B Biointerfaces*, **103** (2013), 238–243.

Geijtenbeek T. B., Gringhuis S. I., *Nat. Rev. Immunol.*, **9** (2009), 465–479.

Gooding J. J., Wibowo R., Liu, Yang W., et al., *J. Am. Chem. Soc.*, **125** (2003), 9006–9007.

Gopalan A. I., Lee K. P., Ragupathy D., Lee S. H., et al., *Biomaterials*, **30** (2009), 5999–6005.

Gordijo C. R., Koulajian K., Shuhendler A. J., Bonifacio L. D., et al., *Adv. Funct. Mater.,* **21** (2011), 73–82.

Gordijo C. R., Shuhendler A. J., Wu X. Y., *Adv. Funct. Mater.,* **20** (2010), 1404–1412.

Gu Z., Aimetti A. A., et al., *ACS Nano,* **7** (2013a), 4194–4201.

Gu Z., Dang T. T., et al., *ACS Nano,* **7** (2013b), 6758–6766.

Guilbault G. G., Lubrano G. J., *Anal. Chim. Acta,* **64** (1973), 439–455.

Halse R., Bonavaud S. M., Armstrong J. L., McCormack J. G., et al., *Diabetes,* **50** (2001), 720–726.

Horwitz M. S., Bradley L. M., Harbertson J., Krahl T., et al., *Nat. Med.,* **4** (1998), 781–785.

Howard B. V., *Ann. N. Y. Acad. Sci.,* **967** (2002), 324–328.

Jin X., Zhang X., Wu Z., Teng D., et al., *Biomacromolecules,* **10** (2009), 1337–1345.

Kang X., Mai Z., Zou X., Cai P., et al., *Anal. Biochem.,* **369** (2007), 71–79.

Kasuga, M., *J. Clin. Invest.,* **116**(7), (2006), 1756–1760.

Katsoyannis P. G., *Diabetes,* **13** (1964), 339–348.

Kauffman K. J., Do C., Sharma S., Gallovic M. D., et al., *ACS Appl. Mater. Interfaces,* **4** (2012), 4149–4155.

Kim H., Kang Y. J., Kang S.Kim K. T., *J. Am. Chem. Soc.,* **134** (2012), 4030–4033.

Kong T., Chen Y., Ye Y., Zhang K., et al., *Sens. Actuat. B Chem.,* **138** (2009), 344–350.

Kumar M. N., Muzzarelli R. A., Muzzarelli C., Sashiwa H., et al., *Chem. Rev.,* **104** (2004), 6017–6084.

Lin J., He C., Zhao Y., Zhang S., *Sens. Actuat. B Chem.,* **137** (2009), 768–773.

Lindall A. W., Jr., Bauer G. E., Dixit P. K., Lazarow A., *J. Cell. Biol.,* **19** (1963), 317–324.

Liu Y., Zhu Y., Zeng Y., Xu F., *Nanoscale Res. Lett.,* **4** (2009), 210–215.

Lorand J. P., Edwards J. O., *J. Org. Chem.,* **24** (1959), 769–774.

Miyake K., Tanaka T., McNeil P. L., *PLoS One,* **2** (2007), e687.

Mo R., Jiang T., Di J., Tai W., et al., *Chem. Soc. Rev.,* **43** (2014), 3595–3629.

Mychaleckyj J. C., Noble J. A., Moonsamy P. V., Carlson J. A., et al., *Clin. Trials,* **7** (2010), S75–S87.

Nakamae K., Miyata T., Jikihara A., Hoffman A. S., *J. Biomater. Sci. Polym. Ed.,* **6** (1994), 79–90.

Pang X., He D., Luo S., Cai Q., *Sens. Actuat B Chem.*, **137** (2009), 134–138.

Patolsky F., Weizmann Y., Willner I., *Angew. Chem. Int. Ed. Engl.*, **43** (2004), 2113–2117.

Preinerstorfer B., Lammerhofer M., Lindner W., *J. Sep. Sci.*, **32** (2009), 1673–1685.

Qi W., Yan X., Fei J., Wang A., et al., *Biomaterials*, **30** (2009), 2799–2806.

Rabkin R., Reaven G. M., Mondon C. E., *Am. J. Physiol.*, **250** (1986), E530–E537.

Rahman M. M., Ahammad A. J., Jin J. H., Ahn S. J., et al., *Sensors (Basel)*, **10** (2010), 4855–4886.

Rajan A. S., Aguilar-Bryan L., Nelson D. A., Yaney G. C., et al., *Diab. Care*, **13** (1990), 340–363.

Ravaine V., Ancla C., Catargi B., *J. Control. Release*, **132** (2008), 2–11.

Saltiel A. R., Kahn C. R., *Nature*, **414** (2001), 799–806.

Schügerl K., Hitzmann B., Jurgens H., Kullick T., et al., *Trends Biotechnol.*, **14** (1996), 21–31.

Seminoff L. A., Gleeson J. M., Zheng J., Olsen G. B., et al., *Int. J. Pharm.*, **54** (1989), 251–257.

Sharon N., Lis H., *Science*, **177** (1972), 949–959.

Shibata H., Heo Y. J., Okitsu T., Matsunaga Y., et al., *Proc. Natl. Acad. Sci. U. S. A.*, **107** (2010), 17894–17898.

Steiner M. S., Duerkop A., Wolfbeis O. S., *Chem. Soc. Rev.*, **40** (2011), 4805–4839.

Sweet I. R., Cook D. L., DeJulio E., Wallen A. R., et al., *Diabetes*, **53** (2004), 401–409.

Tai W., Kasuga, M. (2006). *J. Clin. Invest.*, **116**(7), 1756–1760.

Tai W., Mo R., Di J., Subramanian V., et al., *Biomacromolecules*, **15** (2014), 3495–3502.

Tang B., Cao L., Xu K., Zhuo L., et al., *Chemistry*, **14** (2008), 3637–3644.

Taylor M. J., Tanna S., Taylor P. M., Adams G., *J. Drug. Target*, **3** (1995), 209–216.

Tian K., Prestgard M., Tiwari A., *Mater. Sci. Eng. C Mater. Biol. Appl.*, **41** (2014), 100–118.

Usman Ali S. M., Nur O., Willander M., Danielsson B., *Sens. Actuat. B Chem.*, **145** (2010), 869–874.

Vaz A. F. M., Souza M. P., Vieira L. D., Aguiar J. S., et al., *Radiat. Phys. Chem.*, **85** (2013), 218–226.

Veetil J. V., Jin S., Ye K., *Biosens. Bioelectron.,* **26** (2010), 1650–1655.

Vilarem M. J., Jouanneau J., Le Francois D., Bourrillon R., *Cancer Res.,* **38** (1978), 3960–3965.

Wang J., *Chem. Rev.,* **108** (2008), 814–825.

Wang X. D., Chen H. X., et al., *Biosens. Bioelectron,* **24** (2009a), 3702–3705.

Wang Y., Wei W., et al., *Mater. Sci. Eng. C,* **29** (2009b), 50–54.

Welch C. M., Compton R. G., *Anal. Bioanal. Chem.,* **384** (2006), 601–619.

Whiting D. R., Guariguata L., et al., *Diabetes Res. Clin. Pr.,* **94**(3), (2011), 311–321.

Wen Z., Ci S., Li J., *J. Phys. Chem. C,* **113** (2009), 13482–13487.

Wu P., He Y., Wang H. F., Yan X. P., *Anal. Chem.,* **82** (2010), 1427–1433.

Wu B.-Y., Hou S.-H., Yu M., Qin X., et al., *Mater. Sci. Eng. C,* **29** (2009), 346–349.

Wu S., Huang X., Du X., *Angew. Chem. Int. Ed. Engl.,* **52** (2013), 5580–5584.

Wu Q., Wang L., Yu H., Wang J., et al., *Chem. Rev.,* **111** (2011), 7855–7875.

Xiao Y., Patolsky F., Katz E., Hainfeld J. F., et al., *Science,* **299** (2003), 1877–1881.

Yang J., *Int. J. Biol. Sci.,* **6** (2010), 716–718.

Yang K., She G.-W., Wang H., Ou X.-M., et al., *J. Phys. Chem. C,* **113** (2009), 20169–20172.

Yin R., Tong Z., Yang D., Nie J., *J. Control. Release,* **152 Suppl 1** (2011), e163–e165.

Yoi O. O., Seldin D. C., Spragg J., Pinkus G. S., et al., *Proc. Natl. Acad. Sci. U. S. A.,* **76** (1979), 3612–3616.

Yoon J. W., Onodera T., Notkins A. L., *J. Exp. Med.,* **148** (1978), 1068–1080.

Yu J., Zhang Y., Ye Y., DiSanto R., et al., *Proc. Natl. Acad. Sci. U. S. A.,* **112** (2015), 8260–8265.

Zenkl G., Mayr T., Klimant I., *Macromol. Biosci.,* **8** (2008), 146–152.

Zenkl G., Klimant I., *Microchimica Acta,* **166**(1–2), (2009), 123–131.

Zhao W., Zhang H., He Q., Li Y., et al., *Chem. Commun.,* **47** (2011), 9459–9461.

Chapter 17

Nanostructured Materials for Enzymatic Biofuel Cells

Takanori Tamaki
Tokyo Institute of Technology, Institute of Innovative Research, Laboratory for Chemistry and Life Science, R1-17, 4259 Nagatsuta, Midori-Ku, Yokohama 226-8503, Japan
tamaki.t.aa@m.titech.ac.jp

17.1 Introduction

Enzymatic biofuel cells use enzymes as catalysts to convert the chemical energy of fuels into electricity. Because various nontoxic fuels can be used under moderate conditions, biofuel cells have attracted much attention in recent years as energy sources used near the body. The fuels under consideration are monosaccharides such as glucose and fructose, and alcohols such as ethanol and glycerol. These fuels are oxidized at the anode, and an oxidant such as oxygen is reduced at the cathode. Considering the high intrinsic activity of enzymes, combinations of enzymes and nanostructured materials can enable biofuel cells to power portable devices. Nanostructured materials have a high surface area, and thus, are suitable for immobilizing large amounts of enzymes. However, simply using high-surface-area nanostructured materials with high electron conductivity does not necessarily

Biocatalysis and Nanotechnology
Edited by Peter Grunwald

ISBN 978-981-4613-69-9 (Hardcover), 978-1-315-19660-2 (eBook)
www.panstanford.com

lead to high power density, as summarized in a review published in 2012 (Tamaki, 2012).

Rational design of the electrode is necessary to achieve high power density and high current density. The points to be considered are (1) the amount of enzymes in the electrode, (2) the reaction rate of the enzymes, and (3) the mass transport of fuel or oxidant. Nanostructured materials enable the immobilization of large amounts of enzymes, as discussed above, and thus solve the first point, (1) amount of enzymes in the electrode. The second point, (2) reaction rate of enzymes, is affected by (2-1) how much enzyme retains its native activity in the electrode and (2-2) how efficiently enzymes can transfer electrons to or from the electrodes. Point (2-2) has been addressed from the beginning of research in biofuel cells. When discussing the effectiveness of the electron transfer between enzymes and electrodes, two types of electron transfer need to be considered between enzymes and electrodes: direct electron transfer (DET) and mediated electron transfer (MET). Several enzymes are known to undergo DET to or from the electrode (Cracknell et al., 2008; Ramanavicius and Ramanaviciene, 2009; Shleev et al., 2005). These DET-type enzymes allow a simple electrode design, although the number of DET-type enzymes is limited; examples of such enzymes are [NiFe]- and [FeFe]-hydrogenases, fructose dehydrogenase, and multicopper oxidases such as laccase, bilirubin oxidase, and copper efflux oxidase. For intrinsically non-DET-type enzymes, artificial redox mediators are used to establish MET. A direct covalent linkage of mediators to an electrode surface, or to a polymer forming a redox polymer, is preferred to avoid the leaking of mediators and exclude the requirement of adding mediators in solution. However, the immobilization of mediators reduces their mobility, which decreases the overall electrode reaction rate. For an efficient MET with immobilized mediators, several methods have been established on flat, two-dimensional electrodes. One method involves the reconstitution of an apoenzyme on cofactor–mediator monolayers (Willner et al., 2009). Another involves tethering mediators, such as osmium (Os) complexes, to long spacer arms and then immobilizing enzymes in three-dimensional redox hydrogels consisting of these redox polymers (Heller, 2004; Heller, 2006). Efficient electron transfer obtained using these approaches, in combination with nanostructured materials and efficient mass transport, enables further increases in current density.

The next section introduces nanostructured materials that have been shown to be effective in achieving high-current-density enzymatic biofuel cells. The focus is on current density; another important parameter determining the power density, the open-circuit voltage (OCV), will not be addressed in this chapter because OCV is mostly determined by the components that transfer electrons to or from the electrodes, namely enzymes for DET and mediators for MET.

17.2 Nanostructured Materials Used in High-Current-Density Enzymatic Biofuel Cells

17.2.1 Carbon Black

In conventional polymer electrolyte fuel cells (PEFCs) and direct methanol fuel cells (DMFCs), carbon blacks such as Ketjenblack and Vulcan XC-72, with particle diameters of about 30 nm, have been widely used as a support for platinum nanoparticles, and consequently, determine the structure of the catalyst layer. The catalyst layer possesses two types of pores: The smaller pore, or primary pore, is the space in and between primary particles in an agglomerate, and the larger pore, or secondary pore, is the space between agglomerates (Uchida et al., 1995; Uchida et al., 1996). Secondary pores with sizes from about 40 nm to 1 μm mainly serve as channels for mass transport. Considering that DMFCs with liquid methanol as a fuel have achieved high current densities on the order of 10^2 mA cm^{-2}, the abovementioned problem (3) mass transport of fuels will be circumvented when similar structures are formed by following similar fabrication procedures and using the same types of carbon black.

We next focus on problem (2), reaction rate of the enzymes. In the case of DET-type enzymes, adsorbing the enzymes on the surface of carbon black by immersing the electrode in a solution containing the enzymes achieved high current densities of around 10 mA cm^{-2} (Kamitaka et al., 2007a, 2007b; Tsujimura et al., 2008; Tsujimura et al., 2009; Kontani et al., 2009). An electrode consisting of Ketjenblack with adsorbed fructose dehydrogenase showed fructose-oxidation current densities with a maximum value in the range 7–10 mA cm^{-2} (Kamitaka et al., 2007a). When this bioanode composed of fructose dehydrogenase and Ketjenblack was combined with a biocathode composed of Laccase and carbon aerogel, which has mesopores with an average pore size of 22 nm, to develop a one-compartment biofuel cell, a maximum current density of 2.8 mA cm^{-2} and maximum power density of 850 μW cm^{-2} were obtained at pH 5.0 in a stirred solution (Kamitaka et al., 2007b). A comparison of Ketjenblack with other types of carbon particles, Vulcan XC-72R, carbon nanospheres, and lamp black 101, for DET from fructose dehydrogenase showed that the highest current density per unit weight of carbon particles was obtained at the Ketjenblack-modified electrode. This was attributed to the higher surface area of Ketjenblack, 800 m^2 g^{-1}, compared with the other carbon particles due to its unique hollow structure and small particle size of around 30 nm; Vulcan XC-72R has a similar particle size of about 30 nm and a surface area of 200 m^2 g^{-1}; carbon nanospheres and lamp black 101 have large particle sizes of over 100 nm, with correspondingly low surface areas of around 20 m^2 g^{-1} (Tsujimura et al., 2009). The electrode modified with the adsorbed copper efflux oxidase showed oxygen-reduction current densities as high as 12 mA cm^{-2} at pH 5.0 on a rotating disk electrode at a rotation speed of 10000 rpm (Tsujimura et al., 2008). Ketjenblack-modified carbon paper and copper efflux oxidase were also used for the construction of an air-diffusion biocathode, where oxygen is directly supplied to the electrode from air. Examination of hydrophobic binders, their compositions, and buffer concentrations has enabled current densities as high as 20 mA cm^{-2} to be achieved (Kontani et al., 2009). Further discussion on air-diffusion biocathodes is given in Sections 17.2–17.5.

In the case of MET-type enzymes, redox polymers are incorporated in an electrode consisting of carbon black. Research on enzyme-containing redox polymer films on flat electrodes has revealed that when the thickness of the redox polymer is several to several tens of micrometers, electron transport through the polymer becomes the rate-limiting step because of their low apparent electron diffusion coefficients (Barton et al., 2004). One way to overcome the rate-limiting step is to decrease the electron transport distance in the redox polymer. For this purpose, a short-chain redox polymer was chemically immobilized, or grafted, onto the surface of Ketjenblack, and the surface-modified carbon black was used for the fabrication of a three-dimensional carbon electrode as schematically shown in Fig. 17.1a (Tamaki and Yamaguchi, 2006; Tamaki et al., 2007; Tamaki et al., 2009; Sugiyama et al., 2014). This carbon electrode played a primary role in conducting electrons through the thick electrode, and the grafted redox polymer only needed to transport electrons a short distance of several to several tens of nanometers from the enzyme to the carbon black surface. Thus, the electron transport distance in the redox polymer was reduced by several orders of magnitude, and the rate-limiting step was overcome. The scanning electron microscope (SEM) images of the electrode shown in Fig. 17.1b reveal that the structure is similar to that of PEFC electrodes. A glucose-oxidation current density of 3 mA cm^{-2} was obtained with glucose oxidase and grafted poly(vinylferrocene-*co*-acrylamide) (Tamaki and Yamaguchi, 2006). Graft polymerization of the redox polymer was also applied to hydroquinone for a glucose-oxidizing anode (Tamaki et al., 2007) and to 2,2-azinobis(3-ethylbenzothiazoline-6-sulfonate) for an oxygen-reducing electrode with laccase as the enzyme (Sugiyama et al., 2014). The real surface area of the electrode was 2300 cm^2 per unit projected area when the carbon loading per unit projected area was 1.2 mg cm^{-2}, which was calculated assuming that the double-layer capacitance of the carbon black was 10^{-5} F cm^{-2}. This high surface area enabled the immobilization of a large amount of glucose oxidase, with a surface coverage of about 6×10^{-9} mol cm^{-2} (Tamaki et al., 2010); this value almost agrees with that of a fully packed monolayer of glucose oxidase covering the entire electrode surface. A mathematical model that considered the reaction and diffusion processes in biofuel cell electrodes suggested that this high surface coverage, i.e., the utilization of the entire electrode surface by the enzymes, leads to an increase in the current density of two orders of magnitude to around 10^2 mA cm^{-2} (Tamaki et al., 2009). In other words, only a few percent of the electrode surface was effectively utilized by the glucose oxidase in the electrode. The model calculation also showed that the rate-limiting step of electron diffusion could be overcome by using an electrode with a thin redox polymer layer, even if the redox polymer used has a low apparent electron diffusion coefficient. To investigate the contact between enzymes and mediators, electrochemical measurements were then performed with and without the addition of a free mediator in the solution. The obtained current densities were similar, irrespective of the presence of the free mediator, which suggests that the mediator in the grafted redox polymer made good contact with the glucose oxidase. The remaining point to be considered is (2-1) how much enzymes retain its native activity in the electrode. The activity of adsorbed glucose oxidase was measured by suspending enzyme-coated carbon black in a buffer solution containing glucose to avoid the mass transport limitation of the electron acceptor, oxygen, that occurs in the three-dimensional electrode. The relative activity, defined as the ratio of the specific activities of adsorbed glucose oxidase and free glucose oxidase in solution, was about 10% (Tamaki et al., 2010). These results revealed that the deactivation of glucose oxidase upon physical adsorption on the hydrophobic carbon surface limited the current density, resulting in much lower values than those expected from the results of model calculations. Surface modification of carbon black to make

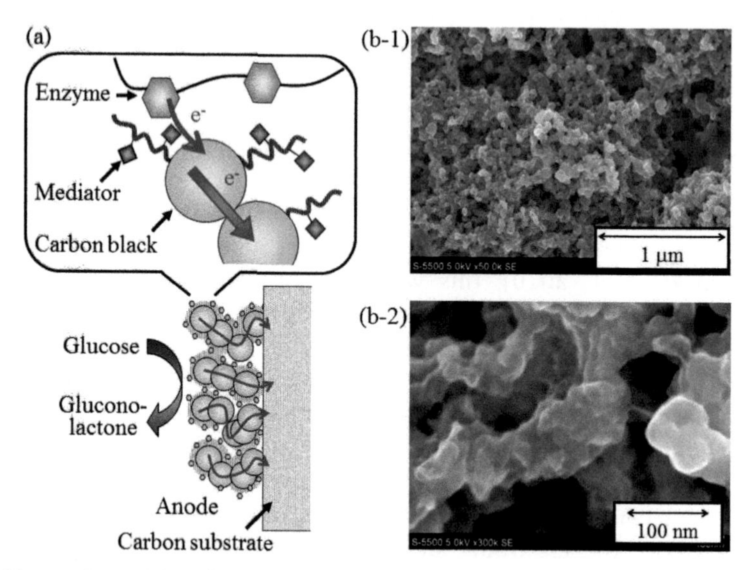

Figure 17.1 (a) Schematic illustration and (b) SEM images of an enzyme-modified electrode comprising redox-polymer-grafted Ketjenblack and glucose oxidase at two magnifications. Reprinted with permission (a) from *Journal of Physical Chemistry B*, 111 (2007), 10312. Copyright 2007, American Chemical Society, and (b) from *Topics in Catalysis*, 55 (2012), 1162. Copyright 2012, Springer.

the surface hydrophilic was shown to be effective in reducing the physical adsorption, and the combination of surface modification and immobilization of glucose oxidase using ammonium sulfate precipitation with cross-linking increased the glucose-oxidation current density to 6 mA cm^{-2}, which is twice as high as the value obtained without the surface treatment (Tamaki et al., 2014). Improving (2-1) the remaining activity of the immobilized enzymes (i.e., those not just physically adsorbed) is necessary to further increase the current density obtained using electrodes made of carbon black.

17.2.2 Carbon Nanotubes

Many researchers have used carbon nanotubes as electrode matrices in enzymatic biofuel cells. However, the random structures formed by casting or dip-coating methods had difficulty achieving current densities higher than 1 mA cm^{-2} when a free mediator was not added to the solution, as previously summarized (Tamaki, 2012). On the other hand, several groups have succeeded in achieving high current densities through rational design of the electrode, as will be introduced in this section.

A multiscale carbon material composed of multi-walled carbon nanotubes grown on carbon paper by chemical vapor deposition was the first glucose-oxidizing electrode that produced a current density above 10 mA cm^{-2}. SEM images of the material at different growth times are shown in Fig. 17.2 (Barton, et al., 2007). A multiscale carbon material coated with a redox hydrogel comprising glucose oxidase, poly(vinylpyridine) (PVP)-[Os(bipyridyl)$_2$Cl], and a cross-linker showed a current density of 22 mA cm^{-2} on a rotating disk electrode at 4000 rpm. The Brunauer–Emmett–Teller (BET) surface area measured by nitrogen physisorption increased by two orders of magnitude to be 82 m^2 g^{-1} after the growth of nanotubes, and the electrochemical capacitance increased to about 90% of that predicted from the BET surface area. The peak pore volume occurred near a pore diameter of 17 nm, with a broad distribution. The glucose-oxidation current density increased tenfold relative to the bare carbon paper. The disproportionate increase in the current density relative to the surface area was ascribed to mass transport limitations and the limited access of enzymes to the nanopores (Barton et al., 2007).

Thus, the improvement in (1) amount of enzymes in the electrode and (3) mass transport of fuels can increase the current density further.

Figure 17.2 SEM images of multiscale carbon materials composed of multi-walled carbon nanotubes grown on a carbon paper at different growth times. Reprinted with permission from *Electrochemical and Solid State Letters*, 10 (2007), B96, [90]. Copyright 2007, The Electrochemical Society.

Engineered porous microwires comprising assembled and oriented carbon nanotubes were shown to be effective in overcoming mass transport limitations while also providing high surface areas. The carbon nanotube microwires were prepared by coagulation spinning of single-walled carbon nanotubes and poly(vinyl alcohol), followed by thermal decomposition of the polymer. The resulting carbon nanotube microwires with diameters of 9.5 μm were highly porous and possessed BET surface areas of about 300 m^2 g^{-1}. SEM images of the cross section of a classical, non-porous carbon fiber and the carbon nanotube microwires are shown in Fig. 17.3. The carbon nanotube fibers were then modified with bilirubin oxidase and poly-*N*-vinylimidazole (PVI)-[Os(4,4′-dichloro-2,2′-bipyridine)$_2$Cl]-*co*-acrylamide. Compared with the conventional carbon fiber with a diameter of 7 μm, the redox peak currents of the carbon nanotube microwires were fivefold higher, despite the same quantity of redox polymer being deposited on both electrode surfaces. The carbon nanotube microwires showed fourfold higher oxygen-reduction current density than the conventional carbon fiber in quiescent solution under air. Under 1 atm oxygen, the current density exceeds 2.5 mA cm^{-2}. Observation of such high current densities in unstirred solutions shows that problem (3), mass transport of oxygen, was efficiently solved by using carbon nanotube microwires. When this biocathode was combined with a bioanode composed of the carbon nanotube microwires, glucose oxidase, and PVP-[Os(*N,N'*-dialkylated-2,2'-bis-imidazol)$_3$] to construct a membrane-less biofuel cell, a maximum power density of 740 μW cm^{-2} was obtained in an unstirred solution at pH 7.2 (Gao et al., 2010).

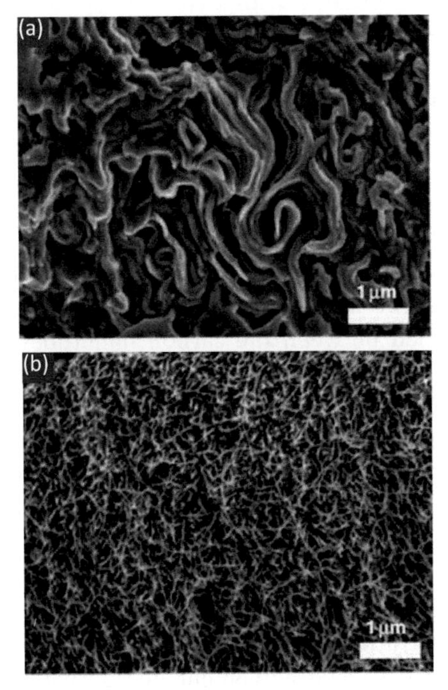

Figure 17.3 SEM images of the cross section of (a) a classical, non-porous carbon fiber and (b) carbon nanotube microwires. Reprinted by permission from Macmillan Publishers Ltd: *Nature Communications*, 1 (2010), 2 doi: 10.1038/ncomms1000. Copyright 2010.

The methods introduced so far involve fabrication of the electrode structures first, followed by immobilization of enzymes in the electrode. Another approach is the regulation of the nanostructure of the electrode in response to the enzyme to be immobilized. This method becomes possible by using a free-standing carbon nanotube forest comprising extremely long single-walled carbon nanotubes with lengths of 1 mm. The pitch of the carbon nanotube forest decreased from 16 to 3.7 nm by liquid-induced shrinkage, which occurs during the drying of liquid-containing as-grown carbon nanotube forests due to the surface tension of the liquid. The use of an enzyme solution as the liquid enabled dynamic entrapment of the enzymes during the shrinkage, as well as in situ regulation of the inter-carbon nanotube pitch to the size of the enzymes, as schematically shown in Fig. 17.4 (Miyake et al., 2011). The carbon nanotube forest was first combined with two DET-type enzymes, fructose dehydrogenase and laccase. The amount of fructose dehydrogenase inside the film was controllable through the enzyme concentration, resulting in different degrees of shrinkage after drying, and increased toward the theoretical limiting value where the void volume of the as-grown carbon nanotube forest is fully occupied by four enzyme molecules with increasing enzyme concentration. The best electrode performance was obtained when the immobilized amount of enzyme was one-quarter of the limiting value, which can be modeled as a linear arrangement of enzyme molecules trapped between the carbon nanotubes. The obtained fructose-oxidation current density was 11 mA cm^{-2} in quiescent solution and 16 mA cm^{-2} in stirred solution at room temperature, without any mediator in the solution. A comparison was made with a conventional random carbon nanotube electrode prepared by a casting method, which showed a low current density of only about 0.4 mA cm^{-2}, suggesting a limited loading of enzymes. The apparent Michaelis–Menten constant, estimated from the currents obtained from the carbon nanotube forest film at various fructose concentrations using the Lineweaver–Burke equation, was around 10 mM. The agreement of this value with that for the reaction in the bulk solution indicates that the nanospace formed by the shrinkage of the carbon nanotube forest can maintain the nature of the enzyme. A biofuel cell with the bioanode and a biocathode composed of laccase and a carbon nanotube forest produced a maximum power density of 1.8 mW cm^{-2} in a stirred solution at pH 5.0 (Miyake et al., 2011). The carbon nanotube forest was also used with an MET-type enzyme, glucose oxidase.

Although there are several reports of DET from glucose oxidase, the flavin adenine dinucleotide cofactor of glucose oxidase is buried in a cleft in the protein so that DET is very sluggish (Cracknell et al., 2008). The carbon nanotube forest also did not work for glucose oxidase without any mediator. The carbon nanotube forest film was then modified, first with a redox polymer, PVI-[Os(bipyridine)$_2$Cl], and then with glucose oxidase. Symmetrical cyclic voltammograms and a proportional dependence of peak currents on scan rate are typical responses of adsorbed redox species. The amount of glucose oxidase in a 20 μm-thick film was 0.86 μg, which corresponds to about 74% of the theoretical value where the enzyme molecules align to form lines in the space between the carbon nanotubes. The obtained glucose-oxidation current densities in the stirred solution were 14.7 mA cm^{-2} at room temperature and 26.7 mA cm^{-2} at 37.5°C. The current densities decreased by half under quiescent conditions. The turnover rate of glucose oxidase was calculated from the current measured at room temperature and the amount of glucose oxidase in the film. The resulting value was 650 s^{-1}, which is comparable to the value of 700 s^{-1} obtained for glucose oxidase in bulk solution containing the natural electron acceptor, oxygen (Yoshino et al., 2013). These results suggest that carbon nanotube forests are effective in solving both problems (1) amount of enzymes in the electrode and (2) reaction rate of enzymes, including (2-1) how much enzymes retain its native activity in the electrode and (2-2) how efficient enzymes can transfer electrons to or from the electrodes.

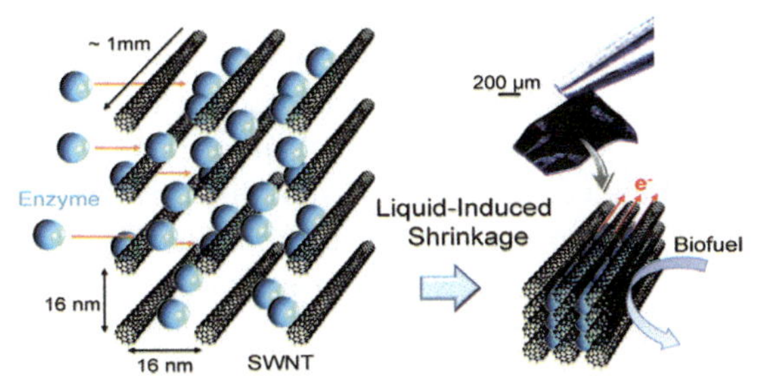

Figure 17.4 Schematic illustration of enzyme entrapment inside a carbon nanotube forest by liquid-induced shrinkage. The photograph shows a free-standing enzyme–carbon nanotube forest ensemble film that can be manipulated with tweezers. Reprinted with permission from *Journal of the American Chemical Society*, 133 (2011), 5129. Copyright 2011, American Chemical Society.

17.2.3 Monolithic Carbon

A new class of nanostructured carbon materials, three-dimensional monolithic carbonaceous foams, has been proposed to provide hierarchical pore structures for both high surface area and fast mass transport. The synthesis was performed by introducing a tetrahydrofuran solution of formophenolic resin into a silica macroporous framework under vacuum, followed by pyrolysis. Photographs and SEM images of the monolithic carbonaceous foams are shown in Fig. 17.5 (Flexer et al., 2010). The monolithic materials have an interconnected hierarchical porosity comprising macropores with sizes centered at 4–5 μm for fast mass transport, and mesopores and micropores to act as anchoring sites for enzyme entrapment. The total macroporous area was 3.4 $m^2 g^{-1}$ and the microporous area was around 523 $m^2 g^{-1}$. A bioanode was fabricated by impregnating the monolithic carbonaceous foams with glucose oxidase, a redox polymer, PVI-[Os(4,4′-dichloro-2,2′-bipyridine)$_2$Cl]-*co*-acrylamide, and a cross-linker. A comparison was made with flat glassy carbon electrodes, upon which the same masses of enzyme and redox polymer were deposited. Cyclic voltammogram analysis obtained in the absence of glucose revealed that the redox hydrogel in the monolithic carbonaceous foams behaved as a thin layer, whereas that on glassy carbon electrode behaved more like a thick layer under semi-infinite diffusion control. The peak area, or redox charge, obtained with the monolithic carbonaceous foams was larger than that obtained with the glassy carbon electrode, although the same quantity of the mediator was deposited on both electrodes. The ratios of the peak areas were 7 at a scan rate of 5 mV s^{-1} and 40 at 100 mV s^{-1}, suggesting that more redox centers are connected to the electrode in the monolithic carbonaceous foams. The glucose-oxidation current density obtained with monolithic carbonaceous foams depends on the rotation rate. However, the value was below that expected from the Levich equation, which suggests that the size and quantity of the macropores were not yet fully optimized. The current density obtained with the monolithic carbonaceous foams was 7.49 mA cm^{-2} at a scan rate of 5 mV s^{-1} and a rotation speed of 10000 rpm, which was 13 times higher than the current density on the glassy carbon electrode (Flexer et al., 2010). The monolithic carbonaceous foams were also applied to a DET-type enzyme, bilirubin oxidase, and an oxygen-reduction current density of 2.1 mA cm^{-2} was obtained at 1000 rpm. The value was 500 times higher than that obtained with bilirubin oxidase adsorbed on a glassy carbon. A rough estimation showed that approximately 20% of a fully packed monolayer of bilirubin oxidase was adsorbed within the porous electrode (Flexer et al., 2011). When a different pathway was employed to synthesize the monolithic carbonaceous foams, the mesopore surface area increased four times and the macroscopic pore volume increased by 12%. The oxygen-reduction current density obtained with bilirubin oxidase increased by 30%. In addition to the increased porosity of the monolithic carbonaceous foams, the redox polymer used with glucose oxidase was changed to PVP-Os(1,1′-dimethyl-2,2′-biimidazole)$_2$-2-[6-methylpyrid-2-yl]imidazole. The maximum glucose-oxidation current density was 19.3 mA cm^{-2} at 0.38 V vs. Ag/AgCl and 2000 rpm, and 15.6 mA cm^{-2} at 0.3 V vs. Ag/AgCl under quiescent conditions. Similar current densities were obtained in the

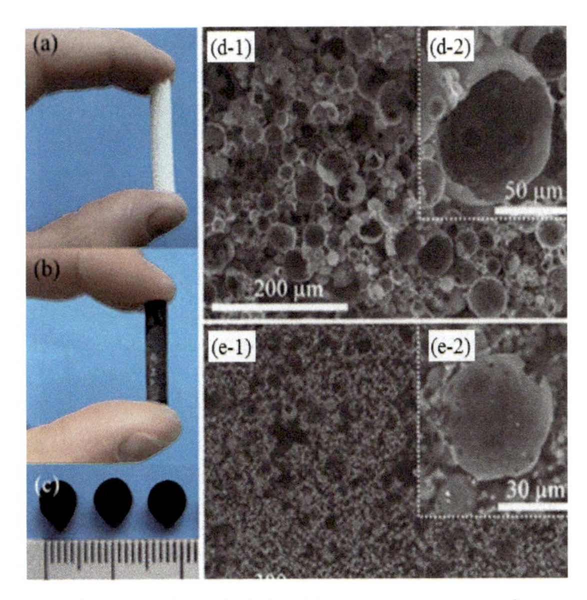

Figure 17.5 Photographs of (a) silica macroporous framework and (b and c) monolithic carbonaceous foams. SEM images of (d) silica macroporous framework and (e) monolithic carbonaceous foams at two magnifications. Reprinted with permission from *Energy & Environmental Science*, 4 (2011), 2097. Copyright 2011, Royal Society of Chemistry.

presence and absence of forced convection, especially below 0.1 V vs. Ag/AgCl, suggesting that (3) the mass transport of fuels is efficient in the material. A membrane-free biofuel cell composed of the bioanode with glucose oxidase and PVP-Os(1,1′-dimethyl-2,2′-biimidazole)$_2$-2-[6-methylpyrid-2-yl]imidazole and the biocathode with bilirubin oxidase generated power densities of 188 μW cm^{-2} at 1000 rpm and 125 μW cm^{-2} under quiescent conditions. The use of PVP-[Os(*N*,*N*′-dialkylated-2,2′-bis-imidazol)$_3$] as the redox polymer in the anode generated a slightly higher power density of 202 μW cm^{-2} at 1000 rpm (Flexer et al., 2013).

Another class of monolithic carbon is carbon cryogels prepared by sol–gel polycondensation of resorcinol and formaldehyde using CO_3^{2-} as a catalyst, followed by freeze-drying and pyrolysis. A change in the molar ratio of resorcinol to the catalyst enabled the control of the pore size in the range 5–40 nm. Cryogels with different pore sizes were used to examine the effect of pore size on the current density obtained with a DET-type enzyme, fructose dehydrogenase. In the experiment, a carbon cryogel block was broken into small particles below 1 μm in diameter to fabricate the electrode. The amount of fructose dehydrogenase adsorbed on the carbon cryogel increased with the pore size, but did not depend on the BET surface area or the pore volume. Furthermore, when the pore size was smaller than the enzyme diameter, fructose dehydrogenase only adsorbed onto the external surface of the carbon cryogels and not inside mesopores. Wider pore sizes of up to about 40 nm resulted in higher current densities (Tsujimura et al., 2010). Further evaluation was performed by fabricating the electrode directly using microcubic monolithic carbon cryogels without breaking them into small particles. Two different carbon cryogels with side lengths of 0.18 and 0.26 mm showed almost identical fructose-oxidation current densities per unit volume, which suggests a uniform distribution of fructose dehydrogenase inside the particles. The catalytic current density per unit projected electrode area was 16 mA cm^{-2} under stirring conditions (Hamano et al., 2012).

17.2.4 Mesoporous Carbon

Another approach to fabricate hierarchical pore structures is to build the macroporous structure using mesoporous carbon materials. A hierarchically structured electrode was fabricated using a combination of magnesium oxide-templated mesoporous carbon (average pore diameter of 38 nm and BET surface area of 580 m^2 g^{-1}) and an electrophoretic deposition method, in which carbon particles dispersed in a solvent are forced to migrate toward an electrode by applying an electric field. The electrode was modified with deglycosylated flavin adenine dinucleotide-dependent glucose dehydrogenase, PVI-[Os(bipyridine)$_2$Cl], and a cross-linker. A schematic illustration and SEM images of the hierarchical bioanode comprising magnesium oxide-templated mesoporous carbon is shown in Fig. 17.6 (Tsujimura et al., 2014). Analysis of cyclic voltammograms revealed that a thin redox-hydrogel layer was formed on the surface of the magnesium oxide-templated mesoporous carbon. A comparison with a flat glassy carbon electrode modified with the same amount of the enzyme and redox polymer showed that the glucose-oxidation current

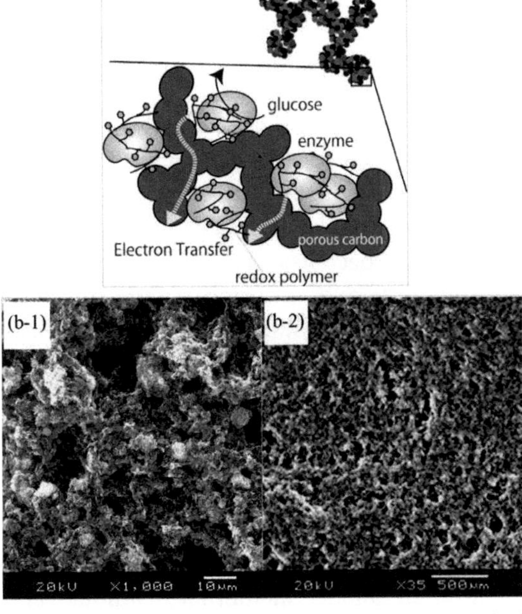

Figure 17.6 (a) Schematic illustration and (b) SEM images of a hierarchical bioanode comprising magnesium oxide-templated mesoporous carbon at two magnifications. Reprinted with permission from *Journal of the American Chemical Society*, 136 (2014), 14432. Copyright 2014, American Chemical Society.

density was 33 times higher for the electrode composed of the magnesium oxide-templated mesoporous carbon. Optimization of the redox-hydrogel loading and the concentrations of buffer and glucose resulted in an extremely high current density of 100 mA cm^{-2} at room temperature and 360 mA cm^{-2} at 55°C, at a rotation speed of 9000 rpm. Without stirring, a current density of 18 mA cm^{-2} was obtained as a steady-state value, and a peak current density of 27 mA cm^{-2} was obtained at 0.3 V vs. Ag/AgCl. The electrode fabricated by a drop-casting technique, which lacked macropores larger than 10 μm, generated a maximum current density of less than 40 mA cm^{-2} at 9000 rpm, suggesting the formation of efficient channels for mass transport by the electrophoretic deposition method (Tsujimura et al., 2014). The extremely high current densities of above 100 mA cm^{-2} obtained in the hierarchically structured electrode composed of magnesium oxide-templated mesoporous carbon reveals that the rational electrode design is actually effective in achieving high current densities.

17.2.5 Gas Diffusion Electrodes

Of the three points to be considered in the rational design of enzymatic biofuel cell electrodes, (3) mass transport of oxygen is difficult to solve solely using nanostructured materials. When oxygen is supplied as dissolved oxygen in aqueous solution, mass transport of oxygen becomes a significant rate-limiting step because the solubility and diffusion coefficient of oxygen are low. A numerical calculation showed that the limited diffusion of oxygen in a dissolved-oxygen electrode can be much improved by employing gas-diffusion electrodes (Barton, 2005). However, the enzyme activity in the gas phase is generally much lower than that in the liquid phase. Thus, the local water activity around the enzyme molecules should be considered in the gas-diffusion electrode. The gas supply of oxygen is experimentally investigated in a membrane–electrode-assembly-type cell, which is a sandwich-type combination of anode, membrane, cathode, and diffusion layers, and is common in conventional PEFCs and DMFCs. The results showed that humidification of the gas stream stabilized the cathode performance to some extent, and that a major limiting factor was likely to be a decrease in the rate of proton transport in the biocathode caused by dehydration (Hudak et al., 2009). Another type of gas-diffusion biocathode is constructed by contacting one side of the electrode with an electrolyte solution and the other side with the gas phase, as shown in Fig. 17.7 (Kontani et al., 2009). A membrane such as cellophane can be located between the electrolyte solution and the cathode to control the supply of solution to the electrode (Sakai et al., 2009). The current densities obtained from gas-diffusion biocathodes were much higher than those obtained from sink-type cells, where oxygen is supplied through a bulk solution phase (Kontani et al., 2009; Sakai et al., 2009). These gas-diffusion biocathodes have been fabricated with carbon black (Hudak et al., 2009; Kontani et al., 2009; Miyake et al., 2011), carbon fiber sheet (Sakai et al., 2009), and carbon nanotubes (Lau et al., 2012; Lalaoui et al., 2015). A balance between hydrophilicity and hydrophobicity should be tuned in the gas-diffusion electrode to achieve efficient mass transport of gas-phase oxygen while retaining enzyme activity.

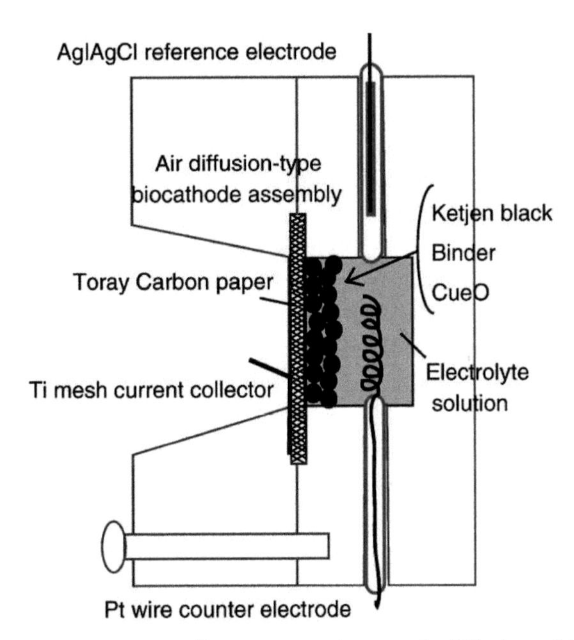

Figure 17.7 Schematic illustration of an air-diffusion biocathode comprising Ketjenblack, a binder, and copper efflux oxidase (CueO) on carbon paper. Reprinted with permission from *Bioelectrochemistry*, 76 (2009), 10. Copyright 2009, Elsevier.

17.3 Other Parameters to be Considered in Enzymatic Biofuel Cells

The previous sections focused on the power density, especially current density, of enzymatic biofuel cells. However, a few other important parameters remain to be considered, which are introduced in this section.

17.3.1 Stability

Stability is another important issue in enzymatic biofuel cells because enzymes can easily lose their activity compared with other catalysts, such as metal nanoparticles. Although the issue of stability has not been mentioned in the previous sections, the immobilization of enzymes in nanostructured materials greatly enhances their stability compared with the immobilization on electrodes without nanostructures.

Carbon nanotube microwires (mentioned in Section 17.2.2) showed higher stability than classical, non-porous carbon fibers with the same loading of redox polymer and enzyme. At a rotation rate of over 2000 rpm, a classical carbon fiber electrode poised at 0.3 V vs. Ag/AgCl lost 80% of its current density in 8 h at 37°C. Under the same conditions, a carbon nanotube microfiber electrode lost only 18% of its initial current density. Thus, the current density after 8 h obtained with carbon nanotube microwires was 10 times higher than that with classical carbon fibers (Gao et al., 2010). The hierarchical electrode composed of magnesium oxide-templated mesoporous carbon (mentioned in Section 17.2.4) also showed enhanced stability compared with a flat glassy carbon electrode. Under continuous operation at 0.5 V vs. Ag/AgCl at room temperature, the hierarchical electrode retained more than 80% of its current density after 7 days, whereas the glassy carbon electrode only retained the same level of current density for 2 days. When the hierarchical electrode was stored in phosphate buffer at 4°C, the catalytic current remained almost unchanged for 220 days, as determined by nearly weekly current measurements (Tsujimura et al., 2014). Although the reasons still need to be clarified, nanostructured materials are effective in improving the stability of enzyme electrodes.

Several other methods to enhance electrode stability have been reported, such as immobilization of enzymes in quaternary ammonium bromide salt-treated Nafion films (Moore et al., 2004; Akers et al., 2005) and genetic engineering of the enzymes (Okuda et al., 2004). These methods, in conjunction with the immobilization of enzymes in nanostructured materials and clarification of the stabilization mechanism, are expected to enhance the durability of enzymatic biofuel cells.

17.3.2 Energy Density

The energy density of fuels should also be considered when enzymatic biofuel cells are to be used to power portable devices. In the research mentioned so far, only one enzyme was employed to oxidize the fuel, and thus, only two electrons were extracted from the fuel. The partial oxidation of fuels results in low energy density per weight or volume of fuel.

The combination of a few or several enzymes has been reported to fully or deeply oxidize fuels by extracting more than two electrons. Methanol was fully oxidized to carbon dioxide using alcohol dehydrogenase, aldehyde dehydrogenase, and formate dehydrogenase (Palmore et al., 1998). A complete oxidation of glycerol was also achieved using a combination of pyrroloquinoline quinone (PQQ)-dependent alcohol dehydrogenase, PQQ-dependent aldehyde dehydrogenase, and oxalate oxidase (Arechederra and Minteer, 2009). A bioanode containing all of the enzymes in the Krebs cycle, which is the main metabolic process in living cells, achieved complete oxidation of pyruvate (Sokic-Lazic and Minteer, 2009). A synthetic catabolic pathway comprising 13 enzymes produced nearly 24 electrons from the glucose unit of maltodextrin (Zhu et al., 2014).

17.4 Conclusion

This chapter summarizes the nanostructured electrode materials used in high-current-density enzymatic biofuel cells. Increasing the current density requires a rational design of the electrodes while considering (1) the amount of enzymes in the electrode, (2) the reaction rate of the enzyme, which is further divided into (2-1) how much enzymes retain their native activity in the electrode and (2-2) how efficiently can enzymes transfer electrons to or from the electrodes, and (3) the mass transport of fuels or oxidant. When the current density is lower than expected, the limiting factor should be clarified to guide modifications intended to increase the current density.

Further investigation into increasing the power density by increasing the current density, stability, and energy density should lead to practical applications of biofuel cells being realized.

References

Akers N. L., Moore C. M., Minteer S. D., *Electrochimica Acta,* **50** (2005), 2521–2525.

Arechederra R. L., Minteer S. D., *Fuel Cells,* **9** (2009), 63–69.

Barton S. C., *Electrochimica Acta,* **50** (2005), 2145–2153.

Barton S. C., Gallaway J., Atanassov P., *Chemical Reviews,* **104** (2004), 4867–4886.

Barton S. C., Sun Y. H., Chandra B., White S., Hone J., *Electrochemical and Solid State Letters,* **10** (2007), B96–B100.

Cracknell J. A., Vincent K. A., Armstrong F. A., *Chemical Reviews,* **108** (2008), 2439–2461.

Flexer V., Brun N., Backov R., Mano N., *Energy & Environmental Science,* **3** (2010), 1302–1306.

Flexer V., Brun N., Courjean O., Backov R., Mano N., *Energy & Environmental Science,* **4** (2011), 2097–2106.

Flexer V., Brun N., Destribats M., Backov R., Mano N., *Physical Chemistry Chemical Physics,* **15** (2013), 6437–6445.

Gao F., Viry L., Maugey M., Poulin P., Mano N., *Nature Communications,* **1** (2010), 2 doi: 10.1038/ncomms1000.

Hamano Y., Tsujimura S., Shirai O., Kano K., *Bioelectrochemistry,* **88** (2012), 114–117.

Heller A., *Physical Chemistry Chemical Physics,* **6** (2004), 209–216.

Heller A., *Current Opinion in Chemical Biology,* **10** (2006), 664–672.

Hudak N. S., Gallaway J. W., Barton S. C., *Journal of the Electrochemical Society,* **156** (2009), B9–B15.

Kamitaka Y., Tsujimura S., Kano K., *Chemistry Letters,* **36** (2007a), 218–219.

Kamitaka Y., Tsujimura S., Setoyama N., Kajino T., Kano K., *Physical Chemistry Chemical Physics,* **9** (2007b), 1793–1801.

Kontani R., Tsujimura S., Kano K., *Bioelectrochemistry,* **76** (2009), 10–13.

Lalaoui N., de Poulpiquet A., Haddad R., Le Goff A., Holzinger M., Gounel S., Mermoux M., Infossi P., Mano N., Lojou E., Cosnier S., *Chemical Communications,* **51** (2015), 7447–7450.

Lau C., Adkins E. R., Ramasamy R. P., Luckarift H. R., Johnson G. R., Atanassov P., *Advanced Energy Materials,* **2** (2012), 162–168.

Miyake T., Haneda K., Nagai N., Yatagawa Y., Onami H., Yoshino S., Abe T., Nishizawa M., *Energy & Environmental Science,* **4** (2011), 5008–5012.

Miyake T., Yoshino S., Yamada T., Hata K., Nishizawa M., *Journal of the American Chemical Society,* **133** (2011), 5129–5134.

Moore C. M., Akers N. L., Hill A. D., Johnson Z. C., Minteer S. D., *Biomacromolecules,* **5** (2004), 1241–1247.

Okuda J., Wakai J., Igarashi S., Sode K., *Analytical Letters,* **37** (2004), 1847–857.

Palmore G. T. R., Bertschy H., Bergens S. H., Whitesides G. M., *Journal of Electroanalytical Chemistry,* **443** (1998), 155–161.

Ramanavicius A., Ramanaviciene A., *Fuel Cells,* **9** (2009), 25–36.

Sakai H., Nakagawa T., Tokita Y., Hatazawa T., Ikeda T., Tsujimura S., Kano K., *Energy & Environmental Science,* **2** (2009), 133–138.

Shleev S., Tkac J., Christenson A., Ruzgas T., Yaropolov A. I., Whittaker J. W., Gorton L., *Biosensors & Bioelectronics,* **20** (2005), 2517–2554.

Sokic-Lazic D., Minteer S. D., *Electrochemical and Solid State Letters,* **12** (2009), F26–F28.

Sugiyama T., Tamaki T., Yamaguchi T., *Journal of Chemical Engineering of Japan,* **47** (2014), 704–710.

Tamaki T., *Topics in Catalysis,* **55** (2012), 1162–1180.

Tamaki T., Hiraide A., Asmat F. B., Ohashi H., Ito T., Yamaguchi T., *Industrial & Engineering Chemistry Research,* **49** (2010), 6394–6398.

Tamaki T., Ito T., Yamaguchi T., *Journal of Physical Chemistry B,* **111** (2007), 10312–10319.

Tamaki T., Ito T., Yamaguchi T., *Fuel Cells,* **9** (2009), 37–43.

Tamaki T., Sugiyama T., Mizoe M., Oshiba Y., Yamaguchi T., *Journal of the Electrochemical Society,* **161** (2014), H3095–H3099.

Tamaki T., Yamaguchi T., *Industrial & Engineering Chemistry Research,* **45** (2006), 3050–3058.

Tsujimura S., Miura Y., Kano K., *Electrochimica Acta,* **53** (2008), 5716–5720.

Tsujimura S., Murata K., Akatsuka W., *Journal of the American Chemical Society,* **136** (2014), 14432–14437.

Tsujimura S., Nishina A., Hamano Y., Kano K., Shiraishi S., *Electrochemistry Communications*, **12** (2010), 446–449.

Tsujimura S., Nishina A., Kamitaka Y., Kano K., *Analytical Chemistry*, **81** (2009), 9383–9387.

Uchida M., Aoyama Y., Eda N., Ohta A., *Journal of the Electrochemical Society*, **142** (1995), 4143–4149.

Uchida M., Fukuoka Y., Sugawara Y., Eda N., Ohta A., *Journal of the Electrochemical Society*, **143** (1996), 2245–2252.

Willner I., Yan Y. M., Willner B., Tel-Vered R., *Fuel Cells*, **9** (2009), 7–24.

Yoshino S., Miyake T., Yamada T., Hata K., Nishizawa M., *Advanced Energy Materials*, **3** (2013), 60–64.

Zhu Z., Tam T. K., Sun F., You C., Zhang Y. H. P., *Nature Communications*, **5** (2014), 3026 doi: 10.1038/ncomms4026.

Chapter 18

Enzymatic Biofuel Cells on Porous Nanostructures

Dan Wen and Alexander Eychmüller

Physical Chemistry, TU Dresden,
Bergstrasse 66b, 01062 Dresden, Germany

alexander.eychmüller@chemie.tu-dresden.de

18.1 Introduction

Biofuel cells (BFCs) belong to a special kind of fuel cells where biocatalysts are employed as the catalysts and biomass as the fuels (Davis and Higson, 2007; Zhao et al., 2009). Compared to conventional fuel cells, BFCs are active in moderate conditions such as room temperature and neutral pH, and therefore they are viewed as a potential green energy technology (Bullen et al., 2006). According to the type of biocatalyst (i.e., living cells and enzymes) to catalyze the oxidation of fuels, the main types of BFCs are defined as microbial BFCs and enzymatic BFCs (EBFCs), respectively. In particular, EBFCs can be formatted into portable power sources and implantable medical devices due to the facile miniaturization and higher power densities they typically possess, in contrast to microbial BFCs (Barton et al., 2004; Leech et al., 2012, Yang et al., 2013). The first EBFC has been pioneered by Yahiro, Lee, and Kimble in 1964, where they utilized

Biocatalysis and Nanotechnology
Edited by Peter Grunwald

ISBN 978-981-4613-69-9 (Hardcover), 978-1-315-19660-2 (eBook)
www.panstanford.com

the electron transfer process of the flavoprotein enzyme system (i.e., glucose oxidase (GOx), D-amino acid oxidase, and yeast alcohol dehydrogenase) as a potential anodic reaction in conjunction with an O_2 cathode (Jahiro et al., 1964). Since then, considerable efforts have been put on EBFCs to gain high power output and good operational stability for their potential applications (Heller, 2004; Zhou and Dong, 2011; Falk et al., 2013b). Some outstanding examples are available in the application field of EBFCs for self-powered biosensors, implantable medical devices, and portable power sources, etc. Katz and Willner invented the concept of self-powered biosensors that utilized EBFC as a biosensor for the fuel (Katz et al., 2001). The first EBFC operating in a living organism was the work of Mano and Heller who implanted their glucose/O_2 cell in a grape (Mano et al., 2003). Sony Cooperation announced an EBFC that can be linked in series and used to power mp3 player (Sakai et al., 2009). Although numerous efforts have been devoted to the development of EBFCs, they are not currently in use outside of the laboratory. This is because still, two critical issues, namely the low power density and the typically short lifetimes, are to be addressed. Both depend on the enzyme stability, the electron transfer rates, the enzyme loadings, and the fuel/oxidant mass transport and have to be improved before EBFCs become really competitive in practical applications (Kim et al., 2006b; Moehlenbrock and Minteer, 2012; Luz et al., 2014).

EBFCs rely on enzymatic bioelectrocatalysis, where electron transfer occurs between the enzyme and the electrode directly (DET) or via a redox mediator (MET). Recent progress in nanobioelectrocatalysis on the basis of novel nanostructures shows great promise to solve the above concerned issues and enables the actual applications (Minteer, 2012; Min and Yoo, 2014). The incorporation of nanomaterials can increase the enzyme loading and facilitate reaction kinetics, thus improving the power density of EBFCs (Cooney et al., 2008b; Kim et al., 2006a; Walcarius et al., 2013). In addition, research efforts have also been made to advance the activity and stability of immobilized enzymes by using nanostructures (Kim and Grate, 2003; Fischback et al, 2008). Among various types of nanostructures, porous substrates with three-dimensional (3D) structure and high porosity are a promising choice. Generally, micro- and mesopores provide high surface areas while macropores guarantee accessibility to the surface (Walcarius, 2013; Liang et al., 2011). In this respect, porous nanostructures have additional remarkable features for the purpose of the EBFCs development. The larger surface area can ensure the immobilization of a large amount of enzymes and corresponding reagents (co-factors and mediators) if needed, and increase the reactive surface area (Sattayasamitsathit et al., 2012; Catalano et al., 2015; Do et al., 2015). The pores are large enough to allow for easy permeation of the electrolyte and oxidant, and they support the fuel transport to the catalyst reaction sites (Naharudin et al., 2014; Chen et al., 2015a). Inspired by the enhanced electron transfer and mass transport abilities, it appears to be reasonable to expect that achievements in porous nanostructured biocatalysts will play a critical role in overcoming the major obstacles for the development of powerful EBFCs (Zhou and Hartmann, 2012; Nardecchia et al., 2013; Tamaki, 2012)].

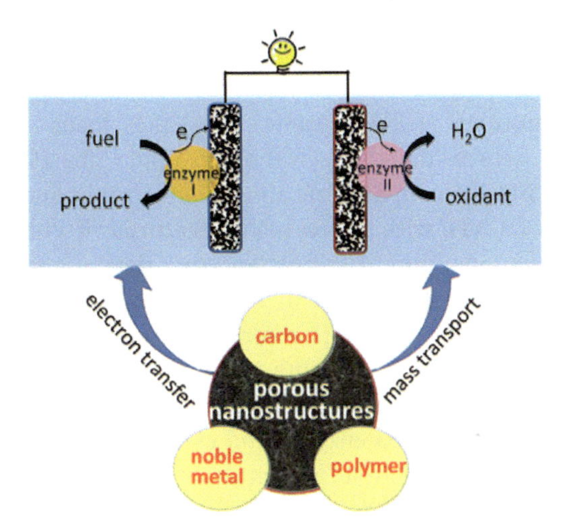

Figure 18.1 Scheme of the EBFC promoted by porous nanostructures from carbon, noble metal, and polymer.

Three parameters, which are commonly used to quantify the performance of EBFCs, are the open circuit voltage (OCV), the maximum power density (P_{max}), and the operational stability. The OCV of an EBFC is determined by the onset potential of the anodic reaction and cathodic reaction; the realization of DET in EBFCs generally leads to high OCV. The P_{max} is determined as a product of the maximum catalytic current density and its corresponding potential, which will benefit from high catalytic currents on both bioelectrodes. Rather than giving a comprehensive review of the topic, this chapter will provide an overview of the recent advances in EBFCs enhanced by porous nanoarchitectures over the past 5 years, from the aspects of their performance and potential applications. It will mainly consider porous structures from carbon materials including carbon nanotubes, graphene, and porous carbon, as well as noble metal and polymeric materials used as electrode materials and enzyme immobilization matrices to develop high performance EBFCs as displayed in Fig. 18.1. In addition, some key issues on how these nanostructured porous media improve the performance of EBFC will be discussed.

18.2 Porous Carbon Nanostructures for EBFCs

18.2.1 Carbon Nanotubes

Carbon nanotubes (CNTs) came into the focus of interest immediately since their discovery by Iijima in 1991 (Iijima, 1991). Specifically, coupling the unique electronic and structural properties of CNTs with the catalytic features of enzymes, thus offers tremendous opportunities for the development of high performance EBFCs (Feng and Ji, 2011; Babadi et al., 2016). Not only they can be used as a support for a high loading and dispersion of enzymes on the electrode due to their ability to assemble in large porous and conductive networks, but also they facilitate electron transfer between enzyme and electrode via a MET and/or DET way. Dong et al. first presented work on the fabrication of a glucose/O_2 EBFC with a pH-dependent power output based on GOx and laccase entrapped CNTs bioelectrodes coupled with a porous carbon matrix (Liu et al., 2005). The P_{max} was around 99.8 (pH 4.0), 14.75 (pH 5.0), 7.94 (pH 6.0), and 2.0 μW cm^{-2} (pH 7.0). And there is no change in the magnitude of cell voltage and current during continuous operation of the EBFC for 3 h. However, a perfluorinated membrane is required to separate the anodic and cathodic compartments due to the free dispersed mediators. Thereafter, Mao and co-authors demonstrated a single-walled CNTs-based one-compartment glucose/O_2 EBFC with a high OCV of 0.8 V (Yan et al., 2006). The CNTs played a role as a support for an anodic biocatalyst (glucose dehydrogenase, GDH) as well as redox mediator for the oxidation of glucose, and also for facilitating the DET of the cathodic catalyst (laccase) for the reduction of O_2 that enables the high OCV. The power output of the cell is only 9.5 μW cm^{-2}, which is probably due to the slow mass transport at the planar electrodes and the slower rate of the DET of laccase compared with the transfer shuttled by redox mediators.

Since then, research on CNTs-based EBFCs has been flourishing to improve the operational performance (Yan et al., 2007; Liu and Dong, 2007; Minteer et al., 2012; Cosnier et al., 2014). Most of the cases involve a simple dispersion of CNTs on planar electrode surfaces, which form quite dense structures and limit the electron transfer and especially the mass transport. This results in the poor power density of the early CNTs-based EBFC configurations. It has been proved that the fabrication of even more porous electrode materials of CNTs is efficient for powerful EBFC design. For example, as a new member of the CNT family, single-walled carbon nanohorn have been employed to build EBFCs with an improved power density and stability (Wen et al., 2011; Wen, Deng et al., 2011). The enhanced performance could be attributed to a considerable capacity of micropores and a little mesoporosity originating from the hexagonal stacking (Inagaki et al., 2004; Murata et al., 2000). In addition, 3D porous CNTs bioelectrodes prepared via a layer-by-layer (LBL) assembly technique were used to construct EBFCs (Deng et al., 2008a; Hyun et al., 2015; Deng et al., 2010). In one case, a polyelectrolyte-GOx/CNTs bioanode via a LBL route showed a power density of 1.34 mW cm^{-2} when assembled into a glucose EBFC (Hyun et al., 2015). In the other case, a fast and simple self-powered biosensor based on glucoe/O_2 EBFC of multilayered CNTs/enzymes bioelectrodes for cyanide detection was developed (Deng et al., 2010).

Engineering CNTs into highly porous nanoarchitectures as the immobilization matrix for enzymes have been proven to be an efficient way to enhance the bioelectrocatalysis and further the performance of EBFCs. For example, Mano et al. demonstrated a high-power EBFC utilizing the newly engineered porous microwires comprising assembled and oriented CNTs, which overcome the limitations of small dimensions and have large specific surface areas (Gao et al., 2010). The poor electron transfer and the slow mass transport of substrates in the biosystem were greatly improved and subsequently the glucose/O_2 cell exhibited an OCV of 0.83 V and P_{max} of 740 μW cm^{-2}. This value is more than tenfold higher than the power density obtained for a carbon fiber EBFC. After 150 h of operation, the cell only lost 20% of its initial power density. Later, the CNT sheets structure was coated by using poly(3,4-ethylenedioxythiophene) via vapor-phase polymerization to produce a highly oriented and porous yarn electrode into a BFC textile. A membrane-free EBFC provided an OCV of 0.70 V and a P_{max} of 2.18 mW cm^{-2}, as shown in Fig. 18.2 (Kwon et al., 2014). Its operation in human serum also generated a high areal power output, as well as a markedly increased lifetime (83% remained after 24 h). With the immobilization of the interconnected enzyme and the redox mediator in a highly conductive, the porous bioelectrode maximized their interaction with the electrolyte and minimized the diffusion distances for fuel and oxidant, thereby enhancing the power density.

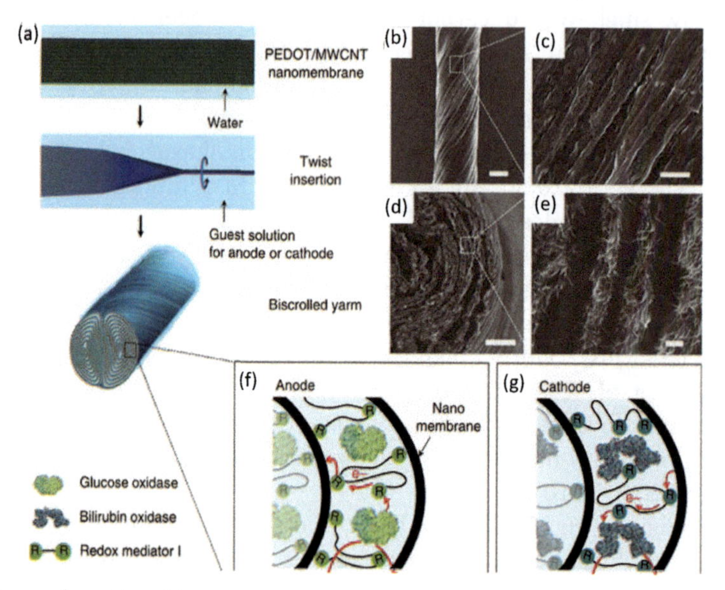

Figure 18.2 The fabrication and structure of biscrolled yarn electrodes for the glucose/O_2 EBFC. The scale bar for (b–e) is 20 μm, 500 nm, 5 μm and 500 nm, respectively. Reproduced with permission (Kwon et al., 2014). Copyright 2014, Nature Publishing Group.

An efficient wiring of enzymes in a porous matrix of CNTs with a high electrical conductivity for the fabrication of glucose EBFCs via mechanical compression was demonstrated by Cosnier group (Zebda et al., 2011; Zebda et al., 2013; Reuillard et al., 2013; Reuillard et al., 2015). This kind of CNT electrodes ensured a good diffusion for the enzyme substrate and the electrical connection of a large amount of entrapped enzymes. Moreover, the compression of the enzyme nanotube mixture likely favored a close proximity at the nanoscale between nanotube and prosthetic sites of the enzymes, leading thus to DET without any loss of activity. The mediator-less EBFC delivered a high power density up to 1.3 mW cm^{-2} and an OCV of 0.95 V. It remained stable for 1 month and displayed 1 mW cm^{-2} power density under physiological conditions. Taking advantage of this improved performance, with respect to both power density and operational stability, the first EBFC implanted in an animal (namely a rat) was developed to generate electric energy from a mammal's body fluids, powering electronic devices like a light-emitting diode (LED) or a digital thermometer (Fig. 18.3). No signs of rejection or inflammation were observed after 110 days of implantation in the rat. Very recently, 1-year stability for a glucose/O_2 EBFC combined with a pH reactivation of the compressed laccase/CNTs biocathode was presented by the same group (Reuillard et al., 2015).

Compared to CNT-based films prepared by other methods, the buckypaper, a self-supporting carbon mat of entangled assemblies (ropes and bundles) of CNTs, is highly porous, easily handling, flexible, mechanically stable and electrically conductive (de Heer et al., 1995; Hussein et al., 2011a). When it was used as a laccase-catalyzed biocathode, a 68-fold higher current density for the O_2 reduction reaction via a DET way was achieved than that from the same CNTs in a non-dispersed agglomerated state as a packed electrode (Hussein et al., 2011b). This is due to the reduced diffusional mass transfer limitation and enhanced electrical conductivity. A freestanding redox buckypaper electrode from multi-walled CNTs for bioelectrocatalytic oxygen reduction via MET reached a maximum current of 2 mA ± 70 μA and an excellent operational stability for two weeks with daily 1 h discharges (Bourourou et al., 2014). Employing the bioelectrodes based on buckypaper modified with PQQ-dependent GDH and/or FAD-dependent fructose dehydrogenase on the anode and with laccase on the cathode, Katz and co-workers developed a series of implanted or portable EBFC devices. They were implanted in a rat or a plant and powered a pacemaker or a wireless transmission system (Castorena-Gonzalez et al., 2013; Southcott et al., 2013; Mac Vittie et al., 2015).

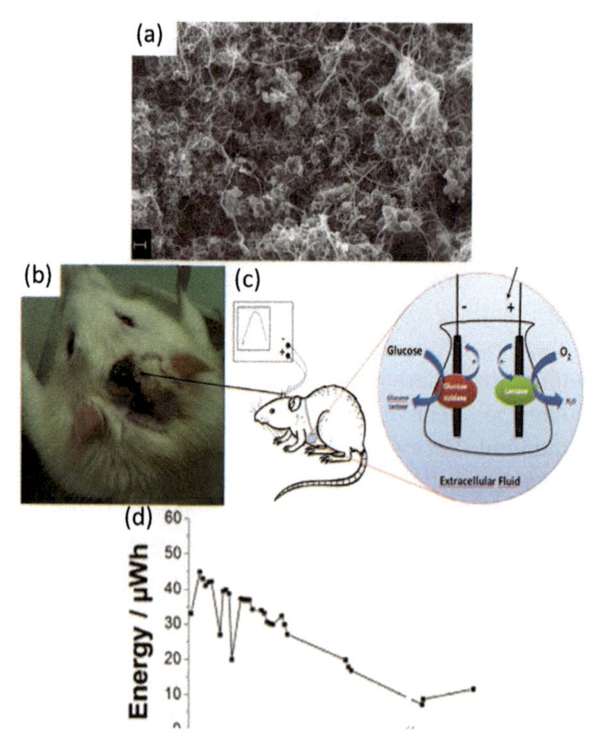

Figure 18.3 (a) SEM micrograph of a cross section of the bioanode. The scale bar corresponds to 200 nm. Reproduced with permission (Zebda et al., 2011). Copyright 2011, *Nature* Publishing Group. (b, c, Schematic representation of a biocompatible EBFC which is implanted in the abdominal cavity of a rat. Reproduced with permission (Zebda et al., 2013). Copyright 2013, Nature Publishing Group. (d) Evolution of the power delivery of the glucose/O_2 EBFC during one year. Reproduced with permission (Reuillard et al., 2015). Copyright 2015, Elsevier Publishing Group.

In addition to the utilization of pristine or simply acid treated CNTs, modification of CNTs is a promising approach to achieve high performance of EBFCs. Covalent functionalization of CNTs with different aromatic groups helps to orient the active sites of multicopper oxidases (MCOs) like laccase and BOD to allow for DET. For example, hydroxylated CNTs were covalently modified with anthracene groups, which led to a significantly enhanced direct electrocatalytic reduction of O_2 by laccase. The EBFCs coupled with a mediated GOx anode and a DET fructose dehydrogenase anode produced a P_{max} of 56.8 and 34.4 μW cm^{-2} respectively (Meredith et al., 2011). Non-covalent methods have also been especially developed for the immobilization of MCOs using pyrene moieties that form π–π interactions with the CNT sidewalls (Lalaoui et al., 2013; Lalaoui et al., 2015; Lalaoui et al., 2016). CNTs modified with oxidative electropolymerization of pyrrole monomers bearing pyrene and *N*-hydroxysuccinimide groups interact with a hydrophobic cavity of laccase for the oriented enzyme immobilization. The resulting laccase-based biocathode showed higher catalytic current (1.85 mA cm^{-2}) and higher stability (50% after one month) for O_2 reduction. Later, a site-specific pyrene-modified laccase and CNTs/Au NPs supramolecular assemblies acted as a biocathode and achieved a highly efficient direct reduction of O_2 at low overpotential accompanied by high catalytic current densities of almost 3 mA cm^{-2}. The modification of CNT surface helps to realize both immobilization and orientation of the enzyme on electrodes to favor DET. Up to now, most of the research has been focused on the MCOs biocathode, while there are few reports on the bioanode DET via anodic biocatalysts orientation.

Nowadays, due to the unique electronic properties and compatibility with biocatalysts, CNTs are the most common nanomaterials for bioelectrode fabrication in EBFCs. The highly porous CNT nanostructures are used to immobilize enzymes by different approaches like physical adsorption, covalent linkage, mechanical compressing, or π-stacking. They dramatically enhanced the enzymatic electrocatalysis through wiring the enzyme active site with the electrode to improve the DET rate or offering a large electrochemically active surface area for immobilizing mediators to improve MET. Coupled with the accelerated mass transport from open pores, the incorporation of porous CNTs into bioelectrodes has been a powerful strategy to enhance the performance of EBFCs and explore their further applications in portable power sources, self-powered electrochemical sensing, implantable medical devices, etc.

18.2.2 Graphene

The wealth of applications and dramatic improvements in the performance of EBFCs by CNTs has guided researches toward other carbon nanomaterials, such as graphene. In the past decade, a success of graphene-based biosensors implies the fact that biocatalysts can be very effectively integrated with this nanomaterial for EBFC fabrication (Liu et al., 2012). Graphene exhibits many functional properties that can be used to facilitate direct or mediated electrical contact between the redox site of the enzyme and electrode surfaces, and finally to develop high-power EBFCs (Ravenna et al., 2015; Filip and Tkac, 2014; Karimi et al., 2015; Jaafar et al., 2015). Li et al. designed an EBFC consisting of co-immobilized graphene-GOx in a silica sol–gel matrix as the anode and graphene-BOD as the cathode (Liu et al., 2010). It showed a P_{max} of about 24.3 μW cm^{-2}, which is nearly two times larger than that of the CNT based system, with the performance lasting for 7 days. In another report, a nanographene platelet-based glucose/O_2 EBFC with GOx as the anodic biocatalysts and the laccase as the cathodic biocatalysts was developed with a P_{max} of ca. 57.8 μW cm^{-2} (Zheng et al., 2010). These two examples point out the successful construction of graphene-based EBFC, however, the power output cannot complete with the most powerful one, which seems to be due to an insufficient 3D structure of graphene surfaces. Compared to its 2D configuration, the 3D assembly of graphene can create larger accessible specific surface areas, interconnected conductive networks, and a superior microenvironment (Chen et al., 2011). The endowed high porosity and diffusion characteristics improved the enzyme bioelectrocatalysis as well as the power density of the EBFCs (Gao and Duan, 2014; Song et al., 2015; Quian and Lu, 2014; Jaafar et al., 2016). A 3D graphene network synthesized with a Ni^{2+}-exchange/KOH activation combination method were used as a substrate for the immobilization of laccase and mediator (Fig. 18.4a). Due to the interconnected network structure and high surface area of the 3D graphene, a high catalytic activity for O_2 reduction was facilitated. A glucose/O_2 EBFC combined with a GOx anode can output a P_{max} of 112 μW cm^{-2} (Zhang et al., 2014).

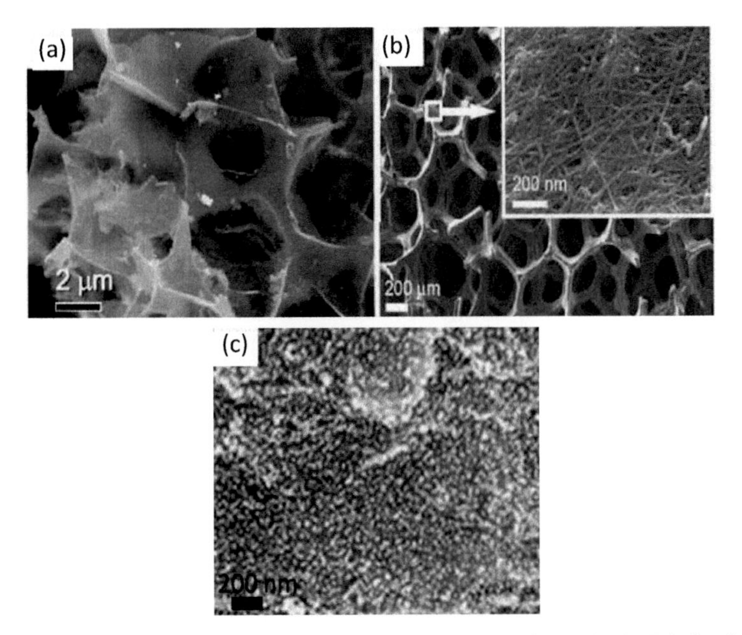

Figure 18.4 SEM images of 3D graphene (a) 3D graphene-SWCNT hybrids (b), and graphene-Au NPs hybrid (c). Reproduced with permission (Zhang et al., 2014; Prasad et al. 2014; Gai et al., 2015). Copyright 2014, American Chemical Society. Copyright 2015, Royal Society of Chemistry.

Interestingly, there are contributions that combine other nano-objects like CNTs and metal nanoparticles (NPs) to construct 3D porous structures of graphene, further enhancing the enzymatic electrode kinetics and improving the power density of EBFCs. With 3D graphene/CNTs cogel electrodes, a membrane/mediator-free rechargeable EBFC was reported. The enhanced power density is attributed to the high available surface area and porosity of the electrode material allowing for high loading of active enzymes and ease of glucose diffusion through the cogel-based electrode (Campbell et al., 2015). In addition, 3D graphene was covered inside out by a dense thin-film network of CNTs with a mesh size comparable to a macromolecule (Fig. 18.4b). A glucose/O_2 EBFC equipped with enzyme-functionalized 3D graphene-CNT hybrid electrodes can nearly attain the theoretical limit of the OCV (~1.2 V) and the highest power density ever reported (2.27 ± 0.11 mW cm^{-2}), together with high stability

(Prasad et al., 2014). Zhang and Zhu et al. reported on graphene–Au NPs hybrids (Fig. 18.4c) to support biocatalysts (Gai et al., 2015; Chen et al., 2015b; Gai et al., 2016). The as-assembled EBFCs reached a P_{max} of 1.96 mW cm^{-2}, which can light up red and yellow LEDs. The cell retained 66% of the output after 70 days.

Up to now, graphene has been applied in EBFCs in a similar way as CNTs, while it is still far from being fully exploited. Taking the advantages of potential merits of graphene, especially commercial benefits, high performance and low cost of the graphene-based EBFCs would be reachable goals.

18.2.3 Porous Carbon Derived 3D Nanostructures

Besides CNTs and graphene, other carbon materials such as carbon paper, carbon fiber, and carbon black can further be manipulated to form 3D porous structures that allow for high enzyme loading and connectivity with an ease of substrate diffusion into the network in the EBFC development (Barton et al., 2007; Tsujimura et al., 2003; Tamaki and Yamaguchi, 2006; Tamaki, 2012; Shitanda et al., 2013). Porous carbon nanostructures with controlled porosity, high surface area and good electrical conductivity draw much attention in the application of EBFCs. The performance characteristics of the recently developed EBFCs based on porous carbon nanoarchitectures are described in the following part.

The use of mesoporous carbon NPs (CNPs, <500 nm in diameter, pore dimensions ~6.3 nm) as a functional material to electrically contact redox proteins with electrodes was introduced by Willner group (Trifonov et al., 2013; 2015). The high surface area and the conducting properties of mesoporous carbon NPs are implemented to design electrically contacted enzyme electrodes for EBFC applications. The mediators were trapped inside the pores by means of redox enzyme caps, resulting in integrated bioelectrocatalytic assemblies. That is, efficient electron transfer was realized with the turnover rate of electrons between the enzymes and the electrodes to be 995 electrons s^{-1} for the GOx-capped ferrocene methanol-loaded CNP anode for the oxidation of glucose and 995 electrons s^{-1} for the BOD-capped 2,2′-azinobis(3-ethylbenzothiazoline-6-sulfonic acid-loaded CNPs as an O_2 reduction cathode, respectively. Thus, a glucose/O_2 biofuel cell yields a power output of ~95 μW cm^{-2}. Japanese researchers demonstrated a stable generation of electricity of a glucose-powered mediated EBFC based on porous carbon particles originating from rice husks through multiple refueling cycles (Fig. 18.5a, Fujita et al., 2014). The bioanode can be refueled continuously for more than 60 cycles at 1.5 mA cm^{-2} without significant potential drops. The whole cells can be repeatedly used to power a portable music player at 1 mW cm^{-3} through 10 refueling cycles. This refuelability is attributed to the immobilized electron transfer mediator and redox enzymes in high concentrations on porous carbon particles, which maintained their electrochemical and enzymatic activities.

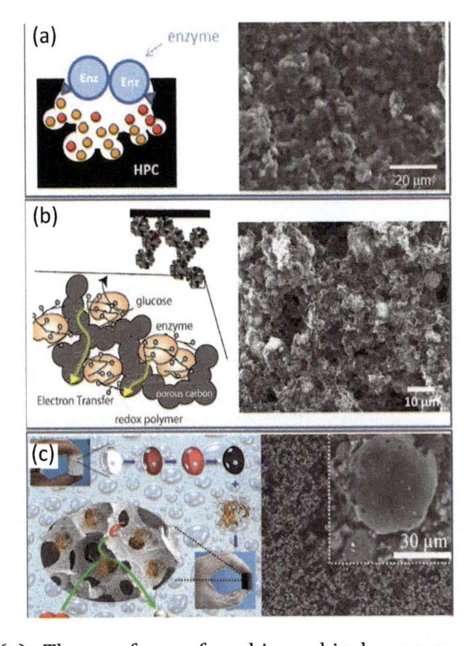

Figure 18.5 (a) The surface of a hierarchical porous carbon (from carbonized rice husks) anode was imaged by SEM. Reproduced with permission (Fujita et al., 2014). Copyright 2014, Nature Publishing Group. (b) Schematic illustration of a hierarchical porous carbon (by the EPD method) based glucose bioanode (left) and its SEM image (right). Reproduced with permission (Tsujimura et al., 2014). Copyright 2014, American Chemical Society. (c) Photos and SEM images of carbonaceous micro/macrocellular foams from hard macrocellular silica Si(HIPE) template. Reproduced with permission (Flexer et al., 2011). Copyright 2011, Royal Society of Chemistry.

A hierarchically structured porous carbon electrode was prepared by the electrophoretic deposition (EPD) of magnesium oxide-templated mesoporous carbon (mean pore diameter of 38 nm) (Tsujiruma et al., 2014). Thanks to macropores (for an efficient mass transport) and mesopores (for high enzyme loading), the resulting glucose bioanode obtained a current density for the oxidation of glucose of 100 mA cm^{-2} at 25°C and pH 7 (Fig. 18.5b). This was 33 times higher than that on a flat carbon electrode at the same biocatalyst composition and loading. The efficient diffusion of the glucose is due to the relatively thin carbon layer (<0.1 mm) and its macroporous structure. Furthermore, the stability of the enzyme electrode was improved by using mesoporous carbon materials, and more than 95% of the initial catalytic current was remained after 220 days.

Mano et al. showed that carbonaceous micro/macrocellular foams can be used for efficient and stable non-specific enzyme entrapment (Flexer et al., 2010, 2011). 3D carbonaceous electrodes with interconnected hierarchical porosity were prepared using Si(HIPE) exotemplating matrices, and were modified with GOx and mediator (i.e., Os polymer). The glucose electrooxidation current was 13-fold larger on the porous electrode than on glassy carbon for the same enzyme loading. When BOD was adsorbed into the porous electrode, the direct reduction of O_2 to water provided a dramatic increase in enzyme loading which allowed for a 500-fold current enhancement and stabilization of the DET current from few hours to several days as compared to conventional flat electrodes.

The compacted electrodes from graphitized mesoporous carbon (GMC) nanopowder were used as an inexpensive method for obtaining a large increase in productive enzyme loading, greatly increasing current densities and stability (Xu and Armstrong, 2013, 2015). Operated under non-explosive H_2-rich air mixtures at 25°C, typical power density from a H_2/O_2 EBFC at a stationary hydrogenase anode and a BOD cathode exceeded 1 mW cm^{-3}. Good prospects for stability were demonstrated by the fact that 90% of the power was retained after continuously working for 24 h, and more than half of the power was retained after one week of non-stop operation. The 3D porous electrodes worked by greatly increasing the catalysts loading (at both the anode and the cathode) and selectively restricting the access of O_2 (relative to H_2) to enzymes embedded in pores at the anode. In addition, there is a report on GOx nanocomposites by entrapping cross-linked GOx aggregates within a GMC network. The GOx-GMC offer a high electrical conductivity and electron transfer rate. Furthermore, they maintained 99% of the initial activity after thermally treatment of at 60°C for 4 h, suggesting the high stability (Garcia-Perez et al., 2016).

18.3 Porous Noble Metal Nanoarchitectures for EBFCs

Due to large specific surface areas, high porosity, intrinsic conductivity, and biocompatibility, porous noble metal (such as Au, Ag, Pt, Pd) nanostructures can yield rapid electron and mass transport pathways for bioelectrocatalysis (Deng et al., 2008b; Zu et al., 2015; Chen et al., 2012). Enzymes conjugated with the nanostructured metal surface inside/outside the pores retain their natural conformation and hence the activity due to the comparable dimensions. In this respect, they are found significant interest for EBFC applications, as they are likely to improve the key issues of lifetime and power density. Even though there are some results on the utilization of nanoporous Au (Hakamada et al., 2012; Salaj-Kosla et al., 2013; Hou et al., 2014; du Toit and Lorenzo, 2015), 3D porous metal nanomaterials via a bottom-up approach of noble metal NPs have been more widely applied in the EBFC configuration. In the following, some of the attractive features of porous noble metal nanoarchitectures from either the NPs assembly or NPs supported on porous templates are discussed.

18.3.1 Porous Template–Assisted Metal NP Nanoarchitectures

Incorporation with porous templates such as carbon, metal, metal oxide, polymer, and even bacteria to assist noble metal NPs, are advantageously exploited for porous noble metal nanoarchitectures in the EBFC development. A DET-based laccase biocathode for O_2 electroreduction at low overpotentials was achieved via a step-by-step covalent attachment of Au NPs to porous graphite electrodes, as shown in Fig. 18.6a. Oriented immobilized laccase molecules that were efficiently wired by the Au NPs showed a very fast DET with a heterogeneous electron transfer rate constant k_0 >> 400 s^{-1} (Guitérez-Sánchez et al., 2012) The same group reported on 3D nanostructured microscale Au electrodes with an electrochemically driven transformation of physically deposited Au NPs modified with suitable biocatalysts (Andoralov et al., 2013; Falk et al., 2013a). A mediator-, cofactor-, and membrane-less EBFC operated in cerebrospinal fluid and in the brain of a rat, producing amounts of electrical power sufficient to drive a self-contained bio-device. It is also capable of generating electrical energy from human lachrymal liquid by utilizing the ascorbate and O_2 naturally present in tears as fuel and oxidant, with a stable current density output of over 0.55 μA cm^{-2} at 0.4 V for 6 h of continuous operation.

A high power and stable membrane-less, mediator-free glucose EBFC with ultrathin configuration was achieved using Ag NP-functionalized hierarchical mesoporous titania thin-film electrodes, where GOx and laccase were immobilized and served as anodic and cathodic catalyst, respectively (Bellino and Soler-Illia, 2014). The mesoporous silica film facilitated the mass transport of products and reactants to/from the electrodes while the Ag NPs enhanced the electrical connection between the biocatalyst and the thin layer electrode. A P_{max} of 602 μW cm^{-2} at 0.68 V was delivered; only about 10% loss in voltage output was observed after continuous operation for 30 h. In addition, with cellulose templated Au and Pt NPs, a 3D origami-based EBFC was demonstrated for self-powered, low-cost, and sensitive biosensing (Wang et al., 2014a; Wang et al., 2014b). Another interesting example is the design of a bioanode consisting of GOx covalently attached to Au NPs that are assembled onto a genetically engineered M13 bacteriophage, a high-surface area template. The resulting "nanomesh" architecture exhibited a DET and achieved a higher current density of 1.2 mA cm^{-2} toward glucose oxidation, compared to most other DET attachment schemes (Blaik et al., 2016).

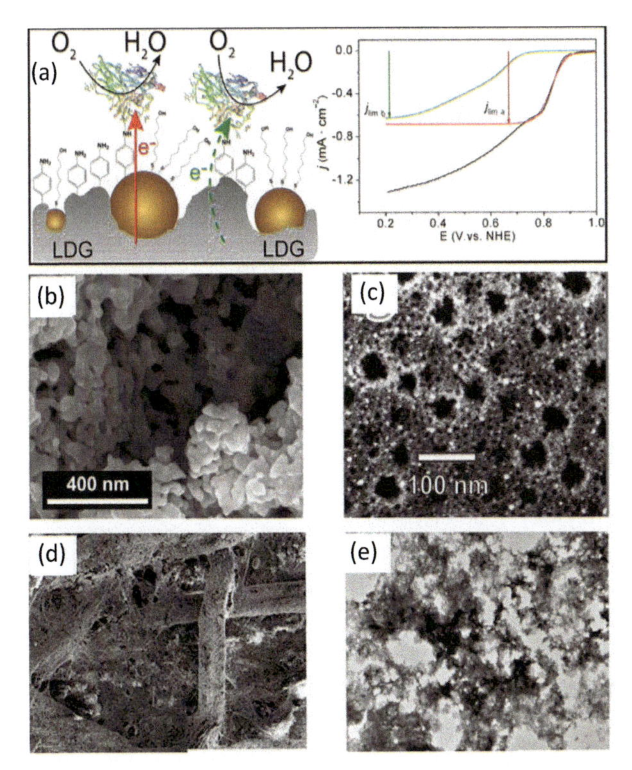

Figure 18.6 (a) Au NPs modified porous low-density graphite (LDG) for DET-based laccase bioelectrodes. Reproduced with permission (Gutiérez-Sánchez et al., 2012). Copyright 2012, American Chemical Society. SEM images of AuNPs modified microelectrodes (b: reproduced with permission (Andoralov et al., 2013). Copyright 2013, Nature Publishing Group), hierarchical mesoporous film after infiltration 701 with Ag NPs (c: reproduced with permission (Bellino and Soler-Illia, 2014) Copyright 2014, Wiley-VCH), AuNPs on the surfaces of the cellulose fibers in PAE after 10 min of growth (d: reproduced with permission (Wang et al., 2014). Copyright 2014, Wiley-VCH), and Au NPs-Coated M13 Bacteriophage (e: reproduced with permission (Blaik et al., 2016). Copyright 2016, American Chemical Society).

18.3.2 Porous Structures from Metal NPs

The use of noble metal NPs as building blocks for 3D porous architectures can be considered as an attractive option for an enzyme loading matrix and current collector in EBFCs. For example, similarly to CNTs, electrostatic LBL assemblies of metal NPs and enzymes have been applied in the construction of the porous bioelectrodes for the EBFC (Deng et al., 2008; Yan et al., 2009). In addition, Willner et al. developed 3D bioelectrodes via electropolymerization of biocatalysts and metal NPs, where both enzyme and the particles were modified with electropolymerizable units, namely thioaniline (Yehezkeli et al., 2011; Lesniewski et al., 2010). The biocathode consisting of a composite of BOD and Pt NPs revealed an effective electrocatalyzed reduction of O_2. GDH and Au NPs were assembled on a roughened Au electrode for the bioelectrocatalyzed oxidation of glucose with a high turnover rate of electrons of ca. k_{et} = 1100 electrons s^{-1}. The accelerated bioelectrocatalysis reduction and oxidation together with the power output were driven by the 3D conductive hybrid enzyme/NPs network.

As a new emerging porous metal nanostructure for bioelectrocatalysis, noble metal aerogels derived from NPs via a controlled destabilization of the NP colloid exhibit high surface areas, through connected porosity distributed from the micro- to the macro- pore size range, and retained metrics of the NPs (Fig. 18.7) (Liu et al., 2015; Wen et al., 2014a; Wen et al., 2014b). The Eychmüller group reported on the controllable synthesis of Pd aerogels with high surface area and porosity by destabilizing colloidal solutions of Pd NPs with variable concentrations of calcium ions. Compared to glassy carbon and Pd NPs, enzyme electrodes based on Pd aerogels co-immobilized with GOx showed much faster electron transfer kinetics of the mediator and mass transport of the mediator at the electrode surface (Wen et al., 2014a). Subsequently, a Pd aerogel-based EBFC was developed, in which the ferrocene (Fc)-coupled Pd aerogel not only provided a 3D porous support for the biocatalyst, but also mediated the bioelectrocatalytic oxidation of glucose as an integrated bioanode. BOD encapsulated into a Pd–Pt alloy aerogel realized the direct electrocatalytic reduction of O_2 at the biocathode (Wen et al., 2014b). Further attempts to enhance the power output of the EBFC by improving the GOx anode performance through the use of mediators with low redox potential are currently under way.

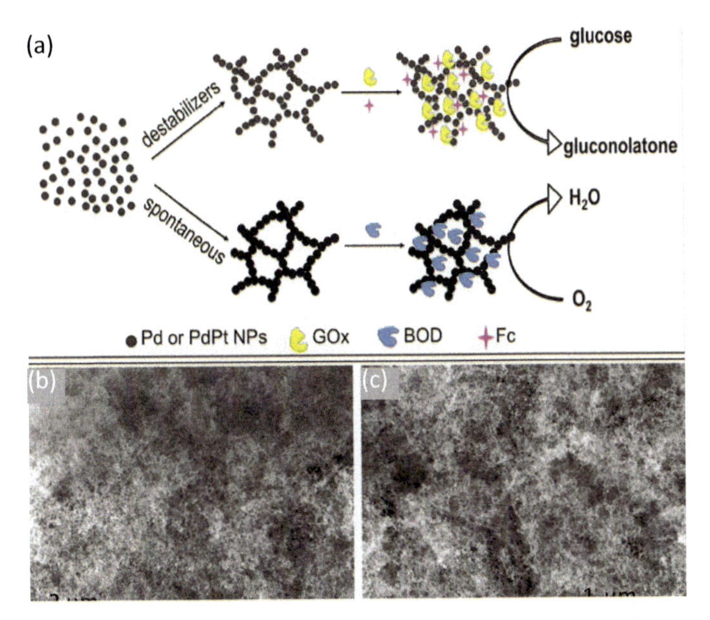

Figure 18.7 (a) Schematic illustration of the Pd-based aerogel formation and corresponding 709 bioelectrodes for the Pd aerogel based EBFC. (b, c) Electron microscopy images of the Pd-Fc and Pd-Pt aerogel. Reproduced with permission (Wen, Liu et al., 2014). Copyright 2014, Wiley-VCH.

18.4 Polymeric Nanostructured Matrix for EBFCs

Porous nanostructures from polymeric materials have also attracted attention in EBFC investigations. Nafion, biopolymers (chitosan, poly(lactic acid), etc.), conducting polymers (polyaniline, polypyrrole, poly(3,4-ethylenedioxythiophene), etc.), and composite polymers have been widely used in EBFCs (Vaghari et al., 2013; Yang et al., 2012; Filip et al., 2014). Rather than the functionalities of polymeric redox mediators or ion-selective membranes, here it mainly focuses on their utilization for the construction of bioelectrodes in EBFCs.

The biocompatible polymer-based 3D electrodes exhibiting multidimensional and multidirectional pore structures, which are abundant, biodegradable, and cost-effective, are possible solutions to improve the performance of EBFCs for further biological application. One of the important examples is the chitosan scaffold which was used to fabricate an enzymatic electrode that oxidized glucose and produced electrical current more effectively than the same electrode made of a chitosan film (Cooney et al., 2008a; Lau et al., 2010). In addition, the rapid oxidation of dopamine or L-noradrenaline by $K_3Fe(CN)_6$ yields a catecholamine polymer with GOx entrapped effectively. Such an enzyme-entrapped polydopamine was applied as a bioanode for EBFC which showed a P_{max} of 1.62 mW cm^{-2} (Chen et al., 2011). Conducting polymers with high electrical conductivity and excellent inherent environmental stability have been used as immobilization matrix for enzymes and mediator compounds (Wang et al., 2011; Latonen et al., 2012). For instance, GOx immobilized in the porous matrix of polyaniline nanofibers improved both activity and stability. Combining an air-breathing cathode, the EBFCs delivered a P_{max} of 292 μW cm^{-2} and remained stable for 2 months of storage and 4 h of operation at 60°C (Kim et al., 2011).

Recently, it has been shown that the integration of nanosized conductive components such as carbons and metals, increase the conductivity of polymers for the facility of electron transfer in EBFCs as displayed in Fig. 18.8 (Liu et al., 2011; Ichi et al., 2014; Dai et al., 2015). The incorporation of chitosan into CNTs achieved a significant increase in longevity and stability of laccase-based biocathodes, namely a stable current response during 2 months. The absence of enzyme inhibition over time could be attributed to the protection of the enzyme by the micro-environment created by the porous 3D-polymer matrix which also prevented the enzyme release in the solution and provided good O_2 diffusion (Ichi et al., 2014). L-3,4-dihydroxyphenylalanine (L-DOPA) was chemically and electrochemically synthesized and used as an in situ enzyme-immobilization matrix for the EBFC application. A GOx-bioanode was prepared through a chemical oxidative polymerization of L-DOPA by Au precursors in the presence of GOx. The enzyme electrode based on this polymer-NP hybrid matrix showed an OCP of ca. 1.0 V and a high P_{max} of 2.62 mW cm^{-2}, when assembled into a glucose/O_2 EBFC (Dai et al., 2015).

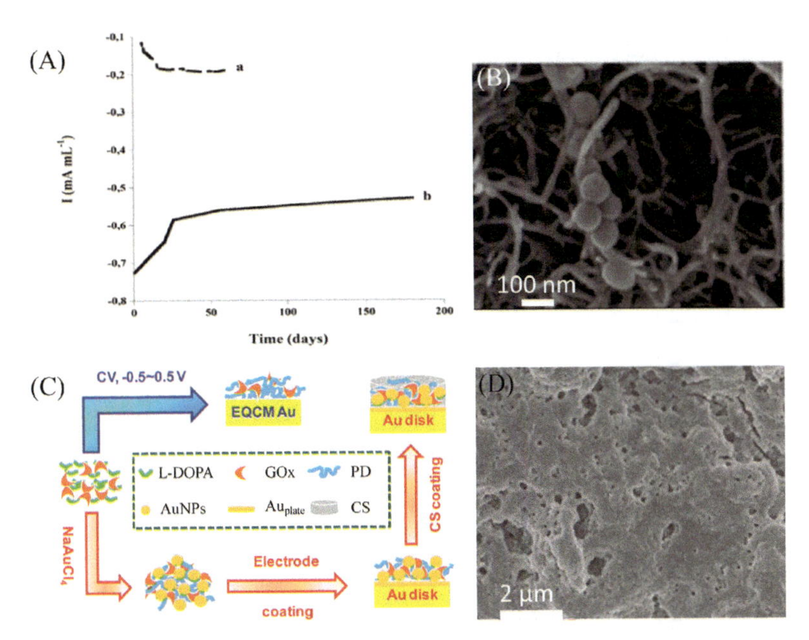

Figure 18.8 (A) Stability of the chitosan–CNTs–laccase cathode under continuous discharge (a) and storage conditions (b). (B) SEM image of a cross section of chitosan–CNT–laccase. Reproduced with permission (Dai et al., 2015). Copyright 2014, Royal Society of Chemistry. (C) Illustration of GOx immobilization into the Au-poly (L-DOPA) matrix as a bioanode. (D) SEM image of the CS/GOx-PDC-AuNPs/Au_{plate}/Au electrode surface. Reproduced with permission (Ichi et al., 2014). Copyright 2015, American Chemical Society.

18.5 Summary

Obviously, the introduction of porous nanostructures in EBFCs enabled the development of highly active and stable enzyme bioelectrodes, and thus improved the performance of the cells (Table 18.1). Due to the novel characteristics like high surface area, good accessibility, and high porosity, porous nanostructured materials with good conductivity have several benefits. First, all such materials play an important role as 3D immobilization matrix, one of the key issues in the EBFC development: (1) increasing the enzyme loading with respect to the amount and efficiency due to the extremely high surface area and comparable

Table 18.1 The operational characteristics of EBFCs based on different porous nanostructures

Porous matrix	Catalysts*	Fuel/oxidant	Operation conditions	OCV (V)	P_{max} ($mW\ cm^{-2}$)	Stability	Refs.
CNTs film via LBL	GDH/Lac	glucose/O_2	PBS (pH 6.5)	0.62	0.30	—	Hyun et al., 2015
CNTs film via LBL	GOx/Pt	Glucose/O_2	PBS (7.4)/air with membrane	—	1.34	—	Deng et al., 2008
CNTs fibers	GOx/BOD	Glucose/O_2	PBS + NaCl (7.2) at 37°C	0.83	0.74	80% after 150 h	Gao et al., 2010
CNT yarns	GOx/BOD	Glucose/O_2	PBS + NaCl (7.2) at 37°C	0.70	2.18	83% after 24 h	Kwon et al., 2014
Compressed CNTs	GOx-Cat/Lac	Glucose/O_2	PBS (7.0)	0.95	1.3	96% after 30 days storage	Zebda et al., 2011
Buckypaper	GDH/Lac	Glucose/O_2	PBS (7.4)	0.6	0.09	—	McVittie et al., 2015
Anthracene-modified CNTs	GOx/Lac	Glucose/O_2	PBS (7.0)/citrite buffer (4.5) with membrane	0.82	0.057	—	Meredith et al., 2011
	FDH/Lac	fructose/O_2	Citrate buffer (pH 4.5)	0.71	0.034		
3D graphene network	GOx/Lac	Glucose/O_2	Acetate buffer (pH 5.0)	0.4	0.121	84% after 72 h	Zhang et al., 2014

Porous matrix	Catalysts*	Fuel/oxidant	Operation conditions	OCV (V)	P_{max} (mW cm^{-2})	Stability	Refs.
Graphene/Au NPs hybrid	FMDH/Lac	Formic acid/O_2	PBS (pH 6.0)	0.95	1.96	80% of OCV after 20 days	Gai et al., 2015
3D Graphene/CNT network	GOx/Lac	Glucose/O_2	Acetic acid buffer (pH 5.0) with membrane	1.2	2.27	80% of OCV after 30 days	Prasad et al., 2014
Mesoporous carbon NPs	GOx-LOx/BOD-Cat	glucose-lactate/O_2-H_2O_2	HEPES buffer (pH 7.0)	—	0.09	—	Trifonov et al., 2015
Porous carbon particle	GDH/BOD	Glucose/O_2	PBS (7.0)	0.6	1	refueled for 60 cycles	Fujita et al., 2014
Compacted mesoporous carbon	Hydrogenase/BOD	H_2/O_2	PBS (6.0)	0.8	0.42	90% after 24 h	Xu and Armstrong., 2013
Au NPs network	GDH/BOD	Glucose/O_2	PBS (7.4)	0.50	0.032	—	Lesniewski et al., 2010
Pd aerogels	GOx/BOD	Glucose/O_2	PBS (7.0)	0.4	0.02	—	Wen et al., 2014
Au NPs on microscale Au	CDH/BOD	Glucose/O_2	PBS (7.4)	~0.7	0.004	90% of OCV after 2 h	Andoralov et al., 2013

(Continued)

Table 18.1 (*Continued*)

Porous matrix	Catalysts*	Fuel/oxidant	Operation conditions	OCV (V)	P_{max} (mW cm^{-2})	Stability	Refs.
Ag NPs on mesoporous TiO_2 films	GOx/Lac	Glucose/O_2	PBS (7.0)	0.91	0.403	92% after 10 days' storage	Bellino, Soler-Illia, 2014
Catecholamine polymers	GOx/Pt	Glucose/O_2	PBS (7.0)/ $KMnO_4$+H_2SO_4 with membrane	1.09	1.62	—	Chen et al., 2011
Porous polyaniline nanofibers	GOx/Pt	Glucose/O_2	PBS (7.4)/air with membrane	—	0.292	stable for 2 months' storage and 4 h of operation at 60°C	Wang et al., 2011
poly(L-3,4-dihydroxyphenylalanine)	GOx/Pt	Glucose/O_2	PBS (7.0)/ $KMnO_4$ + H_2SO_4 with membrane	1.0	2.62	—	Dai et al., 2015

**Abbreviations*: glucose oxidase (GOx), laccase (Lac), bilirubin oxidase (BOD), fructose dehydrogenase (FDH), formate dehydrogenase (FMDH), lactate oxidase (LOx), catalase (Cat), and cellobiose dehydrogenase (CDH).

dimensions from both the size of nanomaterials and the pore size; (2) providing preferred orientation of immobilized enzymes for DET; (3) immobilizing redox mediators, if needed, without leakage for MET. Second, the roughness and dimensionality of porous nanostructures maximize the availability of electron transfer, allowing a decrease in the distance between the active sites and the electrode conductor to improve electron transfer and therefore increase the currents for both DET and MET. Third, the high porosity with the pores being large facilitates the mass transport in the EBFC system in terms of the permeation of electrolyte, fuels, and oxidant. Finally, because of the pore structure and the nano-dimensions, they can improve the stability in two main ways: (1) prevent leaching of the biomolecule into solution to increase the re-usability of the bioelectrodes and (2) provide a better microenvironment for the biomolecule that helps in retaining its structure and function.

Up to now, significant progress has been achieved in EBFCs based on porous nanostructured materials. The power density increased from μW cm^{-2} to mW cm^{-2} and the stability extended from hours to about 1 year. But still such parameters are too low for a performance in commercial applications. The DET would be advantageous to increase the cell OCV and simplify the setup. However, most of the biocatalysts (i.e., GOx) keep their active sites deep in protein shells, which obstruct the direct bioelectrocatalysis of fuels and oxidants. There are some successful examples based on DET (Falk et al., 2012; Zhao et al., 2025), yet they generally yield low current densities. MET with higher bioelectrocatalytic current is limited by the decreased lifetime due to the instability of cycling redox mediators. To make further progress, the combined efforts from materials scientist and biologist are highly desirable. It is critical to develop new designs of materials with high surface areas and good electrical conductivities to improve the bioelectrocatalytic interface and stabilize the biocatalysts, and to optimize the pore structures of the porous matrix to overcome the block of the mass transport. Furthermore, protein engineering and the utilization of enzyme cascades are promising strategies to develop powerful EBFCs. The former allows for the insertion of specific anchoring sites to control and orient the enzyme on the electrode surface and broadens the enzyme working 434 environment with abundant substrates and broader physiochemical conditions (pH, ionic strength, temperature, etc.) (Gueven et al., 2010; Caruana and Howorka, 2010; Holland et al., 2011). The latter dramatically improve the oxidation efficiency of the fuels to increase the current density of the bioanode (Moehlenbrock et al., 2011; Xu and Minteer 2012; Hickey et al., 2014). For example, an enzyme cascade realized the deep or complete oxidation to CO_2 of glucose (the most common anodic substrate) which is generally involving a two-electron reaction in most glucose/O_2 EBFCs (Xu and Minteer, 2012). In this respect, it is interesting to carry out further research on improving the electrochemical performance and stability of the EBFCs along these lines (combination of materials design and enzymology utilization).

Acknowledgments

This work is supported by the AEROCAT project (ERC-2013-ADG 340419) and the DFG project EY16/10-2. D. W. acknowledges the support from the Alexander von Humboldt Foundation.

References

Andoralov V., Falk M., Suyatin D. B., Granmo M., Sotres J., Ludwig R., Popov V. O., Schouenborg J., Blum Z., Shleev S., *Sci. Rep.*, **3** (2013), 3270.

Babadi A. A., Bagheri S., Hamid S. B. A., *Biosens Bioelectron.*, **79** (2016), 850–860.

Barton S. C., Gallaway J., Atanassov P., *Chem. Rev.*, **104** (2004), 4867–4886.

Barton S. C., Sun Y., Chandra B., White S., Hone J., *Electrochem. Solid State Lett.*, **10** (2007), B96–B100.

Bellino M. G., Soler-Illia G. J. A. A., *Small*, **10** (2014), 2834–2839.

Blaik R. A., Lan E., Huang Y., Dunn B., *ACS Nano*, **10** (2016), 324–332.

Bourourou M., Elouarzaki K., Holzinger M., Agnès C., Goff A. L., Reverdy-Bruas N., Chaussy D., Party M., Maaref A., Cosnier S., *Chem. Sci.*, **5** (2014), 2885–2888.

Bullen R. A., Arnot T. C., Lakeman J. B., Walsh F. C., *Biosens. Bioelectron.*, **21** (2006), 2015–2045.

Campbell A. S., Jeong Y. J., Geier S. M., Koepsel R. R., Russell A. J., Islam M. F., *ACS Appl. Mater. Interfaces*, **7** (2015), 4056–4065.

Caruana D. J., Howorka S., *Mol. Biosyst.*, **6** (2010), 1548–1556.

Castorena-Gonzalez J. A., Foote C., MacVittie K., Halámek J., Halámková L., Martinez-Lemus L. A., Katz E., *Electroanalysis*, **25** (2013), 1579–1584.

Catalano P. N., Wolosiuk A., Soler-Illi G. J. A. A., Bellinoa M. G., *Bioelectrochemistry*, **106** (2015), 14–21.

Chen L. Y., Fujita T., Chen M. W., *Electrochim. Acta*, **67** (2012), 1–5.

Chen X. F., Cui D., Wang X. J., Wang X. H., Li W. S., *Biosens. Bioelectron.*, **69** (2015a), 135–141.

Chen Y., Gai P., Zhang J., Zhu J. J., *J. Mater. Chem. A*, **3** (2015b), 11511–11516.

Chen Z., Ren W., Gao L., Liu B., Pei S., Cheng H.-M., *Nat. Mater.*, **10** (2011), 424–428.

Chen C., Wang L. H., Tan Y. M., Qin C., Xie F. Y., Fu Y. C., Xie Q. J., Chen J. H., Yao S. H., *Biosens. Bioelectron.*, **26** (2011), 2311–2316.

Cooney M. J., Lau C., Windmeisser M., Liaw B. Y., Klotzbach T. L., Minteer S. D., *J. Mater. Chem.*, **18** (2008a), 667–674.

Cooney M. J., Svoboda V. C., Lau C. G., Martin G., Minteer S. D., *Energy Environ. Sci.*, **1** (2008b), 320–337.

Cosnier S., Holzinger M., Goff A. L., *Front. Bioeng. Biotechnol.*, **2** (2014), 45.

Dai M. Z., Sun L. G., Chao L., Tan Y. M., Fu Y. C., Chen C., Xie Q. J., *ACS Appl. Mater. Interfaces*, **7** (2015), 10843–10852.

Davis F., Higson S. P. J., *Biosens. Bioelectron.*, **22** (2007), 1224–1235.

de Heer W. A., Bacsa W. S., Chatelain A., Gerfin T., Humphrey-Baker R., Forro L., Ugarte D., *Science*, **268** (1995), 845–847.

Deng L., Chen C., Zhou M., Guo S., Wang E., Dong S., *Anal. Chem.*, **82** (2010), 4283–4287.

Deng L., Shang L., Wang Y., Wang T., Chen H., Dong S., *Electrochem. Commun.*, **10** (2008a), 1012–1015.

Deng L., Wang F., Chen H., Shang L., Wang L., Wang T., Dong S., *Biosens. Bioelectron.*, **24** (2008b), 329–333.

Do T. Q. N., Varničić M., Flassig R. J., Vidaković-Koch T., Sundmacher K., *Bioelectrochemistry*, **106** (2015), 3–13.

du Toit H., Lorenzo M. D., *Biosens. Bioelectron.*, **69** (2015), 199–205.

Falk M., Andoralov V., Silow M., Toscano M. D., Shleev S., *Anal. Chem.*, **85** (2013a), 6342–6348.

Falk M., Blum Z., Shleev S., *Electrochim. Acta*, **82** (2012), 191–202.

Falk M., Narváez Villarrubia C. W., Babanova S., Atanassov P., Shleev S., *ChemPhysChem*, **22** (2013b), 2045–2058.

Feng W., Ji P., *Biotechnol. Adv.*, **29** (2011), 889–895.

Filip J., Monosik R., Tkac J., *Int. J. Electrochem. Sci.*, **9** (2014), 2491–2506.

Filip J., Tkac J., *Electrochim. Acta*, **136** (2014), 340–354.

Fischback M. B., Youn J. K., Zhao X. Y., Wang P., Park H. G., Chang H. N., Kim J., Ha S., *Electroanalysis*, **18** (2008), 2016–2022.

Flexer V., Brun N., Backov R., Mano N., *Energy Environ. Sci.*, **3** (2010), 1302–1306.

Flexer V., Brun N., Courjean O., Backov R., Mano N., *Energy Environ. Sci.*, **4** (2011), 2097–2106.

Fujita S., Yamanoi S., Murata K., Mita H., Samukawa T., Nakagawa T., Sakai H., Tokita Y., *Sci. Rep.*, **4** (2014), 4937.

Gai P., Ji Y., Chen Y., Zhu C., Zhang J., Zhu J. J., *Analyst*, **140** (2015), 1822–1826.

Gai P., Jia Y.-S., Wang W. J., Song R. B., Zhu C., Chen Y., Zhang J., Zhu J. J., *Nano Energy*, **19** (2016), 541–549.

Gao H., Duan H., *Biosens. Bioelectron.*, **65** (2014), 404–419.

Gao F., Viry L., Maugey M., Poulin P., Mano N., *Nat. Commun.*, **1** (2010), 2.

Garcia-Perez T., Hong S.-G., Kim J., Ha S., *Enzyme Microb. Tech.*, **90** (2016), 26–34.

Gueven G., Prodanovic R., Schwaneberg U., *Electroanalysis*, **22** (2010), 765–775.

Gutiérrez-Sánchez C., Pita M., Vaz-Domínguez C., Shleev S., De Lacey A. L., *J. Am. Chem. Soc.*, **134** (2012), 17212–17220.

Hakamada M., Takahashi M., Mabuchi M., *Gold Bull.*, **45** (2012), 9.

Heller A., *Phys. Chem. Chem. Phys*, **6** (2004), 209–216.

Hickey D. P., McCammant M. S., Giroud F., Sigman M. S., Minteer S. D., *J. Am. Chem. Soc.*, **136** (2014), 15917–15920.

Holland J. T., Lau C., Brozik S., Atanassov P., Banta S., *J. Am. Chem. Soc.*, **133** (2011), 19262–19265.

Hou C. T., Yang D. P., Liang B., Liu A. H., *Anal. Chem.*, **86** (2014), 6057–6063.

Hussein L., Rubenwolf S., Stetten F., Urban G., Zengerle R., Krüger M., *Biosens. Bioelectron.*, **26** (2011a), 4133–4138.

Hussein L., Urban G., Krüger M., *Phys. Chem. Chem. Phys.*, **13** (2011b), 5831–5839.

Hyun K. H., Han S. W., Koh W.-G., Kwon Y., *J. Power Sources*, **286** (2015), 197–203.

Ichi S. E., Zebda A., Laaroussi A., Reverdy-Bruas N., Chaussy D., Belgacem M. N., Cinquina P., Martin D. K., *Chem. Commun.*, **50** (2014), 14535–14538.

Iijima S., *Nature*, **354** (1991), 56–58.

Inagaki M., Kaneko K., Nishizawa T., *Carbon*, **42** (2004), 1401–1417.

Jaafar M. M., Ciniciato G. P. M. K., Ibrahim S. A., Phang S. M., Yunus K., Fisher A. C., Iwamoto M., Vengadesh P., *Langmuir*, **31** (2015), 10426–10434.

Karimi A., Othman A., Uzunoglu A., Stanciu L., Andreescu S., *Nanoscale*, **7** (2015), 6909–6923.

Katz E., Bückmann A. F., Willner I., *J. Am. Chem. Soc.*, **123** (2001), 10752–10753.

Kim J., Grate J. W., *Nano Lett.*, **3** (2003), 1219–1222.

Kim J., Grate J. W., Wang P., *Chem. Eng. Sci.*, **61** (2006a), 1017–1026.

Kim J., Jia H., Wang P., *Biotechnol. Adv.*, **24** (2006b), 296–308.

Kim H., Lee I., Kwon Y., Kim B. C., Ha S., Lee J.-H., Kim J., *Biosens. Bioelectron.*, **26** (2011), 3908–3913.

Kwon C. H., Lee S.-H., Choi Y.-B., Lee J. A., Kim S. H., Kim H.-H., Spinks G. M., Wallace G. G., Lima M. D., Kozlov M. E., Baughman R. H., Kim S. J., *Nat. Commun.*, **5** (2014), 3928.

Lalaoui N., Elouarzaki K., Goff A. L., Holzinger M., Cosnier S., *Chem. Commun.*, **49** (2013) 9281–9283.

Lalaoui N., Goff A. L., Holzinger M., Cosnier S., *Chem. Eur. J.*, **21** (2015), 16868–16873.

Lalaoui N., Rousselot-Pailley P., Robert V., Mekmouche Y., Villalonga R., Holzinger M., Cosnier S., Tron T., Goff A. L., *ACS Catal.*, **6** (2016), 1894–1900.

Latonen R.-M., Wang X. J., Sjöberg-Eerola P., Eriksson J.-E., Bergelin M., Bobacka J., *Electrochim. Acta*, **68** (2012), 25–31.

Lau C., Martin G., Minteer S. D., Cooney M. J., *Electroanalysis*, **22** (2010), 793–798.

Leech D., Kavanagh P., Schuhmann W., *Electrochim. Acta*, **84** (2012), 223–234.

Lesniewski A., Paszewski M., Opallo M., *Electrochem. Commun.*, **12** (2010), 435–437.

Liang H. W., Cao X., Zhou F., Cui C.-H., Zhang W. J., Yu S. H., *Adv. Mater.*, **23** (2011), 1467–1471.

Liu C., Alwarappan S., Chen Z., Kong X., Li C.-Z., *Biosens. Bioelectron.*, **25** (2010), 1829–1833.

Liu C., Chen Z., Li C. Z., *IEEE Trans. Nanotechnol.*, **10** (2011), 59–62.

Liu W., Herrmann A.-K., Bigall N. C., Rodriguez P., Wen D., Oezaslan M., Schmidt T. J., Gaponik N., Eychmüller A., *Acc. Chem. Res.*, **48** (2015), 154–162.

Liu Y., Dong S., *Biosens. Bioelectron.*, **23** (2007), 593–597.

Liu Y., Dong X., Chen P., *Chem. Soc. Rev.*, **41** (2012), 2283–2307.

Liu Y., Wang M., Zhao F., Liu B., Dong S., *Chem. Eur. J.*, **11** (2005), 4970–4974.

Luz R. A. S., Pereira A. R., Souza J. C. P., Sales F. C. P. F., Crespilho F. N., *Chemelectrochem*, **1** (2014), 1751–1777.

MacVittie K., Conlon T., Katz E., *Bioelectrochemistry*, **106** (2015), 28–33.

Mano N., Mao F., Heller A., *J. Am. Chem. Soc.*, **125** (2003), 6588–6594.

Meredith M. T., Minson M., Hickey D., Artyushkova K., Glatzhofer D. T., Minteer S. D., *ACS Catal.*, **1** (2011), 1683–1690.

Min K., Yoo Y. J., *Biotechnol. Bioprocess Eng.*, **19** (2014), 553–567.

Minteer S. D., *Top Catal.*, **55** (2012), 1157–1161.

Minteer S. D., Atanassov P., Luckarift H. R., Johnson G. R., *Mater. Today*, **15** (2012), 166–173.

Moehlenbrock M. J., Minteer S. D., *Chem. Soc. Rev.*, **37** (2008), 1188–1196.

Moehlenbrock M. J., Toby T. K., Pelster L. N., Minteer S. D., *ChemCatChem*, **3** (2011), 561–570.

Murata K., Kaneko K., Kokai F., Takahashi K., Yudasaka M., Iijima S., *Chem. Phys. Lett.*, **331** (2000), 14–20.

Nardecchia S., Carriazo D., Ferrer M. L., Gutiérrez M. C., Monte F., *Chem. Soc. Rev.*, **42** (2013), 794–830.

Nasharudin M. N., Kamarudin S. K., Hasran U. A., Masdar M. S., *Inter. J. Hydrogen Energy*, **39** (2014), 1039–1055.

Prasad K. P., Chen Y., Chen P., *ACS Appl. Mater. Interfaces*, **6** (2014), 3387–3393.

Qian L., Lu L., *RSC Adv.*, **4** (2014), 38273–38280.

Ravenna Y., Xia L., Gun J., Mikhaylov A. A., Medvedev A. G., Lev O., Alfonta L., *Anal. Chem.*, **87** (2015) 9567–9571.

Reuillard B., Abreu C., Lalaoui N., Goff A. L., Holzinger M., Ondel O., Buret F., Cosnier S., *Bioelectrochemistry*, **106** (2015), 73–76.

Reuillard B., Goff A. L., Agnes C., Holzinger M., Zebda A., Gondran C., Elouarzaki K., Cosnier S., *Phys. Chem. Chem. Phys.*, **15** (2013), 4892–4896.

Sakai H., Nakagawa T., Tokita Y., Hatazawa T., Ikeda T., Tsujimura S., Kano K., *Energy Environ. Sci.*, **2** (2009), 133–138.

Salaj-Kosla U., Scanlon M. D., Baumeister T., Zahma K., Ludwig R., Conghaile P. Ó., MacAodha D., Leech D., Magner E., *Anal. Bioanal. Chem.*, **405** (2013), 3823–3830.

Sattayasamitsathit S., O'Mahony A. M., Xiao X. Y., Brozik S. M., Washburn C. M., Wheeler D. R., Gao W., Minteer S. D., Cha J., Burckel D. B., Polsky R., Wang J., *J. Mater. Chem.*, **22** (2012), 11950–11956.

Shitanda I., Kato S., Hoshi Y., Itagakia M., Tsujimura S., *Chem. Commun.*, **49** (2013), 11110–11112.

Song Y., Chen C., Wang C., *Nanoscale*, **7** (2015), 7084–7090.

Southcott M., MacVittie K., Halámek J., Halámková L., Jemison W. D., Lobel R., Katz E., *Phys. Chem. Chem. Phys.*, **15** (2013), 6278–6283.

Tamaki T., *Top. Catal.*, **55** (2012), 1162–1180.

Tamaki T., Yamaguchi T., *Ind. Eng. Chem. Res.*, **45** (2006), 3050–3058.

Trifonov A., Herkendell K., Tel-Vered R., Yehezkeli O., Woerner M., Willner I., *ACS Nano*, **7** (2013), 11358–11368.

Trifonov A., Tel-Vered R., Fadeev M., Willner I., *Adv. Energy Mater.*, **5** (2015), 1401853.

Tsujimura S., Kawaharada M., Nakagawa T., Kano K., Ikeda T., *Electrochem. Commun.*, **5** (2003), 138–141.

Tsujimura S., Murata K., Akatsuka W., *J. Am. Chem. Soc.*, **136** (2014), 14432–14437.

Vaghari H., Jafarizadeh-Malmiri H., Berenjian A., Anarjan N., *Sustain. Chem. Process.*, **1**, (2013), 16.

Walcarius A., *Chem. Soc. Rev.*, **42** (2013), 4098–4140.

Walcarius A., Minteer S. D., Wang J., Lin Y., Merkoçi A., *J. Mater. Chem. B*, **1** (2013), 4878–4908.

Wang Y., Ge L., Ma C., Kong Q, Yan M., Ge S., Yu J., *Chem. Eur. J.*, **20** (2014a), 12453–12462.

Wang Y., Ge L., Wang P., Yan M., Yu J., Ge S., *Chem. Commun.*, **50** (2014b), 1947–1949.

Wang X. J., Sjöberg-Eerola P., Immonen K., Bobacka J., Bergelin M., *J. Power Sources*, **196** (2011), 4957–4964.

Wen D., Deng L., Guo S., Dong S., *Anal. Chem.*, **83** (2011), 3968–3972.

Wen D., Herrmann A.-K., Borchardt L., Simon F., Liu W., Kaskel S., Eychmüller A., *J. Am. Chem. Soc.*, **136** (2014a), 2727–2730.

Wen D., Liu W., Herrmann A.-K., Eychmüller A., *Chem. Eur. J.*, **20** (2014b), 4380–4385.

Wen D., Xu X., Dong S., *Energy Environ. Sci.*, **4** (2011), 1358–1363.

Xu L., Armstrong F. A., *Energy Environ. Sci.*, **6** (2013), 2166–2171.

Xu L., Armstrong F. A., *RSC Adv.*, **5** (2015), 3649–3656.

Xu S., Minteer S. D., *ACS Catal.*, **2** (2012), 91–94.

Yahiro A. T., Lee S. M., Kimble D. O., *Biochim. Biophys. Acta*, **88** (1964), 375–383.

Yan Y. M., Baravik I., Tel-Vered R., Willner I., *Adv. Mater.*, **21** (2009), 4275–4279.

Yan Y., Su L., Mao L., *J. Nanosci. Nanotechnol.*, **7** (2007), 1625–1630.

Yan Y., Zheng W., Su L., Mao L., *Adv. Mater.*, **18** (2006), 2639–2643.

Yang J., Ghobadian S., Goodrich P. J., Montazami R., Hashemi N., *Phys. Chem. Chem. Phys.*, **15** (2013), 14147–14161.

Yang X. Y., Tian G., Jiang N., Su B. L., *Energy Environ. Sci.*, **5** (2012), 5540–5563.

Yehezkeli O., Tel-Vered R., Raichlin S., Willner I., *ACS Nano*, **5** (2011), 2385–2391.

Zebda A., Cosnier S., Alcaraz J.-P., Holzinger M., Goff A. L., Gondran C., Boucher F., Giroud F., Gorgy K., Lamraoui H., Cinquin P., *Sci. Rep.*, **3**, (2013), 1516.

Zebda A., Gondran C., Goff A. L., Holzinger M., Cinquin P., Cosnier S., *Nat. Commun.* **2**, (2011), 370.

Zhang Y., Chu M., Yang L., Tan T., Deng W., Ma M., Su X., Xie Q., *ACS Appl. Mater. Interfaces*, **6** (2014), 12808–12814.

Zhao M., Gao Y., Sun J. Y., Gao F., *Anal. Chem.*, **87** (2015), 2615–2622.

Zhao F., Slade R. C. T., Varcoe J. R., *Chem. Soc. Rev.*, **38** (2009), 1926–1939.

Zheng W., Zhao H. Y., Zhang J. X., Zhou H. M., Xu X. X., Zheng Y. F., Wang Y. B., Cheng Y., Jang B. Z., *Electrochem. Commun.*, **12** (2010), 869–871.

Zhou M., Dong S., *Acc. Chem. Res.*, **44** (2011), 1232–1243.

Zhou Z., Hartmann M., *Top. Catal.*, **55** (2012), 1081–1100.

Zhu C., Du D., Eychmüller A., Lin Y., *Chem. Rev.*, **115** (2015), 8896–8943.

Chapter 19

Nanoplasmonic Biosensors

Bruno P. Crulhas, Caroline R. Basso, and Valber A. Pedrosa

Department of Chemistry and Biochemistry, Institute of Bioscience, UNESP-Botucatu, SP 18618-000, Brazil

vpedrosa@ibb.unesp.br

19.1 Introduction

During the last decades, analytical science has provided new possibilities for getting chemical information about objects and systems. As a result, novel methodologies have been developed focusing on low cost, miniaturization, and simplification of existing techniques (Oliveira et al., 2008; Malhotra and Chaubey 2003). Analytical methods for identification and determination quantitative and qualitative properties of chemical and biological compounds are increasingly used in several areas of diagnosis, including diseases caused by viruses and bacteria, environmental monitoring, pathogen detection, antigen–antibody binding, detection of chemicals substances for defense and security systems (Grieshaber et al., 2008; Oliveira et al., 2008).

In this context, biosensors can afford continuous monitoring, in vivo and in real time, thereby using less amount of sample, increasing their employability for the commercial market. The purpose of a biosensor is to produce a signal with a power that

Biocatalysis and Nanotechnology
Edited by Peter Grunwald

ISBN 978-981-4613-69-9 (Hardcover), 978-1-315-19660-2 (eBook)
www.panstanford.com

is proportional in magnitude or frequency to the concentration of analyte. The biosensors can be classified according to the biological active compound (e.g., enzymatic, cellular and immunological) and type of transducer used (e.g., optical, electrochemical, acoustic and colorimetric). Among these several types of biosensor, the optical ones are of particular importance (Fatebello-Filho and Capelato, 1992; Healy et al., 2007; Lojou and Bianco, 2006).

Optical biosensors are based on the detection of changes on absorbing electromagnetic radiation in visible/infrared and between reactants and reaction produced by the measurement of light emission from a luminescent process (Roiter et al., 2012). The measured quantities are absorption, adsorption and desorption, refractive and reflectivity index, fluorescence, phosphorescence and wavelength (Sriram et al., 2015). Because of their extreme sensitivity to variations of active refractive index (RI) and within the surface plasmon decay length (δ_d) for kinetic monitoring, nanoplasmonics based refractometric platforms have been developed. All these properties present advantages such as speed of response, results in real time, good biocompatibility, ease of interaction, low cost, and high sensitivity in end-point detection and also can detect at extremely low concentrations by monitoring the biological phenomena through surface plasmon resonance (SPR) or localized surface plasmon resonance (LSPR) (Tokarev et al., 2008; Mayer and Hafner, 2011; Sriram et al., 2015; Wu et al., 2015).

Surface plasmons are charge density oscillations confined to coinage metal, excited by light at an incident wavelength, resulting in the appearance of intense surface plasmon absorption bands (Tokarev et al., 2010). SPR is based on the sensitivity of surface plasmon polaritons on, for example, a gold film to changes in the refractive index measured as the change in reflected laser light passed through a prism and reflected back from the film. One of the variations of the SPR technique that currently is of high interest due to its enormous application potential is the LSPR technique (Roiter et al., 2012; Lupitskyy et al., 2008).

Unlike SPR, where excited plasmons oscillate along the interface between metal and its dielectric surrounding, LSPRs are confined to nanoparticles of metals such as gold (AuNPs), silver (AgNPs), and platinum (PtNPs) (Lupitskyy et al., 2008;

Wu et al., 2015). Hence this nanoscale-related phenomenon enables detection of molecular interactions near nanoparticle surface through shifts in LSPR spectral peak (Roiter et al., 2012). The dependence of resonance angle into the surface is caused by the interaction of an evanescent electromagnetic field (evanescent wave) with molecules present in a short distance from the particles on surface. This phenomenon itself can be described accurately by classical physics as shown in the following expression for plasma energy E_p:

$$E_p = h\sqrt{\frac{ne^2}{mE_0}},$$

where m is the electron mass, e is the electron charge, E_0 is the permittivity of free space, and n is the electron density.

Suspensions of these noble metal nanoparticles Interact strongly with light producing colored solutions caused by light absorption in UV-Vis range by electron oscillations and photon interaction with conduction band of the metal nanostructures (Mayer and Hafner, 2011). Two properties are particularly characteristic for LSPR; first, electric field is greatest at the surface of metal nanoparticles but decreases rapidly with distance, and furthermore, the sensitivity of LSPR; and second, the particle optical extinction, is maximal at plasmon resonant frequency at visible wavelength for nanoparticulate structures and depends on refractive index of the immediate surrounding. LSPR peaks are detected by spectral scattering measurements on single nanoparticles or spectral extinction measurements in dense film.When spacing between nanoparticles is smaller than its diameter, electromagnetic interaction may occur through coupling mechanism between dipoles causing a color shift toward to red (lower energy) in LSPR spectrum; however when distance between nanoparticles increases 2.5 times their size they behave as isolated species (Murphy et al., 2008; Zhang et al., 2014; Estevez et al., 2014; Sriram et al., 2015).

In particular, LSPR transducer response is very sensitive to changes close to surface, which is a prerequisite for measurement of kinetics interaction between biomolecules. Additional properties of nanoparticles such as shape, size, distribution, shell thickness, dielectric constant of surrounding medium and spatial organization may influence intensity and the band position. The shape of (spherical) nanoparticles, nanorods, and nanostars also interferes significantly with the sensitivity of the experimental refractive index (Tokareva et al., 2006; Nikoobarht and El-Sayed, 2003; Zheng et al., 2014; Langille et al., 2012; Wu et al., 2005). Particles with sharp tips, nanoshells, and nanotriangles produce higher refractive index sensitivities because metal nanoparticles with this format create localized sensing/mode volume of highly enhanced electric field intensity. For analysis of optical properties and extinction at long wavelengths limited by electrostatic dipole present in spherical nanoparticles, the Mie theory (Mie's solution to Maxwell's equation) is used and represented as follows:

$$E(\lambda) = \frac{24\pi N_A a^3 \varepsilon_m^{3/2}}{\lambda \ln(10)} x \frac{\varepsilon_i}{(\varepsilon_r + \chi\varepsilon_m)^2 + \varepsilon_i^2},$$

where $E(\lambda)$ is the extinction (concerning the sum of the scattering and absorption), a is the radius of the metal sphere, λ is the wavelength of the absorbed radiation, m is dielectric constant of the medium, N_A is the particle area density, ε_i is the imaginary part of the dielectric function of the metal particle, ε_r is the real part, and χ is the term that describes the appearance of the nanoparticle (in the case of a sphere this value equals two). In the case of a single nanosphere, its spectrum in a dielectric middle external will depend on factors such as its material (ε_i and ε_r), the dielectric constant the middle (ε_m) and its radius (a).

The wavelength-selective absorption in LSPR may occur with an extremely high molar extinction coefficient of up to 3×10^{11} $M^{-1}cm^{-1}$ due to increasing local electromagnetic field near the nanoparticle surface (Jensen et al., 2000). In the LSPR technique, the free electrons of valence shell present on nanoparticles surface enter into resonance with incident light, remain confined into the surface and start to emit light continuously along the x axis, this process depends strongly on refractive index (RI) of nanoparticles, changes in RI, decay length (δ_d), and nanopattern thickness. The Drude model is used to find functional forms of dependencies between LSPR peak wavelength and dielectric properties of the medium:

$$\varepsilon_1 = 1 - \frac{\omega_p^2}{\omega^2 + y^2}$$

Here y is the damping parameter of the bulk metal and ω_p p is the plasma frequency. From this equation, the expressions for the LSPR frequency ω_{max} and for LSPR peak wavelength λ_{max} can be derived. This model describes electron transport in conductors considering collisions between freely moving electrons of metal nanoparticles and their ionic lattice (Mayer and Hafner, 2011).

The LSPR in metal nanoparticles has been widely employed for analytics, biocatalytic process, analysis of biomolecules, monitoring of local properties of biomaterials and bioanalytical applications (Zheng et al., 2014). The metals commonly reported in literature for LSPR-based sensors and biosensors are gold and silver. Among these, Au nanoparticles have attracted increasing interest in the production of label-free devices for biosensing because their extinction spectrum is highly sensitive to the dielectric constant of the surrounding media (Wu et al., 2015). Biosensors based on Au nanoparticles are easier to handle, are highly stable, can be bioconjugated with various ligands, exhibit chemical stability and resistance to oxidation, and are simple to fabricate. However, AgNPs have higher RI sensitivity, sharper resonances, higher scattering efficiency, narrower plasmon line widths, and sharper plasmon resonance curve (Mayer and Hafner, 2011; Sriram et al., 2015). The nanoparticle surfaces may also be modified by deposition of different materials, for example, by self-assembled monolayer (SAM) formation with alkane thiols of different chain lengths creating a covalent bond. This SAM modification results in a decrease in LSPR sensitivity with increasing distance tantamount to the increasing SAMs chain length indicating the need to differentiate bulk refractive index sensitivity of nanoparticle from its molecular detection sensitivity (Basso et al., 2015).

LSPR-based different types of chemical and biosensors measuring light transmission (T) within the UV-Vis range are known as T-LSPR setups. Just as in the case of LSPR, electrons from the Au or AgNP surfaces will oscillate when excited by a wavelength of incident light inducing photon absorption and scattering sensitive to dielectric properties of surrounding media.

The major advantage of LSPR technique is a low-cost solution for label-free qualitative and quantitative analysis of small molecules in complex samples and is well suited for use in clinical diagnostics (Tokareva et al., 2004; Tokarev et al., 2008; Mayer and Hafner, 2011).

19.2 Applications

19.2.1 Molecular Biosensors: Cell Analysis and Drug Delivery

The study of cellular behavior is generally conducted using population-based studies (e.g., western blot, immunoassay, reverse-transcription, and quantitative Polymerase Chain Reaction (PCR) (Carlo and Lee, 2006; Chattopadhyay et al., 2014). However, such methodologies just provide knowledge about the average group of cells; the diversity within populations as well as heterogeneity from biological environment is completely lost. A wide range of studies revealed a relevant role from heterogeneous sub-populations (Marusyk and Polyak, 2010; Burrell et al., 2013). Therefore, in order to understand how cellular microenvironment and cell dynamics are organized, techniques at single-cell level started to be a promising field for research (Spiller et al. 2010; Bakstad et al., 2012).

LSPR is a powerful tool for biological and chemical sensing and usually relies on following mechanisms: (1) resonant Rayleigh scattering; (2) charge transfer interactions at the surface of nanoparticles (3) nanoparticles aggregation, and (4) changes in the local refractive index. Also, LSPR is widely used to enhance the surface optically related processes such as fluorescence, plasmon resonance energy transfer (PRET), and surface-enhanced Raman spectroscopy (SERS) (Krull and Petryayeta, 2011).

The high sensitivity of LSPR has the advantage that it detects molecular binding events and conformational changes and provides steady-state and kinetic data. LSPR is a well-known platform to investigate a variety of mutual reactions between biological molecules, including short-range changes and can be the basis for sensing molecular interactions near nanoparticle surface allowing direct measurement of the cell environment (Kneipp et al. (2006a); Huang et al., 2008; Breuzard et al.,

2004). LSPR sensors have the advantage of not requiring microfabrication, achieving significant impact in several biological areas, including proteomics, point-of-care diagnostics, and drug discovery (Breuzard et al., 2004; Kelly et al., 2003; Willets and Van Duyne, 2007; Mayer and Hafner, 2011).

Usually, single-cell techniques are utilized to verify gene regulation, protein translocation, cell-to-cell interactions, cell fate (i.e., division, apoptosis), and diseases. Single-cell analysis techniques mostly rely on fluorescence systems to acquire live and real-time image from dynamic cellular processes (Spiller et al., 2010; Bakstad et al., 2012). Fluorescent compounds are cell permeable dyes or fluorescently modified proteins and firefly luciferase reporters. Despite, fluorescence being a robust tool for single-cell analysis, it still has major limitations: (1) sensitivity to photobleaching, which limits real-time experiments during prolonged processes (e.g., cell division and proliferation); (2) fluorescent probes commonly have overlapping emission spectra, which could decrease the number of monitored cellular processes at one time; and (3) fluorescence signal does not contain molecular vibration, limiting its use for studying specific molecular profile inside the cells or even cell-to-cell communication (Kang et al., 2012; Kang et al., 2013; Austin et al., 2013).

19.2.2 Monitoring Cellular Environment: Cell State and Fate

Combination of LSPR techniques and gold nanoparticles (AuNPs) are used for bioimaging applications, with focus in differentiate a diseased cell state from a healthy cell environment (Austin et al., 2014; Dreaden and El-Sayed, 2012). Several research groups have employed AuNPs and LSPR techniques; for example, El-Sayed's group employed enhanced Rayleigh scattering and SERS of AuNPs to identify and selectively destroy human oral cancer cells (Huang et al., 2006; Huang et al., 2007; El-Sayed et al., 2006). Other groups employed SERS and AuNPs for selective labeling and visualization of cancerous cells (in vitro and in vivo) by means of antibody-AuNPs conjugates (Seekell et al., 2011; Kneipp et al. (2006b); He et al., 2008; Aaron et al., 2007; Zavaleta et al., 2013).

Combination of AuNPs with LSPR techniques can provide noninvasive and long-term studies in monitoring cell behavior. AuNPs can be used as a probe for non-specific intracellular targets or as an organelle-specific targeting (Ba et al., 2010; Yang et al., 2012; de la Fuente and Berry, 2005; Ju et al., 2014). Kneipp et al. (2006a) analyzed the local pH in live cells using LSPR and SERS. This methodology avoids photodecomposition of the sensor and allows free selection of the excitation wavelengths. Recently, Kang et al. (2010) showed that AuNPs bioconjugating with specific peptides can be transported to the nuclei of cancer cells. The results show evidence that nanomaterials localized at cell nucleus can specifically affect cellular function causing cytokineses and apoptosis.

On the other hand, for specific organelle targeting, several studies started to focus on the mitochondria, an organelle involved in a variety of cellular functions (e.g., apoptosis, cytosolic Ca^{2+} uptake, and storage, and also regulating ROS). Thus, targeting mitochondria becomes a strategy for disease therapy and probing such an environment could provide valuable information about a variety diseases, such as obesity, Parkinson's, and cancer (Karatas et al. 2009; Zhuang et al., 2014; Wang et al., 2010a).

Ju et al. (2014) used a cytochrome c binding aptamer for successful accumulation of silica-coated AuNPs in mitochondria. Combinations of organelle-targeted AuNPs and LSPR techniques have the potential to reveal new and precise information of organelle response from internal and external stimuli. To achieve single-molecule probing in single cells analysis, higher imaging resolution might be required. The spatial resolution depends not only on imaging optics but also on the size of plasmonic probes. For example, promising results have been obtained by super-resolution imaging of single molecules in living cell using a 5 nm gold nanoprobe and detecting light absorption instead of scattering by photothermal imaging optics (Leduc et al., 2013).

19.2.3 Intracellular Detection: Biomedical Diagnostics and Drug Delivery

Intracellular assays and biosensing based on LSPR are a novel platform. In addition, this method has potential for applications in biomedical diagnostics, drug delivery, and determination of therapeutic efficacy at the cellular level with higher sensitivity, specificity, and extremely low limit of detection (Bakstad et al., 2012; Kneipp et al., 2006b; Mayer and Hafner, 2011). Nanoplasmonic particles started to be used for cell imaging, drug–cell interaction, and photothermal therapy. Differently shaped Au nanostructures have been employed as thermal converters to verify irreversible cell damage: Nanospheres, nanorods, and nanoshells exhibited strongly enhanced absorption in visible and near-infrared region (Thaxton and Mirkin, 2005).

Detection of cancer cells in blood samples was achieved using SERS by combining magnetic nanoparticles coated with epithelial specific antibody (EpCAM) and anti-her2 antibody conjugated with AuNPs (Sha et al., 2008). Briefly, modified magnetic nanoparticles have specific affinity to epithelial cells, while anti-her2-SERS tags can recognize cancer cells by over-expressed her2 receptor with a limit of detection of less than 10 cells mL^{-1}. This method achieved high specificity and sensitivity using whole blood samples. Nie's group modified AuNPs with polyethylene-glycol hydrogel (PEG) and encapsulated Raman reporters for in vivo targeting of tumors. PEG coating provided colloidal stability of AuNPs over a wide range of pH and ionic strength. The coating presented increased hydrodynamic diameter of 20 nm and nanotags were covalently conjugated to a single chain fragment variable (ScFv) antibody binding with high specificity and affinity to the epidermal growth factor (EGF) receptor on tumor cell surfaces (Qian et al., 2008).

Another example for biomedical screening is the use of Ag triangular arrays for the detection of Alzheimer's disease biomarker. In summary, amyloid-beta-derived-diffusible ligand (ADDL) antigen was immobilized on the surface of the nanoparticles and coated with a self-assembled monolayer of MUA (mercaptoundecanoic acid) and 1-octanethiol (Haes et al., 2004). ADDL antigen conjugation was achieved using 1-ethyl-3-(3-methyl-aminopropyl)-carbodiimide hydrochloride (EDC). Finally, binding of specific rabbit polyclonal anti-ADDL IgG antibody was verified and associated with saturation response of LSPR, which provided 18.5 nm shift at a concentration of 100 nM of anti-ADDL, showing higher specificity and lower limit of detection.

Drug delivery systems (DDS) are a promising strategy to ensure selective and specific release of drugs by maintaining or even improving the efficacy. Several DDS have been modeled and AuNPs showed a robust platform due to biocompatibility and easy visualization of successful delivery through optical techniques. Doxorubicin (DOX, chemotherapeutic agent), is a well-known model drug for such applications with inherent fluorescence and when loaded onto AuNPs it adopts a quenched state. Upon successful release, DOX's fluorescence will be restored allowing for real-time drug delivery dynamic measurements to be conducted. Diverse groups have reported successful variations of this DDS format (Song et al., 2012; Wang et al., 2011; Gu et al., 2012).

El-Sayed's group (Austin et al., 2015) performed a DDS in conjunction with strongly enhanced Raman scattering from DOX molecule for monitoring dynamic release of DOX in live human carcinoma cells and found an inverse relationship between DOX's Raman and fluorescence signal, resulting in the disappearance of the DOX Raman spectra with the simultaneous restoration of its fluorescence emission, thus providing a successful drug delivery system. Ock et al. (2012) utilized SERS to monitor glutathione (GSH) anticancer drug release from AuNPs (in vitro and in vivo). Briefly, in this system, 6-mercaptopurine or 6-thioguanine were adsorbed onto the surface of AuNPs and distinct Raman bands from two anticancer agents were monitored before and after external GSH administration. After delivering GSH, Raman bands of both drugs were reduced, indicating their successful desorption from AuNP surface.

Drug uptake by plasma membranes of living cells can also be monitored with high sensitivity (10^{-10} M) by SERS for antitumor drug mitoxantrone (MTX); Kneipp's group researched drug diffusion through cell membrane using AuNPs, AgNPs and nanoaggregates delivering it intracellularly by endocytosis after the incubation of nanoparticles with cells for several hours (Kneipp et al. 2006a, 2006b). Diffusion of anticancer drug by membrane of U87-MG cancer cells was performed by silver-coated silica beads, and it was able to observe kinetics of diffusion (Balint et al., 2010). Shamsaie et al. (2007) reported intracellular growth of AuNPs within MCF10 epithelial cells that were not translocated to endosomes and lysosomes, as confirmed by the absence of Raman bands around 500 cm^{-1} associated with stretching vibrational mode of disulfide bonds that are abundant in lysosomal proteins. An approach to deliver SERS probes to cell nucleus was demonstrated by co-functionalizing nanoparticles with HIV-derived Trans-Activator of Transcription (TAT) peptide along with Raman reporter molecules (Gregas et al., 2010). Electroporation allowed for rapid uptake of nanoparticles by the cells, and AgNPs in cell cytoplasm became highly localized (Lin et al., 2009).

Multi-type bioconjugation and targeting strategies have been established to drug delivery systems and plasmonic nanoprobes could be used in different cell compartment, such as cell membrane, mitochondria, peroxisome, and nucleus. LSPR conjugation and delivery strategies are still desired to expand the scope of plasmonic nanoprobes to target specific intracellular organelles or even specific molecules (Krull and Petryayeva, 2011; Mayer and Hafner 2011).

Recent advance in studying intercellular and intracellular biochemical processes have made important contributions to our understanding of biology in the past several decades and has significant impact on cell imaging and drug delivery. LSPR Technologies, single molecular imaging, and gene regulation have allowed a deeper understanding of cellular functions and mechanisms in drug delivery.

19.2.4 Aptamer-Based Nanoplasmonic Biosensors

In recent years, several research groups started to work with aptamers using different emphases (Song et al., 2008; Keefe et al., 2010; Chen et al., 2011; Iliuk et al., 2011; Kong et al., 2011). Various detection techniques, including fluorescence, surface-plasmon resonance (SPR), surface-enhanced Raman scattering (SERS), and fluorescence resonance-energy transfer (FRET), are used with aptamers. Integration of aptamers with optical nanomaterials (e.g., quantum dots, metal nanoparticles, carbon nanotubes (CNTs), graphene oxide (GO), and silica nanoparticles

(SiNPs) represents some examples of materials that can be used for fabricating aptamer-based biosensors with plasmonic nanostructures.

Plasmonics is a new field with focus on electromagnetic responses of metal nanostructures and interactions with light at the metal–dielectric interface (Atwater and Polman, 2010). When resonant photons are confined within a plasmonic nanoparticle, they excite localized surface-plasmon oscillations giving rise to a strong surface electromagnetic field that propagates around the particle and decays over a distance comparable to the particle's size (Jain et al., 2007). For example, Raman scattering (SERS) is the evaluation of species adsorbed onto particle surface (Mirkin, 2000; Nie and Emory, 1997). The high sensitivity of plasmon spectra to the particle size and local dielectric environment also provides a new methodology for detecting biomolecules, where detection signal is solely based on changes in plasmonic spectra (Anker et al., 2008; Penga and Miller, 2011).

DNA or RNA aptamers have been identified for several targets, such as proteins, peptides, amino acids, antibiotics, small chemicals, viruses, whole cells, organelles, and even metal ions, with high affinity and specificity. In addition, these molecules started to be applied in major biomedical fields, such as diagnostics, therapeutics, and bio-analytics (Yuan et al., 2012).

Aptamer-based biosensors stand out among other biorecognition receptors, such as antibodies and enzymes due their unique properties: high flexibility of structure and convenience in design of their structure, which allow the development of various novel aptasensors with high stability and affinity (Navani and Li, 2006; Gao et al., 2004; Gao et al., 2005). Therefore, major applications of aptamer-based nanoplasmonic biosensors are cell monitoring, cell targeting, and detection of specific biomolecules. Aptamers, as molecular probes with high specificity and selectivity, can readily distinguish between cancerous and healthy cells at molecular level. The combination of aptamers with nanomaterials as signal reporting groups therefore represents a powerful diagnostic tool for the detection of cancer and other diseases in early stage.

19.2.5 Aptamer-Based LSPR Biosensor

The optical properties of noble metal nanoparticles, which exhibit unique extinction spectra, have became an alternative strategies to development of optical biosensors. Interaction between metals and aptamers is a wide field in biosensors applications, particularly for biomedical purposes. Furthermore, optical properties of metal nanostructures have high chemical stability and allow the utilization of LSPR for the construction of sensitive aptamer-biosensors (Kennedy et al., 2011).

For example, Chen et al. (2010) reported prion protein (PrP^c) aptamer-modified AgNPs that could be used as contrast imaging agents for dark-field light scattering and transmission electron microscopy of neuroblastoma SK-N-SH cells. In addition, PrP^c-AgNPs can be internalized into plasma membrane, lysosome and endocytic structure through aptamer-mediated endocytosis.

Aptamer modified AgNPs were used to target and for imaging subcompartments of live cells. Sun et al. (2011) demonstrated a Sgc8c DNA aptamer conjugated to an Ag cluster for targeting the nucleus of CCRP-CEM cells (T lymphoblast cells). Hwang et al. (2010) reported a multi-modal imaging probe consisting of cobalt–ferrite nanoparticles protected by a silica shell for cancer-specific targets. They extended applications to aptamer-functionalized nanomaterials in cellular analysis and delivery for studying several types of cells and different cellular processes. Besides in vitro selection from random nucleic acid pools, the introduction of unnatural nucleotides into nucleic acid pools to improve diversity of functional groups may further enhance the chance to obtain aptamers for additional cellular targets (Hwang 2010).

19.2.6 Aptamer-Based SERS Biosensors

Since the discovery of SERS, it has been shown that Raman scattering cross-section of a molecule can be increased up to 10^7–10^8-fold, being a valuable alternative to fluorescence sensors. Therefore, electromagnetic field hot spots by SERS could result in devices with high sensitivity and specificity for in vitro and in vivo biosensing (Fabris et al., 2007; Braun et al., 2007; Bonham et al., 2007). Theoretical (García-Vidal and Pendry, 1996), and experimental (Orendorff et al., 2006; Maltzahn et al., 2009) results have proved that Raman scattering signal could presumably emanate from large electromagnetic field produced.

Irudayaraj et al. produced a SERS aptasensor based on gold nanorods and gold nanoparticles (AuNR/AuNPs) to detect thrombin in human-blood serum (Wang et al., 2010b). Basically, AuNRs were modified with a thrombin-binding antibody to capture α-thrombin. On the other hand, AuNPs were conjugated with a Raman reporter-labeled aptamer, producing a protein sandwich between the AuNRs and AuNPs for SERS detection. The close interaction between α-thrombin and antibody brought AuNPs and AuNRs created hot spots under laser excitation to produce an enhanced Raman scattering signal. In absence of thrombin, SERS detection did not show enhancement mainly due to missing signal from AuNPs (Cao et al., 2002). Dluhy's group described a highly sensitive SERS method for the detection of influenza viral nucleoproteins by Ag nanorods (Negri et al., 2011). The SERS spectrum of aptamer–nucleoprotein complex provided direct binding between a polyvalent anti-influenza aptamer and nucleoproteins of influenza strains. Kim et al. reported a SERS-based aptamer, using a bifunctional adenosine-sensitive aptâmero (Kim et al., 2010). A SERS hot spot was created between the Au surface and an AuNP attached to aptamer via a biotin-avidin linkage. AuNP was conjugated with a Raman tag molecule 4-aminobenzenethiol (4-ABT) and when target molecule adenosine was added to the system, the SERS spectrum of 4-ABT increased depending on the concentration of the analyte. In addition, AFM imaging confirmed that the mean height of AuNP-bearing aptamer decreased by 5.6 nm, in agreement with the observed SERS intensity change. This strategy may be further developed by using several bifunctional aptamers, each with its own Raman tag for simultaneous detection.

19.3 Future Directions

In this contribution, we highlighted new techniques and application examples of LSPR related to the fabrication of molecular and aptamer-based biosensors. LSPR methodologies have potential to become a powerful tool for sensitivity improvement of biological detections. LSPR offers opportunities to tune and control optical behavior. Combination of LSPR and several types of nanoparticles with different shape, size, and composition can improve the sensitivity of a bioassay or a biosensor; however, many questions concerning the shape features of nanoparticles and their effect on LSPR still remain a matter of debate. Using higher-order structures, such as nanostars, nanoflowers, and nanocylinders, further extends the application potential of LSPR. However, the major limitations are high reproducibility combined with sensitivity, directly impacting quantitative data analysis. Furthermore, variability in the size of NPs and lack of control about spatial positioning and orientation of the biomolecules remain a challenge in the application of LSPR techniques, although LSPR provides reliable information about analytes in close vicinity to a NP surface with high sensitivity and specificity.

Despite the challenges with respect to practical applications of LSPR, the scientific literature based on assays developments continues to grow at an exponential rate. LSPR can provide rapid, easy, and cost-effective experimental designs with at least comparable analytical performance. Moreover, LSPR allows single-nanoparticle detection and probing extremely small volumes, such as single-cell analysis and biomolecules detection. In the near future, it could be an interesting platform for the miniaturization of the point-of-care analysis. Additionally, LSPR improvement, including the integration of microfluidics, will provide a platform for multiplexed analysis, which is crucial for clinical diagnostics and proteomics.

LSPR techniques combined with aptamers (aptamer-functionalized NP-substrates) are mainly limited by difficulties in bioconjugation chemistry. Although still being a challenge in the field of bioanalytical development, several research groups are focused on improving conjugation of aptamers and probes for application in connection with LSPR techniques. Aptamer-conjugated nanoplasmonic biosensors rely basically on the proof-of-concept research, and an evaluation of more complex samples is still in its infancy.

References

Aaron J., Nitin N., Travis K., Kumar S., José-Yacamán M., Coghlan L., Follen M., Richards-Kortum R., Sokolov K., Collier T., Biomedo, **12** (2007), 34007–34011.

Anker J. N., Hall W. P., Lyanderes O., Shan N. C., Zhao J., Van Duyne R. P., *Nat. Mater.*, **7** (2008), 442–453.

Atwater H. A., Polman A., *Nat. Mater.*, **9** (2010) 205–213.

Austin L. A., Kang B., El-Sayed M. A., *J. Am. Chem. Soc.*, **135** (2013), 4688–4691.

Austin L. A., Kang B., El-Sayed M. A., *Nanotoday*, **10** (2015), 542–558.

Austin L. A., Mackey M., Dreaden E., El-Sayed M., *Arch. Toxicol.*, **88** (2014), 1391–1417.

Ba H., Rodríguez-Fernández F., Stefani F., Feldmann J., *Nano Lett.*, **10** (2010), 3006–3012.

Bakstad D., Adamson A., Spiller D. G., White M. R. H., *Curr. Opin. Biotechnol.*, **23** (2012) 103–109.

Balint S., Rao S., Sanchez M. M., Huntosova V., Miskovsky P., Petrov D. J., *Biomed. Opt.*, **15** (2010), 27005–27005.

Basso C. R., Tozato, C. C., Araujo J. P., Pedrosa V. A., *Anal. Method.*, **7** (2015), 2264–2267.

Bonham A. J., Braun G., Pavel I., Moskovits M., Reich N. O., *J. Am. Chem. Soc.*, **129** (2007), 14572–14573.

Braun G., Lee S. J., Dante M., Nguyen T. Q., Moskovits M., Reich N., *J. Am. Chem. Soc.*, **129** (2007), 6378–6379.

Breuzard G., Angiboust J. F., Jeannesson P., Manfait M., Millot J. M., *Biochem. Biophys. Res. Commun.*, **320** (2004), 615–621.

Burrell R. A., McGranahan N., Bartek J., Swanton C., *Nature*, **501** (2013), 338–345.

Cao Y. W. C., Jin R. C., Mirkin C. A., *Science*, **297** (2002), 1536–1540.

Carlo D. D., Lee L. P., *Anal. Chem.*, **78** (2006), 7918–7925.

Chattopadhyay P. K., Gierahn T. M., Roedere M. R., Love J. C., *Nat. Immunol.*, **15** (2014), 128–135.

Chen T., Shukoor M. I., Chen Y., Yuan Q., Zhu Z., Zhao Z., Gulbakan B., Tan W., *Nanoscale*, **31** (2011), 546–556.

Chen L. Q., Xiao S. J., Peng L., Wu T., Ling J., Li Y. F., Huang C. Z., *J Phys. Chem. B*, **114** (2010), 3655–3659.

de la Fuente J. M., Berry C. C., *Bioconjug. Chem.*, **16** (2005), 1176–1180.

Dreaden E. C., El-Sayed M. A., *Acc. Chem. Res.*, **45** (2012), 1854–1865.

El-Sayed I. H., Huang X., El-Sayed M. A., *Cancer Lett.*, **239** (2006) 129–135.

Estevez M. C., Otte M. A., Sepulveda B., Lechuga, L. M., *Anal. Chim. Acta*, **806** (2014), 55–73.

Fabris L., Dante M., Braun G., Lee S. J., Reich N. O., Moskovits M., Nguyen T. Q., Bazan G. C. J., *Am. Chem. Soc.*, **129** (2007), 6086–6087.

Fatebello-filho O., Capelato M. D., *Química Nova*, **15** (1992), 28–39.

Gao X., Cui Y. Y., Levenson R. M., Chung L. W. K., Nie S., *Nature*, **8** (2004), 969–976.

Gao X., Yang L., Petros J. A., Marshall F. F., Simons J. W., Nie S., *Curr. Opin. Biotechnol.*, **16** (2005), 63–72.

García-Vidal F. J., Pendry J. B., *Phys. Rev. Lett.*, **77** (1996), 1163–1166.

Gregas M. K., Scaffidi J. P., Lauly B., Vo-Dinh T., *Appl. Spectrosc.*, **64** (2010), 858–866.

Grieshaber D., Mackenzie R., Voros J., Reimhult E., *Sensors*, **8** (2008), 1400–1458.

Gu Y. J., Cheng J., Man C. W. Y., Wong W. T., Cheng S. H., *Nanomed. Nanotechnol. Biol., Med.*, **8** (2012), 204–211.

Haes A. J., Hall W. P., Chang L., Klein W. L., Van Duyne R. P., *Nano Lett.*, **4** (2004), 1029–1034.

Haes, A. J., Van Duyne, R. P., *Anal. Bioanal. Chem.*, **379** (2004), 920–930.

Haes A. J., Hall W. P., Chang L., Klein W. L., Van Duyne R. P., *Nano Lett.*, **4** (2004), 1029–1034.

He H., Xie C., J. Ren, *Anal. Chem.*, **80** (2008), 5951–5957.

Healy D. A., Hayes C. J., Leonard P., Mckenna L., O'kennedy R., *Trends Biotechnol.*, **25** (2007), 125–131.

Huang X., El-Sayed I. H., Qian W., El-Sayed M. A., *J. Am. Chem. Soc.*, **128** (2006), 2115–2120.

Huang X., El-Sayed I. H., Qian W., El-Sayed M. A., *Nano Lett.*, **7** (2007), 1591–1597.

Huang X. H., Jain P. K., El-Sayed I. H., El-Sayed M. A., *Lasers Med. Sci.*, **23** (2008), 217–228.

Hwang D. W., Ko Y. H., Lee J. H., Kang K., Ryu S. H., Song I. C., Lee D. S., Kim S. A., *J. Nucl. Med.*, **51** (2010), 98–105.

Iliuk A. B., Hu L., Tao W. A., *Anal. Chem.*, **83** (2011), 4440–4452.

Jain P. K., Huang W., El-Sayed M. A., *Nano Lett.*, **7** (2007), 2080–2088.

Jensen T. R., *J. Phys. Chem. B.*, **104** (2000), 10549–10556.

Ju E., Li Z., Liu Z., Ren J., Qu X., *Appl. Mater. Interfaces*, **6** (2014), 4364–4370.

Kang B., Afifi M. M., Austin L. A., El-Sayed M. A., *ACS Nano*, **7** (2013), 7420–7427.

Kang B., Austin L. A., El-Sayed M. A., *Nano Lett.*, **12** (2012), 5369–5375.

Kang B., Mackey M. A., El-Sayed M. A., *J. Am. Chem. Soc.*, **132** (2010), 1517–1519.

Karatas Ö. F., Sezgin E., Aydýn Ö., Culha M., *Colloids Surf. B Biointerfaces*, **71** (2009), 315–318.

Keefe A. D., Pai S., Ellington A., *Nat. Rev. Drug Discov.*, **9** (2010), 537–550.

Kelly K. L., Coronado E., Zhao L. L., Schatz G. C., *J. Phys. Chem. B*, **107** (2003), 668–677.

Kennedy L. C., Bickford L. R., Lewinski N. A., Coughlin A. J., Hu Y., Day E. S., *Small*, **7** (2011), 169–183.

Kim N. H., Lee S. J., Moskovits M., *Nano Lett.*, **10** (2010), 4181–4185.

Kneipp K., Kneipp H., Kneipp J., *Acc. Chem. Res.*, **39** (2006a), 443–450.

Kneipp J., Kneipp H., McLaughlin M., Brown D., Kneipp K., *Nano Lett.*, **6** (2006b), 2225–2231.

Kneipp J., Kneipp H., Wittig B., Kneipp K., *Nano Lett.*, **7** (2007), 2819–2823.

Kong R. M., Zhang X. B., Chen Z., Tan W., *Small*, **7** (2011), 2428–2436.

Krull U. J., Petryayeva E., *Anal. Chem. Acta*, **706** (2011), 8–24.

Langille M. R., Personick M. L., Zhang J., Mirkin C. A., *J. Am. Chem. Soc.*, **134** (2012), 14542–14554.

Leduc C., Si S., Gautier J., Soto-Ribeiro M., Wehrle-Haller B., Gautreau A., Giannone G., Cognet L., Lounis B., *Nano Lett.*, **13** (2013), 1489–1494.

Lin J. Q., Chen R., Feng S. Y., Li Y. Z., Huang Z. F., Xie S. S., Yu Y., Cheng M., Zeng H. S., *Biosens. Bioelectron.*, **25** (2009), 388–394.

Lojou E., Bianco P., *J. Electroceram.*, **16** (2006), 79–91.

Lupitskyy R., Motornov M., Minko S., *Langmuir*, **24** (2008), 8976–8980.

Malhotra B. D., Chaubey A., *Sens. Actuators B*, **91** (2003), 117–127.

Maltzahn G. V., Centrone A., Park J., Ramanathan R., Sailor M. J., Hatton T. A., Bhatia S. N., *Adv. Mater.*, **21** (2009), 3175–3180.

Marusyk A., Polyak K., *Biochim. Biophys. Acta Rev. Cancer*, **1805** (2010), 105–117.

Mayer K. M., Hafner J. H., *Chem. Rev.*, **111** (2011), 3828–3857.

Mirkin C. A., *Inorg. Chem.*, **39** (2000), 2258–2272.

Murphy C. J., Gole A. M., Stone J. W., Sisco P. N., Alkilany A. M., Goldsmith E. C., Baxter S. C., *Acc. Chem. Res.*, **41** (2008) 1721–1730.

Navani N. K., Li Y., *Curr. Opin. Chem. Biol.*, **10** (2006), 272–281.

Negri P., Kage A., Nitsche A., Naumannc D., Dluhy R. A., *Chem. Commun.*, **47** (2011), 8635–8637.

Nie S., Emory S. R., *Science*, **275** (1997), 1102–1106.

Nikoobakht B., El-Sayed M. A., *Chem. Mater.*, **15** (2003), 1957–1962.

Ock K., Jeon W. I., Ganbold E. O., Kim M., Park J., Seo J. H., Cho K., Joo S. W., Lee S. Y., *Anal. Chem.*, **84** (2012), 2172–2178.

Oliveira M. D. L., Correa M. T. S., Coelho L. C. B. B., Diniz F. B., *Colloids Surf., B*, **66** (2008), 13–19.

Orendorff C. J., Gearheart L., Jana N. R., Murphy C. J., *Phys. Chem. Chem. Phys.*, **8** (2006) 165–170.

Penga H. I., Miller B. L., *Analyst*, **136** (2011), 436–447.

Qian X. M., Peng X. H., Ansari D. O., Yin-Goen Q., Chen G. Z., Shin D. M., Yang L., Young A. N., Wang M. D., Nie S. M., *Nat. Biotechnol.*, **26** (2008), 83–90.

Roiter Y., Minko I., Nykypanchuk D., Tokarev I., Minko S., *Nanoscale*, **4** (2012), 284–292.

Seekell K., Crow M. J., Marinakos S., Ostrander J., Chilkoti A., Wax A., *J. Biomed. Opt.*, **16** (2011), 116003-1-116003-12.

Sha M. Y., Xu H. X., Natan M. J., Cromer R., *J. Am. Chem. Soc.*, **130** (2008), 17214–17215.

Shamsaie A., Jonczyk M., Sturgis J., Robinson J. P., Irudayaraj J., *J. Biomed. Opt.*, **12** (2007), 20502–20502.

Song S., Wang L., Li J., Zhao J., Fan C., *Trends Anal. Chem.*, **27** (2008), 108–117.

Song J., Zhou J., Duan H., *J. Am. Chem. Soc.*, **134** (2012), 13458–13469.

Spiller D. G., Wood C. D., Rand D. A., White M. R., *Nature*, **465** (2010), 736–745.

Sriram M., Zong K., Vivekchand S. R. C., Gooding J. J., *Sensors*, **15** (2015), 25774–25792.

Sun Z. P., Wang Y. L., Wei Y. T., Liu R., Zhu H. R., Cui Y. Y., Zhao Y. L., Gao X. Y., *Chem. Commun.*, **47** (2011), 11960–11962.

Thaxton C. S., Mirkin C. A., *Nat. Biotechnol.*, **23** (2005), 681–682.

Tokarev I., Tokareva I., Gopishetty V., Katz E., Minko S., *Adv. Mater.*, **22** (2010), 1412–1416.

Tokarev I., Tokareva, I., Minko, S., *Adv. Mater.*, **20** (2008), 2730–2734.

Tokareva I., Minko S., Fendler J. H., Hutter. E., *J. Am. Chem. Soc.*, **126** (2004), 15950–15951.

Tokareva I., Tokarev I., Minko S., Hutter E., Fendler J. H., *Chem. Commun.*, **31** (2006), 3343–3345.

Wang Y., Lee K., Irudayaraj J., *Chem. Commun.*, **46** (2010a), 613–615.

Wang L., Liu Y., Li W., Jiang Y., Ji X., Wu X., Xu L., Qiu Y., Zhao K., Wei T., Li Y., Zhao Y., Chen C., *Nano Lett.*, **11** (2010b), 772–780.

Wang F., Wang Y. C., Dou S., Xiong M. H., Sun T. M., Wang J., *ACS Nano*, **5** (2011), 3679–3692.

Willets K. A., Van Duyne R. P., *Annu. Rev. Phys. Chem.*, **58** (2007), 267–297.

Wu H. Y., Chu H. C., Kuo T. J., Kuo C. L., Huang M. H., *Chem. Mater.*, **17** (2005), 6447–6451.

Wu C., Zhou X., Wei J., *Nanoscale Res. Lett.*, **10** (2015), 1–6.

Yang S., Ye F., Xing D., *Opt. Express*, **20** (2012), 10370–10375.

Yuan Q., Lu D., Zhang Z., Chen Z., Tan W., *Trend Anal. Chem.*, **39** (2012), 72–86.

Zavaleta C. L., Garai E., Liu J. T. C., Sensarn S., Man-della M. J., Van de Sompel D., Friedland S., Van Dam J., Contag C. H., Gambhir S. S., *Proc. Natl. Acad. Sci. U. S. A.*, **110** (2013), 2288–2297.

Zhang B., Kumar R. B., Dai H. J., Feldman, B. J., *Nat. Med.*, **20** (2014), 948–953.

Zheng Y., Zhong X., Li X., Xia Y., *Part. Syst. Charact.*, **31** (2014), 266–273.

Zhuang Q., Jia H., Du L., Li Y., Chen Z., Huang S., Liu Y., *Biosens. Bioelectron.*, **55** (2014), 76–82.

Chapter 20

Enzyme Biocomputing: Logic Gates and Logic Networks to Interface and Control Materials

Marcos Pita

Instituto de Catálisis y Petroleoquímica (CSIC), C/Marie Curie 2 L10, 28049 Madrid, Spain

marcospita@icp.csic.es

20.1 Introduction

The technological revolution that has taken place in the last century relies on advanced computing systems. Many attribute the first conceived computer to Alan Turing, who proposed in 1937 "the use of hypothetical devices able to perform any conceivable mathematical computing operation if such can be represented as an algorithm" (Turing, 1937). Since the beginning of computers' development, the computing operations used to define the algorithms have relied on the logic operations defined by George Boole in the Nineteenth Century (Boole, 1854). Thereafter computing knowledge and technology have grown exponentially, first adopting transistors and later diving into miniaturization to develop nowadays' computers, which rely on nanotechnology. Still, almost every computing step is performed by sequential Boolean logic gates operations.

Biocatalysis and Nanotechnology
Edited by Peter Grunwald

ISBN 978-981-4613-69-9 (Hardcover), 978-1-315-19660-2 (eBook)
www.panstanford.com

Although conventional computing is completely established in society, there are research efforts towards the development of unconventional computing approaches. These approaches require processing the information in absence of silicon-chip devices (Adamatzky, 2006). One example of unconventional computing is chemical computing. In this case several chemicals mixed in a solution perform Boolean logic operations (Adamatzky, 2005). Chemical reactions including a catalytic process show additional virtues to mimic digital logic computing operations. Catalytic processes may take place in either homogeneous solutions or heterogeneous interfaces. Catalysts can be immobilized on solid supports, which may work simultaneously as transducers for the chemical information processing. Such process can be miniaturized down to single molecule computing operations, which may implement useful algorithms (Sienko et al., 2005). A particularly interesting kind of catalysts are those responsible for controlling and self-regulating living organisms, the biocatalysts. Most biocatalysts are enzymes, proteins specifically designed to catalyze, promote, and regulate biochemical reactions. The first tests of applying biomolecules to computing aimed at processing information through the DNA hybridization assembly, expecting to accomplish massive parallel data processing rather than serial, which might yield faster results for huge-enough combinatorial problems (Adleman, 1994). Enzymes, the most common biocatalysts, have also their place in unconventional computing; a groundbreaking new field has lately become known as enzyme biocomputing (Niazov et al., 2006). Although a computational application of biomolecular systems is no match to conventional computing, enzymes present several advantages to be considered an alternative. Enzymes are very specific to the chemical reactions they catalyze, often interacting only with one particular substrate among many similar ones. An enzyme can identify its specific substrate in a complex mixture of chemicals existing in either a cell or an artificial aqueous solution. These properties can suffice to pursue alternatives to Si-based electronics such as complex biosensing of several interconnected biomarkers with their current level being processed and computed to offer a final binary decision in the form of "yes/no" to some health-related problems. Additionally, the logic bioanalysis can be coupled to drug release or other intervention, yielding new therapeutic possibilities. Therefore, during the last decade, different approaches of using biomolecules to mimic the circuitry of digital processing units have been explored. Such approaches include enzymes as the pillars to mimic digital logic operations (Pita and Katz, 2011). Unconventional computing may find an application's niche where conventional computing may not be effectual, i.e., diagnosis or novel biomedical sensing. Regarding such applications, enzyme reactions permit a complex signal processing and analyze different biomarkers specifically designed for an injury or a medical condition. Current research addresses the design of signal-processing networks that rely on enzymatic reactions to process complex biochemical signals (Pita and Katz, 2011; Wang et al., 2010; Katz et al., 2012a). The challenge and at the same time the most difficult task in biosensing is the simultaneous analysis of biomarkers for multiple pathologies and/or physiological conditions. It should be noted that such biomarkers often interfere with each other or may be non-conclusive for a specific condition analyzed.

This chapter will analyze how enzyme biocomputing paradigms have been developed, how their complexity has grown from single logic operations to networks, the evolution from model to applied systems, and their interface with different materials to achieve applied systems and devices.

20.2 Enzyme Biocomputing

Binary logic operations are the basis of digitalization and computing. In electronics, digital circuits are built on combination of logic gates. There are several logic gates, each one able to make a specific decision. When logic gates are assembled, they form circuits or networks, which perform specific tasks. To define a digital operation we need three elements: the input signals, the operator, and an output (Boole, 1854). Inputs are the initial values provided to the system. The logic operator takes the input values and transforms them. There are different operators that differ in the nature of the transformation they perform; each operator tethers a relationship between each combination of input signals and the specific output given. More common logic operators are AND, OR, XOR, NAND, NOR, NOT, InhA, and InhB. All the results for each digital logic operator can be resumed as a truth table that correlates the output corresponding to each combination of inputs (Table 20.1). Finally, input and output values are digitalized, so they can only take two values: 0 or 1.

Table 20.1 Truth tables for different Boolean logic operations

AND		
Input A	Input B	Output
0	0	0
0	1	0
1	0	0
1	1	1

OR		
Input A	Input B	Output
0	0	0
0	1	1
1	0	1
1	1	1

XOR		
Input A	Input B	Output
0	0	0
0	1	1
1	0	1
1	1	0

NOT	
Input	Output
0	1
1	0

Inh A		
Input A	Input B	Output
0	0	0
0	1	1
1	0	0
1	1	1

Inh B		
Input A	Input B	Output
0	0	0
0	1	0
1	0	1
1	1	1

NAND		
Input A	Input B	Output
0	0	1
0	1	1
1	0	1
1	1	0

NOR		
Input A	Input B	Output
0	0	1
0	1	0
1	0	0
1	1	0

Realization of logic operations with enzyme-catalyzed reactions implies a series of challenges. The first one is to identify each agent in the biochemical reaction with one of the three elements essential to accomplish a digital logic operation: the inputs, the operator, and the output. Selecting suitable chemicals for each element is critical because such selection will determine how the enzyme-logic gate will work. A possible configuration may be to set the chemical substrates as input signals, the enzyme(s) as the operator's machinery and one of the chemical products as output signal. Another possible configuration would be to define enzymes as input signals, and their substrate as the machinery. A third kind of configuration may comprise both enzymes and some of its substrates are input signals, and different substrates as machinery. Due to its catalytic nature, enzymes are not likely to work as output signals.

A second limitation to overcome is how to read the information processed by the enzyme logic gates. This issue can be addressed taking advantage from the chemical and/or physical changes that take place when the enzyme logic gate operates and links such change to a suitable transducing material. A key research field in biocomputing has been the development of materials that can sense signals originated by the biochemical reactions and give a response only when the output coming from the enzyme logic operation commands it. Whereas in electronics, 1 and 0 input and output signals consist of either the appearance or the absence of an electric current or voltage, in enzyme computing such currents are substituted by changes in concentration, oxidation state or other physical properties of the compounds taking part in the enzymatic reaction. Measurable parameters whose values can change over the course of a biochemical reaction are the solution pH, the optical absorbance at a given wavelength, or even temperature.

An additional challenge for enzyme biocomputing is connecting single logic operations to form a network able to process information at a higher degree of complexity. It is very important to carefully design the network to avoid undesired cross-reactivity between different logic operations belonging to the same network, and may yield false information processing.

A final challenge regards digitalization of signals and addition of correcting algorithms to reduce the "chemical noise" and avoid processing errors that may result from the operation of the enzyme logic gates. This is particularly challenging due to the nature of enzymatic reactions' kinetics, generally driven by the Michaelis–Menten model. Following such kinetics the reaction rate increases linearly with substrate concentration during the initial phase, reaching a substrate concentration value where saturation occurs and the reaction rate vs. concentration reaches a plateau. Correlating saturation concentration with the value 1 for an input signal is straightforward; however, for the case of input signal 0, very low substrate concentrations yield a large slope in the response curve, which translates into a noisy output signal that causes noise amplification. Minimizing the spread of this noise has been matter of research (Katz et al., 2010a).

An introduction to enzyme logic gates' realization may start with enzymatic reactions that model single logic operations such as AND, OR, XOR (Strack et al., 2008a). All these enzyme logic gates were designed as follows: the machinery includes the enzyme substrates and a dye precursor, i.e., 2,2′-azino-bis(3-ethylbenzothiazoline-6-sulfonic acid (ABTS), and the enzymes that will carry out the reaction were the input signals, as can be seen in Fig. 20.1:

- AND gate. Machinery: 100 mM glucose + 100 μM ABTS dissolved in an air-equilibrated aqueous solution. Input A: the enzyme glucose oxidase (GOx). The input value 0 is defined as no enzyme added, whereas input value 1 corresponds to 0.15 enzyme activity units. Input B: the enzyme Microperoxidase (MP11). The input value 0 is defined as no enzyme added, whereas input value 1 corresponds to MP11 540 nM. The output signal is defined as the appearance of $ABTS^{+}$, characterized by its light absorption centered at λ = 415 nm which can be monitored with a UV-Vis spectrophotometer. Output 1 occurs when the input values A and B is 1,1. In this case GOx oxidizes the glucose giving gluconic acid and H_2O_2, MP11 takes the appearing H_2O_2 and ABTS yielding $ABTS^{+}$. The other three possible input combinations are 1,0; 0,1; and 0,0; all of them imply the absence of either GOx, MP11 or both of them, therefore forbidding the final oxidation of ABTS.

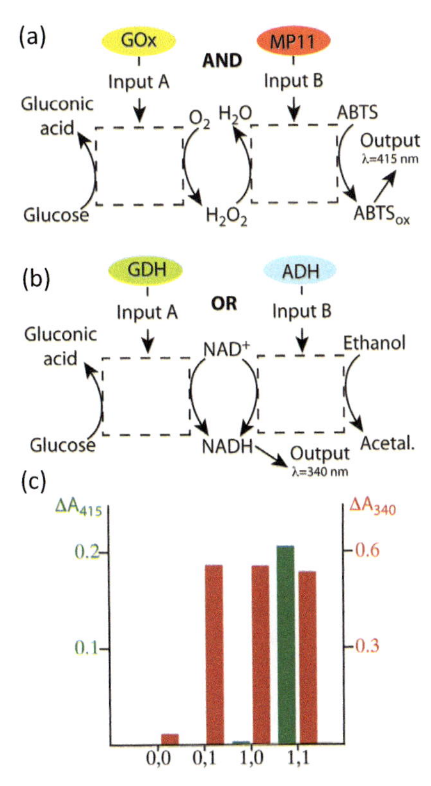

Figure 20.1 (a) Scheme of the biochemical reactions taking place in the AND enzyme logic gate. (b) Scheme of the biochemical reactions taking place in the OR enzyme logic gate. (c) Output results for the AND and the OR gates. Adapted from Strack et al. (2008a).

- OR gate. Machinery: ethanol, glucose, NAD^+. Input A: glucose dehydrogenase (GDH). The input value 0 is defined as no enzyme added, whereas input value 1 corresponds to 1.5 enzyme activity units. Input B: alcohol dehydrogenase (ADH). The input value 0 is defined as no enzyme added, whereas input value 1 corresponds to 3.14 enzyme activity units. The output signal is NADH, characterized by a strong absorption band at λ = 340 nm, which can also be monitored with a UV-Vis spectrophotometer. Adding either GDH or ADH yields the reduction of NAD^+ to NADH and consequently the output is 1. Adding both enzymes yields the same effect on NADH production. 0.0 is the only input combination that yields 0 output.

- XOR gate. Machinery: ethanol, H_2O_2, NADH, and NAD^+. Input A: ADH. The input value 0 is defined as no enzyme added, whereas input value 1 corresponds to 3.14 enzyme activity units. Input B: MP11. The input value 0 is defined as no enzyme added, whereas input value 1 corresponds to 540 nM MP11. Output: absolute value for variation of absorption at λ = 340 nm. Input combination 0,0 does not add any biocatalyst to the machinery, so [NADH] remains constant and there is no change in absorbance, which is an output 0. Input combination 1,0 adds ADH to the system, which yields an increase in [NADH], rising the absorbance at λ = 340 nm, which corresponds to an output 1. Input combination 0,1 adds MP11 to the system, consuming NADH and H_2O_2, therefore decreasing the absorbance at λ = 340 nm, which corresponds to an output 1. Final combination 1,1 triggers two biocatalytic reactions, ADH produces NADH while MP11 consumes it in such an adjustment that [NADH] remains constant, keeping the absorbance at λ = 340 nm constant, which corresponds to an output 0.

It should be noted that to maximize the performance of model enzyme logic gates, the concentrations used were above typical physiological levels and enhance the output responses. One thing in common among these first models was that the biocatalysts comprised the input signals. Many other enzyme logic gates have been designed and developed, such as NAND and NOR (Zhou et al., 2009), considering the universal logic operations because in electronics their combination allows to develop any other logic system. The key to develop NAND and NOR systems was the development of an inverter system, able to transform the output 1 in either an AND or an OR gate into a 0 output and, vice versa, to transform the output 0 in the same AND/OR gates into an output 1. Such an inverter had as input glucose concentration, the output of AND and OR previous gates. Glucose fed the inverter machinery, which comprised the enzymes GOx, MP11 and ADH as well as the chemicals ethanol and NAD^+, in equilibrium with air. Glucose input 0 means that ADH oxidizes ethanol while reduces NAD^+ to NADH, detected at λ = 340 nm as output 1. Glucose input 1 implies its oxidation by GOx, producing H_2O_2. The H_2O_2 is taken by MP11 to oxidize the

NADH produced by ADH enzyme, thereby avoiding the absorption at λ = 340 nm and yielding an output 0.

Single enzyme logic gates may be, as conventional logic gates, connected to each other and perform sequential enzyme-logic operations. It is possible to concatenate three AND enzyme logic gates mimicking a keypad lock operation (Strack et al., 2008b) combining some substrates and enzymes as input signals. The enzymatic keypad lock comprises three AND logic gates, where the output of the first gate works also as input for the second gate and the second output works as input for the third and last gate. The first AND enzyme logic gate was operated by the inputs sucrose and the enzyme invertase (Inv). Input combination **1,1** was the only combination to produce glucose, which is the output **1**. The second AND enzyme logic gate was served with glucose input (coming from the first AND gate) and GOx. The machinery for this operator is the existing O_2 dissolved in the solution. Again only the input combination **1,1** yields the production of H_2O_2, which is the output **1**. The third AND enzyme logic gate is fed with the inputs H_2O_2 and MP11, and the machinery is ABTS. The input combination **1,1** yields the final output, $ABTS_{ox}$, which absorbs radiation at λ = 415 nm. The above-mentioned keypad lock was assembled by preparing multiple solutions, each one resembling a keypad button, where all of them but three were placebos. The three activating solutions, namely A, B and C, corresponded to the input **1,1** for the first logic gate, input B = **1** for the second logic gate and input C = **1** for the third logic gate. When ABC was set to the enzymatic keypad lock, the output resulted **1**. If any of the placebos was added, then an essential chemical was substituted by a non-function solution and the final output was **0**. A final requirement for the keypad lock development was to ensure that the correct order of inputs A, B, and C was the only one granting the output **1**. Such challenge was met by tuning the threshold values for the $ABTS_{ox}$ generated and the reaction time after each input, which was set to 20 min.

Besides enzyme logic gates, other computational systems including a colorimetric-based output has been developed, such as flip-flop S-R memory units (Pita et al., 2009) or signal filters (Privman et al., 2010a). Set-Reset memory units are switchable between **0** and **1**, which are activated by two input signals: set (S) and reset (R). S and R can be applied at two digital levels defined as **0** and **1**. Application of **S = 0** and **R = 0** must preserve the current state of the memory unit independently of its value. **S = 1** should take the system to state **1**, while **R = 1** must take the system to the unit state **0**, regardless of the initial state of the unit. The operation of such memory unit is explained by means of a table that shows the states of the unit before (Q_t) and after (Q_{t+1}) applying different combinations of the signals S and R, (Table 20.2).

Table 20.2 Operation of the Set-Reset flip-flop memory unit for different combinations of Set-Reset signals

SET–RESET FLIP FLOP			
Set (S)	**Reset (R)**	**Initial State, Q_r**	**New State, Q_{t-1}**
0	0	Q_t (0 or 1)	Q_t(0 or 1) no change
0	1	Q_t (0 or 1)	0
1	0	Q_t (0 or 1)	1
1	1	Q_t (0 or 1)	Not Allowed, unstable

The S-R memory unit was mimicked by enzyme-catalyzed reactions using a key reversible enzyme called diaphorase (Fig. 20.2). Diaphorase oxidizes NADH to NAD^+ while reducing an oxidized redox mediator. The memory unit was developed for two reading outputs, a colorimetric one where the mediator used was 2,6-dichloroindophenol (DCIP) and an electrochemical one where the mediator was ferricyanide. The memory machinery included the enzymes diaphorase, GOx, HRP, and ADH together with the substrates DCIP and NAD^+. For the colorimetric case, the SET input was triggered by glucose addition, which starts the already mentioned enzymatic cascade that comprises GOx and HRP, oxidizing DCIP and yielding an absorption band at λ = 600 nm. The RESET input was ethanol, which is oxidized by ADH and simultaneously reduces NAD^+ to NADH. NADH is used by diaphorase to reduce the $DCIP_{ox}$, "erasing" the SET output and turning it into RESET (Pita et al., 2009). The signal filters aim to reduce the chemical noise, which generally is a factor for **0**-kind outputs. They consist of an additional chemical that reacts spontaneously with the output chemical of the enzyme logic gate to be filtered, taking it back to its initial configuration. The strength of the filter can be regulated by the concentration of that additional chemical (Privman et al., 2010a).

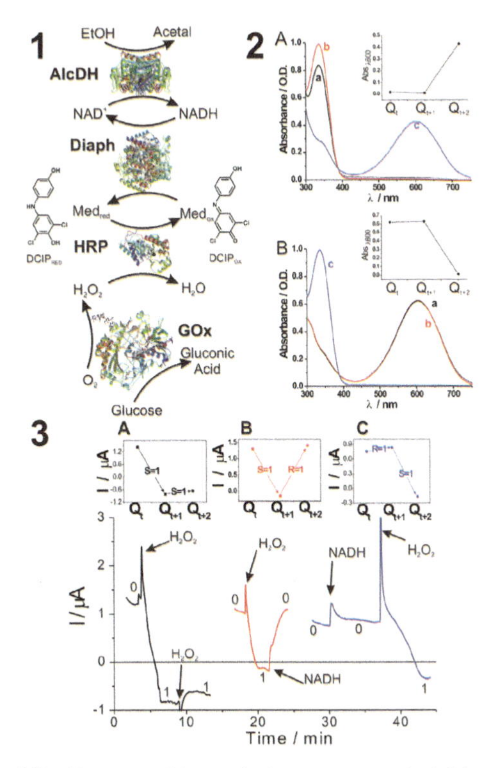

Figure 20.2 (1): Enzyme biocatalytic system mimicking a Set-Reset flip-flop memory operation. The system has a core that operates on DCIP formed by HRP and Diaphorase, and a terminal part formed by ADH and GOx which sets or resets the Diaphorase reduced or oxidized step. (2) Absorbance spectra corresponding to the flip-flop system. (2A) The initial state is reduced (Q_t = 0): (A) the system is pretreated with NADH, no signals applied (Q_t = 0); (B) the system treated with the Reset signal **1** (ethanol), keeping the system in **0** state (Q_{t+1} = 0); (C) The system treated with the set signal **1** (glucose) converting the system to **1** state (Q_{t+2} = 1); Inset: Absorbance at λ = 600 nm at the initial **0** state, after **Reset = 1** and after **Set = 1** signals. (2B) The initial state of the system is oxidized (Q_t = 1): (A) no inputs applied (Q_t = 1); (B) The system treated with the set signal **1** (glucose) preserving the system in **1** state (Q_{t+1} = 1); (C) the system treated with the reset signal **1** (ethanol) converting the system to **0** state (Q_{t+2} = 0). Inset: absorbance at λ = 600 nm at the initial **1** state, after **Set = 1** and after **Reset = 1** signals. (3) Chronoamperometric read out of the state changes in the enzyme-based flip-flop system starting from $\boldsymbol{Q_t}$ **= 0**: (3A) Sequence of **S = 1** and **S = 1** signals; (3B) sequence of **S = 1** and **R = 1** signals; and (3C) sequence of **R = 1** and **S = 1** signals. Copyright ACS 2009. Images reproduced with permission from Pita et al. (2009).

When mimicking Boolean logic operations with enzyme-catalyzed reactions, it is very important to minimize the noise and the possibility to transmit wrong information. Whereas a digital operation can have only two values, either **0** or **1**, enzymatic reactions turn out in the form of continuous measurable signals rather than discrete values. Thresholds are the tools to digitalize such continuous output signals; however, thresholds need to be defined, as they are arbitrary levels that define the maximum possible value which limits the **0** signal and the minimum possible value to reach **1** signal value. Between the two thresholds appears an undefined region that yields a non-determined output for the logic operation. The output of the enzyme logic gates is defined as a physical signal dependent on the presence of a chemical at a certain concentration. Enzymes are very sensitive to their substrates in a very selective way, which produce some noise in the output signal even when the substrate appears at a concentration lower than the threshold value for input value **0**. This effect can hinder the performance of enzyme-logic systems and is worthy to be analyzed. Single (Privman et al., 2008) and concatenated (Privman et al., 2009a) enzyme logic gates, approaches to reduce the noise (Melnikov et al., 2009), and research on enzymatic systems that offer a sigmoidal response (Privman et al., 2009b) have been published.

Most of the studies mentioned up to this point regard the conceptual development of how to drive a biocatalytic reaction to mimic a logic operation; however, the step forward has to go necessarily in only one direction: how to make these logic systems operate on sensing materials. Such achievement yields all application possibilities for enzyme logic gates: multiple biosensing, command of responsive materials, discrimination of normal healthy situations from hazard levels, to correlate biomarkers and make them trigger a specific actuator, etc. Many examples have been described, although the most successful linkage of enzyme logic gates with smart responsive materials is built on one pillar: the control of pH.

20.3 Connecting Enzyme-Logic Gates and Sensor Materials

The output signals of the enzyme logic gates mean a change in the environment that can be coupled with materials specifically designed to sense and act accordingly. There are at least two different output signals suited for this purpose, which may act independently or together: electrochemical signals and pH changes. pH changes performed by single enzyme logic gates were firstly described as AND or OR logic operations including a reset system, Fig. 20.3 (Tokarev et al., 2009). The described AND gate comprised urea, sucrose and Na_2SO_4 in an aqueous solution as machinery with the enzyme GOx as input A, and the enzyme Inv as input B. The system keeps its pH value unless the input combination 1,1 is added to the solution. 1,1 combination initiated a series of biocatalytic oxidative processes, yielding gluconic acid in the further step. Within the range of minutes, pH decreases below 4. The coupled Reset system had the enzyme urease as input signal. Addition of urease degrades the urea present in the system, producing ammonia. This reaction pulls the pH of the working solution back to its original neutral value or even to a slightly alkaline value. The OR gate machinery comprised an aqueous solution including glucose, ethyl butyrate and urea, and GOx and esterase as inputs A, and B, respectively. Any combination different from 0,0 added an enzyme to the system that produces an acid, gluconic acid in the case of GOx or butyric acid in the case of esterase; therefore pH is acidified for inputs 0,1; 1,0 and 1,1. The OR enzyme logic gate was also coupled to the same reset system. These pH-controlling enzyme logic gates were tested against a pH sensitive material, an alginate membrane grafted on an indium-tin oxide (ITO) electrode. The changes in the gel porosity upon pH control were monitored through atomic force microscopy (AFM), and other changes such as the membrane permeability were monitored by impedance spectroscopy; the response of the material to setting and resetting the pH value was confirmed.

One of the finest materials for sensing pH changes ranging from neutral to moderately acidic is polyvinylpyridine (PVP), particularly polyvinyl-2-pyridine (P2VP) or polyvinyl-4-pyridine (P4VP). PVP can be grafted on materials activated with bromosilanes, mainly oxides like silica nanoparticles (Motornov et al., 2008a) or ITO. Nitrogen heteroatoms present in each pyridinyl monomer of the PVP are protonated within the pH range 5.5–4.5. As consequence of the protonation PVP turns from hydrophobic to hydrophilic. This range is compatible with the action range of the above-described AND-OR gates, making them suitable

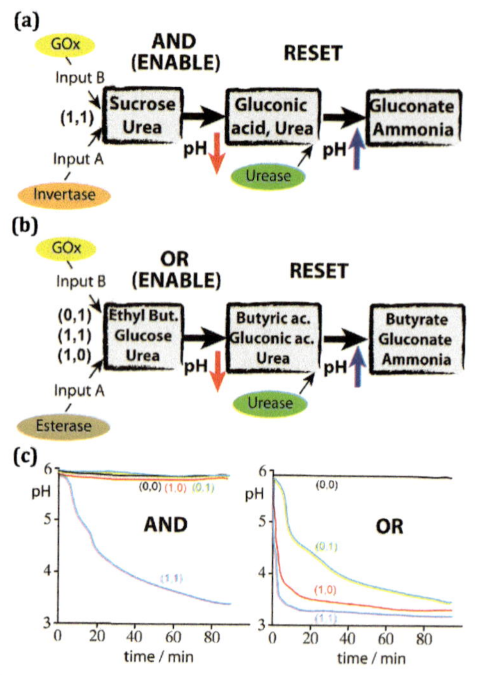

Figure 20.3 (a) Reaction scheme for the biochemical AND logic gate with the enzymes GOx and Invertase used as input signals to activate the gate operation. The absence of enzyme is taken as 0 input signal, the presence of enzyme is taken as 1. The reset function is triggered by urease. (b) Reaction Scheme for the biochemical OR logic gate with the enzymes GOx and esterase used as input signals to activate the gate operation. The absence of enzyme is taken as 0 input signal, the presence of enzyme is taken as 1. The reset function is triggered by urease. (c) Typical evolution of pH upon each input combination gated by the AND (left) and the OR (right) operations. Adapted from Tokarev et al. (2009).

to control the protonation of PVP-modified silica nanoparticles (Fig. 20.4b). The output **0** of the enzyme logic operations means that no acid is generated and pH remains neutral, so the grafted PVP is deprotonated and hydrophobic, keeping a shrunk conformation; therefore, the particles aggregated into clusters of ~3 μm in size. On the other hand, an output **1** of the enzyme logic gates acidified the solution pH, causing the protonation of the PVP, turning into a hydrophilic and swollen configuration; additionally the particles repel and separate from each other. Urease-based reset function forces the system to the initial configuration. Aggregation and disaggregation of the particles was possible to monitor by AFM and dynamic light scattering. Other materials sensitive to pH controlled with these enzyme logic gates were Pickering emulsions (Motornov et al., 2009a), which could shift from water-in-oil emulsion to oil-in-water and therefore turn on (2.5 mA at pH = 4) and off (100 μA at pH 6) the conductivity of the solution.

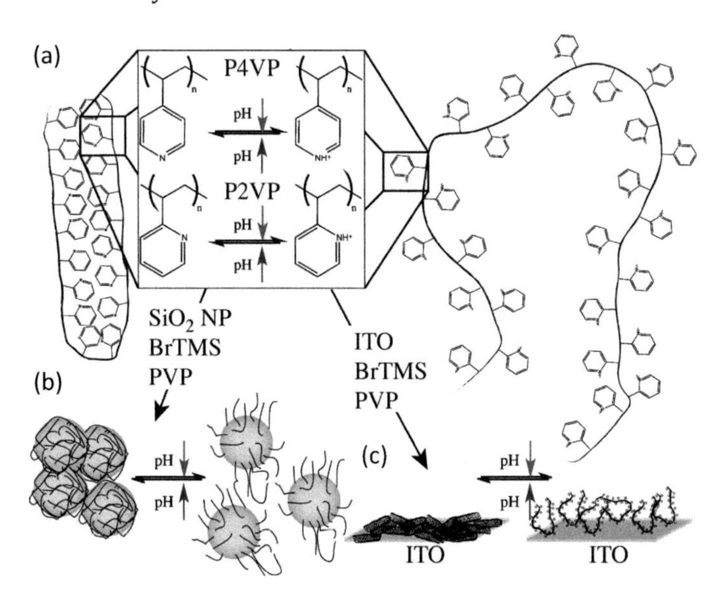

Figure 20.4 (a) Schematic representation of PVP protonation and deprotonation conformations. Central magnification shows the configurations for P2VP and P4VP. (b) PVP grafted on 200 nm silica nanoparticles that aggregate and disaggregate upon pH changes, adapted (Motornov, 2008a). (c) Shrunk and swollen conformations of PVP grafted on an ITO glass surface commanded by pH value (adapted from Tam et al., 2008a).

It can be noted that the enzyme logic gates used to control pH described in Fig. 20.3 use the enzymes as input signals. As enzymes are the biocatalyst of the reactions it is obvious to develop systems where the enzymes are included in the machinery and their substrates are the input signals. Such configuration is coherent with regular chemical engineering processes and enables a transition to flow-analysis systems. This can only be accomplished by immobilizing the enzymes. The first pH-oriented enzyme logic gates where the enzymes were immobilized used two sets of particles: amino-functionalized silica microparticles as enzyme support and carboxylic functionalized gold-shelled magnetic nanoparticles as plasmonic pH sensors (Pita et al., 2008). In this case the biocatalytic process was transformed from a homogeneous one, where the total reaction occurred in a liquid phase, to a heterogeneous one including two solid phases and a liquid phase. The separation of phases allowed preparing a one-pot reaction, where the pH shifted originated at the bottom of the phial and the pH shift output could be magnetically concentrated on the top of the aqueous phase by simply placing an external magnet slightly above the top of the phial. The mercaptopropionate-modified particles were protonated upon a pH decrease given by the enzyme logic gates' operation, decreasing the density of electrons on the gold shell and therefore blue-shifting the Plasmon wavelength band. When the reset system was activated, the mercaptopropionic acid was deprotonated, reversing the Plasmon to its original resonance band. The study was successful thanks to keeping an accurate control of the final acidic pH thus avoiding irreversible aggregation of the nanoparticles.

Another transducer used to reveal pH-controlling enzyme logic gates was a semiconducting silicon chip enhanced with gold nanoparticles functionalized with mercaptopropionic acid (Kramer, 2009). In this case the protonation-deprotonation controlled by the enzyme logic gates caused a reversible change in the capacitance of the Si chips, altering its flat band potential and serving as output signal of the system.

PVP turned into a major role material as interface suitable to transduce biocatalytically generated pH changes into electronic signals. It can be grafted on ITO electrodes (Taylor, 1995; Fig. 20.4c), and loaded with redox organometallic complexes, which can act as mediators in enzymatic reactions (Kenausis et al., 1996). An ITO electrode is a glass slice covered with a thin layer of the tin-doped indium oxide and can be easily functionalized with silane coupling agents (epoxysilane, bromosilane, etc.), thus introducing a functional group that facilitates the grafting of PVP. In particular, an epoxy-terminated functionalization yields a polymer brush configuration, providing a single point polymer-surface bond and the consequent high mobility of the polymer chains. A PVP-ITO electrode demonstrated to switch off the electric current at pH higher than 5.5 while switching it on at pH lower than 4.5 due to its already explained protonation-deprotonation process (Fig. 20.4c). PVP-ITO electrodes have been used to control the signal transmission from hydrophilic redox probes such as ferrocyanide upon the pH commanded by the external enzyme logic gates AND and OR described above (Wang et al., 2009). The pH-transducing electrode was also shown to sense and act according to the output signal given by a complex biocatalytic network consisting of five enzyme logic operations processing 16-input combinations (Privman, M., et al., 2009).

An additional advantage of PVP is the possibility to make its pyridinyl derivatives to act as ligands to metal complexes, becoming redox-active and also charge carriers (Kenausis et al., 1996; Fig. 20.5). A very successful configuration is loading PVP with osmium complexes because once they are attached to the polymer they stand covalently linked, and they can act as redox mediators for many redox enzymes; additionally, the selection of ligands can influence the redox potential, allowing its tuning for an optimal mediated electron transfer. Initial works used a high load of Os complexes pending from the polymer (Kenausis et al., 1996), making a redox polymer suitable for electron transfer between the enzymes and the electrode. However, by decreasing the concentration of Os complex bonded to the PVP, it became possible to switch on and off the electrochemical process upon pH variation: under acidic pH the Os-PVP is hydrophilic, swollen and its flexibility allows the electron transfer. At pH > 5.5 the pyridinyl derivatives deprotonate; the polymer turns hydrophobic, shrinks on the electrode surface, and its rigidity forbids the electron transfer. This on-off process has been shown to work for redox enzymes such as GOx, which used the Os complex as electron acceptor in the absence of oxygen (Tam et al., 2008a). It was also shown to be controlled with pH-output enzyme logic gates (Tam et al., 2008b).

The PVP and Os-PVP modified electrodes show two main application fields: bioelectrochemical sensors and switchable biofuel cells. PVP-modified electrodes have been mainly used to demonstrate the conceptual development of multi-biosensing: enzyme logic gates are designed, interconnected and optimized to process the presence or absence of several biochemicals in such a way that the last step is the one that produces an acid. pH is therefore changed and the electrode is switched on, giving a response. On the other hand, the Os-PVP modified electrode offers a more interesting possibility. The Os complex can act as redox transfer mediator for an enzyme working at the cathode of a biofuel cell. A typical enzyme for biofuel cell cathodes is laccase, a polyphenol oxidase that naturally oxidizes polyphenols by reducing O_2 to H_2O directly, avoiding the formation of H_2O_2, a harmful compound. The Os-PVP polymer can substitute the polyphenols and act as electron donors for the laccase. As the PVP is still a pH-responsive switchable polymer, by adding a pH-controlling set of enzyme logic gates into the cathodic compartment of the biofuel cell it was possible to switch the electric current on and off by the command of the logic gates' input signals. Such a biofuel cell was developed for single enzyme logic gates (Amir et al., 2009) and 4 inputs logic networks (Privman, M., et al., 2009); moreover, it was also demonstrated that the enzymes used to process the logic gates can be cross-linked to antibodies and set specific antigens as input signals (Tam et al., 2009a). Once antibody-antigen plays a role, the panoply of logic operations able to control the process is almost endless.

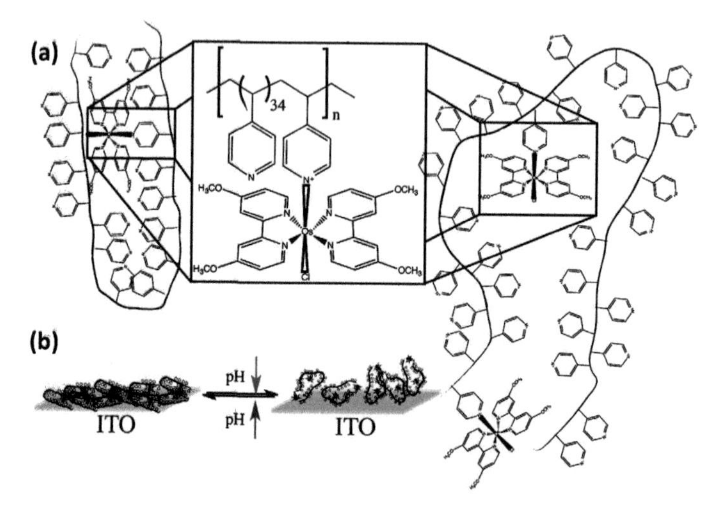

Figure 20.5 (a) Poly-4-vinylpyridine loaded with an average of one osmium complex $Os(dmobpy)_2Cl_2$ each 35 monomers, showing the shrunk (left) and swollen configuration acquired at neutral or acidic pH, respectively. (b) Schematic representation of the grafted Os-PVP on ITO glass electrodes and its configuration change upon pH changes.

20.4 Polymer-Brush-Modified Electrodes with Bioelectrocatalytic Activity Controlled by pH Value

Functionalizing electroactive surfaces with signal-responsive polymers yielded the establishment of a completely new electrode switching behavior (Motornov et al., 2008b; Combellas et al., 2009; Choi et al., 2007). The most common approach relies on surface-attached pH-responsive polyelectrolyte brushes that can reversibly switch between an electrically charged hydrophilic form and a neutral hydrophobic state, with the first one being permeable for oppositely charged redox species and the latter one not permeable for ionic species (Motornov et al., 2008b; Harris et al., 2000; Park et al., 2004). For example, switchable behavior of pH-controlled poly(4-vinyl pyridine) (P4VP)-brush attached to an electrode surface can be followed by impedance spectroscopy demonstrating high and low electron transfer resistance for the hydrophobic (neutral, pH 6) and hydrophilic (protonated, pH 4) states, respectively (Macvittie et al., 2013; Fig. 20.6). The primary

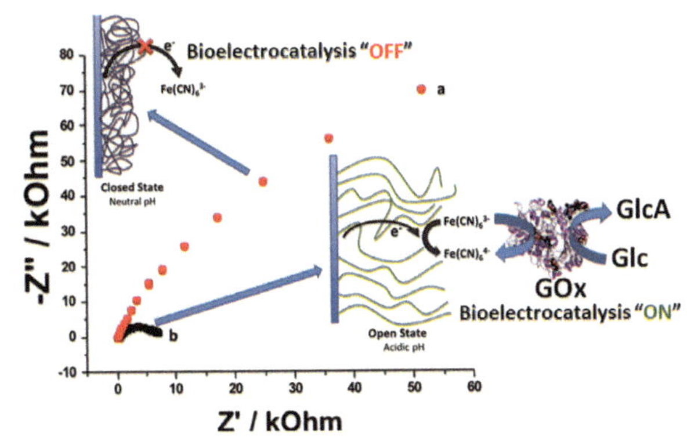

Figure 20.6 Impedance spectra obtained on the "closed", pH 6.0, (a) and "open", pH 4.0, (b) states of the poly(4-vinyl pyridine) (P4VP)-modified electrode. Bias potential −450 mV (vs. a quasi-reference electrode), frequency range 100 mHz–10 kHz. The impedance spectra were recorded in the presence of $[Fe(CN)_6]^{3-}$ (1 mM) and in the absence of GOx and glucose. The bioelectrocatalytic process converting glucose (Glc) to gluconic acid (GlcA) in the process biocatalyzed by GOx and mediated by $[Fe(CN)_6]^{3-}$ was analyzed separately by cyclic voltammetry (not shown in the figure). Adapted from MacVittie et al. (2013), with permission; Copyright American Chemical Society, 2013.

pH-switchable redox reaction of $[Fe(CN)_6]^{3-}$ can be coupled to the oxidation of glucose catalyzed by GOx and mediated by $[Fe(CN)_6]^{3-}$ (Fig. 20.6). Polyelectrolytes such as P4VP can switch on and off their permeability upon local pH-changes generated on the electroactive surface by electrochemical reactions, in other words an applied potential on the electrode can switch on or off the electrode (Tam et al., 2010a; Tam et al., 2010b).

An example is the modification of an ITO electrode surface with a P4VP-brush, which later was utilized to reversibly switch the activity at the interface upon electrochemical signals (Tam et al., 2010a). An applied potential high enough to reduce molecular oxygen electrochemically yielded the consumption of protons at the electrode surface, thus increasing the pH value and causing the deprotonation and re-structuring of the immobilized P4VP-brush, Fig. 20.7A. The initial state at pH = 4.4 for the protonated P4VP-brush allows the anionic $[Fe(CN)_6]^{4-}$ redox species to reach the electrode (Fig. 20.7B, curve a). On the other hand the local pH of 9.1 achieved electrochemically caused the deprotonation of the P4VP. The state reached by the polymer brush under such conditions is shrunken and hydrophobic, forbidding the access of anionic redox species and inhibiting its redox reaction (Fig. 20.7B, curve b). However, local pH changes do not provoke big pH changes in the bulk electrolyte solution. The interface relaxed to its electrochemically active state by turning off the applied potential while homogenizing the cell solution either by stirring or by letting a 10 min diffusion take place and the P4VP reaches the pH equilibrium. The approach described above allowed to "close" the electrode surface by means of a reversible inhibition caused electrochemically (Fig. 20.7B, inset). The opposite electrochemically triggered "opening" process was achieved for an electrode modified with a mixture of polyacrylic acid (PAA) and poly(2-vinyl pyridine) (P2VP) (Tam et al., 2010b). Similarly to the P4VP-only modified electrode example, the pH at the interface value was raised upon electrochemical reduction of O_2 causing the conformational change of the polyelectrolyte from neutral to negatively charged state due to deprotonation of the polyacrylic acid carboxylic residues (Tam et al., 2010b). This caused the electrode switching from neutral to a negatively charged state, granting cationic redox species (e.g., $[Ru(NH_3)_6]^{3+}$) access. This approach may be applied to several interfacial systems, providing a panoply of switchable electrodes equipped with external controllable activity, and may be useful for various bioelectrocatalytic applications, such as biofuel cells or biosensors.

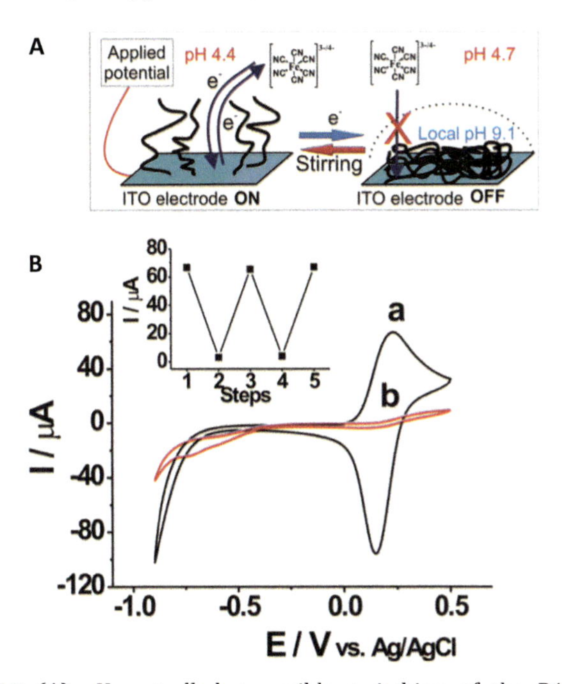

Figure 20.7 (A) pH-controlled reversible switching of the P4VP-brush between "ON" (left) and "OFF" (right) states allowing and restricting the anionic species penetration to the electrode surface, thus activating and inhibiting their redox process. (B) Cyclic voltammograms obtained on the P4VP-brush-modified ITO-electrode in the presence of 0.5 mM $K_4[Fe(CN)_6]$: (a) prior to the application of the potential on the electrode, (b) after application of −0.85 V to the electrode for 20 min. The background electrolyte was composed of 1 mM lactic buffer (pH 4.4) and 100 mM sodium sulfate saturated with air. Inset: The reversible switching of the peak current value upon "closing" the interface by the electrochemical signal and restoring the electrode activity by the solution stirring. Adapted from Tam et al. (2010a), with permission; Copyright American Chemical Society, 2010.

It is possible to go a step further by sophisticating the pH-switchable interfaces upon modifying the electrode with mixed-polymer systems (Motornov et al., 2008b; Motornov et al., 2009b; Tam et al., 2010c). For example, a mixed brush comprising poly(2-vinylpyridine) (P2VP) and polyacrylic acid (PAA) linked to an ITO electrode showed the three possible charged states by controlling the bulk pH: At pH 6, PAA is dissociated and the mixture negatively charged; at pH 3, the P2VP component is protonated and the mixture is positively charged, and at pH = 4.5 the mixture becomes neutral due to the compensation of PAA and P2VP opposite charge values (Motornov et al., 2009b; Tam et al., 2010c; Fig. 20.8A).

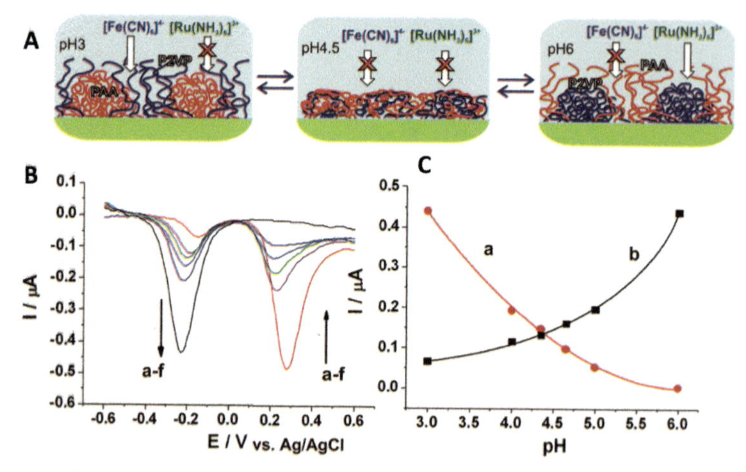

Figure 20.8 (A) The polymer brush permeability for differently charged redox probes controlled by a solution pH value: (left) the positively charged protonated P2VP-domains allow the electrode access for the negatively charged redox species; (middle) the neutral hydrophobic polymer thin-film inhibits the electrode access for all ionic species; (right) the negatively charged dissociated PAA-domains allow the electrode access for the positively charged redox species. (B) The differential pulse voltammograms (DPVs) obtained for the mixed-polymer (P2VP/PAA) brush in the presence of $[Fe(CN)_6]^{4-}$, 0.5 mM, and $[Ru(NH_3)_6]^{3+}$, 0.1 mM, at the variable pH of the solution: (a) 3.0, (b) 4.0, (c) 4.35, (d) 4.65, (e) 5.0 and (f) 6.0. The background solution was composed of 0.1 M phosphate buffer titrated to the specified pH values. (C) The peak current dependences on the pH value for the anionic, $[Fe(CN)_6]^{4-}$, (a) and cationic, $[Ru(NH_3)_6]^{3+}$, (b) species, as derived from the DPVs measured at the variable pH values. Adapted from Tam et al. (2010c), with permission.

The control of the bulk pH allows discriminating different redox species (i.e., $[Fe(CN)_6]^{4-}$ and $[Ru(NH_3)_6]^{3+}$) present in the electrolyte solution depending on their charge sign, which is only allowed if opposite to the polymer brush mixture charge sign. When the polymer mixture is positively charged at pH < 4, the $[Fe(CN)_6]^{4-}$ finds access to the conducting support granted, showing electrochemical redox activity. On the other hand the $[Ru(NH_3)_6]^{3+}$ species can access the electrode only at pH > 5, when the PAA-P2VP mixture brush is negatively charged; when the pH value is between 4 and 5, the PAA-P2VP mixture behaves neutral and is not permeable for either positive or negative redox probes, forbidding the access to the electrode surface and therefore muting any electrochemical reaction. Changing the pH value of the electrolyte solution gradually yields a reversible transition from the electrochemical response of $[Fe(CN)_6]^{4-}$ to the redox event of $[Ru(NH_3)_6]^{3+}$ and turns back to the initial $[Fe(CN)_6]^{4-}$ response (Motornov et al., 2009b; Tam et al., 2010c), Fig. 20.8B–C. This tunable behavior of the switchable-modified interface will allow redirection of bioelectrocatalytic processes from one pathway to another, thus controlling bioelectrocatalytic reactions with external signals.

20.5 Development of Sensors Including Smart Materials Controlled by Enzyme Logic Gates

The synergy of materials that can respond to changes in their local environment and networks of enzymes able to produce such changes can be used to build advanced sensors. The major advantage of such sensors is to include in the answer provided already processed information, which may allow devices with a higher autonomy range as well as simplicity for final potential users. Efficiency in this direction requires enzyme logic networks that work beyond the proof of concept, and can suffice a reliable interface between physiological systems and artificial sensors. In the case of medical applications, the enzyme logic gates have to process the information provided by specific biomarkers related to the monitored medical condition(s), take such biomarkers as input signals, and finally offer processed information as a final binary answer like "yes" or "no." Such answer should take in account each biomarker present as healthy level or pathologic level. Examples have been described for the identification and the analyses of several proteins marking multiple sclerosis

(Margulies et al., 2009). Enzyme logic gates can process simultaneously multiple biomarkers, each of them related with more than one injury but each injury being responsible for only one combination of biomarkers, therefore pointing out which injury is taking place. Such systems should read significant changes in the amount of those specific biomarkers provoked by a physiological reaction upon injury. An early example of using enzyme logic gates to process and detect the biochemical information signaling and differentiate two physical injuries such as a traumatic brain injury and hemorrhagic shock using an AND-Identity biocomputing system was reported (Manesh et al., 2009). The detection and information processing of the biomarkers glucose, lactate and norepinephrine by a logic system comprising the enzymes GDH, HRP, and lactate oxidase (LOx) could differentiate those two injuries.

Research efforts have been driven to the increase of complexity involving biocomputational systems. The evolution has taken two different directions, multiplexing parallel enzyme logic gates or the creation of logic gates' networks. An enzyme logic network has been used to differentiate between traumatic brain injury and a soft tissue injury, analyzing the information given by ten different biomarkers. A high amount of biomarkers raises the reliability of the system when compared with a simpler one (Halamek et al., 2010a). A large number of specific biomarkers can be correlated using several enzymatic steps to increase the effectiveness of the network, although possible interference of chemical intermediates should be avoided. In order to overcome such limitations multiplexing the system in parallel reactions is a promising option. A diagnostic method to analyze simultaneously several biomarkers in parallel has been applied, identifying each of six possible injuries as a binary yes-no answer (Halamek et al., 2010b). The system allowed the identification of six physical injuries by means of enzyme logic AND/NAND gates, processing injury-specific biomarkers. The system reached the development level to be successful under human serum conditions (Zhou et al., 2011). An additional advantage of multiplexing enzyme logic gates for specific health issues is that it offers many possibilities to customize the systems on demand.

The switchable electrodes described in the previous section can also be coupled to enzyme logic gates that process medical status' information. A P4VP-ITO electrode has performed successfully as an electrochemical transducer for a logic AND operation designed to detect liver malfunction (Privman, M., et al., 2011). The logic system's machinery comprised the enzyme GDH and the chemicals alanine, α-ketoglutarate, glucose, and ferrocyanide dissolved in 1.2 mM Tris-buffer solution. The biomarkers for the liver injury are the enzymes alanine transaminase (ALT) and lactate dehydrogenase (LDH), which were defined as the inputs. Typical values for concentrations of ALT and LDH under health conditions were defined as input value 0 whereas the typical values for pathological concentration were defined as 1 (Fig. 20.9A).

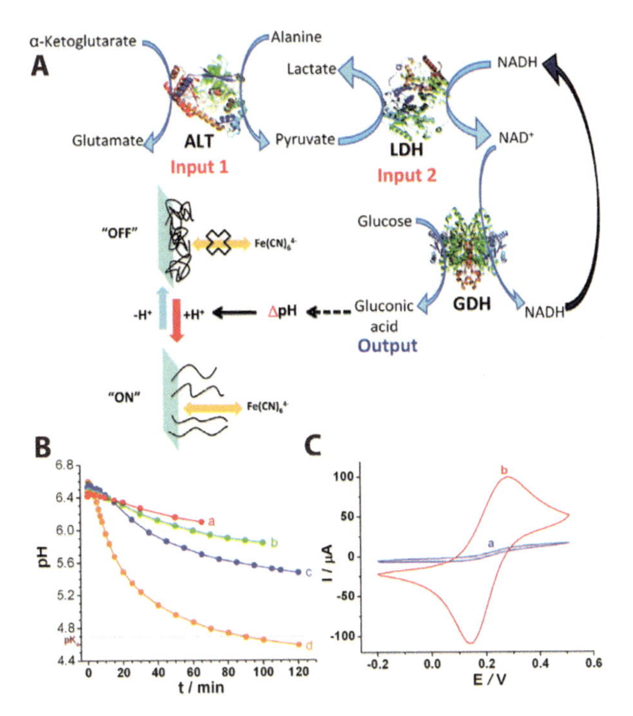

Figure 20.9 (A) Biocatalytic cascade used for the logic processing of the chemical input signals, causing in situ pH changes and activation of the electrode interface. (B) pH changes generated in situ by the biocatalytic cascade activated with various combinations of the two chemical input signals, ALT, LDH: (a) 0,0; (b) 0,1; (c) 1,0; (d) 1,1. The dotted line corresponds to the pK_a value of the PVP brush. (C) Cyclic voltammograms obtained for the P4VP-ITO electrode in (a) the initial OFF state at t = 1 min, pH 6.3, and (b) the ON state enabled by the ALT, LDH input combination 1,1, at t = 80 min, pH 4.75. Reprinted with permission from Privman et al. (2011). Copyright 2011 American Chemical Society.

The output readout was the pH value; constant pH means output 0 whereas pH acidification due to gluconic acid production was output 1. Such pH change activates the P4VP-ITO electrode and an external electrochemical probe, $[Fe(CN)6]^{4-}$, can reach the electrode and provide an electrochemical signal (Figs. 20.9B,C). The four possible input combinations were tested, and the input 1,1 was the only one able to trigger the whole biocatalytic cascade. Input 1, ALT, catalyzes the transformation of α-ketoglutarate into glutamate while producing pyruvate from alanine. Input 2, LDH, catalyzes the reduction of pyruvate to lactate and the oxidation of NADH to NAD^+. When NAD^+ is produced GDH can oxidize glucose to gluconic acid, which is responsible for the decrease in pH that switches ON the electrode. The behavior of the electrode when switched ON or OFF was characterized by cyclic voltammetry and impedance spectroscopy.

20.6 pH-Triggered Disassembly of Biomolecular Complexes on Surfaces Resulting in Electrode Activation

While the systems discussed above demonstrated reversible electrode activation-inhibition for bioelectrocatalytic processes upon pH changes, in some applications (e.g., to trigger drug release; Katz et al., 2015) one-time activation will be sufficient. The system exemplified below (Gamella et al., 2014) represents irreversible disassembly of biomolecular complexes on surfaces due to pH change, resulting in the electrode activation for a bioelectrocatalytic process.

Figure 20.10 (top) illustrates an electrode interface that was pre-assembled aiming at being electrochemically non-active. The electrode was assembled in a step-by-step chemical modification. The basal conductive supporting material was "buckypaper," a 2D material obtained by the compression of multi-wall carbon nanotubes (MWCNTs) (Strack et al., 2011). Glucose dehydrogenase dependent on pyrroloquinoline quinone (PQQ-GDH) was bounded to the MWCNTs by the cross-linker 1-pyrenebutanoic acid succinimidyl ester. This heterofunctional linker binds covalently with the amino residues present in Lys of the protein structure by forming amide bonds while interacting with the MWCNTs via the pyrenyl moiety, stabilized by π-π stacking interactions. Such enzyme linking provides efficient direct electron transfer from the enzyme active center (PQQ) to the MWCNT, offering an electrical current upon glucose oxidation (Katz et al., 2013a). The carboxylic residues of the amino acids belonging to the enzymes were transformed to esters in order to react with amino-functionalized 200 nm diameter silica microparticles. This reaction yielded the attachment of the silica particles to the PQQ-GDH layer on the buckypaper via amide bonds. Afterwards amino groups of the immobilized silica particles were incubated with iminobiotin-NHS ester, thus forming a biotinylated interface. It should be mentioned that iminobiotin is an analog of biotin but with the significant difference, that its avidin affinity is dependent on the pH. For pH values that range from neutral to basic the free base form of iminobiotin shows still a high specific affinity to bind the avidin. On the other hand when the pH is acidic the iminobiotin protonates and its interaction with avidin is hindered (Orr, 1981). This characteristic makes the buckypaper functionalized with further iminobiotin molecules to bind with avidin at pH 7.5. A set of iminobiotin/avidin complexes reacted with a batch of GOx previously functionalized with iminobiotin. This final modification allowed the immobilization of GOx and silica particles forming an affinity complex including an iminobiotin/avidin/iminobiotin bridge. The purpose of using the silica particles was to include a platform with a high surface density that provides a high load of biocatalitically active GOx, whose mission is to intercept all the glucose and to forbid its penetration to the deeper layer on the modified electrode. The assembled electrode was tested by means of electrochemical measurements in a pH 7.5 solution containing 20 mM glucose. The cyclic voltammograms recorded in either the presence or absence of glucose did not show significant differences (Fig. 20.11A, curves a and b, respectively), confirming that the electrode does not give any Faradaic response to the glucose oxidation, although the glucose was biocatalitically oxidized in a process independent from the electrode. Evidence of the process was the formation of hydrogen peroxide as byproduct of GOx biocatalytic activity. The PQQ-GDH was capable of direct electron transfer with the electrode but GOx consumed efficiently all the glucose at the external silica particles' shield layer. The modified electrode was activated by its reaction with an acidic solution of 100 mM citrate buffer at pH 4.5. This environment caused the protonation of iminobiotin and its concomitant decrease of affinity towards avidin, and the latter detachment of the iminobiotin/avidin/iminobiotin complex, removing GOx from the interface (Fig. 20.10, bottom). After the incubation period at acidic pH the electrode was set back to pH = 7.5 and a cyclic voltammogram was recorded, showing a bioelectrocatalytic response comparable to that observed before attaching the silica-GOx glucose shield (Fig. 20.11A, curve c), and confirming the activation of the electrode after eliminating the GOx outer layer. Fig. 20.11B, curve b, shows a negative potential appearing at ca. –80 mM (vs. Ag/AgCl) after the pH change that caused the removal of the avidin complex. It should be noted that the potential remained in the presence of 20 mM glucose if the GOx shield was not removed from the interface (Fig. 20.11B, curve a). The bioelectrocatalytic process activated on the electrode by the disassembly of the biomolecular complex upon the pH change was used for drug release triggering from another connected electrode (Gamella et al., 2014).

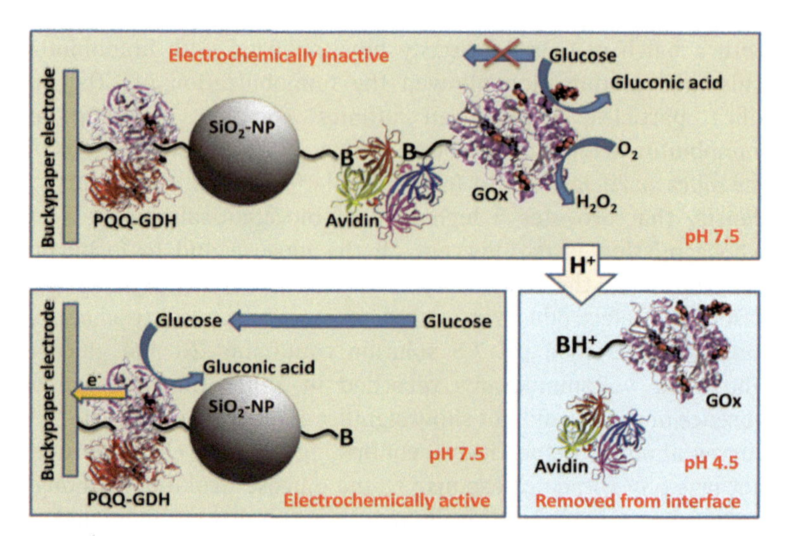

Figure 20.10 Electrochemically inactive (top) and active (bottom) electrode states and the transition from the inactive to active state by removing GOx from the interface upon cleaving the pH-sensitive affinity bridge in the presence of a pH signal. Adapted from Gamella et al. (2014), with permission; Copyright American Chemical Society, 2014.

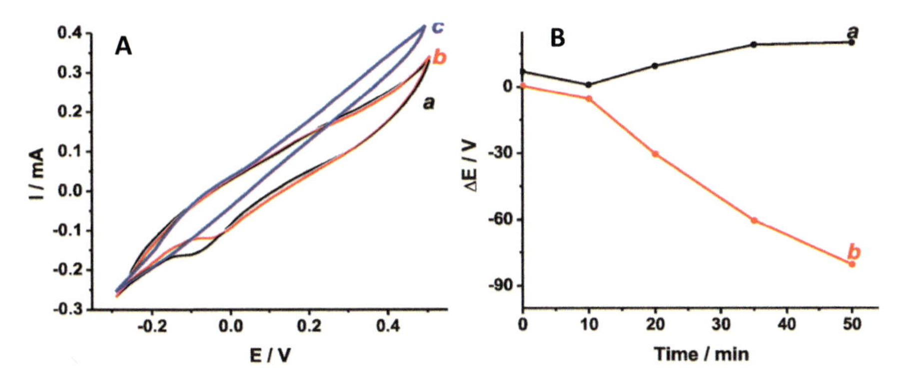

Figure 20.11 (A) The cyclic voltammograms obtained on the modified electrode after its complete assembling including PQQ-GDH and GOx in the absence (a) and presence (b) of 20 mM glucose demonstrating the bioelectrocatalytically inactive state. The pH-induced removal of GOx results in the electrochemically active state demonstrating the bioelectrocatalytic current (c). The cyclic voltammograms were obtained in the background electrolyte solution containing MOPS-buffer (50 mM), Na_2SO_4 (100 mM) and $CaCl_2$ (1 mM) and dissolved O_2, pH 7.5, with the potential scan rate 1 $mV \cdot s^{-1}$. (B) The electrode potential changes upon addition of glucose (20 mM) to the solution: (a) in the presence of GOx on the modified electrode surface, pH 7.5, (b) after pH-change-induced removal of GOx from the electrode interface (the potential dynamics was measured when the pH was returned to the initial value of 7.5). The potential was measured *vs.* an Ag/AgCl reference electrode in a background solution containing MOPS-buffer (50 mM), Na_2SO_4 (100 mM) and $CaCl_2$ (1 mM) and dissolved O_2. Adapted from Gamella et al. (2014), with permission; Copyright American Chemical Society, 2014.

20.7 Conclusions and Perspectives

Extensive work performed in the area of switchable electrode interfaces has led to numerous systems controlled by a large variety of physical and chemical signals (Bocharova et al., 2012; Katz et al., 2013b). Activation–inhibition of (bio)electrocatalytic processes by changing pH values represents particularly successful and important research direction since the pH variation can be coupled to enzyme-catalyzed reactions changing pH in situ (Katz, 2012b; Privman, M. et al., 2009; Privman, M., et al., 2011). Particularly important results have been achieved when the research advances in signal-responsive materials, modified electrodes and biocatalytic reactions were integrated into a new research area. Coupling of "smart" switchable electrodes with sophisticated multi-step biochemical pathways could be envisaged in the continuing research. The biochemically controlled switchable electrodes will operate as an interface between biological and electronic systems in future micro/nano-robotic devices (Katz, 2014).

Switchable electrodes controlled by pH signals, particularly generated in situ by biocatalytic reactions, have allowed "smart" biofuel cells producing electrical power dependent on complex variations of biochemical signals (Katz et al., 2009; Katz, 2010b). Enzyme-biocatalytic (Tam et al., 2009b; Amir et al., 2009) and immune-biorecognition (Tam et al., 2009a) systems have been developed to control performance of switchable biofuel cells (Halamek et al., 2010c). Their applications for self-powered wireless biosensors (Falk et al., 2014) and complex bioelectronics devices e.g., keypad lock systems (Halamek et al., 2010a; Halamek et al., 2010b; Halamek et al., 2010d) have been demonstrated. Further development in this direction resulted in bioelectrocatalytic switchable systems demonstrating novel memristor properties (MacVittie et al., 2014). New emerging materials involving graphene (Parlak et al., 2014) will open additional configurations to accomplish biocatalyzed switchable electronics. The present developments in the area of switchable modified electrodes and future expectations in bioelectronics in general are based on the application of a multi-disciplinary approach which will require contribution from electrochemists, specialists in materials science and unconventional biomolecular computing (Katz et al., 2012b).

References

Adamatzky, A., De Lacy Costello, B., Asai, T. (eds.) (2005), *Reaction-Diffusion Computers*. Elsevier, New York.

Adamatzky, A., Teuscher, C. (eds.) (2006), *From Utopian to Genuine Unconventional Computers*. Luniver Press, Beckington.

Adleman, L. M., *Science*, **266** (1994), 1021–1024.

Amir, L., Tam, T. K., Pita, M., Meijler, M. M., Alfonta, L., Katz, E., *J. Am. Chem. Soc.*, **131** (2009), 826–832.

Bocharova, V., Katz, E., *Chem. Record*, **12** (2012), 114–130.

Boole, G. (1854), *An Investigation of the Laws of Thought*. Macmillan ed.; reprinted with corrections, Dover Publications, New York 1958.

Choi, U. Y., Azzaroni, O., Cheng, O., Zhou, F., Kelby, T., Huck, W. T. S., *Langmuir*, **23** (2007), 10389–10394.

Combellas, C., Kanoufi, F., Sanjuan, S., Slim, C., Tran, Y., *Langmuir*, **25** (2009), 5360–5370.

Falk, M., Alcalde, M., Bartlett, P., De Lacey, A. L., Gorton, L., Gutierrez-Sanchez, C., Haddad, R., Leech, D., Ludwig, R., Magner, E., Mate, D. M., Ó-Conghaile, P., Ortiz, R., Pita, M., Pöller, S., Ruzgas, T., Salaj-Kosla, U., Schuhmann, W., Sebelius, F., Shao, M., Stoica, L., Tilly, J., Toscano, M. D., Vivekananthan, J., Wright, E., Shleev, S., *PLoS ONE*, **9** (2014), e109104.

Gamella, M., Guz, N., Mailloux, S., Pingarron, J. M., Katz, E., *ACS Appl. Mater. Interfaces*, **6** (2014), 13349–13354.

Halamek, J., Bocharova, V., Chinnapareddy, S., Windmiller, J. R., Strack, G., Chuang, M. C., Zhou, J., Santhosh, P., Ramirez, G. V., Arugula, M. A., Wang, J., Katz, E., *Mol. Biosyst.*, **6** (2010a), 2554–2560.

Halamek, J., Tam, T. K., Chinnapareddy, S., Bocharova, V., Katz, E., *J. Phys. Chem. Lett.*, **1** (2010c), 973–977.

Halámek, J., Tam, T. K., Strack, G., Bocharova, V., Pita, M., Katz, E., *Chem. Commun.*, **46** (2010d), 2405–2407.

Halamek, J., Windmiller, J. R., Zhou, J., Chuang, M. C., Santhosh, P., Strack, G., Arugula, M. A., Chinnapareddy, S., Bocharova, V., Wang, J., Katz, E., *Analyst*, **135** (2010b), 2249–2259.

Harris, J. J., Bruening, M. L., *Langmuir*, **16** (2000), 2006–2013.

Katz, E., *Electroanalysis*, **22** (2010b), 744–756.

Katz, E. (ed.) (2012), *Biomolecular Information Processing-From Logic Systems to Smart Sensors and Actuators*, Wiley-VCH, Weinheim, Germany.

Katz, E. (ed.) (2014), *Implantable Bioelectronics: Devices, Materials and Applications*. Wiley-VCH, Weinheim, Germany.

Katz, E., Bocharova, V., Privman, M., *J. Mater. Chem.*, **22** (2012b), 8171–8178.

Katz, E., MacVittie, K., *Energy Environ. Sci.*, **6** (2013a), 2791–2803.

Katz, E., Minko, S., Halamek, J., MacVittie, K., Yancey, K., *Anal. Bioanal. Chem.*, **405** (2013b), 3659–3672.

Katz, E., Pingarron, J. M., Mailloux, S., Guz, N., Gamella Carballo, M., Melman, G., Melman, A., *J. Phys. Chem. Lett.*, **6** (2015), 1340–1347.

Katz, E., Pita, M., *Chem. Eur. J.*, **15** (2009), 12554–12564.

Katz E., Privman, V., *Chem. Soc. Rev.*, **39** (2010a), 1835–1857.

Katz, E., Wang, J., Privman, M., Halamek, J., *Anal. Chem.*, **84** (2012a), 5463–5469.

Kenausis, G., Taylor, C., Katakis, I., Heller, A., *J. Chem. Soc. Faraday Trans.*, **92** (1996), 4131–4136.

Kramer, M., Pita, M., Zhou, J., Ornatska, M., Poghossian, A., Schoning, M. J., Katz, E., *J. Phys. Chem. C*, **113** (2009), 2573–2579.

MacVittie, K., Katz, E., *J. Phys. Chem. C*, **117** (2013), 24943–24947.

MacVittie, K., Katz, E., *Chem. Commun.*, **50** (2014), 4816–4819.

Manesh, K. M., Halamek, J., Pita, M., Zhou, J., Tam, T. K., Santhosh, P., Chuang, M. C., Windmiller, J. R., Abidin, D., Katz, E., Wang, J., *Biosens. Bioelectron.*, **24** (2009), 3569–3574.

Margulies D., Hamilton, A. D., *J. Am. Chem. Soc.*, **131** (2009), 9142–9143.

Melnikov, D., Strack, G., Pita, M., Privman, V., Katz, E., *J. Phys. Chem. B*, **113** (2009), 10472–10479.

Motornov, M., Sheparovych, R., Katz, E., Minko, S., *ACS Nano*, **2** (2008b), 41–52.

Motornov, M., Tam, T. K., Pita, M., Tokarev, I., Katz, E., Minko, S., *Nanotechnology* (2009b) Article 434006.

Motornov, M., Zhou, J., Pita, M., Gopishetty, V., Tokarev, I., Katz, E., Minko, S., *Nano Lett.*, **8** (2008a), 2993–2997.

Motornov, M., Zhou, J., Pita, M., Tokarev, I., Gopishetty, V., Katz, E., Minko, S., *Small*, **5** (2009a), 817–820.

Niazov, T., Baron, R., Katz, E., Lioubashevski, O., Willner, I., *Proc. Natl. Acad. Sci.*, **103** (2006), 17160–17163.

Orr, G. A., *J. Biol. Chem.*, **256** (1981), 761–766.

Park, M. K., Deng, S. X., Advincula, R. C., *J. Am. Chem. Soc.*, **126** (2004), 13723–13731.

Parlak, O., Turner, A. P. F., Tiwari, A., *Adv. Matter.*, **26** (2014), 482–486.

Pita, M., Katz, E. J., *Comput. Theor. Nanosci.*, **8** (2011), 401–408.

Pita, M., Kramer, M., Zhou, J., Poghossian, A., Schoning, M. J., Fernandez, V. M., Katz, E., *ACS Nano*, **2** (2008), 2160–2166.

Pita, M., Strack, G., MacVittie, K., Zhou, J., Katz, E., *J. Phys. Chem. B*, **113** (2009), 16071–16076.

Privman, V., Arugula, M. A., Halamek, J., Pita, M., Katz, E., *J. Phys. Chem. B*, **113** (2009a), 5301–5310.

Privman, V., Halamek, J., Arugula, M. A., Melnikov, D., Bocharova, V., Katz, E., *J. Phys. Chem. B*, **114** (2010), 14103–14109.

Privman, V., Pedrosa, V., Melnikov, D., Pita, M., Simonian, A., Katz, E., *Biosens. Bioelectron.*, **25** (2009b), 695–701.

Privman, V., Strack, G., Solenov, D., Pita, M., Katz, E., *J. Phys. Chem. B*, **112** (2008), 11777–11784.

Privman, M., Tam, T. K., Bocharova, V., Halamek, J., Wang, J., Katz, E., *ACS Appl. Mater. Interfaces*, **3** (2011), 1620–1623.

Privman, M., Tam, T. K., Pita, M., Katz, E., *J. Am. Chem. Soc.*, **131** (2009), 1314–1321.

Sienko, T., Adamatzky, A., Rambidi, N. G., Conrad, M. (eds.) (2005), *Molecular Computing*. MIT Press, Cambridge, Massachusetts.

Strack, G., Luckarift, H. R., Nichols, R., Cozart, K., Katz, E., Johnson, G. R., *Chem. Commun.*, **47** (2011), 7662–7664.

Strack, G., Ornatska, M., Pita, M., Katz, E., *J. Am. Chem. Soc.*, **130** (2008b), 4234–4235.

Strack, G., Pita, M., Ornatska, M., Katz, E., *Chem. Bio. Chem.*, **9** (2008a), 1260–1266.

Tam, T. K., Ornatska, M., Pita, M., Minko, S., Katz, E., *J. Phys. Chem. C*, **112** (2008a), 8438–8445.

Tam, T. K., Pita, M., Motornov, M., Tokarev, I., Minko, S., Katz, E., *Adv. Mater.*, **22** (2010b), 1863–1866.

Tam, T. K., Pita, M., Motornov, M., Tokarev, I., Minko, S., Katz, E., *Electroanalysis*, **22** (2010c), 35–40.

Tam, T. K., Pita, M., Ornatska, M., Katz, E., *Bioelectrochemistry*, **76** (2009b), 4–9.

Tam, T. K., Pita, M., Trotsenko, O., Motornov, M., Tokarev, I., Halamek, J., Minko, S., Katz, E., *Langmuir*, **26** (2010a), 4506–4513.

Tam, T. K., Strack, G., Pita, M., Katz, E., *J. Am. Chem. Soc.*, **131** (2009a), 11670–11671.

Tam, T. K., Zhou, J., Pita, M., Ornatska, M., Minko, S., Katz, E., *J. Am. Chem. Soc.*, **130** (2008b), 10888–10889.

Taylor, C., Kenausis, G., Katakis, I., Heller, A., *J. Electroanal. Chem.*, **396** (1995), 511–515.

Tokarev, I., Gopishetty, V., Zhou, J., Pita, M., Motornov, M., Katz, E., Minko, S., *ACS Appl. Mater. Interfaces*, **1** (2009), 532–536.

Turing, A., *Proc. London Math. Soc.*, **S2-42** (1937), 230–265.

Wang J., Katz, E., *Anal. Bioanal. Chem.*, **398** (2010), 1591–1603.

Wang, X., Zhou, J., Tam, T. K., Katz, E., Pita, M., *Bioelectrochemistry*, **77** (2009), 69–73.

Zhou, J., Arugula, M., Halamek, J., Pita, M., Katz, E., *J. Phys. Chem. B*, **113** (2009), 16065–16070.

Zhou, J., Halamek, J., Bocharova, V., Wang, J., Katz, E., *Talanta*, **83** (2011), 955–959.

Chapter 21

Functional Nano-Bioconjugates for Targeted Cellular Uptake and Specific Nanoparticle–Protein Interactions

Sanjay Mathur, Shaista Ilyas, Laura Wortmann, Jasleen Kaur, and Isabel Gessner

Faculty of Mathematics and Natural Sciences, University of Köln, Department of Chemistry, Greinstrasse 6, D-50939 Köln, Germany

sanjay.mathur@uni-koeln.de

21.1 Introduction

Functional nanostructures for biomedical applications are mainly used in the therapeutic and diagnostic fields as well as biomimetic implant materials. Hereby, the portfolio of applications spans from drug delivery, hyperthermia, bio-imaging (optical imaging (OI), magnetic resonance imaging (MRI), positron emission tomography (PET), and computed tomography (CT)), and cell separation to protein profiling (Lee et al., 2012b). Among commonly used materials, iron oxide nanoparticles (IONPs) are known for their superparamagnetic behavior, gold (Au) and silver (Ag) NPs for plasmonic properties, mesoporous silica particles for their high adsorption capacity and physiological stability (Lee et al., 2012a; Lee et al., 2015; Wegner and Hildebrandt, 2015; Biffi et al., 2015).

Biocatalysis and Nanotechnology
Edited by Peter Grunwald

ISBN 978-981-4613-69-9 (Hardcover), 978-1-315-19660-2 (eBook)
www.panstanford.com

Although all these materials have different application areas, the considerations to tune their in vivo physiological behavior are always driven by assessment of toxicological potential and assimilation effects. Since NPs can be designed for desired application by tailoring their size, surface texture and composition, these factors also influence their pharmacokinetic and pharmacodynamics in biological environment. Therefore, by considering the transfer of a NP-system in biomedicine from the bench to the clinic, the changes from their "chemical identity" in lab toward their "biological identity" in vitro to their "physiological identity" in vivo have to be taken into account (Fig. 21.6) (Walkey and Chan, 2012). Consequently, NPs have to be colloidally stable in high ionic strength liquids over a long period. Moreover, NPs should have certain size ranges to circumvent unwanted clearance by the body defense system and should be able to reach their target by active or passive accumulation. Also in terms of protein corona formation in the presence of proteins, surface functionality and charge should be carefully considered. In this context, functional nano-bioconjugates designed by tagging a targeting drug unit and an imaging moiety (optical or radioactive) to NPs carries great potential for early detection, accurate diagnosis and targeted therapy of various diseases. Converting a pristine (as-synthesized) NP into a bioresponsive vector suited for cellular interactions require a synergy between surface properties and conjugation chemistry as shown in Fig. 21.1.

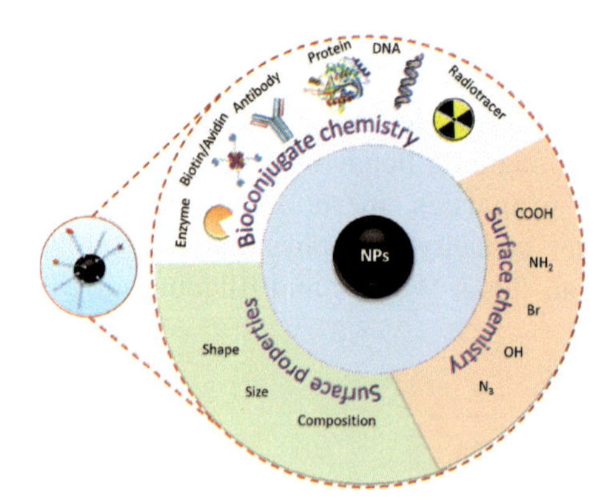

Figure 21.1 A model for a nanocarrier with possible surface chemistries for various biomedical applications.

Furthermore, there is a huge interest in NPs that can be functionalized with various biomolecules such as enzymes, peptides, tumor markers, fluorescent dyes, polymers, drugs and some proteins (Huy et al., 2014; Wu et al., 2015; Imperiale et al., 2015). Such functionalized probes play a key role in number of applications including carriers for drug delivery, specific labeling, bio-sensing and work as tools for biomedical research (Conde et al., 2014). Among various biomolecules, proteins hold a dominant position within biomedical studies, by offering multiple functions and serve as building blocks for enzymes, antibodies and hormones used in different cell activities.

Therefore, efforts to develop NP–protein conjugates become increasingly important due to their potential applications as therapeutics and diagnostics agents (Conde et al., 2014). However, the nature of the desired and undesired NP–protein interaction and effect of ligand conjugation on the structural properties of proteins are the special areas of ongoing research and explained in more detail under Section 21.3.

To obtain NP–biomolecule conjugates, several synthetic methods have been developed (Di Marco et al., 2010; Turcheniuk et al., 2013). Besides electrostatic or noncovalent interactions, more specific covalent-bonding strategies have been recently demonstrated. Two of the most commonly used approaches are carbodiimide coupling and click chemistry. Comparative analysis of carbodiimide coupling over click chemistry is well discussed in literature and was investigated in detail by Bolley et al., (2013), who functionalized the same IONPs with hydroxymethylenebisphosphonate as surface active ligand bearing a carboxylic acid or alkyne function. Afterwards, they coupled the fluorophore Rh123, the stabilizing agent polyethylene glycol (PEG) and the targeting unit cRGD with either an amine or acid linker *via* carbodiimide and click chemistry coupling, respectively. The performed MRI measurements and biological stability studies revealed that click functionalized NPs showed significant better grafting and targeting affinity compared to NPs obtained by carbodiimide coupling (Bolley et al., 2013).

This is in accordance with numerous other examples of click reactions reported for the preparation and functionalization of polymeric micelles and NPs, liposomes and polymersomes, capsules, microspheres, metal silica NPs, carbon nanotubes and fullerenes (Bolley et al., 2013; Moses and Moorhouse, 2007; Sperling and Parak, 2010).

Since its introduction by Sharpless and co-workers, the term "click chemistry" is known as one of the most versatile and modular approaches to couple two reactive partners in a facile, selective, reliable and high yield reaction under mild conditions (Kolb et al., 2001). Meanwhile, click chemistry has become one of the most common and reliable methods to link molecules in a covalent fashion and is finding ever-increasing applications in a variety of disciplines including the chemistry of nanomaterials, chemical biology, drug delivery, and medicinal chemistry (Binder, 2008; Hou et al., 2012; Kolb and Sharpless, 2003; Neibert et al., 2013). Typical examples of click chemistry reactions include cycloaddition reactions, such as the 1,3-dipolar family, and hetero Diels–Alder reactions, nucleophilic ring-opening reactions of epoxides and aziridines, carbonyl reactions such as formation of hydrazones, Michael additions, and cycloaddition reactions (Jùrgensen, 2000; Adolfsson et al., 1999; Kolb et al., 1994).

During the last few decades, hard-matter nanosystems, e.g., Au and other noble metal, IONPs, quantum dots, and silicon dioxide (SiO_2) NPs, have been extensively investigated because of their unique physical and chemical properties, which strongly depend on their size and shape. Although these nanosystems are of great interest for researchers in different disciplines, they have been the focus of special attention in the biomedical field, where they have found application as drug carriers, labeling agents, vectors for gene therapy and imaging agents in magnetic resonance imaging (Wang et al., 2012a).

Within this chapter, we aim to provide a concise overview and scope of click chemistry and their application in nano-bioconjugated vectors. The focus lies on Au and IONPs as most investigated examples for highly functional protein-conjugated nanoprobes.

21.2 Introduction to the "Click" Cycloaddition

The Cu(I)-catalyzed variant of the Huisgen 1,3-dipolar cycloaddition of azides and alkynes (CuAAC), which has become increasingly popular to afford 1,2,3-triazoles (Hein et al., 2008), owes its usefulness in part to the ease with which azides and alkynes can be introduced into a molecule and their relative stability under a variety of conditions. In addition, CuAAC reaction can be performed in a variety of solvents such as water, ethanol or tert-butyl alcohol, etc. (Tornøe et al., 2002). Other advantages of CuAAC reaction include mild reaction conditions, high yield and efficiency under physiological conditions, and its chemo-selectivity, which allows labeling of functional biomolecules such as peptides, proteins, nucleic acids, polysaccharides, etc. (Rostovtsev et al., 2002). However, the original Huisgen 1,3-cycloaddition reaction, i.e., the reaction of unactivated azides and alkynes proceeds slowly in the absence of a catalyst and generally requires elevated temperatures and long reaction time and yields a mixture of 1,4- and 1,5-triazoles, rendering this reaction unsuitable for most biomedical applications (Fig. 21.2).

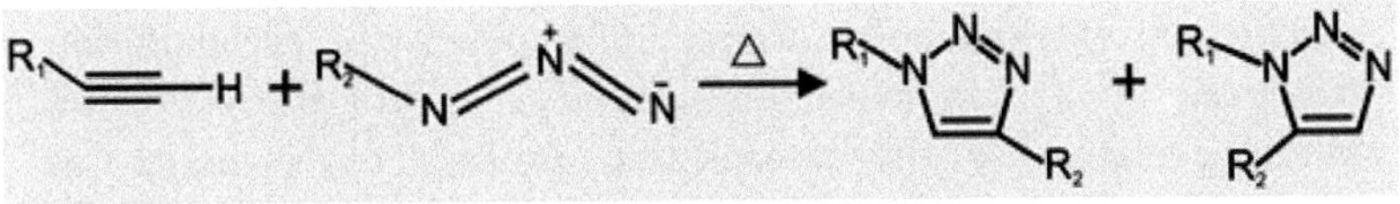

Figure 21.2 The thermal 1,3-cycloaddition reaction of terminal alkynes with azides.

A copper-catalyzed variant on the other hand that follows a different mechanism can be conducted under aqueous conditions, even at room temperature and allows the regio-specific synthesis of the 1,4-disubstituted isomers (Fig. 21.3). The rate of CuAAC is increased by a factor of 10^7 relative to the thermal process, making it conveniently fast at and below room temperature. This high-yield reaction tolerates a variety of functional groups and affords the 1,2,3-triazole product with minimal work-up and purification (Himo et al., 2005). Further interest in this reaction stems from the interesting biological activity of 1,2,3-triazoles and advantageous properties such as high chemical stability (generally inert to severe hydrolytic, oxidizing, and reducing conditions, even at high temperature), strong dipole moment (4.8–5.6 Debye), aromatic character, and hydrogen bond accepting ability (Hein and Fokin, 2010). Thus, the triazole ring can productively interact in several ways with biological molecules and organic and inorganic surfaces and materials, serving as a hydrolytically stable replacement for the amide bond.

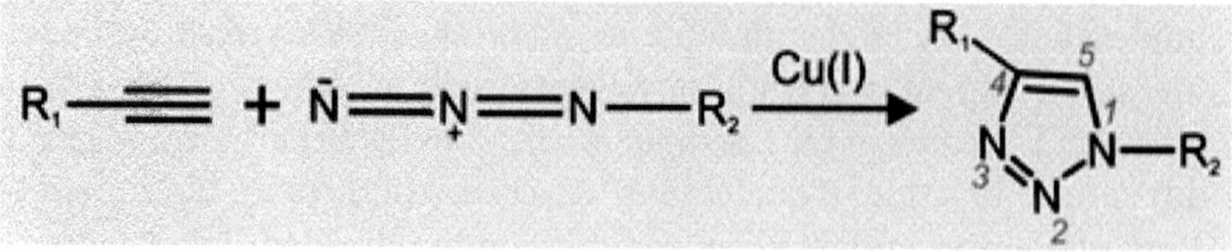

Figure 21.3 Copper(I)-catalyzed cycloaddition reaction of terminal alkynes with azides.

21.2.1 Mechanism of Cu(I)-Catalyzed Alkyne–Azide

Although the thermal dipolar cycloaddition of azides and alkynes occurs through a concerted mechanism, DFT calculations on monomeric copper acetylide complexes indicate that the concerted mechanism is strongly disfavored relative to a stepwise mechanism (Meldal and Tornøe, 2008). Stepwise cycloaddition catalyzed by a monomeric Cu(I) species lowers the activation barrier relative to the uncatalyzed process by as much as 11 kcal/mol, which is sufficient to explain the incredible rate enhancement observed under Cu(I) catalysis. The stepwise catalytic cycle (Fig. 21.4) begins with formation of a Cu(I) acetylide species *via* the π complex (2). Calculations also indicate that copper coordination lowers the pK_a of the alkyne C–H by up to 9.8 pH units, thus making deprotonation in aqueous systems possible without the addition of a base. Following the formation of the active copper acetylide species, azide displacement of one ligand generates a copper acetylide-azide complex, such as the dicopper species (3). Complexation of the azide activates it toward nucleophilic attack of acetylide carbon C(4) at N(3) of the azide, generating metallocycle (4). Consistent with this mechanism, experimental results indicate that electron-with drawing substituents on the alkyne accelerate Cu(I)-catalyzed alkyne-azide coupling. This metallocycle positions the bound azide group properly for subsequent ring contraction to form a triazole-copper derivative (5). Protonation of triazole-copper derivative followed by dissociation of the product (6) ends the reaction and regenerates the catalyst.

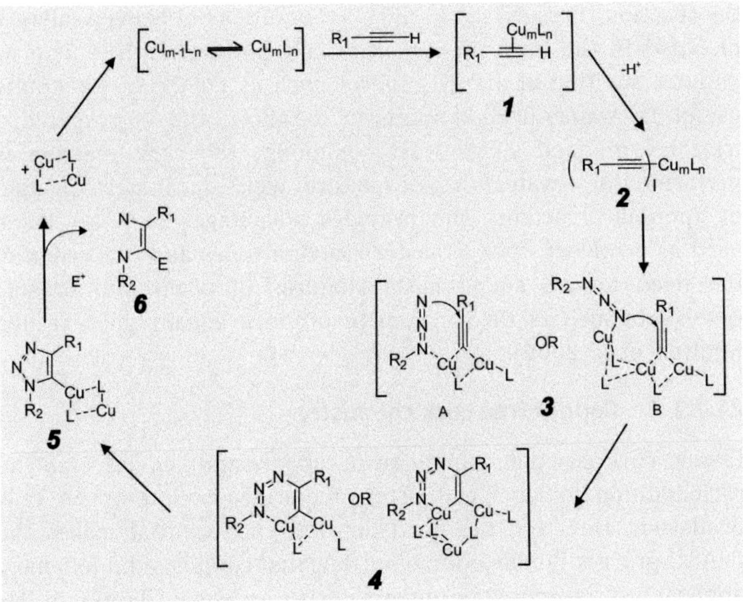

Figure 21.4 Outline of plausible mechanisms for the Cu(I) catalyzed reaction between organic azides and terminal alkynes (redrawn from Meldal and Tornøe, 2008).

21.2.1.1 Catalysts and solvents

In situ generation of Cu(I) from Cu(II) salts such as Cu(II) sulfate pentahydrate occurs by comproportionation with copper metal or by reduction and has the advantage of not requiring inert atmospheres despite the instability of the Cu(I) oxidation state in the presence of oxygen (Wang et al., 2012b). Although both reduction and comproportionation have a wide scope and tolerate many organic functional groups, copper-metal comproportionation is generally limited to special applications, such as biological systems, that preclude the use of most reducing agents. This preference for reduction likely results from a combination of the longer reaction times required for comproportionation and the simpler workup of the reduction method. Oxidation of copper metal provides another method for generating the Cu(I) catalyst for triazole formation. Copper(I) salts (iodide, bromide, chloride, acetate) and coordination complexes such as $[Cu(CH_3CN)_4]PF_6$ and $[Cu(CH_3CN)_4]OTf$ have also been commonly employed in the reaction (Presolski et al., 2011). Conditions utilizing a mixture of Cu(0) in the form of wire, turnings, powder, or NPs with or without addition of a Cu(II) source such as $CuSO_4$ are also quite useful in water/alcohol mixtures to afford the corresponding triazoles in good yield. Most commonly, the click reaction is performed in a water/alcohol mixture, which facilitates solvation of lipophilic reactants and provides advantages of water being used as a solvent, such as faster reaction times and preclusion of the need for an added base. Mixtures of water and organic cosolvents such as DMSO seem to produce equally good results (Molteni et al., 2006).

21.2.1.2 Copper-free click chemistry

Classic click reactions comprise a copper-catalyzed azide-alkyne cycloaddition to label and detect molecules of interest in cells or tissues. However, the need for a copper catalyst makes the CuAAC process nonideal for bioorthogonal coupling and for many applications in inorganic/nanomaterials science (Baskin et al., 2007). The metal-free strain-promoted azide-alkyne cycloaddition (SPAAC) takes advantage of ground-state destabilization (strain) to accelerate triazole formation under ambient or physiological conditions (Codelli et al., 2008; Agard et al., 2004). The reaction of azides with strained alkynes, such as cyclooctynes, relieves that burden by readily forming a triazole product without a toxic catalyst (Fig. 21.5). Such reactions, in addition to other cycloadditions, are now being referred to as Cu-free click reactions. Current SPAAC tools are based on the cyclooctyne (OCT) platform which are extremely reactive toward azides. All cyclooctyne derivatives can be designed by alteration of the reactivity through modulation of strain or electronics parameters. A selection of these Cu-free click reagents is shown in Fig. 21.5.

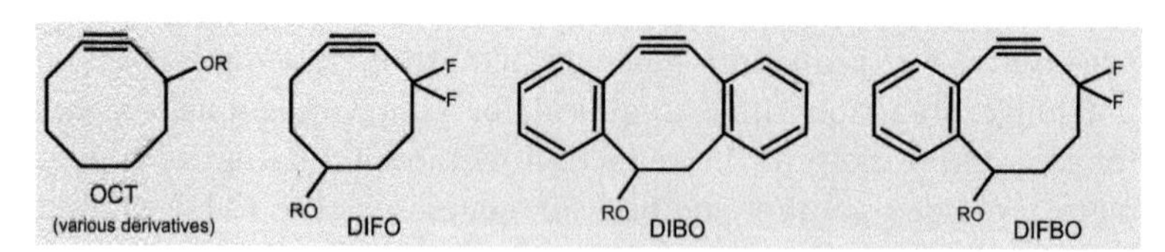

Figure 21.5 Selected strain-promoted azide-alkyne cycloaddition (SPAAC) reagents based on cyclooctyne (OCT) and alternative alkene platforms for metal-free click coupling.

The strain in this eight-membered ring allows the reaction with azide-modified molecules to occur in the absence of catalysts or extreme temperatures or solvents, enabling the study of the surface of live cells, and preventing copper-induced damage of fluorescent proteins such as GFP in fixed and permeabilized cells. Click dibenzocyclooctyne (DIBO) alkyne reagents are compatible with copper-free click chemistry reactions that can capture or detect azide-substituted targets. The key to copper-free click chemistry is the DIBO core of the Click-DIBO alkyne reagent. It was shown that the rate of SPAAC reaction using cyclooctyne derivatives can be enhanced through attachment of fluorine atoms at the propargylic position in its structure. Monofluorinated cyclooctyne, difluorinated cyclooctyne (DIFO) and dibenzocyclooctyne (DIBO) react faster with organic azides as compared to OCT (Jewett et al., 2010; Berkel et al., 2008; Lallana et al., 2012).

21.2.2 Formation of Protein Corona on NPs Surface

When NPs are injected into the blood stream, they might interact with organic molecules such as lipids, proteins, nucleic acids even metabolites owning to their small size and large surface to mass ratio. As a result, biological molecules immediately start to cover their surface and form a dense protein layer called "corona" (Fig. 21.6). Overall, the newly formed NP-complexes differ significantly from parent NPs and play pivot role for NPs interactions with biological systems by influencing their bio-distribution profile at the intracellular level.

Bhattacharya et al., (2014) documented an interesting report on the shifting identity and fate of particles in biological systems. Upon entry into the living system, NPs acquire a different biological identity due to adsorption or removal of proteins. The new NP–protein conjugates impart a direct effect on cellular recognition and cytotoxic profile through intra- or extracellular dissolution of NPs during biomedical applications. Depending on the size, structure, and surface chemistry, NPs usually observe a hard and soft protein corona when encountering biological fluids. It is known that hard corona formation mainly depends on the interaction of NPs with the surrounding proteins available in the biological media. With the passage of time, the dominating protein-protein interactions give rise to a short-lived soft protein layer at the NPs surface.

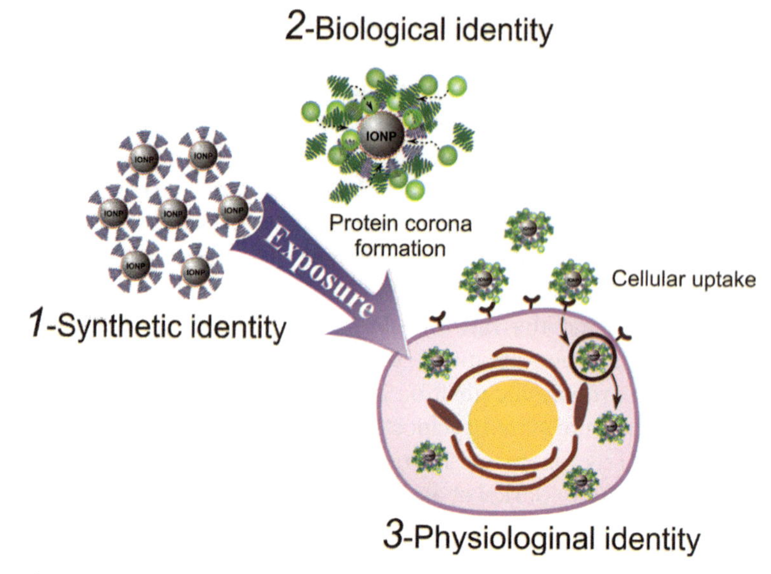

Figure 21.6 Schematic description of NP interaction with their environment. Starting from "synthetic identity" in lab, to the "biological identity in vitro and physiological reaction in vivo significantly affected by the NP-protein interaction and the formation of protein corona (redrawn from Walkey and Chan, 2012).

Treuel et al., (2010) showed the influence of physiocochemical properties of protein corona on cellular uptake of NPs. They modified human serum albumin (HSA) on NPs to investigate the results for protein corona formation with 5 nm dihydrolipoic acid-coated quantum dots (DHLA-QDs). HSA modified by succinic anhydride (HSA_{suc}) showed a threefold decreased binding affinity while HSA modified by ethylenediamine (HSA_{am}) offered 1000-fold enhanced affinity toward the NPs. Results showed that HSA_{suc} formed thicker protein adsorption (8.1 nm) than native HSA (3.3 nm), whereas the HSA_{am} corona (4.6 nm) is only slightly thicker. The charge distributions of both HSA constructs brought remarkable changes of the binding affinity of the protein. Pronounced variations in the kinetics and internalizations of NPs indicated that even small physiochemical changes of the protein corona may affect biological environments completely (Treuel et al., 2010).

As various forces are involved in the process of adsorption it becomes challenging to define a set of general rules about protein interactions for different types of NPs. Based on various reports, certain observations can be made to get a better overview of the NP–protein interactions. The material composition, hydrophobicity, hydrophilicity, surface chemistry, pH (ion composition), temperature and many other factors strongly influence the protein adsorption (Vladkova et al., 2010). The effect of these factors on NP–protein interactions is well documented and their consideration can enable us to produce the paramount NP-hybrid system.

21.2.2.1 Aggregation and colloidal stability of nanostructures

The colloidal stability of particles strongly depends on their size, concentration, and surface charge and leads to conformational changes to the adsorbed protein. Agglomerated NPs present uneven surfaces that cause structural changes in protein conformation and consequently increase the rate of nonspecific interactions (Nel et al., 2009). One way of controlling aggregation of NPs can be the use of different stabilizing agents. The biopolymers commonly used for stabilizing NP–protein conjugates may include dextran, polyvinyl pyrrolidone (PVP) and poly(ethyleneglycol) (PEG), poly(vinyl) alcohol (PVA), poly(ethyleneimine) (PEI), and chitosan (Teekamp et al., 2015). Huy et al., (2014) recently presented the synthesis of chitosan-coated IONP-protein conjugate for pathogen (*Vibrio cholerae*) separation. After binding with a specific antibody, NPs conjugates showed stability and allowed separation of pathogen from water samples at a low concentration of 10 cfu mL^{-1}. This approach can be used in parallel with new diagnostic tests for faster detection of pathogens (bacteria and viruses).

21.2.2.2 Hydrophobicity and surface charge

Hydrophobic surfaces favor protein adsorption and denature them with a greater extent than hydrophilic surfaces (Ashby et al., 2014). Hydrophobic molecules may have more protein binding sites, which are due to the result of clustering of hydrophobic polymer chains. Hydrophobicity of polystyrene NPs increases the rate of protein adsorption from blood plasma (Tao et al., 2016). Similarly, higher amounts of albumin are adsorbed on the surface of hydrophobic and negatively charged particles. During adsorption on NPs, proteins may undergo structural changes, which are thermodynamically favored if the NPs are charged or hydrophobic. The conformation changes in proteins are irreversible after desorption (Deitcher et al., 2010). QDs linked with mercaptoundecanoic acid denature and inactivate the enzyme chymotrypsin while PEG-coated quantum dots presented less loss for the same by absorbing the enzyme (Walling et al., 2009). This shielding behavior of PEG is also shown for other PEG-modified NPs, which are more stable at high salt concentrations and in biological environments, as they show less nonspecific binding to proteins and cells (Gref et al., 2000; Liu et al., 2007). This results in longer blood circulation times by controlling the density of PEG on NP surface.

The nonspecific interactions between NP–protein conjugates influence the route and efficiency of NPs uptake based on the mode of functionalization. Lynch et al., (2009) indicated that the structural modification of proteins can be influenced by the surface properties of positive, negative, and neutral ligands on Au NPs. They showed that proteins denature in the presence of charged ligands, either positive or negative, but neutral ligands keep the natural structure of the proteins. While Brennan et al., (2006) showed the role of negatively charged molecules on the adsorption of lipase. They covalently prepared a hybrid material consisting of active lipase enzyme to PEGylated Au NPs. Fluorometric lipase activity assays showed seven fully active lipase molecules to each Au NP. The higher activity of the product was possible due to the covalent attachment of negatively charged lipase molecules to Au NPs. It was stated that increase in negative charge controls the effect of increased hydrodynamic radius of the NPs to reduce nonspecific interactions. Calatayud et al., (2014) showed the role of Fe_3O_4 NPs with positively (polyethyleneimine, PEI-MNPs) and negatively (polyacrylic acid, PAA-MNPs) charged ligands on the adsorption of proteins and cellular uptake. Results showed that adsorption of proteins is time dependent and delivers NP–protein clusters depending on the availability of free functional groups at the surface of the polymer. After few minutes of incubation, the NP–protein adsorption showed an increase of fivefold in the hydrodynamic radius. After 24 h incubation, NP–protein aggregates showed further increase in the hydrodynamic radius of PAA-MNPs (~1500 nm) and PEI-MNPs (~3000 nm). The final properties of these cluster revealed distinguishable effects on NP uptake. In another report, Kralj et al., (2012) presented the functionalization of fluorescent SiO_2 NPs with positive and negative surface charges at physiological pH. The uptake of the NPs was strongly affected by the absorption of the proteins from the culture medium.

21.2.2.3 Particle size and curvature

The influence of particles size and therefore curvature cannot be ignored as it implants an influence on protein adsorption. In general there is a trend observed that smaller sized NPs result in a higher curvature showing less NP–protein interaction caused by the smaller available surface area (Walkey and Chan, 2012). In a report, Au NPs (12 nm) had shown greater affinity to bind fibrinogen proteins as compared to smaller (7 nm) Au NPs (Deng et al., 2012). In another report 100 nm SiO_2 NPs induced greater loss to the structure and activity for the lysozyme as compared to ultra-small 4 nm SiO_2 particles (Vertegel et al., 2004). The protein adsorption on NP gives a new biological identity to the nanomaterials, which may modulate the subsequent cellular uptake, internalization mechanism and also the subcellular distribution. In an interesting article, Shang et al., (2014) reviewed the effect of NP size on active and passive cellular uptake for different particles. They indicated an optimal core size, in the range of 30–50 nm for maximum cellular uptake in different cell lines for mesoporous SiO_2, Au, and Fe_3O_4 NPs. Mironava et al., (2010) performed uptake studies which indicated that NPs might follow different penetration pathways into cells due to different size range. They showed that Au NPs (45 nm) entered the cells *via* clathrin-mediated endocytosis, while the smaller NPs (13 nm) penetrated *via* phagocytosis. Similarly, in another report, Rejman et al., (2004) investigated the penetration pathway of fluorescent latex beads (different size ranges) with B16 cell lines. The ultra small NPs showed their penetration into the cells *via* phagocytosis. The NPs < 200 nm in size exhibited cell internalization by clathrin-coated pits, while the NPs > 200 nm observed caveolae-mediated internalization. Regardless of enormous efforts in this area, it still remains thought-provoking to reliably correlate a specific cellular uptake with NP size.

21.2.3 Controlled Attachment of Proteins to NP Surfaces

The design of effective NP–protein conjugates involves the retention of protein structure and function and is a challenging task. To attach proteins to the NPs surface in a controlled manner, covalent and noncovalent conjugations are usually used (Fig. 21.7). Electrostatic adsorption is based on the noncovalent interactions between the NPs and proteins while direct methods depend on the covalent attachment of the ligands to the NPs. Both methods have their strengths and limitations and provide functional hybrid systems with interesting applications in biotechnology, biomedicine and catalysis. The nanomaterial conjugation (covalent or noncovalent) to proteins is a complex interface due to nonspecific adsorption of proteins to the NP surface. The unwanted adsorption occurs due to noncovalent binding of some amino acids with molecules available on the particle surface and leads to irreversible denaturation of the proteins (Lee et al., 2012b).

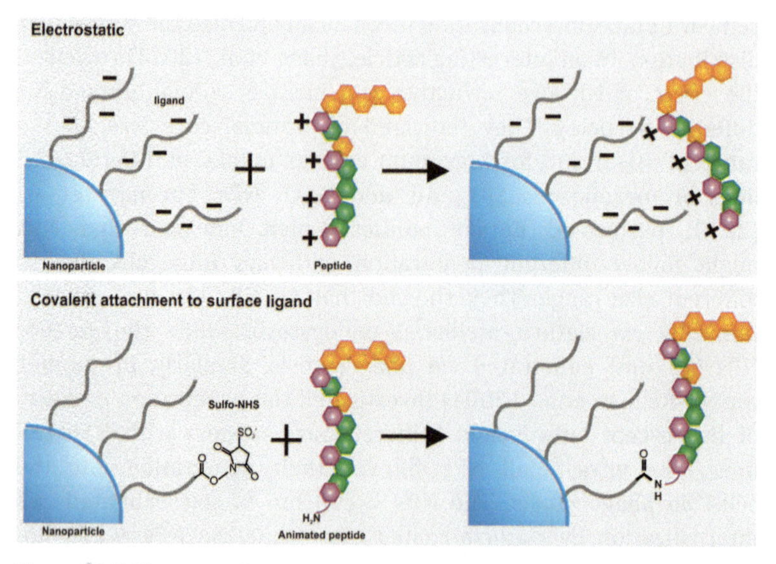

Figure 21.7 Electrostatic interactions, opposite charges on the surface of the NP and the peptide are used to mediate charge–charge-based NP-peptide assembly, covalent chemical attachment, EDC-based coupling of amines to carboxyls or NHS- and maleimide-mediated conjugation to amines and thiols (redrawn from Sapsford et al., 2013).

This is particularly essential when considering differential binding sites of physiologically active proteins on the NPs. At this stage, it becomes challenging to precisely determine the physical and chemical NP–protein interface due to the existence of multiple interactions. However, recent studies *via* precise control through the surface chemistry influence on specificity and uptake of NPs in biological systems. Table 21.1 shows different ligands for the bioconjugation of proteins at NPs through selected chemistries. Immobilization of ligands, provides certain reactive groups which can be used for subsequent bioconjugation on NPs or proteins through selected chemical reactions.

Table 21.1 Recent advances in bioconjugation of proteins, listed by material type, size, biological medium and the type of protein (Saptarshi et al., 2013)

Overview of proteins at NPs, surface chemistries and size				
Nano material	Size range (nm)	Surface chemistry	Biological medium	Protein
Iron oxide	5–20, 50–200	NH_2, COOH, Dextran, PEG-NH_2, PEG-COOH	Human plasma, Bovine serum albumin	Transferrin, Albumin, Fibrinogen chains
Gold	15–20	P(AA)-coated polymer	Human plasma, Bovine serum albumin	Albumin, Fibrinogen chains, Apo lipoprotein, Tissue development protein
Silica	8–25	Silanol	Human plasma	Transferrin, Albumin, Cellular and serum proteins Fibrinogen

21.2.3.1 Non-covalent attachment of proteins to NPs

In terms of electrostatic adsorption, "naked" or ligand-coated NPs are used to generate NP–protein conjugates. The ligands can be attached to the NPs based on specific optimized parameters such as steric control on pH or polarity of the reaction. The ligand facilitates the generation of positively or negatively charged NPs that can be attached with oppositely charged proteins. The classical example is the strong interaction between biotin and streptavidin due to its high binding affinity with the order of

Kd = 10^{-14} M (Chirra et al., 2011). The high affinity of the biotin-streptavidin binding makes this system an attractive model for studying protein–ligand interactions. Mattoussi et al., (2000) first reported the electrostatic interactions between negatively charged lipoic acid capped QDs and a positively charged recombinant protein.

In electrostatic adsorption, the proteins may interact with various orientations of NPs owing to inherent nonspecific interactions, which also result in the retention of the native protein structure. Studies have demonstrated that an appropriate NP surface chemistry can make specific interactions successful including retention of the protein conformation. For example, in case of L-phenylalanine functionalized NPs, cytochrome "c" binds on the face of the protein while in case of L-aspartic acid functionalized nanoprobes, the cytochrome "c" attaches to the cationic side of the protein (Bayraktar et al., 2007). By changing the charge of NP ligand, a specific face of the protein had been targeted. To reduce nonspecific interaction, oligo (ethylene glycol) can be an ideal choice as recently Au NPs protected with oligo molecules reduced nonspecific binding of proteins producing a stable system with NP–protein conjugates (Schollbach et al., 2014). Using a similar approach, Mukhopadhyay et al., (2012) published the interactions of bare NPs with horse heart protein cytochrome c, which lead to reduction of the protein while PEG-coated magnetite NPs did not show similar behavior.

21.2.3.2 Covalent attachment of proteins to NPs

Compared to electrostatically adsorbed proteins, covalent attachment of proteins on NPs provide greater control over stability, reactivity and aggregation state of NPs in complex biological media. The covalent affinity of NP–protein conjugates mainly depends on the interactions of protein side chains (e.g., glutamate, lysine) to the NPs, surface conjugation and functionalization of NPs (Cedervall et al., 2007). The surface fuctionalization by direct conjugation of proteins to NPs with thiol groups (–SH) or amine-carboxylate, EDC/NHS reaction using carboxylate-NPs and amino groups on protein surface or click reaction using alkyne-modified protein and azide-NPs has provided better control over NPs aggregation and retention of protein structure (Veronese and

Pasut, 2005). The use of appropriate ligands such as PEG for conjugating proteins to NPs has given greater understanding in terms of protein structure and activity. This approach has been used by Day et al., (2010) to produce antibody-conjugated gold–gold sulfide NPs for imaging and therapy of breast cancer. The NPs were conjugated with anti-HER2 and anti-IgG antibodies and orthopyridyl-disulfide-poly (ethyleneglycol)-*N*-hydroxysuccinimide (OPSS-PEG-NHS) was used as a linker. These PEG-antibodies were produced as a result of a stable amide bond formation between primary amines on the antibody and carboxyl groups on the PEG molecule. The influence of different conjugation strategies on the conformation and functions of Gαi1, a subunit of heterotrimeric G-proteins, to Au NP was investigated by Singh et al., (2013). Both kinds of conjugation methods influenced the intrinsic Gαi1 GTPase function affecting the kinetics of GTP hydrolysis in opposite modes. Noncovalent conjugation showed inhibitory effect on GTPase function while covalent conjugation dramatically accelerated it.

21.2.4 Analysis of NP–Protein Conjugates

21.2.4.1 Purification and isolation of proteins

When NP–protein conjugation is achieved, it becomes essential to purify the NP–protein conjugate from unbound protein and free NPs. This step is crucial to achieve the selective and specific binding of the NP–protein conjugate to another molecule it encounters. Isolation and purification of the desired NPs-conjugate has been achieved by a variety of approaches such as, gel filtration, ion exchange, spin column, HPLC and gel electrophoresis (Aggarwal et al., 2009; Welton et al., 2015). Among these methods, gel electrophoresis has been used as a pillar of molecular biology to separate NPs from their conjugates, protein, DNA, RNA based on their different sizes and shapes (Hanuaer et al., 2007). Recently, Hu et al., (2015) successfully used western blot electrophoresis for the separation of proteins based on the length of the polypeptides in NPs samples. The identified proteins were further transferred to nitrocellulose membrane and stained with antibodies specific to the target protein. HPLC has also been effective in separating NP–protein conjugates from nonconjugated proteins and free NPs. Aubin-Tam M-E et al., (2005) attached the aminoethanethiol ligand at the surface of Au NPs. The ligand based NPs were further conjugated to yeast cytochrome "c." Elution curves in the absorption spectra showed isolated fragments, which indicate presence of target protein conjugates in the sample. In case of Fe_3O_4 NPs, an external magnet has been useful to isolate NP–protein conjugates from unbound proteins. However, this is not practicable for purifying NP–protein conjugates in labeling experiments. In such experiments, practice of external magnet transports labeled and unlabeled particles together. To circumvent the mixture of both NPs linked to protein and free NPs, Ilyas et al., (2013) from our working group simplified this process by pulling out the protein of interest from the leaf proteome using a specific antibody against the target protein.

21.2.4.2 Quantification of proteins

The determination of protein concentration in an aqueous sample is significant for electrophoresis, protein purification, and research applications, ranging from diagnosis of diseases to determination of enzyme activity (Noble and Bailey, 2009). There are several well established assays which provide information how to analyze protein concentrations using dye based absorbance measurements (Bradford, colorimetric, Lowry, and bi-cinchoninic acid (BCA) assays), UV protein spectroscopy measurements and the fluorescent dye based assays (Rimkus et al., 2011; Gessner et al., 2006). Each quantification method has its own advantages and limitations according to its suitable application area. In addition, conjugation of NPs with different proteins can be studied with certain analytical techniques such as fluorescence spectroscopy. This spectroscopy has been used to measure the interactions of NPs with single purified protein and showed sensitivity at nano-molar quantities of NPs (Rocker et al., 2009). Mass spectrometry (MS) is traditionally used to quantify amounts of different proteins bound to NP surfaces (Rahman et al., 2013). Additionally, flow fractionation and light scattering can be used for particle size distribution in solution (Ratanathanawongs Williams and Lee, 2006).

21.3 Examples of Highly Functional NP–Protein Conjugates

21.3.1 IONPs: Synthesis and Surface Activation

IONPs are one of the most studied examples in nanomedicine due to their outstanding physicochemical properties and also relevance in most scientific fields. Caused by the high stability of both Fe^{2+} and Fe^{3+} over a wide pH range several oxides, hydroxides and oxyhydroxides containing Fe^{2+}, Fe^{3+} or both can be isolated (Navrotsky et al., 2008). Besides hematite (α-Fe_2O_3) and maghemite (γ-Fe_2O_3), the most common used iron oxide phases are the mixed valent magnetite (Fe_3O_4) and the nonstoichiometric Wüstite (FeO_x) (Hu et al., 2007; Lv et al., 2011). The most thermodynamically stable phase hematite is mostly used for catalysis, chemical sensors, pigments and for energy generation and storage because of the high chemical stability, suitable band gap and antiferromagnetic behavior. Various morphologies of hematite-NPs have been synthesized such as spindles, rods, wires, tubes, cubes and spheres (Jia et al., 2005; Hu et al., 2007; Hu and Yu, 2008; Lv et al., 2011).

The iron oxide phases with a higher net magnetic moment, namely magnetite and maghemite, are commonly used for biomedical application including imaging, hyperthermia, drug delivery or magnetic force-based gene delivery (Lee et al., 2015). It is well known that size and composition of magnetic NPs influence their physical properties and thus have a huge impact toward their capability for hyperthermia and as MRI contrast agent (Jun et al., 2005; Lee and Hyeon, 2012). In this context, it is crucial not only to tailor the size and morphology of the magnetic NPs, but also to control their phase and composition. Several methods such as thermal decomposition, co-precipitation and solvothermal synthesis have been established to obtain NPs with the required phase, size and morphology (Wu et al., 2008; Laurent et al., 2008).

Traditional and commercially available iron oxide based NPs are commonly produced by the co-precipitation method (Lu et al., 2007; Laurent et al., 2008). Thereby, the formation of magnetite from a water based Fe^{2+}/Fe^{3+} salt solution is forced by the addition of base under inert condition. A basic understanding of this method was recently given by Ahn et al., (2012), who investigated the phase formation of magnetite by a phase transition from akaganeite to goethite followed by a topotactic transition *via* hematite and maghemite to magnetite. Furthermore, the formation pathway is dependent on the local ammonia concentration leading to ferrous hydroxide primer at high local pH enforcing a transition to magnetite *via* lepidocrocite. They proposed a reaction scheme in which Fe^{2+} and Fe^{3+} have separate, but interrelated pathways toward the magnetite phase (Fig. 21.8).

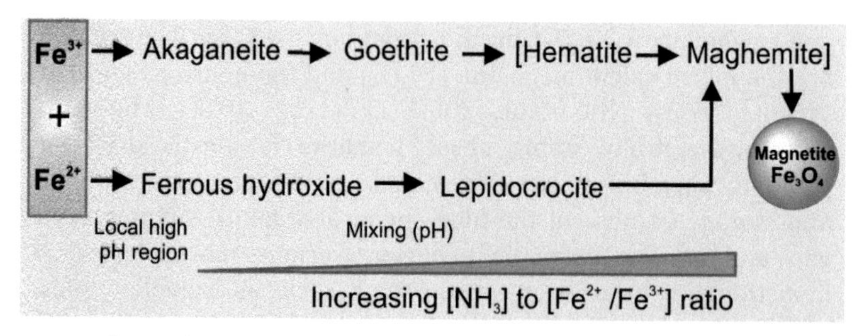

Figure 21.8 Schematic outline of the proposed reaction pathway for the formation of magnetite NP by co-precipitation method. Redrawn from Ahn et al., (2012).

The main challenge of this method is the control over the reaction parameters as several factors such as the type of precursor salt, the reaction temperature, pH value and ionic strength of the media can influence the building mechanism and therefore the formation of polydisperse particles with uncontrolled morphology and wide size distribution, is commonly observed. Moreover, once the particles are built, they are very sensitive toward oxidation by air and dissolution by acidic environment. Therefore, particle stabilization is absolutely necessary.

To improve the control over size, shape, and crystallinity of the NPs, the thermal decomposition method of organometallic compounds is often used. Here, high-boiling organic solvents which contain stabilizing surfactants such as fatty acids, oleic acid, and hexadecylamine are commonly used. Generally, the ratio of organometallic compound, surfactant and solvent together with reaction time and temperature are crucial for the control over size and shape of the particles. Park et al., (2004) reported an ultra-large-scale synthesis of monodisperse IONPs by using iron chloride and sodium oleate that produced an iron oleate complex which was in situ decomposed in a high boiling solvent at temperatures between 240 and 320°C. They were able to synthesize particles with controlled size (in a range from 5 to 22 nm) and morphology (spherical/cube). In a similar approach monodisperse iron particles in a range from 6 to 15 nm by using iron pentacarbonyl as precursor were synthesized whereby add-on oxidation was desired to lead to the magnetite phase (Park et al., 2005).

Cavelius et al., (2012) precisely controlled the phase composition of IONPs (Fe, FeO, Fe_3O_4, α-Fe_2O_3) by adjusting the synthesis conditions of thermal decomposition of iron(II) and iron(III) oxalates in high-boiling solvents. By addition of trimethylamine-*N*-oxide monodisperse hematite particles in the size range of 6–25 nm were obtained using oleylamine and oleic acid acting as capping ligands.

One huge drawback of this method is the outcome of hydrophobic particles, which makes a partial or complete ligand exchange necessary for yielding hydrophilic particles. To overcome an additional functionalization process an easy one pot hydrothermal synthesis can be used, which yields highly crystalline and hydrophilic NPs. The simple reaction is carried out in a sealed vessel, i.e., an autoclave, which is heated beyond or below the boiling point of the solvent. During the process, a supercritical fluid is created. The result is a solvent with high viscosity and surface tension lack which causes easier solution of chemical compounds. Even if the reaction is run under the boiling temperature of the solvent, the solubility and the reactivity of metal salts and complexes increase, preferred by faster running kinetic processes and shorter diffusion ways. Thereof lower reaction temperatures and times result unlike in classical solid phase reactions and also the formation of thermodynamic metastable phases is possible. This and the high crystalline state of the product are the great advantages of this synthetic method. Wang et al. recently reported the synthesis of 10 nm Fe_3O_4 NPs using oxidized dextran as a coating agent. The yielded particles were obtained by a one-pot solvothermal method in water and showed strong magnetization, low cytotoxicity and a high T_2-weighted MR image signal intensity (r_2 = 250.5 m/Ms), 3.7

times larger than the commercial product (Huang et al., 2012). Recently, our working group reported the synthesis of ultra-small superparamagnetic Fe_3O_4 (5 nm) NPs using vitamin C as a chemical reduction as well as capping agent in the oxidized form. Phantom experiments on the contrast agent revealed an enhanced r_2/r_1 ratio of 36.4 (r_1 = 5 m/Ms and r_2 = 182 m/Ms) when compared to the clinically approved agents (Xiao et al., 2011). We further extended this approach toward green tea (GT) and GT quinones (epicatechin (EC) and epigallocatechin-3-gallate (EGCG)), which can act similar to Vitamin C both as effective reduction und capping agent. GT capped Fe_3O_4-NP showed a higher r_1/r_2 value compared to EGCG capped Fe_3O_4 (209.6 compared to 95.21). Furthermore, in vitro studies revealed the good biocompatibility of the samples as well as in vivo studies showed the preferred accumulation of the NP in the tumor cells (Xiao et al., 2015).

The dispersibility and stability of IONPs is vital for their application in the biological environment (Amstad et al., 2011; Turcheniuk et al., 2013). Without any stabilizing agent (dispersant) on the nanoparticle surface, the particle will agglomerate and precipitate out of the solution due to the strong attractive van der Waals interactions and high magnetic potential as well as their possible interplay with biological molecules from the environment. To stabilize the NPs and protect them toward the physiological environment a thoughtful choice of the stabilization agent is required.

Hereby two classes of dispersants are distinguished namely high weight polymers (M_w ≥ 10 kDa) and small molecules (Amstad et al., 2011). High weight polymers physically adsorb to the NP surface *via* functional groups (e.g., hydroxyl, amine, carboxylic acid) due to electrostatic and hydrophobic interaction as well as hydrogen bonding (Muthiah et al., 2013; Amstad et al., 2011). This class of polymers involves dextran, chitosan, poly(vinyl)amine (PVA), poly(acrylic acid) (PAA), poly(ethyleneglycol) (PEG) and poly(ethylene imine) (PEI). Nowadays most commercially available MRI contrast agents are coated with those natural carbohydrate polymers (Turcheniuk et al., 2013).

However, their limitation is the lack of stability due to the weak electrostatic interaction and resulting reversible physisorption. Thereby, weak protecting of the NP core toward biomolecules, which could cause toxic effects and drastically decrease blood circulation times in vivo are observed (Muthiah et al., 2013; Amstad et al., 2011; Turcheniuk et al., 2013; Liu et al., 2013).

To overcome these drawbacks, small molecules can be applied to the surface built up by an anchor, which should have a high affinity toward the iron oxide core, a spacer, providing high colloidal stability and protection against the biological environment, and an optional surface functionality. By choosing each of the building blocks carefully, an optimal dispersant for the desired field of application is produced (Muthiah et al., 2013; Amstad et al., 2011; Turcheniuk et al., 2013). Hereby the hydrodynamic radius of the particles can be tuned as well as the well-defined assembly of the dispersant toward the NP surface leads to a controlled presentation of the surface functionalities. This is particularly important, because NP size, stability, dispersant shell thickness and surface functionalities are the crucial factors for their applicability in biological environment (Jun et al., 2008).

There are several anchors used for iron oxide stabilization such as dopamine and its derivatives, phosphonates, carboxyl groups and silanes (Muthiah et al., 2013; Amstad et al., 2011; Turcheniuk et al., 2013). The most commonly used anchor groups are summarized in Fig. 21.9. Irreversible linkage of the anchor toward the NP surface is required as well as a high adsorption and a low desorption rate. Besides carboxylic acid, phosphates and phosphonates, catechol-based ligands have been paid much attention during the last decades. Their use was inspired by the living organism where catechols are used for the fixation of metals and surface adhesion. L-3,4-dihydroxyphenylalanine (DOPA)/Fe^{3+} and dopamine/Fe^{3+} complexes were intensively studied regarding their structure and electronic interactions so that the interplay between the metal centre and the ligand is well understood (Yuen et al., 2012). Nevertheless, recent reports showed a lack in stability of PEG-dopamine-coated NPs as well as surface corrosion of IONPs after replacement of oleic acid by dopamine (Shultz et al., 2007). This reflects the low affinity toward dopamine and the iron oxide atoms present on the nanoparticle surface. Due to their electron withdrawing effect, nitrocatechols show a better affinity and stabilizing effect toward the iron oxide surface. In a direct comparison of PEG stabilized IONPs using either DOPA/dopamine or nitro-DOPA/nitro-dopamine, the PEG-nitro-DOPA and PEG-nitro-dopamine showed much higher stability (Amstad et al., 2009). In line with this, Gillich et al., (2013) substantiate the suitability of nitrocatechol as a single-foot irreversible binding anchor for iron oxides. They determined the influenced of linear versus dendritic polymer shell architecture using nitrocatechol as an anchor group. The dendron-stabilized NPs showed a higher colloidal stability particularly with regard to conventional brushed systems. Furthermore, they had a relatively low overall hydrodynamic size, a reversible temperature-induced aggregation behavior and a substantially reduced shell thickness.

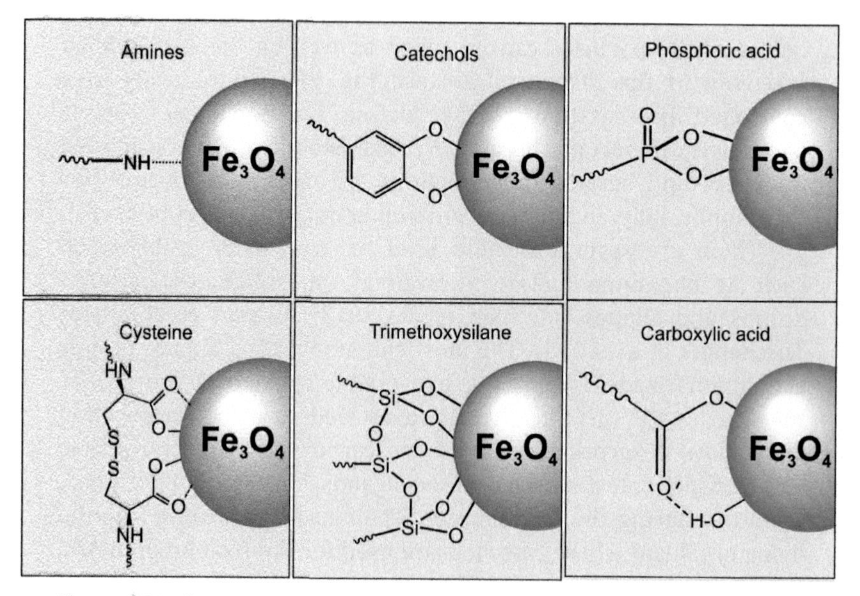

Figure 21.9 Summary of commonly used anchor units of stabilization agents for magnetite nanoparticles.

In addition, a suitable spacer has also great influence toward the NP stability, their job is to keep the cores so far apart that neither magnetic nor van der Waals interactions dominate the system. As soon as the two cores draw closer, they are restricted by the volume of the polymer shell. The spacer also enables the incorporation of further modalities such as targeting ligands and a second imaging modality.

Another way to ensure stability and provide tuned surface functionalities is the encapsulation of IONPs within a silica shell (Wortmann et al., 2014; El-Gamel et al., 2011; Wu et al., 2008).

The silica shell protects the iron oxide core before environmental influences, is itself physiological inert and provides high colloidal stability. Beyond that, those core-shell structures can contain additional features such as tuneable mesopores for drug-delivery and fluorophore-storage. Their well-defined and tuneable structures (i.e., size, morphology, and porosity) and reconcilable surface chemistry make them interesting candidates for (optical) imaging contrast agents and as drug delivery systems. Especially as they show photo-physical stability, biocompatibility and high colloidal stability. Secondly, the surface can be easily modified by well-known siloxane chemistry with antibodies, aptamers, and polymers.

For the add-on functionalization using click chemistry, surface stabilizing agents which are already bearing the desired alkyne or azide function can be introduced *via* ligand exchange, as shown in Fig. 21.10.

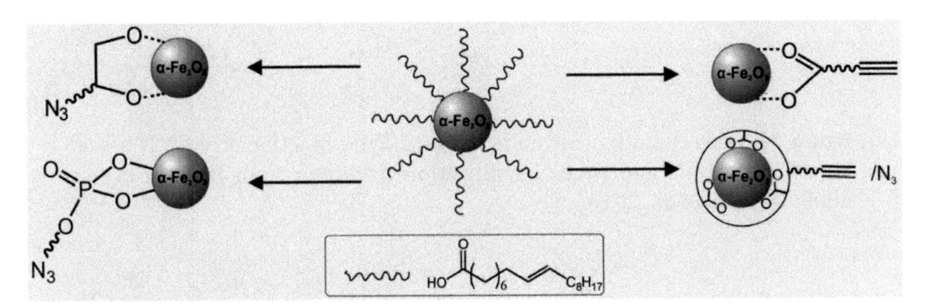

Figure 21.10 Preparation of the azide or alkyne-functionalized Fe_3O_4 *via* the ligand exchange method (redrawn from Li and Binder, 2011).

Furthermore, surface functionalities can be introduced in situ by direct synthetic strategies on the NPs surface as reported by Hayashi et al., (2009). In this reaction, iron(III) 3-allylacetylacetonate was hydrolyzed using hydrazine to obtain allyl functionalized magnetite NPs which were then further converted into Br-functionalized Fe_3O_4. Azide groups were finally incorporated by a salt metathesis reaction (Fig. 21.11, upper reaction). Moreover, starting from allyl-Fe_3O_4 NPs, direct attachment of cysteine was possible by a thiol-ene click reaction (Fig. 21.11, lower reaction) (Hayashi et al., 2010).

In another report, Schätz et al., (2009) demonstrated the surface incorporation of azide groups on SiO_2-coated magnetite

NPs. Two different pathways were proven successful, whereby 3-azidopropyltriethoxysilane was either post-grafted on the surface or directly employed during the particle synthesis (Fig. 21.12).

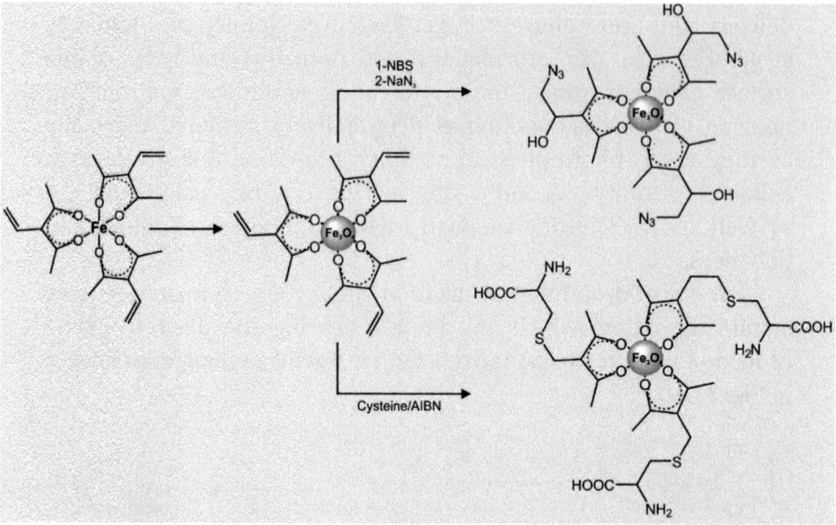

Figure 21.11 Preparation of the azide or alkyne-functionalized MNPs *via* in situ hydrolysis and ligand modification (redrawn from Hayashi et al., 2009; Hayashi et al., 2010).

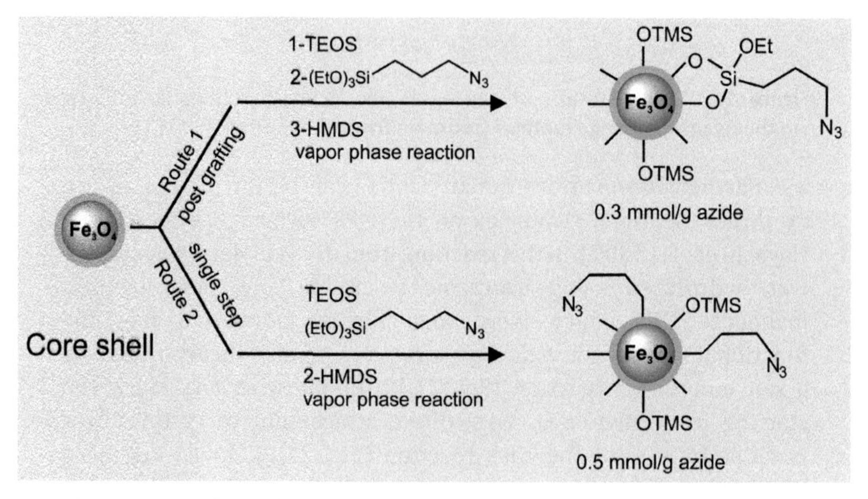

Figure 21.12 Azide-functionalized magnetite@silica NPs following a post-grafting (left) and a single-step (right) protocol (redrawn from Schätz et al., 2009).

21.3.1.1 Separation and selective conjugation of proteins from a proteome

In biomedical research, separation of biological moieties such as proteins and DNA from their native environment is prerequisite for their analysis and characterization. In this condition, inherent magnetism in Fe_3O_4 NPs offers idealism to remove physioadsorbed species and debris from the nano-bioconjugate to enhance the specificity of the application. In our working group, Ilyas et al., (2013) successfully used this approach for the selective conjugation and separation of Cys-protease from crude leaf proteome based on click-functionalized magnetite NPs. In this work microwave synthesized superparamagnetic iron oxide NPs (SPIONs, ca. 6 nm) were conjugated with different biomolecules (biotin and cys protease inhibitor) *via* click chemistry protocols. The functional NPs were further used for the conjugation and separation of proteins from a crude proteome. The detection of attached biotin with streptavidin-HRP and fluorescent Rh conjugates provided a model system for the justification of this approach (Fig. 21.13a–c). Furthermore, capturing proteases on the functionalized SPIONs significantly simplified the process and showed selectivity and reproducibility, which leads to potential for target-specific application. Figure 21.14 shows the steps involved for functionalization and magnetic separation of particles.

To get quantitative prospect for the interaction between the NPs and protein, biotin@SPIONs and streptavidin, a concentration course assay was performed using fluorescent streptavidin. Results showed a capture limit of 0.55–1.65 μg protein/100 μg particles, which provides basis for the concentration assessment between NPs and the protein of interest (Fig. 21.13c–d).

To examine whether E-64@SPIONs can selectively capture cysteine proteases, a leaf extract of the model plant *Arabidopsis thaliana* with E-64@SPIONs and control beads (NPs without E-64) were incubated and the captured proteins were analyzed with an antibody against RD21, an abundant cysteine protease in *Arabidopsis* leaf extract. After antibody binding, the beads were washed and boiled in SDS sample buffer and eluted proteins were separated on a protein gel to analyze the protein blot using a secondary antibody that detects the RD21 antibody. The RD21 antibody was only detected using E-64@SPIONs and not by the control beads (Fig. 21.15). Pre-incubation of leaf proteomes with an excess of E-64 before adding the SPIONs prevented the capture of RD21 on the NPs, thereby demonstrating that capture was specific for cys proteases. E-64 contains reactive epoxy groups, which specifically conjugated with proteins of interest thereby giving us signals for the detection in western blots.

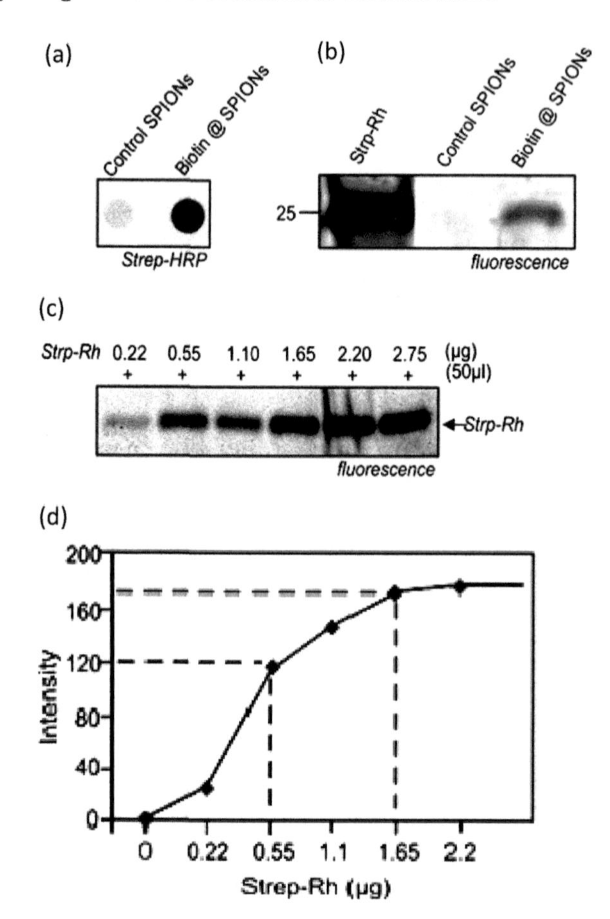

Figure 21.13 Biotin@SPIONs capture streptavidin-HRP and rhodamine. (a) Biotin@SPIONs was detected with streptavidin-HRP and chemiluminescence. (b) Captured streptavidin-rhodamine was detected with SDS-PAGE electrophoresis. (c) Concentration course of streptavidin-rhodamine using Biotin@SPIONs (100 µg). Each mixture of Biotin@SPIONs and streptavidin-rhodamine was incubated for 1 h, maximum emission intensities at 580 nm were recorded after each addition of streptavidin rhodamine solution. (d) Saturation binding curve determined by quantification of luminescence intensity (reprinted from Ilyas et al., 2013).

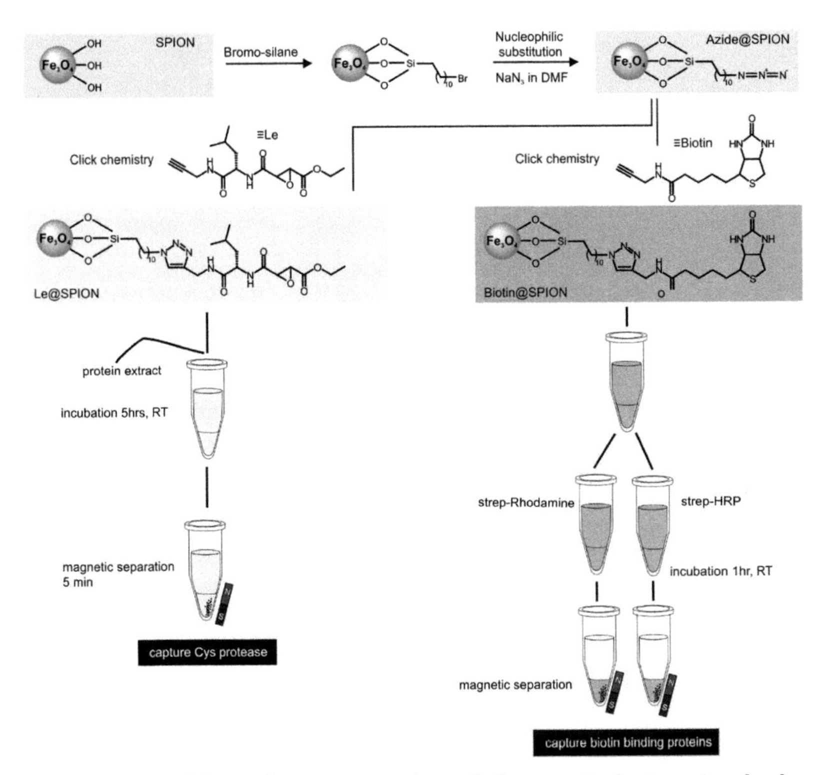

Figure 21.14 Schematic representation of the practical steps involved. Microwave-assisted synthesis, click chemistry, and functionalization of SPIONs to capture biotin binding protein and Cys protease in plant proteome. Structure of the alkyne-biotin probe (upper right), structure of the Cys protease inhibitor E-64 containing a leucine-epoxide reactive group provides specificity for PLPCs (redrawn from Ilyas et al., 2013).

In this system, the protease capturing *via* functionalized SPIONs provides an excellent alternative to handle extensive protein work as approach offers significant advantage especially when multiple proteins need to be studied. The functionalized NPs enabled specific and efficient way of identifying entire proteome through specific binding and thereby circumventing the multi-step copurification use in conventional approach. This conjugation method therefore can be extended to other proteins and NPs, for the sensitive isolation of protein from blood samples such as serum and plasma. In addition, biotin functionalized SPIONs are suitable for binding with streptavidin bearing modified proteins for nanomedicinal applications.

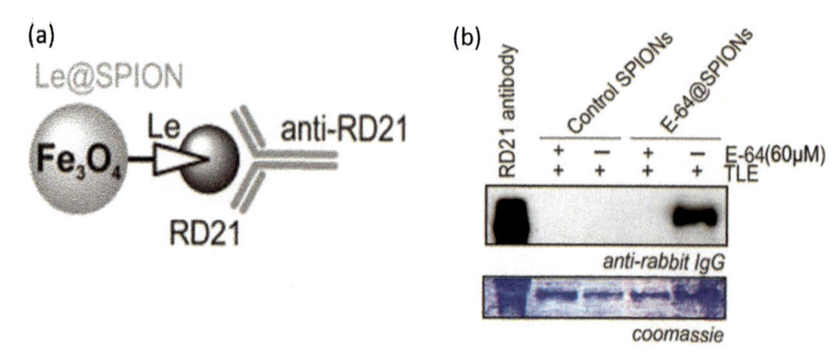

Figure 21.15 E-64@SPIONs capture Cys protease RD21 from plant proteome. (a) Schematic representation of capturing and detecting, Cys protease RD21 using Le@SPIONs (leucine-epoxide E-64). (b) Detection of RD21 antibody captured on E64@SPIONs incubated with leaf proteome. The eluted proteins were separated on protein gel, analyzed on protein blot using a secondary antibody linked to horseradish peroxidase (HRP) that detects the RD21 antibody. The RD21 was only detected using E-64@SPIONs but not control SPIONs. Lower panel represents the loading control. TLE represents total leaf extract (reprinted from Ilyas et al., 2013).

Another interesting report for protein separation and purification based on click chemistry and NPs was provided by Jian et al., (2012). They showed Fe_3O_4@SiO_2 and their attachment with Cu2+-IDA (iminodiacetic acid, indicated by purple in Fig. 21.16) as an affinity ligand *via* Cu(I) click reaction. The NPs were used for the removal of the highly abundant protein, hemoglobin (Hb), from BSA and HSA. Fe_3O_4@SiO_2-Cu^{2+}–IDA conjugate showed high adsorption capacity and specificity toward Hb protein suggesting promising affinity material in the removal of abundant proteins in proteomic analysis (Fig. 21.16).

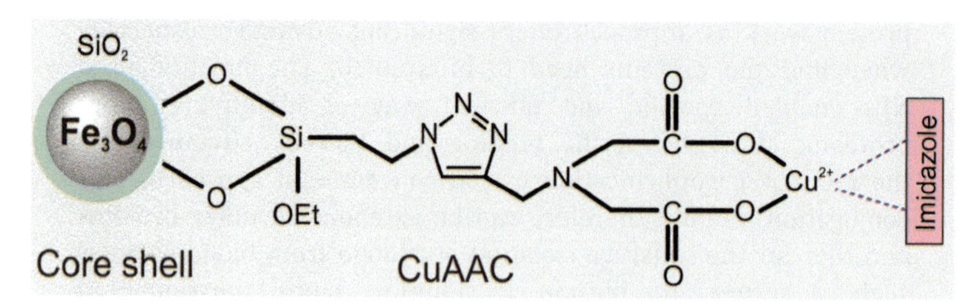

Figure 21.16 IDA-Cu functionalized IONPs *via* the CuAAC reaction (redrawn from Jian et al., 2012).

Von Maltzahn et al., (2008) were able to successfully investigate the utility of peptide conjugated NPs for tumor targeting based on a click reaction. The amine moiety was immobilized on dextran-Fe_3O_4 NPs. The targeting peptides (cyclic LyP-1) were specifically linked to azide functionalized NPs which offered excellent stability during the circulation determined by intravenous administration. The conjugated NPs were successfully penetrated the tumor interstitium and offered specific binding to p32-expressing tumor cells in vivo (Fig. 21.17).

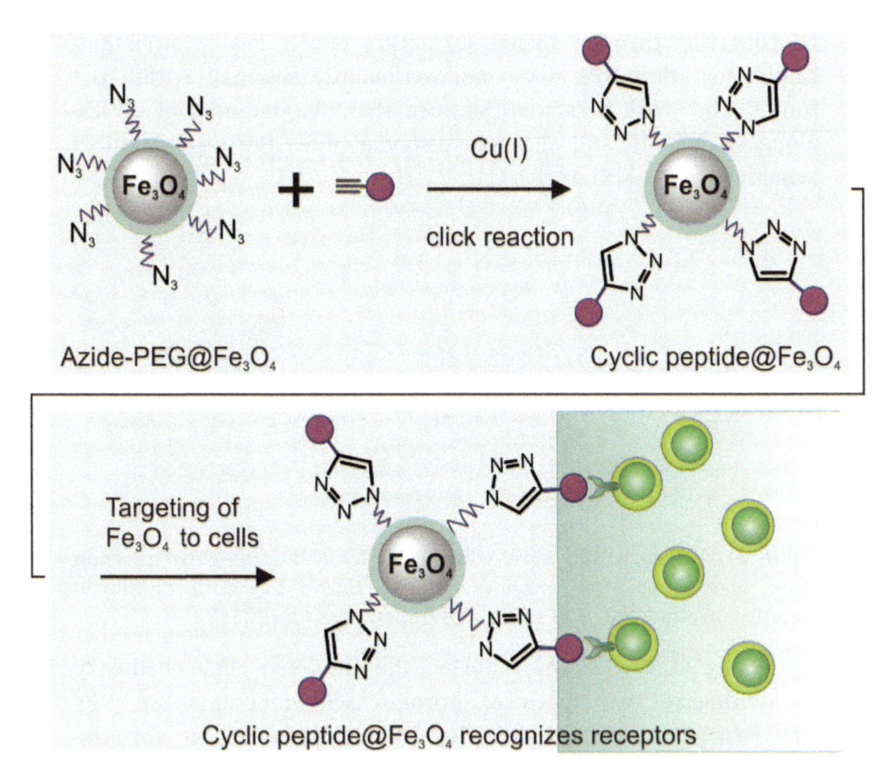

Figure 21.17 Fluorescent, IONPs with azido-PEG groups. Conjugation of cyclic targeting peptides (purple) bearing pendant alkynes to azido-PEG NPs *via* click reaction allows specific targeting of the NPs to cells expressing the receptor (green) (redrawn from von Maltzahn, Geoffrey, et al., 2008).

Similarly, Gallo et al., (2014) reported CXCR4 (peptides) and matrix metalloproteinase (MMP, enzyme) responsive IONPs for in vivo and in vitro assays. The structure of the targeting ligand (cyclopentapeptide) contains peptide sequences cleavable by MMP enzyme overexpressed in tumor. The NPs with alkyne or azide moieties were immobilized by PEG molecules and further immobilized with a peptide targeting ligand by copper-free [3+2] cycloaddition reaction following cleavage by MMP enzyme. These constructs were tested in in vitro studies where T_2 signal enhancements of 160% were measured when particles were incubated with cells overexpressing MMP and CXCR4. The tumor bearing mice showed the signal enhancing ability with the NP-construct through cluster formation which can be attributed to the fact that NPs were more efficiently retained within the tumor. The work presented showed that the potential of CXCR4 along with MMP and click chemistry enables the production of sensitive cancer MRI in vivo (Fig. 21.18).

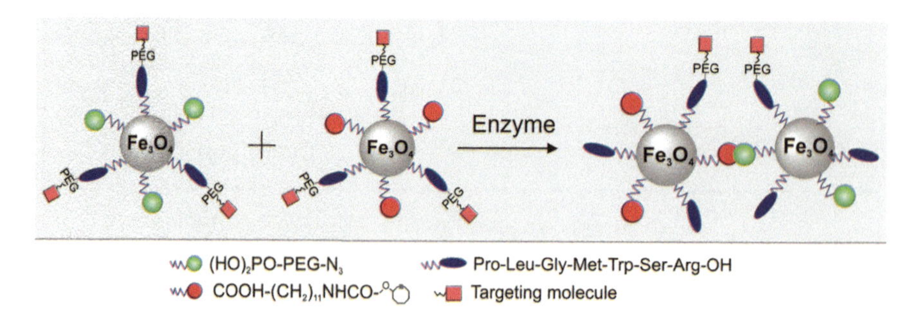

Figure 21.18 Two IONPs were synthesized for a bioorthogonal reaction after cleavage by MMP enzymes, which exposes the azide or alkyne moieties on either set of NPs (adapted from Gallo et al., 2014).

Zhang et al., (2014) reported the thiol-ene click chemistry to synthesize two types of boronic acid functionalized NPs [(I) Fe_3O_4@MPS@PBA and (II) Fe_3O_4@SH@AAPBA)] for the efficient enrichment of glycoproteins. Alkene-or thiol-coated Fe_3O_4 (Fe_3O_4@MPS and Fe_3O_4@SH) were generated by sol–gel reaction with a 3-methacryloyloxypropyltrimethoxylsilane (MPS) or 3-mercaptopropyltriethoxysilane (MPTES). The carbon-carbon double bonds/thiol groups on the surface of the NPs offer clickable sites to react with 4-mercaptophenylboronic acid (4-MPBA) or 3-acrylamidophenylboronic acid (AAPBA) during the subsequent click reaction (Fig. 21.19).

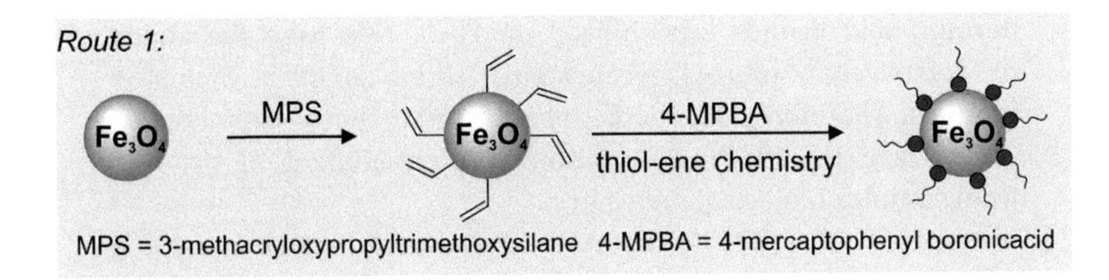

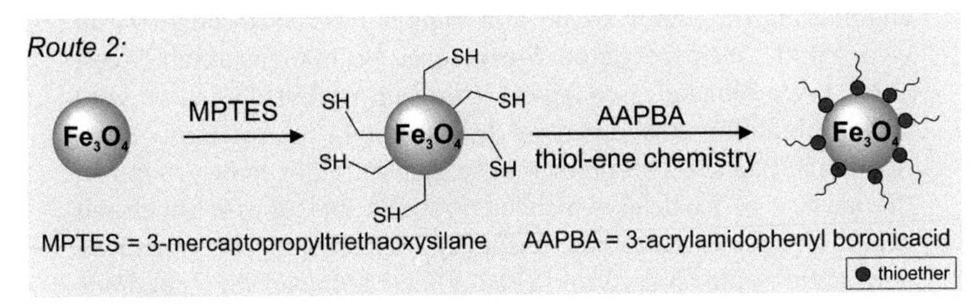

Figure 21.19 Schematic illustration of the two synthesis routes of boronic acid modified Fe_3O_4 NPs *via* thiol-ene click reaction (adapted from Zhang et al., 2014).

Steady-state binding evaluated the selectivity and efficiency of protein separation by click functionalized NPs toward glycoproteins ovalbumin (OVA), transferrin (Trf), non-glycoproteins lysozyme (Lyz) and horse heart cytochrome "c" (Cyt C). For the adsorption and separation of glycoproteins and non-glycoproteins, functionalized NPs were incubated in protein samples with different concentrations and magnetically separated and washed with PBS buffer to remove unbound proteins. Subsequently, the click-products were added into an acidic eluted solution to release the proteins. Click-functionalized NPs offer greater specific binding to glycoproteins (OVA &Trf) than non-glycoproteins (Lyz and Cyt C). The Fe_3O_4@SH@AAPBA NPs presented better binding capacity toward OVA and Trf than Fe_3O_4@MPS@PBA NPs which may be due to the higher grafting density of boronic acid ligands *via* thiol–ene click reaction. To further illustrate the applicability of click functionalized Fe_3O_4 NPs, the enrichment of glycoprotein from egg white samples was performed. Here, click functionalized NPs were incubated in 200-fold or 400-fold dilutions of egg white in PBS, then gently and magnetically separated and washed and immersed in acidic eluted solution. The results confirmed that boronic acid ligands immobilized on Fe_3O_4 NPs have the ability to selectively capture glycoproteins from complex biological samples. Therefore, this methodology can be successfully applied for surface modification, isolation and enrichment of proteins from complex biological molecules.

Lee et al., (2014) reported a successful tumor-targeting efficiency of functionalized NPs *via* site-specific metabolic glyco-engineering on tumor tissue and copper-free click chemistry in vivo. Firstly, tetraacetylated *N*-azidoacetyl-D-mannosamine (Ac_4M anNA$_z$) was loaded into glycol chitosan particles (CNPs) with an amphiphilic structure *via* hydrophobic interactions. Here, Ac_4ManNA$_z$ acts as precursor for the generation of azide groups on the surface of particles which intravenous injection of generated azide functions on tumor tissue specifically by the enhanced permeation and retention (EPR) effect followed by metabolic glycol engineering. During a second intravenous injection, these chemical groups acted as receptors and presented enhanced tumor-targeting ability of drug-containing NPs by copper-free click chemistry in vivo. Overall results showed that there are significantly more binding molecules on the surface of tumor cells. The azide groups as small receptors can be expressed on the cell surface in large quantities regardless of the types. In addition to Fe_3O_4, other phases of iron oxide such as γ-Fe_2O_3 have also been investigated for surface modification by click reactions, although only very limited numbers of literature are available regarding click chemistry combined with protein conjugation compared to magnetite. White et al., (2006) reported the attachment of two types of ligands (phosphonic acid-azide and carboxylic acid-alkyne) on γ-Fe_2O_3 NPs *via* click chemistry. The results demonstrated successful utilization of click chemistry for the attachment of "universal ligands" which were well dispersed in a range of solvents.

As stated in the introduction, Bolley et al., (2013) published superparamagnetic fluorescent γ-Fe_2O_3 NPs targeting $\alpha v\beta 3$ integrins based on carbodiimide coupling and click chemistry for the elaboration of multimodal imaging contrast agents. In their initial step, NPs were functionalized with hydroxymethyl-enebisphonates bearing carboxylic acid or alkyne functionality. In the next step, these reactive sites were used for the covalent attachment of PEG, fluorophore rhodamine (dyes) and $\alpha v\beta 3$ integrin targeting peptide cRGD (Fig. 21.20).

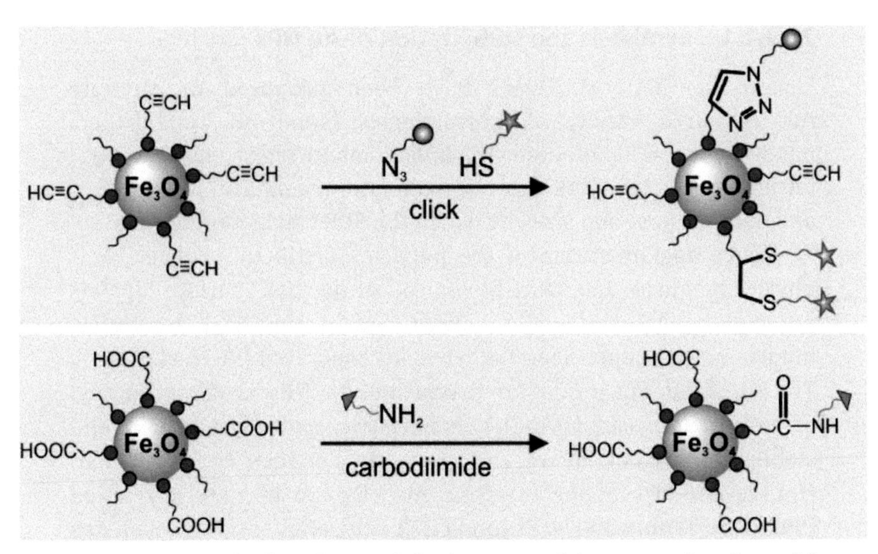

Figure 21.20 Carbodiimide vs. click chemistry elaboration of multimodal imaging contrast agents (redrawn from Bolley et al., 2013).

During surface characterization, they compared the results of carbodiimide and click chemistry. The grafting of active targeting moiety enhanced targeting affinity to integrin in click chemistry as well as the bio-stability and MRI properties. Fluorescence data showed the similar trends obtained by click chemistry conjugation.

21.3.2 Au NPs

Nobel metal NPs like gold and silver with controlled geometry, optical properties and surface chemistry are under intensive research for biomedical applications due to their SPR and presumed chemical inertness. As Au NPs have always fascinated the human beings, there is a long history of medically used colloidal Au NPs back to the Middle Ages in Europe. Nowadays, research focuses in diagnostics, therapy and immunology (Dykman and Khlebtsov, 2012). Their main fields of application are genomics, biosensors, immunoassays, photothermolysis of cancer cells and tumors, targeted drug and antigen delivery and optical bio-imaging of cells and tissues with state-of-the-art nanophotonic detection systems (Dreaden et al., 2012).

21.3.2.1 Synthesis and stabilization of Au NPs

Several synthetic strategies have been proposed to generate Au NPs in a variety of morphologies (spherical, rod shaped, nano-shells with tunable thickness, nano-cages, hollow NPs, tetrahedral, octahedral, cubic, icosahedral, triangular bipyramides, and nanoprisms) and sizes by which the SPR can be tuned.

Since agglomeration of the particles distinctly changes their physicochemical and also biological properties, surface ligands are commonly employed to either obtain electrical or sterical hindrance and hence keep the particles separated (Yu et al., 2011). The most famous route for producing Au NPs is the reduction of chloroauric acid ($HAuCl_4$) using citric acid as reduction and stabilization agent. By varying the ratio between citric acid and $HAuCl_4$, the size of the resulting Au NPs can be precisely tuned (Shan and Tenhu, 2007). Beyond citric acid, amines and phosphates can be also used as stabilization agents, but the most stable bond is formed between gold and thiolate groups following the HSAB principle. This includes ligand molecules bearing a –SH or S–S group. In this context, Brust et al., (1994) developed a well-known synthesis whereby Au NPs with a mean size between 1–3 nm are formed in a two-phase system in the presence of alkanethiol. Several other methods have been established to synthesize monodisperse Au NPs, including microemulsion, sonochemistry, photochemistry, and radiolysis (Zhou et al., 2009). Considering biomedical application, water dispersibility is of crucial importance, but often implicates the need for a ligand exchange reaction in order to obtain desired surface functionalities. In this context, the in situ preparation of water dispersible Au NPs has been reported, whereby $HAuCl_4$ is reduced by $NaBH_4$ in the presence of glutathione (GSH) (Schaaff et al., 1998; Negishi et al., 2004). Furthermore, in recent studies of our workgroup, the hydrolytic decomposition of molecular precursors for the formation of ligand-free Au NPs of 5.5 nm size has been developed (Zopes et al., 2013; Stein et al., 2015). This simplifies the subsequent surface decoration with various ligands including biomolecules. Additionally, biological molecules can bind to Au NPs in a non-specific manner. This is advantageous for introducing a specific functionality but also has to be taken into account for in vitro and in vivo applications as it could cause unwanted interactions.

As stated earlier, one way to prevent this, is the surface functionalization with PEG-chains which prolong blood circulation times and uptake by the RES.

21.3.2.2 Click-functionalized Au NPs

Terminal azides or alkynes have to be present at the NP surface in order to modify Au NPs *via* the CuAAC reaction. A general synthetic strategy for azide functionalization of NPs is shown in Fig. 21.21, *via* a ligand exchange method. A variety of alkyne and azide functional ligands have been used to replace alkanethiol ligands and hence introduce the desired azide or alkyne functionality (Li and Binder, 2011). The CuAAC reaction has also been performed on the surfaces of Au NPs under microwave conditions, which allowed the reaction time to be kept under 10 min. By use of short reaction time, particle aggregation appeared to be negligible in the $CuSO_4$-sodium ascorbate catalytic system with dioxane-*t*-BuOH-H_2O (1:1:0.5) as solvent, enabling many alkyne-functionalized molecules to be clicked to the Au NPs in yields of 78–100% (Li and Binder, 2011).

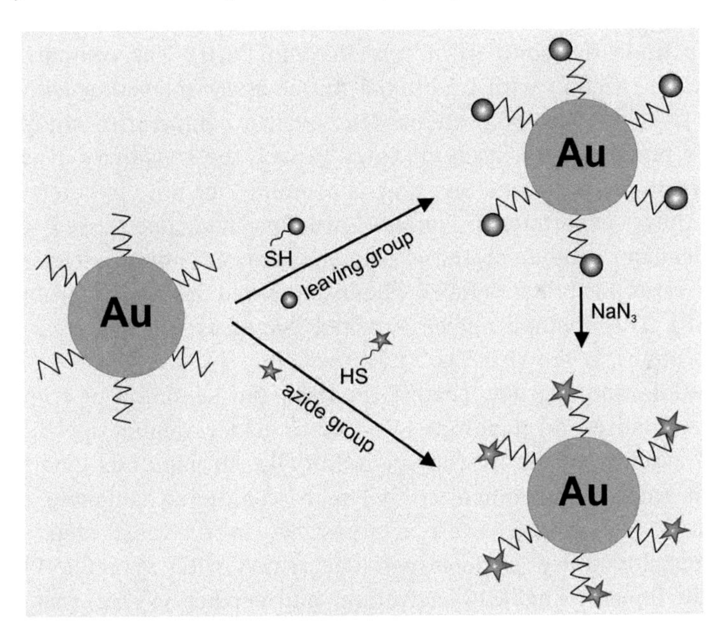

Figure 21.21 Synthesis of azide-functionalized Au NPs *via* direct ligand exchange approach (adapted from Li and Binder, 2011).

Furthermore, Zhu et al., (2012) recently published the detection and quantification of proteins with a click chemistry-based assay by two types of Au NPs, each modified with thiol terminated azide or an alkyne functionalized group. In the process, a functionalized Au NP-based colorimetric assay has been used to quantify the total protein concentration. Both Au NPs (azide-and alkyne-functionalized) were mixed equally to obtain homogeneous dispersions of NPs. In a separate reaction, proteins were mixed in Biuret reagent consistent of potassium hydroxide and hydrated copper(II) sulfate together with potassium sodium tartrate. In the next step, equal amounts of functionalized Au-NPs were mixed in the above-mentioned solution. Cu(II) is reduced to Cu(I) by proteins in the alkaline solution.

The absorbance spectra corresponding to each bottle in Fig. 21.22 (left) also validated the aggregation of Au NPs Fig. 21.22 (right) and revealed that the strong alkaline solution prevented the aggregation of functionalized Au NPs. The proteins were detected with bathocuproine disulfonic acid (BCDSA), a Cu(I) chelator and inhibitor of the CuAAC reaction. When Cu(I) was sequestered by BCDSA, the mixed Au NPs solution stayed red even upon the addition of proteins and Cu(II). The comparison of click reaction with traditional biuret assay showed sensitivity for the detection of proteins. Moreover, a comparative study of click reaction with Bradford assay proved the sensitivity of click reaction (accurate and less time consuming) for protein detection. In these experiments, various proteins had been used and molecular weights of these proteins showed minor effects on the reproducibility of click chemistry-based assay for proteins (CAP). This method allows a naked eye-assay for the presence of proteins, as shown in Fig. 21.22.

Gole and Murphy, (2007) reported the synthesis of trypsin functionalized Au nanorods in order to achieve higher specificity and stability of the NP-conjugate. Initially, Au nanorods modified with carboxyl-functionalized polymer (containing sulfonate and maleic acid groups) were synthesized. In the next step, the carboxyl-modified polymer was conjugated with an amine-PEG-azide linker using EDC activation, and further clicked (copper catalyzed 1,3-dipolar cycloaddition reaction) with alkyne-labeled trypsin. In the above system, the maleic acid groups participated in the amide bond formation whereas sulfonate groups offered electrostatic stabilization to the nanorods and prevents their aggregation in aqueous media. Trypsin labeling was achieved by its lysine residues with EDC activated 4-pentynoic acid. The trypsin-Au NP conjugates offered purity, stability, and specific activity which is threefold higher than other conjugates designed by electrostatic adsorption or EDC coupling between the Au NPs and lysine residues. Trypsin quantification was achieved by using Bradford assay that involved the binding of dye (Coomassie Brilliant blue) to amino acid residues of the trypsin.

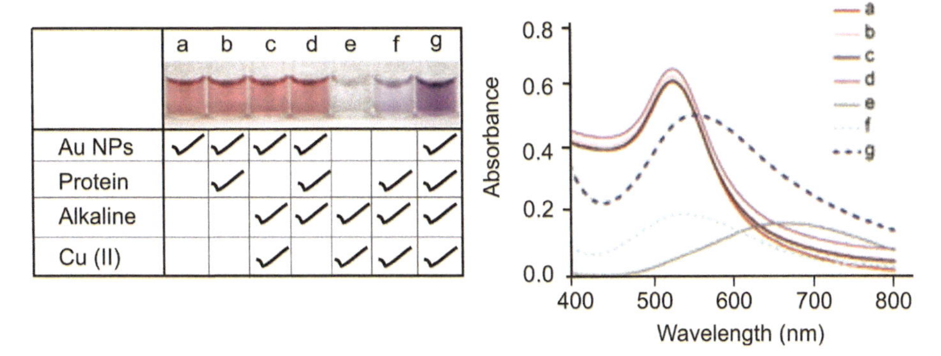

Figure 21.22 Parts a–g show the essential compounds of CAP at room temperature (left), UV–vis spectra of the solutions from parts a–g (right). Alkaline solution is the biuret reagent. Reprinted with permission from Zhu et al., (2012) Copyright 2012, American Chemical Society.

Kim et al., (2010) reported the production of bio-luminescent nano-sensors for the detection of protease activity based on the Au-luciferase conjugate. In this work the alkyne-modified 5 nm Au NPs were attached to azide-modified recombinant luciferase (Luc8-pep-N3) *via* a Cu(I) catalyzed [3 + 2] cycloaddition reaction. The click reaction offered an efficient conjugation between NPs and the luciferase protein. The intein-mediated ligation enabled this protein to be site-specifically conjugated on the Au NP. In UV-vis analysis, the final product (Luc8-pep-Au NPs) displayed a slightly red-shifted spectrum (520 nm) because of their surface modification compared to unconjugated NPs (Au-alkyne) with a strong absorption at 516 nm. Bradford assay showed that conjugation offered 4.7 proteins per Au NPs by displacing its monolayer with 2-mercaptoethanol and quantifying the luciferase concentration. The results indicated that the site-specific orientation of luciferase on NPs is critical for detecting protease activity.

Only recently, Hu et al., (2015) published a study about core–shell gold glyconanoparticles which were prepared *via* a copper-free click chemistry and used for the selective detection of lectins by a colorimetric assay. Core-shell NPs where thereby produced by a SPAAC between azido galactoside and a lipid cyclooctyne which was followed by the self-assembly of the resulting galactolipid on PEG functionalized Au NPs. The plant lectin protein peanut agglutinin (PNA) was then selectively detected in the presence of other proteins *via* UV-vis and plasmonic resonance scattering spectroscopy.

21.4 Conclusion

Given their nanoscopic dimensions, ultra-small particles (<100 nm) are in general well suited for interactions with the cells; however the current challenge of the nanomedicine is to transform inorganic nanoparticles into signal-generating vectors that demand attachment of different imaging modalities, targeting ligands, and therapeutic payloads onto a single nanoparticle carrier. In this context, click chemistry provides a promising platform for enhancing biomedicinal capabilities through selective conjugation of macro-biomolecules such as proteins, antibodies to nanoparticulate carriers. An effective bioconjugation strategy based on click chemistry reactions can lead to new biochemical protocols with faster detection rates and higher reliability at low concentrations.

The interaction of proteins and nanoparticles play a vital role to gain a comprehensive understanding of changing biological identity of engineered nanoparticles when present in physiological conditions. Although the synthetic identity (composition, toxicity) of nanoparticles is not entirely diluted in vivo systems, the adsorption of proteins and other biomolecules can drastically affect their cellular uptake or attachment. Important challenges pertinent to the biomedical application of nanoparticles includes the limited understanding of specific chemical interactions between nanoparticles and proteins. Given the fact that structure and function of proteins can be influenced by a number of parameters including the chemical environment (pH, ionic strength) and surface properties of nanoparticles, synthetic protocols providing covalent immobilization of proteins and other biomolecules on nanoparticulate carriers can improve their application potential in biomedical analysis. The complex structures of the proteins and their comparably larger sizes than nanoparticles hamper the characterization process and so far only few analytical studies clarifying the nano-bio interaction and attempting a quantification of surface attached probe groups are reported in the literature. Nevertheless, the recently gained results presented in this chapter demonstrate the ongoing interest in nanoparticle–protein interactions and display the promising opportunities of protein nanoconjugates for a vast variety of biomedical applications. Especially iron oxide and gold nanoparticles have strongly been investigated in terms of bioconjugation strategies using click chemistry which offer a simple and efficient way to covalently attach proteins to nanoparticle surfaces. Targeting cancer cells with functional nanoprobes possessing a targeting drug unit and an imaging moiety carries great potential for early detection, accurate diagnosis, and targeted therapy of various diseases. In summary, click chemistry combined with functional nanoparticles is a promising avenue in developing new biochemical assays with improved signal transformation and facile signal readout for diagnostic investigations.

References

Adolfsson H., Converso A., Sharpless K. B., *Tetrahedron Lett.,* **40** (1999), 3991–3994.

Agard N. J., Prescher J. A., Bertozzi C. R., *JACS Commun.,* **126** (2004), 15046–5047.

Aggarwal P., Hall J. B., McLeland C. B., Dobrovolskaia M. A., McNeil S. E., *Adv. Drug Deliv. Rev.,* **61** (2009), 428–437.

Ahn T., Kim J. H., Yang H. M., Lee J. W., Kim J. D., *J. Phys. Chem. C,* **116** (2012), 6069–6076.

Amstad E., Gillich T., Bilecka I., Textor M., Reimhult E., *Nano Lett.,* **9** (2009), 4042–4048.

Amstad E., Textor M., Reimhult E., *Nanoscale,* **3** (2011), 2819–2843.

Ashby J., Pan S., Zhong W., *ACS Appl. Mater. Interfaces,* **6** (2014), 15412–15419.

Aubin-Tam M.-E., Hamad-Schifferli K., *Langmuir,* **21** (2005), 12080–12084.

Baskin J. M., Prescher J. A., Laughlin S. T., Agard N. J., Chang P. V., Miller I. A, Lo A., Codelli J. A., Bertozzi C. R., *Proc. Natl. Acad. Sci. U. S. A.,* **104** (2007), 16793–16797.

Bayraktar H., You C. C., Rotello V. M., Knapp M. J., *J. Am. Chem. Soc.,* **129** (2007), 2732–2733.

Berkel S. S., Van, Dirks A. T. J., Meeuwissen S. A., Pingen D. L. L., Boerman O. C., Laverman P., Delft F. L. Van, Cornelissen J. J. L. M., Rutjes F. P. J. T., *ChemBioChem,* **9** (2008), 1805–1815.

Bhattacharya K., Farcal L., Fadeel B., *MRS Bull.,* **39** (2014), 970–975.

Biffi S., Voltan R., Rampazzo E., Prodi L., Zauli G., Secchiero P., *Expert Opin. Drug Deliv. Taylor Fr.,* **12** (2015), 1837–1849.

Binder W. H., *Macromol. Rapid Commun.,* **29** (2008), 951.

Bolley J., Guenin E., Lievre N., Lecouvey M., Soussan M., Lalatonne Y., Motte L., *Langmuir,* **29** (2013), 14639–14647.

Brennan J. L., Hatzakis N. S., Tshikhudo T. R., Dirvianskyte N., Razumas V., Patkar S., Vind J., et al., *Bioconjug. Chem.,* **17** (2006), 1373–1375.

Brust M., Walker M., Bethell D., Schiffrin D. J., Whyman R., *Chem. Commun.,* (1994), 801–802.

Calatayud M. P., Sanz B., Raffa V., Riggio C., Ibarra M. R., Goya G. F., *Biomaterials,* **35** (2014), 6389–6399.

Cavelius C., Moh K., Mathur S., *Cryst. Growth Des.,* **12** (2012), 5948–5955.

Cedervall T., Lynch I., Lindman S., Berggård T., Thulin E., Nilsson H., Dawson K. A., Linse S., *Proc. Natl. Acad. Sci. U. S. A.,* **104** (2007), 2050–2055.

Chirra H. D., Sexton T., Biswal D., Hersh L. B., Hilt J. Z., *Acta Biomater.,* **7** (2011), 2865–2872.

Codelli J. A., Baskin J. M., Agard N. J., Bertozzi C. R., *J. Am. Chem. Soc.,* **17** (2008), 11486–11493.

Conde J., Dias J. T., Grazú V., Moros M., Baptista P. V., la Fuente J. M. de, *Front. Chem.,* **2** (2014), 1–27.

Day E. S., Bickford L. R., Slater J. H., Riggall N. S., Drezek R. A., West J. L., *Int. J. Nanomed.,* **5** (2010), 445–454.

Deitcher R. W., O'Connell J. P., Fernandez E. J., *J. Chromatogr. A,* **1217** (2010), 5571–5583.

Deng Z. J., Liang M., Toth I., Monteiro M. J., Minchin R. F., *ACS Nano,* **6** (2012), 8962–8969.

Dreaden E. C., Alkilany A. M., Huang X., Murphy C. J., El-Sayed M. A., *Chem. Soc. Rev.,* **41** (2012), 2740.

Di Marco M., Razak K. A., Aziz A. A., Devaux C., Borghi E., Levy L., Sadun C., *Int. J. Nanomedicine,* **5** (2010), 37–49.

Dykman L., Khlebtsov N., *Chem. Soc. Rev.,* **41** (2012), 2256–82.

El-Gamel N. E. A., Wortmann L., Arroub K., Mathur S., *Chem. Commun.,* **47** (2011), 10076–10078.

Gallo J., Kamaly N., Lavdas I., Stevens E., Nguyen Q. De, Wylezinska-Arridge M., Aboagye E. O., Long N. J., *Angew. Chem. Int. Ed.,* **53** (2014), 9550–9554.

Gessner A., Paulke B., Mu R. H., *Pharmazie,* **61** (2006), 293–297.

Gillich T., Acikgöz C., Isa L., Schlürer A. D., Spencer N. D., Textor M., **7** (2013), 316–329.

Gole A., Murphy C. J., *Langmuir,* **24** (2007), 266–272.

Gref R., Lück M., Quellec P., Marchand M., Dellacherie E., Harnisch S., Blunk T., Müller R. H., *Colloids Surf. B,* **18** (2000), 301.

Hanauer M., Pierrat S., Zins I., Lotz A., Sönnichsen C., *Nano Lett.,* **7** (2007), 2881–2885.

Hayashi K., Moriya M., Sakamoto W., Yogo T., *Chem. Mater.,* **21** (2009), 1318–1325.

Hayashi K., Ono K., Suzuki H., Sawada M., Moriya M., *Chem. Mater.,* **22** (2010), 3768–3772.

Hein J. E., Fokin V. V., *Chem. Soc. Rev.,* **39** (2010), 1302–1315.

Hein C., Liu X.-M., Wang D., *Pharm. Res.,* **25** (2008), 2216–2230.

Himo F., Lovell T., Hilgraf R., Rostovtsev V. V., Noodleman L., Sharpless K. B., Fokin V. V, *J. Am. Chem. Soc.,* **127** (2005), 210–216.

Hou J., Liu X., Shen J., Zhao G., George P., *Expert Opin. Drug Discov.,* **7** (2012), 489–501.

Hu X. L., Jin H. Y., He X. P., James T. D., Chen G. R., Long Y. T., *ACS Appl. Mater. Interfaces,* **7** (2015), 1874–1878.

Hu B. X., Yu J. C., *Adv. Funct. Mater.,* **18** (2008), 880–887.

Hu B. X., Yu J. C., Gong J., Li Q., Li G., *Adv. Mater.,* **19** (2007), 2324–2329.

Huang S., Yan W., Hu G., Wang L., *J. Phys. Chem. C,* **116** (2012), 20558–20563.

Huy T. Q., Chung P. Van, Thuy N. T., Blanco Andujar C., Thanh N. T. K., *Faraday Discuss.,* **175** (2014), 73–82.

Ilyas S., Ilyas M., Hoorn R. A. L. Van Der, Mathur S., *ACS Nano,* **7** (2013), 9655–9663.

Imperiale J. C., Nejamkin P., Sole M. J. del, Lanusse C. E., Sosnik A., *Biomaterials,* **37** (2015), 383–394.

Jewett J. C., Sletten E. M., Bertozzi C. R., *JACS Commun.,* **132** (2010), 3688–3690.

Jia C.-J., Sun L.-D., Yan Z.-G., You L.-P., Luo F., Han X.-D., Pang Y.-C., Zhang Z., Yan C.-H., *Angew. Chem. Int. Ed.,* **44** (2005), 4328–4333.

Jian G., Liu Y., He X., Chen L., Zhang Y., *Nanoscale,* **4** (2012), 6336.

Jun Y., Huh Y., Choi J., Lee J., Song H., Kim S., Yoon S., et al., *J. Am. Chem. Soc.,* **127** (2005), 5732–5733.

Jun Y., Lee J., Cheon J., *Angew. Chem. Int. Ed.,* **47** (2008), 5122–5135.

Jùrgensen K. A., *Angew. Chem. Int. Ed.,* **39** (2000), 3558–3588.

Kim Y.-P., Daniel W. L., Xia Z., Xie H., Mirkin C. A., Rao J., *Chem. Commun. (Camb).,* **46** (2010), 76–78.

Kolb H. C., Finn M. G., Sharpless K. B., *Angew. Chem. Int. Ed. Engl.,* **40** (2001), 2004–2021.

Kolb H. C., Sharpless K. B., *Drug Discov. Today,* **8** (2003), 1128–1137.

Kolb H. C., Vannieuwenhze M. S., Sharpless K. B., *Chem. Rev.,* **94** (1994), 2483–2547.

Kralj S., Rojnik M., Romih R., Jagodič M., Kos J., Makovec D., *J. Nanoparticle Res.,* **14** (2012), 1151.

Lallana E., Sousa-Herves A., Fernandez-Trillo F., Riguera R., Fernandez-Megia E., *Pharm. Res.,* **29** (2012), 1–34.

Laurent S., Forge D., Port M., Roch A., Robic C., Elst L. Vander, Muller R. N., *Chem. Rev.,* **108** (2008), 2064–2110.

Lee N., Hyeon T., *Chem. Soc. Rev.,* **41** (2012), 2575–2589.

Lee S., Koo H., Na J. H., Han S. J., Min H. S., Lee S. J., Kim S. H., et al., *ACS Nano,* **8** (2014), 2048–2063.

Lee D.-E., Koo H., Sun I.-C., Ryu J. H., Kim K., Kwon I. C., *Chem. Soc. Rev.,* **41** (2012a), 2656–72.

Lee W., Loo C., Van K. L., Zavgorodniy A. V., Rohanizadeh R., *J. R. Soc. Interface,* **9** (2012b), 918–927.

Lee N., Yoo D., Ling D., Cho M. H., Hyeon T., Cheon J., *Chem. Rev.,* **115** (2015), 10637–10689.

Li N., Binder W. H., *J. Mater. Chem.,* **21** (2011), 16717–16734.

Liu G., Gao J., Ai H., Chen X., *Small,* **9** (2013), 1533–1545.

Liu Y., Shipton M. K., Ryan J., Kaufman E. D., Franzen S., Feldheim D. L., *Anal. Chem.*, **79** (2007), 2221–2229.

Lu A. H., Salabas E. L., Schüth F., *Angew. Chem. Int. Ed.*, **46** (2007), 1222–1244.

Lv B., Xu Y., Wua D., Sun Y., *CrystEngComm*, **13** (2011), 7293–7298.

Lynch I., Salvati A., Dawson K. A., *Nat. Nanotechnol.*, **4** (2009), 546–547.

Maltzahn G. von, Ren Y., Park J.-H., Min D.-H., Kotamraju V. R., Jayakumar J., Fogal V., Sailor M. J., Ruoslahti E., Bhatia S. N., *Bioconjug. Chem.*, **19** (2008), 1570–1578.

Mattoussi H., Mauro J. M., Goldman E. R., Anderson G. P., Sundar V. C., Mikulec F. V., Bawendi M. G., *J. Am. Chem. Soc.*, **122** (2000), 12142–12150.

Meldal M., Tornøe C. W., *Chem. Rev.*, **108** (2008), 2952–3015.

Mironava T., Hadjiargyrou M., Simon M., Jurukovski V., Rafailovich M. H., *Nanotoxicology*, **4** (2010), 120–137.

Molteni G., Bianchi C. L., Marinoni G., Ponti A., *New J. Chem.*, **30** (2006), 1137–1139.

Moses J. E., Moorhouse A. D., *Chem. Soc. Rev.*, **36** (2007), 1249–1262.

Mukhopadhyay A., Joshi N., Chattopadhyay K., De G., *ACS Appl. Mater. Interfaces*, **4** (2012), 142–149.

Muthiah M., Park I. K., Cho C. S., *Biotechnol. Adv.*, **31** (2013), 1224–1236.

Navrotsky A., Mazeina L., Majzlan J., *Science*, **319** (2008), 1635–1638.

Negishi Y., Takasugi Y., Sato S., Yao H., Kimura K., Tsukuda T., *JACS Commun.*, **126** (2004), 6518–6519.

Neibert K., Gosein V., Sharma A., Khan M., Whitehead M. A., Maysinger D., Kakkar A., *Mol. Pharm.*, **10** (2013), 2502–2508.

Nel A. E., Mädler L., Velegol D., Xia T., Hoek E. M. V., Somasundaran P., Klaessig F., Castranova V., Thompson M., *Nat. Mater.*, **8** (2009), 543–557.

Noble J. E., Bailey M. J. A. (2009) *Quantitation of Protein*. Elsevier Inc.

Park J., An K., Hwang Y., Park J., Noh H., Kim J., Park J., Hwang N., Hyeon T., *Nat. Mater.*, **3** (2004), 891–895.

Park J., Lee E., Hwang N.-M., Kang M., Kim S. C., Hwang Y., Park J.-G., et al., *Angew. Chem. Int. Ed. Engl.*, **44** (2005), 2872–2877.

Presolski S. L., Hong V. P., Finn M. G., *Curr. Protoc. Chem. Biol.*, **3** (2011), 153–162.

Rahman M., Laurent S., Tawil N., Yahia L., Mahmoudi M., Nanoparticle and protein corona, in *Protein-Nanoparticle Interact.*, Vol. 15 (2013), pp. 21–45.

Ratanathanawongs Williams S. K., Lee D., *J. Sep. Sci.*, **29** (2006), 1720–1732.

Rejman J., Oberle V., Zuhorn I. S., Hoekstra D., *Biochem. J.*, **377** (2004), 159–169.

Rimkus G., Bremer-Streck S., Grüttner C., Kaiser W. A., Hilger I., *Contrast Media Mol. Imaging*, **6** (2011), 119–125.

Rocker C., Potzl M., Zhang F., Parak W. J., Nienhaus G. U., *Nat. Nanotechnol.*, **4** (2009), 577–580.

Rostovtsev V. V., Green L. G., Fokin V. V., Sharpless K. B., *Angew. Chem. Int. Ed. Engl.*, **41** (2002), 2596–2599.

Sapsford K. E., Algar W. R., Berti L., Gemmill K. B., Casey B. J., Oh E., Stewart M. H., Medintz I. L., *Chem. Rev.*, **113** (2013), 1904–2074.

Saptarshi S. R., Duschl A., Lopata A. L., Duschl A., Lopata A. L., *J. Nanobiotechnology*, **11** (2013), 26.

Schaaff T. G., Knight G., Shafigullin M. N., Borkman R. F., Whetten R. L., *J. Phys. Chem. B*, **102** (1998), 10643–10646.

Schätz A., Hager M., Reiser O., *Adv. Funct. Mater.*, **19** (2009), 2109–2115.

Schollbach M., Zhang F., Roosen-runge F., Skoda M. W. A., Jacobs R. M. J., Schreiber F., *J. Colloid Interface Sci.*, **426** (2014), 31–38.

Shan J., Tenhu H., *Chem. Commun. (Camb).*, **44** (2007), 4580–98.

Shang L., Nienhaus K., Nienhaus G. U., *J. Nanobiotechnol.*, **12** (2014), 1–11.

Shultz M. D., Ulises Reveles J., Khanna S. N., Carpenter E. E., *J. Am. Chem. Soc.*, **129** (2007), 2482–2487.

Singh V., Nair S. P. N., Aradhyam G. K., *J. Nanobiotechnol.*, **11** (2013), 7.

Sperling R. A., Parak W. J., *Phil. Trans. R. Soc. A*, **368** (2010), 1333–1383.

Stein B., Zopes D., Schmudde M., Schneider R., Mohsen A., Goroncy C., Mathur S., Graf C., *Faraday Discuss.*, **181** (2015), 85–102.

Tao X., Jin S., Wu D., Ling K., Yuan L., Lin P., Xie Y., Yang X., *Nanomaterials*, **6** (2016), 1–14.

Teekamp N., Duque L. F., Frijlink H. W., Hinrichs W. L. J., Olinga P., *Expert Opin. Drug Deliv.*, **12** (2015), 1311–1331.

Tornøe C. W., Christensen C., Meldal M., *J. Org. Chem.*, **67** (2002), 3057–3064.

Treuel L., Malissek M., Gebauer J. S., Zellner R., *ChemPhysChem*, **11** (2010), 3093–3099.

Turcheniuk K., Tarasevych A. V., Kukhar V. P., Boukherroub R., Szunerits S., *Nanoscale*, **5** (2013), 10729–10752.

Vladkova T. G., A review. *Int. J. Polym. Sci.*, (2010), 296094:1–296094:22.

von Maltzahn G., Ren Y., Park J.-H., Min D.-H., Kotamraju V. R., Jayakumar J., Fogal V., Sailor M. J., Ruoslahti E., Bhatia, S. N., *Bioconjug. Chem.,* **19** (2008), 1570–1578.

Veronese F. M., Pasut G., *Drug Discov. Today,* **10** (2005), 1451–1458.

Vertegel A. A., Siegel R. W., Dordick J. S., *Langmuir,* **20** (2004), 6800–6807.

Walkey C. D., Chan W. C. W., *Chem. Soc. Rev.,* **41** (2012), 2780–2799.

Walling M. A., Novak J. A., Shepard J. R. E., *Int. J. Mol. Sci.,* **10** (2009), 441–491.

Wang A. Z., Langer R., Farokhzad O. C., *Annu. Rev. Med.,* **63** (2012a), 185–198.

Wang D., Zhao M., Liu X., Chen Y., Li N., Chen B., *Org. Biomol. Chem.,* **10** (2012b), 229–231.

Wegner K. D., Hildebrandt N., *Chem. Soc. Rev.,* **44** (2015), 4792–834.

Welton J. L., Webber J. P., Botos L.-A., Jones M., Clayton A., *J. Extracell. Vesicles,* **4,** 10.3402/jev.v4.27269 (2015).

White M. A., Johnson J. A., Koberstein J. T., Turro N. J., *JACS Commun.,* **128** (2006), 11356–11357.

Wortmann L., Ilyas S., Niznansky D., Valldor M., Arroub K., Berger N., Rahme K., Holmes J., Mathur S., *ACS Appl. Mater. Interfaces,* **6** (2014), 16631–16642.

Wu X., Chen J., Wu M., Zhao J. X., *Theranostics,* **5** (2015), 322–344.

Wu W., He Q., Jiang C., *Nanoscale Res. Lett.,* **3** (2008), 397–415.

Xiao L., Li J., Brougham D. F., Fox E. K., Feliu N., Bushmelev A., Schmidt A., et al., *ACS Nano,* **5** (2011), 6315–6324.

Xiao L., Mertens M., Wortmann L., Kremer S., Valldor M., Lammers T., Kiessling F., Mathur S., *ACS Appl. Mater. Interfaces,* **7** (2015), 6530–6540.

Yu M., Zhou C., Liu J., Hankins J. D., Zheng J., *J. Am. Chem. Soc.,* **133** (2011), 11014–11017.

Yuen A. K. L., Hutton G. A., Masters A. F., Maschmeyer T., *Dalt. Trans.,* **41** (2012), 2545–2559.

Zhang S., He X., Chen L., Zhang Y., *New J. Chem.,* **38** (2014), 4212–4218.

Zhou J., Ralston J., Sedev R., Beattie D. A., *J. Colloid Interface Sci.,* **331** (2009), 251–262.

Zhu K., Zhang Y., He S., Chen W., Shen J., Wang Z., Jiang X., *Anal. Chem.,* **84** (2012), 4267–4270.

Zopes D., Stein B., Mathur S., Graf C., *Langmuir,* **29** (2013), 11217–11226.

Chapter 22

Cell-Free Expression Based Microarrays: Applications and Future Prospects

Apurva Atak and Sanjeeva Srivastava

Department of Biosciences and Bioengineering, Indian Institute of Technology Bombay, Powai, Mumbai 400076, Maharashtra, India

sanjeeva@iitb.ac.in

22.1 Introduction

22.1.1 Introduction to Microarrays

During the technological outburst of the last century, proteomics is one branch of "omics" that has evolved with the advances in high-throughput, multiplexed analytical platforms. The genetic information that is generated by certain genes that are either expressed or in an active state is carried by the mRNA. Several mRNA corresponding to those genes are produced during this process of transcription, which functions to carry out this information from the nucleus of a cell to the cytoplasm for the initiation of protein synthesis. While these mRNA act as excellent markers to understand the various alterations taking place in the genome, they are very labile and tend to degrade rapidly; therefore, it is necessary to convert it into a more stable complementary DNA

Biocatalysis and Nanotechnology
Edited by Peter Grunwald

ISBN 978-981-4613-69-9 (Hardcover), 978-1-315-19660-2 (eBook)
www.panstanford.com

(cDNA) form. The basic principle underlying the microarray technology is the binding of the complementary sequences to each other. The restriction endonucleases act to cut the DNA molecules and markers that facilitate detection attached to these DNA fragments. Once these DNA fragments react with probes on the DNA chip, the binding of the target DNA fragments along with their complementary sequences takes place on DNA probes, and the unbound fraction is washed away. A laser beam facilitates the excitation of the fluorophore and emitted light allows the detection of the target DNA fragments. The data is recorded on a computer to generate a pattern based on the fluorescence emission and DNA identification (Govindarajan et al., 2012). After the completion of the human genome project, the focus has shifted on protein research using proteomic tools. In 2003, Mingyue He and Michael Taussig published a protocol for the generation of protein microarrays from PCR amplified DNA product using a cell-free expression system. The PCR amplified DNA was incubated with rabbit reticulocyte lysate, which resulted in production of tagged proteins depending on the construct. The expressed proteins were immobilized in situ via its tag. They also used the arrays developed using this method to study the protein function, interactions and for identification of mutants (He and Taussig, 2003). This method was further modified by other researchers to overcome the shortcomings. Another approach is known as the Multiple Spotting technique, where, DNA is firstly spotted onto the solid surface, followed by spotting of in vitro transcription and translation reaction mixture exactly on the first spot. The synthesized proteins are immobilized onto the surface and detected by specific tag (Angenendt et al., 2006). The introduction of protein microarrays has made it possible to screen a large number of proteins in a time and cost-effective manner for applications relating to the analysis of protein–protein interaction, protein expression profiles, and biomarker discovery for several disease conditions (Bertone and Snyder, 2005). Tissue microarray is a recent addition to the microarray technology to overcome the shortcomings of the DNA and protein microarray platforms. This platform is majorly used to validate targets identified from discovery phase screening (Jawhar, 2009). In this chapter, we describe various protein microarray platforms, with special emphasis on cell-free expression based microarrays.

Various applications of cell-free expression microarrays and the future prospect of this technology have also been discussed.

22.1.2 Types of Microarrays

22.1.2.1 DNA microarrays

The introduction of DNA microarray platforms revolutionized the potential of carrying out DNA and/or RNA hybridization analysis in micro-miniaturized high-throughput manner. The DNA microarrays, also commonly referred to as the DNA chip or biochip, are collection of microscopic DNA probes (DNA spots/DNA oligos) that are attached on a solid surface. These are commonly used to simultaneously quantify a large number of genes or multiple genotypic regions at genome-wide scale, which are detected by means of fluorescence labeling, silver labeling, or chemiluminescence labeled targets to determine the expression levels of the detected nucleic acids. The DNA microarrays function on the core principle of hybridization between two DNA strands (Bednar, 2000). DNA microarrays revolutionized gene expression analysis and motivated the protein biologists to explore chip-based platforms for protein expression analysis.

22.1.2.2 Protein microarrays

Although protein microarrays have followed DNA microarray field in 2000, various types of protein microarray platforms have been generated in this short time. There are three types of protein microarrays, which are most commonly used in practice for studying the various biochemical activities of proteins. These can be categorized as analytical microarrays, functional microarrays and reverse phase protein microarrays (RPPA).

The analytical microarrays are majorly used to analyze a complex mixture of proteins so as to measure their binding affinities, binding specificities, and in order to determine the levels of protein expression from a complex mixture. The analytical microarray platform generally employs the application of a library of antibodies, or aptamers, which are traditionally arrayed on a glass slide. The entire array is then probed with a protein solution and further detected by a detection system to analyze several parameters in a high-throughput manner (Bertone and Snyder,

2005). The antibody based analytical microarrays are most commonly used for the differential expression analysis of diseased tissue/cell lysate in a clinical setting (Sreekumar et al., 2001).

The application of full-length functional proteins or protein domains in an array is what distinguishes the functional protein microarray platforms from that of the analytical microarrays. These full-length functional proteins are used to study the biochemical changes taking place in the entire proteome in a single high-throughput rapid experiment. Furthermore, the functional protein microarrays are commonly used to study protein–protein, protein–DNA, protein–RNA, protein–phospholipid, lectin–cell, lectin–glycan, and protein–small-molecule interaction studies (Kodadek, 2001; Zhu et al., 2001; Hall et al., 2004; Hall et al., 2007; Hu et al., 2011; Zhu and Qian, 2012).

Based on a similar background to the analytical microarrays wherein the cells are isolated from various tissues of interest, RPPA have gained popularity. The RPPA takes into account the lysis of cells or tissue of interest, and then these lysates are arrayed on nitrocellulose slides by using a contact pin microarrayer. These slides are then probed with antibodies against the target protein to allow detection of protein expression. This technique majorly employs chemiluminescent, fluorescent, or colorimetric measurements to allow precise detection. Herein, the reference peptides printed on the slides permit the quantification of target proteins. The RPPA technology is widely used in the detection of altered proteins, specific post-translational modifications, and dysfunctional protein pathways to not only facilitate the detection of rapid and stable progression of a disease but also to define specific therapeutic interventions to allow treatment of the disease in the most effective manner (Speer et al., 2005; Hall et al., 2007). The introduction of protein microarray platforms allowed for the ease of performing gene expression analysis in a high-throughput manner. These advancements have also revolutionized our understanding and studies on functional biochemical changes and interaction among several proteins. Additionally, the development of the RPPA technology further made it possible to study not only differentially expressed proteins but also post-translational modifications.

22.1.2.3 Tissue microarrays

Tissue microarrays aid in the high-throughput analysis of gene expression in multiple tissue samples in addition to allowing parallel molecular profiling of samples at DNA, RNA, and protein levels. The introduction of the tissue microarray platform has opened up new avenues for pathologists to perform large-scale analyses in a time saving and low cost manner, using techniques like immunohistochemistry or fluorescence in situ hybridization (FISH). However, what differs in tissue microarray platform from the DNA microarrays is that this technique employs the use of small histological sections from clinical tissue samples instead of having cDNA or oligonucleotide samples (Jawhar, 2009). The tissue microarrays comprise paraffin blocks constructed from clinical biopsy samples and are thereafter re-embedded onto a microarray block at defined coordinates followed by staining the area of interest with hematoxylin and eosin (Wang et al., 2002; Jawhar, 2009). Following this, the tissue microarray is arrayed on an empty paraffin block on specifically assigned coordinates. This arraying process can be repeated several times to generate a final tissue microarray block. Upon construction of the final tissue microarray block, several sections can be cut and prepared for subsequent morphological or molecular analysis (Shergill et al., 2004). The constant advancement in the microarray technologies have not only made it easier to carry out gene expression analysis from DNA or RNA hybridized samples, lysed cells or tissues of interest, but have also revolutionized the studies of post-translational modifications, dysfunctional disease pathways, and proteins interacting to several other partners like proteins, DNA, RNA, lectins, and small molecules. Furthermore, application of the tissue microarray platform demonstrated its potential in a clinical setting, allowing pathologists to carry out high-throughput analyses in a cost-effective and high-throughput manner.

22.1.3 Cell-Free Synthesis–Based Microarrays

The limiting factor of the cell-based expression systems was production and functional maintenance of a diverse class of proteins. In addition to this, the requirement of several in vivo expression systems for recombinant protein production following its purification was a time consuming process. In this regard, cell-free protein synthesis has proved to be a valuable tool in understanding the translation of mRNAs into functional polypeptides ever since it was first reported (Nirenberg and Matthaei, 1961). With the discipline of proteomics constantly evolving and with the accumulation of technical advances in the field, cell-free protein synthesis has also been used to enhance our understanding on antibiotic drug discovery and has proven to be a valued tool for small-scale expression of toxic products. Recent technical advances have been aimed at increasing productiveness by implementing several enhancements to the configuration, energetics, and robustness of the reactions. Perhaps the most exciting output of this technique is the ability to easily manipulate both-the reaction components and conditions, in order to make the process of protein synthesis predominantly amenable to automation and miniaturization thereby allowing multiple applications to the discipline of protein arrays, and evolution and real-time labeling of several protein samples.

While the cell-free protein synthesis systems have been developed from several organisms, systems derived from the extracts of *Escherichia coli*, wheat germ, and rabbit reticulocyte lysates remain the most commonly used. The cell-free protein synthesis systems facilitate the production of protein directly from a PCR fragment or mRNA template, eliminating the need for cloning into *E. Coli*, thereby providing ease to synthesize protein in a high-throughput manner. While the advantages of the cell-free synthesis systems clearly outweighs that of the cell-based systems, cell-free synthesis systems have also been demonstrated to express multiple templates to facilitate the production of a large number of proteins in a single reaction. Another note-worthy mention with regard to the achievements of the cell-free synthesis systems over the cell-based synthesis systems is that these systems generate soluble and functional proteins thereby facilitating a rapid way to functional protein analysis. Not only are the cell-free systems capable of producing proteins that are generally intolerable (like toxic and unstable proteins) to live cells under physiological conditions, they also allow generation of specific proteins in site specific manner to allow translation by efficiently incorporating the non-natural or chemically-modified amino acids thereby generating novel molecules for proteomic applications (Jackson et al., 2004). Among various approaches widely used in the cell-free synthesis based microarrays, the nucleic acid programmable protein array (NAPPA), DNA array to protein array (DAPA), and protein in situ array (PISA) have gained major interest for various biological applications.

22.1.3.1 Nucleic acid programmable protein array

In this approach (Fig. 22.1), the proteins are synthesized directly from the template DNA to the surface of the array. The nascent protein is then captured by an affinity reagent (Ramachandran et al., 2008) thereby avoiding the concern of displaying the protein on a solid surface to facilitate the detection of highly variable biochemical properties such as oligomerization states, post-translational modifications, protein stability, binding affinities and specificities in a multiplex manner.

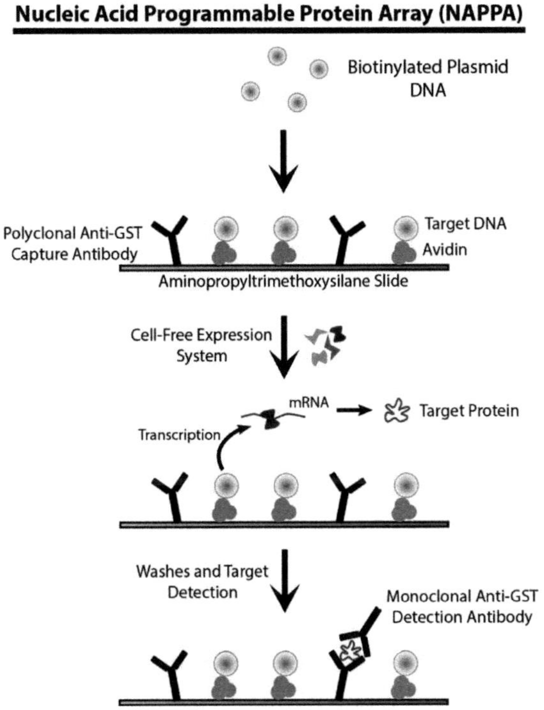

Figure 22.1 Schematic representation of nucleic acid programmable protein array (NAPPA) technology.

This approach was designed and developed in 2004 by LaBaer's laboratory at Harvard University based on the addition of cDNA clones into expression plasmids and also adding a transcriptional promoter and an in-frame polypeptide capture tag. This powerful platform posed several advantages over the typical protein microarrays such as (i) glycerol stock production of clones facilitated the long-term maintenance of gene integrity; (ii) due to the verification of clone sequence, high fidelity was ensured; and (iii) insertion of clones into plasmids permitted the incorporation of additional tags and antibiotic resistance genes for specific selection (Díez et al., 2015).

Ramachandran et al. used biotinylated plasmids encoding desired proteins containing a GST-tag. These biotinylated plasmids were then spotted with avidin, and anti-GST antibody (for protein immobilization at a later stage) onto aminopropyltrimethoxysilane slides to facilitate plasmid binding. Using the NAPPA approach, they were able to array 29 different human DNA replication initiation proteins by co-expressing all the 29 different proteins in solutions together on array to enable the detection of protein–protein interactions. Their study revealed 110 interactions with 63 novel interactions (Ramachandran et al., 2004). The NAPPA approach demonstrated the cost-effective and high-throughput manner of in situ protein production at protein microarray levels without bias for any given class of protein whether transcription factors, membrane proteins or kinases.

22.1.3.2 DNA array to protein array

DAPA was introduced by He et al. in 2007 and is characterized by face-to-face assembly between a slide containing the DNA templates for the proteins and a pre-coated slide containing the protein-capturing reagent. This novel method allowed the generation of several protein microarray copies from the same DNA microarray template. A cell-free extract soaked permeable membrane was used to place the cell-free system in between the two Ni-NTA coated glass slides in order to allow transcription/translation (He et al., 2008; Stoevesandt et al., 2011). The DAPA method (Fig. 22.2) was the first successful method that had the capability to make multiple copies of protein arrays from a single DNA microarray. However, one of the major concerns here was that the thickness of the used membrane limited the minimal spotting distance and its non-homogeneity was directly projected into the generated protein pattern.

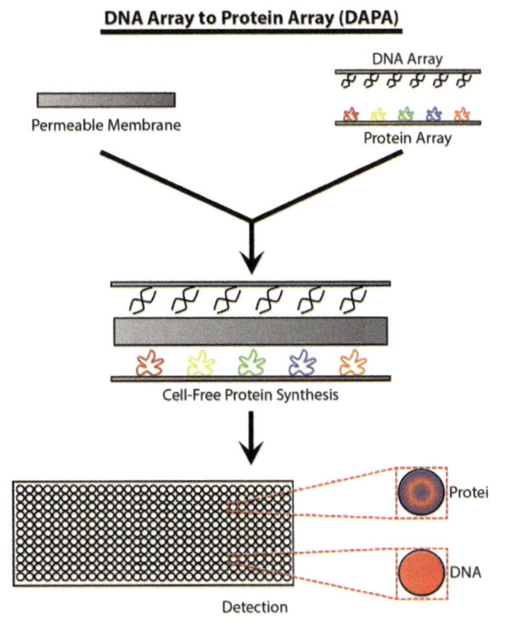

Figure 22.2 Schematic representation of DNA array to protein array (DAPA) technology.

Later in 2012, Stoevesandt attempted to fix the issue by designing a membrane-free DAPA system. In this approach, they separated the glass slide with the DNA array and the NTA capture surface with a parafilm. Cell-free expression mixture was used to fill the generated gap. While this simple solution did circumvent membrane abrasion as seen in the original DAPA setup, issues relating to any tilt between the DNA and protein capture slides led to the formation of a gradient in spot size on the protein copy due to the rather undefined filling process (Stoevesand, et al., 2011).

As an additional improvement to this platform, in 2013, Schmidt et al. investigated the influence of various other supporting coats like Ni-NTA, epoxy, 3D-epoxy, and polyethylene glycol methacrylate (PEGMA) to obtain higher yields of the protein and spot morphology optimization. Further, by application of a tag-specific capture antibody on a surface coating that is protein repellent in nature, they were able to improve the specificity of protein capturing and obtained numerous expressed proteins analogous to clinical protein arrays (Schmidt et al., 2013). The DAPA approach was not only successful in the generation of copies of protein microarrays from the same DNA microarray template but also in carrying out studies on transcription and translation as and when required.

22.1.3.3 Protein in situ array

PISA is a technique which allows the one step rapid generation of protein arrays directly from DNA templates by using cell-free protein expression and simultaneous immobilizations at a surface in situ. In this technique (Fig. 22.3), depending on the source of the genetic material, primers for individual genes or gene fragments are designed and the genes or the gene fragments are produced by PCR or RT-PCR. Cell-free protein synthesis using coupled transcription and translation in order to produce a double $(His)_6$-tagged protein is employed and the reaction is carried out on a Ni-NTA-coated surface, which facilitates the adhesion of the synthesized proteins thereby resulting in the generation of PISA.

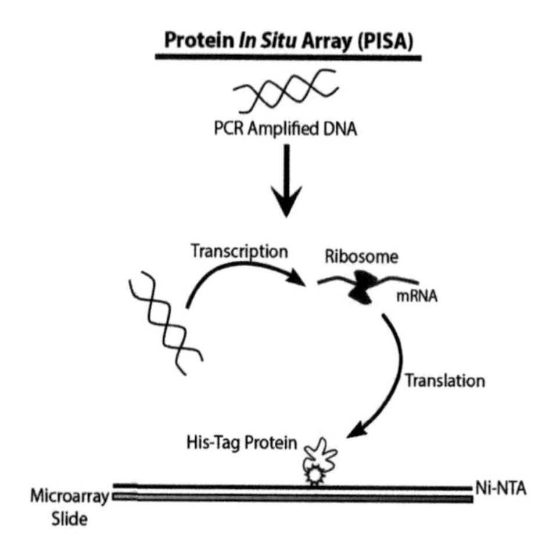

Figure 22.3 Schematic representation of protein in situ array (PISA) technology.

He et al. first demonstrated the application of this method by functional immobilization of antibody fragments. In order to model their approach, they made use of a human anti-progesterone antibody fragment of known specificity. They also showed that this antibody, when expressed and immobilized, retains its initial antigen binding specificity. Further, they arrayed cloned antibody fragments from a library and carried out a screen with labeled antigens, distinguishing binders from non-binders in a reproducible manner (He and Taussig, 2001). One of the most important advantages of this technique is that the proteins that are generated by cell-free synthesis are usually soluble and functional, and the use of PCR-generated DNA enables rapid production of proteins or domains, where the cloned material is not available solely on the basis of genome information.

22.1.3.4 HaloTag technology

The HaloTag technology (Fig. 22.4) mainly comprises two elements: the HaloTag protein and the Halo ligand. The HaloTag can be genetically fused to any protein of interest, and the Halo ligands are a variety of organic molecules that enable irreversible binding to the HaloTag protein. The HaloTag is a 33 kDa-engineered derivative of a bacterial hydrolase. The HaloTag approach enabled the oriented capture of proteins in order to avoid loss of function. This technique was based on the primary objective to ensure firm and oriented immobilization of the expressed proteins on the array surface by covalent binding between the HaloTag-fused protein and the HaloTag ligand. Not only did this covalent binding prevent the loss of protein function, but material loss during the washing steps was also prevented. The direct interaction of protein and ligand also helped in quantification of protein. The HaloTag technique has previously been used to study the protein–protein interactions of five well-characterized proteins pairs and verified stability of three enzyme arrays (Nath et al., 2008). Furthermore, studies pertaining to the validation of autoantibodies generated against denatured proteins commonly employed the NAPPA platform using GST tagged proteins. However, when subjected to denaturation in vitro, the proteins were washed off from the surface. Therefore, Wang et al made use of HaloTag chemistry, which covalently binds the proteins with HaloTag (expressed in vitro using in vitro transcription and translation) and its ligand and could withstand the harsh denaturation conditions. Thus, HaloTag based arrays expressing proteins in its native conformation as well as in denatured conditions were used to study the changes in the autoantibody profile generated against them in order to study the type of epitope present on those proteins. They reported the changes in some of the autoantibodies when compared with those of the denatured counterparts. These alterations were attributed to the difference in epitopes of some of the proteins, which are either exposed or unfolded upon denaturation of the protein (Wang et al., 2013).

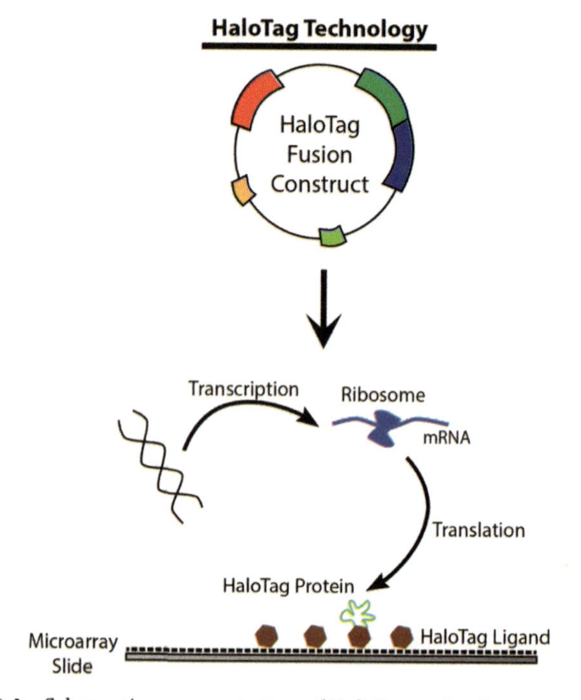

Figure 22.4 Schematic representation of HaloTag technology.

22.1.4 Label-Free Detection Techniques for Protein Arrays

Traditional protein microarray system employed label-based detection strategies; however, to overcome the limitations of labels, which may introduce artifacts and false signals; label-free strategies have been used for detection of protein arrays in a label-free manner, wherein the intrinsic properties such as mass and dielectric property of the probing molecules can be measured. Applications of label-free techniques, which include surface plasmon resonance (SPR), Surface plasmon resonance imaging (SPRi), carbon nanotubes (CNTs), micro-electromechanical cantilevers have been increasingly used for microarray applications (Ray et al., 2010; Chandra et al., 2011). The introduction of label-free protein array strategies have also helped in determining the real-time reaction kinetics of biomolecular interactions by eliminating the false interaction by tagged molecules (Ji et al., 2008). However, the expensive fabrication procedures, morphological inconsistencies of sample spots, and inadequate understanding of biosensors often limit their practice (Ray et al., 2010). In addition to this, the concerns of specificity and false positives due to non-specific binding to the immobilized proteins and/or adsorption to the sensor surface remains to be a worry for majority of the label-free assay systems (Ramachandran et al., 2005). However, the extensive applications of label-free detection strategies have successfully demonstrated elimination of false interaction of tagged molecules.

22.2 Applications of Cell-Free Synthesis–Based Microarrays in Basic and Clinical Research

Ever since the fame attained by success of DNA microarrays, protein microarray have been trying to repeat the same success. The traditional cell-based methods employed to generate these protein microarrays were considered a strenuous and time-consuming process, wherein technical challenges concerning to the purity of the protein, preservation of protein folding, its functionality during purification and immobilization steps, and shorter shelf-life and stability were amongst the major setbacks.

The introduction of cell-free synthesis based microarray systems eliminated the hassle of cloning and expressing proteins in host cells by making use of template DNA for direct synthesis of protein expression in vitro in the presence of crude cell lysates harvested from growing cells containing the necessary means for successful transcription and translation of the expression system. While cell-free based microarray expression systems not only facilitate the direct conversion of genetic readouts to their functional proteins but also enable the parallel synthesis of numerous proteins in a high-throughput reaction making the technique suitable for both functional and structural proteomics.

22.2.1 Protein Interaction Studies

Over the last decade, efforts aimed at the synthesis of complex proteins containing multiple disulphide bonds have increased. However, the achievement of this in a cell-free system requires the hydrophobic regions of the target protein to be shielded from one another in order to provide proper natural chemical environment, encourage the formation of disulphide bonds, and promote their isomerization (Carlson et al., 2012). Several studies have been reported wherein the identification of novel protein interactions by use of tagged peptides have been achieved. In 2004, a pioneering study by Ramachandran et al. used NAPPA to immobilize and express 29 proteins, which demonstrated functional initiation of DNA replication. Their findings resulted in detection of 63 novel protein interactions amongst the total 110 proteins identified (Ramachandran et al., 2004). Following this encouraging discovery in 2008, they also applied the use of a high-density NAPPA technique to study and identify binary interactions wherein, their array revealed 449 previously untested genes amongst the 647 unique genes identified by arraying the query protein with its corresponding cDNA clone with protein specific antibodies to facilitate detection (Ramachandran et al., 2008). In 2009, another pioneering study by Hurst et al. demonstrated the potential of the HaloTag technology in the detection of protein–protein interactions by use of rabbit reticulocyte lysate (RRL) cell-free system (Hurst et al., 2009). The evolution of NAPPA and the HaloTag platforms have generated options to study novel protein–protein interaction in cell-free systems thereby making it possible to shield the hydrophobic regions of the target proteins and providing the necessary natural environment.

22.2.2 Biomarker Discovery and Autoantibody Profiling

Cell-free synthesis based microarray systems have demonstrated their potential in biomarker discovery and autoantibody profiling as well as applications related to the pharmaceutical industry. The high-throughput readout capacity of proteomic screens facilitates the analysis of a large number of patient tissue and serum samples suffering from diseases such as cancer. While identification of biomarkers allow for early detection of the diseased state, individual screens can appear an arduous task. The use of protein microarrays accelerated the screening process by rapidly investigating thousands of proteins in a single run. Protein microarray technologies also enable detection of antibody-antigen interactions either by use of a fluorescently labeled antigen or application of secondary labeled antibodies to detect antigen binding (Chandra and Srivastava, 2010).

Detection of tumor specific antibodies has also been achievable using cell-free NAPPA microarray by using sera from cancer patients. A report by Anderson et al. suggests that when they probed sera from breast cancer patients on a protein microarray having p53 antibodies, they were able to detect antibody binding by using HRP-conjugated antihuman IgG. In order to confirm the specificity of their approach, they also performed ELISA to confirm their findings and found that they were in agreement with the findings of the protein microarray. Additionally this screening also confirmed that p53 antibodies are not only related to tumor burden and p53 mutation but they also noted a decline in the antibody concentration after neoadjuvant chemotherapy (Anderson et al., 2008).

In order to study the profile of serum proteins differentially expressed in bladder cancer, antibody array fabricated with 254 antibodies were probed with serum to study the levels of respective proteins in 37 bladder tumor patients and 58 control subjects. These arrays classified the bladder cancer patients from control group with a classification rate of 93.7%. Another set of antibody array consisting of 144 antibodies was generated, which included antibodies that were detected in at least 50% of the first set of arrays. These arrays were probed with serum, which revealed a significant alteration of a serum protein c-met, which was shown to be associated with the grade of the tumor, pathological stage and survival prediction. Further, c-met expression levels in various grades of bladder cancer were validated using immunohistochemistry and enzyme immunoassays (Sanchez-Carbayo et al., 2006).

Protein microarrays have also been used for short-listing the candidate antigens for Q fever caused by *Coxiella burnetii* in humans, which could help in developing the kits for diagnosis of Q fever by using recombinant antigens. Robert Heinzen's group used *E. coli* based IVTT to translate the ORFs for 1988 *C. burnetii* proteins. The crude IVTT lysates were spotted onto nitrocellulose membrane, which was later incubated with serum from patients suffering from Q fever at early and late phase or those vaccinated against *C. burnetii* with Q-Vax. Fifty antigens, which were able to elicit antigen specific antibody response was shortlisted for validation by ELISA, which could help in development of diagnostic test kits (Beare et al., 2008). In order to detect immunodominant antigens of *Francisella tularensis*, which is the causative agent of tularemia, serum from the BALB/c mice immunized with killed *Francisella tularensis*, immune complexes and CpG adjuvants was probed against immunodominant antigens in *F. tularensis* and 48 novel immunoreactive antigens were identified. Most of these antigens were found to be membrane proteins, which were considered as potential targets for designing the non-living vaccines (Eyles et al., 2007).

In another study using IVTT based protein expression method, 50 potential vaccine candidates for *Anaplasma marginale* were screened for their immunogenicity to activate the CD4+ T-lymphocytes. These potential vaccine candidates were expressed using *E. coli* cell-free IVTT method and purified using antibody (anti_His/FLAG tag) based bead-affinity purification method. Various proteins of *A. Marginale* like MSP2, MSP3 VirB10, OMP4, OMP9, EF-Tu, Ana29, and OMA87 after processing and presentation by APCs (antigen presenting cells) consistently activated CD4+ T-lymphocytes even at low concentrations. This method could be useful in the rapid expression and identification of potential vaccine candidates which can stimulate CD4+ T-cells to generate memory T-cells to confer protection against the pathogen (Lopez et al., 2008). In order to identify the potential protein targets, that will be useful for the preparation/designing of vaccine against *P. falciparum* infections, Doolan et al. expressed 250 *P. falciparum* proteins using *E. coli* cell-free IVTT method and directly printed on a chip. Serum samples from the subjects infected with pathogen and vaccinated with the attenuated organism were probed for the identification of strong immunogens.

Seventy-two highly reactive *P. falciparum* antigens were identified, which could be potential targets for the development of vaccines (Doolan et al., 2008).

Nucleic acid-programmable protein arrays fabricated with 262 outer membrane proteins and exported *P. aeruginosa* PAO1 proteins were used to study the humoral immune response in 22 cystic fibrosis and 16 acutely infected patients with *P. aeruginosa*. Twelve proteins which could be promising targets for vaccine development were identified against which adaptive immunity was developed in the patients and the results were validated using ELISA (Montor et al., 2009).

Type-I diabetes is an autoimmune disorder, in which the pancreatic β-cells are destroyed by our own immune system, thus resulting in the insulin deficiency. In type-I diabetes antibodies against the islet antigens insulin (IAA), glutamate decarboxylase (GADA), and the protein tyrosine phosphatase-like protein IA-2 (IA-2A) are known to be present. Achenbach et al. studied the relationship between the risk of diabetes with antibody titers, epitope specificity and IgG subclass to classify the patients into different groups with the risk of diabetes ranging from 7% to 89% in 5 years using protein A/G radiobinding assays (Achenbach et al., 2004). In another study, serum autoantibodies were screened in 124 T1D patients and 95 healthy subjects against ~6000 human proteins using NAPPA method. A validation set consisting of 46 T1D patients and healthy controls were used for validating the results from the preliminary screening study. Twenty-six novel autoantibody markers along with the known T1D associated autoantigens (IA-2A) were identified in the T1D patient's sera. NAPPA technology using whole human proteome has helped in the identification of autoantigens in various autoimmune diseases including type-1 diabetes (Miersch et al., 2013). High-density HD-NAPPA, which are superior to the standard NAPPA chips, have exhibited higher signal to noise ratio. DNA coding for 761 viral antigens (with GST tag) from 25 different viruses were printed on the chips to screen for antibodies against these viral antigens in juvenile idiopathic arthritis (JIA) and type 1 diabetes (T1D). Both JIA and T1D patients showed high seroreactivity to the herpes simplex virus antigens with respect to the control samples. However, only T1D patient's serum showed high seroreactivity to the EBV viral antigens, but not the JIA patients. This HD-NAPPA technology could be useful in understanding the role of different pathogens in autoimmune disorders (Bian et al., 2015).

Autoantibody profiles in plasma and synovial fluid of 10 JIA patients were analyzed using NAPPA platform, which showed a strong correlation between the autoantibodies identified in both the biological fluids. Each chip consisted of 768 human genes against which the autoantibodies were screened. Unsupervised cluster analysis segregated the JIA patients into low antibody (6 patients) and high antibody (4 patients) level groups. Autoantibody study using NAPPA platform could yield reliable markers for disease diagnosis, clinical outcomes, and understanding the disease etiology (Gibson et al., 2012).

Davies et al. developed a high-throughput PCR recombination cloning and expression technology, where large numbers of proteins were synthesized using *E. coli* cell-free in vitro transcription/translation system and printed on the chip to screen the humoral immune response against large number of infectious diseases. They printed 185 vaccinia virus proteins on a chip and screened for immune response in human, primates and mice immunized with vaccinia virus and identified vaccinia specific antibodies in more than 20 human sera. This is a rapid and less laborious method than the conventional protein purification methods to find potential targets for vaccine development (Davies et al., 2005).

A suspension bead array platform developed by Wong et al. employed the proteins/antigens having a GST or FLAG-tag which were produced using cell-free in vitro transcription/translation system and captured onto anti-GST/FLAG tag containing beads. Serum samples to be probed were added to the array and the human IgG were detected using standard detection methods. Seventy-one of 72 GST tagged proteins (98.6%) were successfully expressed and detected. Bead array studies conducted for a viral protein EBNA-1 in human serum was found to be very reproducible with intra-assay variation of 3–8% and inter-assay variation of 5%. This method could be used for rapid detection of antibodies produced against pathogen/tumor/self antigens for diagnostic purpose (Wong et al., 2009). In summary, the introduction of high-throughput proteomic screens have helped in the realization of carrying out otherwise time consuming screens to identify the diseased state from patient tissue or serum sample achievable in a short time frame to aid in rapid diagnosis.

22.2.3 Small-Molecule Studies and Other Applications

The induction of the immune system to evoke an immune response against the pathogen is crucial in inflammatory response. In this light, the alterations in expression of genes was studied using reverse transcriptase in vitro translation kit on oligonucleotide arrays, which targeted 32381 mouse genes. A macrophage cell line was used for the study wherein the cells were treated with lipopolysaccharide and non-treated cells were used as controls. The gene expression profiles were studied in both the cases, and the genes which showed more than 4-fold alterations were shortlisted for gene enrichment analysis. The study was aimed at studying the lipopolysaccharide target genes in order to study the alterations arising from treatment by medicinal herbs extracts (Huang et al., 2006). The NAPPA platform was also put to use in order to study the chemical protein acetylation by adding radiolabeled 14C-acetyl-CoA to the translation mixture. Acetylation is thought to take place chemically majorly in mitochondria and enzymatically outside the mitochondria due abundance of Acetyl CoA inside the mitochondria. The human proteins undergoing chemical acetylation were screened and subjected to mass spectrometric analysis. They reported the presence of a basic residue contributing toward chemical acetylation of proteins. However, the proteins identified from their study involved proteins located outside the mitochondria but which are known to be acetylated in vivo. Thus, they reported that chemical acetylation is not restricted to mitochondrial region only (Olia et al., 2015).

In vitro expression arrays could be used for the detection of toxins, which interfere with protein expression. Mei et al. designed a chip consisting of miniaturized wells, in which, the in vitro transcription and translation takes place. These in vitro protein expression chips were used to analyze the efficacy of tetracycline and cycloheximide on inhibition of protein synthesis using *E. coli* expression and rabbit reticulocyte expression system. Tetracycline showed significant inhibition of protein synthesis in *E. coli* expression system, but not in the rabbit reticulocyte expression system. On the contrary, cycloheximide showed significant inhibition of protein expression in rabbit reticulocyte expression system, but not in the *E. coli* expression system. These results are in accordance with the known fact that tetracycline is more effective against bacteria than cycloheximide. These protein expression chips could be useful in the screening of a large number of drug molecules to identify potential drugs which are effective against only bacteria but not eukaryotic system (Mei et al., 2005). The cell-free systems have demonstrated remarkable potential for the screening of small-molecule compounds arising from various sources in addition to providing opportunities to study epigenetic modifications that are more commonly seen in an in vivo setting thereby opening new avenues that can be easily studied by use of high-throughput proteomic screens.

22.3 Future Prospects

Remarkable technological advancement has taken place over the last decade that has allowed the establishment of cell-free based microarray platforms. Majority of the advantages offered by the cell-free expression systems are mainly accomplished by highly productive batch-fed configurations. Despite the improvements that have led to the attainment of protein concentration as high as milligram per milliliter of reaction, significant room for further improvements remains.

Substantial research is yet to be accomplished in order to not just simplify the current detection methods but also to enhance the sensitivity of detection. Furthermore, improvements to eliminate cross-reactivity and to minimize reagent consumption also need to be achieved. One particular discipline where the cell-free technology can make significant impact is with regard to protein folding since a relatively high portion of proteins acquired by in vitro and in vivo systems are generally insoluble or misfolded. While the addition of detergents or chaperones to the reaction mixture tend to have a positive effect at times, other complementary approaches like use of hybrid systems that comprise lysates from different sources, including those from archaea, may provide a more robust fold setting (Katzen et al., 2005).

Efforts toward finding a highly efficient energy regeneration systems is one of the key issues toward making this technological approach more cost-effective. Further, as the applications of the cell-free synthesis systems continue to grow, research with this technology can also be focused toward the discipline of RNA silencing due to the less-restrictive biochemical framework of this approach for studying the RNAi machinery or validating siRNAs (Novina and Sharp, 2004).

It, however, goes without saying that the cell-free expression system is a powerful, flexible, and an ever expanding technology with the ability to manipulate any reaction condition, thereby generating novel applications wherein the novelty of applications is only limited to one's creativity. While from cell biologists to biochemists have all benefitted from the cell-free expression systems, the high-throughput proteomics applications have greatly benefitted from cell-free technologies.

References

Achenbach P., Warncke K., Reiter J., Naserke H. E., Williams A. J., Bingley P. J., Bonifacio E., and Ziegler A. G., *Diabetes,* **53** (2004), 384–392.

Anderson K. S., Ramachandran N., Wong J., Raphael J. V., Hainsworth E., Demirkan G., Cramer D., Aronzon D., Hodi F. S., Harris L., Logvinenko T., and LaBaer J., *J. Proteome Res.,* **7** (2008), 1490–1499.

Angenendt P., Kreutzberger J., Glokler J., and Hoheisel J. D., *Mol. Cell Proteomics,* **5** (2006), 1658–1666.

Beare P. A., Chen C., Bouman T., Pablo J., Unal B., Cockrell D. C., Brown W. C., Barbian K. D., Porcella S. F., Samuel J. E., Felgner P. L., and Heinzen R. A., *Clin. Vaccine Immunol.,* **15** (2008), 1771–1779.

Bednar M., *Med. Sci. Monit.,* **6** (2000), 796–800.

Bertone P., and Snyder M., *FEBS J.,* **272** (2005), 5400–5411.

Bian X., Wiktor P., Kahn P., Brunner A., Khela A., Karthikeyan K., Barker K., Yu X., Magee M., Wasserfall C. H., Gibson D., Rooney M. E., Qiu J., and LaBaer J., *Proteomics,* **15** (2015), 2136–2145.

Carlson E. D., Gan R., Hodgman C. E., and Jewett M. C., *Biotechnol. Adv.,* **30** (2012), 1185–1194.

Chandra H., Reddy P. J., and Srivastava S., *Expert Rev. Proteomics,* **8** (2011), 61–79.

Chandra H., and Srivastava S., *Proteomics,* **10** (2010), 717–730.

Davies D. H., Liang X., Hernandez J. E., Randall A., Hirst S., Mu Y., Romero K. M., Nguyen T. T., Kalantari-Dehaghi M., Crotty S., Baldi P., Villarreal L. P., and Felgner P. L., *Proc. Natl. Acad. Sci. U. S. A.,* **102** (2005), 547–552.

Díez P., González-González M., Lourido L., Dégano R., Ibarrola N., Casado-Vela J., LaBaer J., and Fuentes M., *Microarrays,* **4** (2015), 214.

Doolan D. L., Mu Y., Unal B., Sundaresh S., Hirst S., Valdez C., Randall A., Molina D., Liang X., Freilich D. A., Oloo J. A., Blair P. L., Aguiar J. C., Baldi P., Davies D. H., and Felgner P. L., *Proteomics,* **8** (2008), 4680–4694.

Eyles J. E., Unal B., Hartley M. G., Newstead S. L., Flick-Smith H., Prior J. L., Oyston P. C., Randall A., Mu Y., Hirst S., Molina D. M., Davies D. H., Milne T., Griffin K. F., Baldi P., Titball R. W., and Felgner P. L., *Proteomics,* **7** (2007), 2172–2183.

Gibson D. S., Qiu J., Mendoza E. A., Barker K., Rooney M. E., and LaBaer J., *Arthritis Res. Ther.,* **14** (2012), R77.

Govindarajan R., Duraiyan J., Kaliyappan K., and Palanisamy M., *J. Pharm. Bioallied Sci.,* **4** (2012), S310–S312.

Hall D. A., Ptacek J., and Snyder M., *Mech. Ageing Dev.,* **128** (2007), 161–167.

Hall D. A., Zhu H., Zhu X., Royce T., Gerstein M., and Snyder M., *Science,* **306** (2004), 482–484.

He M., Stoevesandt O., Palmer E. A., Khan F., Ericsson O., and Taussig M. J., *Nat. Methods,* **5** (2008), 175–177.

He M., and Taussig M. J., *Nucleic Acids Res.,* **29** (2001), E73–E73.

He M., and Taussig M. J., *J. Immunol. Methods,* **274** (2003), 265–270.

Hu S., Xie Z., Qian J., Blackshaw S., and Zhu H., *Wiley Interdiscip. Rev. Syst. Biol. Med.,* **3** (2011), 255–268.

Huang H., Park C. K., Ryu J. Y., Chang E. J., Lee Y., Kang S. S., and Kim H. H., *Arch. Pharm. Res.,* **29** (2006), 890–897.

Hurst R., Hook B., Slater M. R., Hartnett J., Storts D. R., and Nath N., *Anal. Biochem.,* **392** (2009), 45–53.

Jackson A. M., Boutell J., Cooley N., and He M., *Brief Funct. Genomic Proteomic,* **2** (2004), 308–319.

Jawhar N. M., *Ann. Saudi Med.,* **29** (2009), 123–127.

Ji J., O'Connell J. G., Carter D. J., and Larson D. N., *Anal. Chem.,* **80** (2008), 2491–2498.

Katzen F., Chang G., and Kudlicki W., *Trends Biotechnol.,* **23** (2005), 150–156.

Kodadek T., *Chem. Biol.,* **8** (2001), 105–115.

Lopez J. E., Beare P. A., Heinzen R. A., Norimine J., Lahmers K. K., Palmer G. H., and Brown W. C., *J. Immunol. Methods,* **332** (2008), 129–141.

Mei Q., Fredrickson C. K., Jin S., and Fan Z. H., *Anal. Chem.,* **77** (2005), 5494–5500.

Miersch S., Bian X., Wallstrom G., Sibani S., Logvinenko T., Wasserfall C. H., Schatz D., Atkinson M., Qiu J., and LaBaer J., *J. Proteomics,* **94** (2013), 486–496.

Montor W. R., Huang J., Hu Y., Hainsworth E., Lynch S., Kronish J. W., Ordonez C. L., Logvinenko T., Lory S., and LaBaer J., *Infect. Immun.,* **77** (2009), 4877–4886.

Nath N., Hurst R., Hook B., Meisenheimer P., Zhao K. Q., Nassif N., Bulleit R. F., and Storts D. R., *J Proteome Res.,* **7** (2008), 4475–4482.

Nirenberg M. W., and Matthaei J. H., *Proc. Natl. Acad. Sci. U. S. A.,* **47** (1961), 1588–1602.

Novina C. D., and Sharp P. A., *Nature,* **430** (2004), 161–164.

Olia A. S., Barker K., McCullough C. E., Tang H. Y., Speicher D. W., Qiu J., LaBaer J., and Marmorstein R., *ACS Chem. Biol.,* **10** (2015), 2034–2047.

Ramachandran N., Hainsworth E., Bhullar B., Eisenstein S., Rosen B., Lau A. Y., Walter J. C., and LaBaer J., *Science,* **305** (2004), 86–90.

Ramachandran N., Larson D. N., Stark P. R., Hainsworth E., and LaBaer J., *FEBS J.,* **272** (2005), 5412–5425.

Ramachandran N., Raphael J. V., Hainsworth E., Demirkan G., Fuentes M. G., Rolfs A., Hu Y., and LaBaer J., *Nat. Methods,* **5** (2008), 535–538.

Ray S., Mehta G., and Srivastava S., *Proteomics,* **10** (2010), 731–748.

Sanchez-Carbayo M., Socci N. D., Lozano J. J., Haab B. B., and Cordon-Cardo C., *Am. J. Pathol.,* **168** (2006), 93–103.

Schmidt R., Cook E. A., Kastelic D., Taussig M. J., and Stoevesandt O., *J. Proteomics,* **88** (2013), 141–148.

Shergill I. S., Shergill N. K., Arya M., and Patel H. R., *Curr. Med. Res. Opin.,* **20** (2004), 707–712.

Speer R., Wulfkuhle J. D., Liotta L. A., and Petricoin E. F., 3rd, *Curr. Opin. Mol. Ther.,* **7** (2005), 240–245.

Sreekumar A., Nyati M. K., Varambally S., Barrette T. R., Ghosh D., Lawrence T. S., and Chinnaiyan A. M., *Cancer Res.,* **61** (2001), 7585–7593.

Stoevesandt O., Taussig M. J., and He M., *Methods Mol. Biol.,* **785** (2011), 265–276.

Wang J., Barker K., Steel J., Park J., Saul J., Festa F., Wallstrom G., Yu X., Bian X., Anderson K. S., Figueroa J. D., LaBaer J., and Qiu J., *Proteomics Clin. Appl.,* **7** (2013), 378–383.

Wang H., Wang H., Zhang W., and Fuller G. N., *Brain Pathol.,* **12** (2002), 95–107.

Wong J., Sibani S., Lokko N. N., LaBaer J., and Anderson K. S., *J. Immunol. Methods,* **350** (2009), 171–182.

Zhu H., Bilgin M., Bangham R., Hall D., Casamayor A., Bertone P., Lan N., Jansen R., Bidlingmaier S., Houfek T., Mitchell T., Miller P., Dean R. A., Gerstein M., and Snyder M., *Science,* **293** (2001), 2101–2105.

Zhu H., and Qian J., *Adv. Genet.,* **79** (2012), 123–155.

Chapter 23

Overview of the Current Knowledge and Challenges Associated with Human Exposure to Nanomaterials

Ali Kermanizadeh,[a] Kim Jantzen,[a] Astrid Skovmand,[a] Ana C. D. Gouveia,[a] Nicklas R. Jacobsen,[b] Vicki Stone,[c] and Martin J. D. Clift[d]

[a]*University of Copenhagen, Department of Public Health, Section of Environmental Health, Copenhagen, Denmark*
[b]*National Research Centre for the Working Environment, Copenhagen, Denmark*
[c]*Heriot Watt University, School of Life Sciences, Nano Safety Research Group, Edinburgh, United Kingdom*
[d]*In Vitro Toxicology Group, Institute of Life Sciences, Swansea University Medical School, Singleton Park Campus, Swansea, Wales SA2 8PP, United Kingdom*

m.j.d.clift@swansea.ac.uk

23.1 Introduction

There has been significant progress in the field of nanotoxicology, which has coincided with the increasing advancements made in the field of nanotechnology (Donaldson et al., 2006; Hoet et al., 2004; Maynard et al., 2006; 2007). Due to the increased use and diversity of nanomaterials (NM) there is an imperative need to deduce whether NM exposure could result in the onset of adverse human health effects. In order to address such health

Biocatalysis and Nanotechnology
Edited by Peter Grunwald

ISBN 978-981-4613-69-9 (Hardcover), 978-1-315-19660-2 (eBook)
www.panstanford.com

concerns, research into the effects of engineered NMs has received considerable interest over the past 20 years. This research has provided important information that has enabled the progression and realisation of some of proposed advantages of nanotechnology (Maynard et al., 2006; 2007; Oberdorster et al., 2005). Currently, despite certain advances in the field of nanotoxicology, the testing strategies for the determination of the NM-related biological impact(s) remain insufficient (Oberdorster et al., 2005).

This chapter will focus upon current knowledge available regarding NM-induced adverse effects towards human health. The chapter outlines outstanding issues associated with the field of nanotoxicology in order to offer suggestions that might improve experimental design towards future research. The chapter predominantly focuses on observations in animals, although a few human studies do exist (i.e. Apostoli et al., 1994; Brown et al., 2002; Jones et al., 2015; Nemmar et al., 2002; Stradling et al., 2002). These are limited in number and quality mainly due to the fact that it is often technically difficult to assess translocation of NMs in relevant and realistic human exposure scenarios (hence, they will receive no further attention hereafter).

23.2 Exposure Routes of NMs into and within the Human Body

With the ever-increasing sources and proposed applications of NMs, humans are continuously and inevitably exposed to NMs of all shapes and sizes. The skin, airways and the gastrointestinal tract (GIT) can be considered to be in direct contact with the external environment, so it is not surprising to find that all three systems are primary exposure sites for NMs.

The availability of a substance to an organism determines the dose internalised which then influences the "toxicokinetics", which describe the uptake, transport, metabolism, sequestration to different compartments and finally elimination of a substance from the organism. These parameters are essential, since the toxicity of a substance is dependent on which organs or cell types are exposed, and how long the substance remains at the site of exposure. Therefore, the potential for NMs to accumulate and exhibit toxicity at distal sites is greatly influenced by the route of exposure.

Some NMs have been shown to translocate to secondary organs situated at a distance from the original point of entry (Gieser et al., 2014; Lipka et al., 2010). It is also important to note, that in certain scenarios (e.g. medicine), NMs are intended to be directly injected into the human systemic circulation, and therefore are prone to both primary and secondary organ exposure. For a succinct and holistic summary of the NM human exposure routes, and their potential subsequent translocation to secondary organs, please refer to Oberdorster et al. (2007). It is important to note however, that the route of exposure is (again) crucial with respect to the amount of NMs reaching distal organs. For example, NMs reaching the central nervous system following exposure via the epidermis will be orders of magnitude lower before reaching the brain (Oberdorster et al., 2009).

23.3 Pulmonary

In terms of human exposure, the predominant focus, historically, of the NM toxicology field has been towards understanding the pulmonary effects of NMs. This is founded upon the respiratory system being identified as the key organ for research following occupational dust and particle matter (PM) exposure (Clift et al. 2011). Although the cardiovascular system was also highlighted, the respiratory system was considered the primary target organ, and thus research focussed predominantly upon the human lung (Clift et al., 2011). From this core of research, the "ultrafine hypothesis" was derived (Seaton et al., 1995). Epidemiological studies showed that the ultrafine component of PM_{10} influenced lung function, increasing morbidity and mortality adversely compared to exposure to the fine particle component (Peters et al., 1997). These findings, put in context with the historical understanding of how particles (and fibres (e.g. asbestos) (Donaldson et al., 2010)) interacted with the human lung, clearly showed this organ to be of considerable interest and importance to the (nano)particle toxicology community.

It is understood that inhaled materials deposit in different sites of the respiratory tract via either, or combination of, interception, impaction, sedimentation or diffusion (Geiser and Kreyling, 2010). NMs may enter cells by diffusion—it has been speculated that fibrous materials can physically pierce cell membranes (Murphy et al., 2012). In addition, active, energy dependent, cellular uptake of NMs is also important. The method by which the materials are transported through the lung is predominately influenced by endocytosis, and potentially by transcytosis across epithelial and endothelial cells (Chuang et al., 2014; Papp et al., 2008). Endocytosis is very much associated with single NM passage across the epithelium in the alveolar region and can occur by receptor-mediated mechanisms, clathrin-mediated mechanisms and by formation of caveolae. Yet, when considering the more representative NM-lung cell interaction of NM aggregates/agglomerates, it is understood that macrophages play a key role in re-distribution of NMs in the lung, and beyond, through transporting material into the thoracic lymph nodes (Konduru et al., 2014; Wang et al., 2012).

Thus, the alveolar region is, arguably, the most important region of interest for inhaled particles (and fibres), and therefore also for NM. Cells representative of this region (e.g. macrophages, epithelial cells, endothelial cells, dendritic cells, fibroblasts) of the human lung therefore, have been the foundation for a plethora of studies conducted in order to deduce the potential biological hazard associated with NM inhalatory exposure. It is also important to note that whilst many approaches towards elucidating the NM-lung cell interaction have been undertaken, including monoculture analysis, bi-culture, and further multi-cellular systems (e.g. coculture systems) (Rothen-Rutishauser et al., 2008), there have also been copious in vivo investigations conducted (Kim et al., 2014; Landsiedel et al., 2014). Whilst in the latter sections of this chapter, specific examples of studies are given, it is beyond the scope to adequately summarise all of the NM-lung cell/tissue analysis that has occurred since the early 1990s. In this respect, please refer to Bakand et al. (2012) for a succinct overview of the inhalatory exposure effects of NMs. Notably the study of Ferin et al. (1992) showed that titanium dioxide NMs were able to create an enhanced and prolonged inflammatory response, compared to their non-nanosized particle counterpart. Such findings arguably incited the research known today as "nanotoxicology" (Donaldson et al., 2004), but more importantly discussed the concept that not only size, but also number, mass dose and surface area were key characteristics potentially driving the biological effects noted. Surface area was considered key beyond others, with further research in the early 2000s reporting that the surface area of different NMs was able to drive the surface reactivity, and therefore the biological (mammalian cell) impact observed (Duffin et al., 2007). More recently research has shown that the surface properties of each NM and the subsequent protein coatings are also extremely important in their interaction with the pulmonary and immune cells (Johnston et al., 2012; Mailander et al., 2009).

In addition to the interaction with the alveolar barrier in the human lung, a comprehension of the potential for translocation of NMs into the bloodstream has also been gained (Choi et al., 2010). It is well known through the work of Kreyling and colleagues, (e.g. Kreyling et al., 2009; 2010) that NMs may enter the blood and are transported to a range of secondary target organs, including the liver, kidneys, lymph nodes, spleen, heart and the brain (Bai et al., 2014; Mailander et al., 2009; Sadauskas et al., 2009; Zhang et al., 2010). The impact of NMs upon the major organs noted here is further discussed in the next sections; however, it is important to note that much of the research into how NMs interact with other organs other than the lung is still in its relative infancy.

23.4 Extra-Pulmonary Biokinetics and Potential Adverse Effects

23.4.1 Cardiovascular System

The impact of particles on the cardiovascular system has been of great interest due to the realisation that exposure to PM_{10} was associated with numerous cardiovascular complications and increased mortality in exposed individuals (Schwartz, 2004). Therefore, there is a need to explore the mechanisms for the increased risk of cardiovascular diseases upon exposure to air pollution particulates. Furthermore, it is imperative to explore whether the mentioned PM-induced adverse effects can correlate to engineered NMs (Stone et al., 2016). There have been suggestions that exposure to certain NMs may result in progression to atherosclerosis, altered platelet function and detrimental effects on the tissues of both the myocardium and vascular vessel walls (at a cellular level the effects include oxidative stress and related inflammation) (Mills et al., 2009). Additionally, a positive correlation between the exposure to PM and the development of cardiovascular disorders such as myocardial infarction, hypertension, atherosclerosis, heart rate variability, thrombosis and coronary heart disease has been demonstrated (Sheng et al., 2013). In a recent investigation the exposure of C57BL/6 mice post intravenous (IV) exposure to a panel of five different NMs including a titanium dioxide (TiO_2) anatase (38 nm), TiO_2 rutile (67 nm), silica (SiO_2) (47 nm), zinc oxide (ZnO) (150 nm) and silver (Ag) (15 nm) was carried out. The authors noted small effects on haemodynamics (SiO_2 and Ag NMs caused elevated blood pressure in arterioles, but the mean arterial pressure remained unaltered), although found that TiO_2 anatase induced thrombosis formation and in vitro platelet aggregation. These findings indicate that cardiovascular effects are not confined to NMs originating from ambient air pollution and may rely on the chemical composition and crystallinity of the materials (Haberl et al., 2015). Furthermore, the intratracheal instillation (IT) of hamsters at doses of 5, 50 or 500 μg/animal of the standard reference material SRM1650 (diesel exhaust model particle—18–30 nm) resulted in both venous and arterial thrombosis formation. This finding was concomitant with non-cytotoxic levels of pulmonary inflammation (increased neutrophils and histamine levels in bronchoalveolar lavage fluid measured 1 h post exposure). The study also showed that platelet activation was increased in blood from the animals dosed with 50 μg of NMs. Collectively, the data indicated that exposure to diesel exhaust particles cause pulmonary inflammation with subsequent effects on platelet-related thrombi formation; albeit the study addresses acute effects of a single bolus diesel exhaust particle exposure, and not the chronic, continuous dose exposure which is arguably more physiologically relevant (Nemmar et al., 2002). The chronic effects of exposure to ambient air materials (103 $\mu g/m^3$, ~80 nm), whole diesel exhaust (436 $\mu g/m^3$, ~78 nm) and filtered diesel exhaust gas (particle free) was investigated in an Apolipoprotein E *(ApoE)-/-* mice model (animals prone to development of hyperlipidaemia, hypercholesterolemia and atherosclerosis). The animals were exposed in an inhalation chamber for 5 h/day and 4 days/week for a total duration of up to 5 months. The data demonstrated increased levels of phenylephrine induced vasoconstriction as well as vascular cell adhesion molecule 1 (VCAM-1) levels in sera. In addition, plaque progression was increased in the exposed animal groups (ambient air significantly contributed towards plaque progression compared to whole diesel exhaust only). In another recent investigation, the composition of protein coronas on amorphous SiO_2 NMs of varying sizes (~10, 16 and 55 nm) suspended in human whole-blood was investigated. The authors noted a total of 125 proteins identified on the circulating NMs. The study demonstrates the importance of the protein corona (defined as proteins adsorbed onto the particles' surface) formation on NMs circulating in the blood and the influence of this on the toxicity and eventual fate of NMs (Tenzer et al., 2011). In a bio-distribution study, Wistar rats were exposed to two gold (Au) NMs with different sizes (1.4 nm and 18 nm) via the IV route (26.57 μg). It was demonstrated that 24 h post exposure between 0.1–0.2% of the injected dose were retained in the heart for the smaller and larger materials respectively (Semmler-Behnke et al., 2008).

As an example of an in vitro investigation on the cardiovascular system the adverse effects of SiO_2 (13 nm) on murine macrophage cell line (J774A.1) was carried out. First, it was found that the materials were engulfed in greater quantities with increased NM concentrations (in part co-localises with lipid droplets). The study further demonstrated that exposure to the SiO_2 NM exposure was associated with increased cytotoxicity, intracellular reactive oxygen species (ROS) production and accumulation of triglycerides and cholesterol which is regarded as pivotal to foam cell formation—a hallmark of atherosclerotic progression (Petrick et al., 2014). The endothelium is extremely important in regulating the tone of the vascular system (Vesterdal et al., 2012). Thus, human umbilical vein endothelial cell (HUVEC) line is often utilised for analysing endothelial cells function in vitro settings. These cells were exposed to SiO_2 NMs at a concentration range of 25–100 μg/ml for up to 24 h. It was noted that exposure was associated with a concentration and time dependent toxicity, inhibition of anti-oxidant activity and intracellular ROS generation (indicative of oxidative damage) (Duan et al., 2013). In a similar investigation HUVEC cells were exposed to carbon black (Printex 90) (14 nm) at concentrations of 0.1, 10, 50 or 100 μg/ml for up to 24 h. The data showed an increase in ROS production which was associated with cell membrane damage. Furthermore, a concentration-dependent up regulation of intercellular adhesion molecule 1 (ICAM-1) and VCAM-1 on the surface of HUVEC cells was noted. In the same study the vasomotor function was assessed using segments of aorta and mesenteric arteries from C57BL/6 mice. The aorta was exposed to the NMs for 30 min resulting in a reduced pressure–diameter relationship (decreased both acetylcholine and sodium nitroprusside responses as well as receptor-dependent vasoconstriction caused by phenylephrine) (Vesterdal et al., 2012).

In summary, adverse effects in terms of cardiovascular effects are invariably only observed in experimental animals given relatively high doses, often by instillation into the lungs or the direct injection into the blood. However, several toxicological studies have demonstrated that combustion and model NMs can gain access to the blood following inhalation or instillation and can enhance experimental thrombosis. Although largely simplified, it is also possible to state that cardiovascular accumulation is size-dependent following IV or intraperitoneal (IP) exposure. It has also been suggested that that individuals suffering from a myocardial infarction might have a higher permeation and retention of NMs (most likely due to abnormal or disrupted vasculature) (Dvir et al., 2011). Airway exposure to air pollution particles and nanomaterials is associated with similar effects on atherosclerosis progression, augmented vasoconstriction and blunted vasorelaxation responses in arteries. Importantly, the data does show that certain NMs, including TiO_2, carbon black and carbon nanotubes can have similar hazards to the vascular system as combustion-derived PM.

23.4.2 Central Nervous System

It is widely believed that NMs can gain entry to the central nervous system (CNS) after inhalation either via the sensory nerve endings in the sinuses or via the blood–brain barrier (BBB) after systemic distribution. Due to the similarities in size of NMs to odorous molecules, they can deposit in the olfactory mucosa, move towards the olfactory bulb via nerve axons and ultimately reach the CNS (Oberdorster et al., 2009). The intranasal uptake of NMs into the olfactory rods via the olfactory mucosa followed by their translocation into the olfactory bulb, crossing the synapses of the olfactory glomerulus and finally into mitral cell dendrites has been demonstrated for carbon (36 nm) NMs (Oberdorster et al., 2004) and MnO (30 nm) (Elder et al., 2006). The BBB is a highly specialised structure which separates blood from the cerebrospinal fluid, limiting the entry of substances into the brain. One speculation as to how this occurs is the formation of free radicals and increased oxidative stress which may disrupt normal BBB function (Sharma et al., 2012). Hence, a number of pathologies, including hypertension and allergic encephalomyelitis have been associated with increased permeability of the BBB to NMs in experimental designs (Siddiqi et al., 2012).

In a recent study the effects of NM size in potential neurotoxicity was investigated in male Sprague-Dawley rats exposed to three differently sized metal NMs (Ag, Cu (copper) and Al (aluminium)—20–30, 50–60 and 120–150 nm respectively). The materials were suspended in Tween 80 and administered via the IP route (50 mg/kg) once daily for one week. Following the exposure, serum protein bound tracers (Evans blue and radioactive iodine) were utilised to detect any excess leakage of serum proteins into the brains microenvironment (indicative of BBB dysfunction). The data indicated a significant inverse correlation between particle size and inhibition of the BBB. This was associated with increased damage to neurons, glial and myelin cells with decreasing NM size. Up-regulation of neuronal nitric oxide synthase (NOS) linked to formation of nitric oxide was also increased following exposure to the smaller NMs. Finally, the smaller Ag and Cu caused much greater BBB dysfunction compared to the Al NMs. Ag exposure caused greater neuronal damage and NOS up-regulation followed by Cu and Al (Sharma et al., 2013). It is not surprising to find different NMs have different degrees of neurotoxic effects, mainly due to their inherent physiochemical properties. However, the effects of different materials will not be discussed in detail here with the exception of hinting at the potential increased adverse effects associated with soluble NMs. There is no doubt that dose is crucial in any NM-induced adverse effects in the CNS. In a recent study CD-1 female mice were exposed to Cu NMs (23.5 nm) via intranasal instillation at three different concentrations (low dose (1 mg/kg), a middle dose (10 mg/kg) and a high dose (40 mg/kg)) every other day for 15 or 21 days. The data demonstrated that brain homeostasis in different regions to be highly sensitive to NM dose administered. An example of this was an increase in Homovanillic acid (dopamine metabolite and a biomarker for dopamine dysfunction indicating several neurological disorders) in the hippocampus after a medium and high dose, decreases in the cerebral cortex after a low dose and increased in the cerebellum after a high dose but decreases in the low and medium dose. Similarly, nitric oxide (NO) levels increased in the striatum after a low and medium dose but decreased in the olfactory region following exposure to the high dose. The brain is a sensitive and complex organ with any change in neurotransmitter activity may be indicative of a toxic effect (Zhang et al., 2012). In another study, CD-1 mice exposed to iron oxide NMs (21 nm) via the intranasal route at low and repeated doses (130 μg) resulted in elevated levels of anti-oxidants and NOS; while the glutathione (GSH) levels significantly decreased in the olfactory bulb and the hippocampus. Transmission electron microscopy (TEM) analysis demonstrated some ultra-structural alterations in the nerve cells, including dendrite degeneration, membranous structure disruption and lysosome increase in the olfactory bulb, slight dilation in the rough endoplasmic reticulum and lysosome increase in the hippocampus (Wang et al., 2009). As a further example of in vitro NM-induced CNS effects, exposure of PC12 cells to differing concentrations of single-walled carbon nanotubes (SWCNT) (W: 1–2 nm, L: 2 μm) resulted in a concentration dependent decrease in cell viability, cell-cycle arrest, induction of apoptosis and intracellular anti-oxidant depletion (Wang et al., 2011).

In summary, such a review of (some of) the available experimental data demonstrates that protein coating and functionalisation might enhance the NMs ability to translocate across the BBB. The NMs that enter the blood stream will be coated with numerous human plasma proteins and might be recognised as endogenous protein by endothelial cell receptors (Mickler et al., 2012). Conversely, the NM surface charges have been shown to alter blood–brain integrity (Jiang et al., 2009) and should be considered towards undertanding the impact and distribuiton of NMs upon and in the brain.

23.4.3 Gastrointestinal Tract

Ingestion of NMs can potentially occur directly from food, water or orally administered medicines (Card et al., 2011). In addition, retrograde transfer of NMs by mucocilliary clearance from the lungs may result in the ingestion of materials. It is believed that the vast majority of ingested NMs are rapidly passed through the GIT and lost via the faeces (Papp et al., 2008). However, the harsh environment of the stomach, with its low pH and digestive enzymes might alter the characteristics of NMs and therefore their subsequent fate in the human body. Surface properties of NMs play an important role in their translocation from the GIT. It has been suggested that charged materials exhibit poor bio-availability due to electrostatic repulsion and mucus entrapment (Hoet et al., 2004). It is also reported that NM uptake in the GIT occurs by the microfold cells (M cells) that cover the Peyer's patches. This is similar to antigen delivery to dendritic cells or lymphocytes via transcytosis, as well as the transport through epithelial cells via endocytosis, persorption through gaps from shredding at villous tips and ineffective tight junctions (Møller et al., 2012). Once in sub-mucosal tissue, NMs are capable of entering the lymphatics and the blood capillaries. In a study in which Fischer rats which were orally exposed to 14C-labelled C_{60} fullerenes, around 98% of the ingested material were cleared in the faeces within two days and the final 2% were renally excreted (Yamago et al., 1995). However, the exposure of F344 rats orally dosed with Ag NM (56 nm) at 500 mg/kg for 90 days resulted in alkaline phosphatase (ALP) and cholesterol changes in the blood indicative of slight liver damage. In addition, histopathologic examination revealed a higher incidence of bile-duct hyperplasia, with or without necrosis, fibrosis, and/ or pigmentation in the treated animals. There was also a dose-dependent accumulation of Ag in all tissues examined. Finally, a significant accumulation of Ag was noted in the liver and the kidneys (Kim et al., 2010). In another study Fischer rats were exposed to a single intragastric dose of C_{60} (0.7 nm) or SWCNT (~2 nm (width) and < 1 μm (length) (0.064 and 0.64 mg/kg/bw suspended in saline or corn oil)). The authors showed that both doses of the SWCNT and C_{60} increased 8-oxo-2-deoxyguanosine (8-oxodG) levels in the liver. The suspension of materials in saline or corn oil had no significant effect on the genotoxicity of NMs in the liver. Finally, an increased mRNA expression of 8-oxoguanine DNA glycosylase in the liver was noted following C_{60} treatment, although no significant increase in DNA repair activity was detected. The data from this study indicated that low dose oral exposure to C_{60} and SWCNT NMs can result in direct genotoxic effects in the liver (Folkmann et al., 2009). However, the same NMs were not able to induce genotoxicity in colonic mucosa cells. It can be argued that from an evolutionary perspective, enterocytes may be less susceptible to toxic effects of exogenous materials, compared to other cell types in the human body. Another explanation could be related to the age of the rats used in the study (sacrificed at 9 weeks of age); it has previously been shown that uptake of 1 μm labelled polystyrene particles was approximately nine times higher in adult (5 months) rats compared to young (6–8 weeks) rats (Seifert et al., 1996). Another study in oral exposure of Sprague-Dawley rats to SiO_2 (12 nm), iron oxide (60 nm) and Ag (11 nm) NMs at doses of up to 2000 mg/kg/bw following a single or repeated exposure (daily for 13 weeks) was carried out. Here, no adverse effects were noted in terms of haematological or histological changes within the GIT, despite signs of inflammation in hepatic tissues (Yun et al., 2015).

Overall, it appears that in vivo NM induced adverse effects in the GIT is rare; however, numerous in vitro studies have noted toxicity in cells originating from the GIT (the most popular cell type utilised being the Caco-2 cell line—human cell line derived from colon cancer). In one such study, Caco-2 cells were exposed to two differently sized SiO_2 NMs (15 and 55 nm, amorphous) resulted in increased cytotoxicity, genotoxicity, intracellular ROS generation and interleukin (IL) 8 secretion; with all adverse effects more significant following exposure to the smaller of the NMs (Tarantini et al., 2015). Similarly, exposure of Caco-2 cells to

SiO_2 (14 nm), (TiO_2) (<10, 20–80 and 40–300 nm) and ZnO (10–20 nm) NMs resulted in varying degrees of cytotoxicity and genotoxicity. Interestingly, neither the genotoxic nor the cytotoxic potential of these NMs seemed to be affected by either size or surface area, which emphasises that NM hazard identification could be impacted by additional parameters (i.e. PAH-content, metal content, combustion method) (Gerloff et al., 2009). A note of caution is required, however, as a direct comparison of in vitro and in vivo data is often difficult. Currently, there is little evidence which shows that ingestion of NMs cause direct adverse effects to the gastrointestinal tract in vivo. However, hazard identification in the GIT is fairly complex and will be discussed in more detail in the following sections. It is abundantly clear that translocation following oral exposure is very possible with the liver highlighted as a potentially vulnerable organ for any potential toxicity.

23.4.4 Hepatic System

It has been demonstrated that NMs in blood have the tendency to accumulate in the liver in large quantities (Almeida et al., 2011; Balasubramanian et al., 2013; Kreyling et al., 2014; Sadauskas et al., 2009). For blood-borne materials, the phagocytic Kupffer cells (hepatic resident macrophages) can be considered as a key clearance system with subsequent potential for accumulation of dose in the liver (Chen et al., 2012; Liu et al., 2014). The reticuloendothelial system in the liver is exposed to all NMs absorbed from the GIT into the cardiovascular system due to that fact that all blood exiting the GIT does so in the hepatic portal vein that directly perfuses the liver. One of the livers main functions is the removal and neutralisation of any potential pathogens that enter the body from the GIT via Kupfter cells (Maemura et al., 2005).

An interesting study investigating the fate and bio-distribution of Au NMs (1.4 and 18 nm) via different routes of exposure utilised the NMs at doses of (26.5 μg for the 1.4 nm NMs and 2.7 μg for the 18 nm NMs) via the IV or IT instillation route in a Wistar rat model. Within 24 h of IV exposure ~94% of the 18 nm and ~48% of 1.4 nm was found in the liver. A relatively small amount (0.5% of the injected dose) was excreted via the hepatobiliary system into the faeces (suggesting uptake and processing via hepatocytes). Following IT exposure, however, 25% of the administered dose was removed from the lungs by mucociliary clearance to the GIT and excreted via the faeces. It was demonstrated that the larger Au NMs were retained in the lungs; however, the 1.4 nm NMs were found in significant quantities in numerous secondary organs including the liver (0.7 ± 0.3% of total dose). The study concluded that small NMs can pass the air/blood barrier of the lungs after IT instillation, with subsequent distribution to secondary organs (Semmler-Behnke et al., 2008).

Further investigation has shown the adverse effects of a panel of 10 engineered NMs (5 TiO_2, 2 ZnO, 2 multi-walled carbon nanotubes (MWCNT) and a Ag) was investigated in a number of hepatic in vitro and in vivo models. The study of the NM-induced cytotoxicity demonstrated that the Ag NMs elicited the greatest level of cytotoxicity (lethal concentration (LC_{50}—2 μg/cm^2)) to a hepatocyte cell line (C3A) after a 24 h exposure. The Ag was followed by the non-functionalised ZnO (LC_{50}—7.5 μg/cm^2) and coated ZnO (LC_{50}—15 μg/cm^2) NMs with respect to cytotoxicity. The LC_{50} was not attained following exposure to any of the other seven engineered NMs (up to 80 μg/cm^2). Next, it was shown that all NMs significantly increased IL8 production from the cells. The ZnO and Ag NMs induced IL8 production at lower treatment concentrations than the TiO_2 and MWCNT NMs. Meanwhile no significant change in tumour necrosis factor (TNF)-α, IL6 or C-reactive protein levels were detected. The urea and albumin production were measured as a means of accessing normal liver function. Liver function was unaffected following acute sub-lethal NM exposure with the exception of the two ZnO NMs at higher concentrations which significantly decreased albumin production. A concentration-dependent decrease in the cellular total glutathione content occurred following exposure of the C3A cells to Ag, ZnO and MWCNT NMs. Intracellular ROS levels were also measured and showed to increase significantly following exposure of the C3A to the low toxicity NMs (MWCNT and TiO_2). The anti-oxidant Trolox had a preventative impact on the cytotoxic effect of NMs, whilst decreased the NM-induced IL8 production after exposure to all NMs, apart from Ag NMs. Finally, following 4 h exposure of the C3A cells to sub-lethal levels of the

NMs, the largest amount of DNA damage was induced by two of the TiO_2 NMs. As before, it was noted that the hepatocytes actively internalised all NMs. This was interesting as the hepatocytes are not specialised phagocytic cells (Fig. 23.1 (Kermanizadeh et al., 2012; 2013a)).

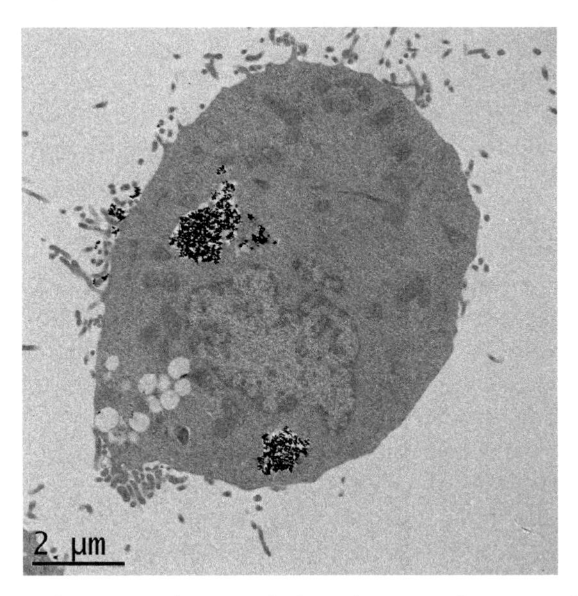

Figure 23.1 Conventional transmission electron microscopy image of the interaction between C3A liver cells (a derivative of the HepG2 liver cell-line) and 10 nm rutile titanium dioxide nanomaterials after 24 h exposure at 0.01 mg/mL.

In addition, similar findings were noted following exposure to the same panel of engineered NMs to primary human hepatocytes (Kermanizadeh et al., 2013b). The final in vitro model was the use of a 3D human liver micro tissue model (incorporated primary hepatocytes, Kupffer cells and endothelial cells) which allowed for the investigation the toxicological effects of the panel of NMs following single or multiple exposure regimes. It was noted that a LC_{50} was not reached for any of the NMs at any of the time-points or concentrations investigated. However, interestingly, the multiple exposures to low concentrations of NMs (in particular for the Ag and ZnO NMs) resulted in increased cytotoxicity over time which can be attributed to the internalisation and accumulation of materials within the cells. This is extremely important as accumulation of NMs might contribute towards their potential human health hazard. Similarly, the repeated NM exposure caused significant damaging potential to the liver tissue compared to the single bolus dose exposure. In this sense, the adverse biological effect were noted to be significantly heightened following treatment with Ag and ZnO NMs, in terms of their ability to cause cytokine secretion, lipid peroxidation and genotoxicity (Kermanizadeh et al., 2014a).

In the first in vivo trials utilising the same NMs the total liver weight was analysed 24 h post IT NM administration. A significant dose-dependent decrease in liver weight was noted after treatment with the ZnO and Ag NMs. The instillation of the Ag resulted in high 8-isoprostane values in the organ indicating hepatic lipid peroxidation. Additionally, significant dose-dependent decreases in glutathione levels were detected after exposure to ZnO and Ag NMs. Ag NMs was the most potent in reducing anti-oxidant levels followed by the two ZnO NMs. The Ag induced GSH depletion was linked to the presence of NMs/ions in the organ itself (high-resolution inductively coupled plasma mass spectrometry) (Gosens et al., 2015). A wide array of the same NMs induced a neutrophil influx into the liver as early as 6 h post IV exposure (myeloperoxidase assay and immunohistochemistry). It was noted that the neutrophils were only involved in the initial phase(s) of the immune response against the NMs. The analysis of mRNA expression in mice livers showed alterations in, at the gene level, of *complement component 3*, *IL6*, *IL10*, *CXCL2* and *ICAM-1* (most notable for ZnO and Ag NMs). Overall, the data suggested that the low doses of NMs were not sufficient to cause any long-term inflammation in the liver. Any changes that were observed after 24 h post exposure, in terms of leukocyte infiltration and changes in gene expression related to inflammation, oxidative stress and apoptosis resolved after 72 h post exposure (Kermanizadeh et al., 2013c). Finally, the role of Kupffer cells in the overall inflammatory response in the organ was assessed following IV exposure of the Ag NMs. The cytokine expression was measured post NM treatment in terms of IL2, IL4, TNF-α, IFN-γ and IL10 released from the organ, with significant up-regulation of TNF-α and IL10. For livers in which the macrophage population was specifically targeted and destroyed this cytokine profile was significantly decreased. The data indicated a potentially important role for Kupffer cells in the anti-inflammatory response and suggested that tolerance to the Ag NMs is favoured over a fully activated immune response (Kermanizadeh et al., 2014b).

In summary, it appears that the route of exposure is crucial in the quantity of NMs reaching the liver. It is believed that NMs introduced into the system via the IV route results in NMs reaching the liver in large quantities with the levels of this translocation to the hepatic system considerably less following exposure via the lungs (these levels also greatly vary between inhalation and instillation exposures). The cytotoxicity of NMs might be influenced by their solubility—it has been shown that materials like Ag and ZnO are potentially more damaging to the organ. Finally, it seems apparent that smaller materials remain in the circulation for longer which influences their accumulation and retention in the liver.

23.4.5 Renal System

The kidneys are principally responsible for the removal of metabolic waste such as urea and ammonia. However, it is believed that other "waste" products, such as NMs, could also be excreted through urine (Eisner et al., 2012). The kidneys receive blood from the renal arteries which branch directly from the dorsal aorta. Despite their relatively small size, the kidneys receive approximately 20% of the entire cardiac output (Roberts et al., 2000), making the organs highly susceptible to all xenobiotics (L'Azou et al., 2008). In theory both the glomerular structures (during plasma filtration) and tubular epithelial cells may be exposed to NMs. Since the major function of the kidneys is to eliminate a variety of potentially harmful substances (including the potential excretion of NMs), these organs are extremely important targets for investigation with regard to nanomaterial exposure and hazard. However, despite this, research into NM-induced toxicity to the organ is sparse. The size of a nanomaterial is a key characteristic that determines its filtration and excretion. It seems obvious, but retention in the organs might be associated with potential and possible damage to the organs. In a set of trials it was attempted to define the renal filtration threshold for NMs by using quantum dots with different hydrodynamic diameters and surface charges. The quantum dots (5–15 nm) used had a cadmium selenide core and a zinc sulfide shell and were coated with anionic, cationic, zwitterionic or neutral molecules. Sprague-Dailey male rats and CD-1 male mice where administered a dose of 10 pmol/g of body weight via IV injection. It was noted that the zwitterionic or neutral organic coatings prevented adsorption of serum proteins, which otherwise increased hydrodynamic diameter and prevented renal excretion. A final hydrodynamic diameter < 5.5 nm resulted in rapid and efficient urinary excretion and elimination of quantum dots (QD)s from the body (Choi et al., 2007). This seems sensible that in principle excretion into the urine from the bloodstream applies to small, water soluble molecules, therefore, large protein bound materials might not pass out though the glomerulus (Girolami et al., 1989). In addition, the ability of anionic and cationic metal-containing quantum dots to adsorb serum proteins and their subsequent retention in the lungs, liver, and spleen is of great importance (Choi et al., 2007). It can therefore be hypothesised that small NMs may enter the kidney and be readily excreted via the urine whilst, on the other hand larger materials may reach and accumulate in the kidneys.

This hypothesis was notably highlighted by a further study which found 12.5 nm (diameter) Au NMs to accumulate in the kidney after an 8 day IP exposure (40, 200 and 400 μg/kg/day) using a male C57/BL6 mice model. No adverse effects to the kidneys post exposure were reported (Lasagna-Reeves et al., 2010). In similar set of trials, TiO_2 NMs suspended in a quasi-synthetic polymer (208–330 nm (diameter)) were found to have a dose-dependent significant accumulation in the kidneys after 90 consecutive days of exposure (2.5, 5 and 10 mg/kg/bw). The TiO_2 NMs caused nephric inflammation, cell necrosis and dysfunction and an increase in the secretion of inflammatory cytokines (Gui et al., 2011).

In an in vitro study the toxicity of a panel of 10 engineered NMs (NMs (five TiO_2, two ZnO, two MWCNT and a Ag)) was investigated in a proximal tubular cell line (HK-2). It was found the two ZnO NMs (LC_{50}—2.5 μg/cm^2) and Ag NMs (LC_{50}—10 μg/cm^2) were highly toxic to the cells. The LC_{50} was not attained in the presence of any of the other engineered materials (up to 80 μg/cm^2). All NMs significantly increased IL8 and IL6 production from the exposed renal cells. Meanwhile no significant change in TNF-α or MCP-1 was detectable. There was a significant increase in ROS following 24 h exposure with the Ag and the two ZnO NMs (sub-lethal concentrations) while no change was observed with any of the other NMs. Finally, genotoxicity measured at sub-lethal concentrations showed a small but significant increase in DNA damage following exposure to 7 of the 10 NMs investigated (with the exception of one of the ZnO and the two TiO_2 NMs) (Kermanizadeh et al., 2013d).

In summary, it is extremely difficult to form any definitive understanding regarding NM induced nephrotoxicology mainly due to limited amount literature published. Presently greater than 35 studies highlight the kidneys as being important for accumulation of NMs (Kermanizadeh et al., 2015); however, in the majority of these studies, the potential toxicity of NMs in question was not investigated.

23.5 Summary and Conclusions

NMs are potentially, and realistically going to be increasingly utilised in everyday products thus highlightling the inevitability of their exposure to humans via a variety of scenarios (e.g. occupational and consumer). Despite the great advantages that these NMs offer, their toxicodynamics are currently not fully understood. From the available literature it appears that the distribution, retention and activity of NMs throughout the body is a real possibility and dependent on material biokinetics that hinge on the physicochemical properties of the NMs and their between the bodys' cellular and molecular components (Hagens et al., 2007; Kreyling et al., 2013; Li et al., 2010).

It is important to note that certain NMs are not insoluble and unchangeable in living systems. It is probable that some NMs (i.e. Au, CB, MWCNT and TiO_2) can be considered as insoluble. These NMs can be taken up from the circulation by phagocytic cells and persist for longer periods primarily in the liver and the spleen. In addition circulating insoluble NMs (smaller than 10 nm) can be excreted by glomerular filtration in the kidneys. However, other NMs may be transformed and dissolved in the body and this will affect their bio-distribution and toxicity. As an example, amorphous silica seems to dissolve in tissues over time (Braakhuis et al., 2014). Additionally, dissolution of ZnO NMs might be important for their rapid distribution from the lung (Konduru et al., 2014). Cadmium (Cd)-based QDs and Ag are likely to be oxidised in physiological environments (Jacobsen et al., 2009; Loeschner et al., 2011; van der Zande et al., 2012). The Cd core of the QDs is often toxic—this has to be taken into account in the interpretation of QD bio-distribution and NM-mediated effects (Sanwlani et al., 2014; Silva et al., 2014; Zhan et al., 2014). Therefore, it is important to understand the nature of NMs and distinguish between material types (virtually insoluble, slowly and rapidly soluble NMs) in order to gain a better understanding of bio-durability as well as their ability to translocate and induce toxicity in secondary organs.

It is crucial that great attention is paid in the selection of biologically relevant concentrations/doses for both in vitro and in vivo studies—even more imperative in terms of biokinetics of NMs. From the literature, the selection of higher effective dose or over-loading the system could lead to enhanced toxicity at the port of entry (and potentially translocation to secondary organs), which might not necessarily be a relevant biological response (Donaldson et al., 2013; Johnston et al., 2013; Landsiedel et al., 2012).

In general, in vivo in vitro extrapolation (IVIVE) is extremely difficult, and currently is a major hurdle within the field, especially considering the recent imposed need to move away from in vivo experimentation (Worth et al., 2014). One of the key reasons for this is that in reality, comparisons between the systems are rarely like for like, i.e. cytotoxicity in an in vitro system is not equivalent to inflammation in vivo. Further limitations exist in that organs are never comprised of only a single cell type, and cross-talk, or cellular interplay between different cell types and different organs is essential for responding to a toxic challenge in vivo. More sophisticated, complex in vitro models such as three-dimensional cultures (Clift et al., 2014), which have been shown to elicit a different effect compared to monocultures. Tissue slices and fluidic models have also been developed in order to improve in vitro risk assessment, with a view to reducing, refining and replacing animal studies (Burden et al., 2016).

Overall, it is evident that there are a greater number of IT instillation studies which show systemic effects in comparison to the more physiologically relevant inhalation exposure. The IT instillation studies mostly appear to show some similarities, with translocation of NMs and accumulation in a number of secondary organs. The liver especially, is pin-pointed as being highly susceptible to damage following IT pulmonary exposure, but as expected, the extent of toxicity varies between different NMs. This is likely due to a host of reasons from variability in size, chemical composition and surface properties which all influence how the NMs behave in vivo and their interactions with biological systems. However, a note of caution is required in forming conclusions from these studies—primarily as inhalation and instillation models are not always comparable. IT exposure (as well as pharyngeal aspiration and nasal instillation) are of limited relevance as a replacement/substitute for realistic exposure scenarios mainly as the deposition patterns of materials in the lung can vary between the methods and the dose-rate for instillation is very different to inhalation (due to high local concentration of NMs). These differences could indicate that adverse effects observed following IT, nasal instillations or pharyngeal aspiration might not necessarily be seen following an inhalation study. Similarly a large number of studies, primarily those using instillation and aspiration, have administered NMs in suspensions. This could introduce further non-physiological observations as these NMs will have altered surface characteristics, agglomeration and deposition patterns differing from their airborne counterparts (Balasubramanian et al., 2013). The implementation of suitable standardised protocols to ensure homogenous distribution of NMs with different surface properties in test substances is still an issue that requires attention. Furthermore, comparing effects of inhaled NMs across rodent test systems and between rodent test systems and humans is a key obstacle for the evaluation of the safety of NMs. In order for the quantitative data from the rodent studies to become useful in terms of NM quantitative risk assessment (performed using statistical and/or biologically based models) a judicious approach for extrapolating the data from rodent to humans is required. In addition, the differences in lung geometry, physiology and the characteristics of ventilation can give rise to differences in the regional deposition of materials in the lung in these species. It is commonly accepted that the variances in regional lung tissue doses cannot currently be measured experimentally (Asgharian et al., 2014). Since rodents are considerably smaller than humans and breathe at significant lower flow rates, there is a size-selective bias in material entry to the respiratory systems of the different species (not all materials that enter the human nasal passages on inhalation are capable of nasal uptake of rodents) (Kuehl et al., 2012). It is also important to note that unlike other laboratory animals and humans, rats appear to be more susceptible to overload-related effects in the lungs due to impaired macrophage-mediated alveolar clearance (Oberdorster, 2002) and might represent a worst-case scenario for NM-induced chronic effects in the lungs. To address (at least partially) some of these issues, computational modelling of dose deposition (i.e. multi path particle deposition model (MPPD)) can be utilised to extrapolate doses and deposition patterns from animals to humans (Asgharian et al., 2014; Kuempel et al., 2006). It is important to state that these models are far from perfect and are governed by a number of assumptions and simplifications which may differ from actual physiological exposures. These shortcomings are discussed in detail elsewhere and will be not deliberated here (Asgharian et al., 2014).

On a similar topic it is crucial to state that issues concerning the extrapolation of data from rodents to humans also exist for other routes of exposure. As an example, the majority of studies about dermal exposure focus on effects/translocation following topical application. However, several problems need to be addressed before the interpretation of the data. First, it is often difficult to define the actual absorbed dose of NMs into the skin layers which are further complicated by the interspecies differences (especially in terms of the density of the hair follicles, thickness of stratum corneum and total skin and the amount of skin lipids) in permeability of the organ (Labouta et al., 2013; Magnusson et al., 2001). As for oral exposure, doses are principally selected based on concentrations of oral uptake of nanoparticle-based food additives or ingredients of cosmetic articles occurring at low doses over long periods of time or according to the

Organisation for Economic Cooperation and Development (OECD) test guidelines.

Another key aspect affecting the biokinetics of NMs is the degree of protein binding to the surface of an NM (Landsiedel et al., 2012). This ties in with the concept of a NM "corona", or the ability of NMs to absorb biomolecules onto their surface (particularly proteins, therefore often termed the protein corona). Similarly, some NMs are being purposefully coated or functionalised. The formation of a corona or a coating will have a significant influence on the surface properties of NMs and can alter a number of physicochemical characteristics such as hydrodynamic size, charge, aggregation and agglomeration behaviour (Saptarshi et al., 2013). All of these modifications will impact the biokinetic characteristics and the way NMs interact with a biological system, and has been shown to affect the degree of cellular uptake (Chithrani et al., 2007; Treuel et al., 2013). This topic has been extensively covered in a recent review (e.g. Gunawan et al., 2014).

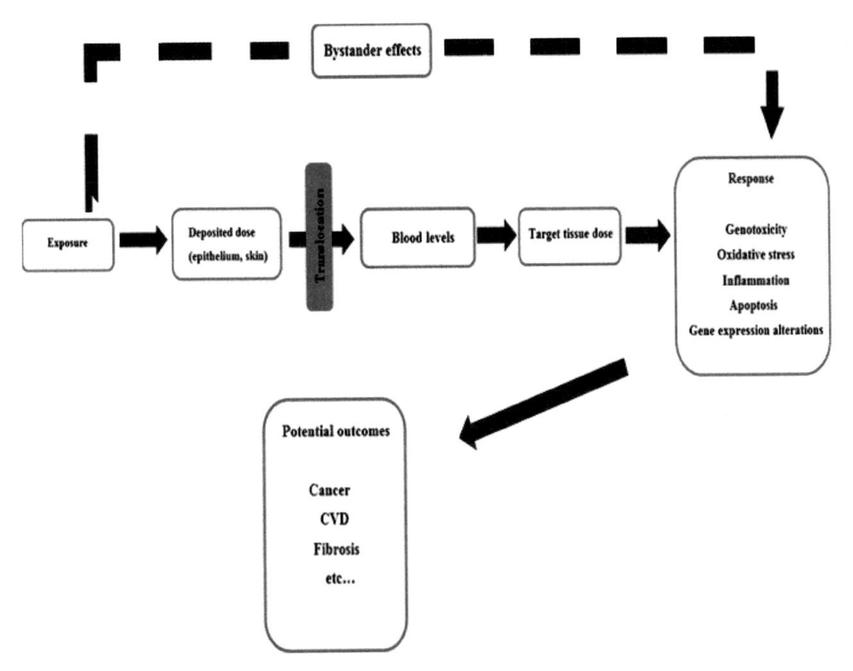

Figure 23.2 Schematic representation of the potential NM-induced human health outcomes and the important milestones investigated (adapted from Kermanizadeh, et al., 2015).

Finally, it has been proposed that NM-induced distal effects could result from molecular signalling cascades originating at the primary site of exposure, which in turn initiate downstream responses in secondary organs. For example, pulmonary inflammation and oxidative stress following NM exposure may result in the release of cytokines, chemokines and long-lived reactive oxygen species into systemic circulation (Magari et al., 2001). Although various mechanisms (Fig. 23.2) have been proposed for the induction of systemic effects by NMs, the biological processes and the underlying molecular pathways resulting in these systemic toxicities remain unclear and require further investigation (Bourdon et al., 2012; Zhao et al., 2011).

23.6 Current and Future Challenges

The quality of the nanotoxicology research (in particular with regard to the characterisation of materials used) has been heavily criticised recently with suggestions at the need for international harmonisation of the physicochemical characteristics of NMs (Krug, 2014). Although characterisation of materials is essential in any nanotoxicological study—the necessity for its regulation seems excessive. The characterisation of any material is always dependent on the toxicological context of the investigation and can differ from study to study. We believe it is the responsibility of the individual researcher (as well as their colleagues via the peer review process) to ensure the quality of the research remains of the highest possible standard. This does not mean that the field would not benefit from some degree of stringency; for example, the standardisation of the most widely used methodologies would be extremely beneficial and ensure that any differences in the observed responses are due to the materials and not the experimental variations. The comet assay for genotoxicity testing is a good example of an assay that is widely used in nanotoxicology with clear existing guidelines for in vivo testing (Møller et al., 2015). Due to expansion of the field of nanotoxicology and the escalating number of the published studies on the topic, it continues to be difficult to publish negative data. NMs are not equally harmful—from a toxicological point of view, it is not logical that only positive effects (i.e. harmful) are published, while the negative data (instances where no toxic/adverse effects are attributed to a material) is largely ignored. In order for the field of nanotoxicology to progress, it is crucial that both positive and negative outcomes are regarded with equal prominence in well-designed experiments. This is also very important for safe-by-design and legislative purposes.

In order to build a solid knowledge base, it is crucial to identify the hazards associated with NM exposure both in vitro and in vivo settings. Furthermore, from the literature it is evident that this is an area of nanotoxicology that still suffers—there is a lack of studies in which realistic, relevant exposure scenarios have been employed that can be used for hazard/risk assessment purposes. This being said, it is important to distinguish between proof of concept studies and those focusing on NM hazard assessment. In order to understand the mechanisms behind toxicity the use of higher dose/concentration might be required and must be judged on a case-to-case basis and justified within the context of the study.

Ordinarily, the risks of materials are assessed through consideration of exposure data and hazard information within risk assessments but this is currently impossible for NMs due to a lack of both exposure and hazard data within the available literature. Risk assessments for human health usually require exposure data and establishment of no observable effect levels (NOELs) that are extrapolated from animal investigations (to the human situation) using uncertainty factors (Aschberger et al., 2011). This risk assessment with respect to the life cycle of NMs is often difficult for numerous reasons. First, NMs are often transformed, e.g. by agglomeration or development of coatings, which may impact their biological effect after uptake (Cai et al., 2014; Kuhn et al., 2014). In addition, the information about the level of exposure to NM is often disintegrated, with the data from most studies unsuitable for the estimation of the exposed dose (as discussed above), and since majority of the predictive exposure models are mass-based, this becomes a major problem. Furthermore, to estimate the impact of a health risk, an accurate representation of the number of potential exposed population is needed. This number strongly relates to the individuals (in the manufacture and/or use) exposed to the nano-based products—with accurate information on this also lacking. The assessment of hazard in vulnerable groups (i.e. infants, elderly and individuals with pre-existing medical conditions) is another area which is currently largely non-existent. Finally, as repeatedly touched upon the lack of appropriate NM hazard assessment studies is also problematic. As exemplified, most of the parameters of the NM risk assessment are currently far from ideal. Hence, to date, only a few reference doses or health-based limit values have been proposed for NM use.

A large body of toxicology data is now being generated for NMs, using a wide variety of models, protocols and end-points. It is clear that not all NMs are equally toxic and these disparities are to a large extent based upon their physical and chemical properties and experimental set-ups and variations. The data generated will hopefully provide the opportunity for a variety of different analyses to be made, ranging from the relatively simple, such as the toxic effects of a particular NM type, to relatively complex issues such as the generation of models to allow structure activity relationships with the overall aim of having the knowledge base to allow for the prediction of the toxicity of new NMs. However, this, in all reality, is still years away; therefore, in the shorter term there is the need for further relevant models to try to predict risk of relevant low doses and appropriate route of exposure of materials, dispersed using relevant physiologically methods (Burden et al., 2016).

Finally, considering all these aspects discussed above, it remains to be determined whether chronic NM exposure could lead to accumulation of sufficiently biologically relevant doses that could initiate and contribute to the progression of disease in humans. By deducing such information, it will be possible to fully elucidate the potential human health hazard posed by NM.

Abbreviations

8-oxo-2-deoxyguanosine: 8-oxodG
Al: aluminium
Alkaline phosphatase: ALP
Apolipoprotein E: ApoE
Blood–brain barrier: BBB

Cadmium: Cd
Central nervous system–CNS
Cu: copper
Gastrointestinal tract: GIT
Glutathione: GSH
Gold: Au
Human umbilical vein endothelial cell: HUVEC
In vivo in vitro extrapolation: IVIVE
Intercellular adhesion molecule 1: ICAM-1
Interleukin: IL
Intraperitoneal: IP
Intratracheal instillation: IT
Intravenous: IV
Lethal concentration: LC
Manganese oxide: MnO
Microfold cells: M cells
Monocyte chemoattractant protein 1: MCP-1
Multi path particle deposition model: MPPD
Multi-walled carbon nanotubes: MWCNT
Nanomaterials: NM
Nitric oxide synthase: NOS
Organisation for economic cooperation and development: OECD
Particle matter: PM
Polycyclic aromatic hydrocarbons: PAH
Quantum dots: QD
Reactive oxygen species: ROS
Silica: SiO_2
Silver: Ag
Single-walled carbon nanotubes: SWCNT
Titanium dioxide: TiO_2
Tumour necrosis factor: TNF
Vascular cell adhesion molecule 1: VCAM-1
Zinc oxide: ZnO

References

Almeida J. P., Chen A. L., Foster A., Drezek R., *Nanomedicine,* **6** (2011), 815–835.

Apostoli P., Porru S., Alessio L., *Science of Total Environment,* **150** (1994), 129–132.

Aschberger K., Micheletti C., Sokull-Kluttgen B., Christensen F. M., *Environment International,* **37** (2011), 1143–1156.

Asgharian B., Price O. T., Oldham M., Chen L. C., Saunders E. L., Gordon T., Mikheev V. B., Minard K. R., Teeguarden T. J., *Inhalation Toxicology,* **26** (2014), 829–842.

Bai R., Zhang L., Liu Y., Li B., Wang L., Wang P., Autrup H., Beer C., Chen C., *Toxicology Letters,* **226** (2014), 70–80.

Bakand S., Hayes A., Dechsakulthorn F., *Inhalation Toxicology,* **24** (2012), 125–135.

Balasubramanian S. K., Poh K. W., Ong C. N., Kreyling W. G., Ong W. Y., Yu L. E., *Biomaterials,* **34** (2013), 5439–5452.

Bourdon J. A., Saber A. T., Jacobsen N. R., Jensen K. A., Madsen A. M., Lamson J. S., Wallin H., Møller P., Loft S., Yauk C. L., Vogel U. B., *Particle and Fibre Toxicology,* **9** (2012), 5.

Braakhuis H. M., Park M. V. D. Z., Gisen I., de Jong W. H., Cassee F. R., *Particle and Fibre Toxicology,* **11** (2014), 18.

Brown J. S., Zeman K. L., Bennett W. D., *American Journal of Respiratory Care and Critical Medicine,* **166** (2002), 1240–1247.

Burden N., Aschberger K., Chaudhry Q., Clift M. J. D., Doak S., Fowler P., Johnston H., Landsiedel R., Rowland J., Stone, V., *Nano Today,* (2016), http://dx.doi.org/10.1016/j.nantod.2016.06.007.

Cai H., Yao P., *Colloids and Surfaces B: Biointerfaces,* **123** (2014), 900–906.

Card J. W., Jonaitis T. S., Tafazoli S., Magnuson B. A., *Critical Reviews in Toxicology,* **41** (2011), 22–49.

Chen Y. P., Dai Z. H., Liu P. C., Chuu J. J., Lee K. Y., Lee S. L., Chen Y. J., *The Chinese Journal of Physiology,* **55** (2012), 331–336.

Chithrani B. D., Chan W. C., *Nano Letters,* **7** (2007), 1542–1550.

Choi H. S., Ashitate Y., Lee J. H., Kim S. H., Matsui A., Insin N., Bawendi M. G., Semmler-Behnke M., Frangioni J. V., Tsuda A., *Nature Biotechnology,* **28** (2010), 1300–1303.

Choi H. S., Liu W., Misra P., Tanaka E., Zimmer J. P., Ipe B. I., Bawendi M. G., Frangioni J. V., *Nature Biotechnology,* **25** (2007), 1165–1170.

Chuang H. C., Juan H. T., Chang C. N., Yan Y. H., Yuan T. H., Wang J. S., Chen H. C., Hwang Y. H., Lee C. H., Cheng T. J., *Nanotoxicology,* **8** (2014), 593–604.

Clift M. J. D., Endes C., Vanhecke D., Wick P., Gehr P., Schins R. P. F., Petri-Fink A., Rothen-Rutishauser B., *Toxicological Sciences,* **137** (2014), 55–64.

Clift M. J. D., Gehr P., Rothen-Rutishauser B., *Archives of Toxicology*, **85** (2011), 723–731.

Donaldson K., Aitken R., Tran L., Stone V., Duffin R., Forrest G., Alexander A., *Toxicological Sciences*, **92** (2006), 5–22.

Donaldson K., Murphy F. A., Duffin R., Poland C. A., *Particle and Fibre Toxicology*, **7** (2010), 5.

Donaldson K., Schinwald A., Murphy F., Cho W. S., Duffin R., Tran L., Poland C., *Accounts of Chemical Research*, **46** (2013), 723–732.

Donaldson K., Stone V., Tran C. L., Kreyling W., Borm P. J. A., *Occupational Environmental Medicine*, **61** (2004), 727–728.

Duan J., Yu Y., Li Y., Yu Y., Sun Z., *Biomaterials*, **34** (2013), 5853–5662.

Duffin R., Tran L., Brown D., Stone V., *Inhalation Toxicology*, **19** (2007), 849–856.

Dvir T., Bauer M., Schroeder A., Tsui J. H., Anderson D. G., Langer R., Liao R., Kohane D. S., *Nano Letters*, **11** (2011), 4411–4414.

Eisner C., Ow H., Yang T. X., Jia Z. J., Dimitriadis E., Li L. L., Wang K., Briggs J., Levine M., SchnermannJ., Espey M. G., *Journal of Applied Physiology*, **112** (2012), 681–687.

Elder A., Gelein R., Silva V., Feikert T., Opanashuk L., Carter J., Potter R., Maynard A., Ito Y., Finkelstein J., Oberdorter G., *Environmental Health Perspectives*, **114** (2006), 1172–1178.

Ferin J., Oberdoster G., *Acta Astronautica*, **27** (1992), 257–259.

Folkmann J. K., Risom L., Jacobsen N. R., Wallin H., Loft S., Møller P., *Environmental Health Perspectives*, **117** (2009), 703–708.

Geiser M., Kreyling W. G., *Particle and Fibre Toxicology*, **7** (2010), 2.

Geiser M., Stoeger T., Casaulta M., Chen S., Semmler-Behnke M., Bolle I., Takenaka S., Kreyling W. G., Schulz H., *Particle and Fibre Toxicology*, **11** (2014), 19.

Gerloff K., Albrecht C., Boots A. W., Forster I., Schins R. P. F., *Nanotoxicology*, **3** (2009), 355–364.

Girolami J. P., Orfila C., Pecher C., Cabosboutot G., Bascands J. L., Moatti J. P., Adam A., Colle A., *Biological Chemistry*, **370** (1989), 1305–1313.

Gosens I., Kermanizadeh A., Jacobsen N. R., Lenz A. G., Bokkers B., de Jong W. H., Krystek P., Tran L., Stone V., Wallin H., Stoeger T., Cassee F. R., *Plos One*, **10** (2015), e0126934.

Gui S., Zhang Z., Zheng L., Cui Y., Liu X., Li N., Sang X., Sun Q., Gao G., Cheng Z., Cheng J., Wang L., Tang M., Hong F., *Journal of Hazardous Materials*, **195** (2011), 365–370.

Gunawan C., Lim M., Marquis C. P., Amal R., *Journal of Materials Chemistry*, **2** (2014), 2060–2083.

Haberl N., Hirn S., Holzer M., Zuchtriegel G., Rehberg M., Krombach F., *Nanotoxicology*, **9** (2015), 963–971.

Hagens W. I., Oomen A. G., de Jong W. H., Cassee F. R., Sips A. J. A. M., *Regulatory Toxicology and Pharmacology*, **49** (2007), 217–229.

Hoet P. H. M., Hohlfeld I. B., Salata O., *Journal of Nanobiotechnology*, **2** (2004), 12–27.

Jacobsen N. R., Møller P., Jensen K. A., Vogel U., Ladefoged O., Loft S., Wallin H., *Particle and Fibre Toxicology*, **6** (2009), 2.

Jiang J., Oberdorster G., Biswas P., *Journal of Nanoparticle Research*, **11** (2009), 77–89.

Johnston H., Brown D., Kermanizadeh A., Gubbins E., Stone V., *Journal of Controlled Release*, **164** (2012), 307–313.

Johnston H., Pojana G., Zuin S., Jacobsen N. R., Møller P., Loft S., Semmler-Behnke M., McGuiness C., Balharry D., Marcomini A., Wallin H., Kreyling W., Donaldson K., Tran L., Stone V., *Critical Reviews in Toxicology*, **43** (2013), 1–20.

Jones K., Morton J., Smith I., Jurkschat K., Harding A. H., Evans G., *Toxicology Letters*, **233** (2015), 95–101.

Kermanizadeh A., Balharry D., Wallin H., Loft S., Møller P., *Critical Reviews in Toxicology*, **45** (2015), 837–872.

Kermanizadeh A., Brown D. M., Hutchison G., Stone V., *Journal of Nanomedicine and Nanotechnology*, **4** (2013c), 157.

Kermanizadeh A., Chauché C., Balharry D., Brown D. M., Kanase N., Boczkowski J., Lanone S., Stone V., *Nanotoxicology*, **8** (2014b), 149–154.

Kermanizadeh A., Gaiser B. K., Hutchison G. R., Stone V., *Particle and Fibre Toxicology*, **9** (2012), 28.

Kermanizadeh A., Gaiser B. K., Ward M. B., Stone V., *Nanotoxicology*, **7** (2013b), 1255–1271.

Kermanizadeh A., Løhr M., Roursgaard M., Messner S., Gunness P., Kelm J. M., Møller P., Stone V., Loft S., *Particle and Fibre Toxicology*, **11** (2014a), 56.

Kermanizadeh A., Pojana G., Gaiser B. K., Birkedal R., Bilaničová D., Wallin H., Jensen K. A., Sellergren B., Hutchison G. R., Marcomini A., Stone V., *Nanotoxicology*, **7** (2013a), 301–313.

Kermanizadeh A., Vranic S., Boland S., Moreau K., Squiban A. B., Gaiser B. K., Andrzejczuk L. A., Stone V., *BMC Nephrology*, **14** (2013d), 96.

Kim Y. H., Boykin E., Stevens T., Lavrich K., Gilmour M. I., *Journal of Nanobiotechnology*, **12** (2014), 47.

Kim Y. S., Song M. Y., Park J. D., Song K. S., Ryu H. R., Chung Y. H., Chang H. K., Lee K. H., Kelman B. J., Jwang I. K., Yu I. J., *Particle and Fibre Toxicology*, **7** (2010), 20.

Konduru N., Murdaugh K., Sotiriou G., Donaghey T., Demokritou P., Brain J., Molina R., *Particle and Fibre Toxicology*, **11** (2014), 44.

Kreyling W. G., Hirn S., Møller W., Schleh C., Wenk A., Celik G., Lipka J., Schaffler M., Haberl N., Johnston B. D., Sperling R., Schmid G., Simon U., Parak W. J., Semmler-Behnke M., *ACS Nano*, **8** (2014), 222–233.

Kreyling W. G., Hirn S., Schleh C., *Nature Biotechnology*, **28** (2010), 1275–1276.

Kreyling W. G., Semmler-Behnke M., Seitz J., Scymczak W., Wenk A., Mayer P., Takenaka S., Oberdorster G., *Inhalation Toxicology*, **21** (2009), 55–60.

Kreyling W. G., Semmler-Behnke M., Takenaka S., Moller W., *Accounts in Chemical Research*, **46** (2013), 714–722.

Krug H. F., *Angewandte Chemie*, **53** (2014), 12304–12319.

Kuehl P. J., Anderson T. L., Candelaria G., Gershman B., Harlin K., Hesterman J. Y., Holmes T., Hoppin J., Norenberg J. P., Yu H., McDonald J. D., *Inhalation Toxicology*, **24** (2012), 27–35.

Kuempel E. D., Tran C. L., Castranova V., Bailer A. J., *Inhalation Toxicology*, **18** (2006), 717–724.

Kuhn D. A., Vanhecke D., Michen B., Blank F., Gehr P., Petri-Fink A., Rothen-Rutishauser B., *Beilstein Journal of Nanotechnology*, **5** (2014), 1625–1636.

Labouta H. I., Schneider M., *Nanotechnology Biology and Medicine*, **9** (2013), 39–54.

Landsiedel R., Fabian R., Ma-Hock L., van Ravenzwaay B., Wohlleben W., Wiench K., Oesch F., *Archives of Toxicology*, **86** (2012), 1021–1060.

Landsiedel R., Saur U. G., Ma-Hock L., Schnekenburger J., Wiemann M., *Nanomedicine*, **9** (2014), 2557–2585.

Lasagna-Reeves C., Gonzalez-Romero D., Barria M. A., Olmedo I., Clos A., Sadagopa Ramanujam V. M., Urayama A., Vergara L., Kogan M. J., Soto C., *Biochemical and Biophysical Research Communications*, **393** (2010), 649–655.

L'Azou B., Jorly J., On D., Sellier E., Moisan F., Fluery-Feith J., Cambar J., Brochard P., Ohayon-Courtes C., *Particle and Fibre Toxicology*, **5** (2008), 22.

Li M., Al-Jamal K. T., Kostarelos K., Reinke J., *ACS Nano*, **4** (2010), 6303–6017.

Lipka J., Semmler-Behnke M., Sperling R. A., Wenk A., Takenaka S., Schleh C., Kissel T., Parak W. J., Kreyling W. G., *Biomaterials*, **31** (2010), 6574–6581.

Liu T., Choi H., Zhou R., Chen I. W., *Plos One*, **9** (2014), e103576.

Loeschner K., Hadrup N., Qvortrup K., Larsen A., Gao X., Vogel U., Mortensen A., Lam H. R., Larsen E. H., *Particle and Fibre Toxicology*, **8** (2011), 18.

Maemura K., Zheng Q. Z., Wada T., Ozaki M., Takao S., Aikou T., Bulkey G. B., Klein A. S., Sun Z. L., *Immunology and Cell Biology*, **83** (2005), 336–343.

Magari S. R., Hauser R., Schwartz J., Williams P. L., Smith T. J., Christiani D. C., *Circulation*, **9** (2001), 986–991.

Magnusson B. M., Walters K. A., Roberts M. S., *Advanced Drug Delivery Reviews*, **50** (2001), 205–227.

Mailander V., Landfester K., *Biomacromolecules*, **10** (2009), 2379–2400.

Maynard A. D., *The Annals of Occupational Hygiene*, **51**(2007), 1–12.

Maynard A. D., Aitkin R. J., Butz T., Colvin V., Donaldson K., Oberdorster G., Philbert M. A., Ryan J., Seaton A., Stone V., Tinkle S., Tran L., Walker N. J., Warheit D. B., *Nature*, **444** (2006), 267–269.

Mickler F. M., Mockl L., Ruthardt N., Ogris M., Wagner E., Brauchle C., *Nano Letters*, **12** (2012), 3417–3423.

Mills N. L., Donaldson K., Hadoke P. W., Boon N. A., MacNee W., Cassee F. R., Sandström T., Blomberg A., Newby D. E., *Nature Clinical Practice Cardiovascular Medicine*, **6** (2009), 36–44.

Møller P., Folkmann J. K., Danielsen P. H., Jantzen K., Loft S., *Current Molecular Medicine*, **12** (2012), 732–745.

Møller P., Hemmingsen J. G., Jensen D. M., Danielsen P. H., Karottki D. G., Jantzen K., Roursgaard M., Cao Y., Kermanizadeh A., Klingberg H., Christophersen D. V., Hersoug L. G., Loft S., *Mutagenesis*, **30** (2015), 67–83.

Murphy F. A., Schinwald A., Poland C. A., Donaldson K., *Particle and Fibre Toxicology*, **9** (2012), 8.

Nemmar A., Hoet P. H. M., Vanquickenborne B., Dinsdale D., Thomeer M., Hoylaerts M. F., Vanbilloen H., Mortelmans L., Nemery B., *Circulation*, **105** (2002), 411–414.

Oberdorster G., *Inhalation Toxicology,* **14** (2002), 29–56.

Oberdorster G., Stone V., Donaldson K., *Nanotoxicology,* **1**, (2007), 2–25.

Oberdorster G., Elder A., Rinderknecht A., *Journal Nanoscience and Nanotechnology,* **9** (2009), 4996–5007.

Oberdorster G., Maynard A., Donaldson K., Castranova V., Fitzpatrick J., Ausman K., Carter J., Karn B., Kreyling W. G., Lai D., Olin S., Monteiro-Riviere N., Warheit D., Yang H., *Particle and Fibre Toxicology,* **2** (2005), 8.

Oberdorster G., Sharp Z., Atudorei V., Elder A., Gelein R., Kreyling W., Cox C., *Inhalation Toxicology,* **16** (2004), 437–450.

Papp T., Schiffmann D., Weiss D., Castronova V., Vallyathan V., Rahman Q., *Nanotoxicology,* **2** (2008), 9–27.

Peters A., Wichmann H. E., Tuch T., Heinrich J., Heyder J., *American. Journal of Respiratory and Critical Care Medicine,* **155** (1997), 1376–1383.

Petrick L., Rosenblat M., Paland N., Aviram M., *Environmental Toxicology,* ahead of print (2014), 1–11.

Roberts R. G., Louden J. D., Goodship T. H. J., *Nephrology Dialysis Transplantation,* **15** (2000), 1906–1908.

Rothen-Rutishauser B., Blank F., Ch Muehlfeld, Gehr, P., *Expert Opinion on Drug Metabolism & Toxicology,* **4**, (2008), 1075–1089.

Sadauskas E., Jacobson N. R., Danscher G., Soltenberg M., Larsen A., Kreyling W., Wallin H., *Chemistry Central Journal,* **3** (2009), 16–23.

Sanwlani S., Rawat K., Pal M., Bohidar H. B., Verma A. K., *Journal of Nanoparticle Research,* **16** (2014), 2382.

Saptarshi S. R., Duschl A., Lopata A. L., *Journal of Nanobiotechnology,* **19** (2013), 11–26.

Schwartz J., *Occupational and Environmental Medicine,* **61** (2004), 956–961.

Seaton A., MacNee W., Donaldson K., Godden D., *Lancet,* **345** (1995), 176–178.

Seifert J., Haraszti B., Sass W., *Journal of Anatomy,* **189** (1996), 483–486.

Semmler-Behnke M., Kreyling W. G., Lipka J., Fertsch S., Wenk A., Takenaka S., Schmid G., Brandau W., *Small,* **4** (2008), 2108–2011.

Sharma A., Muresanu D. F., Patnaik R., Sharma H. S., *Molecular Neurobiology,* **48** (2013), 368–379.

Sharma H. K., Sharma A., *CNS and Neurological Disorders: Drug Targets,* **11** (2012), 65–80.

Sheng L., Wang X., Sang X., Ze Y., Zhao X., Liu D., Gui S., Sun Q., Cheng Z., Hu R., Wang L., Hong F., *Journal of Biomedical Materials Research Part A,* **101** (2013), 3238–3246.

Siddiqi N. J., Abdelhaim M. A. K., El-Ansary A. K., Alhomida A. S., Ong W. Y., *Journal of Neuroinflammation,* **9** (2012), 123.

Silva A. C. A., Silva M. J. B., da Luz F. A. C., Silva D. P., de Dues S. L. V., Dantes N. O., *Nano Letters,* **14** (2014), 5452–5457.

Stone V., Miller M. R., Clift M. J. D., Elder A., Mills N. L., Moller P., Schins R. P. F., Vogel U., Kreyling W. G., Jensen K. A., Kuhlbusch T. A. J., Schwarze P. E., Hoet P., Pietroiusti A., De Vizcaya-Ruiz A., Baeza-Squiban A., Tran L., Cassee F. R., *Environmental Health Perspectives,* (2016), http://dx.doi.org/10.1289/EHP424.

Stradling N., Etherington G., Hodgson A., Bailey M. R., Hodgson S., Pellow P., Shutt A. L., Birchall A., Rance E., Newton D., Fifield K., *Journal of Radioanalytical and Nuclear Chemistry,* **252** (2002), 315–325.

Tarantini A., Lanceleur R., Mourot A., Lavault M. T., Casterou G., Jarry G., Hogeveen K., Fessard V., *Toxicology in Vitro,* **29** (2015), 398–407.

Tenzer S., Docter D., Rosfa S., Wlodarski A., Kuharey J., Rekik A., Kanauer S. K., Bantz C., Nawroth T., Bier C., Sirirattanapan J., Mann W., Treuel L., Zellner R., Maskos M., Schild H., Stauber R. H., *ACS Nano,* **5** (2011), 7155–7167.

Treul L., Jiang X., Nienhaus G. U., *Journal of Royal Society Interface the Royal Society,* **10** (2013), 20120939.

van der Zande M., Vandebriel R. J., Van Doren E., Kramer E., Herrera Rivera Z., Serrano-Rojero C. S., Gremmer E. R., Mast J., Peters R. J., Hollman P. C. H., Hendriksen P. J. M., Marvin H. J. P., Peijnenburg A. A. C. M., Bouwmeester H., *ACS Nano,* **6** (2012), 7427–7442.

Vesterdal L. K., Mikkelsen L., Folkmann J. K., Sheykhzade M., Cao Y., Roursgaard M., Loft S., Møller P., *Toxicology Letters,* **214** (2012), 19–26.

Wang B., Feng W., Zhu M., Wang Y., Wang M., Gu Y., Ouyang H., Wang H., Li M., Zhao Y., Chai Z., Wang H., *Journal of Nanoparticle Research,* **11** (2009), 41–53.

Wang J., Sun P., Bao Y., Liu J., An L., *Toxicology in Vitro,* **25** (2011), 242–250.

Wang Y. X., Wang D. W., Zhu X. M., Zhao F., Leung K. C., *Quantitative Imaging in Medicine and Surgery,* **2** (2012), 53–56.

Worth A., Barroso J., Bremer S., Burton J., Casati S., Coecke R., Corvi R., Desprez B., Dumont C., Gouliarmou M., Grapel R., Griesinger C.,

Halder M., Roi J., Kienzler A., Madia F., Munn S., Nepelska M., Paini A., Price A., Prieto P., Rolaki A., Schaffer M., Triebe M., Whelan M., Wittwehr C., Zuang V., *JRC Science and Policy Report,* **26797** (2014), 1–475.

Yamago S., Tokuyama H., Nakamura E., Kikuchi K., Kananishi S., Sueki K., Nakahara H., Enomoto S., Ambe F., *Chemistry and Biology,* **2** (1995), 385–389.

Yun J. W., Kim S. H., You J. R., Kim W. H., Jang J. J., Min S. K., Kim H. C., Chung D. H., Jeong J., Kang B. C., Che J. H., *Journal of Applied Toxicology,* **35** (2015), 681–693.

Ze Y., Sheng L., Zhao X., Ze X., Wang X., Zhou Q., Liu J., Yuan Y., Gui S., Sang X., Sun Q., Hong J., Yu X., Wang L., Li B., Hong F., *Journal of Hazardous Materials,* **264** (2014), 219–229.

Zhan Q. L., Tang M., *Biological Trace Element Research,* **161** (2014), 3–12.

Zhang L., Bai R., Liu Y., Meng L., Li B., Wang L., Xu L., Le Guyader L., Chen C., *Nanotoxicology,* **6** (2012), 562–575.

Zhang X., Yin J., Kang C., Li J., Zhu Y., Li W., Huang Q., Zhu Z., *Toxicology Letters,* **198** (2010), 237–243.

Zhao F., Zhao Y., Liu Y., Chang X., Chen C., Zhao Y., *Small,* **7** (2011), 1322–1337.

Index